Gerhard Einsele

Sedimentary Basins

Evolution, Facies, and Sediment Budget

With 269 Figures

Springer-Verlag
Berlin Heidelberg New York
London Paris Tokyo
Hong Kong Barcelona
Budapest

Prof. Dr. GERHARD EINSELE
Geologisches Institut
Universität Tübingen
Sigwartstraße 10
7400 Tübingen, Germany

ISBN 3-540-54743-6 Springer-Verlag Berlin Heidelberg New York
ISBN 0-387-54743-6 Springer-Verlag New York Berlin Heidelberg

Library of Congress Cataloging-in-Publication Data. Einsele, Gerhard. Sedimentary basins: evolution, facies, and sediment budget; with 269 figures / Gerhard Einsele. p. cm. Includes bibliographical references and index. ISBN 3-540-54743-6 (Springer-Verlag Berlin). – ISBN 0-387-54743-6 (Springer-Verlag New York) 1. Sedimentary basins. 2. Sedimentation and deposition. 3. Facies (Geology) I. Title. QE571.E36 1992 551.3 – dc20 91-41002

© Springer-Verlag Berlin Heidelberg 1992
Printed in Germany

Typesetting: Camera ready by author
Printing: Beltz, Hemsbach
Binding: J. Schäffer, Grünstadt
32/3145-5 4 3 2 1 0 – Printed on acid-free paper

Preface

The modern geological sciences are characterized by extraordinarily rapid progress, as well as by the development and application of numerous new and refined methods, most of them handling an enormous amount of data available from all the continents and oceans.

Given this state of affairs, it seams inevitable that many students and professionals tend to become experts in relatively narrow fields and thereby are in danger of losing a broad view of current knowledge. The abundance of new books and symposium volumes testifies to this trend toward specialization. However, many geologic processes are complex and result from the interaction of many, seemingly unrelated, individual factors. This signifies that we still need generalists who have the broad overview and are able to evaluate the great variety of factors and processes controlling a geologic system, such as a sedimentary basin. In addition, this also means that cooperation with other disciplines in the natural sciences and engineering is increasingly important.

Modern text books providing this broad overview of the earth sciences are rare. Some are written by several authors together to make sure that all topics are treated properly. When individual authors write a book, they run the risk of creating a text that is less balanced, because they cannot avoid indulging their own preferences for specific topics and field examples. However, this disadvantage can be compensated for by the fact that just one author can produce a more concise and uniform text and include appropriate cross references.

In this one-author book I have tried to put much information into a considerable number of figures. Once the reader has acquired some basic knowledge and has read the text, he should be able to recall it easily by looking at these composite illustrations, for example the various facies models demonstrating both the depositional environment and the resulting vertical sequences. Many of these figures may be regarded as a kind of summary of the chapter in question; therefore, I did not give written summaries at the end of each chapter as found in many other textbooks.

This book addresses both qualitative and quantitative aspects of basin analysis, including topics such as various flux rates, diagenesis, and fluid flow, in the context of plate tectonics and sedimentary geology. Tectonic subsidence and uplift are prerequisites for basin formation and terrigenous sediment supply, but sedimentary processes in a basin are governed by other factors, including water circulation and recycling of nutrients, sediment transport, deposition, and redistribution. The sedimentary facies of a basin are largely controlled by the interrelationship between subsidence, sedimentation rate, and relative sea level change. Basinal sediment budgets are a topic which has been rarely treated in textbooks. Large-scale processes, facies associations, and especially sedimentary sequences are stressed in the book, rather than small-scale sedimentary structures, texture, petrographic characteristics, or detailed descriptions of biogenic sediment components and trace fossils. The latter phenomena are sufficiently described in a number of modern books. Finally, brief sections address the application of basic knowledge to exploration for hydrocarbons, coal, minerals, and deep groundwater.

This book is written for advanced students and professionals who require a comparatively straightforward, elementary treatment of sedimentary basin processes and evolution. The reader should already be familiar with general geology and geologic principles and have some basic knowledge of sedimentology. Quantitative aspects are described by simple equations and idealized examples. The book emphasizes broad, large-scale features of sedimentary basins and their facies associations. It provides only a limited number of case studies, which are chosen mostly from Europe and North America, but from other continents as well. Many experts will probably find that their specific topics are not treated thoroughly enough and that important datails have been omitted. Others may criticize that not all publications relevant to their fields are cited in the reference list. I would be grateful if these colleagues were to inform me when important points are missing or not treated properly.

All books bear the personal stamp of the author. In keeping with this unavoidable tradition, this book is influenced by my experience in studies of both modern marine sediments and ancient sedimentary rocks on land. Furthermore, I have done some work on mass physical properties and the mechanical behavior of soft and overconsolidated sediments, as well as in the area of groundwater behavior. This volume is based partially on courses which I have taught for many years, as well as on an intensive literature study, particularly of papers and topical volumes published during the past three to four years. Nevertheless, I am afraid and even certain that I have missed a great number of important publications, especially those written in languages I cannot read, such as Russian, Chinese, Japanese, and others. I apologize for these omissions, but then, one person is no longer able to evaluate the enormous literature which is published today, even in a limited field of geology.

Finally, I wish to express my thanks to a number of colleagues who reviewed specific chapters of this book and provided me with invaluable comments: Thomas Aigner (Tübingen), Erwin Appel (Tübingen), Robin Bathurst (Liverpool), R. Langbein (Greifswald), Stefan Kempe (Hamburg), Hanspeter Luterbacher (Tübingen), Ulrich von Rad (Hannover), Werner Ricken (Tübingen), Rüdiger Stein (Bremerhaven), Jobst Wendt (Tübingen), Jan Veizer (Bochum), and Andreas Wetzel (Basel). Nevertheless, only I can be taken to task for any shortcomings or errors in this text.

Linda Hobert and Susanne Borchert reviewed the English text and helped clarify many points. Hermann Vollmer produced most of the text's figures. Wolfgang Engel and Susanne Fink of Springer-Verlag, among others, assisted me in many ways in the production of this volume. All this help is gratefully acknowledged.

Last but not least I wish to thank my wife Ruth and my family, who with great fortitude tolerated my incessant work on weekends and in the evenings and did their best to spare me from the usual house chores and many other obligations which I should have taken care of.

Gerhard Einsele

Tübingen, November 1991

Contents

Part I
Types of Sedimentary Basins

1 Basin Classification and Depositional Environments (Overview)

1.1 Introduction

Sedimentary basins are, in a very broad sense, all those areas in which sediments can accumulate to considerable thickness and be preserved for long geological time periods. In addition, there also exist areas of long-persisting denudation, as well as regions where erosional and depositional processes more or less neutralize each other (creating what is known as non-deposition or omission).

In plan view sedimentary basins can have numerous different shapes; they may be approximately circular or, more frequently, elongate depressions, troughs, or embayments, but often they may have quite irregular boundaries. As will be shown later, even areas without any topographic depression, such as alluvial plains, may act as sediment traps. The size of sedimentary basins is highly variable, though they are usually at least 100 km long and tens of km wide.

We can distinguish between (1) active sedimentary basins still accumulating sediments, (2) inactive, but little deformed sedimentary basins showing more or less their original shape and sedimentary fill, and (3) strongly deformed and incomplete former sedimentary basins, where the original fill has been partly lost to erosion, for example in a mountain belt.

As many workers have pointed out, the regional deposition of sediments, non-deposition, or denudation of older rocks are controlled mainly by tectonic movements. Hence, most of the recent attempts to classify sedimentary basins have been based on global and regional tectonic concepts which will be briefly discussed below. In spite of obvious advantages, however, this approach has some serious shortcomings if it is not supplemented by additional criteria. One ought always bear in mind that the characteristics of sediments filling a basin of a certain tectonic type are predominantly controlled by other factors and can be extremely variable. With few exceptions (also discussed later), there is hardly such a phenomenon as a "tectonic sedimentary facies". For example, the broad concept of "geosynclinal sediments", often postulated in the past, was more misleading than helpful.

In addition to tectonic movements in the basinal area itself, sedimentary processes and facies are controlled by the paleogeography of the regions around the basin (peri-basin morphology and climate, rock types and tectonic activity in the source area), the depositional environment, the evolution of sediment-producing organisms, etc. Many sedimentologists therefore prefer a classification scheme based mainly on criteria which can be recognized in the field, i.e., the facies concept and the definition of the depositional environment (fluvial sediments, shelf deposits etc.). A further approach is the subdivision of sediments into important lithologic groups, such as siliciclastic sediments of various granulometries and composition, carbonate rocks, evaporites, etc. Having established the facies, succession, and geometries of such lithologic groups, one can proceed to define the tectonic nature of the basin investigated.

In this book an attempt is made to combine some principal points of these different classification systems and to show the interaction between tectonic and environmental characteristics of depositional areas.

1.2 Tectonic Basin Classification

Basin-generating tectonics is the most important prerequisite for the accumulation of sediments. Therefore, a tectonic basin classification system should be briefly introduced at the beginning of this chapter. Such a basin classification must be in accordance with the modern concept of global plate tectonics and hence will differ from older classifications and terminology.

In recent years, several authors have summarized our current knowledge on the interaction of plate tectonics and sedimentation (e.g., Dickinson in Dickinson and Yarborough 1976; Kingston et al. 1983; Miall 1984; Mitchell and Reading 1986; Foster and Beaumont 1987; Klein 1987; Perrodon 1988) and proposed basin classification systems. Although basically identical, these systems differ somewhat and do not use exactly the same terms. In this text we essentially use the system described by Mitchell and Reading, but add some minor modifications.

The different types of sedimentary basins can be grouped into seven categories, which in turn may be subdivided into two to four special basin types (Table 1.1 and Figs. 1.1 through 1.3):

Table 1.1. Tectonic basin classification.
(After Kingston et al. 1983; Mitchell and Reading 1986)

Basin category	Special basin type or synomym(s)	Underlying crust	Style of tectonics	Basin characteristics
Continental or interior sag basins	Epicontinental basins, infra-cratonic basins	Continental	Divergence	Large areas, slow subsidence
Continental or interior fracture basins	Graben structures, rift valleys and rift zones, aulacogens	Continental	Divergence	Relatively narrow basins, fault-bounded, rapid subsidence during early rifting
Basins on passive continental margins, margin sag basins	Tensional-rifted basins, tension-sheared basins, sunk margin basins	Transitional	Divergence + shear	Asymmetric basins partly outbuilding of sediment, moderate to low subsidence during later stages
Oceanic sag basins	Nascent ocean basin (growing oceanic basin)	Oceanic	Divergence	Large, asymmetric, slow subsidence
Basins related to subduction	Deep-sea trenches	Oceanic	Convergence	Partly asymmetric, greatly varying depth and subsidence
	Forearc basins, backarc basins, interarc basins	Transitional, oceanic	Dominantly divergence	
Basins related to collision	Remnant basins	Oceanic	Convergence	Activated subsidence due to rapid sedimentary loading
	Foreland basins (peripheral), retroarc basins (intramontane), broken foreland basins,	Continental	Crustal flexuring, local convergence or transform motions	Asymmetric basins, trend to increasing subsidence, uplift and subsidence
	Terrane-related basins	Oceanic		Similar to backarc basins
Strike-slip/ wrench basins	Pull-apart basins (transtensional) and transpressional basins	Continental and/or oceanic	Transform motion, ± divergence or convergence	Relatively small, elongate, rapid subsidence

- **Continental or interior sag basins** (Fig. 1.1a). Basins on continental crust are commonly generated by divergent plate motions and resulting extensional structures and thermal effects (cf. Chap. 8.1). In the case of large interior sag basins, however, major fault systems forming the boundaries of the depositional area or a central rift zone may be absent. Subsidence occurs predominantly in response to moderate crustal thinning or to a slightly higher density of the underlying crust in comparison to neighboring areas. In addition, slow thermal decay after a heating event and sedimentary loading can promote and maintain further subsidence for a long time (Chap. 8.1). Alternatively, it was recently suggested that long-term subsidence of intracratonic basins may be related to a decrease of the mantle heat flow above a "cold spot", i.e., to abnormal cooling (Ziegler 1989). In general, rates of subsidence are low in this geodynamic setting (cf. Chap. 12.3).

- **Continental graben structures and rift zones** form narrow elongate basins bounded by large faults (Fig. 1.1b and c). Their cross sections may be symmetric or asymmetric (e.g., half-grabens, see Chap. 11.4 and 12.1). If the underlying mantle is relatively hot, the lithosphere may expand and show updoming prior to or during the incipient phase of rifting. Substantial thinning of the crust by attenuation, which is often accompanied by the upstreaming of basaltic magma, thus forming transitional crust, causes rapid subsidence in the rift zone. Subsequent thermal contraction due to cooling and high sedimentary loading enable continuing subsidence and therefore the deposition of thick sedimentary infillings.

- **Failed rifts and aulacogens** (Fig. 1.1c). If divergent plate motion comes to an end before the moving blocks are separated by accretion of new oceanic crust, the rift zone is referred to as "failed". A certain type of such failed rifts is an aulacogen. Aulacogens represent the failed arm of a triple junction of a rift zone, where two arms continue their development to form an oceanic basin. Aulacogen floors consist of oceanic or transitional crust and allow the deposition of thick sedimentary sequences over relatively long time periods. Basins similar to aulacogens may also be initiated during the closure of an ocean and during orogenies.

- **Passive margin basins** (Fig. 1.1d). The initial stage of a true oceanic basin setting (or a proto-oceanic rift system) is established when two divergent continents separate and new oceanic crust forms in the intervening space. This does not necessarily mean that such a basin type fills with oceanic sediments, but it does imply that the central basin floor lies at least 2 to 3 km below sea level. When such a basin widens due to continued divergent plate motions and accretion of oceanic crust (drifting stage), its infilling with sediments lags more and more behind ocean spreading. Consequently, the sediments are deposited predominantly at the two continental margins of the growing ocean basin. The marginal "basins" developing on top of thinned continental crust are commonly not bordered by morphological highs and represent asymmetric depositional areas. Their underlying crust increasingly thins seaward; hence subsidence tends to become greater and faster in this direction (Chapt. 8.4). Here, sediments commonly build up in the form of a prism (Fig. 1.1d and Chap. 12.2). Some of these marginal basins may be affected and bordered by transform motions (tension-sheared basins). In a sediment-starved environment, subsided transitional crust can create deep plateaus (sunk basins). In general, subsidence of these marginal basins tends to decrease with passing time, unless it is reactivated by heavy sediment loads.

- **Oceanic sag basins or nascent ocean basins** occupy the area between a mid-oceanic ridge, including its rise, and the outer edge of the transitional crust along a passive continental margin (Fig. 1.1f). They commonly accumulate deep-sea fan or basin plain sediments. Due to the advanced cooling of the aging oceanic crust, subsidence is usually low, unless it is activated by thick sedimentary loading near the continental margin. Fault-bounded basins of limited extent are common in conjunction with the growth of mid-oceanic ridges (Fig. 1.1e).

- **Basins related to subduction.** Another group of basins is dominated by convergent plate motions and orogenic deformation. Basins related to the development of subduction complexes along island arcs or active continental margins include deep-sea trenches, forearc basins, backarc basins (Fig. 1.2a and b), and smaller slope basins and intra-arc basins. *Deep-sea trench floors* are composed of descending oceanic crust. Therefore, some of them represent the deepest elongate basins present on the globe. In areas of very high sediment influx from the neighboring continent, however, they are for the most part fil-

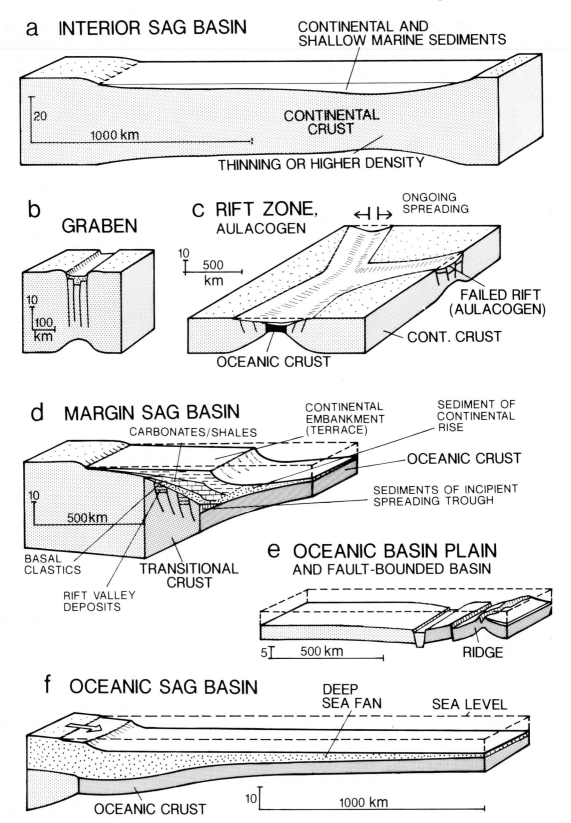

Fig. 1.1a–f. Tectonic basin classification for continental, marginal, and oceanic basins. See text for explanation. (After Dickinson and Yarborough 1976; Kingston et al. 1983; Mitchell and Reading 1986)

led up and morphologically resemble a continental rise. Deep-sea trenches commonly do not subside as do many other basin types. In fact, they tend to maintain their depth which is controlled mainly by the subduction mechanism, as well as by the volume and geometry of the accretionary sediment wedge on their landward side (Chap. 12.5.2). *Forearc basins* occur between the trench slope break of the accretionary wedge and the magmatic front of the arc. The substratum beneath the center of such basins usually consists of transitional or trapped oceanic crust older than the magmatic arc and the accretionary subduction complex (Chap. 12.5.3). Rates of subsidence and sedimentation tend to vary, but may frequently be high. Subsequent deformation of the sedimentary fill is not as intensive as in the accretionary wedge.

Backarc or interarc basins form by rifting and ocean spreading either landward of an island arc, or between two island arcs which originate from the splitting apart of an older arc system (Fig. 1.2a). The evolution of these basins resembles that of normal ocean basins between divergent plate motions. Their sedimentary fill frequently reflects magmatic activity in the arc region.

- Terrane-related basins are situated between micro-continents consisting at least in part of continental crust (Nur and Ben-Avraham 1983) and larger continental blocks. The substratum of these basins is usually oceanic crust. They may be bordered by a subduction zone and thus be associated with either basins related to subduction or collision.

- Basins related to collision. Partial collision of continents with irregular shapes and boundaries which do not fit each other leads to zones of crustal overthrusting and, along strike, to areas where one or more oceanic basins of reduced size still persist (Fig. 1.2c). These *remnant basins* tend to collect large volumes of sediment from nearby rising areas and to undergo substantial synsedimentary deformation (convergence, also often accompanied by strike-slip motions). *Foreland basins,* and *peripheral basins* in front of a fold-thrust belt, are formed by depressing and flexuring the continental crust ("A-subduction", after Ampferer, Alpine-type) under the load of the overthrust mountain belt (Fig. 1.2c and Fig. 1.3a). The extension of these asymmetric basins tends to increase with time, but a resulting large influx of clastic sediments from the rising mountain range often keeps

pace with subsidence (Chap. 12.6). As a result of the collision of two continental crusts, the overriding plate may be affected by "continental escape", leading to extensional graben structures or rifts perpendicular to the strike of the fold-thrust belt (Fig. 1.2c).

Retroarc or intramontane basins (Fig. 1.2b) occur in the hinterland of an arc orogen ("B-subduction" zone). They may affect relatively large areas on continental crust. Limited subsidence appears to be caused mainly by tectonic loading in a backarc fold-thrust belt.

Pannonian-type basins originate from post-orogenic divergence between two fold-thrust zones (Fig. 1.3a). They are usually associated with an A-subduction zone and are floored by thinning continental or transitional crust.

During crustal collision, some foreland (and retroarc) basins can get broken up into separate smaller blocks, whereby strike-slip motions may also play a role (Fig. 1.2c). Some of the blocks are affected by uplift, others by subsidence, forming basinal depressions. The mechanics of such *tilted block basins* were studied, for example, in the Wyoming Province of the Rocky Mountain foreland (McQueen and Beaumont 1989). So-called *Chinese-type basins* (Bally and Snelson 1980) result from block faulting in the hinterland of a continent-continent collision. They are not directly associated with an A-subduction margin, but it appears unnecessary to classify them as a special new basin type (Hsü 1989).

- Strike-slip and wrench basins (Fig. 1.3b and c): Transform motions may be associated either with a tensional component (transtensional) or with a compressional component (transpressional). Transtensional fault systems locally cause crustal thinning and therefore create narrow, elongate pull-apart basins (Chap. 12.8). If they evolve on continental crust, continuing transform motion may lead to crustal separation perpendicular to the transform faults and initiate accretion of new oceanic crust in limited spreading centers. Until this development occurs, the rate of subsidence is usually high. Transpressional systems generate wrench basins of limited size and endurance. Their compressional component can be inferred from wrench faults and fold belts of limited extent (Fig. 1.3c).

In order to identify these various basin categories, one must know the nature of the underlying crust as well as the type of former plate movement involved during basin formation, i.e., divergence or convergence. Even in the case of transform movement, either some

a SUBDUCTION-RELATED BASINS, INTRAOCEANIC

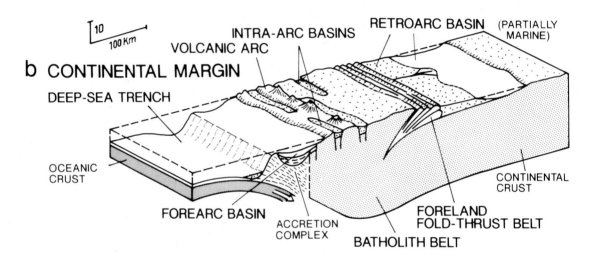

b CONTINENTAL MARGIN

COLLISION-RELATED BASINS

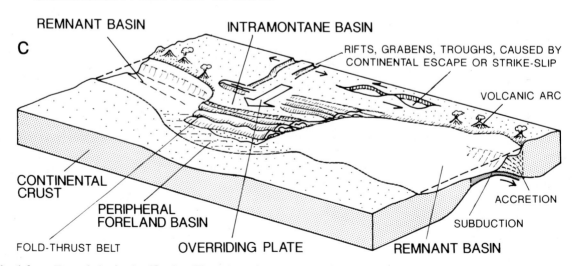

Fig. 1.2a-c. Tectonic basin classification (Fig. 1.1 continued). Subduction and collision-related basins (remnant basin). See text for explanation

a COLLISION-RELATED BASINS

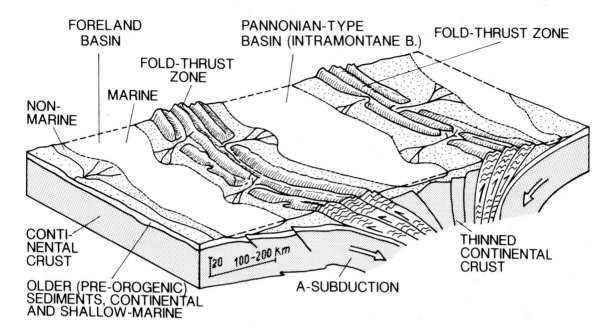

b STRIKE SLIP/WRENCH BASINS

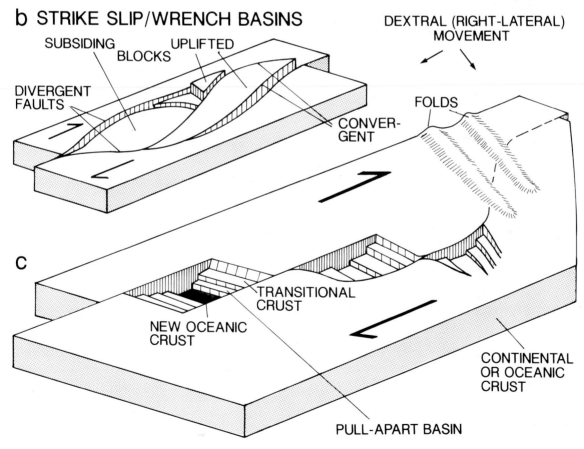

Fig. 1.3a-c. Tectonic basin classification (Fig. 1.2 continued). Collision-related basins and strike-slip/wrench basins. See text for explanation

divergence or convergence must take place. Small angles of convergence show up as wrenching or fold belts, and small angles of divergence appear as normal faulting or sagging.

One should bear in mind that all these basin types represent proto-types of tectonically controlled basins. They offer a starting point for the study and evaluation of basins, but there are no type basins which can be used as a complete model for any other basin (Burchfiel and Royden 1988). Even within a single broad tectonic setting, the development of smaller individual basins may display great variation. As soon as basins are analyzed in greater detail, the broad tectonic basin classification listed above becomes less useful. In addition, over long time periods, a sedimentary basin may evolve from one basin type into another (polyhistory basins) and thus exhibit a complex tectonic and depositional history (Chap. 12.9).

1.3 Pre-, Syn-, and Post-Depositional Basins

Principally, tectonic movements and sedimentary processes can interact in three different ways. These are used to distinguish between different types of sedimentary basins (Fig. 1.4; Selley 1985a):

- **Post-depositional basins.** The deposition of sediments largely predate tectonic movements forming a basin structure. Hence, there is no or little relationship between the transport, distribution, and facies of these sediments and the later evolved basin structure (Fig. 1.4a). However, some relationship between the syn-depositional subsidence phase and the subsequent basin-forming process cannot be excluded.

- **Syn-depositional basins.** Sediment accumulation is affected by syn-depositional tectonic movements, e.g., differential subsidence (Fig. 1.4b). If the sedimentation rate is always high enough to compensate for subsidence, the direction of transport and the sedimentary facies remain unchanged, but the thickness of the sediment in certain time slices varies. In Fig. 1.4b the sediment thickness increases toward the center of the basin. In this case, the basin structure is syn-depositional, but there was hardly a syn-depositional morphological basin controlling the sedimentary facies

of the basin. If sedimentation is too slow to fill up the subsiding area, a morphological basin will develop. Then, the distribution and facies of the succeeding sediments will be affected by the morphology of the deepening basin (transition to the situation shown in Fig. 1.4c).

- **Pre-depositional basins.** Rapid tectonic movements predate significant sediment accumulation and create a morphological basin, which is filled later by post-tectonic sediments (Fig. 1.4c). The water depth in the basin decreases with time, although some syn-depositional subsidence due to sediment loading is likely (Chap. 8.1). Sediment transport as well as vertical and lateral facies development are substantially influenced by the basin morphology.

Of course, there are transitions between these simplified basin types and, as we shall see later (Chap. 12), certain basins may show a complex history and therefore contain pretectonic as well as syn-tectonic or post-tectonic sediments.

1.4 Basin Morphology and Depositional Environments

General Aspects

The geometry of an ultimate basin fill is controlled mainly by basin-forming tectonic processes, but the *morphology of a basin* defined by the sediment surface is the product of the interplay between tectonic movements and sedimentation. Therefore, as already mentioned, a purely tectonic classification of sedimentary basins is not sufficient for characterizing depositional areas. It is true that a sedimentary basin in a particular tectonic setting also often undergoes a specific developmental or subsidence history (Chaps. 8 and 12), but its morphology, including water depth, may be controlled largely by other factors, such as varying influx and distribution of sediment from terrigenous sources (Chap. 11).

For example, a fluvial depositional system can develop and persist for considerable time on top of subsiding crust in various tectonic settings (Miall 1981). Fluvial deposits are known from continental graben structures, passive continental margins, foreland basins,

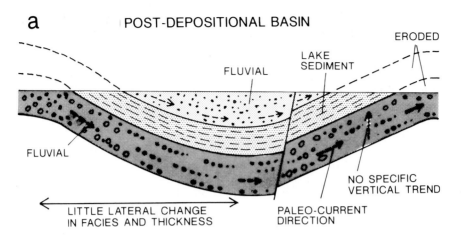

a **POST-DEPOSITIONAL BASIN**

FLUVIAL

LAKE
SEDIMENT

ERODED

FLUVIAL

NO SPECIFIC
VERTICAL TREND

LITTLE LATERAL CHANGE
IN FACIES AND THICKNESS

PALEO-CURRENT
DIRECTION

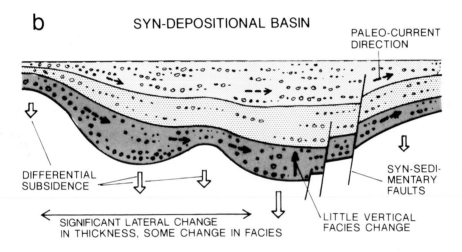

b **SYN-DEPOSITIONAL BASIN**

PALEO-CURRENT
DIRECTION

DIFFERENTIAL
SUBSIDENCE

SYN-SEDI-
MENTARY
FAULTS

SIGNIFICANT LATERAL CHANGE
IN THICKNESS, SOME CHANGE IN FACIES

LITTLE VERTICAL
FACIES CHANGE

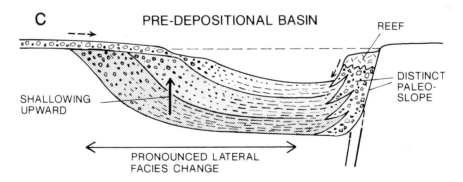

c **PRE-DEPOSITIONAL BASIN**

REEF

SHALLOWING
UPWARD

DISTINCT
PALEO-
SLOPE

PRONOUNCED LATERAL
FACIES CHANGE

Fig. 1.4. a Post-depositional basin created by tectonic
movements after the deposition of sheet-like fluvial and
lake sediments; younger syn-tectonic basin fill is removed
by subsequent erosion. **b** Syn-depositional tectonic movements control varying thicknesses of fluvial and shallow-
marine sediments and generate a basin-fill structure, although a morphological basin barely existed. **c** Rapid, pre-
depositional tectonics creates a deep morphological basin
which is later filled up by post-tectonic sediments. The
geometry of the former basin can be derived from transport
directions and facies distribution

forearc and backarc basins, pull-apart basins, etc. Fluvial sediments accumulate as long as rivers reach the depositional area and supply enough material to keep the subsiding basin filled. Although the basin-forming processes and subsidence histories of these examples differ fundamentally from each other, the sedimentary facies of their basin fills display no or only minor differences. In order to distinguish between these varying tectonic settings, one has to take into account the geometry of the entire basin fill, as well as vertical and lateral facies changes over long distances, including paleocurrent directions and other criteria. Syndepositional tectonic movements manifested by variations in thickness, small disconformities, or faults dying out upward (cf. Fig. 1.4b) may indicate the nature of the tectonic processes involved.

The *erosional base level* and sediment distribution within a basin are additional important factors modifying basin morphology and thus the development of special sedimentary facies. This situation is demonstrated in the elementary model of Fig. 1.5. In a fluvial environment, sediments cannot accumulate higher than the base level and gradient of the stream. If there is more influx of material into the depositional system than necessary for compensation of subsidence, the sediment surplus will be carried farther downslope into lakes or the sea. This signifies that the level up to which a basin can be filled with sediments may depend on the geographic position of the basin in relation to the erosional base. In Tibet, for example, the floors of present-day fluvial basins (intramontane basins and graben structures) are elevated higher than 3000 m in comparison to the coastal fluvial plains elsewhere.

The morphology of water-filled basins may significantly change as a result of depositional processes. Lakes and low-energy basins frequently show a prograding deltaic facies, causing pronounced basinward outbuilding of sediment (Chaps. 2.1.1, 2.2.2, 2.5.1, 3.4.1). Consequently, the areal distribution of the finer-grained sediment in the deeper basin portions decreases with time, although the initial, tectonically controlled basin configuration persists. By contrast, high-energy basins are little influenced by sediment outbuilding (Fig. 1.5). For example, terrigenous sediments transported into high-energy shelf seas tend to be reworked and swept into deeper water by wave action and bottom currents, except for some local seaward migration of the shoreline. Even on deep submarine slopes and in the deep sea, there is no general outbuilding or upbuilding of sediments, because gravity mass movements and deep bottom currents redistribute large quantities of material.

These few examples demonstrate that the most appropriate classification scheme for sedimentary basins depends primarily on the objectives of the study. If tectonic structure and evolution of a region are the main topics, then basin fill geometry and subsidence history derived from the thickness of stratigraphic units (Chap. 8.4) are of primary importance. If, on the other hand, the depositional environment, sedimentary facies, and paleogeographic reconstructions are of primary interest, then the basin classification used should not be strictly tectonic. Such a classification should also take into account changes in

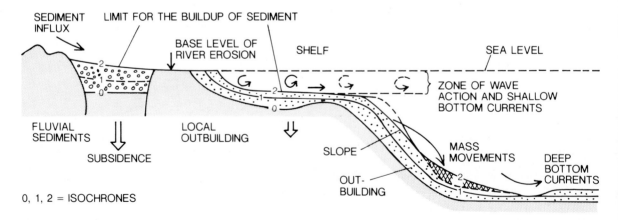

Fig. 1.5. Base level of erosion, hydrodynamic regime in the sea, and gravity mass movements as limiting factors controlling upbuilding and outbuilding of sediments. Note that the model may be modified by sea level changes

basin morphology caused by depositional processes, the chemical and hydrodynamic regimes of the basin, and peri-basin characteristics such as the size and nature of the drainage areas on nearby land.

Many workers distinguish between recent and ancient examples of depositional environments (e.g., Davis 1983; Reading 1986a), because the interpretation of paleoenvironments from the fossil record is subject to greater uncertainties. Furthermore, the methods of investigation and the possibilities of observing certain physical and biological sedimentary structures differ between soft sediments and lithified rocks. Soft material, for example, is suitable for the determination of primary grain size distribution, which in the case of lithified rocks is frequently problematic. On the other hand, any kind of structure is commonly much better visible in ancient rocks than in soft sands and muds. The surface of recent sediments on land and under water can be well observed, but in many cases, for example in fluvial environments, such temporary surfaces are rarely preserved in the sedimentary record. By contrast, indurated beds alternating with weaker material frequently show excellently preserved lower and upper bedding planes with trace fossils, various marks, and imbrication phenomena which are difficult to observe in soft sediments. Diagenesis may, however, also obscure primary bedding features. In addition, there are special sediments in the past, particularly far back in the Earth's history, for which no present-day analogies are known. Such environments are mentioned in Chapter 6.5.

In spite of such various problems between recent and ancient sediments, the depositional environments of both groups are treated jointly in this book, except for some special deposits. After a brief overview in this chapter, the most important groups of depositional environments are described in simplified facies models in Chapters 2 through 6.

Depositional Environments (Overview)

On the surface of our present-day globe, on land and below the sea, hundreds of depositional areas are known which meet the definition of sedimentary basins as described in Chapter 1.1. If we add to this list medium to large ancient sedimentary basins whose fill is still largely preserved, we have some thousand sedimentary basins. Taking into account this large number and the many factors controlling

a sedimentary environment, it appears at first glance that an enormous number of differing depositional environments should exist. This is in fact the case, but nevertheless it is possible to subdivide this great quantity into a limited number of distinct groups which have many characteristics in common.

Such depositional environment models have been extensively described in several textbooks (e.g., Reineck and Singh 1980; Blatt et al. 1980; Scholle and Spearing 1982; Davis 1983; Walker 1984a; Selley 1985a and b; Reading 1986a), and single groups of environments have been dealt with repeatedly in special publications, memoirs, short course notes, etc.

In Figure 1.6 the various types of sedimentary basins are predominantly classified according to their depositional environment and basin morphology. However, peri-basin geomorphology and climate also play a role. One can distinguish between several principal groups, for example:

- Continental (fluvial, glacial, eolian),
  lacustrine, and deltaic environments.
- Adjacent sea basins and epicontinental
  seas of varying salinity.
- Marine depositional areas of normal
  salinity.

As an alternative, a group of "transitional" environments may be defined between continental and marine environments (e.g., Davis 1983). This group includes marine deltas, intertidal environments, coastal lagoons, estuaries, and barrier island systems (cf. Chap. 3). In Part II of this book, a more diversified classification is used with the following main groups:

- Continental sediments.
- Coastal and shallow sea sediments
  (including carbonates).
- Sediments of adjacent seas and estuaries.
- Oceanic sediments.
- Special sediments and environments.

In addition, a chapter deals with depositional rhythms and cyclic sequences which may occur in all groups of depositional environments. In the following, a few general principles for the sedimentary fill of various basins are briefly discussed.

The *fluvial environment* is controlled by its erosional base level as well as by the sediment supply from more elevated regions sufficient

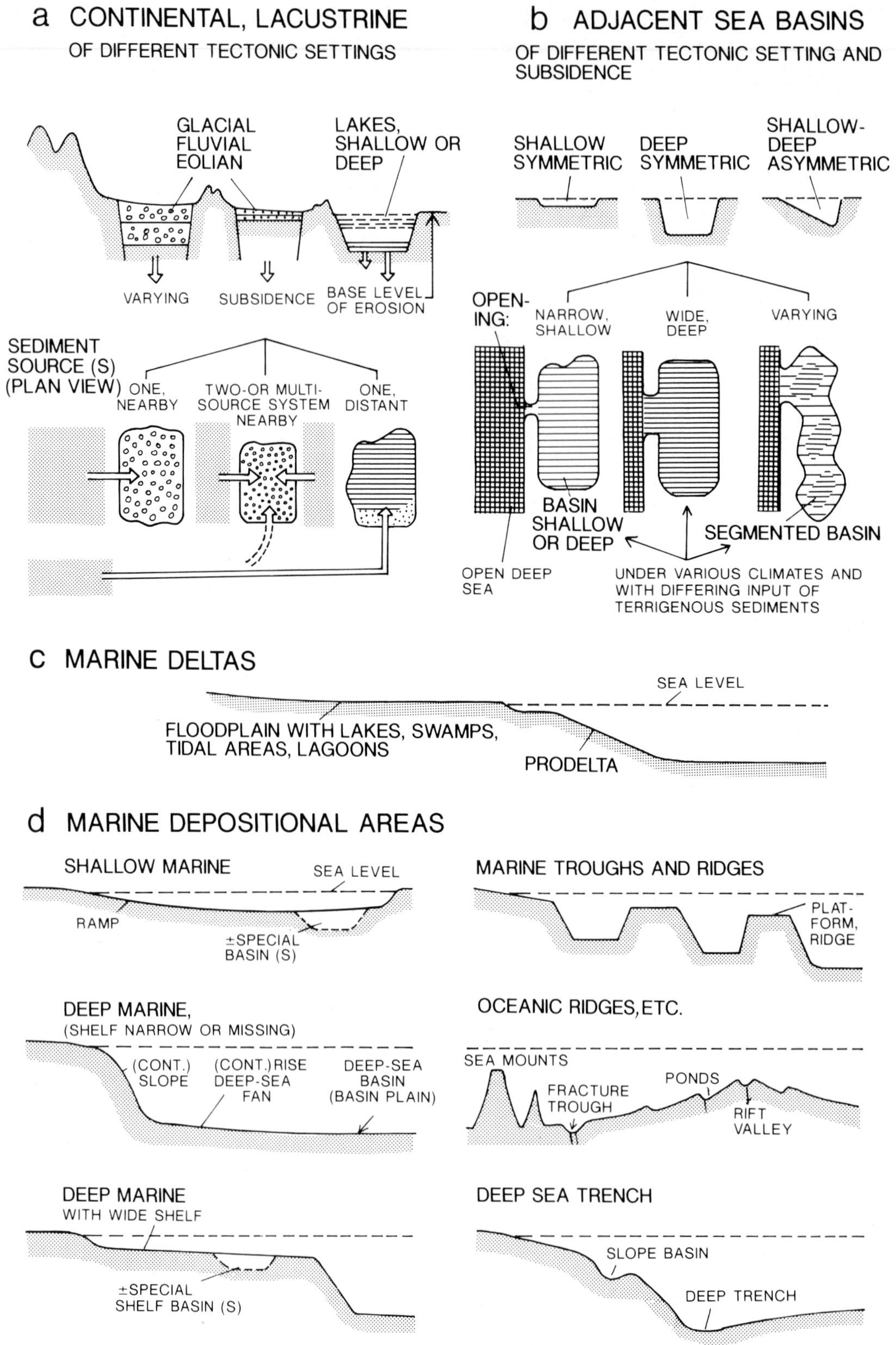

a CONTINENTAL, LACUSTRINE
OF DIFFERENT TECTONIC SETTINGS

b ADJACENT SEA BASINS
OF DIFFERENT TECTONIC SETTING AND SUBSIDENCE

GLACIAL FLUVIAL EOLIAN   LAKES, SHALLOW OR DEEP

SHALLOW SYMMETRIC   DEEP SYMMETRIC   SHALLOW-DEEP ASYMMETRIC

VARYING   SUBSIDENCE   BASE LEVEL OF EROSION

SEDIMENT SOURCE (S) (PLAN VIEW)   ONE, NEARBY   TWO-OR MULTI-SOURCE SYSTEM NEARBY   ONE, DISTANT

OPENING: NARROW, SHALLOW   WIDE, DEEP   VARYING

BASIN SHALLOW OR DEEP   SEGMENTED BASIN

OPEN DEEP SEA

UNDER VARIOUS CLIMATES AND WITH DIFFERING INPUT OF TERRIGENOUS SEDIMENTS

c MARINE DELTAS

SEA LEVEL

FLOODPLAIN WITH LAKES, SWAMPS, TIDAL AREAS, LAGOONS

PRODELTA

d MARINE DEPOSITIONAL AREAS

SHALLOW MARINE   SEA LEVEL

RAMP

±SPECIAL BASIN (S)

MARINE TROUGHS AND RIDGES

PLATFORM, RIDGE

DEEP MARINE, (SHELF NARROW OR MISSING)

(CONT.) SLOPE   (CONT.) RISE DEEP-SEA FAN   DEEP-SEA BASIN (BASIN PLAIN)

OCEANIC RIDGES, ETC.

SEA MOUNTS   PONDS   FRACTURE TROUGH   RIFT VALLEY

DEEP MARINE WITH WIDE SHELF

±SPECIAL SHELF BASIN (S)

DEEP SEA TRENCH

SLOPE BASIN

DEEP TRENCH

Fig. 1.6a-d. Overview of depositional environments, based primarily on basin morphology and peri-basin characteristics. All basins, particularly those on land (**a**) or adjacent to continents (**b** and **c**), are strongly affected by variations in terrigenous input under differing conditions of climate and relief. **d** Various marine basins

to compensate for subsidence in different tectonic settings (Fig. 1.6a). Under these circumstances, the river gradient and thus a more or less constant average net transport direction can be maintained for rather long time periods. A topographic depression, i.e., a syndepositional morphological basin (Fig. 1.4b) can only develop when fluvial transport lags behind basin subsidence. This clear relationship between gradient and transport direction is somewhat modified in the *glacial and eolian environments*. Subglacial abrasion often leads to erosional depressions, over-deepened valleys, and ice-filled troughs, which are later filled with water creating short-lived lakes. Similarly, eolian deflation can generate local depressions in the land surface which, if the groundwater table rises, may be transformed into salt pans. However, such erosional features are normally filled up again with sediments within a short time span. On the other hand, eolian sand can accumulate large "sand seas" reaching elevations well above the surrounding landscape. In addition, wind-blown sand can migrate into different directions, partially up-slope.

The influence of peri-basin morphology on *fluvial-lacustrine sedimentation* is described in Figure 1.6a. Terrigenous material entering the basin may come either from one or several nearby sources, or, solely or in addition, from a distant source. Consequently, deposition will be either texturally immature or markedly mature and display either a fairly uniform or complex composition. In addition, the climate in the source area(s) exerts a strong influence (Chap. 2.2.4). Where sediment accumulation cannot compensate for subsidence, long persisting, deepening lakes or shallow seas evolve (see below).

Marine deltas represent a transitional, highly variable depositional environment between continental and marine conditions (Fig. 1.6c). The subaerial part of such a delta is controlled by fluvial and possibly lacustrine processes, whereas its coastal and subaqueous regions are dominated by the hydrodynamic and chemical properties of the sea. Large terrigenous sediment supply causes prograding of the deltaic complex toward the sea; high sedimentation rates and subsidence enhanced by the sediment load enable the formation of thick, widely extended deltaic sequences. Marine delta complexes provide a particularly good example of depositional environments which are controlled predominantly by exogenic factors (Chap. 3.4).

Adjacent sea basins and *epicontinental seas* are connected with the open sea and therefore exchange basin water with normal ocean water (Fig. 1.6b). The extent of this water exchange and thus the salinity of the basin water strongly depend on the width and depth of the opening to the ocean. In humid regions, adjacent basins with a limited opening tend to develop brackish conditions, while arid basins frequently become more saline than normal sea water. Adjacent basins and epicontinental basins on continental crust are commonly shallow, but basins on oceanic or mixed crust may also be deep. All these basins may show either symmetric or asymmetric cross sections, and they may represent either simple morphological features or basins subdivided by shallow swells into several subbasins (segmented basins). In the latter case, markedly differing depositional subenvironments have to be taken into account. Most of these adjacent basins are still strongly influenced by the climate and relief of peri-basin land regions, which control the influx of terrigenous material from local sources. In addition, more distant provenances may contribute to the sediment fill. In summary, adjacent basins may exhibit a particularly great variety of facies (Chap. 4).

The *shallow sea and continental shelf sediments* are still considerably affected by processes operating in neighboring land regions, which generally provide sufficient material to keep these basins shallow. Strong waves, and surface and bottom currents usually tend to distribute the local influx of terrigenous sediment over large areas. Especially in shallow water, the high-energy, sediment-transporting systems prevent the deposition of fine-grained materials, partially including sands. Therefore, such areas often persist over long time periods without being filled up to sea level. This is also true for widely extended shallow-marine basins, as long as excess sediment volume (in relation to space provided by subsidence) can be stored in special depressions (Fig. 1.6d) or be swept into a neighboring deeper ocean basin. The margin of such basins is commonly characterized by a kind of ramp morphology.

Deeper marine basins are usually bordered by a shelf zone of varying width followed by a wide and normally gentle slope (continental slope, Fig. 1.6d). The foot of the slope in deep water (continental rise) is still gently inclined basinward; it is built up to a large extent by redeposited material derived directly from the slope (slope apron) or by sediments funnelled by submarine valleys and canyons into the deep sea (deep-sea fans). The terms *continen-*

tal slope and *continental rise* are commonly used to describe corresponding features of the present-day passive, Atlantic-type continental margins. These terms, however, imply a plate-tectonic interpretation.

Deep-sea basins or basin plains are the deepest parts of marine environments except for the special features of deep-sea trenches (see below).

Large volumes of terrigenous material can also be collected by the troughs in a *submarine horst and graben topography* bordering the continent. Similarly, deep sea trenches at the foot of relatively steep slopes and slope basins are sites of preferential sediment accumulation (Fig. 1d). Thick, ancient flysch sequences are mostly interpreted as depositions in such basins. Less important sediment accumulation features are small basins, called "ponds", which occur along oceanic ridges, and infillings of narrow troughs due to fracturing of the oceanic crust.

The thin, frequently incomplete sedimentary records on the tops of *submarine ridges, platforms, and seamounts* strongly contrast with all other marine sediments. These deposits are mostly biogenic or chemically precipitated and usually contain only very small proportions of terrigenous or volcaniclastic materials. Although such limited sediment accumulations can hardly be referred to as basin fills, they do constitute an important and diagnostically significant part of larger marine depositional environments.

The direct influence of tectonic basin evolution on sedimentary facies is only evident in areas, where tectonic movements are rapid and nonuniform, such as at the basin margins, or where sediment accumulation lags far behind subsidence. This situation is common in *continental rift and pull-apart basins* during their early stages of evolution, in subduction-related settings, in remnant and foreland basins, and in deep marine environments along oceanic ridges or transform faults far away from large land masses. These problems are further discussed in Chapter 12.

**Some General Trends
for Sediment Accumulation and Facies**

From the previous discussion one can draw some general, straightforward rules for the sediment accumulation and facies in various depositional environments:

- The influence of a terrigenous sediment source on the basin fill decreases from high-relief continental environments, to lowlands and shallow seas, to the deep sea.
- Similarly, the sedimentation rate tends to decrease from highland continental basins to the central parts of large oceanic basins.
- Chemical sediments of some extent commonly form in lowlands and special portions of adjacent shallow seas, but rarely in the other depositional environments.
- Basins with low sedimentation rates tend to accumulate sediments relatively rich in biogenic components. Such basins may persist for long time periods and are therefore often markedly affected by synsedimentary tectonic movements.
- The sedimentary facies of many basin fills do not reflect tectonic basin evolution and specific structural elements. Only in some basin types and/or during the most rapid phase of basin evolution do tectonic movements directly control sedimentary facies. However, the geometry of basin fills, sedimentation rates, and syn- and post-depositional deformations characterize the tectonic style and evolution of the basin considered.

Facies Architecture

The principal characteristics of the various depositional environments include features on different scales. These range from large-scale phenomena, relevant to the facies distribution in the total basin, to micro-scale properties which are studied in a single rock specimen. As Allen (1983) and Miall (1985) have pointed out, the sedimentary basin fill often displays a certain type of stratigraphic architecture, i.e., larger units are built up by a number of smaller, basic units. In single outcrops, generally only the smaller scale units can be observed, which are often not sufficiently diagnostic for the recognition of the true nature of the total basin fill or a large part of it.

The brief summaries for the common depositional environments presented in Chapters 2 through 6 are largely based on these principles. They preferentially show field and outcrop phenomena and how these fit into a larger scale facies model. Micro-scale features and processes are only described in special cases.

Part II
Depositional Systems and Facies Models

2 Continental Sediments

2.1 Glacial Deposits of Lowlands and Shallow Seas

2.1.1 Continental Glacial Deposits

Introduction

Glaciers produce and effect numerous types of both continental and marine sediments. From present-day glaciated and periglacial regions we know a great variety of features characteristic of glacial environments, including many minor structures and patterns which can be observed at or near the land surface. Therefore it is barely possible to summarize all these phenomena in a few simple facies models.

However, if we focus only on glacial sediments of a high geological preservation potential, we can neglect glacial landforms and sediments in highly elevated mountain ranges or periglacial phenomena such as patterned ground, gelifluction, and other gravity mass movements on subaerial slopes. Then the remaining environments to be considered are glaciated lowlands and predominantly shallow seas. In order to understand the sedimentary processes of these selected environments, a general knowledge of glacial phenomena is, of course, necessary, a topic which is treated in many special articles and textbooks (e.g., Eyles 1984; Eyles and Miall 1984; Edwards 1986; Anderson and Molnia 1989; Brodzikowski and van Loon 1991). The following simplified facies models therefore present some information only on those glacial sedimentary facies which have a relatively good chance of being preserved over long geological time.

Continental Ice Sheet Deposits

A continental ice sheet, advancing over areas of low relief, carries debris from its usually more elevated source area and takes up further material by basal erosion. If the base of the glacier is cold and frozen to the ground, more basal debris is generally accumulated than by "temperate" glaciers, which have a wet base sliding relatively easily over the substrate.

The debris left behind after the ice melts (Fig. 2.1a and b) is usually poorly sorted and unstratified and, if its origin is uncertain, described as *diamict* (when unlithified also as *diamicton* or *mixtum*). Such sediments typically contain boulders and large clasts floating in a finer grained matrix of silt and sand. Clays derived from weathered soils are absent (due to the cold climate) or play only a minor role. If the debris clearly originates from glacial action, the terms *till* (for unlithified material) and *tillite* (for lithified material) are common; recently the term *orthotill* (in contrast to *paratill* which has undergone mass movement or is sedimented through a column of water) was recommended (e.g., Shaw 1985). Debris accumulated at the base of glaciers is termed *basal till* or *lodgement till*, because part of the material is trapped by obstacles in the substrate. "Cold-based" glaciers can cause deformation of the underlying bedrock (glaciotectonic structures) and rip up large slabs or plates from the substrate by shearing. One can also observe thrust faults and small-scale folds within the lodgement till. Nearly horizontal or low-angle shear zones within the till cause some crude stratification and imbrication of coarser components. At wet-based ice margins, these phenomena are even better deve-

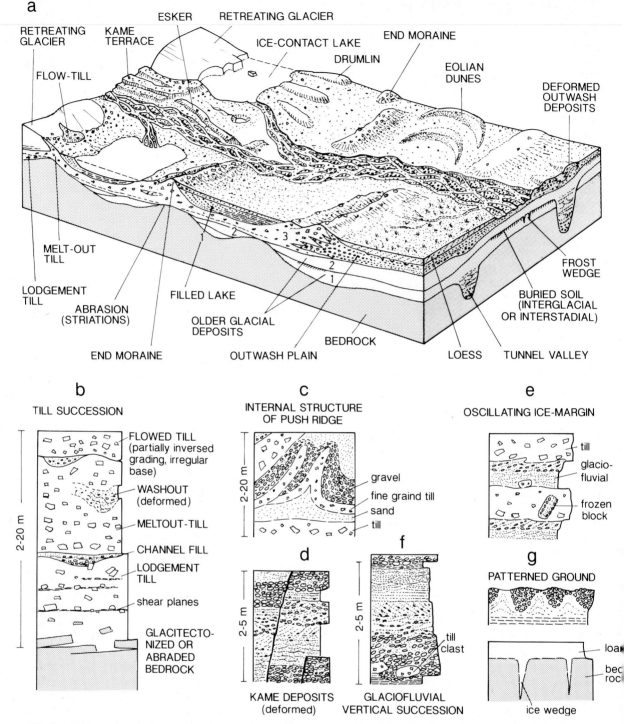

Fig. 2.1. a Landforms and deposits of continental ice sheets. *1, 2, 3*, succession of three glacial periods alternating with warmer interglacial or interstadial times allowing the growth of vegetation and the formation of soil. Note that older glacial deposits and interglacial soils are only partially preserved. **b** Typical sections in continental glacial deposits including periglacial patterned ground and ice wedges. c-g see text. (Based on several sources, e.g., Wagenbreth and Steiner 1982; Eyles 1984; Edwards 1986)

loped. The particles of the relatively thin basal debris layer often undergo intense abrasion (*ice-faceted clasts*) their long axes tending to become aligned parallel to the direction of ice flow. Continued lodgement of clasts on top of the substrate may lead to the formation of lenticular beds of coarse grained diamict.

Melting of glacier ice both at its surface and/or its base leads to the release and accumulation of debris in the form of *melt-out till* (Fig. 2.1a and b). This process usually occurs where the entire ice sheet, or at least its basal part, is stagnant. Hence melt-out till, if not overridden by a later ice advance, is not internally deformed. Its clasts may also be striated, but for the most part melt-out till is less affected by basal or internal abrasion and therefore tends to be coarser grained than lodgement till. Melt-out till may also contain frozen blocks of unlithified substrate which are incorporated into the basal zone of a glacier. Clast orientation is preferentially parallel to the direction of ice flow, and a considerable number of clasts dip up glacier. Due to running melt-water, this type of till also contains locally minor areas of stratified sorted beds. If these are deposited on top of till still holding some ice, they later become contorted or intensely faulted.

Melt-out till accumulating on top of the ice or at the slopes of moraines is frequently redeposited by mass flows (Fig. 2.1). Such *flowtills, flowed tills* or *flow (flowed) diamicts* are difficult to discriminate from debris flow deposits of other environments (Chap. 5.4.1) unless they contain striated clasts or ice-cemented blocks, or unless they show deformation characteristics of ice contact. As intercalations of a glacial sedimentary sequence, such flow tills as well as grain flows originating from glaciofluvial outwash are easier to identify. At standing ice margins, melt-out till and supraglacial outwash may form *end moraines* of considerable height and length. These moraines are frequently overridden and deformed by advancing ice (Fig. 2.1c).

Special morphological features developd in the subglacial environment are *drumlins, eskers,* and *tunnel valleys.* Whereas eskers and drumlins form narrow ridges or elongate hills on the land surface after the ice has melted, tunnel valleys may cut deeply into the underlying bedrock. Tunnel valleys probably result from subglacial meltwater erosion, where the channels are completely filled with water and therefore do not need a continuous gradient. In northern Germany such tunnel valleys locally reach depths of 400 to 500 m below the present sea level. They are cut into weakly indurated Neogene sandstones, have steep side walls, and are infilled by glacial outwash diamict and finer grained water-deposited sediments. They are known only from subsurface investigations. Such deep tunnel valleys have a very high preservation potential in the geologic record, although it may be difficult to identify them as such when only small portions of the former tunnel system are exposed by erosion.

Glaciofluvial Sediments

Transitional facies between till and glaciofluvial sediments are *glacial outwash*, primarily deposited either *supraglacially* on top of the ice *(e.g., kame terraces)*, or *englacially* in large ice cracks or tunnels at the base of the ice *(esker).* These sediments may show sharp lateral facies changes into other glacial deposits, because the meltwater streams were originally bounded by ice.

Glaciofluvial sediments in the *proglacial* region usually show the same characteristics as deposits of alluvial fans and less sinuous braided rivers (Chap. 2.2). Large valley glaciers and some continental ice sheets release enormous quantities of coarse- and fine-grained debris which is transported by meltwater into the proglacial lowlands. Along the ice margin, some of the material is deposited in outwash fans, and further away in *outwash plains (sandur, Fig. 2.1a)*. Outwash deposits of alpine glaciers are mostly rich in gravel, whereas outwash plains of continental ice sheets (for example, those of the Scandinavian Pleistocene ice sheet) consist predominantly of sand. This results not only from the general downstream decrease in grain size, but also from the lower proportion of coarse-grained material delivered by the continental ice mass. In any case, meltwater streams, with their high bedload, tend to accumulate the coarser grained portion of their load and may therefore generate comparatively thick glaciofluvial sequences. Distinctive features of the glacial environment, in contrast to normal fluvial conditions, are deformation structures formed by post-depositional melting of buried ice (Fig. 2.1 b and d), and the occurrence of "frozen blocks" of till or sand, which were carried and finally deposited in this state (Fig. 2.1e and f).

A special feature of proglacial meltwater sediments are the deposits of catastrophic floods, the so-called *jökulhlaups*, which were described from Iceland and other presently or formerly glaciated regions (e.g., Maizels 1989). These floods result form sudden drainage of ice-dammed lakes, from landslides on to glaciers, or from subglacial volcanic eruptions. Jökulhlaup flows may contain very high suspended sediment loads (hyperconcentrated flows with up to 35 % sediment concentration) consisting predominantly of silt, sand, and gravel. These flows predominantly form thick, more or less homogeneous, poorly structured massive beds.

As long as the outwash plains are devoid of vegetation, sand and silt sized particles are easily blown away by strong winds. The sand can be deposited nearby in the form of dunes; the silt is carried farther away and accumulates as *loess*, often in areas where tundra vegetation is then able to grow (Chap. 2.3.5). On surfaces where neither accumulation nor erosion takes place for some time, we often observe the development of *patterned ground, ice wedges*, and other phenomena typical of the periglacial zone (Fig. 2.1g). The accumulations of large rivers draining a glacial/periglacial region may provide good evidence of alternating colder and warmer climates (Fig. 2.2). Cold periods are in-

dicated by patterned ground (cryoturbation), ice wedges, and the deposition of loess; warmer periods enable the growth of more and more varied vegetation and thus the formation of soil. Paleosols are important indicators of former environmental conditions. Loess profiles with intercalated soils deposited far away from the ice margin in a region of little erosion (Fig. 2.2) may provide better evidence of the climatic history of a glaciated area than the more directly deposited glacial sediments themselves. If such loess is transformed by diagenesis into a redbed sequence (Chap. 6.3), its true origin may be difficult to recognize.

Proglacial Lacustrine Deposits

In glacial or proglacial environments, lakes (and their rapid filling by clastic sediments) are very common. They are generated by the erosion of bedrock in valleys or flat foreland areas by ice, by the buildup of moraines or glacial outwash as dams, by the melting of buried ice or, in the case of large ice caps and inland ice, by the depression of the continental crust due to isostasy.

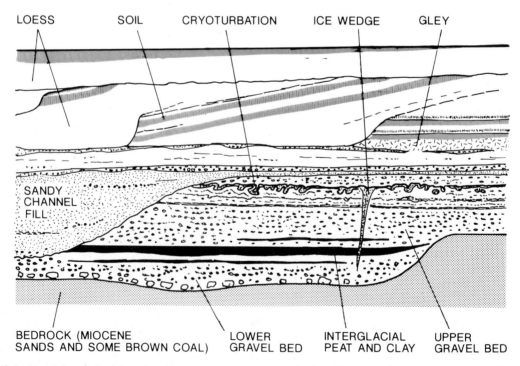

LOESS   SOIL   CRYOTURBATION   ICE WEDGE   GLEY

SANDY CHANNEL FILL

BEDROCK (MIOCENE   LOWER   INTERGLACIAL   UPPER
SANDS AND SOME BROWN COAL)   GRAVEL BED   PEAT AND CLAY   GRAVEL BED

Fig. 2.2. Example of fluvial sediments and loess with intercalated peat layers and soils indicating periods of warmer climate. Apart from loess, periglacial conditions are also shown by a cryoturbated horizon and ice wedging.

Coal pit in Pleistocene/Miocene deposits of the Lower Rhine Valley, northern Ville, near Cologne, Germany. (After Woldstedt and Duphorn 1974)

We are not dealing here with very large glacially formed lakes like the Baltic Sea or the Great Lakes in North America, which persist for a long period after the retreat of the inland ice, but only with smaller glacier-fed lakes. These latter ones receive abundant meltwater, are filled within a short time period, and show typical *glaciolacustrine deposits* (Smith and Ashley 1985). They may be bordered by a calving glacier *(ice-contact lakes)* or be located many kilometers downstream from a glacier *(noncontact glacier-fed lakes).*

Ice-contact lakes usually are small and highly variable, including changing water levels caused by repeated filling and draining. They are characterized by subaqueous outwash deposits emerging from the mouth of an ice tunnel (Fig. 2.3). The rapidly deposited coarse grained material near the tunnel mouth (containing boulders and gravel) grades laterally into sand and silt. It may rest on lodgement till and be interbedded with ablation till, flowtill, and finer grained lake sediments. Small-scale mass flow phenomena and postdepositional collapse structures due to the melting of buried ice are common. The best evidence of ice-contact lakes are *dropstones* derived from floating ice.

Glacier-fed, more distal lakes (Fig. 2.4) are better known and provide more stable depositional conditions than ice-contact lakes, which are often overridden by advancing ice. Meltwater streams drop their coarse grained bed

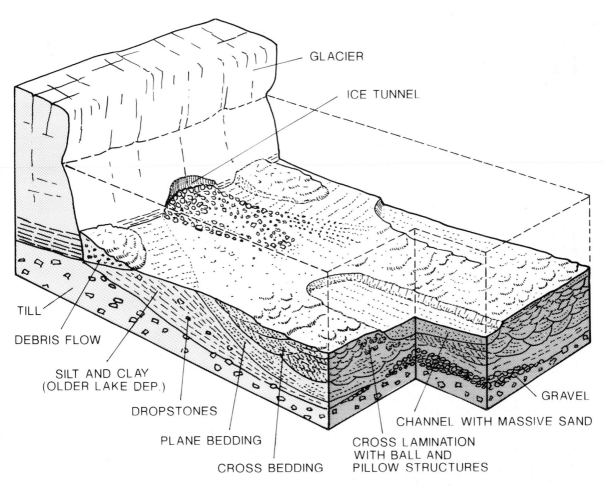

Fig. 2.3. Reconstruction of subaqueous outwash fan in ice-contact lake. Clast-supported boulder gravel at the mouth of the ice tunnel passes laterally and vertically into horizontally laminated, planar and trough cross-bedded sand, and finally into climbing ripple-drift lamination. The fan may be cut by steep-sided "channels" caused by mass flows. The lost material can be replaced by massive or stratified sand. Due to rapid deposition some sand layers are loosely packed and therefore affected by dewatering and/or load casting. (After Shaw 1985)

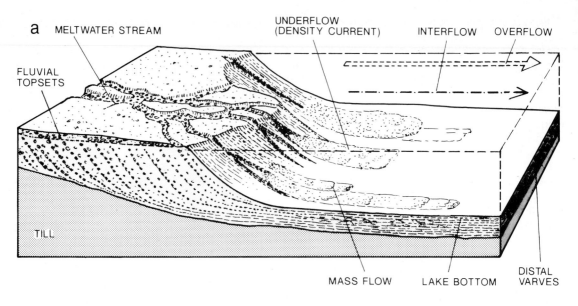

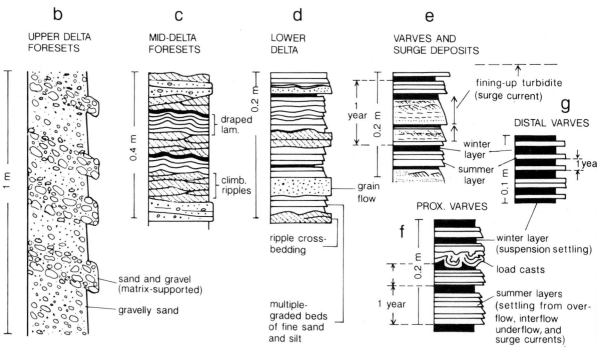

Fig. 2.4. a Sedimentation in a proglacial (non-ice contact), meltwater-fed lake. Note the marked facies change from the upper to the lower delta foresets (**b, c, d**), as well as the different types of annual varves (**e, f, g**) which are partially obscured by event deposits, i.e., distinct surge deposits (lake turbidites) and successions of very thin, graded beds of silt and fine sand. (After Smith and Ashley 1985)

load in the form of a classical lake delta, which can be built out rather rapidly into the lake. The sediment is distributed over the delta front by avalanching, fallout from suspension, river water underflowing the standing lake water, and gravity mass movements (Fig. 2.4a). Upper-delta foresets, if they have not been affected by wave action, show steeply inclined coarse-grained beds (Fig. 2.4b). Their grain size may alternate from gravel-dominated to sandy layers, due to the fluctuating competence of the meltwater stream. Individual foresets reflect special storm events during the summer time rather than annual meltwater conditions. *Mid-delta foresets* are less inclined and finer grained than the *upper-delta foresets*. Caused by high fallout of fine sand from suspension, they frequently show climbing ripple or ripple drift sequences and draped lamination (Fig. 2.4c).

The *lower-delta foresets* consist of rhythmic ripple cross-laminated fine sands and graded beds of fine sand and coarse silt (Fig. 2.4d). These layers are mainly formed by undercurrents or surging currents during the summer time. They are often overlain by somewhat darker, millimeter-thick layers of fine silt and clay deposited during the long winter period. These proximal and irregular *varves* grade into thinner (<1 cm) and better developed *rhythmites* in the distal lake basin (Fig. 2.4). The coarser laminae deposited from summer underflows frequently consist of several sublaminae, indicating a series of events during one summer season (Fig. 2.4f). The rest of the year is represented by a thin, fining upward silt-clay lamina. Whereas these rhythmites may contain a high proportion of randomly occurring, slump-generated, very thin, silty or sandy *turbidites* (Fig. 2.4e), there are also rhythmites which are hardly affected by such depositional events.

These varves are thinner, more regularly bedded, and devoid of any current structures (Fig. 2.4g). Their material is delivered by surface currents (overflows) or by currents crossing the lake at middepth (interflows). The *varves* are produced mainly by vertical settling of fine grained particles and therefore distinctly reflect summer (light laminae) and winter (dark laminae) conditions. The regular bedding of these types of lake bottom sediments may be disturbed by different processes: mass flows, load casting (represented, for example, by small-scale ball-and-pillow structures), and even by burrowing organisms (bioturbation). Overriding ice can promote compaction of the clayey bottomsets and deform some of the lake deposits.

2.1.2 Glaciomarine Sediments

Glaciomarine sediments or sedimentary sequences contain facies types which indicate the direct influence of glaciers (Molnia 1983; Eyles et al. 1985; Grobe 1987; Anderson and Molnia 1989). However, this clear relationship cannot be described by a simple facies model. The net growth of ice on the continents and the advance of ice toward the sea during a cold period is accompanied by sea level fall and vice versa. Furthermore, loading of the crust by thick continental ice sheets causes isostatic subsidence and, after melting of the ice, uplift. However, both sea level fluctuation and isostatic adjustment of the crust do not act synchronously, the latter lagging behind the sea level changes.

If, for example, in a nonglaciated area the global sea level has reached its eustatic minimum level, a glaciated shelf sea may still subside due to the isostatic effect of the ice-load and therefore experience a transgressive sea. When the global eustatic sea level has again reached its maximum, the coaststline of a formerly glaciated area may still continue to rise, causing a regression of the sea.

In short, glaciomarine sedimentary sequences are strongly and complexly affected by migrating coastlines over long distances.

Simultaneously, the continental and lacustrine glacial associations can migrate back and forth a shallow sea and alternate with glaciomarine and normal marine facies types. During low sea level stands, the glaciers may advance over the emerging bottom of a shelf sea and deposit a till-outwash-periglacial-lacustrine facies association on top of marine beds (Fig. 2.6a). The subsequent transgressing sea, in conjunction with a warmer climate, will in general cause a drastic retreat of the ice margin, drown the tills and their associated continental and lacustrine sediments, and enable the deposition of subaqueous glaciomarine sediments. As long as the sea is in contact with the ice margin, subglacial meltwaters can deposit submarine ice-contact fans (Fig. 2.3a and 2.5a) and release their suspended loads as plumes into the sea. Floating ice can drop its solid material over large areas. Its fine and coarse grained rain of particles forms *massive or stratified diamicts* (Fig. 2.5b, c, and d), which locally may be affected by bottom currents, reworked by storm waves, or redeposited as grain flows, mud flows and, in deeper water, as turbidity currents (Fig. 2.5e, f, and g). This facies association is highly variable, depending on the load of the ice, the configuration and hydraulic regime of the marginal sea, and many other factors. The most distinctive features of glaciomarine sediments are dropstones in stratified marine muds which also contain some fossils and frequently display bioturbation.

In some modern and ancient glaciomarine environments, a special carbonate mineral, i.e., calcium carbonate hexahydrate (ikaite) or pseudomorphs after ikaite were found, which

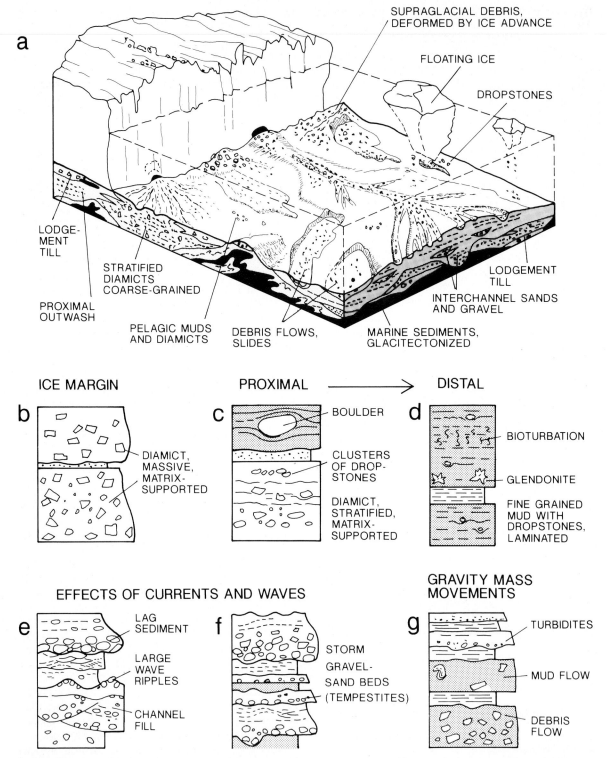

Fig. 2.5. Glaciomarine ice-contact environment (**a**) and the resulting sediment types (**b** through **g**). The advancing ice-margin can disturb the underlying, relatively soft marine sediments and release most of its debris load via tunnels as submarine outwash fans. Suspended material is widely distributed and deposited as mud. Supraglacial debris drops along the ice-margin or is carried by icebergs into the sea. Proximal massive submarine diamicts (**b**) grade seaward into stratified diamicts (**c**) and rhythmic laminated muds with dropstones (**d**). Reworking and redeposition of these facies types leads to lag deposits and large ripples in coarse-grained sands (**e**), proximal tempestites (**f**), or different types of gravity mass movements (**g**). (Based on different sources, including Eyles and Miall 1984)

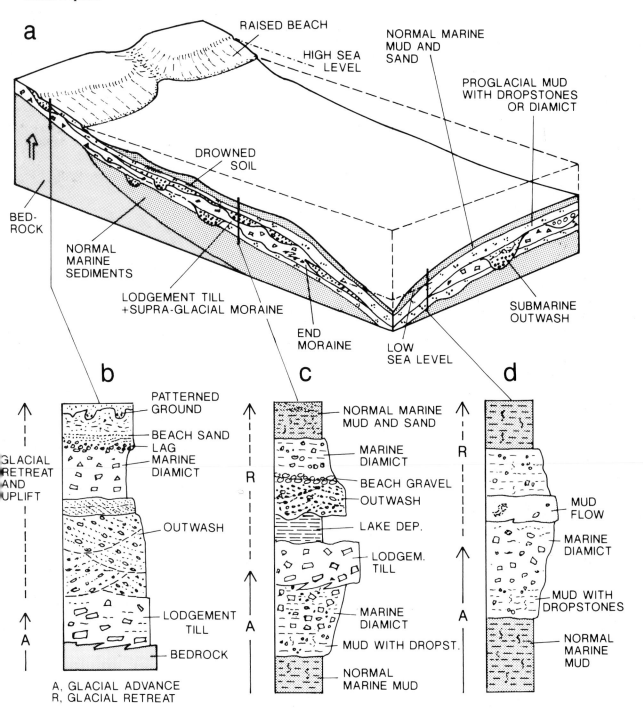

Fig. 2.6. a Glacial deposits on continental shelf, originating from one advance and subsequent retreat of ice, accompanied by lowering and raising sea level, as well as some isostatic subsidence and following uplift due to the growing and waning ice load. **b, c, d** Simplified sequences of one glacial cycle: b Asymmetric cycle along the coast with raised beach, glaciomarine diamict representing only a short intervall. **c** Mid-shelf cycle showing several marine and non-marine subsections. d Fully marine symmetric cycle where the maximum advance of ice is represented by relatively coarse-grained diamict. (Based on several sources, e.g., Boulton and Deynoux 1981; Eyles 1984; Edwards 1986)

are referred to as *glendonites* (Suess et al. 1982; Shearman and Smith 1985; Kemper 1987). This mineral occurs in morgenstern-like or stellate aggregates, several centimeters in size, and indicates sub-zero temperature conditions in the sediments at or somewhat below the sea floor, where these minerals are formed from organic-rich interstitial solutions.

During periods in which no ice reaches the coast of the sea, outwash streams of nearby glaciers deposit their bed load as *proglacial fans and deltas* along the coast. Their suspended load can be distributed over large areas of the sea, especially if flocculation of clay particles and their settling as aggregates is delayed, which in general is caused by mixing of the inflowing fresh water with sea water. In this case, however, the marine sediments carry no direct evidence of glacial influence.

Thus, repeated advance and retreat of the ice sheets across shelf seas can in principle lead to a cyclic sequence of alternating (continental) glacial, glaciomarine and normal marine sediments (Fig. 2.6b, c, and d). Due to glacial erosion and marine current activity, however, often only part of such a complete succession will be preserved.

In regions of very cold climate and sufficient ice accumulation (as today in the Antarctic), large areas of the sea can be covered with comparatively "clean" thick *shelf ice*. Since most of the debris in these floating ice sheets is located near their base and dropped below the ice due to subglacial ablation, the ice margins and floating icebergs are poor in debris and therefore leave behind little evidence of ice action. Neogene and Quaternary proglacial marine sediments in this environment are often rich in diatoms, bioturbated, and contain few dropstones. On the other hand, large and "dirty" valley glaciers reaching the sea can provide great amounts of siliciclastic debris and accumulate glaciomarine sequences several kilometers in thickness in subsiding areas, as for example in the Neogene in the Gulf of Alaska (Powell and Molnia 1989) or on the Barents shelf (Vorren et al. 1989). Icebergs originating from such dirty glaciers frequently carry their debris load over long distances, and hence are the source of dropstones found in pelagic and hemipelagic sediments in large ocean basins far away from glaciated regions. In addition, productivity of biogenic carbonate and opal may significantly vary from glacial to interglacial periods (Henrich 1990).

Final Remarks

Glacial sediments are extremely variable both in texture and composition, because they are deposited in greatly differing subenvironments. Their material may be derived from many sources, some of them extrabasinal (thereby delivering *erratic blocks*), others intrabasinal. Distinctive features of glacial sediments are characteristic deformation structures and dropstones in lacustrine and marine deposits, as well as relics of periglacial structures.

Tills and tillites may be confused with mud flows or debris flows of non-glacial origin, if certain details of their structure or their facies associations with other sediment types are not sufficiently evaluated.

2.2 Fluvial Sediments, Alluvial Fans, and Fan Deltas

2.2.1 Bed Forms, Sedimentary Structures, and Facies Elements

Introduction

For many past decades, fluvial sediments did not receive much attention from sedimentologists, because their origin and interpretation appeared essentially clear in comparison to the deposits of many other environments. Also, fluvial sediments do not form a very large part of the depositional record. In the last 10 to 20 years, however, a considerable number of geologists and sedimentologists have viewed both recent and ancient fluvial sediments as interesting objects of study. This new impetus was initiated not only by academic scientists, but also by many workers interested in fluvial deposits as reservoirs for hydrocarbons and groundwater.

The body of new, detailed knowledge gained in this field has been summarized in several specialized volumes (e.g., Miall 1978, 1980, and 1981; Scholle and Spearing 1982; Collinson and Lewin 1983; Galloway and Hobday 1983; Rust and Koster 1984; Flores et al. 1985; Collinson 1986a; Ethridge et al. 1987). In the context of this book, only a brief and therefore incomplete overview can be included. Mineralogical and petrographic aspects of fluvial deposits, including their provenance from various rock types, can also not be treated in any detail (see, e.g., Davis 1983; Pettijohn et al. 1987; Füchtbauer 1988; Bosellini et al. 1989).

Minor and Medium-Sized Bed Forms and Internal Sedimentary Structures

Based on direct observations in present-day streams and on experimental studies in flumes, the relationship between hydrodynamic regime, erosion, sediment transport, and accumulation is fairly well established. This allows the prediction of the behavior of particles of different size, density, and shape, as well as the formation of minor bed forms, such as small current ripples and larger megaripples or dunes, including their internal sedimentary structures. Some of this knowledge is summarized in Fig. 2.7; a further treatment of this topic is found in several text books (e.g., Blatt et al. 1980; Leeder 1982; Allen 1982; Collinson and Lewin 1983: Davis 1983; Collinson 1986a).

With the aid of the criteria documented in Fig. 2.7, one can, for example, distinguish minor bed forms and internal structures generated in a *lower flow regime*. Gravel is transported only if the current reaches a certain critical velocity (around 100 cm/s). In slower currents, gravel forms a *lag deposit,* which protects the underlying sand from erosion. Stronger currents with velocities higher than the critical value for gravel can

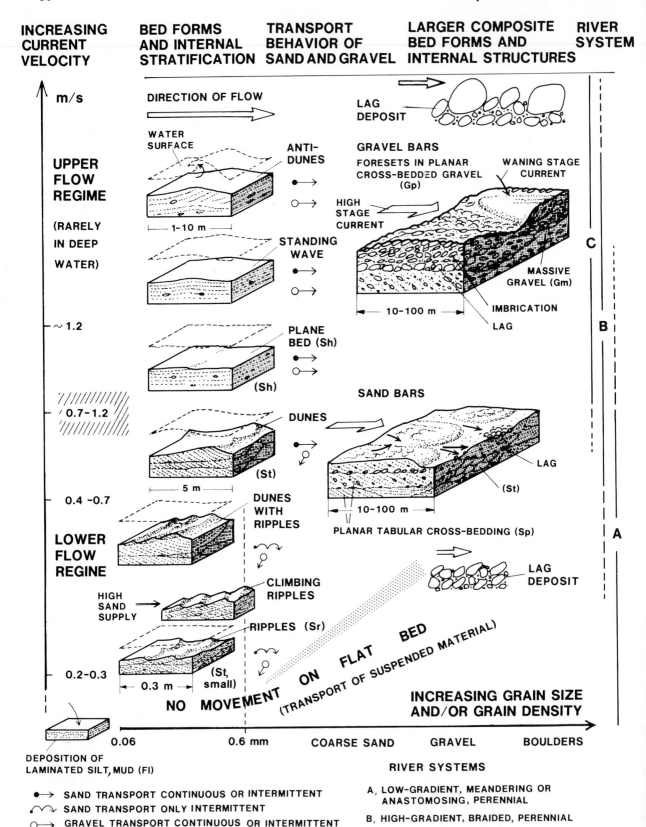

INCREASING CURRENT VELOCITY   **BED FORMS AND INTERNAL STRATIFICATION**   **TRANSPORT BEHAVIOR OF SAND AND GRAVEL**   **LARGER COMPOSITE BED FORMS AND INTERNAL STRUCTURES**   **RIVER SYSTEM**

m/s

DIRECTION OF FLOW

LAG DEPOSIT

WATER SURFACE

ANTI-DUNES

GRAVEL BARS

UPPER FLOW REGIME

(RARELY IN DEEP WATER)

1-10 m

STANDING WAVE

FORESETS IN PLANAR CROSS-BEDDED GRAVEL (Gp)

WANING STAGE CURRENT

HIGH STAGE CURRENT

MASSIVE GRAVEL (Gm)

IMBRICATION

LAG

C

10-100 m

~ 1.2

PLANE BED (Sh)

(Sh)

B

0.7-1.2

DUNES

SAND BARS

(St)

5 m

0.4 -0.7

DUNES WITH RIPPLES

LAG

(St)

10-100 m

PLANAR TABULAR CROSS-BEDDING (Sp)

A

LOWER FLOW REGIME

HIGH SAND SUPPLY

CLIMBING RIPPLES

LAG DEPOSIT

RIPPLES (Sr)

(St, small)

0.2-0.3

0.3 m

NO MOVEMENT ON FLAT BED
(TRANSPORT OF SUSPENDED MATERIAL)

INCREASING GRAIN SIZE AND/OR GRAIN DENSITY

0.06   0.6 mm   COARSE SAND   GRAVEL   BOULDERS

DEPOSITION OF LAMINATED SILT, MUD (Fl)

RIVER SYSTEMS

•→   SAND TRANSPORT CONTINUOUS OR INTERMITTENT

〰→   SAND TRANSPORT ONLY INTERMITTENT

○→   GRAVEL TRANSPORT CONTINUOUS OR INTERMITTENT

↻   GRAVEL ROLLS INTO SCOUR

A, LOW-GRADIENT, MEANDERING OR ANASTOMOSING, PERENNIAL

B, HIGH-GRADIENT, BRAIDED, PERENNIAL

C, HIGH-GRADIENT, EPHEMERAL

transport sand and gravel simultaneously. They form either *gravelly sands* or, if most of the sand has been sorted out due to faster sand transport, *clast-supported gravel beds*. Furthermore, there may be vertical sediment aggradation, for example on top of rippled beds, or downstream migration of the river bed load (Fig. 2.7). Rapid vertical aggradation can occur only if sediment input into the system is high. this may lead to *climbing ripples and ripple drift cross-stratification. Sand waves and antidunes* form in *the upper flow* regime. They are more or less destructive features and are associated with sediment loss rather than with aggradation of fluvial material. Sand waves and antidunes are rarely preserved because they are frequently truncated or completely reworked by subsequent "normal" current action.

Coarse particles such as gravel increase in *roundness* downstream if they originate from angular rock debris and not from older sedimentary rocks with pre-rounded material. They often form characteristic *fabrics* (Fig. 2.7) in which flat pieces dip upstream, with their long axes perpendicular to the flow direction. Less common is the orientation of gravel and pebbles with their long axes parallel to flow. This type of fabric appears to be restricted to high energy flow conditions.

Both criteria, the downstream growth of cross-bedding and the imbrication of gravel, indicate the *current direction* which was responsible for the formation of the bed. Variations of these directions from a general trend are one of the means of discriminating between different fluvial systems in the ancient record (see below).

Although much is known about the physics of sediment transport in rivers and the flow conditions under which minor sedimentary structures are formed (see, e.g., Allen 1982;

Ashley 1990), the *large-scale, three-dimensional depositional structures* of fluvial systems are difficult to reconstruct. However, the latter are the most important diagnostic features for discriminating between various fluvial systems in the ancient record. Many of the common minor bed forms and corresponding internal structures occur in all fluvial environments, but in varying proportions. This is so, because all river systems are characterized by frequently changing water stages and a wide range of current velocities. For example, small sand ripples of various shape, larger dunes, various types of cross-bedding, and flat lamination with or without gravel are observed in nearly all river systems. Consequently, these small-scale structures alone have only limited diagnostic value. Rather, it is the vertical or lateral succession of such structures, or the proportion of specific structures in the total sediment body, which can be used to identify the overall depositional environment.

To some extent, these limitations also control the *grain size distribution* in river deposits. Both small and large streams may, for example, develop a braided river system, whether they are sand or gravel-dominated. In both cases, the transport capacity of channel flow is intermittently sufficient to move gravel as bed load. Hence, the proportion of gravel in such fluvial deposits reflects rock types, relief, and weathering conditions in the source area rather than in the depositional environment. On the other hand, the grain size distribution available in the source area, including jointed hard rocks, has only limited influence on the grain size distribution of river deposits (Walger 1964; Ibbeken 1983). Even after a short transport distance, river sediments show a characteristic frequency curve resembling a log-normal distribution, while the material in the source area may significantly deviate from this distribution.

Fig. 2.7. Relationship between current velocity (hydraulic regime), grain size, minor and medium-sized bed forms, and internal sedimentary structures of fluvial deposits (overview). (Drawn after different sources, e.g., Harms and Fahnestock 1965; Walker and Cant 1984; Collinson 1986a). The boundary between lower and upper flow regime (in terms of mean current velocity) strongly depends on water depth. The larger bed forms and their internal structures result from fluctuating water stages and current velocities, and therefore cannot be attributed to certain flow conditions. Small ripple forms develop only in fine to medium sands (< 0.6 mm grain diameter). For combined,

more or less continuous transport of sand and gravel, current velocities higher than 70 to 120 cm/s (upper flow regime) are needed. The resulting beds are either horizontally stratified sands with some matrix-supported gravel, or planar cross-bedded gravelly sands (for symbols see Table 4.1). Gravel exposed to currents which are only capable of eroding and transporting sand roll into developing scours. If all sandy material is eroded, gravel may form lag deposits, which protect underlying finer material from further reworking. In different river systems (*A* through *C*), certain flow conditions and bed forms prevail

In contrast to the transport of individual particles and the formation of small-scale sedimentary structures, which can be simulated in laboratory flumes, medium-sized *sand bars or gravel bars* can be studied only in present-day rivers. They move during floods, have different shapes (transverse, longitudinal, linguoid bars), and are modified and reorganized by repeated changes between lower and higher water stages in the channel system. As a result, the bars frequently show composite internal structures (Fig. 2.7), which may be associated with both upper and lower flow conditions. They may contain lag deposits, gravel beds and sand beds.

The terminology used in Figs. 2.9 and 2.10 was proposed by Miall (1978) as well as Rust and Koster (1984). The most important lithofacies types are listed in Table 2.1.

Clast or framework-supported gravel (Gm, Gt, and partially Gp) originates from high energy flow transporting and accumulating coarse bedload and keeping sand and finer material in suspension. The remaining pore space between the gravel is usually infiltrated later by sand, when the flow velocity has decreased. The *matrix-supported gravel* results either from debris flows (Gms), or from the simultaneous transport of sand and gravel in a river with a high flow regime (Gp).

Table 2.1. Small-scale facies or bed types in fluvial sediments

Facies code	Description	Interpretation
Gms	Massive matrix (sand and mud) supported gravel	Debris flow deposit
Gm	Massive or crudely bedded gravel	Longitudinal bars, lag deposits, sieve deposits
Gt	Trough cross-bedded, clast-supported gravel	Minor channel fills
Gp	Planar cross-bedded gravel and/or matrix-supported gravel	Linguid bars or deltaic growth
Sh	Horizontally stratified sand	Upper flow regime
St	Trough cross-stratified sand	Lower flow regime
Sp	Planar cross-stratified sand	Transverse bars, lower flow regime
Sr	Ripple marks and small-scale cross stratification [a]	Lower flow regime
Fm	Massive, fine sandy mud or mud	Overbank or drape deposits
Fl	Laminated or cross-laminated fine sand, silt or mud	Overbank or waning flood deposits
P	Pedogenic concretions (carbonate)	Soil formation

[a] Description somewhat changed by the author

Large-Scale Phenomena and Types of Fluvial Systems

The most distinctive features of fluvial systems are large-scale phenomena, such as the size and geometry of channels, their sinuosity and ability to migrate, associated compound bars, and the occurrence of more or less extended overbank deposits, natural levees, etc. The factors controlling these large-scale structures are more complex than those mentioned for the small-scale and medium-scale phenomena. In the case of meandering channels, it is a general rule that the wave length and width of the meanders grow with increasing river discharge and slope of the river valley or fluvial plain. Braided stream systems with channels of low sinuosity tend to form in environments of large sediment supply and relatively steep gradients.

In present-day fluvial environments, the channel system and its sinuosity are the most striking, diagnostic features. With the aid of these criteria, one can distinguish between several types of fluvial systems, although there are no sharp boundaries between these depositional environments (Fig. 2.8):

- *Alluvial fans and fan deltas* (bed-load channels).
- *Braided rivers and braidplains* (bed-load-channels).
- *Meandering river systems* (mixed-load and suspended-load channels).
- *Anastomosed river systems* (predominantly suspended-load channels).

Another classification (already included above) is based on the mode of sediment transport in the channel systems (e.g., Schumm 1981; Galloway 1981):

1. **Bed-load channels** typically have steep gradients, high width/depth ratios (>40), and channel patterns of low sinuosity. These channels tend to migrate laterally, and channel fills are coarse-grained and contain little suspended-load material. The total alluvium of bedload systems consists predominantly of channel and channel-flank deposits (Fig. 2.9). Silty and muddy floodplain deposits play a subordinate role.

2. **Mixed-load fluvial systems** preserve a higher percentage of floodplain deposits which consist of silts, muds, and (locally) backswamp carbonaceous muds and clays. Their channels are more stable and have a lower width/depth ratio than those of the bed-load system; they are mainly filled with sand and minor proportions of silt and clay. These channel fills are flanked by levee sands and silts and crevasse-splay sands (Fig. 2.9), thus raising the overall sand content to about 20 to 40 %.

3. **Suspended-load fluvial systems** are characterized by high-sinuosity single channels of great stability and low width/depth ratio (<10). Their gradient and stream power are usually low. Their channel fills contain a high proportion of silts and muds. Silty or muddy levee deposits are well developed, and flood basin deposits consist predominantly of overbank fines or, in humid climates, of backswamp and lacustrine muds.

Some authors also separate *gravel-dominated systems* from *sandy fluvial systems* (e.g., Walker and Cant 1984), because grain size is an important indicator of relief (that is, tectonic activity) as well as climate in the source area. High relief and arid or periglacial/paraglacial conditions favor the production of coarse-grained materials and the primary input into the different fluvial transport systems. In addition, the resistivity of source rocks exposed to weathering and erosion processes also plays an important role (Chap. 9).

In ancient fluvial sediments, it is usually very difficult or impossible to reconstruct the former paleo-channel systems and their sinuosity, unless there are many very large outcrops. If such exposures are missing, it is of some use to analyse the proportions of bed-load and suspended-load in the channel fills and in the total flood basin to determine the mode of sediment transport. However, this has its limitations, and consequently many workers have attempted to improve the facies analysis of ancient fluvial environments through other means. The simplest approach is to subdivide the sedimentary fill of a fluvial basin into channel sediments and finer-grained overbank or floodplain deposits. This can easily be done in systems where the difference in grain sizes between these two sub-environments is sufficiently large. A general trend in the relationship between these two principal facies types is shown in Fig. 2.8. Thus, for example different types of sandstone reservoir geometries can be distinguished (Fig. 2.9a). One should use caution. however, when deducing subsurface channel distribution and geometry from the present-day, active channel type (Fig. 2.9b).

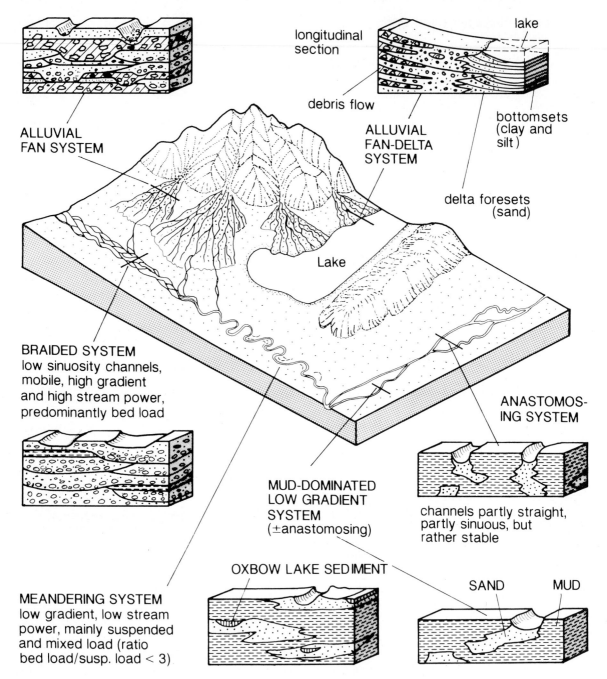

Fig. 2.8. Principal types of fluvial systems and generalized characteristics of their cross sections (vertical scale exaggerated)

Basic Facies Elements and Architecture of Fluvial Systems

In order to describe and reconstruct modern and ancient fluvial systems in more detail than that above, Allen (1983) and Miall (1985) have proposed subdividing fluvial deposits into eight *basic architectural elements* (Fig. 2.10). These elements differ, however, in dimension and rank as structural units forming the total fluvial sediment body. Single elements are bounded by bed contacts of different order, the most prominent ones being erosion surfaces, such as those at the bases of channels.

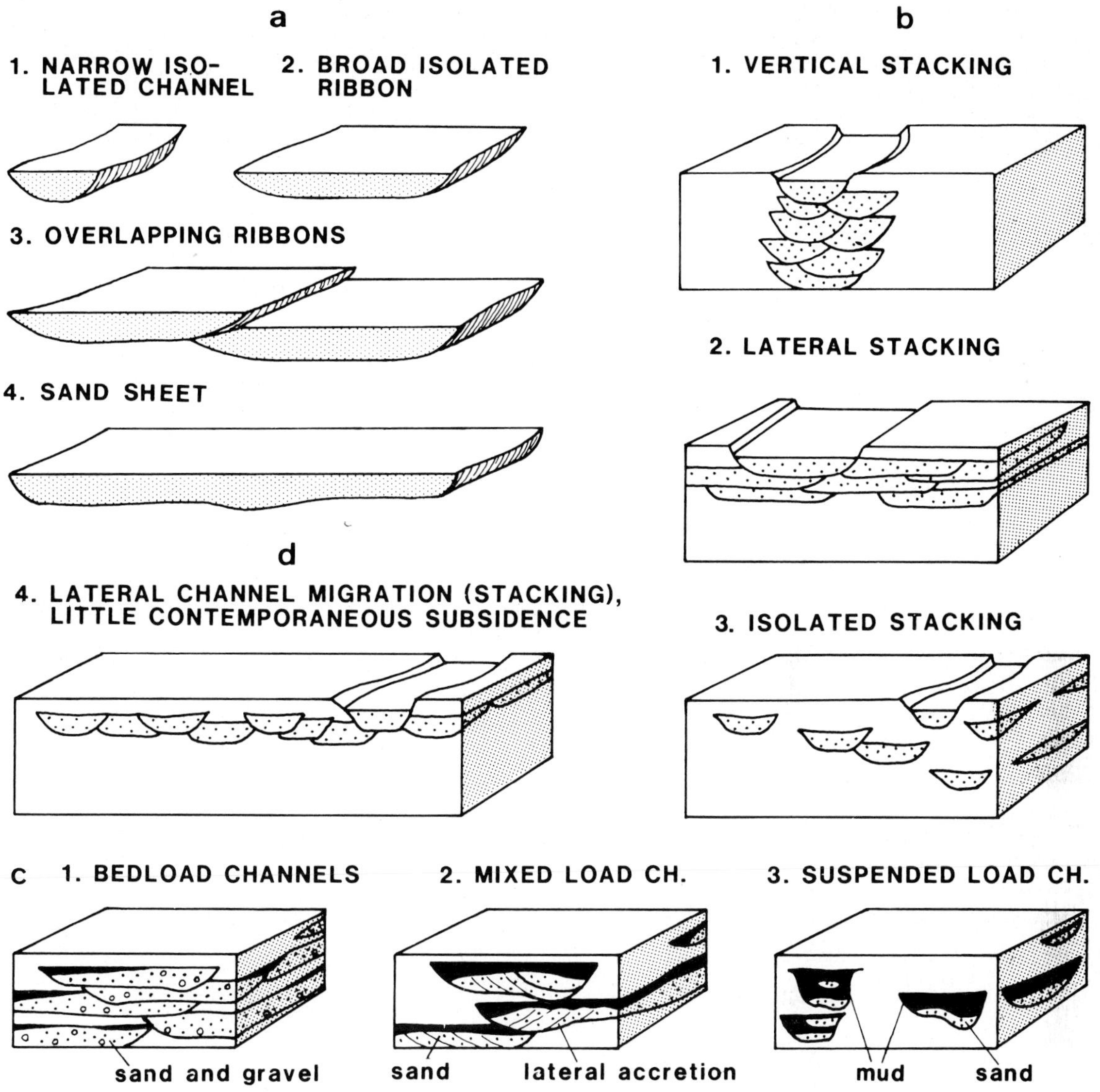

a

1. NARROW ISO-LATED CHANNEL   **2. BROAD ISOLATED RIBBON**

3. OVERLAPPING RIBBONS

4. SAND SHEET

b

1. VERTICAL STACKING

2. LATERAL STACKING

3. ISOLATED STACKING

d

4. LATERAL CHANNEL MIGRATION (STACKING), LITTLE CONTEMPORANEOUS SUBSIDENCE

c   **1. BEDLOAD CHANNELS**   **2. MIXED LOAD CH.**   **3. SUSPENDED LOAD CH.**

sand and gravel   sand   lateral accretion   mud   sand

Fig. 2.9a-c. Principal types of sand and sandstone reservoir geometries generated by channel fills in fluvial systems. **a** Single sandstone bodies. **b** Different types of stacked channel sands. Note that the active channels visible at the surface (e.g., cases *b1*, *b3*, and *b4*, showing hardly any difference) do not necessarily indicate the nature of the buried channel complex; *b1* is associated with rapid subsidence, *b4* with little subsidence. (Ethridge 1985; Miall 1985, modified). **c** Channel fills of bed-load, mixed-load, and suspended-load rivers. (Based on Galloway 1985)

Large elements may be composed of several smaller elements. For example, channel fills can consist of up to five subordinate units. It is, however, the smaller elements which we can recognize in outcrops of limited size. Therefore, an approach from smaller to larger architectural elements is recommended for the study of such compound structures seen in many fluvial sediments. The physical processes generating such compound deposits are not discussed in this approach.

Fluvial deposits may also alternate with eolian sediments when sand is blown out from dry, neighboring alluvial plains, channels, or land surfaces (Chap. 2.3). Commonly, such eolian intercalations only form thin beds, frequently on top of coarser lag deposits, because they are partially eroded by subsequent floods. Modern examples of intercalated fluvial/eolian deposits have been described from several regions, for example from the Great Sand Dunes in Colorado (Fryberger et al.

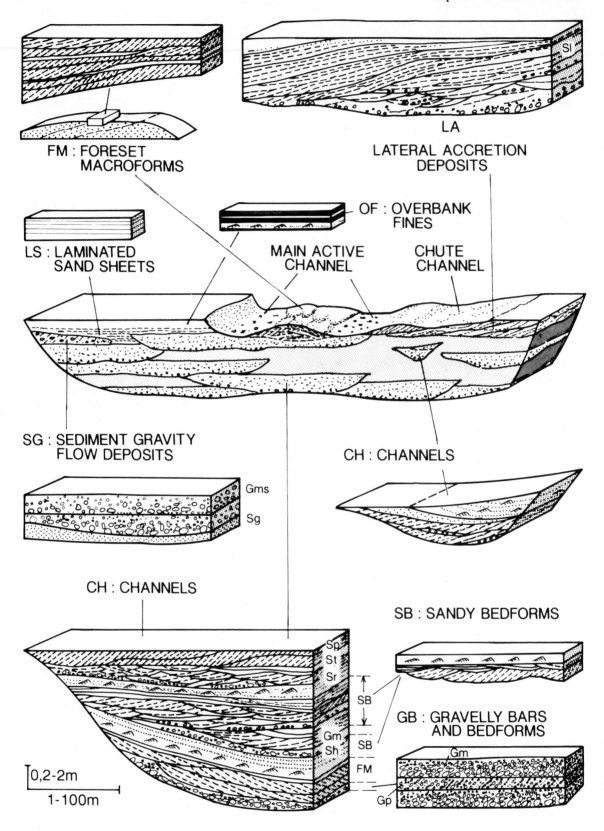

FM : FORESET MACROFORMS

LA
LATERAL ACCRETION DEPOSITS

OF : OVERBANK FINES

LS : LAMINATED SAND SHEETS

MAIN ACTIVE CHANNEL

CHUTE CHANNEL

SG : SEDIMENT GRAVITY FLOW DEPOSITS

CH : CHANNELS

Gms
Sg

CH : CHANNELS

SB : SANDY BEDFORMS

GB : GRAVELLY BARS AND BEDFORMS

Sp
St
Sr
SB
Gm
Sh
SB
FM
Gm
Gp

0,2-2m
1-100m

1979); ancient examples include parts of the predominantly fluvial Triassic Buntsandstein in Europe (e.g., Mader 1985; Marzo 1986).

2.2.2 Alluvial Fans and Fan Deltas

Alluvial Fans

Alluvial fans are cone-shaped piles of sediment formed at the foot of highlands where streams confined by narrow valleys emerge into an adjacent lowland (Figs. 2.8 and 2.11; Bull 1977; Heward 1977; Gloppen and Steel 1981; Nilsen 1982). A series of overlapping alluvial fans generates a clastic wedge. Fan deltas are alluvial fans that have built into a lake or the sea. The proximal facies of alluvial fans and the subaerial part of fan deltas are essentially the same, but the subaqueous sediments of fan deltas differ strongly from those of their subaerial counterparts. There are also some differences between alluvial fans in arid and humid climates. Alluvial fans in arid to semi-arid regions have been frequently described, whereas fans in humid regions have been rarely studied.

Sedimentation on alluvial fans begins where the streams leave their confined valleys and lose some of their transport efficiency. Basically, alluvial fans are composed of two types of sediment: *stream deposits* and *sediment gravity flows* (Fig. 2.11a). Current-transported sediments usually predominate. They are deposited either from ephemeral or perennial water flow in the channel system or, after extreme rain storms, from *sheet-floods* inundating large parts of the alluvial fan. Sometimes, gravel is concentrated locally to form *sieve deposits* (coarse gravel and boulders devoid of finer-grained matrix). From time to time, large debris flows with a muddy-sandy matrix reach the proximal and mid-fan area and bury part of the pre-existing, radiating channel system. At their lower end, such debris flows terminate in characteristic lobes, and they often concentrate large boulders and gravel at their outer margin, forming *levees*. Later, new channels cut into the mass flow deposits and rework and redistribute great proportions of their material.

Many alluvial fans are only a few kilometers long, but some can reach a length of more than 50 km. The *stream gradients* of fans in arid to semi-arid climates often decrease from 1 to 3 degrees at the head of the fan to 0.1 to 0.5 degrees at its base. Similarly, grain size decreases down fan, and roundness of gravel increases.

In *humid regions*, alluvial fans have lower gradients and are dominated by stream processes with marked seasonal variations in run-off. The mid and lower fan areas are vegetated and therefore less susceptible to reworking. They are cut by a limited number of active, narrow channels. The sedimentary processes in such fans, particularly those of humid tropical regions, are only poorly known.

One of the largest present-day examples of this type is the Kosi alluvial fan in India, draining the area around Mount Everest (Wells and Dorr 1987). This fan has a length of more than 150 km, but a very gentle slope (mean value 0.04°). River flow varies greatly due to the monsoonal climate. The channels are partly braided, and partly meandering and anastomosed, but there are also many abandoned channels forming oxbow lakes. The dominant grain size is sand and finer-grained material; sheetfloods and mud flows are absent. Minor primary sedimentary struc-

---

Fig. 2.10. Basic architectural elements of fluvial deposits according to Miall (1985), somewhat changed, varying scale. Note that there are simple and more complex elements consisting of several simple units. *OF* Overbank fines, sheet-like geometry, predominantly vertical aggradation of lithofacies *Fl*, mud or silt with thin lenses or laminae of silt to fine sand, commonly showing ripple cross-lamination; *LS* Laminated sand sheets, up to several meters thick, produced by flash floods, Sh lithofacies, and other sandy bedforms. *SB* Sand bedforms, including *Sh, St, Sp*, and *Sr* lithofacies (see Table 4.1 for explanation of symbols). Typical examples: fields of dunes and transverse bars, crevasse channel and crevasse splay deposits (Fig. 2.16). *GB* Gravelly bars and bedforms comprising lithofacies *Gm* and *Gp*, frequently alternating with *SB* and *MF* (in proximal regions); *FM* foreset macroforms of the active main channels, i.e., large compound bar forms, consisting of several co-sets of presumably upper flow regime bed forms; predominantly smaller-scale element in *SB*. *LA* Lateral accretion deposits (including point bar deposits), with variable internal geometry and lithofacies, consisting of different smaller-scale elements, for example *GB* (at the base) and *SB*, gently dipping surfaces toward the main channel. *CH* Channel fills of different size and geometry (Fig. 2.9). The total fill may show several episodes of erosion and refilling (*multi-storey fills*), and each part of such a fill can consist of smaller-scale elements such as *GB, FM, SB, OF* or *LS, SB, OF*, which often display fining-upward successions, but also large-scale elements such as FM and LA play a role. *MF* Mass flow deposits, mainly *Gms* lithofacies, frequently associated with *GB*. Author's note: the elements *FM* and *LA* are difficult to define clearly

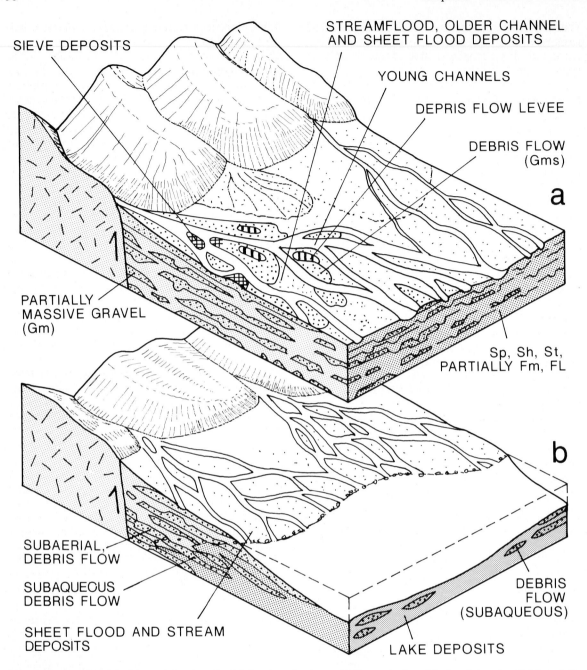

SIEVE DEPOSITS

STREAMFLOOD, OLDER CHANNEL
AND SHEET FLOOD DEPOSITS

YOUNG CHANNELS

DEPRIS FLOW LEVEE

DEBRIS FLOW
(Gms)

a

PARTIALLY
MASSIVE GRAVEL
(Gm)

Sp, Sh, St,
PARTIALLY Fm, FL

b

SUBAERIAL,
DEBRIS FLOW

SUBAQUEOUS
DEBRIS FLOW

SHEET FLOOD AND STREAM
DEPOSITS

DEBRIS
FLOW
(SUBAQUEOUS)

LAKE DEPOSITS

Fig. 2.11. Simplified facies models of **a** alluvial fan (proximal to mid fan region) and **b** fan delta. See Table 2.1 for explanation of symbols

tures are frequently obscured and obliterated by vegetation and bioturbation. The maximum thickness of these fan deposits is about 900 m.

The Kosi fan is thought to be analogous to the depositional environment of fluvial molasse accumulating in the fore-deep of a high mountain range.

In all alluvial fan deposits, *current directions* deduced from sedimentary structures show radial flow patterns from the fan head down fan (Collinson and Thompson 1982; Walker and Cant 1984; Pettijohn et al. 1987). Due to their braided stream system, current directions measured in a limited area display little variation (in comparison to meandering systems). Vertical sequence profiles may be quite irregular without showing a particular trend because they are controlled by several factors.

Flood events and debris flows are stochastic processes with greatly varying recurrence intervals. Nevertheless, they generate small-scale repeated successions, which are frequently a few meters thick. On the other hand, processes outside of the depositional area (extrabasinal factors) control long-term trends. Such factors include the amplification or reactivation of relief in the hinterland (tectonic control, see e.g. North et al. 1989), climatic changes affecting weathering conditions and erosion in the drainage area, and the base level of fans entering a lake.

For example, *coarsening-upward sequences* (Fig. 2.12a) may reflect fan growth during continuous faulting, i.e., uplift of the source area or subsidence of the fan region. The fan then progrades toward the lowland. *Fining-upward sequences* are generated if a short phase of faulting is followed by retreat of the scarp front and lowering of relief in the highlands (Fig. 2.12b). In paraglacial environments, the same phenomena can be generated in a different way: Periods of glacial advance lead to coarsening-upward successions, whereas fining-upward sequences result from glacial retreat.

Modern alluvial fans are associated with the source area containing the deposits of a confined stream, as well as with talus or colluvium from the adjacent slopes. In general, these facies types are not preserved in ancient examples. Down-slope, alluvial fans grade into other alluvial deposits, mostly those of a braided river system flowing perpendicular or at high angle to the outbuilding fans (Fig. 2.8). The lower fan sediments, consisting predominantly of sand and mud, may feed wind-blown sand transport systems or be associated with eolian sand (Fig. 2.21). On abandoned fan surfaces, vegetation starts to grow and soil forming processes take place. Distal fans may reach lacustrine environments, for example, playa lakes in arid regions (Chap. 2.5).

Paleosols and the interfingering of fan deposits with characteristic other facies types, such as playa sediments, tillites, and glaciolacustrine sediments, permit the recognition of their paleoenvironment. Ancient fan deposits occasionally contain plant relics, traces of burrowing organisms, and other fossils which may be useful in identification.

Fan Deltas

Coastal alluvial fans prograding into a lake or into the sea form fan deltas (Figs. 2.11b and 2.13). As soon as the streams, carrying a high bed-load, reach the standing water body, they drop their coarse material at the shore face and in prodelta foresets (Chap. 2.5). The intensity of reworking, sorting, and redeposition, as well as the transport of material along the shoreline, depend on the wave energy and, in marine environments, on the tidal range. In the case of lakes and protected embayments, fan progradation is little influenced by these processes. Gravel and sand accumulate at the mouths of streams until they become unstable from time to time and move as subaqueous debris flows into deeper water (cf. Chap. 5.4.1). There, they alternate with muddy lake or marine deposits. On high-energy coasts, some of the coarse material dropped at the river mouth is transported alongshore adjacent beaches where it forms distinctive beach gravel (cf. Chap. 3.1); some sand and gravel is swept by storms into deeper water. It is not possible to describe these greatly varying conditions by one simple facies model (also see Nemec and Steel 1988).

However, two to three common processes affecting fan deltas should be briefly mentioned. Figure 2.13 depicts an alluvial fan or braided river entering an active tectonic graben filled by a lake. The normal facies association in such a basin fill is a succession of topsets, foresets, and bottom sets. Lateral switching of the river mouth, however, complicates this simple facies pattern and leads to repeated displacements of the site where the coarsest material is deposited. Thus, in conjunction with ongoing subsidence, several coarsening-upward sequences may be generated on top of each other (Fig. 2.13). Mass flow deposits, and silty and muddy turbidites (Chaps. 2.5.1 and 5.4.2) reflect the prograding delta front in more distal regions. This model also applies to proglacial lakes fed by meltwater streams (Chap. 2.1).

Fluctuations in the water level in lakes or the sea affect both the subaerial and subaqueous facies of fan deltas (Fig. 2.14). They are reflected by fluvial terraces and coarse-grained river mouth deposits at varying elevations. A lowering of the lake or sea level causes the emergence of delta foresets and the subsequent cutting of fluvial channels into the foresets. These channels are commonly filled later with fluvial deposits.

The thick Oligocene-Miocene conglomerates of the famous Meteora monasteries in Greece represent coarse-grained foresets of a fan delta with intercalated channel fills (Ori and Roveri 1987) which may be interpreted as incised

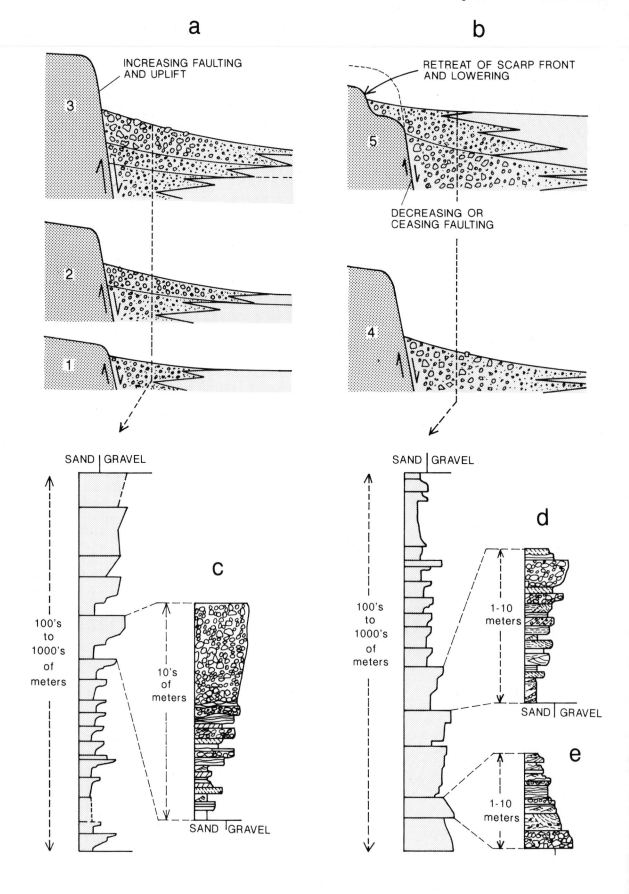

a

b

INCREASING FAULTING
AND UPLIFT

3

2

1

RETREAT OF SCARP FRONT
AND LOWERING

5

DECREASING OR
CEASING FAULTING

4

SAND | GRAVEL

c

100's
to
1000's
of
meters

10's
of
meters

SAND | GRAVEL

SAND | GRAVEL

d

1-10
meters

SAND | GRAVEL

e

1-10
meters

100's
to
1000's
of
meters

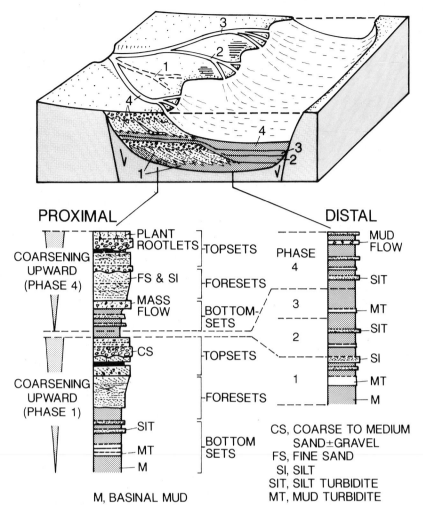

Fig. 2.13. Facies pattern of topsets, foresets, and bottomsets of a lake affected by switching river mouth and subsiding lake bottom. Note several phases of delta prograding, coarsening-upward sequences, and gravity mass movements reflecting the approaching delta front

Fig. 2.12. Idealized vertical sequences in alluvial fan deposits and their possible cause. **a** Large-scale coarsening-upward sequence due to continuous faulting and fan progradation (Stages *1* through *3*). **b** Large-scale fining-upward sequence caused by retreat of scarp front and lowering of relief in source area (Stages *4* and *5*), or lateral shifting and abandonment of fan (not shown). **c** and **d** Small-scale coarsening-upward cycles due to the prograding of individual fan lobes. **e** Small-scale fining-upward cycle with channelized base generated by bar processes or filling of braided channel. (After Ethridge 1985)

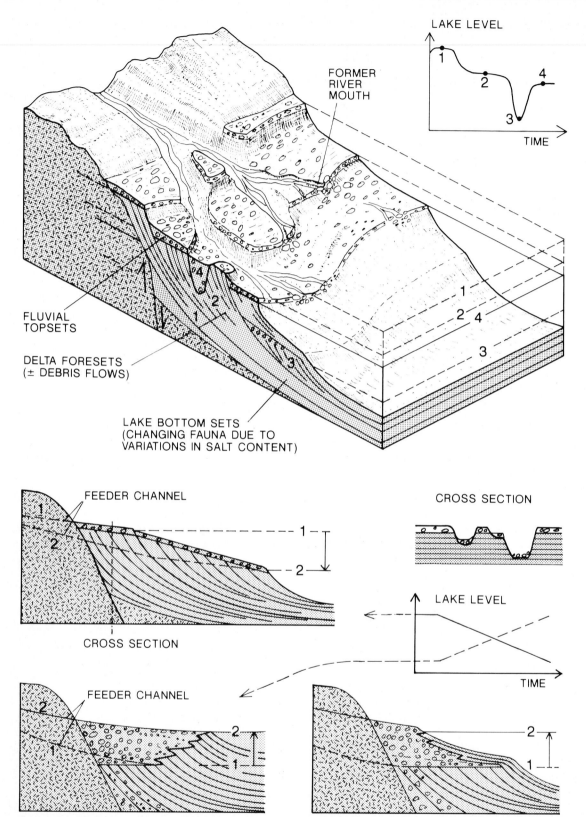

Fig. 2.14. Fluvial terraces, river mouth bars, and lake sediments affected by fluctuations in lake level (Stages 1 through 4). (Based on author's own observations in Lake Burdur area, Turkey; also see Price and Scott 1991)

valley fills during low sea level stands (compare Chapter 7.4).

A fluctuation water level also causes characteristic vertical and lateral facies changes in lake deposits (Fig. 2.14) and shallow-marine sediments.

2.2.3 Various River Systems and Their Sediments

Braided Rivers and Braidplains

If we define *sinuosity* as the ratio between channel length and valley length, braided systems have a low sinuosity of the order of 1.1 to 1.2, while meandering channels reach values >1.5 (Fig. 2.8). Braided streams usually consist of several individual channels separated by bars and islands and therefore form a wide, shallow stream bed. Braided rivers develop near areas of high relief, which deliver relatively large amounts of debris, gravel and sand into the fluvial system. A prominent modern example is the Brahmaputra River in Bangladesh, draining large areas in the High Himalayas (Bristow 1987). The recent sedimentation of this river produces a sheet-like sand body approximately 20 km wide and 40 m thick. The channels can be subdivided into larger and smaller ones and have a very high width/depth ratio of about 50:1 to as much as 500:1. Medium-sized channels may move laterally at rates up to 1 km/a (a = year). Continual reworking within the channel belt and occasional channel switching generates complex internal structures within the sand body. The most important depositional process is lateral accretion, followed by downstream and upstream accretion.

Glaciofluvial outwash rivers also belong to this low-sinuosity, multiple-channel category. Both the valley gradient and the stream power are relatively high, the latter often intermittently so. From all these characteristics, it can be inferred that braided systems are bed load-dominated, i.e., they carry and deposit chiefly gravel and sands. Therefore, they consist predominantly of channel and channel-flank deposits, while silty and muddy floodplain facies are subordinate. Downstream, they often display a progressive decrease in grain size, as well as in bed forms and internal sedimentary structures.

In the *proximal reaches* of such multiple-channel bed-load rivers (Miall 1985), *lon-gitudinal gravel bars* are characteristic channel bed forms (Fig. 2.15a). They develop as a result of intermittent clast accretion under flow passing obliquely over the bar. Therefore, the bars migrate not only downstream parallel to the current, but also laterally. The resulting planar sheets of gravel show indistinct, crude horizontal stratification. The interstices in the gravel are usually later filled with sand during low water periods, but in this system sandy beds are relatively rare in the proximal zone, unless there is little gravel available in the source area. Sandy beds develop best at somewhat higher topographic elevations within the braided system next to the active channel system. Here, minor channels active during floods may drop their bed load during waning flow.

A typical vertical sequence in such a proximal braided system is shown in Figure 2.15b. The dominant sediments are *multistory gravel units,* originating from channel bars. Many bars are somewhat graded. Interbedded with the gravels are thin lenses of sand representing deposition in abandoned channels or sand wedges at the edge of bars. In places, one can observe repeated successions of fining-upward gravel-sand sequences 1 to 2 m thick, but in general it is difficult to identify the bottoms of former channels and the geometry of their fill. The reason is that the underlying and neighboring sediments also consist largely of gravel.

Downstream, the predominantly gravelly beds grade into beds consisting partly of smaller pebbles and sand. In the lower, more active channels, bar gravels dominate, whereas sands and pebbly sands are common at higher topographic elevations. In rarely flooded areas, some silt and mud may be deposited and preserved. In the total assemblage, the gravel content varies between 10 and 70 %. As a result of downstream and laterally migrating sand and gravel bars, planar and trough cross-bedding are the most important internal sedimentary structures. Both the lateral migration and sudden abandonment of channels due to avulsion cause *fining-upward channel fill sequences* a few meters thick (Fig. 2.15c). Such *autocyclic sequences* (Chaps. 7.1 and 7.4) are considered the most distinctive characteristic feature of this type of braided river deposit.

In *distal reaches* of braided fluvial plains, runoff is distributed by numerous shallow channels. The gradient and stream power of the system are lower than in proximal areas, and sand is the dominant grain size deposited. Typical present-day examples of this situation

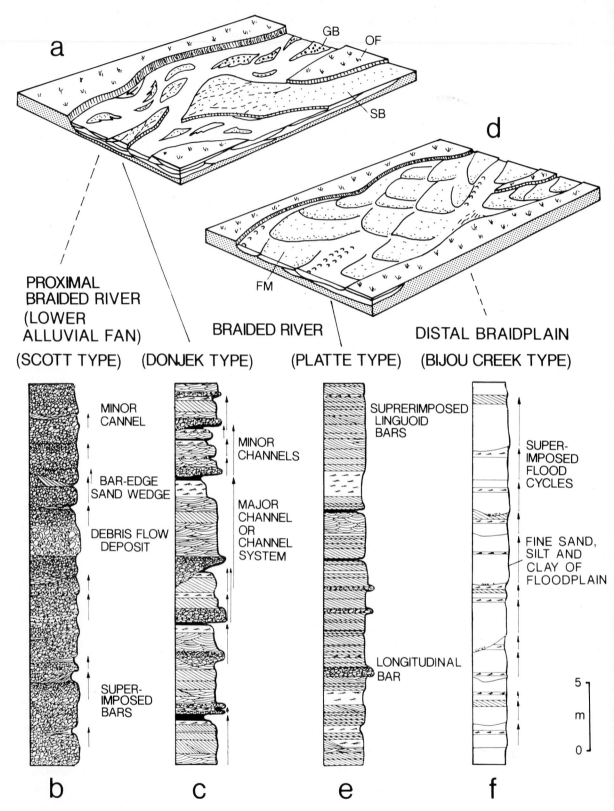

Fig. 2.15a-f. Braided river systems. **a-c** Proximal to middle reaches, gravel-dominated (**b**), or sand-dominated (**c**) with minor proportion of gravel. **d-f** Distal, sand-dominated system with wide channels and flat, linguoid sand bars (**d** and **e**), or wide floodplain rarely inundated by flash floods (**f**). (After Miall 1985)

Table 2.2. Main characteristics of selected braided river systems. (Some authors identify more types, but there are no sharp boundaries between the different subenvironments.) (After Miall 1985)

Subenvironment, general description	Percentage of gravel (%)	Prevailing lithofacies and internal structure	Minor lithofacies[a]	Name of type (Miall 1985)
Proximal braided river deposits (including lower alluvial fans), gravel-dominated	>50	Massive gravel	Gp, Gt Sp, St Sr, Fl Fm	Scott
Braided river deposits (middle reaches), rich in gravel	10-70	Planar and trough cross-bedded gravel and sand, massive gravel	Sh, Sr Fl, Fm	Donjek
Mainly sandy		Trough and planar cross-bedded sand	Sh, Sr Sl, Gm Fl, Fm	South Saskatchewan
Distal braided river deposits, sand-dominated	<10 Mainly sand	Trough and planar cross-bedded sand[b]	Sh, Sr Gm, Fl Fm	Platte
Silt and mud-dominated	Silt and mud	Massive and fine laminated silt, mud and sand	St	Slims
Sandy river plains subject to flash floods	Mainly sand	Horizontal and low angle cross-bedded sands	Sp Sr	Bijou creek

[a] For explanation of symbols, see Table 2.1
[b] Fining-upward channel fills cause distinct minor sedimentary cycles

are the *distal outwash plains* of glaciated areas *(sandar)*, but the same type also occurs in non-glacial river systems. The channel sands form large *linguoid bars* (Fig. 2.15d), sand waves, and smaller bed forms such as dunes and ripples (see Fig. 2.7). In vertical sections, planar cross-bedding prevails, whereas trough cross-bedding is less common (Fig. 2.15e). Gravel may occur in small lenses, and *overbank fines* are occasionally intercalated. Fluvial cyclicity due to the filling and abandonment of channels is much less pronounced than in the medium reaches of this depositional system. In the most distal parts of outwash rivers, or in regions which receive mainly fin2e-grained silts and muds from the hinterlands, the river system is characterized by a low gradient, as well as by low-relief channels and bars. The main sediment types are massive, laminated and cross-laminated sandy silts.

Sand-dominated distal braidplain environments are somewhat different from the braided systems described so far. Such braidplains appear to be rather common in arid regions. Here, runoff and sediment accumulation are controlled by rare flash floods over a broad river plain with very shallow channels. Flow may be either channelized or occur as sheet floods. During waning flow, most of the sand is deposited essentialla in sheets, displaying horizontal lamination resulting from upper flow regime conditions. Such sands often grade vertically into finer and small-scale trough cross-bedded layers (Fig. 2.15f).

The most characteristic features of the different subenvironments of braided river systems are summarized in Table 2.2.

Figure 2.17a shows a generalized large-scale cross-section of a Cenozoic alluvial coastal plain in Texas (Galloway 1981), which is composed of several contemporaneous, individual fluvial systems. The axes of these systems coincide with the centers of the broad sand belts, while the regions between the sand bodies are occupied by the sediments of floodplains, smaller channels and lakes. The character of the individual fluvial deposits varies between braided bed-load and meandering mixed-load systems. The core of a sandy braided river deposit consists of wide, amalgamated, vertically stacked, sandy and conglomeratic channel fills and finer-grained, sandy-silty sheet splay units (Fig. 2.17b). These grade laterally into interbedded floodplain deposits punctuated locally by isolated channel fills.

Fluvial sheet sandstones similar to these in Texas have been described from several regions, for example from the Eocene in the South Pyrenees, Spain (Marzo et al. 1988).

Meandering Rivers

Meandering river systems develop one principal, relatively narrow channel of high sinuosity (>1.5) and are dominated by mixed load or predominantly suspended load. Their overall sand content often averages 20 to 40 %. If meandering rivers are associated with a wide floodplain, the channel sediments may be restricted to a comparatively narrow zone within the flood basin where they form a meander belt (Fig. 2.16).

The different architectural elements of the fluvial sediments shown in Fig. 2.10 can best be observed in meandering systems. In a sinuous channel segment, one can distinguish the following morphological features and depositional subenvironments (Fig. 2.16):

- Channels and channel fills.
- Point bars and lateral accretion complexes.
- Chute bars.
- Channel plugs (oxbow lakes).
- Levee and crevasse splay deposits.
- Alluvial flooplain deposits.

The **channel floor** is usually covered by lag sediments consisting of the coarsest material transported by the river during peak flood. This channel lag may also contain mud clasts or blocks eroded from the banks. Lag sands and gravel usually accumulate between scour pools and form flat, elongate bars displaying either imbrication of gravel or crudely laminated and planar cross-bedded gravelly sand.

Point bars accumulate on the inner sides of river bends, while on the outer side material from the bank is eroded. In this way, the curvature of the meander tends to become increa-

singly exaggerated until the river produces short-cuts, leaving behind abandoned channel segments (oxbow lakes, Fig. 2.16). Most of the point bar material is eroded from the upstream channel banks. It is deposited in areas of lower velocity turbulence. Because sediment moves up and out of the channel onto the bar, cross sections of point bars often show *fining-upward sequences*, with sands on top of channel lags (Fig. 2.16). Similarly, the internal structures grade from horizontal bedding (upper flow regime) to large-scale and small scale trough cross-bedding (lower flow regime). The most distinctive feature of point bars is *lateral accretion (low-angle, "epsilon" cross-bedding)* which may be visible at the surface by the development of ridge-and-swale topography (Fig. 2.16). The swales can be filled with mud, and older portions of the point bar covered by levee sands and silts or floodplain sediment.

This idealized point bar architecture is often modified in nature by *chute channels* cutting into the point bars and producing chute bars, small terraces associated with different water stages in the main channel, or other irregularities. The resulting internal structures of such a *lateral accretion complex*, a term which is preferred by some experts on fluvial sediments, is less regular and more variable than the major architectural units of the fluvial deposition shown in Figs. 2.10 and 2.16.

Chute bars result from chute channels which direct part of the river flow across the surface of a point bar during flood stage. In this way relatively coarse-grained bed-load material can be deposited as lag or chute bar on the eroded top of a lateral accretion complex (Fig. 2.16). While channel lag deposits are common on the upstream part of the chute, the downstream part is often characterized by imbricated pebble sheets and large-scale planar and trough cross-bedded sands.

Channel plugs are infillings of cutoff meander segments (oxbow lakes). Since the further influx of bed load is terminated, the abandoned channel segments are slowly filled with fine-grained material washed in from the neighboring floodplain. In humid climates, organic matter (including peat) may accumulate in the lake or swamp.

Levee and crevasse splay deposits. Many meandering channels are accompanied by flat ridges or dams sloping away from the channel into the floodplain. These levees are built up during moderate floods which just reach the elevation of the channel bank or ridge. Due to decreasing flow velocity, sand is deposited

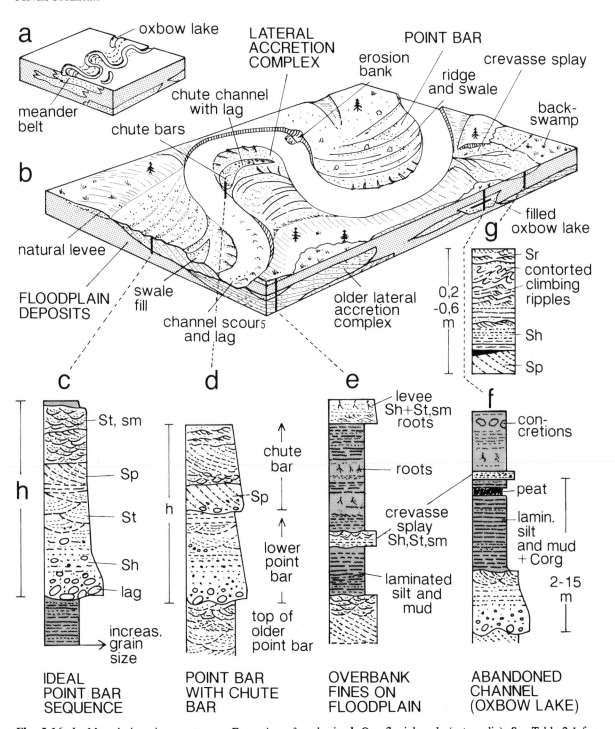

Fig. 2.16a-h. Meandering river system. **a** Formation of sandy meander belt within a flood basin. **b** Different subenvironments of meandering channel. **c-g** Characteristic vertical sections of the youngest sediments of the flood basin. **h** One fluvial cycle (autocyclic). See Table 2.1 for explanation of symbols; *sm* small-scale. (Based on different sources, e.g., Walker and Cant 1984; Galloway 1985; Miall 1985)

along the channel banks, grading into silt somewhat farther away. Locally, channel water may spill over the levees into the flood-plain, forming *crevasse splays*. The fallout of sand and silt usually extends farther into the floodplains than the levees, but such crevasse splays can also contribute to the buildup of the levees. The prevailing internal structures of these sand sheets may resemble those of thin sandy turbidites, showing some grading, horizontal lamination and small-scale ripple cross-bedding, including climbing ripples and occasional convolute or contorted cross-bedding. These structures are, however, often mashed or destroyed by the roots of vegetation. Whereas distal crevasse splays become interbedded with floodplain deposits, levee sands often tend to be reworked by subsequent channel migration.

Floodplain deposits accumulate during rare inundations. They consist predominantly of suspended load, i.e., silt and mud, though fine sand may also be present in areas where the peak flood currents are sufficiently strong to transport this grain size. The deposits from individual large floods reach thicknesses of only a few millimeters or, locally, a few centimeters. Such thin beds may be either somewhat graded, or internally finely laminated or cross-bedded. A series of flood layers can show distinct lamination. Floodplains may be wetlands and backswamps or areas of dessication and *calcrete* development. Therefore, primary sedimentary structures are often destroyed by vegetation, mud cracking, salt precipitation, concretion formation (such as carbonate nodules or caliche), and other soil forming processes.

Fluvial red beds. Originally gray or brown flood deposits may become red through time, if they contain little or no organic matter, as for example in arid and semi-arid regions. In this case, brownish iron hydroxides deposited with the original sediment can be later transformed into hematite and thus produce very fine-grained, evenly dispersed red pigment (cf. Chap. 6.3). Another mechanism generating red beds is the weathering of iron-bearing silicate minerals within the sediment during shallow-burial diagenesis (Walker 1967; van Houten 1973; Füchtbauer 1988). Comparatively unstable minerals, such as biotite, hornblende, and smectite can release iron which, under oxic conditions, slowly forms hematite.

Swamps. In humid climate and on very gently sloping alluvial plains, the groundwater level can reach the surface. As a result, *back-swamps* and lakes occupy parts of the interchannel area (Fig. 2.18c). These depressions are filled either with sediments delivered by crevasse splays and peak floods forming flat, deltaic sequences and well-laminated fine muds, or they accumulate organic detritus and peat. Numerous *coal deposits*, with individual coal seams usually limited in extent and thickness, formed in alluvial plains (e.g., Flores 1981; Galloway and Hobday 1983; Flores et al. 1985; Lyons and Alpern 1989). Many of these ancient floodplains were situated not only near the paleo-coastline of a former sea, but also in more elevated fluvial environments, including intermontane basins. In addition to the non-marine fauna in the accompanying sediments, abundant siderite concretions often indicate a fresh or brackish water depositional environment (see Chap. 2.5).

Large-Scale Facies Associations

The large-scale vertical and lateral facies interrelationship between channel fills and levee and crevasse splay sands, on the one hand, and floodplain and lake deposits, on the other, is indicated in Fig. 2.17c. In large-scale cross-sections of a meandering system, the interchannel deposits predominate. Under the load of younger sediments, these fluvial sediments are affected by differential compaction (Chap. 13.2), leading, for example, to considerable deformation of fine-grained floodplain deposits and coal seams near the channel fills, which themselves are little compacted. Because compaction begins early in interchannel areas, peat and subsequent coal seams may become thicker there than near the channels, where they frequently split up into several thin bands before pinching out (Fig. 2.17c).

We can conclude that floodplain deposits are the most important facies type in meandering fluvial systems, if the river system is not constrained by nearby valley slopes. Apart from this exception, floodplain deposits make up more than half of the total floodbasin sediments. The rate of their vertical aggradation is a major factor controlling the total fluvial system. High rates are associated with abundant supplies of suspended load and substantial subsidence. They tend to generate vertical stacking of the channel deposits (point bars, etc., including relatively good preservation of levee deposits, see Figs. 2.9 and 2.17). The contrary situation, however, leads to pronounced channel migration, lateral stacking of channel deposits, and hence more reworking

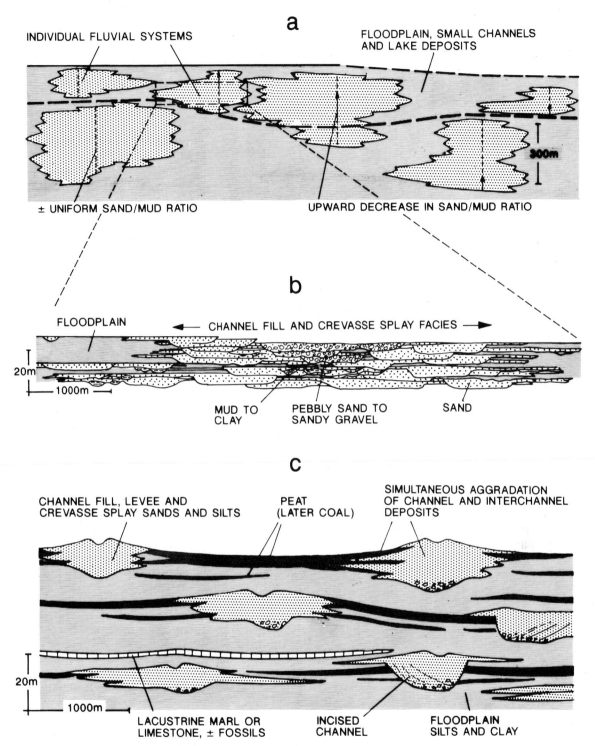

Fig. 2.17a-c. Macro-scale cross sections of fluvial systems. **a** Generalized cross section of Cenozoic coastal plain, Texas, demonstrating the principal large-scale depositional elements. **b** Section of **a**, bed-load river deposits forming a sand belt several km wide and about 20 m thick (**a** and **b** after Galloway 1981). **c** Idealized channel and interchannel-backswamp-lake deposits of a mixed-load meandering system. Channel sands are accompanied by levee and crevasse splay sands; floodplain deposits make up the majority of the basin fill. Coal, carbonaceous shales and some freshwater limestones accumulate in poorly drained backswamps or lakes. Aggradation of channel fills and floodplain deposits occurs either simultaneously (rapidly subsiding flood basin), or the channels cut into pre-exixting fluvial sediments. (Based on Flores 1981; Galloway and Hobday 1983)

and removal of the levee and floodplain sediments.

The sedimentation rate of floodplain deposits in terms of vertical upbuilding varies greatly. Some values mentioned in the literature are of the order of one to several meters per thousand years (also see Chap. 10.2).

Anastomosed Rivers

In contrast to the braided and meandering river systems, anastomosed rivers and their deposits are known only from a few recent and ancient examples (Smith 1983; Walker and Cant 1984). Such systems appear to result from an elevating base level at the downstream end of the river course. This in turn creates a low gradient and causes the overbank deposition of fine-grained material.

Anastomosed rivers develop an inter-connected network of straight to sinuous channels, which are relatively narrow and deep (Fig. 2.18). Because their banks are built up of fine-grained sediment and are usually covered with vegetation, channels and islands are remarkably stable. The channels are filled with sand and gravel. Since vertical aggradation proceeds comparatively rapidly, rather thick and narrow ribbons of channel sand can form, bounded by the sandy silts of the levees. On the river plain or in the wetlands between the channels, laminated silty clays and clayey silts accumulate, or some other facies types such as backswamp deposits and peats containing large amounts of organic matter and iron sulfides may develop. Modern examples of anastomosed rivers have been described from western Canada and central Australia (Rust and Legun 1983).

In the ancient record, one can expect that anastomosed river systems are often associated with *coal seams* generated in low-gradient alluvial plains. Dense vegetation favors soil formation and hampers erosion and thus reduces the supply of sand and coarser grained material to the river system; low river gradients and relatively stable levees enable widely extended backswamps. However, it appears that a generally accepted facies model for this type of river system has not yet been developed.

Mud-Dominated Low-Gradient Rivers

In areas of very low relief, seasonally or perennially hot, dry climates, and deeply weathered clayey soils, fluvial sediments tend to become mud-dominated. Such conditions are found in the present-day arid to semi-arid Lake Eyre Basin in Australia, where the deposits of the Cooper and Diamantina Rivers, with their braided or meandering channel system infillings, can consist predominantly of muddy material (Rust 1981; Rust and Legun 1983; Rust and Nanson 1989).

In these cases, the mud occurs in sand-sized *pedogenic aggregates,* which originate from repeated drying and wetting at or near the land surface, and partially from the bottoms of shallow dry lakes or their accompanying wind-blown clay dunes (see Chap. 2.3.3). The clay content of these aggregates may surpass 50 %, the rest being silt and sand-sized particles. As long as the clay aggregates are moist and exposed to the air, they stick together. After drying, they form microcracks and can be eroded by running water. In rivers the clay aggregates are kept separate and are transported like sand or small pebbles as bed (or traction) load over long distances. The *mud aggregates* are remarkably durable, even under upper flow conditions. Finally, they form porous structures similar to sand bars. Thus, not only are overbank flood deposits (resulting largely from the suspended load) fine-grained, but also the deposits of point bars or lateral accretionary complexes. Even the channel fills may consist predominantly of mud. After burial and compaction, the aggregate texture of these muds can be obscured or entirely destroyed. Hence, such mud-dominated river systems are difficult to recognize in the ancient record.

This may be the reason why only a limited number of ancient examples of mud-dominated fluvial deposits are known (Rust and Nanson 1989). It is likely that a number of red and green claystones were deposited in this manner. To identify them, careful study of their texture and sedimentary structures is necessary.

Muddy river bed loads can also originate from the weathering and erosion of overconsolidated claystones and mudstones in various climates (cf. Chap. 9.2). In these cases, sand and gravel-size lithoclasts can constitute a major part of the bed load, but are commonly commonly accompanied by normal coarse-grained material.

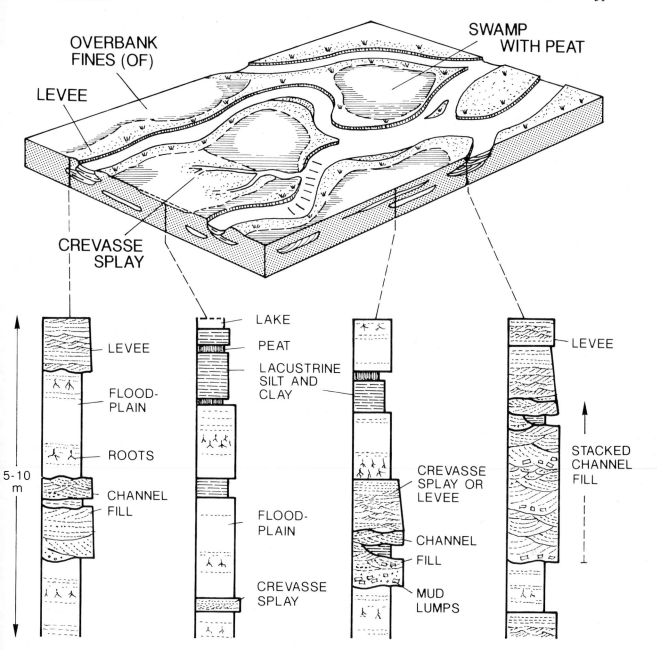

Fig. 2.18. Anastomosing fluvial system with low to high sinuosity and branching channels. Channel sediments form isolated ribbon sand bodies, often accompanied by fine sandy to silty levee deposits. Lateral accretion deposits play a minor role. Crevasse channels and crevasse splays are common. Interchannel areas accumulate overbank fines (floodplain deposits) or shallow lacustrine muds and peat. Note the great differences between the vertical profiles! (Based on Smith 1983; Miall 1985)

2.2.4 Large-Scale Lateral and Vertical Evolution of Fluvial Systems

Fluvial sediments of some thickness accumulate and are preserved only in subsiding regions (relative to a more or less fixed erosional base, such as sea level). The gradient of the main channel of a fluvial system tends to decrease systematically downstream. Thus, a complete idealized system would include an alluvial fan (steepest gradient) evolving to a braided stream, then to a meandering and possibly anastomosing system (lowest gradient, Fig. 2.19). As part of this evolution, the bulk of the sediments becomes finer, and floodplainplain deposits predominate over channel fills.

However, rock types in the hinterland and changes in climate also strongly control both the type and grain size distribution of fluvial systems. Humid climates and abundant vegetation generally reduce mechanical erosion and promote the formation of fine clay minerals (Chap. 9.1). In addition, the mineralogical composition of fluvial sands is affected by

changing climate. In this respect, the so-called *first-cycle* fluvial sandstones derived from nearby crystalline rocks are of particular interest. If such sandstones have not been subjected to subsequent significant metamorphism, the degree of their compositional maturity as expressed in quartz/feldspar/rock fragment percentages (e.g., Pettijohn et al. 1987; Girty et al. 1988) may be used as an indicator of *paleoclimate* (e.g., Basu 1985; Suttner and Dutta 1986). Provided that the petrographic composition of the source rocks was more or less constant over a long period of time, high feldspar and rock fragment

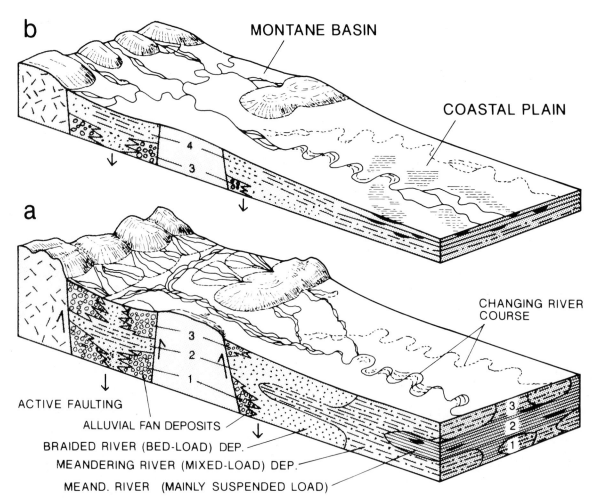

Fig. 2.19a,b. Evolution (Stages *1* through *4*) of depositional fluvial systems. **a** Active faulting maintains high relief during either semi-arid (Stages *1* and *3*) or humid climates (Stage *2*). Stages *1* and *3* are characterized by extensive alluvial fans in the montane basin, as well as braided bed-load rivers and meandering mixed-load rivers on the coastal plain. Stage *2* displays a predominance of mixed-load and suspended-load meandering rivers on the

coastal plain and fewer fan deposits. **b** Waning tectonic activity, lowered relief, and widespread vegetation (Stage *4*) reduce the significance of braided streams and lead to extensive fine-grained floodplain deposits of meandering and anastomosing systems, including interchannel lake deposits and peat. Note the various vertical trends at different locations

contents reflect an arid paleoclimate, while a relative enrichment in chemically stable quartz indicates warm and wet paleoenvironmental conditions. High relief in the source area diminishes the influence of climate on the composition of sands, unless they have been exposed to extended weathering during alluvial storage (Johnsson 1990). Other criteria, such as the varieties of quartz present, may serve to identify different source areas.

An idealized example in Figure 2.19 demonstrates large-scale lateral and vertical facies changes in a subsiding *montane basin* and on a sediment-aggrading coastal plain, reflecting the tectonic and climatic history of the area. The montane basin generally receives sediments from several small and medium local sources, indicated by strongly differing transport directions, and material of possibly greatly varying composition. In contrast, *coastal plains* are fed by a limited number of large point sources, distributing their granulometrically and compositionally mixed sedi-

ment load in a fan-like fashion. This is accomplished by frequent changes in river course (avulsion) over the plain with the aim of evenly aggrade the depositional surface. The variation in mean transport directions of such a radiating distribution system, however, is much less than that of a montane basin. Humid phases tend to provide more fine-grained, quartz-rich material to the fluvial system (Fig. 2.19, Stage 2). Particularly in combination with lowered relief and decreasing stream gradient, such conditions favor mixed-load and predominantly suspended-load meandering and anastomosing systems (Fig. 2.19b). High groundwater levels may generate interchannel swamps and lakes.

There are several other tectonic and depositional scenarios which may also explain the various vertical and lateral, large-scale facies successions of recent and ancient fluvial systems. The above-mentioned general rules may be used to identify the nature of fluvial systems not discussed in this text.

2.3 Eolian Sediments

2.3.1 Introduction

Wind-blown dust and sand contribute to the composition of various types of sediments which ultimately accumulate in fluvial or sub-aquatic environments. Thus, for example, flood plains of desert rivers, coastal sediments, and deep-sea muds may contain significant proportions of air-borne dust or sand, but sedimentary structures and faunal characteristics of these environments only indicate the final mode of dust or sand deposition. Eolian sediments, however, are deposited under the direct influence of wind without the influence of running or standing water. They still exhibit primary features typical of eolian environments. Eolian sediments have been understood by geologists for a long time, because their accumulation can be observed in the present-day deserts. Numerous articles and most textbooks present more or less detailed descriptions of these sediments (e.g., Glennie 1970; Bigarella 1972; Cooke and Warren 1973; Hunter 1977 and 1981; McKee 1979; Warren 1979; Blatt et al. 1980; Galloway and Hobday

1983; Chorley et al. 1984; Collinson 1986b; Frostick 1987; Pettijohn et al. 1987).

The medium of transport of eolian sediments is the atmosphere. Planets or moons without an atmosphere are devoid of eolian processes and sediment, but even a very thin atmosphere such as that on Mars is sufficient to produce sand dunes. Deserts constitute about 20 % of the present-day land surface of the Earth, but only 20 to 45 % of the area classified as arid is covered by "sand seas" (ergs) (Lancaster 1990). The major part of deserts comprises highlands, stony plains, alluvial fans, and playas. Deserts range from low-latitude, hot and arid regions within the zones of trade winds and monsoons to cold and dry polar regions. The western margins of continents affected by cold ocean currents and land-locked continental interiors, particularly on the leeward side of the principle wind directions, may acquire the characteristics of deserts. Beaches with high sand supply can cause dune fields of limited extent under various climatic conditions.

In the geologic past, the general principles governing the formation of deserts were the same as today, but the proportion and regional distribution of deserts on the land surface were frequently quite different from the present situation. During the cold phases of the Pleistocene, the climatic belts and thus also the large subtropical deserts shifted poleward and gained in extent (e.g., Sarnthein 1978; Leinen and Sarnthein 1989). Continental drift and the changing configuration of the oceans and land masses significantly affected global climate and thus also the areal extent of deserts and proportion of eolian sediments in the geological record. In the Permo-Triassic, for example, when the older continents had created one supercontinent (Pangea), arid climates and eolian sediments played a much greater role than today. Since eolian sediment transport is favored by the lack or absence of vegetation, the evolution of land plants during the Earth's history has to be taken into consideration. Prior to the Devonian, plant life was limited, and also in the pre-Upper Cretaceous, soil protection by dense roots of angiosperms did not yet exist. Colder climate and the advance of glaciers during glacial periods led to a reduction in land surface covered by vegetation. For all these reasons, the contribution of eolian processes to sediment accumulation and basin filling can have been been much greater in the past than today.

Winds are an effective agent for taking up and transporting fine to medium-grained, unconsolidated materials, if the sediment particles do not adhere to each other, for example by moisture. Hence, the land surface must be *dry and not be protected by vegetation*. Fine silt is whirled up and kept in *suspension* as long as the turbulence of the air is sufficiently high to counteract its settling by gravity (Fig. 2.20). Thus, silt may travel from desert areas over distances of tens to several hundreds to a thousand kilometers and reach vegetated land. The larger and heavier sand grains move near the land surface by *saltation* and therefore migrate much more slowly than dust. Both dust and sand can, in contrast to fluvial transport, also migrate upslope, and their movement is not channelized by valleys. Particularly dust can easily overcome hilly and mountainous regions and settle far away from its source area. Thus, eolian sediments are less topography-controlled than most other types of depositions. They record the dominant wind pattern of the region and time period studied.

2.3.2 Eolian Sands

Source, Composition, Color, Texture, and Migration of Sand

Sand Sources

Various sources supply wind-blown sand:

- Fluvial sands of alluvial fans, ephemeral streams and flood plains which fall dry part of the year or longer periods.
- Coastal sands of mixed origin including biogenic components.
- Weathering of older sandstones including former dune sands.
- Weathering of plutonic and metamorphic rocks.
- Sand-size silt-clay aggregates including salt particles from playa lakes.

Composition of Sand

Because eolian sands may be derived from different sources, their composition varies considerably, a fact which was pointed out by several authors (e.g., Solle 1966; Glennie 1970; McKee 1979). Many present-day sand dunes are, however, mineralogically fairly mature, i.e., they represent quartz arenites with minor amounts of feldspar and other weatherable silicate phases. The thin, platy minerals of mica are usually rare in eolian sands, because they behave differently during air transport than the other, more compact mineral grains. At first glance, the predominance of quartz is somewhat surprising because chemical weathering in arid regions is limited (Chap. 9.1) and, therefore, cannot account for the advanced maturity of most of the desert sands. The reason for the minor importance of immature sands is that many dune sands are not derived from nearby sources of the desert region itself, but are transported by rivers from areas of intensive weathering into the desert. In other cases, the sands are recycled from older, more or less mature sandstones. Coastal sand dunes composed of carbonate grains or special dune fields consisting of gypsum grains are fed by nearby primary sources. Their potential to be preserved in the geologic record or to become recycled is limited.

Color of Eolian Sands

Many recent sand dunes are yellow, light-brown or reddish in color which is caused by iron hydroxide impurities, for example limonite on the surface of the sand grains. Ancient eolian sandstones are commonly red-colored. The hematite pigmentation of these rocks usually results from the decomposition of unstable, iron-bearing minerals during diagenesis (Walker 1979; also see Chap. 6.3). The *reddening* commonly increases with the age of the sands. In the Australian dune fields, the dunes farthest away from the source area exhibit the most intensive reddening (Wopfner and Twidale 1988).

Transport and Texture of Eolian Sands

The capacity of winds to transport solid particles of a density similar to that of quartz (2.65 g/cm^3) is mostly limited to fine and medium sand (grain diameter 0.125 to 0.5

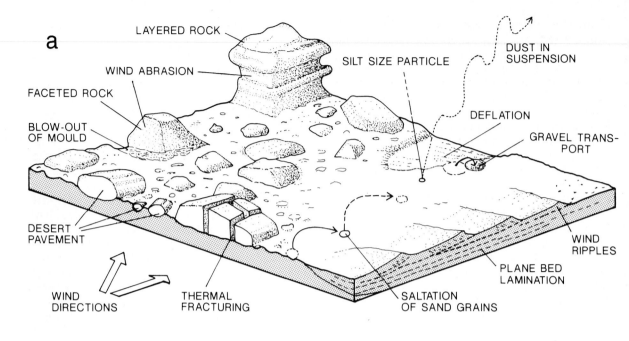

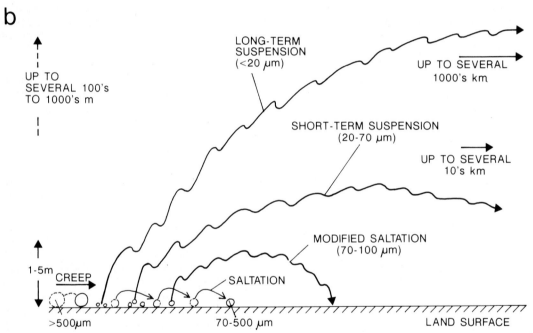

Fig. 2.20. a Various modes of eolian transport of silt, sand, and gravel, specific features of desert surfaces including deflation and wind abrasion, and bed forms on flat surface with limited sand accumulation (e.g., in interdune areas, as shown in Fig. 2.21). **b** Modes and distance of eolian particle transport by moderate storms (wind velocities of 10 to 20 cm/s) in relation to grain sizes. (After Pye 1987)

mm). These size fractions predominate in coastal sand dunes (mainly fine sand) and inland sand seas (fine to medium sands). Only very strong storms (wind velocity on the order of 20 to 30 m/s) can slowly move pebbles and gravel by creep. These particles are too coarse to be lifted and therefore roll along the surface where they suffer impacts by saltating sand grains. Silt size particles, on the other hand, are transported very fast (Fig. 2.20;

Chap. 2.3.4). Hence, wind is very effective in producing and accumulating *well sorted sands*.

Sorting of sand in eolian dunes and particularly in interdune areas may, however, also be poor. Each individual storm event can move a certain population of grain sizes (Folk 1971). Thus, sand beds originating from a number of storms of varying intensity tend to be less sorted than individual sand laminae. Relatively fine-grained and very well sorted sand is typically found in coastal dunes, consisting of pre-sorted beach sand, and on the crests of large inland dunes. Coarser and less sorted sand is characteristic of dune flanks and interdune areas.

Due to saltation transport and continued grain impacts, eolian sands become better *rounded* than sands accumulating in any other environment, but in many sand dunes not all of the grains are well rounded. Generally, smaller grains remain angular longer than larger ones. The *grain surface* of young eolian sands commonly displays frosting and small pits caused by grain impacts. Older eolian sands often lose this distinctive criterion due to diagenetic overgrowth and cementation of the pore space (Chap. 13.3).

Wind-blown sands travel with varying velocities. Sand blown over a flat and smooth surface migrates fast. As soon as the sand has to overcome obstacles of some size and elevation, it tends to settle on the leeside. The *migration of total dune bodies* is much slower than that of single grains. It is a function of the dune size and sand volume. Small dunes such as isolated barchans (see below) move relatively fast (several tens of meters per year), while large sand dunes and sand seas are fairly stable.

In the trade wind belt of the western Sahara, for example, the position of barchan dunes was repeatedly photographed from the air, because a railway transporting iron ore from inland to the coast was endangered by sand dunes. Isolated dunes migrate with a mean velocity of approximately $c = 300/H$ (m/a), if the height H of the dunes is measured in meters (mean value of H = 9 m, Sarnthein and Walger 1974). In a cross-section perpendicular to the dominating wind direction, a sand volume of approximately 200 m^3/a per 1 km width is transported in the form of barchan dunes seaward to the West African coast. Sand drift on flat bottom is, however, significantly more effective for transporting sand (on the order of 50 000 m^3/a per km width) in this region.

These values indicate that in this area large volumes of sand reach the coast line and the sea (see below). The giant draas of large sand

seas, on the other hand, migrate at rates as slow as 1 to 2 cm/a.

Bed Forms and Sedimentary Structures of Inland Dunes

Bed Forms

Accumulating eolian sand generates specific *bed forms* which are found more or less in all deserts. As long as the climate is sufficiently arid (precipitation less than 100 to 150 mm/a) vegetation on sand is virtually absent and the eolian bed forms are mobile. Dry, sandy bed forms move, but more or less maintain their shape. The barchan dunes with their two horns pointing downwind (Fig. 2.21a) represent a typical example of this behavior. They lose sand on the windward side and add about the same amount of sand on the leeside. The dune migrates by this process, but only slowly changes its shape and volume.

Eolian bed forms occur in various shapes and sizes. Here, only the most important types can be mentioned. The smallest features are *wind ripples* on flat ground (Fig. 2.20a). They display a high length/height ratio (ripple index ≥ 15) and commonly consist of medium to coarse sand, because the finer sand is sorted out and accumulated in larger dunes. The largest grains of wind ripples, possibly including pebble-size grains and heavy minerals, are concentrated along ripple crests. Cross lamination is frequently indistinct. These ripples may alternate with plane or low-angle bedded sand forming thin, sheet-like sand bodies around the margins of dune fields or interdune areas (sea below). Climbing ripples which occur on the inclined surfaces of all types of large dunes are widespread (Fig. 2.21a). They consist of fine sand, and their direction of migration (perpendicular to the ripple crests) frequently deviates from the dominant wind direction controlling the movement of larger dunes.

The *large dunes* tend to be oriented either parallel (longitudinal) or transverse to the prevailing wind direction (Fig. 2.21 a). Many forms, such as the *barchan dunes*, however, exhibit transverse (the center of the barchans) and longitudinal components (the horns). The less elevated horns migrate faster than the higher center of the dune. Isolated barchans develop in areas of limited sand supply at the edge of larger sand seas. They may be arranged in longitudinal or oblique chains. As a result of growing size and and diminishing

a

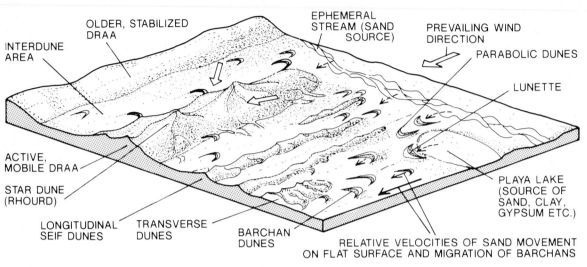

OLDER, STABILIZED DRAA

EPHEMERAL STREAM (SAND SOURCE)

PREVAILING WIND DIRECTION

PARABOLIC DUNES

INTERDUNE AREA

LUNETTE

ACTIVE, MOBILE DRAA

PLAYA LAKE (SOURCE OF SAND, CLAY, GYPSUM ETC.)

STAR DUNE (RHOURD)

LONGITUDINAL SEIF DUNES

TRANSVERSE DUNES

BARCHAN DUNES

RELATIVE VELOCITIES OF SAND MOVEMENT ON FLAT SURFACE AND MIGRATION OF BARCHANS

b

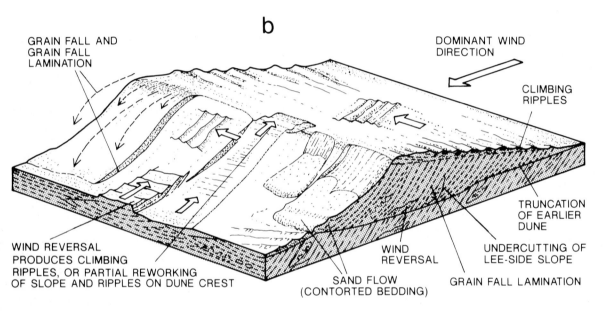

GRAIN FALL AND GRAIN FALL LAMINATION

DOMINANT WIND DIRECTION

CLIMBING RIPPLES

TRUNCATION OF EARLIER DUNE

WIND REVERSAL PRODUCES CLIMBING RIPPLES, OR PARTIAL REWORKING OF SLOPE AND RIPPLES ON DUNE CREST

WIND REVERSAL

UNDERCUTTING OF LEE-SIDE SLOPE

SAND FLOW (CONTORTED BEDDING)

GRAIN FALL LAMINATION

c   d

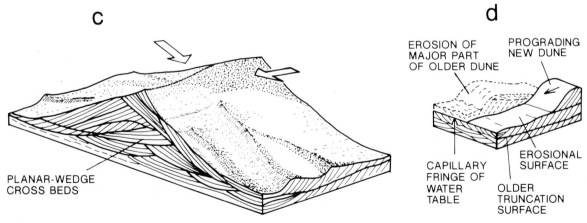

PLANAR-WEDGE CROSS BEDS

EROSION OF MAJOR PART OF OLDER DUNE

PROGRADING NEW DUNE

CAPILLARY FRINGE OF WATER TABLE

EROSIONAL SURFACE

OLDER TRUNCATION SURFACE

space between the barchan dunes, these may pass into more straight-crested transverse dunes. *Parabolic dunes* are mainly known from coastal regions and periglacial environments. In contrast to the barchans, the lower arms of the parabolic dunes tend to stick to the moist and/or vegetated ground, while the higher, central part of the dune migrates or is partially blown away.

Longitudinal sand ridges or seif dunes are the most widespread form in the present-day deserts, for example in Australia, but presumably their preservation potential in the ancient record is limited. Seif dunes develop parallel or subparallel to the dominant wind or the resultant of two wind directions (Fig. 22.21a). They grow downwind and tend to be spaced at equal intervals (often 100 to 300 m). Part of their sand may be derived from their interdune area from where it is transported by side winds or so-called helicoidal flow processes to the dune. Seif dunes typically reach several tens of kilometers in length and are 10 to 30 m high.

The largest eolian sand bodies are *draas*, i.e., elongate sandy hills of considerable height (up to several hundreds of meters) and wide spacing (some kilometers). They are typical representatives of giant *sand seas*. Because draas constitute huge volumes of sand, they require long time periods to form (on the order of at least 0.1 Ma) and move very slowly. Draas may be regarded as complex dunes resulting from the coalescence and growing together of other dune types. A special feature of draas are the so-called star dunes or rhourds (Fig. 2.21a) with high central peaks and radiating arms reflecting multidirectional wind patterns.

Sedimentary Structures

Eolian sands display various large and small-scale sedimentary structures (e.g., McKee 1979; Collinson and Thompson 1982). The most conspicuous features are large-scale pla-

nar cross beds which characterize transverse dunes and the central parts of barchans (Fig. 2.21b), but also occur in the other dune types. The cross beds develop on the lee-side of the dune crests and dip at angles around 30 degrees. In the case of transverse and barchan dunes, their strike is approximately perpendicular to the dominant wind direction. The foresets of the planar cross beds result from two different processes (Fig. 2.21b):

- *Grain fall* of sand particles blown over the dune crest blankets the lee slope. The angle of dip of these mostly thin, grain-fall laminae is 28 degrees or less.
- *Avalanching sand flows* on steeper slopes create thicker, irregular foresets. This process is favored by occasional rain falls providing an additional load to the normally dry sand. Due to its moisture content, the avalanching sand gains some cohesion and may break up into small blocks or display contortions. These features are distinctive for eolian sands, because loose sand under water cannot develop such structures. The slip planes of the sand flows are inclined up to about 35 degrees.
- Several *minor features* may be superimposed on the dominant, large-scale cross beds (Fig. 2.21b). Climbing ripples caused by various wind directions migrate with the dominant wind to the dune crest and provide sand for grain fall lamination and sand flow foresets. Additional climbing ripples result from aerodynamic eddies behind the dune crest or from wind reversals. They move upward on the lower lee-side slope of the dune and create a characteristic cross lamination which is frequently preserved in ancient eolian sandstones. Further climbing ripples of differing orientation can develop on both the windward and lee-side slope of the dune. Overall, climbing ripple stratification can distinctly contribute to the structures of large-scale bed forms. Strong wind reversals may blow away sand on the lee-slope and allow it to accumulate on the gentler windward side of the dune.

Fig. 2.21. a Most important dune forms in relation to prevailing wind directions. From right to left increase in sand supply and volume of sand accumulation, but decrease in the velocity of dune migration. Playa lake porvides, besides sand, sand-size clay aggregates, carbonate, and evaporites such as gypsum which make up part of the neighboring dunes. (Based on various sources, e.g., Cooke and Warren 1973; McKee 1979). **b** Large transverse or barchan dune mainly displaying unidirectional grain fall lamination and sand flow cross bedding. These large-scale bedding phenomena are superimposed by small-scale climbing ripple lamination caused by various subordinate, temporal wind directions. (Based on Hunter 1981, strongly modified). **c** Hypothetical cross section of longitudinal dune (after McKee 1979). **d** Truncation of older dune at capillary fringe of groundwater table and prograding new dune over erosion surface

An *interdune area* which is overridden by transverse or barchan dunes thus typically exhibits the following vertical succession of structures (from top to bottom):

- Climbing ripple lamination and plane bed lamination near the dune crest.
- Large-sclae foresets of grain-fall lamination and sand flow.
- Wind ripple lamination and parallel to subparallel laminae sticking to the flat ground due to adhesion caused by soil moisture and sparse vegetation (adhesion ripples, adhesion laminae).
- In places: lag sediment (desert pavement) where sand and dust is blown away.

This *dune sequence* is commonly incomplete in the ancient record for two reasons: (1) A migrating dune can truncate its own deposits such that only the lowermost part of the dune is preserved. (2) Large dune fields may be eroded by wind action down to the capillary fringe of the (climatically fluctuating) groundwater table, where the sand is stabilized by adhesion (Fig. 2.21d). In this case, widely extended, approximately horizontal truncation surfaces are generated. Particularly the dip of the preserved large-scale cross beds is used to determine paleo-wind directions and to identify the former dune type. Transverse dunes show the least scatter of wind directions, whereas longitudinal dunes tend to display a bimodal distribution of paleo-wind pattern (see below).

The *sedimentary structures of longitudinal dunes and draas* are complex as a result of more than one dominant wind direction and the coalescence of different basic dune types to these sand bodies. Our knowledge about the structures of compound dunes is limited, because it is difficult to study these phenomena in recent examples consisting of dry loose sand. Longitudinal dunes commonly exhibit two or three large-scale cross beds dipping laterally from the dune crest (Fig. 2.21b). Frequently, the upper part of these foresets is truncated by subsequent lateral wind action. In this manner, typical wedge-shaped planar cross beds are generated. In addition, these dunes may develop all the minor features previously mentioned for barchan and transverse dunes. However, only the lowermost portions of the dune, if any, are normally preserved. The huge draas and star dunes are assumed to produce particularly thick sets of cross beds (up to several tens of meters in height) which are also known from ancient eolian sandstones.

Finally it should be mentioned that eolian sands may display the imprints of rain drops and various biogenic structures. Burrows and traces of small animals, particularly arthropods, and molds of plant roots are common in present-day examples and were frequently found in ancient eolian sandstones. Even footprints of large animals were found in such rocks (McKee 1979).

Coastal Dunes

Coastal sand dunes normally do not reach the same large extent as many inland sand seas. They develop along both arid and humid coast lines. Their sand is derived from the beach zone and blown inland by prevailing onshore winds (Fig. 2.22a). In humid regions, the interdune and backdune areas are covered by vegetation and the inland migrating sand is soon trapped and stabilized by moist soils and plants. In arid regions, the coastal sand can travel far inland, produce a broad dune belt, and contribute to the formation of large, interior sand seas.

The *dune types* of coastal sands are principally the same as those of the interior deserts. Many coasts are characterized by a narrow longitudinal sand ridge of considerable thickness (mostly up to several tens of meters). Where the coastal sand ridge is blown away, transverse, barchanoid, and parabolic dunes may follow farther inland, but it is hardly possible to establish a general rule for the succession of special dune types. In the more arid zones, barchan dune fields originating from the coastal sands may pass landward into large transverse and longitudinal dunes. The internal structures of coastal dunes resemble those of inland desert dunes, but tend to be still more complex. They may be affected by storm wave erosion and flooding of their interdune areas, and locally they interfinger with fluvial, lake and lagoonal deposits. Root horizons from vegetation and generation of soils may alter the primary structures and facies of the sands. Dunes along arid coasts often alternate with coastal sabkhas and may prograde over supratidal algal mats and salt crusts.

Coastal dune sands frequently contain considerable amounts of bioclastic carbonate or may consist entirely of carbonate sand derived from various marine organisms. Such sands are readily stabilized due to the dissolution and reprecipitation of carbonate near their surface (McKee and Ward 1983). This *early lithification* of carbonate-bearing eolian dunes is

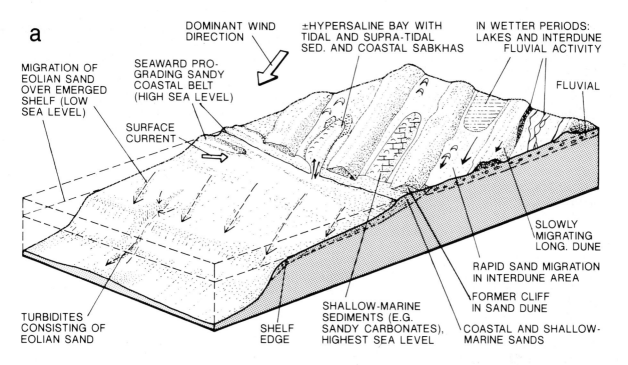

a

MIGRATION OF
EOLIAN SAND
OVER EMERGED
SHELF (LOW
SEA LEVEL)

DOMINANT WIND
DIRECTION

±HYPERSALINE BAY WITH
TIDAL AND SUPRA-TIDAL
SED. AND COASTAL SABKHAS

IN WETTER PERIODS:
LAKES AND INTERDUNE
FLUVIAL ACTIVITY

FLUVIAL

SEAWARD PRO-
GRADING SANDY
COASTAL BELT
(HIGH SEA LEVEL)

SURFACE
CURRENT

SLOWLY
MIGRATING
LONG. DUNE

RAPID SAND MIGRATION
IN INTERDUNE AREA

FORMER CLIFF
IN SAND DUNE

TURBIDITES
CONSISTING OF
EOLIAN SAND

SHELF
EDGE

SHALLOW-MARINE
SEDIMENTS (E.G.
SANDY CARBONATES),
HIGHEST SEA LEVEL

COASTAL AND SHALLOW-
MARINE SANDS

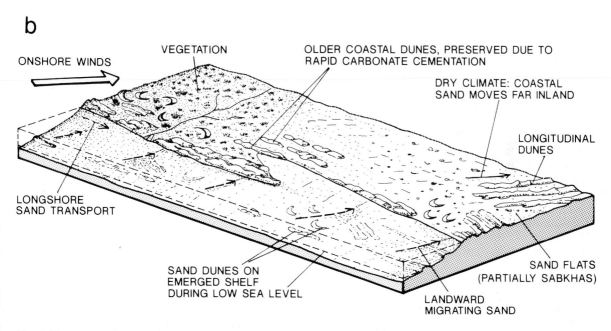

b

ONSHORE WINDS

VEGETATION

OLDER COASTAL DUNES, PRESERVED DUE TO
RAPID CARBONATE CEMENTATION

DRY CLIMATE: COASTAL
SAND MOVES FAR INLAND

LONGITUDINAL
DUNES

LONGSHORE
SAND TRANSPORT

SAND FLATS
(PARTIALLY SABKHAS)

SAND DUNES ON
EMERGED SHELF
DURING LOW SEA LEVEL

LANDWARD
MIGRATING SAND

Fig. 2.22. a Interaction between seaward migrating eolian sands and marine processes including relative sea level changes. The development of lakes or fluvial systems in interdune areas is also indicated. (Mainly based on Sarnthein and Diester-Haass 1977; Schwarz et al. 1975; Einsele et al. 1977). **b** Various types of coastal dunes developing on coasts with prevailing onshore winds. In arid regions, the coastal sand may migrate far inland to feed interior sand seas, while the sand is stabilized by vegetation in the more humid zones. During low sea level, shelf sand can contribute to the buildup of coastal dunes. See text for further explanation

known, for example, from the southern coast
of the Mediterranean, from Florida and the
Bahamas, or from southern Australia, where
several chains of Quaternary dunes are aligned
parallel to a prograding coast. The older the
dunes, the more they are cemented by carbonate and thus gain a high preservation potential
and are protected from truncation, as shown
in Fig. 2.21d.

**Large-Scale Facies Associations
with Eolian Sands**

Interaction with Fluvial and Playa Sediments

Continental eolian sands may alternate laterally and vertically with deposits of marginal
desert and fluvial environments. Sand dunes
often overlie alluvial fans, fluvial deposits, or
playa sediments and frequently pass laterally
into such sediments deposited in the interdune
or marginal dune areas (e.g., Langford 1989;
Langford and Chan 1989). Fluvial flow in
interdune areas erodes dunes, but may also
contribute to the preservation of basal portions of the sand dunes when both fluvial and
eolian aggradation occur successively. Then,
fluvial channel and overbank-interdune deposits may form elongate, lenticular bodies in
between eolian dunes. Flooding of interdune
areas can create ephemeral lakes. The lateral
interfingering of eolian sands and interdune
calcareous or siliceous lake sediments in Pleistocene and older sediments is mostly ascribed
to a change to a wetter climate (Fig. 2.22a).
Then the mobile sand dunes are also stabilized
to some extent by vegetation and soil formation. During the subsequent geologic development, however, the higher portions of the
dunes tend to become eroded by fluvial action.
Thus, in a normally subsiding continental
basin, the thickness of the eolian sands preserved is limited.

Thickness of Eolian Sands

The thickness of eolian sands in the modern
large sand seas commonly attains only several
tens to a few hundreds of meters. The average
sand thickness of widely extended longitudinal
dune fields is usually on the order of only 10
to 20 m, for example in Australia. One has to
consider, however, that the Australian dunes
accumulated in the relatively short time period
of approximately 30 ka (Wasson 1986; Wopfner and Twidale 1988). These and many other

modern occurrences of eolian sand are affected significantly by the shift of climatic
belts during the Pleistocene. These shifts diminished rather than enhanced the accumulation of large volumes of eolian sand, because
great proportions of the older sands were repeatedly redistributed by fluvial processes and
deposited in other environments, including
coastal areas.

Some ancient examples of eolian sediments
exhibit very thick sequences of wind-blown
sand. In Europe, Permian sandstones (Rotliegendes) mostly have a thickness up to a few
hundreds of meters, but locally they reach
more than 1000 m, probably due to accentuated synsedimentary subsidence. The Mesozoic Navajo sandstone of North America and
eolian sandstones in China also attain several
hundred meters in thickness. Longer persisting
dry climate, sufficient subsidence, and the
inability of plants to grow in desert environments in the early Paleozoic favored the accumulation of eolian sands.

Interaction with Shallow-Marine Sediments

Eolian sands may also interact with coastal and
shallow-marine sediments. In the case of dominant offshore winds as, for example, along
the western margin of the Sahara desert in
Africa, sand dunes, drifting sand, and dust are
blown into the Atlantic by trade winds (Fig.
2.22a). The sands are redistributed by waves,
longshore and tidal currents, and generate a
seaward prograding sandy coastal belt. The
interdune area of seaward migrating, longitudinal dunes may be occupied for some
time by hypersaline marine bays or lagoons.
Tidal and supratidal flats bordering coastal
sabkhas are characterized by sediments containing high proportions of eolian sand, including quartz-sandy algal mats and stromatolites (Schwarz et al. 1975). Short-period, high
sea levels, flooding the interdune areas, may
lead to the deposition of thin, shallow-marine,
calcareous sands in between large dunes (Fig.
2.22a). A particularly high, long-persisting sea
level, however, will cause dune erosion and
liquefied sand flows, and shift the entire
coastline landward. Then the basal marine bed
may consist of a massive, fine grained sandstone derived from the pre-existing, planated
eolian dunes, as for example observed in the
upper Jurassic of Utah, North America (Eschner and Kocurek 1986). The preservation of
the dune topography may be better with a
very rapid, low-energy transgression, when

the dunes were already cemented by evaporites or carbonate.

During sea level fall, the coastline moves toward the shelf edge, enabling the eolian sand to migrate over the emerged shelf. If the sand reaches the head of submarine canyons, it is trapped, easily liquefied, and transported by turbidity currents into deeper water (Sarnthein and Diester-Haass 1977; Eschner and Kocurek 1986). The present-day shelf sediments off west Africa, which accumulated since the Holocene transgression, contain high proportions of eolian sand and dust (Einsele et al. 1977). A situation similar to the Pleistocene marine sediments on the west African shelf is also assumed for Permian marine sandstones and siltstones in Texas and New Mexico (Fischer and Sarnthein 1988).

Distinctive Features of Eolian Sandstones (Summary)

The characteristic features of modern and ancient eolian sands and sandstones are summarized as follows (see also, e.g., Blatt et al. 1980; Galloway and Hobday 1983; Collinson 1986b):

Observations in single outcrops:
- Large-scale, 1 to more than 10 m high sets of steep, truncated tabular cross beds and planar-wedge cross beds are the most striking features in field exposures. The cross bed sets are either vertically stacked without muddy intercalations (large dunes), or they alternate with thin, flat-bedded or rippled interdune deposits. The dips of cross beds frequently indicate varying paleo-wind directions and wind reversals.
- Minor distinctive features include avalanching and brecciation of individual cross laminae, and an abundance of climbing ripples with a high length/height ratio, imprints of rain drops, desert pavement, faceted gravel, and other ventifacts.
- Ocurrence of plant roots, burrows and tracks of animals living in the desert or in coastal dune environments.
- In some regions: Intercalations of volcanic ash falls without any indication of fluvial transport as, for example, observed in Permian eolian sandstones of Argentina (Limarino and Spaletti 1986).

Lateral and vertical facies transitions:
- Occurrence of broad, widely extended deflation surfaces.

- Alternation with and lateral transition into playa and sabkha deposits, fluvial and alluvial fan systems.
- Great areal extent of interior sand seas, but more elongate geometry of coastal dunes.
- Lateral transition of coastal dunes or seaward migrating inland dunes to the shore zone and shallow-marine environments.

Texture and composition:
- Mostly fine to medium-grained sand, frequently well sorted, fairly to well rounded, frosted grain surfaces of modern eolian sands.
- Mineralogically immature to fairly mature, depending on the source and recycling of sand.
- Scarcety of mica; silt and clay are found only in the interdune area.
- Yellowish, brown, or red staining.

In spite of this considerable number of criteria, thin eolian sand layers of limited extent, intercalated in fluvial or other deposits, are frequently overlooked and sometimes difficult to identify.

2.3.3 Clay Dunes

Clay dunes or *lunettes* occur along the margin of seasonally drying shallow lakes in semi-arid climates, for example in North Africa (Tunesia) and South Australia. Mud from the lake bed is either entrained by waves and redeposited in emergent shoals (Lees and Cook 1991), or the drying mud and salt in the lake crinkles and forms silt and sand-sized pellets by aggregation (Bowler 1973; Pye 1987). These consist of clay, silt, some sand, and carbonate and/or gypsum and other salts. The pellets are blown away from the lake surface to the downwind lake margin. The larger *clay pellets* usually accumulate close to the lake in the form of flat, parabolic dunes; finer pellets are dispersed more widely. After rainfall, the clay dunes become stabilized as a result of cohesion between the wet clay and silt particles. Hence, no further migration of these dunes can take place, but they may grow larger and higher in subsequent dry seasons. Specific internal structures are normally absent in these dunes. They may become vegetated and repeatedly develop soils as they grow. Some examples show carbonate nodules originating from near-surface carbonate dissolution and reprecipitation similar to the formation of caliche. Some ancient, massive, red mudstones, for example in the Keuper (Upper Triassic) of central Europe, are believed to represent ancient clay dunes formed in a slowly subsiding, semi-arid basin with frequently switching shallow ephemeral lakes. In addition, the deposition of eolian dust (see below) may have contributed to the formation of such poorly defined deposits.

2.3.4 Eolian Dust, Loess

Transport, Texture, Source, and Composition

Similarly to sand-size particles, dust particles are entrained into the air by drag and aerodynamic uplift exerted by the turbulent wind current. The *threshold wind velocities* to move both fine sand and silt-size dust are on the order of 0.1 to 0.3 m/s (Morales 1979; Pye 1987). Storms capable of entraining and transporting dust commonly occur several times each year. The deflation surface, however, must be dry and the dust particles should not adhere to each other by cohesion. Even low moisture contents or low concentrations of cementing salts can significantly raise this threshold velocity and thus prevent dust transport. Furthermore, the roughness of the surface plays an important role. A dense desert pavement resulting from the removal of sand and dust finally terminates further deflation.

While there is little difference in the threshold velocities for the entrainment of fine sand and dust, the modes of *particle transport and distance of migration* of these grain size fractions differ significantly (Fig. 2.20b). The finer the dust particles are, the higher and longer can they be transported in the atmosphere. Global wind systems may carry fine-grained dust over distances of several thousand kilometers and distribute it widely.

Dust mainly consists of *silt-size particles* smaller than about 20 μm (Pye 1987); larger grains settle quickly back to the ground when the turbulence of strong wind decreases. Far-traveled dust particles are smaller than 10 μm and many are even smaller than 2 μm. Such dust may remain in the atmosphere for weeks and then largely accumulate in the oceans. Loess which is commonly transported shorter distances is composed mainly of particles in the range from 10 to 50 μm.

The dust particles may be derived from several *different sources*. Mechanical and chemical weathering provide silt and clay-size material. Grinding of solid rocks by glacial action and fluvial transport can produce substantial amounts of silt-size particles. In desert regions, silt is generated by eolian abrasion of rocks during saltation of sand grains (Fig. 2.20a). Finally, volcanoes can eject large amounts of silt-size material (Chap. 2.4). The relative importance of these various mechanisms to produce dust particles differs from area to area, reflecting the effects of climate, relief, lithology, and geomorphic history (Pye 1987). In regions of cold climate, glacial and fluvial abrasion as well as frost action are most important. Mountain regions with their high rate of mechanical erosion (Chap. 9.2) generally provide more silt-size material than low-relief areas.

The *mineralogical composition* of dust also varies in relation to the source area. Many, relatively coarse grained, local dusts are rich in quartz, feldspar, and carbonate minerals. They reflect the composition of nearby source rocks, such as quartz sandstones and carbonate sequences. With increasing distance from the source, however, textural and mineralogical sorting of the dust particles takes place. Consequently, far-traveled dusts tend to be enriched in fine-grained micas and clay minerals and also may contain organic matter such as pollen grains, fungal spores, seeds, etc. With the aid of such components, the source of the dust may be determined (e.g., Sirocko and Sarnthein 1989).

Dust Deposition

Eolian dust is deposited when the velocity and turbulence of the dust storm wane or the dust particles are washed out by precipitation. Another possibility is the coalescence of smaller particles to larger aggregates which can no longer remain in suspension. The deposition of dust is accelerated by the presence of vegetation which creates *surface roughness* and thus impedes air flow and subsequent resuspension (Yaalon and Dan 1974). Settling dust can also accumulate on the downwind side of hills and other topographic highs, or in depressions; moist surfaces trap the dust grains.

Measured present-day *rates of dust deposition* on land range between 10 to about 200 g/m^2 per year, corresponding to sedimentation rates of approximately 0.5 to 10 cm/ka. Locally and discontinuously, the sedimentation rate of dust must have been higher in the past, for example for Pleistocene loess with rates up to several tens of cm/ka (see below). The contribution of dust to oceanic sediments is less known, although this topic has been studied recently by several workers (see, e.g., Chamley 1989; Leinen and Sarnthein 1989; Weaver 1989). Dust deposition rates in the ocean reported in the literature vary between 0.1 and more than 10 g/m^2 per year. Taking into account the very low sedimentation rates of deep-sea sediments (in large areas 1 cm/ka, Chaps. 5.3 and 10.2), one can draw the conclusion that eolian dust should make up a significant proportion of marine sediments in regions, where offshore winds from deserts reach the open ocean. The carbonate-free fraction of early Cretaceous to late Miocene pelagic to hemipelagic sediments in the North Atlantic consists predominantly of eolian silt and clay (Lever and McCave 1983). In the eastern Atlantic, northwestern Pacific, and northern Indian ocean, eolian dust may comprise more than 50 % of the total modern sediment. Such sediments mainly consist of fine-grained quartz and clay minerals characteristic of their source areas. During the last 8000 years, the dust flux into the Arabian Sea reached nearly the same volume as the suspended load of the Indus river (Sirocko and Sarnthein 1989). It appears that dust deposition in the oceans increased in the Neogene in conjunction with a long-term climatic change.

Pleistocene Loess

Unconsolidated silty dust sediments of Pleistocene or regionally also somewhat older age, which are deposited on land are termed *loess*. Unweathered and nonredeposited loess is homogeneous, non- or weakly stratified and highly porous (Pye 1987). It consists predominantly of quartz, feldspar, mica, clay minerals, and carbonate grains in varying proportions and is mostly buff or yellowish in color. The principle grain size of loess ranges from 20 to 40 μm, but loess can also contain minor proportions of fine sand (≥ 63 μm) and clay (≤ 4 μm). Loess exhibits a wide areal distribution on the present land surface, is agriculturally important, and frequently used for dating Pleistocene processes and sediments.

At least two principally different *source areas of loess* are distinguished. In Europe and North America, most loess is derived from glaciated areas and accumulated under cold and relatively dry conditions (*cold loess*) with steppe vegetation not very far from the Pleistocene glaciers. Most of the silt-size material was blown out of widely distributed, partially abandoned meltwater streams.

By contrast, the so-called *desert loess* is derived from arid regions. This is, for example, true of most of the Chinese loess which covers a large area and attains locally more than 300 m in thickness (Liu Tungsheng 1988). According to paleomagnetic dating (Heller and Liu Tungsheng 1984) loess accumulation in China began 2.4 Ma B.P. and took place discontinuously in relatively cool, dry periods. During warmer and wetter phases, dust transport from the desert areas to the west slowed down and soils formed on top of the loess deposits. The average accumulation rates of loess in Asia and Europe varied between 2 and about 25 cm/ka, but in fact they were much higher during the cold stages when loess was actually deposited. Around the time of the last glacial maximum, loess accumulated at a rate of 0.5 to 3 m/ka (Pye 1987).

Loess is easily eroded by running water and fluvial action. The Yellow River (Huang Ho) in China and some other rivers of eastern Asia carry extremely high loads of suspended material, mostly derived from loess, into the sea. Here, they cause unusually high sedimentation rates of siliciclastic material and the rapid shallowing of coastal seas with prograding shorelines.

Pre-Quaternary loess deposits have rarely been reported. Considering the widespread occurrence of Quaternary loess, covering as much as 5 to 10 % of the present land surface of the earth, lithified loess (*loessite*) should be preserved more frequently in the ancient record than known so far. Possibly the thickest loessite (about 500 m) was described from Upper Carboniferous to Lower Permian mixed fluvial and eolian red beds in northwestern Colorado (Johnson 1989). Part of this sequence is characterized by homogeneous, structureless sandy siltstones interpreted as loessite. This view as supported by intercalations of indistinct paleosols, the lateral gradation of these beds into fluvial deposits, and the inferred paleogeographic setting of the sedimentary basin. Other occurrences of loessite were described from some Precambrian to Neogene deposits in North America and Norway.

2.4 Volcaniclastic Sediments (Tephra Deposits)

2.4.1 General Aspects and Terms

Hardly a depositional system exists in which volcaniclastic beds or some reworked pyroclastic components in other sediments are absent. In fact, in several of the basin types described in Chapter 1.2, volcaniclastic deposits play a significant part in the total basin fill. The contribution of volcaniclastic material to the total volume of sediments in various basin types may be as high as about 25 %. For this reason, a brief review of volcaniclastic-producing phenomena is necessary in the context of this book. Chapin and Elston (1979), Fisher and Schmincke (1984, 1990), Cas and Wright (1987), Schmincke (1988), Schmincke and Bogaard (1991) describe volcaniclastic sediments and their genesis in more detail.

The term *volcaniclastic sediments* or *tephra deposits* refers to all types of volcanic fragments regardless of grain size, grain shape, composition, origin, and depositional process. Volcaniclastic fragments are subdivided into the following groups:

- *Juvenile fragments*, comprising two subgroups:

1. *Pyroclasts* generated during explosive volcanic eruptions caused or dominated by degassing of magma.
2. *Hydroclasts* formed during eruptions due to the contact of hot magma with external water (phreatomagmatic eruptions).
- *Accessory or cognate fragments* are derived from older, co-magmatic rocks.
- *Accidental fragments* originate from underlying rocks of any composition.

The latter two types of fragments are collectively called *lithic fragments or lithoclasts*. Based on the grain size of volcanic fragments, the following terms are used:

- Blocks and bombs (diameter $\geq$ 64 mm).
- Lapilli (2-64 mm).
- Volcanic ash ($\leq$ 2 mm).
- Volcanic dust ($\leq$ 0.063 mm).

The most important group of fragments dealt with here are *juvenile fragments*. They represent chilled samples of the erupted magma and indicate the nature of the volcanic eruption. Depending on the state of magma prior to eruption, these fragments may consist of volcanic glass, or they are partially crystallized. Magmatic explosive eruptions of basalt and basaltic andesite are dominated by lapilli-sized pyroclasts which show many *vesiculae* caused by the degassing magma. Dark colored pyroclast accumulations of a basic to intermediate composition are called *scoria*. Because of the fluid nature of the erupting basaltic magma, the shape of larger fragments may be affected by their flight through the air and their impact on the ground surface (angular or drop-like form, ropy or stringy surface). More viscous, silicic to intermediate magmas (e.g., phonolites and trachytes) commonly produce highly vesicular pyroclasts *(pumice)*. These consist of volcanic glass, but may also contain crystals formed in the magma chamber prior to eruption. Pumice clasts are usually light in color and vary in grain size and shape. Some are less dense than water and hence float.

Phreatomagmatically fragmented juvenile clasts (hydroclasts) tend to be more blocky and less vesicular than pyroclasts. For angular glass fragments the term *shard* is frequently used.

Tephra deposits can be grouped into three genetic types according to their mode of transport and deposition:

- Fallout deposits.
- Flow deposits.
- Surge deposits.

These primary accumulations of volcanic activity are frequently reworked both on land and in the sea due to considerable relief commonly created by volcanism. Slopes unprotected by soils and vegetation or, under water, by early diagenetic processes favor the erosion and redistribution of primary volcanic material. This is incorporated into other sediments as *epiclasts,* causing significant changes in the composition of these rocks.

2.4.2 Tephra Deposits on Land and Below the Sea

Various Tephra Deposits on Land

Fallout tephra deposits result from pyroclastic and phreatomagmatic eruptions. Their material is ejected from a volcanic vent. Hot tephra and gas produce a buoyant plume rising high into the atmosphere (Fig. 2.23a). After a first phase of radial expansion, the ash cloud is directed downwind and may spread over distances of several hundreds to thousands of kilometers. With increasing distance from the source, the fallout layer becomes systematically thinner, finer grained, and better sorted. *Scoria fall deposits* are composed largely of basaltic to andesitic magma. *Pumice fall deposits* originate from highly viscous andesitic, to rhyolitic, phonolitic magmas. They represent the widely dispersed classic *plinian type of ash cloud fallout.*

Ash-fall deposits form under varying conditions including phreatomagmatic eruptions and co-ignimbrite processes (see below). They consist of lapilli and finer grained material. The pyroclasts and hydroclasts settle not only through air, but also through lake water and sea water. Thus, they generate widespread thin ash layers which are often well preserved in

depressions on land, in lakes, and in parts of the sea. Bentonite layers *(tonsteins)* in marine and lake sediments as well as in coal-bearing sequences are mostly derived from former volcanic ash. In coastal areas and in shallow seas, however, thin fallout beds are commonly reworked and mixed with normal marine sediments. In addition, intensive bioturbation frequently masks the occurrence of thin fallout layers in subaqueous environments. Ash layers on hill slopes may be incorporated into periglacial processes and be used to date normal debris covers. Mixed shallow-water sediments and thicker accumulations of tephra on the shelf may be transported by turbidity currents into deeper water and form ash turbidites which sometimes show "double grading" (Fig. 2.23a).

Pyroclastic flows mostly originate from the collapse of overloaded eruption columns. They are hot, partially fluidized gas-solid mixtures with high particle concentrations causing the flow to move downslope and to fill topographic depressions and pre-existing valleys (Fig. 2.23b). If the flow enters a lake or the sea, a second *"co-ignimbrite"* ash cloud is generated which may lead to widespread fallout on top of and beyond the pyroclastic flow deposits. Subaqueous pyroclastic flow material may mix with water and generate tephra-dominated mass flows. Evidence for the hot emplacement of pyroclastic flow deposits is, among other criteria, the occurrence of carbonized wood and welded tephra. Massive, welded silicic and pumiceous pyroclastic flows are referred to as *ignimbrites.*

An idealized section of a pyroclastic flow unit may begin with thin, stratified surge deposits (see below), followed by the main body of the pyroclastic flow or ignimbrite, and end with volcanic ash from fallout (Fig. 2.23c). Ignimbrites are generally poorly sorted, but may show some indistinct layering due to changes in grain size of the welded primary pyroclasts. Some grading of the basal zone and inverse grading of the higher part of the flow unit are frequently observed. The fallout deposit at the top consists of thin, stratified, graded ash layers. Stacked flow units may somewhat differ in grain size, color, and composition.

Three main types of pyroclastic flow deposits are distinguishable (Fig. 2.23c):

1. *Block- and ash-flow deposits* have an ash matrix and contain large blocks of the same magma type.

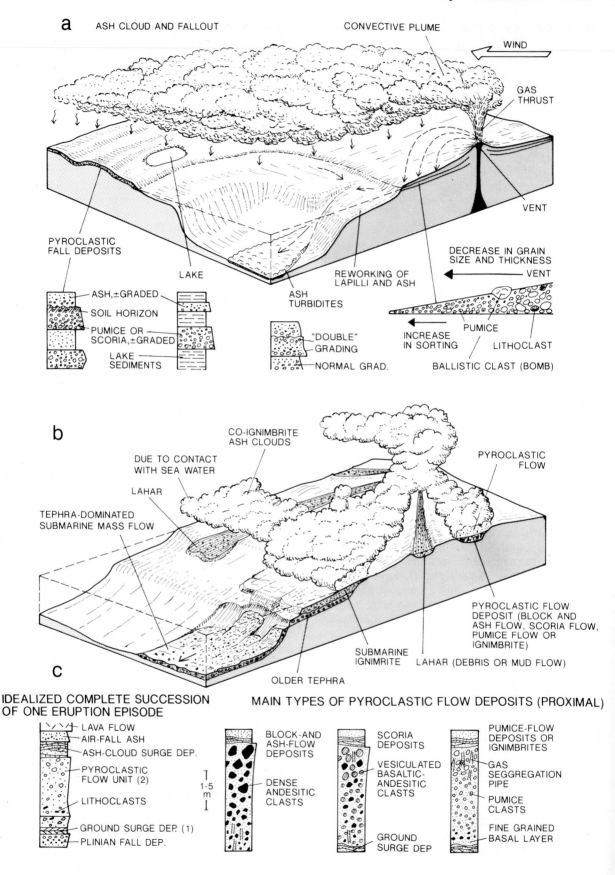

a   ASH CLOUD AND FALLOUT   CONVECTIVE PLUME

WIND

GAS THRUST

VENT

PYROCLASTIC FALL DEPOSITS

LAKE

REWORKING OF LAPILLI AND ASH

ASH TURBIDITES

DECREASE IN GRAIN SIZE AND THICKNESS

VENT

ASH, ±GRADED
SOIL HORIZON
PUMICE OR SCORIA, ±GRADED
LAKE SEDIMENTS

"DOUBLE" GRADING

NORMAL GRAD.

INCREASE IN SORTING

PUMICE

BALLISTIC CLAST (BOMB)

LITHOCLAST

b   CO-IGNIMBRITE ASH CLOUDS

DUE TO CONTACT WITH SEA WATER

LAHAR

PYROCLASTIC FLOW

TEPHRA-DOMINATED SUBMARINE MASS FLOW

PYROCLASTIC FLOW DEPOSIT (BLOCK AND ASH FLOW, SCORIA FLOW, PUMICE FLOW OR IGNIMBRITE)

OLDER TEPHRA

SUBMARINE IGNIMRITE

LAHAR (DEBRIS OR MUD FLOW)

c
IDEALIZED COMPLETE SUCCESSION OF ONE ERUPTION EPISODE

MAIN TYPES OF PYROCLASTIC FLOW DEPOSITS (PROXIMAL)

LAVA FLOW
AIR-FALL ASH
ASH-CLOUD SURGE DEP.
PYROCLASTIC FLOW UNIT (2)
LITHOCLASTS
GROUND SURGE DEP. (1)
PLINIAN FALL DEP.

1-5 m

BLOCK-AND ASH-FLOW DEPOSITS

DENSE ANDESITIC CLASTS

SCORIA DEPOSITS

VESICULATED BASALTIC-ANDESITIC CLASTS

GROUND SURGE DEP

PUMICE-FLOW DEPOSITS OR IGNIMBRITES

GAS SEGGREGATION PIPE

PUMICE CLASTS

FINE GRAINED BASAL LAYER

2. *Scoria flow deposits* consist of basaltic to andesitic ash, lapilli, and larger clasts.

3. *Pumice-flow deposits* or ignimbrites are poorly sorted, normally welded, massive tephra layers as described above.

Volcanic mud flows and debris flows (lahars) are generated in the same way as other gravity mass flows (Chap. 5.4), i.e., under normal temperature with the aid of water. They are, however, similar to pyroclastic flow deposits in composition and structure. In fact, they may form the distal facies of pyroclastic flows. Lahars tend to show a more polymict composition, better rounded clasts, and higher proportions of clay-sized matrix than pyroclastic flows. They form either during eruptions or independently of specific volcanic activities. They comprise not only a large proportion of the proximal facies on the lower slope of explosive volcanoes, but frequently travel distances of 10 to 200 km. Lahars also transport volcanic fragments to the coast and thus provide the source for volcaniclastic submarine debris flows and turbidites.

Pyroclastic surges are high-velocity, low-density turbulent flows which are caused by various mechanisms. Most important are *base surges* initiated by phreatomagmatic eruptions (interaction of magma with external water). These can create a collar-like cloud near the Earth's surface expanding radially in all directions (Fig. 2.24a) as also observed in nuclear explosions. Such a debris-laden base surge may reach velocities up to 100 m/s and shatter all trees and other objects many kilometers away from the locus of eruption. *Phreatomagmatic base surges* are commonly "wet" and have a low temperature. Other types of surges are associated with pyroclastic flows *(ground surges)* and the collapse of an eruption column *(ash-cloud surges)*.

Base surge deposits are commonly thin and irregular. They mantle to some extent the ground surface, but tend to reach greater thicknesses in topographic depressions (Fig. 2.24a). They predominantly consist of poorly sorted sand to gravel-size particles of various composition; juvenile clasts often make up only a small proportion of the total sediment, while older tephra and lithoclasts constitute the major part. The sedimentary structures of base surge deposits at any given locality are unidirectional. They include wavy planar lamination, low-angle cross bedding, climbing dunes and antidunes, and chute-and-pool structures. Characteristic features are also low-angle truncations and steeply inclined laminae draping obstacles on the stoss side (Fig. 2.24b and c). The occurrence of base surge deposits is mostly limited to a few kilometers from the source, after which they grade into laminated fallout layers.

**Tephra Deposits Derived
from Submarine Volcanic Activity**

Submarine volcaniclastic sediments are widespread and have a high preservation potential in the rock record. For those derived from volcanic eruptions on land, generation and mode of deposition was already mentioned in the previous section. The nature and products of volcanic activity under the sea are, however, much less known and must be interpreted from their results rather than from direct observation. A further problem is the differentiation of primary from redistributed (epiclastic) volcanigenic products.

The model in Fig. 2.24d is an attempt to describe two characteristic stages of a submarine eruption. During the most *active phase* of the volcano, large volumes of tephra are ejected high into the overlying water body. Part of the pyroclastic material settles back to the sea floor to form a subaqueous pyroclastic flow. Other proportions, such as highly vesicular pumice, may rise to the water surface and float until they come to rest along a coast. Fine-grained ash can remain in suspension for considerable time and thus be widely dispersed by ocean currents. When the volcanic activity slows down, the ejected material cannot maintain a steady pyroclastic flow. As

---

Fig. 2.23. a Plinian volcanic eruption with gas thrust and wind-driven convective plume. Tephra deposits include ballistic bombs and blocks, lapilli, and volcanic ash. Note that ash-cloud derived fall deposits uniformly drape the landscape (mantle bedding), form distinct layers in lakes, but are commonly reworked and redistributed in coastal areas and shallow seas.

b Collapsing eruption column leads to pyroclastic flow and the formation of ignimbrites and co-ignimbrite ash clouds with widespread fallout. **c** Idealized sections of pyroclastic flow deposits. (After Sheridan 1979; Cas and Wright 1987; Schmincke and Bogaard 1991). See text for further explanation

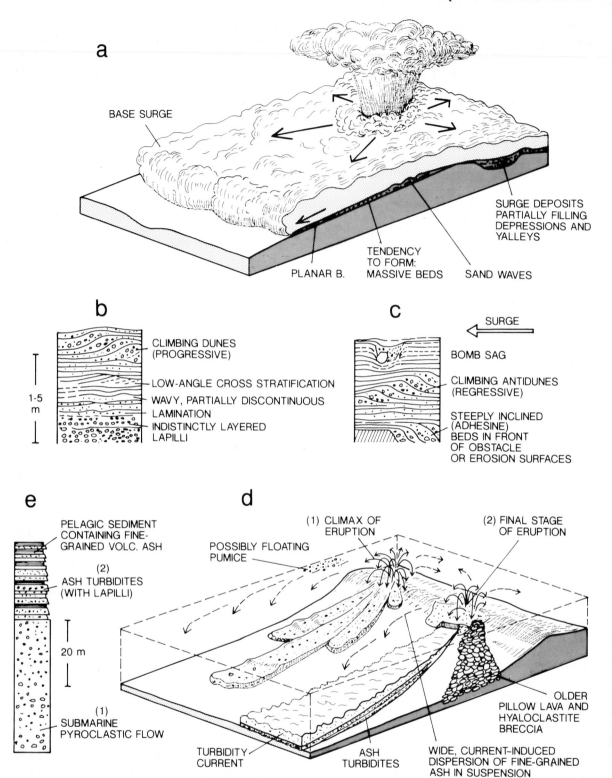

Fig. 2.24. a Base surge and partially topography-controlled surge deposits. **b** and **c** Characteristic sedimentary structures of base surge deposits. **d** Model for submarine eruptions; stage *1* Large volumes of ejecta generate subaqueous pyroclastic flows; stage *2* Decreasing amount of erupting material causes intermittent turbidity currents. **e** Idealized compound section (left) reflects maximum and waning stages of submarine eruptions. (After Wohletz and Sheridan 1979; Allen 1982; Cas and Wright 1987)

a result, it accumulates near the volcanic vent until the slope of the stored tephra fails. The sliding mass frequently evolves into a turbidity current, by which the volcanic material is intermittently transported into deeper water and deposited as *ash turbidites*. Towards the end of the eruptive phase, the turbidity currents become less frequent and carry finer ash. In addition, the pelagic settling intervals may contain some fine-grained ash which was distributed by normal oceanic currents.

Divers have directly observed submarine lava flows forming pillow lava. However, it is not clear whether submarine eruptions can generate hot pyroclastic flows and welded tephra (ignimbrites). Welding in pyroclastic flow can occur on island volcanoes and may continue when the flow enters the sea.

Deep-sea ash layers commonly exhibit a sharp base and a more gradational top. They are moderately affected by bioturbation. They originate from both subaerial and submarine volcanic eruptions and occur several hundreds of kilometers away from the source areas. Wind-driven ash fallen on the sea surface may additionally be transported by ocean currents. Bottom currents rework and redistribute ash accumulated on the sea floor. Ash settled on top of submarine highs can be carried by gravity mass movements into deeper water. Due to this variety of transport machanisms, the areal distribution and thicknesses of marine ash layers is very irregular. In spite of this, deep-sea ash layers provide excellent indicators of former explosive volcanism, and they permit high resolution dating of marine sediments.

Transport Distances of Various Tephra

As already pointed out in the previous sections, volcaniclastic material can be widely distributed and deposited far from its source. The approximate range of primary transport distances for various tephra is indicated in Fig. 2.25 a. Although these values greatly vary in relation to the size of the volcano, the magnitude of eruption, the type of magma, the relief of the land surface or sea bottom around the eruption center, and other factors, they clearly demonstrate the potential, significant contribution of volcanic processes to the filling of sedimentary basins. In addition to primary transport, loose volcaniclastic material is frequently reworked and redeposited by fluvial, eolian, and marine processes and can thus principally be transported over

unlimited distances. However, much of this redistributed, fine-grained and mostly altered volcanic material is mixed with "normal" siliciclastic sediments and remains unidentified.

Volcaniclastic sediments are easily recognized only up to a certain limit in the field. This limit is set, for example, by a minimum thickness of an ash layer, the occurrence of characteristic primary sedimentary structures (Fig. 2.24), or a high, easily visible proportion of reworked pyroclastic material in various types of sandstones and conglomerates, debris flows and turbidites. The examples in Figs. 2.25b and 2.26a demonstrate how characteristics of continental tephra deposits change with increasing distance from the source. Most of the typical volcaniclastic beds pinch out after some distance and are replaced by normal fluvial sandstones and mudstones with decreasing proportions of reworked pyroclastic material (see also below).

Depositional Environment and Facies of Tephra

The volcaniclastic facies can be defined in relation to distance from the source, the depositional environment, or to the primary petrologic-chemical composition (Fisher and Schmincke 1984, 1990). In this section, the depositional environment of the rocks is of particular interest, but other approaches such as the petrofacies of sandstones (based on the occurrence of special mineral phases and their ratios) are also important tools in recognizing and evaluating volcaniclastic rocks. In settings where the chemical characteristics of magma change with time, the chemical composition of concurrently produced volcanic rocks may permit a stratigraphic correlation of rock types which vary in other characteristics.

Continental Tephra Deposits

In the continental realm, proximal tephra deposits near the source tend to fill valleys on the slopes and form coarse-grained alluvial fans at the base of the volcanic structures (Fig. 2.25b). At greater distances from the source (intermediate source, medial), predominantly gravel- and sand-size volcaniclastics alternate with the deposits of braided streams and/or are mixed with these fluvial beds. In distal depositional sites far from the source, sand- and silt-size volcaniclastics are increasingly

a

PRIMARY TRANSPORT DISTANCES FROM SOURCE

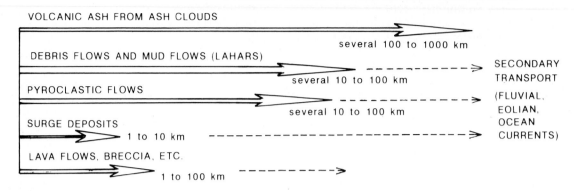

b

CONTINENTAL AND LACUSTRINE ENVIRONMENTS

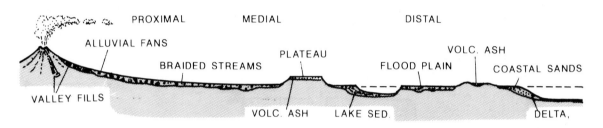

c

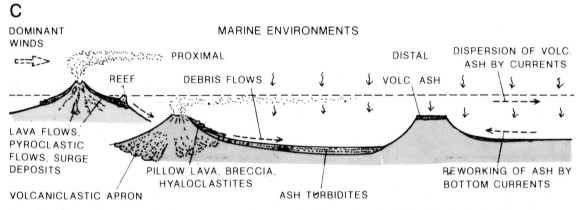

Fig. 2.25. a Frequent ranges of primary transport distances of various tephra. **b**, Volcaniclastics in continental environments in relation to distance from source. **c** Same as **b**, but in marine environments.

incorporated into the deposits of fluvial plains or swept into lakes. Finally, volcaniclastic material can also reach a marine delta and the coast. The areal distribution of primary volcaniclastic beds is more dispersed over the landscape than the linearly accumulating, purely fluvial sediments. Debris flows, sheet floods, and so-called *hyperconcentrated flows* with a relatively high proportion of suspended material tend to generate broad sheets of

sand- and gravel-size pyroclasts extending laterally beyond the range of normal fluvial channels. Airfall ash commonly covers even larger areas, but it is easily redeposited by fluvial activity. If ash accumulates on isolated plateaus which receive no other sediments and undergo little erosion, airfall deposits may build up sequences of some thickness.

Figure 2.26 illustrates *proximal to distal facies changes* in a Miocene *fluvial to lacus-*

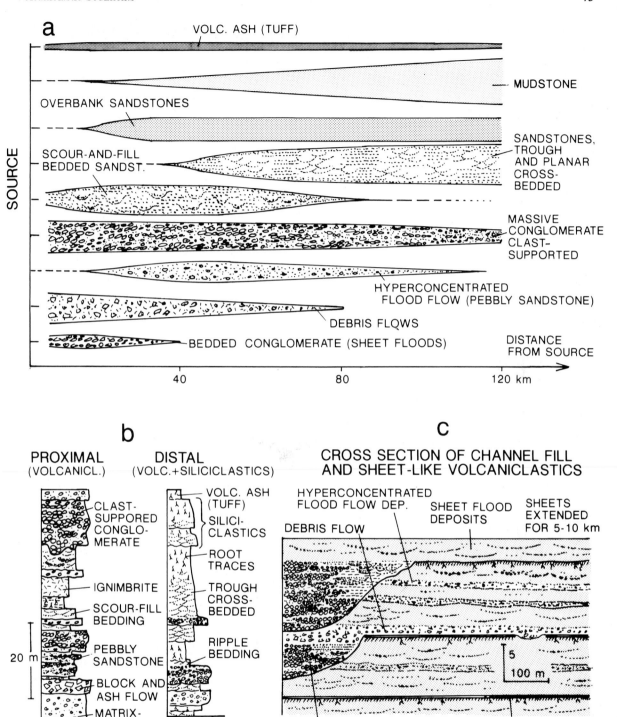

Fig. 2.26. a Relative abundance (thicknesses of bars) of continental volcaniclastic facies in relation to distance from the source, Miocene, Washington. **b** Characteristic sections of **a** in the proximal and distal range.

c Cross section perpendicular to flow direction showing syn-eruption, sheet-like volcaniclastics and siliciclastic to mixed valley fills generated in inter-eruption periods. (After Smith 1988)

trine environment in Washington which was episodically affected by volcanic eruptions (Smith 1988).

In a time span of 5 Ma, a 350 m thick, dacitic volcaniclastic/fluvial sequence accumulated which can be traced as far as 120 km to the east of the volcanic centers. In the proximal depositional regime, volcaniclastic sediments predominate, including debris flows, bedded conglomerates produced by sheet floods, and crudely stratified, normally graded deposits from hyperconcentrated flows. In addition, scour-and-fill bedded sandstones with pebble lenses indicate transport conditions related to volcanic processes. These sedimentary structures are characterized in transverse section by broad, low-angle onlapping stratification and are also thought to result from sheet floods. All of these volcaniclastic-dominated facies types pinch out at distances between 40 and 80 km from the source (Fig. 2.26a), whereas thin volcanic ash layers remain unaffected.

By contrast, the siliciclastic-dominated beds, whose material is mainly derived from various older rocks, display either little change (clast-supported conglomerates) with increasing distance from the paleovolcanoes, or their proportion in the total basin fill increases distally (trough and planar bedded channel sandstones, overbank sandstones, and mudstones). The fine-grained material consists mostly of altered and reworked volcaniclastics.

The interrelationship between extensive paleosols, volcanic ash layers (tuffs), and deposits from sheet floods and debris flows, on the one hand, and laterally restricted fills of concurrently incised channels, on the other hand, is shown in Fig. 2.26c. Channels and soils were formed in the relatively long inter-eruption periods, while rapidly deposited syn-eruption sediments covered larger areas and partially modified the paleo-drainage pattern.

Marine Tephra Deposits

The distribution and facies of volcaniclastic sediments in marine environments is more complex than on land. Besides subaerial processes, subaqueous gravity mass movements, surface and bottom currents play a part here (Fig. 2.25c). Around volcanic islands and seamounts, thick and widely extended volcanic aprons are built up which consist of lava flows, breccias, pyroclastic flows, hyaloclastites, and debris flows. At greater distances from the source, debris flows, mud flows, and ash turbidites accumulate to great thicknesses and alternate with pelagic marine sediments. Submarine plateaus may receive airfall ash and water-suspended fine-grained ash distributed by surface currents. Bottom currents can redistribute the tephra layers in various marine environments including the deep sea (cf. Chap. 5).

2.4.3 Volcaniclastic Sediments in Basins of Various Tectonic Settings

The contribution of primary (juvenile) and reworked (epiclastic) volcanic material in the total fill of various sedimentary basins depends on the tectonic setting of the basin (Chap. 1.2). Furthermore, a specific magma type erupting during volcanic activity may be restricted to a certain stage in basin evolution, for example to the initial stage of a continental rift zone. In this section, only a few major basin types are briefly mentioned in which volcaniclastics play an important part. Fisher and Schmincke (1984, 1990) and Cas and Wright (1987) treat this topic in more detail.

- **Intraplate continental volcanism** is associated with an extensional state of the crust allowing mantle magma to ascend to the surface. This situation corresponds with the early stage of a continental sag basin or aulacogen which does not evolve into an advanced rift zone (sea below). The nature of the volcanism is mostly basaltic, but more acid types also occur. Characteristic features of this volcanism are extensive, thick lava sheets (plateau basalts), valley fill lavas, and many cinder cones. In a continental realm, fluvial and lake sediments may be interspersed with basaltic volcanics.

- **Continental rift zones** with initial up-doming, as for example the East African Rift (Chap. 12.1), may display widespread basic, but in places also more silicic flood volcanism. Such lavas are erupted from large, laterally continuous fractures in the crust. The columnar jointed, massive lava flows may be accompanied by ropy *(pahoehoe)* and completely fragmented, *rough (aa) lava*. The preservation potential of these features is limited, but their weathered and reworked products may play a considerable role in the early filling of the subsequent graben zone. With the onset of extension and rifting, a petrologically diverse volcanism can be observed in limited areas of the axial graben. It may range from mafic to silicic rocks of more or less alkaline nature. Pyroclastic flows and ignimbrites are common, and their eroded material is incorporated into the predominating fluvial and lacustrine sediments of this stage. Later, in an advanced rifting stage with prevailing marine sediments (cf. Chap. 12.1), the influence of volcanism in the area of extended continental crust diminishes. In the newly developed spreading center, oceanic tholeiitic basalts are formed.

- **Intraplate oceanic volcanism** occurs on older oceanic crust. Typical representatives are the volcanic island-sea mount chains in the Pacific (Chap. 12.5.4). The basaltic volcanoes are built up of pillow lavas, *hyaloclastites* (fine-grained sideromelane glass shards), pyroclastic material, and massive lava. They eject large volumes of volcanic debris into the atmosphere and into the sea. The volcaniclastic material may mix with reef detritus, shallow-water carbonate, and pelagic sediments.

- **Mid-oceanic ridge volcanism** leads to the accretion of new oceanic crust. This mechanism yields the highest production rate of volcanic rocks, but only a minor proportion of the ascending magma is widely dispersed as volcaniclastic material. The basaltic magma extrudes from fissures in the median valley and produces abundant pillow basalts, other lava flows, volcanic breccias, and pyroclastic deposits. These volcanic rocks are characteristically associated with pelagic sediments such as radiolarian chert, siliceous green and red clay, or limestones. Metalliferous sediments of limited extent can accumulate near hydrothermal vents.

- **Island arc and continental margin arc volcanism** associated with an active subduction zone (Chap. 1.2, Fig. 1.2 and Chap. 12.5) is most effective in producing large volumes of volcaniclastic materials. The volcanoes are lined up in a narrow, at least several hundred kilometers long zone parallel to a deep-sea trench. The *magma production rate* is a function of the subduction rate and possibly enhanced by concurrent spreading of a backarc basin. In the case of an island arc, the volcanoes may be partially submerged or emerged above sea level. Their volcanic products are mainly basaltic or andesitic, but locally more silicic differentiates including ignimbrites have been observed. Arc volcanism along Anden-type continental margins exhibits high proportions of intermediate and silicic, largely calc-alkaline material. They may also eject large volumes of ignimbrites.

The 1980 eruption of Mount St. Helens in Washington is a medium-scale example of this type of arc volcanism. Due to westerly winds, the pyroclastic fallout was dispersed about 1000 km to the east of the volcano. At a distance of 300 km east of the source, the ash fall deposit was 2 cm in thickness (Lipman and Mullineaux 1981). Contemporaneous debris flows and mud flows, originating on the summit and the flanks of the volcano, travelled as far as 130 km and filled valleys and pre-existing lakes.

Estimates on prehistoric volcanic eruptions partially yield much higher volumes and transport distances of the pyroclastic material. Single explosive eruptions can eject 10 to 10^3 km^3 of magma and disperse volcanic ash more than 1000 km from the source (Kukal 1990: Schmincke and Bogaard 1991).

Forearc basins sediments (Chap. 12.5.3) range from continental to shallow-marine and deep-marine. In the case of an island arc, they tend to be rich in volcaniclastic material of basaltic to andesitic character. Forearc basins associated with continental margin arcs, on the other hand, accumulate more acidic to intermediate, calc-alkaline tephra. Deep forearc basins typically display abundant debris flows and thick turbidite sequences rich in volcaniclastics.

Backarc basins (Chap. 12.5.4) commonly provide less deep depositional environments ranging from fluvial, lacustrine to shallow-marine. If the dominant winds are directed toward the continent, backarc basins can accumulate high volumes of fallout deposits from ash clouds over the entire basin. Lava flows, pyroclastic flows, and debris flows, however, enter the basin from the side of the island arc and thus create an asymmetric basin fill. The backarc basins east of the Andes in South America provide a prominent example of thick continental volcaniclastic sequences derived from a volcanic arc to the west.

Tephra Volume and Time Span of Volcanic Activity

Island arc volcanism can persist for considerable time periods in relation to the time span needed for subduction of an ocean basin.

Given an ocean of 1000 km width and a subduction rate of 5 cm/a, the closure of the ocean and hence the volcanic activity will last for 20 Ma. The *volume* of ejected *volcanic material* depends on the proportion of the subducted crust which is molten and extruded at the surface. If a plate 1 km long (parallel to the subduction zone), 1000 km wide, and 6 km thick, is subducted and only 1 % of the volume of this rock body is transformed into volcaniclastic material, a sediment volume of approximately 60 km^3 results. Evenly distributed over a basin of 120 km width and 1 km length, the volcaniclastic material can form a layer 500 m thick in a time span of 20 Ma.

This simple and purely theoretical calculation demonstrates that subduction-related volcanism can produce indeed very large amounts of volcaniclastic material. Nakamura (1974, cited in Fisher and Schmincke 1984, 1990) estimated even higher production rates of volcanic material along island arcs (global total 0.75 km^3/a). According to his data, at least 10 % of the subducting oceanic plate may ascend to the surface in the form of volcanic material.

2.4.4 Alteration, Diagenesis, and Metamorphism of Volcaniclastic Rocks

The major part of volcanic minerals and particularly glassy materials are thermodynamically unstable under normal temperatures and pressures near the Earth's surface. In addition, volcaniclastic sediments have, in contrast to lava flows, high initial porosities and permeabilities for circulating water. As a result, volcaniclastic sediments are easily altered by pedogenesis as well as under increasing burial depth by diagenesis and metamorphism (cf. Chap. 13.3).

These processes are complex and cannot be treated here (see, e.g., Hughes 1983; Fisher and Schmincke 1984, 1990). Near the Earth's surface and on the sea floor, the unstable components take up water and are transformed into clay minerals (e.g., smectites, phyllipsite, and others), or they form various zeolites. With increasing burial depth, mechanical compaction and cementation reduce the pore space and create dense rocks. Chlorite and kaolinite may replace the early formed clay minerals, but later most water-bearing minerals are dehydrated and transformed into more compact mineral phases such as feldspars (e.g., albite), epidote, iron oxides, and also calcite.

For these and other reasons, the appearance (facies) of *ancient volcaniclastic rocks* in outcrops as well as their petrographic characteristics may deviate significantly from those of young counterparts. In a strongly altered state, it is hardly possible to discriminate between juvenile components and epiclastic volcanic material. Without some experience and the usage of special methods, the correct identification and genetic interpretation of such rocks is difficult. Guidelines for the classification of lithified and metamorphic volcaniclastic rocks are given by Cas and Wright (1987).

The frequently used terms *spilite* and *keratophyre* designate degraded metamorphic rocks of volcanic origin (Hughes 1983). These rocks have re-equilibrated under low-temperature conditions and are recrystallized without showing much deformation. The term spilite comprises rocks of predominantly basaltic composition and texture, while keratophyres represent rocks of intermediate to acid composition. In both cases, the initial mineral phases are replaced by a mineralogy corresponding more or less to the greenschist facies.

The metamorphic rocks of so-called greenstone belts in Precambrian shields also consist, to a great part, of former volcanic flows and volcaniclastic sediments.

2.5 Lake Sediments

2.5.1 Different Lake Systems and Their Sediments

General Aspects

Where ancient lake deposits are well exposed, their origin as lacustrine sediments can usually be recognized at first glance. Many lake deposits are particularly well horizontally bedded or laminated and, in addition, show frequent vertical changes in lithology and color. Fresh-water carbonates, thin dolomitic beds, iron oxides or siderite concretions, oil shales, and evaporites (commonly differing from those of the marine realm) characterize certain stages and types of lacustrine environments. One or several of these rock types are found in many lake deposits. Features characteristic of strong wave and current action are rare, although storm-wave erosion is known from large lakes. True tidal currents are absent, but wind stress may repeatedly drive water against a shore and create "wind-tidal flats".

Faunal remains usually comprise only a limited number of species and are often restricted to certain beds. They may indicate a varying chemical environment from fresh to brackish and hypersaline water. *Fresh-water lakes* usually contain less than 1 g/l dissolved constituents, *brackish water lakes* 1-5 g/l, and *salt lakes* more than 5 g/l. Highly concentrated brines reach ion concentrations of 200 to 300 g/l and more. The pH of lake waters may vary from about 4 to 10. Lake deposits are frequently associated with fluvial sediments and fossil soils (paleosols).

From these criteria one can conclude that lake deposits display some distinctive features, but also include a great variety of facies types and vertical sequences which cannot be described and explained here in detail. A more thorough treatment of this topic is given, for example, by Kukal (1971), Dean and Fouch (1983), Eugster and Kelts (1983), Allen and Collinson (1986) and several special publications mentioned below.

Lakes originate from different exogenetic and endogenetic geological processes. They occur in areas of crustal subsidence such as rift zones, continental sag basins, and foreland basins (cf. Chap. 1.2) which cannot be inundated by sea water. They are also very common in glaciated regions (Chap. 2.1) and in some arid to semi-arid lowlands where wind-induced deflation leads to morphological depressions. Most of the present-day lakes are shaped by glaciers and therefore provide poor analogs for the interpretation of ancient lacustrine systems, which were predominantly controlled by tectonic movements. Furthermore, only few modern lakes have an areal extent and accumulate sediment thicknesses similar to those of prominent fossil counterparts.

Lakes are often referred to as "clearing basins" of rivers. The total bedload and most of the suspended load of entering rivers settle out in the lake. Lakes with high influx of detrital clastics are short-lived, as for example ice-contact lakes, proglacial lakes (Chap. 2.1.1), and many lakes found in graben structures and rift zones bordered by mountain ranges. Small and shallow lakes of this type may become filled within a few thousand years or less and therefore rarely experience long-term drastic climatic changes. In contrast, large and deep lakes, and particularly lakes in areas of long-lasting crustal subsidence and low terrigenous influx, can persist during considerable geological time periods. It is usually this latter type of lake which displays the sensitivity of lacustrine sediments to changing climatic conditions indicated by pronounced, repeated vertical facies changes. Such variations affect either the lake area itself or its drainage area, or both of them.

Lake Systems (Overview)

The great diversity of lake basins and their sedimentary fill makes it difficult to summarize our knowledge in a simple classification scheme (Horie 1978; Allen and Collinsen 1986). Apart from the factors mentioned above, we have to consider physical, hydrological, biological, and chemical processes which play a great part, particularly in long-persisting lakes. In contrast to the open sea, waves and wind-driven currents are generally much less important in lakes for the distribution and reworking of sediments. These processes operate to some extent along the lake shores but normally do not affect the lake bottom, unless the lake is very shallow and large.

For these reasons, lake waters tend to become *stratified* either permanently or seasonally (Fig. 2.27b1). Permanent stratification is characteristic of tropical lakes (temperature stratification, *oligomictic lakes*) and lakes in which the bottom water is more saline than surface water *(meromictic lakes)*. In both cases, the bottom water *(hypolimnion)* can become completely stagnant and be depleted in oxygen and nutrients, which leads to restricted benthic life and the preservation of organic matter produced in the near-surface water *(epilimnion)*. If such stratified conditions are maintained for a long time period, i.e., for at least several thousands of years, laminated muds rich in organic matter *(sapro-*

pel) can accumulate. Hence, lacustrine black shales in the ancient record indicate either a warm climate with small seasonal variations in temperature and a lake basin with limited inflow and outflow of river water, or a completely closed lake system (see below).

In regions of temperate climate, the surficial lake water is warmer during the summer and less dense than the deeper water; hence, lakes with limited inflow and outflow are stratified. However, during the winter the surface water cools and reaches the same or a higher density as the bottom water. Consequently, the total lake water body can be turned over once or two times a year and mixed *(monomictic* or *dimictic* lakes, Fig. 2.27a). In this case, a permanently stagnant hypolimnion with fully anoxic conditions cannot develop. Similarly, a major river crossing a lake will cause underflow (Fig. 2.27a) if its water is cool and comparatively dense. Then the total lake water body including the hypolimnion is well supplied with oxygen and enables a fairly high seasonal biogenic production (e.g., diatoms, different types of algae, etc).

Lakes with marked seasonal changes in their physical, chemical, and biological conditions tend to produce fine *annual varves* which are a few tenths to a few millimeters thick (Fig. 2.27a1). Such varves are well developed and preserved in lakes with little benthic bottom life where influx of terrigenous material is not too high and irregular. Such conditions are fairly common in regions of humid climate where permanent, sufficient inflow of river water keeps the lake filled and enables outflow and thus a hydrologically *open system*. In this case, water chemistry, the areal extension of the lake, its relatively narrow beach zone, etc. (i.e., the entire depositional system) are kept more or less stable. Most of these open lakes are short-lived fresh-water systems which are dominated proximally by river-derived clastic sediments and distally by mixtures of fine-grained clastics, carbonate, and biogenic silica.

Fig. 2.27a-c. Overview of different lake systems. **a** Open lake system with inflow and outflow; the lake water chemistry is river-dominated. **a1** Well oxygenated, oligotrophic lake (poor in nutrients). **a2** Chain of connected lakes, at the downstream end open or closed; arid to semi-arid regions show a tendency to increased salinity and fractioning of evaporites from the higher to the lower lakes. **b** Closed lake systems without outflow, concentration of dissolved species increasing with lifetime of lake. **b1** stable stratification of lake water and eutrophic conditions with oxy-gen-deficient bottom water (also feasible for a). **b2** Perennial lake, sufficiently deep with enough inflow to maintain a permanent water fill. **b3** Ephemeral lake, usually dry or reduced to small shallow pond, seasonally or irregularly flooded. **b4** Ephemeral or permanently dry lake with groundwater table below lake floor, and downward leaching of salts. **c** Open sea-lagoon-lake system, controlled mainly by subsurface sea water and groundwater flow; regional and seasonal variations in water chemistry (Coorong model, Australia). For further explanation see text

a "OPEN" LAKE SYSTEMS

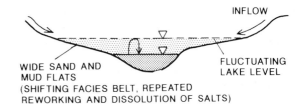

OUTFLOW   OVERFLOW   INFLOW

DENSITY OVERTURN   INTERFLOW
UNDERFLOW (THIN TURBIDITES)

WATER FRESH OR SLIGHTLY BRACKISH,
RIVER-DOMINATED CLASTIC SEDIMENTS,
TENDENCY TO:

a1 OLIGOTROPH

LITTLE NUTRIENTS

NO PERMANENT STRATIFICATION   BOTTOM LIFE
WELL OXYGENATED

ANNUAL VARVES OR POOR LAMINATION,
LITTLE ORG. MATTER, PARTLY IRON
HYDROXIDES, TURBIDITES

b2 PERENNIAL LAKES

FLUCTUATING LAKE LEVEL

PERMANENT WATER BODY, BUT CHANGING SALINITY

BEDDED CARBONATES AND/OR
THICK EVAPORITES ALTERNATING
WITH FINE-GRAINED CLASTICS
OR BLACK SHALES

a2 LAKE CHAIN
(IN SEMIARID TO ARID REGION)

INCREASING SALINITY

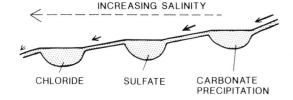

CHLORIDE   SULFATE   CARBONATE PRECIPITATION

b "CLOSED" LAKE SYSTEMS

INFLOW

WIDE SAND AND MUD FLATS
(SHIFTING FACIES BELT, REPEATED
REWORKING AND DISSOLUTION OF SALTS)

FLUCTUATING LAKE LEVEL

b1 EUTROPH

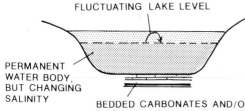

EPILIMNION (WELL OXYGENATED)

± STABLE STRATIFICATION   MESOLIMNION
BOTTOM LIFE SPARSE OR ABSENT   HYPOLIMNION, OXYGEN MINIMUM ZONE

DISTINCT LAMINATION, RICH IN ORG.
MATTER, ± SIDERITE CONCRETIONS

b3 EPHEMERAL LAKES
(PLAYA, INLAND SABKHA)

MUD CRACKS, CRUSTS OF MG-CALCITE,   FRESHWATER
DOLOMITE, GYPSUM, OR HALITE   CARBONATE (SPRING)

EVAPORATIVE PUMPING   CENTRAL THIN SALT
FROM HIGH WATER TABLE   DEPOSITION, ALTERNATING
WITH CLASTICS

b4 WATER TABLE BELOW LAKE FLOOR

CLAY DUNES (WIND-BLOWN CLAY PELLETS)

DEFLATION AND LEACHING OF SALTS

GROUNDWATER BRINE

c SEA-LAGOON-LAKE SYSTEM

CHANGING WATER CHEMISTRY

SEA

GROUNDWATER

DIFFERENT CARBONATES (INCL. DOLOMITE)

Hydrologically *closed lake systems* have no outflow (Fig. 2.27b). They are common in regions of warm, semi-arid climate. If evaporation of lake water is about equal to inflow, a permanent water body can be maintained *(perennial lakes,* Fig. 2.27b1 and b2). However, as a result of minor climatic variations in the range of 10 to one 100 years, the water level of such lakes may fluctuate considerably and affect a wide beach and near-shore zone (sand flats or mud flats, Fig. 2.27b and b1) by repeated emergence and inundation.

The water level of the well-known Great Salt Lake in North America, for example, varied by about 10 m in the Holocene, but in the late Pleistocene (Lake Bonneville stage) it stood about 300 m higher than the present level (Currey 1990).

During periods of high water level, the chemistry of lake water may change markedly by dilution of highly mineralized water or redissolution of salts precipitated during emergence in the near-shore zone. Deeper lakes tend to evolve stable stratified conditions with oxygen-depleted bottom water.

During longer time periods (thousands of years) the concentration of highly soluble species in the lake water tends to increase and finally to lead to the precipitation of salts. *Ephemeral* or *playa lakes* (inland sabkhas) are shallow and fall dry most of the time, apart from a central pond containing highly concentrated brines (Fig. 2.27 b3). By "evaporative pumping", i.e., ascending capillary water from a shallow groundwater table, they can precipitate salts at the surface. If the groundwater table drops too deep for capillary water to reach the surface, salts of the playa or salt pan are leached and transported into the underground (Fig. 2.27b4, also see below).

Finally, several open lake systems may be combined to form a *lake chain* (Fig. 2.27a2), where the last (lowermost) lake has either an outflow or is closed. In arid to semi-arid regions, such lake chains may display a systematic change in water chemistry and thus also in their precipitated salts. Similarly, the chemistry of lakes near the ocean may be influenced by both seaward flowing groundwater and landward intruding sea water (Fig. 2.27c). This situation can create special chemical environments for the formation of coastal carbonates including dolomite (Chap. 3.3.2).

Open Lake Systems in Regions of Humid Climate

Lakes in regions of temperate, humid climate can accumulate four main types of sediments (Dean 1981):

- Detrital clastic material (primarily siliciclastics and reworked carbonates).
- Autochthonous biogenic and bio-induced carbonate.
- Autochthonous biogenic silica.
- Sediments rich in organic matter.

Lakes Dominated by Detrital Clastics

The detrital clastic components of lake sediments reflect the relief, climate, and the rock types present in the drainage area of the lake (Chap. 9). Proglacial lakes (Chap. 2.1.1) and many lakes in mountainous areas are dominated by terrigenous clastic sediments including detrital carbonate. Due to a high sedimentation rate, the autochthonous sediment components are strongly diluted, or they are produced in quantities too low to gain a significant influence. Sedimentation in such lakes is largely controlled by physical processes distributing both the bedload and supended load of the entering river(s) over the lake (see, e.g., Sturm and Matter 1978). The river bedload builds a delta out into the lake which usually has a lobate or birds-foot form and typically consists of fluvial topsets, lake foresets, and bottomsets (forming a classical "Gilbert-type" delta). However, where rivers discharge into very shallow lakes, less regular deltaic sediments may accumulate.

In deeper lakes, a great part of the sandy river bedload is transported from oversteepened delta slopes into deeper water either directly by *underflows* during river floods, or by gravity mass movements (evolving into turbidity currents, see Chap. 5.4.2). Thus, the principal sediment types on the bottom of such lakes are silty-sandy, comparatively thin turbidites alternating with fine-grained mud which settles slowly from suspension. The sediment-laden flows of large rivers can also form subaqueous channels with levees on the delta slope extending into the deeper lake (Fig. 2.28 a) as, for example, observed in Lake Geneva.

If the river water with its suspended load during floods is less dense than the bottom water of the lake, the supended matter is distributed by *overflow* (surface currents) or *interflow* over the lake (Fig. 2.27a). Rhythmic sequences resulting from these processes range from sandy, mostly thin-bedded proximal turbidites to varve-type distal successions. The recurrence time of turbidite events is seldom controlled by the seasons, but rather by irregular storms occurring several times per year and/or rare larger events. Lake slopes at some distance away from river deltas receive only fine-grained material from overflows and intermediate flows (apart from autochthonous sediments, see below). Part of these fine-grained slope sediments is transported by slumps and mud flows into the deeper lake.

The sedimentation rates of lakes receiving water from large to medium-sized rivers draining nearby mountainous regions are very high. The following orders of magnitude for the different zones of a detrital, terrigenous-dominated lake are characteristic (also see Wright et al. 1980; Hsü and Kelts 1984; Sun Shuncai 1988):

- Prodelta area: several tens to hundreds of meters per thousand years (ka).
- Lake center: 3 to 10 m/ka.
- Lake slopes: 1 to 3 m/ka.

These values demonstrate that such lakes have a short life time and that prodelta and proximal bottom sediments make up the bulk of their sediment fill. This typically shows a distinct coarsening-upward sequence from silty, varve-type bottom sets to sandy prodelta foresets and fluvial sands and gravel (Fig. 2.28b).

Lakes with Significant Proportions of Biogenic Sediments

Many lakes or parts of them exhibit a transitional stage between over-supplied and sediment-starved conditions (Fig. 2.28c through e). The contribution of biogenic skeletal components to lake sediments is strongly influenced by the chemistry of lake water and nutrient supply. Waters rich in earth alkali ions, particularly calcium, favor the growth of lime-secreting organisms, whereas lakes poor in calcium ions but relatively rich in dissolved silica and nutrients (such as phosphorus and nitrate) are characterized by a relative enrichment of biogenic opaline silica. Usually both biogenic carbonate and opaline silica are found in lake sediments, but in widely varying fractions of the total sediment.

In so-called *hard-water lakes* which, at least seasonally, are oversaturated with respect to calcium carbonate, calcareous shells of organisms play a significant part in the littoral zone, even in lakes which are well supplied with detrital clastics (Fig. 2.28 c). Charophytes and bluegreen algae extract CO_2 from lake water for photosynthesis and thus cause precipitation of calcium carbonate at their surface. Successive layers of calcite coatings form different types of crusts and nodules (oncoids), associated with reworked, sand-sized carbonate.

During the warm season calcium carbonate is also precipitated chemically in the form of tiny micritic crystals of low-Mg calcite. Then the epilimnion becomes markedly oversaturated as a result of increased temperature and CO_2 consumption by phytoplankton (Fig. 2.28d). The tiny calcite particles are partially redissolved in the hypolimnion which is undersaturated with respect to calcium carbonate, due to the release of CO_2 from decaying organic matter and a lower pH. The rest of the micritic calcite (lutite) generates a thin, light-colored layer in the profundal lake zone which is later (in fall and winter time) overlain by clayey material and settling organic debris (Kelts and Hsü 1978). In spring a high production of diatoms may contribute to the formation of such *nonglacial annual varves*. However, these varves on the lake bottom can only be preserved when benthic life is absent or very limited.

As demonstrated in Fig. 2.28e, the oxygen content of the hypolimnion (as well as terrigenous input) may vary from lake to lake or in the same lake with time. The bottom sediments of such lakes therefore may show transitional facies between bioturbated marly silts and clays, marls, and light-colored, highly calcareous muds (seekreide), or between sediments poor and rich in organic matter. Muds containing considerable amounts of organic detritus but still showing some bioturbation are called *gyttja*. Well laminated, organic-rich muds are referred to as *sapropel,* which frequently contains more than 10 % organic carbon (also see below).

Soft-water lakes poor in calcium but sufficiently rich in nutrients, particularly phosphorus and nitrate, may accumulate sediments rich in opaline silica derived largely from diatom frustules. Such lakes are common in regions of cool climate. A prominent example is the deep, relatively old Lake Baikal in Siberia, which exists since the Miocene and has accumulated about 2000 m of sediments. The sedimentation rate of such lakes is comparatively low (on the order of 10 cm/ka). Arctic lakes, which are not associated with glaciers,

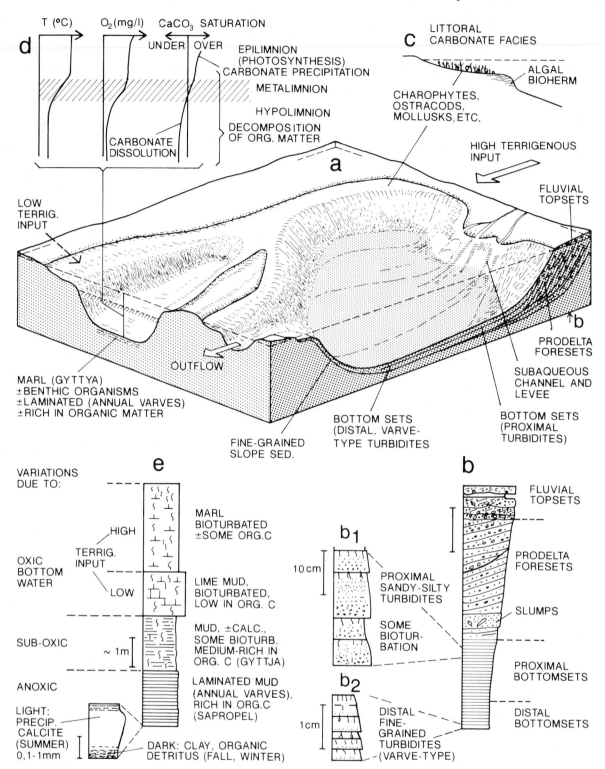

Fig. 2.28. a Scheme of open lake system with high terrigenous input into main basin and low input in adjacent hard-water lake. **b** Coarsening-upward sequence of detrital clastics-dominated basin fill with (**b1**) proximal and (**b2**) distal (varve-like) turbidites. **c** Littoral carbonate facies influenced little by terrigenous material. **d** Some charac-teristics of the epi- und hypolimnion of hard-water lakes allowing calcite precipitation in summer time (based on Dean 1981). **e** Variations in profundal (bottom) sediments of hard-water lake due to changing terrigenous input and oxygenation of bottom water. Note the occurrence of annual varves

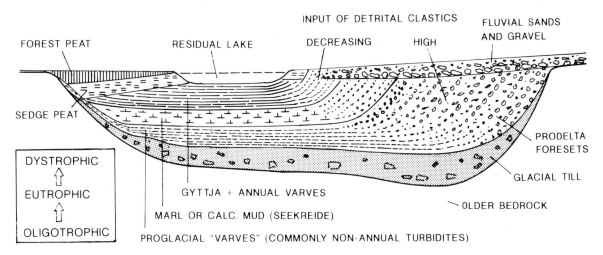

Fig. 2.29. Generalized scheme of the post-glacial sediment fill of a glacier-shaped lake. (After Dean 1981). It is assumed that the input of detrital clastics decreases with time and finally ends

commonly are poor in nutrients and organic production and often receive little detrital material (Kipphut 1988). Lakes in arid polar regions may even precipitate salts such as hydrated sodium sulfate and calcium chloride (Müller 1988). Shallow, poorly drained lakes with the growth of peat tend to become acidic and develop reducing conditions.

Post-Glacial Lake Evolution

The **post-glacial development** of a glacier-shaped lake in humid temperate climate is shown in Fig. 2.29. The overall tendency is the evolution from a relatively deep, large, water-filled basin to a small pond and finally dry land. At the beginning of this development, the lake floor may be coverd by thick glacial till (as, e.g., observed in several lakes along the northern rim of the Alps). The first lake sediments after the retreat of ice are sandy prodelta foresets and silty *glacial varves* deposited by turbidity currents in the profundal zone. These sediments and their relatively high sedimentation rates reflect intensive mechanical erosion and reworking of bedrock by ice action and meltwater in the drainage area. Vegetation is absent or sparse in the surrounding land and, due to the still unfavorable climatic conditions and low nutrient supply, the organic productivity of the lake is also very low at this stage. Consequently, the rapidly accumulating clastic sediments are poor in organic matter. In a second stage, vegetation starts to spread out over the land and reduces

the input of detrital clastics into the lake. Simultaneously, dissolution of carbonates and chemical weathering of silicates in the drainage area increase and thus provide the lake with calcium, silica, and nutrients.

Once calcite has reached saturation, the lake precipitates calcium carbonate in summer time as described above and accumulates, depending on the terrigenous input, a sediment more or less rich in carbonate (marly clay, marl, seekreide). This sediment is still fairly poor in organic matter. As soon as a dense vegetation cover is established, more plant remains are swept into the lake, and the fertility of the lake increases as a result of higher nutrient supply and recirculation of nutrients from decaying organic matter within the lake. Thus, the lake tends to become *eutrophic* with a high production of phytoplankton and algae in the epliminion and an oxygen deficiency in the hypoliminion. The resulting sediment is a gyttja or sapropel, which still may contain a considerable proportion of calcium carbonate.

The relatively high lacustrine sedimentation rate, as compared to marine sediments, favors the preservation of organic matter (Chap. 10.3.3). As the lake shallows, the littoral zone of rooted aquatic vegetation begins to grow outward into the lake, the algal gyttjy around the lake margins is replaced by *sedge peat* (Fig. 2.29). Peat provides a stable substrate for the growth of higher plants, the remains of which generate forest peat. If the main river is diverted from the lake, the lake water becomes poor in nutrients and acidic (i.e., *dystrophic)* as a result of decaying plants with

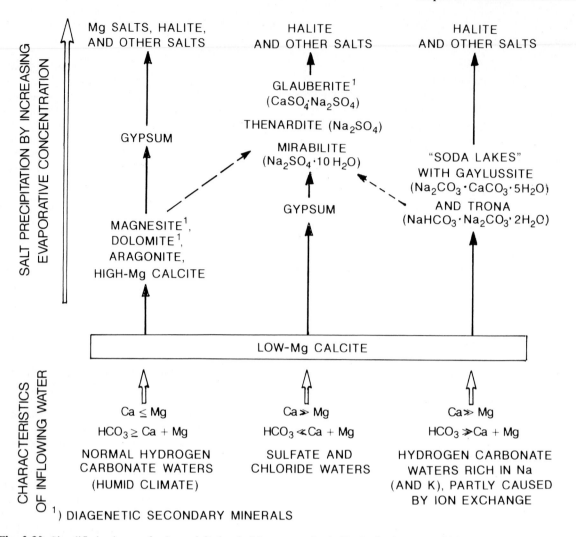

Fig. 2.30. Simplified scheme of salt precipitation in lakes due to increasing evaporative concentration. The principal lines of evolution are controlled by the composition of inflowing water; minor deviations from these trends are omitted. *Vertical sequences* of lake sediments reflect succession of salt precipitation. (After Eugster and Hardie 1978)

the generation of humic substances. Consequently, the life of the lake ends with a peat layer or fluvial deposits on top of prodelta foresets.

Closed Lake Systems in Semi-Arid to Arid Climate

Overview

Salt lakes can show a variety of phenomena depending on the characteristics of their drainage areas (e.g., volcanic versus sedimentary rocks) or their stage of evolution. The most important types of salt deposits in lakes and their possible vertical successions are summarized in Fig. 2.30. Regardless of the composition of inflowing waters, the first mineral which is precipitated chemically or biochemically due to an increasing salt concentration, is normal, low-Mg calcite. Then, in relation to the chemistry of inflowing water, three different lines of development can be distinguished:

– The inflowing water is of the common, earth alkali-hydrogen carbonate type which is characteristic for many rivers draining carbonate-bearing sedimentary rocks and plutonic rocks in highlands. Prograding evaporation leads to the precipitation of earth alkali carbonates (*carbonate lakes*) and later to gypsum and other evaporites.

- A second line of salt deposits starts with waters comparatively rich in sulfate and/or chloride, as well as calcium. Then calcite precipitation is followed by gypsum and Na_2SO_4 minerals including glauberite (Fig. 2.30, *sulfate lakes*). The final stage is again characterized by halite and other highly soluble minerals.

- The prerequisite for a third series of salts is lake water rich in hydrogen carbonate and alkali ions. Such waters may come from drainage areas rich in volcanic rocks or from regions where Na is enriched by cation exchange in the subsurface. After a *soda lake* stage with the accumulation of sodium carbonate minerals, again halite and other highly soluble minerals precipitate.

The *residual brines*, i.e., the brines after the precipitation of halite, may contain rare elements, such as K, Li, U, P, B, F, Br, and J. This is in particular the case, when the drainage area of the lake is rich in volcanic rocks and primary minerals undergoing first-cycle weathering. In some present-day salt lakes, some of these elements are of economic interest.

In ancient lake deposits, the highly soluble constituents are commonly missing, because they were dissolved by circulating groundwater or dissipated by diffusion into neighboring rocks. Furthermore, some of the primary salt minerals may be replaced by secondary, diagenetic minerals (e.g., dolomite, glauberite, and gaylussite). In addition, some authigenic silicate minerals may form in conjunction with salt deposits (e.g., zeolites).

Alkaline brines can dissolve considerable amounts of silica. If the lake water is diluted and the pH drops to lower values, bedded chert or chert nodules are precipitated, as observed for example in Lake Magadi in the East African rift zone (Eugster and Hardie 1978; cf. Chap. 12.1.2).

Closed lake systems have in common that their fluctuating water level causes exposure and considerable reworking of their nearshore sediments. They exhibit widely shifting facies belts, and their deltas move basinward during lowstands. Sheet floods may repeatedly inundate large areas and leave behind thin, graded beds.

Carbonate Lakes

Precipitation of calcite ($CaCO_3$) and later also of gypsum ($CaSO_4 \cdot 2H_2O$) leads to an increasing Mg/Ca ratio in the lake water. This, in conjunction with a high pH (about 9) and high alkali ion concentrations, favors the formation of Mg-rich carbonates such as high-Mg calcite and, in extreme cases, proto-dolomite and hydro-magnesite. During early diagenesis, dolomite ($CaCO_3 \cdot MgCO_3$) and sometimes even magnesite ($MgCO_3$) are generated (Müller et al. 1972). In many modern to Pleistocene lakes, dolomite may have formed as primary mineral (Last 1990).

Due to a balance between inflow and evaporation, the level of such lakes may be kept more or less constant, but the ion concentration of the water increases with time (Fig. 2.31). Then the resulting sedimentary sequences in the lake center show an evolution from calcareous to dolomitic marls, which may be rich in organic matter, and finally to evaporites (Fig. 2.31 e). If such a long-term trend is superposed by shorter term climatic variations, the resulting sequence frequently displays rhythmic or cyclic phenomena, for example marls alternating with carbonate layers, or kerogen-rich marls alternating with claystones or marls less rich in organic matter (Fig. 2.31f).

An alternative to this model is a slowly dropping lake level leading likewise to an increase in ion concentration of the shrinking water body. Consequently, the highest elevated littoral zone is characterized by biogenic and bio-induced carbonate deposited under non-saline conditions (Fig. 2.31b). This deposit may alternate with fluvial sands, beach sands, or alluvial fans. In stages 2 and 3 of the lake evolution, the littoral zone migrates basinward and rests on sediments deposited during a higher lake level (Fig. 2.31c and d). Thus, shallowing-upward sequences are generated which also indicate an upward increase in salinity. In a cross-section from the marginal, littoral zone to the center of the lake (Fig. 2.31, b to d), the carbonate facies changes from that of a nonsaline to a highly saline lake. Marginal, chemically precipitated micritic carbonate mud is often exposed to the air and then may develop desiccation cracks or tepees (Chap. 3.2.2).

Although *siderite concretions* are rare in the sediments of present-day lakes, they appear to be rather common in some ancient lake and brackish water deposits. Siderite can form diagenetically under reducing conditions

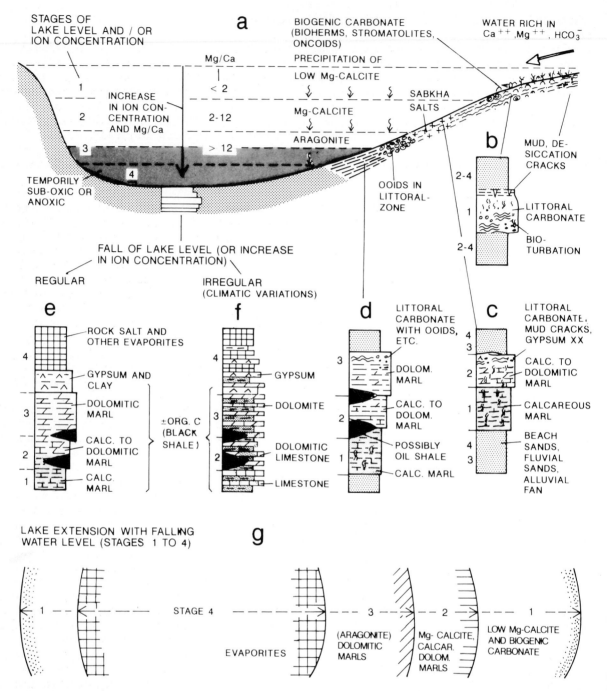

Fig. 2.31. a Scheme of carbonate-dominated, more or less closed lacustrine depositional system where either (1) the lake level maintains its elevation but ion concentration increases with time, or (2) ion concentration grows with falling lake level. In both cases, the Mg/Ca ratio increases, and primary carbonate precipitation proceeds from low-Mg calcite to aragonite. It may be followed by gypsum and other evaporites. During early diagenesis high-Mg calcite and aragonite are frequently converted into dolomite.
b, c, d Incomplete marginal lake sections resulting from falling lake level and prograding fluvial sediments. **e** and **f** Complete sections in lake center, **e** without, and **f** with influence of short-term climatic variations. (Based on different sources, e.g., Müller et al. 1972; Kelts and Hsü 1978; Eugster and Kelts 1983)

where the calcium and sulfide contents of pore waters are low and, in addition, the Fe/Ca mol ratio relatively high (≥ 0.4; Füchtbauer and Richter 1988). Since the water of many lakes contains much less sulfate than sea water, the source for generating iron sulfide with the aid of sulfate reducing bacteria is limited. Therefore, iron carbonate has a better chance of being formed in lake sediments than in marine sediments, provided iron-bearing minerals can release sufficient iron.

Sulfate Lakes, Soda Lakes, Chloride Lakes

After precipitation of gypsum and possibly other sulfate minerals, the remaining lake water contains only salts of very high solubility. In order to precipitate these components, i.e., predominantly Na, Mg, Cl, and SO_4, a further drastic increase in the total ion concentration of the lake water is required. The nature and succession of salts deposited after this process is controlled by the chemistry of the lake water (cf. Fig. 2.30). The most prominent salt occurring in most of the lakes at this stage is halite (NaCl) (Fig. 2.32a). It may be accompanied by Mg salts and some other compounds which are rarely preserved in the ancient record. Saline waters rich in sodium and sulfate precipitate thenardite (Na_2SO_4), mirabilite ($Na_2SO_4 \cdot 10H_2O$), and glauberite ($CaSO_4 \cdot Na_2SO_4$) before halite is formed. Soda lakes precipitate natron ($Na_2CO_3 \cdot 10H_2O$) and trona ($NaHCO_3 \cdot Na_2CO_3 \cdot 2H_2O$) (Fig. 2.32a2 and a3).

As a result of increasing aridity these lakes may develop the following sequence (from top to bottom):

- Trona and gaylussite ($Na_2CO_3 \cdot CaCO_3 \cdot 5H_2O$, formed diagenetically).
- Dolomite and gaylussite (in summer).
- Calcite and dolomite.

Prior to the precipitation of trona, the ion concentration of the brine must be very high. Therefore, the comparatively low evaporation loss of one year can cause precipitation of a fairly thick trona layer on the order of 5 mm. Similarly, annual layers of halite (rock salt) reach a thickness on the order of 1 cm (cf. Chap. 6.4.2).

Playa Lakes (Inland Sabkhas)

Whereas perennial, long-persisting salt lakes can accumulate rather thick and pure salt deposits (Fig. 2.27b2), ephemeral playa lakes usually leave behind only thin salt layers alternating with clastic material. These playa lakes are dry most of the time, with a salt pan in their center (Figs. 2.27b3 and 2.32b). Their marginal zone is commonly characterized by wide sand and mud flats, which may develop thin salt crusts (consisting, e.g., of proto-dolomite on the outer flats and gypsum and halite on the inner flats). In addition, this zone shows desiccation cracks and locally also some sparse vegetation and burrowing organisms (Fig. 2.32b1). Well laminated *algal mats* are fairly common on the mud flats, but some centimeters or decimeters below the surface they are mostly destroyed.

As long as the capillary fringe of the groundwater table reaches the lake floor, salts such as halite, gypsum, and sodium sulfate crystallize interstitially from the ascending groundwater. This process is referred to as "evaporative pumping". It leads to the formation of nodules or irregular layers of these salts below the surface of sand and mud flats. The flats and the central salt pan are occasionally inundated by sheet floods. Then part of the salt is redissolved and later reprecipitated in the salt pan on top of silty and clayey layers resulting from the flood event. Reworked carbonate crusts and algal mats may locally form *intraclast breccias or flat pebble conglomerates*.

The salts dissolved during floods and concentrated in depressions having no outflow may also be leached by infiltrating rain water and enhance the *mineralization of groundwater*. This is the case when the groundwater level as well as its capillary fringe drop below the lake floor, as for example reported from a number of Quaternary lakes in southeastern Australia (Bowler and Teller 1986; Teller and Last 1990). Then vegetation can begin to grow and form some soil on the lake floor. Figure 2.33 shows the interaction between a fluctuating groundwater level and the sedimentary processes in such a lake. It can be seen that the lake sediments respond very sensitively to slight changes in the hydrologic regime of the lake and thus also to minor variations in climate. They frequently exhibit minor sedimentary cycles of widely differing thickness and duration. If leaching of salts and deflation are involved, such cycles tend to become asymmetric (Fig. 2.33). It is interesting to note that in this case not the salt deposits, but the soil horizon corresponds with the lowermost groundwater table and therefore most likely with the dryest period. For that reason, salt

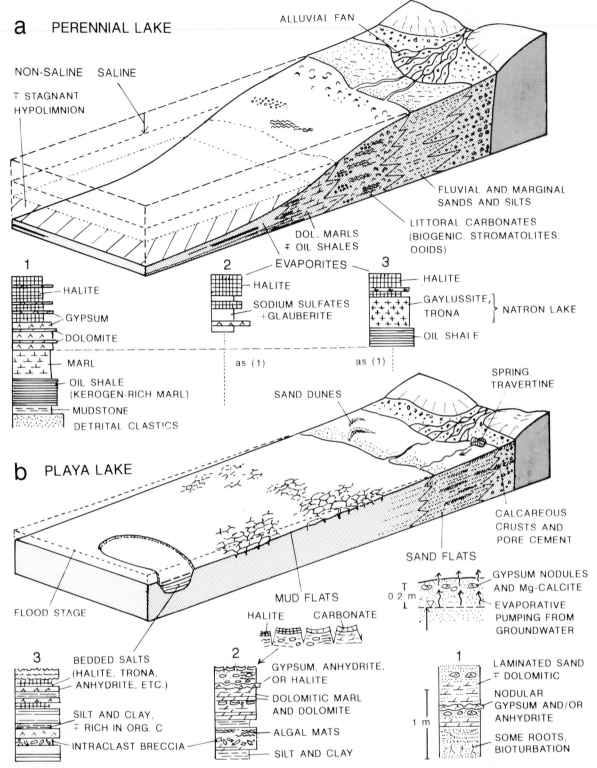

Fig. 2.32a,b. Generalized facies models for **a** perennial, predominantly saline lakes and **b** ephemeral playa lakes (continental sabkhas). **a1-a3** Idealized sequences of central lake: *1* carbonate-sulfate-chloride lake; *2* sulfate-chloride lake; *3* soda-chloride lake. **b1-b3** Sequences of *1* sand flats, *2* mud flats, and *3* of residual playa lake. Note that part of the salts precipitated on the surfaces of *1* and *2* is leached during subsequent flood or later, after burial, within sediment. For further explanation see text

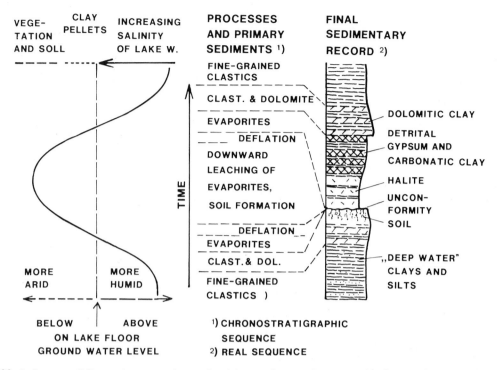

Fig. 2.33. Influence of fluctuating groundwater level in playa lake sediments. The deepest groundwater level, associated with the dryest period, allows downward leaching of evaporites, growth of vegetation, and soil formation. Wet-dry climatic cycle tends to generate an asymmetric sedimentary succession. (After Bowler and Teller 1986)

layers or soil horizons of distant lakes should be correlated with caution, if their hydrological regimes are not equal (Bowler et al. 1986).

Drying lakes with silty-clayey sediments, including some salts, also favor the formation of "clay pellets" on the lake floor which are blown away by wind to form clay dunes (or lunettes, cf. Chap. 2.3.4) along the rim of the lake (Fig. 2.27b4). If a lake is exposed to this deflation process for some time, its floor is lowered and its sedimentary record interrupted.

Black Shale Deposition in Lakes

In oligotrophic lakes, nutrient supply and thus also organic productivity in the epilimnion are relatively low. The oxygen demand for decomposition of organic matter is therefore limited and can usually be balanced by oxygen supply (Demaison and Moore 1980). However, if both nutrient supply and organic productivity are high and cannot be compensated for by oxygen supply, eutrophic lakes develop. In these cases, oxygen demand for the destruction of organic matter within the water column exceeds oxygen supply provided by inflowing water or lake water overturn.

Oxygen supply to the bottom water of lakes is usually sufficient in regions of temperate, humid climate. Here, seasonal overturning of the lake takes place and, due to a positive water balance leading to outflow, cold, well-oxygenated river inflow sinks to the bottom, crosses the lake, and thus improves the oxygen supply of the hypolimnion. For these reasons, all the present-day, large, temperate lakes of the northern hemisphere are oxic in their natural state, for example the Great Lakes of North America, numerous smaller lakes of the Alpine region in Europe, and even the 1600 m deep Lake Baikal in Siberia. As a result of waste water injection, however, some of these lakes developed anoxic bottom waters (e.g., Lake Erie, Lake Zürich).

In contrast, lakes in zones of warm climate, which reach some depth (about 100 m and more), tend to develop a *stable stratification* throughout the year. (A prominant exception from this rule is the large but shallow Lake Victoria in the East African rift zone.) In addition, warm water lakes dissolve less oxygen than cold water lakes. Therefore, oxygen supply to the hypolimnion is very limited and thus favors the development of anoxic bottom waters. Under these conditions, as already mentioned earlier, both the open freshwater lakes and the closed saline lake systems can produce oil shales with high contents of organic matter. Of course, in both cases a low input of detrital clastics (or a limited precipitation of salts) is required for the formation of sediments rich in organic matter.

In fresh-water lakes of tropical and subtropical regions, black shale deposition coincides with periods of high nutrient supply and fertility. A well-known example of this lake type is the 1500 m deep Lake Tanganyika in the East African rift zone (Degens et al. 1971; Cohen 1989). Its drainage area is semi-humid, and the lake has an outflow. Most of the river load settles in the upstream Lake Kivu. Littoral sediments of Lake Tanganyika are rich in carbonate including ooid shoals and beach rock. Anoxic conditions with some hydrogen sulfide in the water prevail below a water depth of about 200 m. Sediments deposited in the shallower part of the lake contain 1-2 % organic carbon, whereas the laminated anoxic sediments in deeper water reach 7-11 % organic C mainly derived from diatoms. The sedimentation rate of the basinal muds is about 0.5 m/ka; interbedded silty-sandy turbidites are common.

Even highly saline waters are not devoid of organisms (Larsen 1980). Although only a few species exist, these can grow in large quantities due to the abundance of nutrients and a high temperature. Bluegreen algae and bacteria in particular, but also some planktonic organisms, copepods, nematods, crustacea, and higher plants manage to live under these conditions and produce and decompose organic matter (Eugster and Hardie 1978). Recent measurements of primary productivity, for example in the Great Salt Lake, Utah, have shown that salt lakes represent some of the most productice ecosystems (Eugster 1985). Phytoplankton productivity as well as the growth rate of algal mats can be very high. Because most of this production is destroyed by grazing higher organisms and microbial decomposition, only a small fraction of this production is eventually deposited together with the inorganic sediment fraction (cf. Chap. 10.3.3). However, this fraction is sufficient for the formation of organic-rich oil shales prior to the precipitation of salts. Even within evaporite sequences, black layers rich in organic carbon and some pyrite are encountered.

Organic matter is generally preserved better in lake sediments than in marine deposits for the following reasons:

1. The sedimentation rate in lakes is usually much higher (often by one to two orders of magnitude) than in open marine environments and thus protects part of the organic matter from decomposition on the lake floor.
2. In contrast to marine sediments, a low sulfate content in the pore water of many lake sediments prevents strong activity of sulfate reducing bacteria that simultaneously consume organic matter. Degradation of organic matter must therefore be accomplished mainly by the less efficient methane fermentation (Demaison and Moore 1980, also see Chap. 14.1).

For the same reason, sulfide supply is commonly limited and cannot precipitate all the released iron as pyrite; consequently siderite nodules can form (as, e.g., found in the former lake stage of the Black Sea; Degens and Stoffers 1980). Anoxic lake sediments are usually also rich in carbonate.

Under changing climate, such lake sediments may display vertical successions alternating between (Fig. 2.32):

- Kerogen-rich beds, coinciding with the more humid phase and high lake level.
- Evaporitic beds, less rich in organic matter, associated with the more arid phase.

Successions of this type are common in many ancient lake deposits, for example in the Triassic and in the Eocene Green River Formation of North America (see below).

2.5.2 Recent and Ancient Examples of Lake Sediments

Many modern and ancient lake sediments have been described in detail. Apart from the examples mentioned above, including the Great Salt Lake in Utah, the Caspain Sea, the Dead Sea, Lake Chad in North Africa, and Lake Eyre in Australia provide particularly interesting present-day examples and are therefore briefly mentioned here:

Caspian Sea. This modern, closed, brackish lake (13 g/l dissolved species) is dominated by detrital clastics, because two large rivers (the Volga river from the north and the Kura river from the Caucasus ranges in the west) flow into the basin. However, the Caspian Sea also exhibits a considerable production of autochthonous (biogenic and bio-induced) carbonate. Its littoral zone and shallow northern part are therefore characterized by the deposition of sediments rich in carbonate, including shell-beds and oolites.

In the 800 to 1000 m deep central and southern part of the lake, fine-grained marls accumulate with a very high sedimentation rate on the order of 1 to 10 m/ka (1 to 10 km/Ma). Because these sediments are both rich in organic matter and are buried rapidly, they enable hydrocarbon generation and oil production from Pliocene to Quaternary strata in the Baku area. Irregularly occurring gas eruptions and numerous mud volcanoes testify to the fact that oil and gas generating processes, as well as differential compaction of these young sediments, are still in operation (also see Chap. 14.2).

Another, interesting feature of the Caspian Sea described in many textbooks (e.g., Sonnenfeld 1984; Müller 1988) is a large, shallow lagoon on its eastern margin, the Kara Bogas Gol. As a result of very high evaporation in the warm, arid climate, the water level of the lagoon drops below that of the Caspian Sea and thus causes continuous inflow of brackish water. The ion concentration of this water increases from the inlet to the inner part of the lagoon and hence leads to the precipitation of carbonate, gypsum, glauberite and halit in a lateral succession (cf. Chap. 6.4.1).

The **Great Salt Lake** in Utah is known for its fluctuating water level and very wide sand and mud flats during lowstand. The lake has a salt concentration about four to seven times as high as normal sea water, but the composition of the dissolved salts is similar to that of sea water. The lake water is oversaturated with respect to carbonate; gypsum and more soluble salts are precipitated only during lowstands in special marginal parts of the lake. In shallow water, ooids, primarily composed of aragonite, some Mg-calcite, and dolomite are common. In the central part of the perennial lake, muds are deposited which are more or less rich in organic matter (Spencer et al. 1984).

The **Dead Sea** is a salt lake occupying the deepest depression (about 400 m below sea level) in the young Jordan rift graben. The lake is about 400 m deep, but below or adjacent to the Dead Sea, more than 3000 m thick salt deposits have been discovered (Neev

and Emery 1966). The lake is fed by the river Jordan which partially drains regions with young volcanic rocks. In addition small springs nearby the lake discharge high-concentrated salt water originating from the leaching of older marine salt deposits. The salt content of the lake (about 300 g/l) is eight to nine times higher than that of normal sea water, but its composition differs considerably from sea water. NaCl makes up only on third of the total salt content, whereas $MgCl_2$ reaches about one half. The brine is remarkably rich in Ca, K, and Br (4.6 g/l), but poor in sulfate. The stable pycnocline, observed years ago between the somewhat lighter surface water and the heavier, rather old water of the hypolimnion, appears to be gone (Beyth 1980).

The Dead Sea is not completely barren of life; it contains bluegreen algae, some planktic microorganisms, and several types of bacteria. At present aragonite, gypsum, and some halite precipitate, but most of the gypsum is dissolved in the anoxic hypoliminion by bacterial sulfate reduction. The generation of hydrogen sulfide leads to the formation of iron sulfide. Thus, dark, mainly calcareous mud with some gypsum, halite, and organic matter is deposited. Precipitation of halite in the form of "salt reefs" is taking place in the shallow southern part of the lake, where "end brines" from evaporation pans mix with Dead Sea brines (Beyth 1980).

Lake Chad in North Africa (cf. Fig. 9.10) and Lake Eyre in Central Australia are playa lakes draining very large, tectonically stable areas. During relatively long dry periods, the widely extended sand and mud flats are exposed to wind action and develop large fields of eolian dunes (Eugster and Hardie 1978). Lake Chad is mainly fed by the river Chari draining the highlands of Cameroun. If the lake level rises, the interdunal depressions are filled with water. The present lake water is rather fresh, favoring the deposition of clayey muds, but some of the small interdunal lakes were converted into salt pans exhibiting crusts of trona, natron, halite, and gypsum. Other pans precipitate thenardite, halite, and other salt minerals indicating that most of the time a uniform brine does not exist.

Lake Eyre in Australia is inundated by floods in quite irregular time intervals on the order of 10 years and more. The mean annual runoff of the Lake Eyre basin is only 3.5 mm; thus the lake falls dry more than 90 % of the time (Kotwicki and Isdale 1991). Its drainage region comprises an area 140 times greater than that of the lake floor. Potential evaporation exceeds precipitation (average of 220 mm/a in the entire drainage area) by a factor of about 12, and in the lake itself, i.e., in the arid continental core, by a factor of 30 (Bowler 1986). The shape of the lake is partially controlled by wind action. On the downwind margin the ephemeral basin is bordered by transverse dunes which are built up by material blown out from the dry lake floor. On the upwind side the lake exhibits a cliffed margin which tends to migrate landward. Irregular deflation on the lake floor may leave behind islands within the lake. During the long dry intervals, the mudflats dry sufficiently to permit the efflorescence of salts at or somewhat below the surface.

The nature of the salts which can be preserved is controlled mainly by the groundwater chemistry. If it is saturated with respect to calcium sulfate, gypsum precipitates (gypsum-dune building phase). Later, as a result of a prolonged warm and arid period, the groundwater becomes saturated with respect to halite (halite-saturation phase). Lake Eyre, as well as a number of other Australian ephemeral lakes, thus show a development from gypsum to halite deposition and vice versa.

Ancient lake deposits can be recognized using the criteria discussed above. Some of the most distinctive features are (1) the absence of marine fauna, (2) the occurrence of salt minerals which cannot be derived from normal sea water, and (3) successions with marked lithological variations and pronounced cyclicity.

Well studied, frequently cited examples of ancient lake sediments are those of the classic **Newark supergroup** (Triassic-Jurassic) in eastern North America (van Houten 1964; Gore 1989) and the marl-carbonate-gypsum-halite successions of the shallow **Triassic (Keuper) basin** in Central and Western Europe (e.g., Schröder 1982), parts of which were repeatedly filled by prograding birdsfoot deltas or wider alluvial plains (Wurster 1964).

Of great economic interest is the occurrence of widespread oil shales and trona beds in the **Eocene Green River Formation** in North America (Eugster and Hardie 1978; Eugster 1985). The oil shales of the small **Lake Messel** in Southern Germany were deposited at about the same time. They are famous for their excellently preserved fauna, particularly vertebrates (Koenigswald and Michaelis 1984; Schaal and Ziegler 1988). The well-known **Miocene Ries impact crater** in Southern Germany was once occupied by a playa lake where algal bioherms, dolomite, gypsum, and also some oil shales were deposited (Wolff and Füchtbauer 1976). Later, it evolved into a fresh-water lake and finally fell dry.

More examples of lacustrine sediments deposited in Phanerozoic time are described by Picard and High (1981), Allen and Collinson (1986), and Talbot and Kelts (1989).

2.5.3 Specific Features of Lakes and Lake Sediments

Sedimentation Rates and Lifetime of Lakes

As already mentioned above, most of the postglacial open lake systems fed by rivers draining mountainous regions show very *high sedimentation rates* on the order of 1 to 10 m/ka and more. Such lakes, particularly shallow ones, have only a comparatively short lifetime. In the Caspian Sea about 1 m/ka is deposited, which signifies that Quaternary lake sediments reach a thickness of more than 1 km. Such high sedimentation rates are capable of compensating even for rapid subsidence (Chap. 8) and thus lead to a fast filling of subsiding lake basins, including closed lake systems in arid and semi-arid zones.

Provided the ion concentration of perennial lakes or that of groundwater below an ephemeral lake has already reached saturation with respect to principal salts, further small evaporation losses can cause a rapid precipitation of salts (several meters to tens of meters per 1000 years). Using the data for chemical (and mechanical) denudation discussed in Chapter 9, one can estimate the time necessary for the accumulation of salts (and detrital clastics) in a lake basin representing a certain fraction of the total drainage area (cf. Chap. 11.2).

However, some modern and ancient lakes are also characterized by rather *low sedimentation rates* (less than 0.1 m/ka). Large lakes of lowland regions belong to this category when their ratio of drainage area to lake area is low. They receive only limited quantities of predominantly fine-grained, clastic material which is distributed over a large lake area. Consequently, the lake is filled up slowly. Such lakes can persist for long time periods, provided their floor subsides sufficiently. Then, different carbonates may constitute a significant sediment component. The widespread European Keuper marls, for example, alternating with dolomitic limestone beds (cf. Chap. 2.5.2), belong to this group. The sediments of long-persisting lakes commonly reflect climatic and hydrologic changes in their drainage area and may experience transitions from an open to a closed drainage system, including marked variations in lake extent and salinity.

Remarks to Specific Sediment Successions and Lake Deposits

Sediment Successions

Lake sediments and particularly their evaporites are extremely variable. In response to changes from dryer to wetter climate and vice versa, lake deposits show characteristic successions. They may display rhythmic or cyclic sequences (cf. Chaps. 7.1 through 7.3). It is not possible to demonstrate all these phenomena in a few facies models. For a number of large, long-persisting individual lakes, composite facies models are required which combine aspects of both the open and the closed lake systems (e.g., Sullivan 1985; Gore 1989). Arid lake basins close to the sea may receive influx from both fresh-water rivers and the ocean, resulting in a complicated system which precipitates various carbonates and evaporites (e.g., Decima et al. 1988).

Isotopes Indicating Paleoclimate

Ancient shorelines and sediments of closed lakes are very sensitive indicators of paleoclimates. The paleoenvironmental interpretation of such lakes can be refined considerably by stable isotope studies (Oberhänsli and Allen 1987; Talbot and Kelts 1989). The isotopic signature of lake carbonate, for example, reflects the river input and climate of the drainage area rather than the situation in the lake itself. A relative enrichment of the heavy carbon isotope ^{13}C indicates that plant material was strongly affected by bacterial activity and that the lake floor possibly emerged. Isotope studies also reveal that primary lake sediments, such as sulfates and carbonates, may undergo significant diagenetic changes (e.g., Decima et al. 1988).

Subsurface Brines and Diagenesis

In addition, it should be borne in mind that salts, which precipitated on or directly below the surface of present-day sand and mud flats, are mostly not preserved in deeper cores. Nodules of gypsum and thin salt layers may dissolve completely in the subsurface and thus enhance the ion concentration of groundwater. Later, the cavities produced by salt dissolution are frequently filled by secondary calcite or quartz. Salts are preserved in the subsurface if the surrounding groundwater has reached saturation for these salts. Furthermore, most of the preserved salts undergo significant diagenetic changes within the sediment. Some are transformed to new salt minerals (Eugster and Hardie 1978; Chambre syndicale 1980), and porewaters rich in silica may lead to the formation of authigenic silicate minerals such as zeolites.

Residual Brines and Mineral Deposits

After precipitation of halite and/or other, highly soluble principal salts, the residual brine may become relatively rich in rare elements (Chap. 2.5.1) and hence be of economic interest in some places. As a result of mineral precipitation, oxidation of pyrite, and decomposition of organic matter, brines can also become acidic. Then they can take up considerable amounts of heavy metals by dissolving oxide coatings and/or by oxidation of metal-rich black shale (Eugster 1985). Even alkaline brines may carry heavy metals. These metals can precipitate as sulfides along a reducing front within or in coherence with the lake sediments and thus form ore bodies containing significant amounts of Cu, Zn, Pb, and Co. Acidic groundwater brines rich in ferrous iron precipitate iron hydroxides when they ascend to the floor of well oxidized, shallow, or dry lakes.

3 Coastal and Shallow Sea Sediments (Including Carbonates)

3.1 Beach and Shoreface Sediments

3.1.1 Coastal Processes, Beach and Shoreface Sands

Waves, Wave-Generated Currents, and Coastal Processes

Beach processes and, in a broader sense also coastal and shelf processes, are controlled predominantly by wind waves, tidal waves, and wave-generated currents. In this first chapter on coastal and shelf sediments we shall briefly discuss wind waves and wind-generated currents of the coastal zone (wave-dominated shorelines). Detailed descriptions of these processes have been given in several books, recently for example by Carter (1988). Tidal effects and other oceanic currents will be referred to in Chapters 3.2 and 3.3.

Waves are the result of mechanical energy transfer from the wind onto the water surface. The large surface of an ocean can absorb much more energy from the atmosphere than that of a small lake, i.e., areas with a long "wind-fetch" can create much larger waves than water bodies of limited length parellel to the wind direction.

Primarily, waves do not transport water and therefore do not induce currents. This is a secondary effect taking place in shallow water where "deep-water waves" are transformed into "shallow-water waves" and finally dissipate.

Deep-water waves only affect a specific layer of the water mass. The thickness of this uppermost layer is approximately L/2, where L is the *wave length* (Fig. 3.1a). The depth of L/2 below the water surface is also referred to as the *wave base* or the depth up to which wave-induced currents exert a significant influence on the sea floor. Between the water surface and the wave base, the water particles perform circular (orbital) movements. At the surface, the diameter of these circles is equal to the wave height H, and the time T, necessary for a water particle to finish one rotation, is the *wave period*. This is the same period of time measured at a fixed point from one passing wave crest to the next one. Below the sea surface the diameter of the rotating water particles is reduced and their *orbital velocity* slows down until this movement almost completely ceases at the wave base. Although the water affected by deep-water waves remains more or less in place, the migrating waves transport energy, for example from localized storm centers to remote coastlines.

As soon as the water depth is less than the wave base (L/2), the deep-water waves start to be transformed into *shallow-water waves*. A significant change of the wave characteristics takes place, if the water depth is as shallow as one-sixth to one-eighth of the wave length. Thereafter, wave length and speed of wave propagation decrease, but their energy is packed into a smaller area and their height increases (Fig. 3.1). Simultaneously, the formerly circular motion of single water particles is transformed into an elliptic and, near the seabed, into a bidirectional oscillating movement. Approaching the shoreface and the beach the waves become shorter, asymmetrical, and their height increases until they collapse in the breaker and surf zone at a water depth of approximately 4/3 H (H = wave height in deep water; Fig. 3.1).

If a wave front migrates obliquely toward the coast, its landward section "touches the bottom" and is affected by wave-length re-

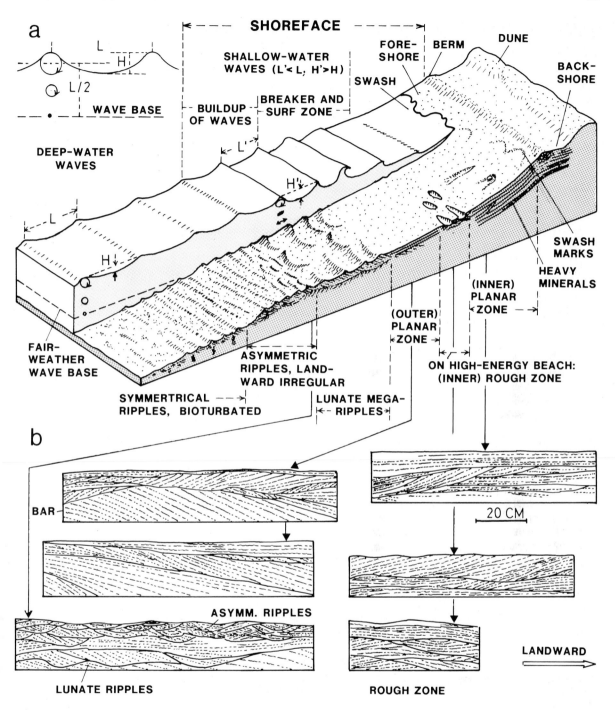

Fig. 3.1. a Transition from deep-water waves to shallow-water waves and swash (backwash) in the shoreface zone, as well as associated bedforms and internal sedimentary structures. No offshore sand bars present; not to scale. For further explanation see text. **b** More complex structures due to shifting of the facies zones under changing wave characteristics. (After Clifton et al. 1971, and others)

duction earlier than the other part. As a result, the waves undergo *refraction* and change their course toward the land (Fig. 3.2a). In this way the wave energy is concentrated at headlands, where it promotes strong erosion, exhibited by cliffs and a wave-cut platform slightly below mean sea level (Fig. 3.2c). In an advanced, and at times inactive state of cliff erosion, the beach zone may consist mainly of fallen blocks and gravel, displaying a characteristic imbrication.

In bays protected by headlands, the wave energy is reduced by refraction, and instead of erosion, deposition of sand along the beach prevails. Both processes, strong wave attack at the headlands and prograding beaches in the bays, favor a development from (usually young) tectonically controlled irregular coastlines to straightened (old) coastlines of tectonically stable areas (Fig. 3.2a and b).

Shoaling waves periodically transport water toward the beach. The water runs up over the

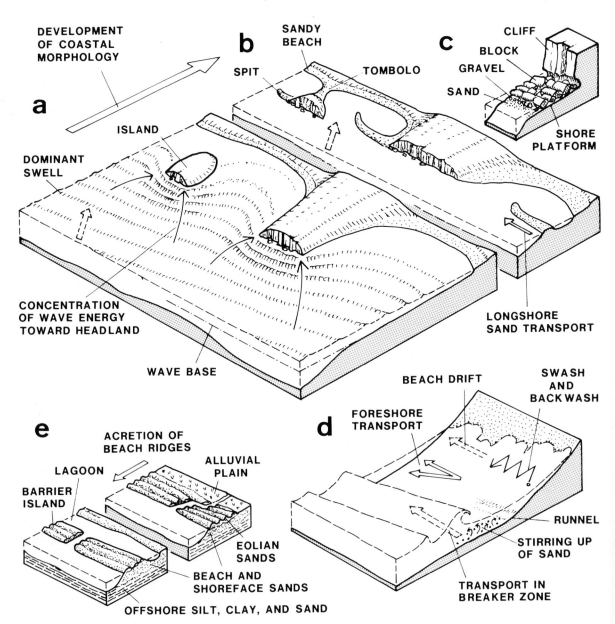

Fig. 3.2. a,b Two stages of coastal headland erosion due to wave refraction and longshore sand transport. c Wave-cut platform and cliff erosion. d Sand movement in the foreshore zone. e Beach ridges and barrier islands resulting from longshore sand transport

foreshore as *swash* and returns due to gravity as *backwash* and bottom current to the sea. If waves approach the coastline obliquely, part of the water masses spilled onto the beach does not flow back immediately, but is directed parallel to the shoreline. This occurs everywhere along the shore and thus the longshore current is reinforced until it reaches rather high velocities in the order of 1 m/s. The preferred line of movement is along a specific depression between the foreshore zone and, if present, the inner sand bar (Fig. 3.2b and d).

Further along the coast, the longshore currents return as *rip currents* back into deeper water (Fig. 3.3a). They can produce erosive channels in the breaker zone and carry sand and particularly finer grained material into the lower shoreface and offshore zone where the rip currents disperse. In the breaker and surf zone much sand and sometimes even gravel is mobilized. This material can be transported by longshore currents in considerable quantities within a narrow belt landward of the breaker zone. Most of this sediment is not carried in suspension, but as bedload in numerous small steps, and partly by a "sawtooth movement" (beach-drifting) caused by alternating swash and backwash, directed obliquely toward the shore (Fig. 3.2d).

Longshore sand transport distributes the sand provided by entering rivers or cliff erosion, and also in some cases by eolian dunes. The migrating sand creates prominent coastal features such as hooks and spits behind headlands or tombolos connecting former islands with the main coast (Fig. 3.2b) and is also a main factor in the formation of broad beach-ridge strand plains *(chenier plains,* see below).

Beach and Shoreface Sands and Their Budget

The Beach-Shoreface Zone

The complicated and permanently changing hydraulic regime of the nearshore zones is reflected by the beach-shoreface profile and its sediments. There is a specific dynamic equilibrium between this profile, the grain-size of sediments, and the wave type. Apart from special features like sand bars, the slope of the beach-shoreface profile usually diminishes from foreshore to deeper water. The mean angle of this slope is low along shorelines subjected to steep high-energy waves (high H/L ratio), because a high energy input is dissipated most efficiently by a wide, flat beach profile. In contrast, low energy input

under flat waves enables a rather steep beach gradient. Furthermore, the presence of coarse-grained sand or gravel increases the slope angle of the beach-shoreface profile. These slope angles therefore often vary between 0.2° (high-energy waves, very fine sand) and approximately 10° (low-energy waves, coarse sand).

A gentle slope profile appears to be especially interesting, because under these conditions a broad coastal sand belt can develop. A further consequence of these rules is the fact that many coastlines exhibit a seasonal change in the beach profile. Due to a higher proportion of steep waves in winter time, the *winter beach* tends to be lower and its slope gentler than the *summer beach*. The lost sand is usually stored in submerged *sand bars* parallel to the coast at some meters depth below mean sea level (Fig. 3.2a) and later, under a subdued wave regime, slowly swept back onto the beaches. Loss of beach sand to deeper water is also caused by single storm events (see below).

Bed Forms and Sedimentary Structures

The bedforms and internal sedimentary structures along the beach-shoreface profile are shown in Fig. 3.1a for fairweather wave conditions. They reflect the transformation of deep-water waves (oscillatory flow) to shoaling waves, generating land-directed flow and return flow into the foreshore zone under upper flow regime conditions.

At depths near, or below the fairweather wave base (often 10 to 20 m), long-crested symmetrical wave ripples, produced earlier by rare storm waves, are bioturbated during normal fairweather conditions. Landward, these inactive ripples pass into active ripples which become increasingly asymmetric, irregular, and short-crested. Larger, lunate-shaped megaripples are frequently observed in the breaker zone. All these bedforms are associated with small-scale or larger scale cross-bedding which is predominantly oriented toward the land. In the foreshore zone the most characteristic feature is parallel to low-angle cross-bedding dipping seaward.

The upper face of beach and foreshore sand may display distinctive swash marks including fine shell debris, as well as small bones and teeth, backwash rills and kolk marks around gravel or shells, and sometimes rhomboid ripple marks formed under a very thin cover of running water.

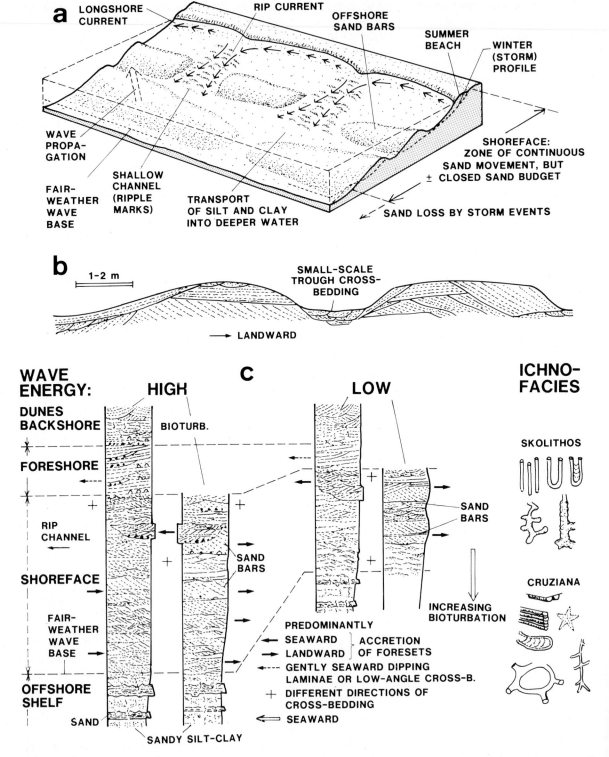

Fig. 3.3. a Summer and winter (storm) beach profile, offshore sand bars, longshore and rip currents. For further explanation see text. **b** Sedimentary structures of offshore sand bars, generalized. (From various sources, e.g., Werner 1963; Davis et al. 1972). Note the prevailing landward dip of foresets. **c** Complete synthetic vertical sections of the shoreface-beach zone of high- and low-energy shorelines, partly with rip channels and offshore bars. Note the changing orientation of foresets in different types of cross-bedding. (Based on Fig. 3.1, Elliott 1986a, and others; ichnofacies after Ekdale et al. 1984)

Foreshore and backshore sands frequently contain dark laminae or layers consisting of *heavy mineral concentrations*. In some places these *placer deposits* are of economic interest, if they contain significant quantities of certain minerals (e.g., rutile, zircon, ilmenite, and monazite).

High-Energy Shorelines

The general tendency in the lateral sequence of bedforms and sedimentary textures can be modified on high-energy coasts or on shorefaces which exhibit one or several shore-parallel *sand ridges* or *offshore bars* (barred shorelines). On a nonbarred high-energy shoreface, two zones with predominantly planar bedding occur, an inner (upper) foreshore planar zone and an outer (lower) planar zone within the breaking waves (Fig. 3.1). In between the two planar zones the sea bed becomes rough due to dotted and irregular erosional and depositional features.

Offshore sand bars are complex structures, because they often migrate in different directions, either landward, seaward, or parallel to the shoreline. Prominent internal structures are landward-dipping large-scale foresets and more or less horizontal lamination, but small-scale cross-bedding is also common (Fig. 3.3b).

Gravelly shoreface deposits are characterized by large, asymmetric gravel wave ripples (wave length frequently between 100 and 150 cm, height 10 to 20 cm), which may form at water depths down to more than 6 m (Hart and Plint 1989). Gravel forms decimeter-thick, massive or cross-bedded layers alternating with thinner sand beds. Pebble fabric and cross-bed orientation indicate predominantly alongshore sediment transport.

Shoreline Migration
and Vertical Sediment Successions

The facies zones of the shoreface migrate back and forth permanently in response to the momentarily active wave regime. As a result, vertical sequences record frequent variations between neighboring facies zones, described above. They may include the internal structures of sand bars and rip channel fillings (Figs. 3.1b and 3.3c).

Seaward prograding shorelines generally show a tendency for the sands to coarsen upward. Simultaneously, bioturbation structures are less readily preserved due to the permanent erosion and redeposition of sand in the upper shoreface zone. Transgressive shoreface sediments display the opposite vertical development with a fining-upward sequence.

Gently sloping beaches create a wide belt of shoreface sands, whereas steeply sloping shorelines generally only produce a narrow belt of coastal sands. Then the transitional facies between deeper water and the swash-backwash zone may be largely absent.

The Sand Budget of the Beach-Shoreface Zone

An aspect of some interest is the sand budget of the total beach-shoreface zone of a particular area. It is fed by river input, cliff erosion, and longshore sand transport, and it is reduced by longshore transport to other areas, as well as by losses to coastal eolian sand dunes. Sand is also transported into deeper water via submarine canyons beginning in the shoreface zone and by storm events producing tempestites seaward of the shoreface (see below and Chap. 5.4).

Apart from these long-term gains and losses, the sand budget of a given shoreline section appears to remain fairly stable under a defined wave regime. Although the beach-shoreface sands are permanently in motion and often migrate landward and seaward with the seasons, their volume remains approximately constant, at least for a limited, geologically relevant time period. This may also be the case during rising and falling sea levels. In this instance, a major part of the coastal sand belt can migrate either landward or seaward without leaving behind many relics in deeper water or in emerged coastal areas.

On the other hand, observations and measurements of several present-day coastlines have revealed that *longshore currents* are capable of transporting enormous volumes of sand (up to several hundred thousands of cubic meters per year). For this reason, sand accretion in sections of coastline protected against strong wave attack is common and may proceed comparatively quickly. In some cases, it is possible to show that a shoreline consisting of sandy beach ridges has migrated roughly 1 m/a seaward during the last few thousands of years. Accretion of coastal sand may take place directly on the mainland coast or in front of barrier islands (Fig. 3.2e). High lateral sand supply may generate a wide beach and shoreface sand belt within a geologically short time period.

3.1.2 Storms and Storm Deposits (Tempestites)

In storm-dominated shallow-marine environments, the shoreface processes described so far, as well as processes in deeper regions of the inner and possibly outer shelves, are strongly affected and modified by rare storm events. Storms can deposit sandy material in the supratidal zone (supratidal storm layers, Chap. 3.2.2), as well as generate special non-channelized flow conditions in deeper water. These frequently produce characteristic sheet-like sand and mud beds *(tempestites)* of considerable lateral extent (reviewed, e.g., by Allen 1982; Johnson and Baldwin 1986; Morton 1988; Nummedal 1991; Seilacher and Aigner 1991). These are briefly described here.

Storm Generation

Strong and laterally extensive storms develop under special climatic and geographic conditions. *Tropical storms* (hurricanes) are initiated in low-latitude regions within the trade wind belt (cf. Fig. 5.1e), where the surface temperature of the sea is in excess of 26.5 °C. They travel westward and are deflected toward the poles by Coriolis forces (Chap. 5.2), in a manner similar to surface ocean currents. *Extratropical storms* are common in zones of mid-latitudinal, eastward-directed atmospheric circulation along polar fronts. They migrate eastward, are more consistent in their direction, and can reach at least the same or even greater magnitudes than tropical storms.

In viewing the geologic record, one can assume that tropical storms prevailed during warm periods exhibiting gentle temperature gradients between the poles and the equator, whereas extratropical storms were more frequent and important during cooler periods with steep temperature gradients.

A third mechanism producing landward-directed storms are the *monsoons*. They are created when large landmasses establish intense summer lows and winter highs. The summer season is characterized by strong onshore winds carrying large quantities of moisture inland. The best present-day example is the monsoon belt of Asia along the margin of the Indian and Pacific Oceans. Monsoonal circulation can also influence the tracks and landfall locations of tropical and extratropical storms.

Storm Action

Storms not only generate large waves with high amplitudes, great lengths, and deep wave bases, but can also drive a net mass flux of water toward or away from the coast. The set-up of water along a wide section of the coast during a storm flood is enhanced by onshore winds, low barometric pressure, abundant rainfall, converging shorelines, and broad, shallow shelves. Such *storm surge elevations* are particularly effective in producing bottom currents and transporting sand into deeper water (Fig. 3.4a). In the northern hemisphere, shore-parallel, or obliquely onshore-directed, wind-driven surface currents are deflected to the right by Coriolis forces and may cause and maintain such water set-up for several hours or days. Since this additional water can neither flow back as surface currents nor flow away as longshore currents, a seaward pressure gradient is established in the water column, leading to downwelling and return bottom flows. These are also deflected by Coriolis forces and therefore flow obliquely away from the coastline.

Currents modified in this way by Coriolis forces, pressure gradients, and bottom topography are referred to as *geostrophic currents*. Seaward-directed bottom flows may be fortified by increased density of the currents, caused by suspended material stirred up from the shoreface. In regions with high tides, geostrophic flows can be augmented by tidal ebb currents.

Storm waves and storm-induced geostrophic currents operate simultaneously and therefore cause a hydrodynamic system which is referred to as *combined flow* (Harms et al. 1982; Snedden et al. 1988). Back-and-forth oscillation of the ground wave is superimposed on the quasi-steady bottom current (Fig. 3.5). In such a flow regime the net shear stress imparted on the sea floor may erode and move sediment during one half of the wave stroke, but be insufficient to do so during the other half.

The relative magnitude of wave- and current-induced shear stresses differs from the shoreface to the inner and outer shelves. The wave-induced component rapidly increases landward and may cause erosional features more or less normal to the shore, whereas the current-induced bed shear stress grows at a much lower rate in this direction. Hence, the shoreface and parts of the inner shelf are dominated by wave-induced oscillatory shear during storms, while the deeper environments

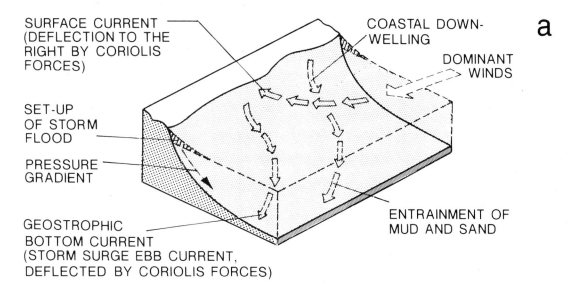

SURFACE CURRENT
(DEFLECTION TO THE
RIGHT BY CORIOLIS
FORCES)

COASTAL DOWN-
WELLING

DOMINANT
WINDS

SET-UP
OF STORM
FLOOD

PRESSURE
GRADIENT

GEOSTROPHIC
BOTTOM CURRENT
(STORM SURGE EBB CURRENT,
DEFLECTED BY CORIOLIS FORCES)

ENTRAINMENT OF
MUD AND SAND

a

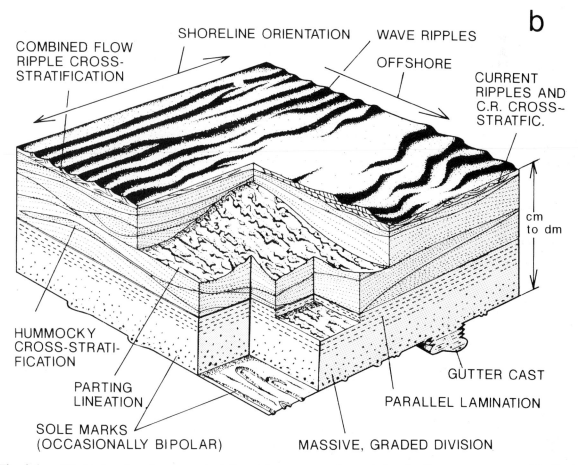

b

COMBINED FLOW
RIPPLE CROSS-
STRATIFICATION

SHORELINE ORIENTATION

WAVE RIPPLES

OFFSHORE

CURRENT
RIPPLES AND
C.R. CROSS-
STRATFIC.

cm
to dm

HUMMOCKY
CROSS-STRATI-
FICATION

PARTING
LINEATION

SOLE MARKS
(OCCASIONALLY BIPOLAR)

GUTTER CAST

PARALLEL LAMINATION

MASSIVE, GRADED DIVISION

Fig. 3.4. a Wind-induced surface current, deflected by Coriolis forces, set-up of storm flood, and generation of geostrophic bottom current. (After Walker 1984b). **b** Cur-rent- and wave-induced sedimentary structures of idealized sandy tempestite and their relationship to the shoreline. (After Leckie and Krystinik 1989)

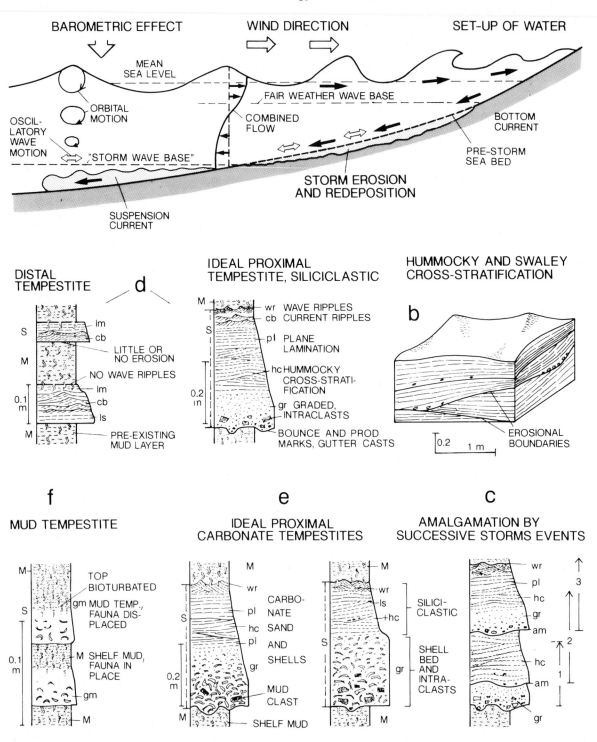

Fig. 3.5. a Hydraulic regime during storm in a section across the shoreface and inner shelf, simplified. Coastal set-up of water and basinward bottom current superimposed by oscillating wave motion (combined flow conditions). **b** Shoreface sands displaying hummocky and swaley cross-stratification, when muddy interbeds are missing. **c** Amal-gamation of proximal tempestites. **d** Siliciclastic proximal and distal sandy tempestites. **e** Proximal carbonate or mix-ed carbonate/siliciclastic tempestites. **f** Distal mud tempes-tites. (Based on Dott and Bourgeois 1982; Walker et al. 1983; Aigner 1985, and others)

are controlled mainly by steady, obliquely offshore or almost shore-parallel currents (Duke 1990).

Storm-Induced Bed Forms and Sedimentary Structures

Storm-generated bed forms show a distinct trend from the beach into deeper water (cf. Chap. 3.1.1 and Fig. 3.1):

- *Beach*. Storms frequently erode seaweed and shells of fauna living in the sediment and accumulate them on the beach, where they form linear beds parallel to the shoreline.
- The *upper shoreface* (surf zone) displays megaripples oriented parallel to the shoreline, as well as flat swash lamination on the offshore bar crests. The resulting internal structures are dominated by trough cross-stratification and planar or low-angle swash laminae. Somewhat channelized longshore currents and rip currents form additional sedimentary structures of differing orientation (Chap. 3.1.1).
- The *middle shoreface* is still within the range of fairweather waves and thus essentially controlled by wave action. High-stress oscillatory flow, such as that created during storms, generates both flat and swaley bed forms with nearly horizontal or low-angle swaley cross-stratification (Fig. 3.5). Due to the fact that the sediments of this zone are under continuous agitation, the deposition of mud is prevented, and bottom life is limited to filter-feeding infauna. The sedimentary record of rare, heavy storms is usually destroyed on the upper and middle shoreface.
- On the *lower shoreface* (5 to 20 m water depth), the stress imparted by geostrophic currents reaches about the same magnitude as storm-wave induced oscillating stress. Hence, for a short time period, both flow components may stir up sand and mud, which are redeposited as graded beds at the same location or nearby. The typical bed form of this combined flow regime is thought to be *hummocky cross-stratification* (HCS or S_{hc}, see below and Fig. 3.5). Laboratory experiments, however, have shown that this bed form can also be produced by purely oscillatory flow (Southard et al. 1990). Hummocky cross-stratification appears to form only in fine-grained sand and may be associated with wave-formed, coarse-grained ripples (Leckie 1988). Deceleration of storm flows over slight topographic highs and down their lee sides may cause the coarser sand

fraction to settle first, while the mud and some fine sand are carried farther offshore. Individual, graded beds resulting from a storm event are called *tempestites*. In the long intervals between large storms, the lower shoreface can be intensely bioturbated, but sufficiently thick tempestites are burrowed only on their tops.
- On the *inner shelf* and parts of the outer shelf, particularly in high-energy environments, storm-induced combined flow may still generate hummocky stratified sand layers or thin, graded sandy beds resulting from laterally flowing suspension currents. If the suspension originates from storms and the sandy layers alternate with shelf muds, the storm-induced beds represent *distal tempestites*. As the storm currents subside, wave-driven reworking may form oscillation ripples on top of the HCS-stratified or graded beds. Muddy interbeds are produced either by storms which have moved inland and eroded fine-grained material, or by the slow and repeated deposition of suspended river load.
- At *greater water depths* on the outer shelf, the current component of the combined storm flow becomes dominant, leading to current-rippled fine sand and silt beds. Since only the largest storms can affect the sea floor in this depth zone, storm events are followed by long periods of quiescence. These allow intensive bioturbation and thus frequently obliterate the storm record. However, if an extremely heavy storm hits the deeper sea bottom, the resulting tempestite may be relatively thick and widespread and thus have a good chance of being preserved and buried under younger, "normal" sediment. The time interval between such storm events may be several hundreds or thousands of years.

In this way, numerous successive storm events may build up a more or less rhythmic tempestite-shale sequence over a certain geologic time period (see below).

Recently, it was lively debated whether or not the shore-normal current directions deduced from tool marks and gutter casts at the base of many ancient tempestites were generated by geostrophic currents (Leckie and Krystinik 1989; Higgs 1990; Hart et al. 1990; Duke 1990). Geostrophic currents should run obliquely or almost parallel to the coast due to Coriolis forces. It appears that shore-normal current directions measured at the base of storm sand beds reflect the dominant influence of wave-induced near-bottom flow during the storm peak, whereas internal structures such as cross bedding frequently deviate from this direction. They result in large part from the geostrophic current component of combined flow. Shore-parallel rippled tops may be caused by subsequent waves approaching the shore-line at nearly right angles. In addition, some graded sand beds on the shelf may also be generated by river-fed density underflows.

Characteristics of Tempestites

The typical tempestite is a sand bed produced on the lower shoreface and in somewhat deeper water, where during fair weather lateral sand transport is not an important factor. Here, we can distinguish between an erosional and a depositional phase in the formation of a storm bed, as with the standard sand turbidite (Chap. 5.4.2). In some cases, erosional depth can be determined from casts of animal burrows with known geometry, and which were shaped just prior to the storm event (Fig. 3.6).

Erosional features due to the combined flow conditions are unidirectional, or *bipolar* if the oscillatory flow component of the wave action dominates. Therefore, bipolar or even multi-directional sole marks, such as bounce and prod marks and *gutter casts* (Fig. 3.5), are of particular diagnostic value. Frequently, small channels with coarse grained infillings are also observed. They are usually oriented more or less perpendicular to the shoreline, but may also display other directions and thus indicate a specific predominant wind-storm system (Aigner 1985).

Still, during peak storm flow, the deposition of lag sediments takes place, forming the base of the storm layer. It often contains mud clasts, disarticulated or broken shells of molluscs and other organisms, and occasionally gravel and small rock fragments or reworked algal structures, as known from Precambrian and Cambrian *flat pebble conglomerates* (Fig. 3.6).

When the storm wanes, the finer grained sandy material stirred up during the storm peak phase starts to settle according to the hydraulic properties of the individual particles or aggregates. The resulting bed is therefore distinctly *graded* and may consist predominantly of either siliciclastic terrigenous material, or mainly or entirely of coarser grained shell debris, mostly carbonate. This sublayer is followed or may be replaced by a parallel-laminated and/or low-angle cross-stratified section displaying the typical *hummocky cross-stratification* (Fig. 3.5). This structure is usually well developed in the zone of significant combined flow, i.e., above the storm wave base, where high-energy flow conditions with a strong oscillatory component (orbital velocities ≥ 0.5 m/s) develop. The medium to

large-scale hummocks exhibit erosional faces and an undulose or sharp-crested top surface. Their laminae vary in thickness and dip direction, but basinward downlap of the basal sets is common, because hummocks tend to migrate with the offshore-directed bottom flow. At greater water depths, hummocky cross-stratification becomes less distinct and is more or less replaced by parallel lamination.

The uppermost sandy part of an ideal sandy tempestite is deposited during the final stage of a waning storm and consists of ripple cross-bedding. The top of the storm bed frequently displays *oscillatory ripple marks* oriented approximately parallel to the shoreline (Fig. 3.5). It may be succeeded by storm-reworked mud, if this material is not carried away by bottom currents.

Sedimentary Structures and Amalgamation

A complete tempestite shows the following sedimentary structures (from top to bottom, Fig. 3.5):

- Redeposited shelf mud (muddy tail of combined flow) S_m)
- Wave ripples and wave ripple cross-stratification   (S_{wr})
- Low-angle hummocky cross-stratification   (S_{hc})
- Parallel lamination and current ripple cross-stratification   (S_{pl})
- Graded layer with basal lag deposit (S_{gr})
- Erosional base with sole marks (in places bipolar or multi-directional) and casts of animal burrows.
- Normal shelf mud, intensely bioturbated   (M)

The symbols proposed here are similar to those used for turbidites (Chap. 5.4.2); S signifies "storm beds".

Such idealized, complete tempestites are produced only in certain water depths. In shallower water, the graded subdivision, S_{gr}, is often reduced or missing, and large-scale hummocky cross-stratification, S_{hc}, and swaley stratification are most characteristic. Furthermore, there is evidence of scours and small channels, often filled with coarse-grained material and/or large shells. The uppermost subdivisions, S_{wr} and S_m, are rarely realized in this area, because the finer grained material is sorted out and carried away.

A common phenomenon in the proximal zone is *amalgamation* (am) or *cannibalism*

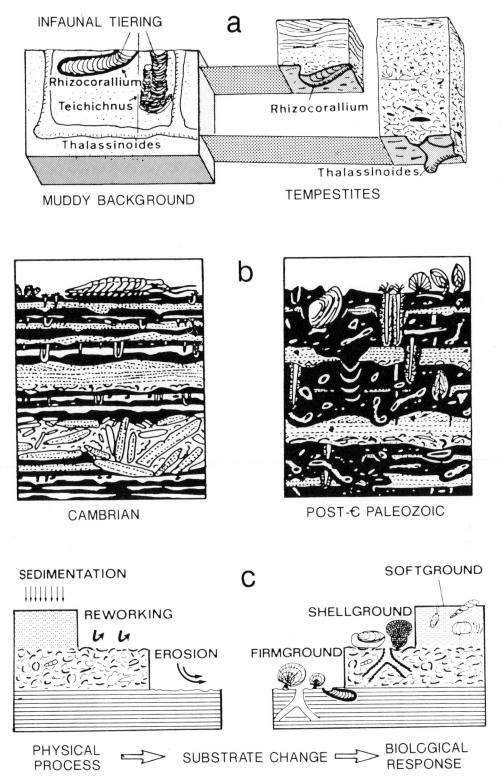

Fig. 3.6. a Infaunal tiering in trace fossil associations, indicating minimal depth of subsequent storm erosion (from Aigner 1985). **b** Increasing diversity of infauna and epifauna, affecting storm beds and their subsequent degree of obliteration, from Cambrian to post-Cambrian times. Note that older, thin tempestites and flat pebble conglomerates are less bioturbated and therefore better preserved than younger ones. (Sepkoski 1982). **c** Biological response to substrate change caused by different levels of storm erosion. (Aigner 1985)

(Figs. 3.5 and 3.7). These terms signify that either thin tempestites resulting from previous average storm events, or the upper parts of pre-existing thick tempestites are reworked by storms of particularly high erosional capacity. Thus, the material of older tempestites is incorporated into new ones. This process may occur repeatedly, until finally a very big storm event produces a thick bed, the base of which can no longer be reworked by successive storms. Proceeding from deeper to shallower water, amalgamation and repeated reworking become increasingly important. Multiple reworking, in turn, will abrade and break up mechanically unstable particles; even some chemical effects cannot be excluded, for example the accelerated dissolution of carbonate in an environment unfavorable for carbonate preservation. Consequently, amalgamation leads to *increasingly "mature" lag deposits*. Thin, delicate skeletal remains are broken and ground into sand and silt, and are partially redeposited elsewhere or, in some cases, dissolved.

Skeletal remains of vertebrates, particularly teeth (but also coprolites which have been phosphatized prior to reworking), may be concentrated in the basal layer of storm beds affected by amalgamation. Thus, certain types of *bonebeds, shell coquinas, grainstones, packstones, and placer deposits* may all result from amalgamation in proximal storm-dominated environments. No sharp boundary can be drawn between the characteristics of beds caused by repeated amalgamation and "condensed" layers, which also frequently reflect several episodes of reworking. However, condensed layers commonly represent a much longer time period than amalgamated tempestites and include long intervals of non-deposition or omission (also see Chap. 7.4).

Due to wind and current directions that vary from storm to storm and even during individual storm events, some of the storm-eroded material at a certain locality may be lost, while other material is gained. Therefore, the new tempestite may be thicker or thinner than the eroded bed and consist of material from different sources, some being autochthonous, some allochthonous.

Composition and Infauna

As far as the grain size distribution and composition of tempestites are concerned, a wide range is possible, from quartzose sandy types with hardly any fossil remains to calcareous or partially siliceous types. The latter may frequently form wackestones or packstones that contain rather large components. For example, many of the building stones of the famous pyramids in Egypt come from Eocene nummulitic limestones, which were formed as storm deposits (Aigner 1982).

As mentioned above, storm erosion commonly disturbs and removes *pre-event epibenthic and shallow infaunal populations*. This is especially obvious when muddy sediments at water depths below the shoreface are so affected. The remains of such fauna are incorporated into the overlying event deposit or transported into deeper or shallower water. The exhumed and partly washed out burrows are filled with sand, forming characteristic casts at the base of the storm bed (Fig. 3.6a). *Burrow-fill sedimentation* has also been observed after a hurricane struck the carbonate platform of Caicos in the West Indies (Wanless et al. 1988). Here, the pre-existing, mounded, approximately 2 m deep bottom was flattened, the seagrass cover partially removed, and *Callianassa* burrows filled with coarse molluscan and peloidal-skeletal packstone, but a continuous storm deposit did not accumulate.

The *post-event community* recolonizes either the top of the tempestite, or erosional surfaces which were not covered by the storm bed (Fig. 3.6c), for example in somewhat elevated proximal regions. Because of the change in substrate, the new epibenthic fauna and ichnofauna may differ in character and species content from the background community of the shelf muds. The firmer, coarser substrates attract oysters, brachiopods, crinoids, stromatolites, and firm-ground burrowers of Glossifungites association. In distal zones, where only thin tempestites are deposited and reworking by successive storms is rare or insignificant, burrowing by post-event benthic organisms can markedly overprint or completely obscure storm event stratification (Fig. 3.7a).

Because the diversity of such cummunities and their ability to live below the sediment surface generally increased during the Earth's history, tempestites of Precambrian to Cambrian age had a better chance of not being destroyed by bioturbation than post-Cambrian storm beds (Fig. 3.6b).

Proximal-Distal Trends

The proximal-distal tempestite trends discussed above are summarized in Figures 3.5 and

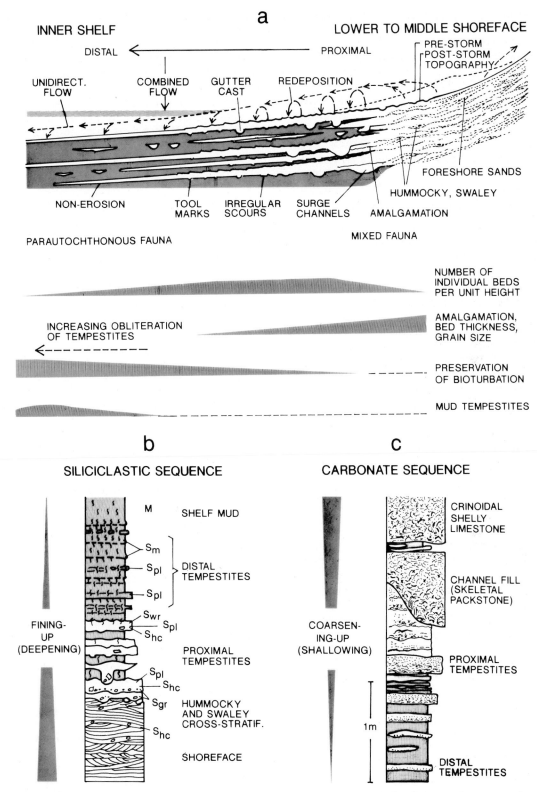

Fig. 3.7. a Hypothetical cross section of storm deposits from shoreface to inner shelf with proximal-distal trends. **b** Idealized deepening (thinning) or fining upward siliciclastic tempestite sequence. **c** Coarsening (thickening) or shallowing upward calcareous tempestite sequence (number of beds reduced). (Mainly based on Dott and Bourgeois 1982; Aigner 1985; Johnson and Baldwin 1986)

3.7. The nearshore zone of swaley and large-scale hummocky cross-stratification is usually not included in the term tempestite. Relatively thick and often amalgamated tempestites begin when muddy intercalations are present. However, tempestites tend to change laterally in thickness and frequently pinch out. Further seaward, i.e., more distally, the number of individual tempestites over a certain time span first tends to increase, because amalgamation becomes rarer, but it then decreases due to the limited distance which the suspended load of weaker storms can travel. Distal tempestites are thin and fine-grained, show the same inorganic sedimentary structures as distal turbidites (but differ in their faunal characteristics), and may end up as mud tempestites. Distinctive features useful in the discrimination of tempestites and turbidites are summarized by Einsele and Seilacher (1991).

In high-energy shelf seas, *distal tempestites* may occur at water depths up to and in excess of 50 m. They show greater discontinuity than do proximal ones and are traceable, for example in the North Sea, over tens of kilometers in water depths up to 30 m (Aigner and Reineck 1982). Widely extensive, fairly thick, sandy tempestite sequences require a substantial continuous source of sand, which is usually provided by river deltas or rapidly eroding sandy coastal cliffs. In mud-dominated shallow environments, mud tempestites may be common. They are, however, difficult to recognize and therefore have been frequently overlooked.

Frequency of Tempestite Events

Storms are represented in the geologic record by tempestites less frequently than they hit present-day coastal zones. In modern seas, a 100-year storm appears to be an exceptionally large storm event (Nummedal 1991). Modern storm layers in seaward prograding sediments on the inner shelf (20 to 30 m of water depth) of northeast Japan formed over time intervals of 20 to 100 years (Saito 1989a). However, for many ancient tempestite sequences we have to assume average sedimentation (or subsidence) rates of approximately 20 to 100 m/Ma. If 1 m of vertical section contains two to five well developed tempestites, the time interval between two tempestites ranges from about 1000 years to possibly $\geq$10 000 years. This surprisingly long recurrence time suggests that even prominent storm beds are wiped out or obscured by subsequent very rare, extremely large storm events. The previous action of weaker storms may be inferred from the occurrence of mechanically abraded, coarse particles and/or mixing of materials from different sources in a preserved tempestite bed.

Tempestite Sequences in Basin Analysis

As for turbidite sequences (Chap. 5.4.2), the occurrence of tempestites and their proximal and distal facies provides a useful tool in basin analysis (e.g., Aigner 1985). Tempestite sequences may reflect three different basin evolution scenarios:

1. *Steady-state conditions* in a foreshore-shelf environment, i.e., the average sedimentation rate more or less compensates for subsidence. In this case, comparatively thick sequences of alternating tempestites and shelf muds can develop, and the paleo-water depth at a certain location within the basin will remain about constant.

2. *Deepening basin*, with an average sedimentation rate lower than subsidence. In this case the vertical sequence will display a trend from thick, coarse-grained, proximal tempestites to thin, fine-grained ones, and finally end up with mud tempestites or purely autochthonous shelf muds (Fig. 3.7b). Usually, such a transition zone will not be very thick, because the sandy tempestites vanish at relatively shallow water depths (mostly between 20 to 50 m).

3. *Shallowing basin*, with a sedimentation rate higher than subsidence. The resulting tempestite sequence coarsens (thickens) upward (Fig. 3.7c), and the transition zone from shelf muds to a siliciclastic or bioclastic foreshore and beach environment will also be of limited thickness, as in case (2). Such regressive conditions favor the amalgamation of storm beds (Morton 1988).

Hence, tempestites can be distinguished from turbidites not only on the basis of certain sedimentary structures (particularly hummocky cross-stratification and wave ripples) and faunal characteristics of individual beds, but often by rapid changes in their vertical sequences.

3.2 Sediments of Tidal Flats and Barrier-Island Complexes

3.2.1 Tidal-Influenced Environments and Sediments

Tides, Tidal Ranges, and Tidal Currents

Generation of Tides

Tidal processes operate along the coasts of all large oceans and are described in many books and special articles (e.g., King 1972; Dietrich 1975; Davis 1978; Pethick 1984; Carter 1988; de Boer et al. 1989). Tides and tidal waves are caused by the attraction of the Moon and, to a lesser degree, by the Sun. On the side of the Earth facing the Moon, as well as on the side opposite to it, the water level of the ocean is raised, while perpendicular to this line the water level is depressed (Fig. 3.8a). Due to the counter-clockwise rotation of the earth, the piled-up water mass dragged along by the Moon travels clockwise until it hits a con-tinental margin and its shore, where it is forced to flow back toward the center of the ocean basin. This process is repeated once or twice daily (diurnal and semi-diurnal tides) and its intensity fluctuates in 2-week periods from high spring tides (with the Moon and Sun working in the same direction) to low neap tides and vice versa.

Apart from the gravitational effect of the Moon and the Sun, the magnitude of the tides also depends on the period with which the water mass of an ocean, an enclosed sea, or a bay tends to oscillate in its basin. If this period corresponds approximately with the periodically changing gravity forces of the Moon and Sun, the tidal range is substantially enhanced.

Tidal Waves, Tidal Ranges, and Tidal Currents

Tidal waves (as used here, not the erroneous synonym of storm surges or tsunamis) have a very long wave length and therefore in some respects behave, even in deep ocean basins, like wind-generated waves in shallow water (Chap. 3.1). In deep water the tides are hardly noticeable, but when a tidal wave approaches the continental slope and shallow water, the migrating water masses are piled up and often bundled to considerable height (Fig. 3.8b and c).

Tides of great amplitude only develop on the coasts of very large water bodies, i.e., at the margins of the oceans. Lakes and smaller sea basins separated from the open ocean only show very small tidal effects (Fig. 3.8d). Even the Mediterranean does not have tides greater than 10 to 20 cm, because its connection with the open sea is too narrow to allow tidal waves to enter from the Atlantic into the adjacent Mediterranean basin. For other reasons, the tides along the coasts of most oceanic islands, as well as along elongated continental coast lines behind narrow or missing shelves, are also limited (<1 m).

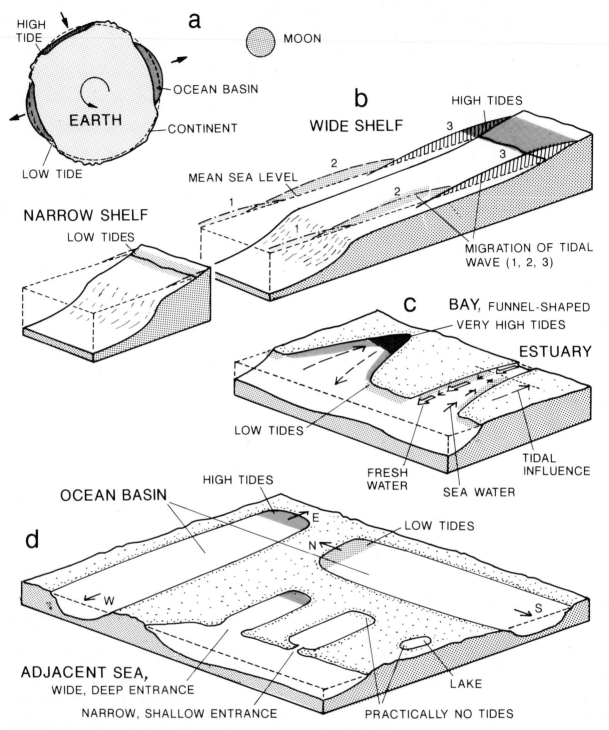

Fig. 3.8. a Simplified view of a rotating slice of the Earth beside a stationary Moon to explain oceanic tides. b Landward migration of tidal wave leading to an increase in tidal range from deep to shallow water, as well as influence of shelf width on tidal range. c Influence of coastal morphology on tidal range and tidal effect in an estuary. d Influence of the orientation of elongated ocean basins on tidal range and tides in adjacent basins

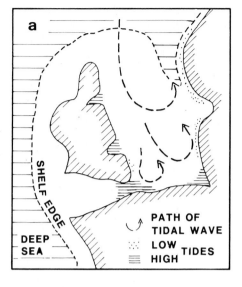

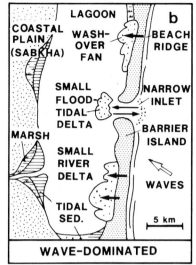

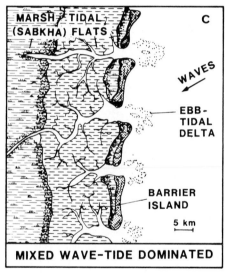

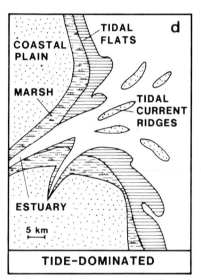

Fig. 3.9. a Simplified scheme of the development of several tidal systems in an epicontinental sea similar to that of the present-day North Sea. **b, c, d** Typical examples of microtidal, mesotidal, and macrotidal coastlines. (After Hayes 1980; Reineck 1984)

High tides and consequently *strong tidal currents* are generated under one or both of the following conditions:

- Wide shelf or continental sea (Fig. 3.8b),
- Large bay with funnel-shaped opening (Fig. 3.8c).

On the continental shelves of the present-day oceans, tides higher than 2 to 3 m usually occur in regions where the shelf is at least 100 to 200 km wide. The tidal range is further enhanced, if tidal waves behind a shallow sea are funneled into a bay with a wide opening (producing tidal ranges up to 10 to 15 m). East-west extended bays, fjords, or adjacent seas give rise to higher tides than north-south extended seaways (Fig. 3.8d). Adjacent basins display high to very high tides only when their openings are wide and deep, such as, for example, the Gulf of California.

These general rules about tidal ranges can locally be modified in many ways, particularly along irregularly shaped coastlines and in epicontinental seas. An example of tides reinforced by resonance and modified by the topography of the basin as well as Coriolis forces (see Chap. 5.2) is the North Sea (Fig. 3.9a).

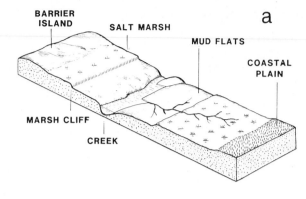

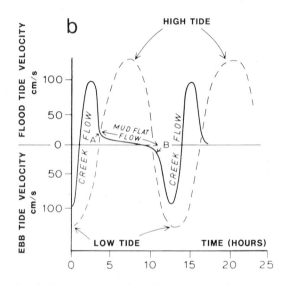

Fig. 3.10. a Cross section of a creek-mudflat-saltmarsh system at the North Sea coast. **b** Changing current velocities *(solid line)* of single water particles during a tidal cycle *(broken line)*. Peak velocities occur only in the creek at water levels below mean water stand (0) during the first half of the flood and the second half of the ebb period. At water levels above about 0, water flows with low velocities over the mudflats (between points *A* and *B*). (After Pethick 1984)

In this simplified figure, the entering tidal wave from the North Atlantic forms three more or less separately working tidal systems. Tidal ranges are high only at the western and southern margins of the epicontinental basin, and very low on the eastern side. From this and other examples we may conclude that while the recognition of ancient tidal deposits is very useful for paleogeographic reconstructions, tidal ranges can vary considerably, even within the same basin.

Tides also penetrate into the river mouths, particularly if these are wide and form es-

tuaries (Figs. 3.8c and 3.9d). When freshwater discharge is limited, salt water can intrude as bottom water far upstream into the estuary. With abundant fresh water flow, the salt water is hampered from moving upriver and therefore forms a "salt wedge" near the mouth of the estuary (Fig. 3.8c and Chap. 4.2, Fig. 4.1).

The incoming tidal water is called *flood*, the outgoing *ebb*. The highest water level reached at a certain point is the *high water line*, the lowest point the *low water*. The difference between mean high and mean low water is the "normal" or *mean tidal range*. According to their elevation with respect to the mean high and mean low water line, the tidal environment is subdivided into three zones:

- Supratidal zone above the mean high water line.
- Intertidal zone between the mean high and mean low water lines.
- Subtidal zone below the mean low water line.

The higher the tidal range, the more water has to be transported from deep water into the coastal zone and back again during one tidal cycle. Consequently, high tides are associated with strong tidal currents flowing onshore and offshore. They often reach velocities of 1 to 2 m/s, with local values of up to 4 to 8 m/s. In tidal flats their maximum velocities commonly develop in the middle of the flood or ebb period, when the rise or fall of sea level is most pronounced (Fig. 3.10b). At other locations, for example in special tidal channels or at the heads of estuaries, the tidal current maxima and minima may occur earlier or later in relation to the tidal cycle.

Tidal Ranges, Coastal Morphology, and Hydrographic Regime

There is a distinct relationship between tidal amplitude and coastal morphology (Table 3.1 and Fig. 3.9b-d; Hayes 1980; Reineck 1984; Elliott 1986a).

In addition, coastal morphology and sedimentation in tidal flats are strongly influenced by wind-generated waves. Therefore it has become common for sedimentologists to subdivide coastal areas into three subenvironments (e.g., Elliott 1986a):

- *Wave-dominated shorelines:* beaches, micro tidal barrier islands and cheniers,
- *Mixed wave-tide influenced shorelines:*

Table 3.1. Tidal-influenced environments

	Tidal range	Coastal morphology
- Microtidal	< 2 m	Long barrier islands, few inlets
- Mesotidal	2-4 m	Short barrier islands with numerous inlets, ebb and flood tidal deltas
- Macrotidal	> 4 m	Small or missing islands, estuaries with subtidal ridges

mesotidal barrier islands with tidal inlets and ebb and flood tidal deltas, and
- *Tide-dominated shorelines:* tidal flats, estuaries and associated sand ridges.

This subdivision is based on the assumption that wind waves are the dominant coastal process when the tidal range is less than 2 m. Features such as beaches, sand spits, and long barrier islands are controlled mainly by wind-generated waves. In contrast, tidal ranges in excess of 4 m are the dominant influence on coastal areas and their sediments, although wind waves also play some part. In this chapter, both tide-dominated and mixed wave-tide influenced environments will be discussed.

General Characteristics of Tidal Sediments

Modern tidal sediments commonly consist of (from higher to lower elevated zones):

- Mudflats.
- Mixed sand-mudflats.
- Sandflats.

Furthermore, they display various characteristics which can be easily recognized in ancient sedimentary rocks. Tidal sediments are therefore of outstanding interest for paleoenvironmental and paleogeographic reconstructions. Besides numerous special articles, several textbooks present summeries and some special books deal entirely with this topic (e.g., Ginsburg 1975; Friedman and Sanders 1978; Reineck and Singh 1980; Klein 1985c; Elliott 1986a; Boer et al. 1988).

Provenance and Distribution of Tidal Sediments

Tidal flat sediments may have different sources: input by rivers, cliff erosion, and reworked material from the offshore sea bottom. The proportion of mudflats, mixed sand-mudflats, and sandflats also depends on the availability of mud and sand in the source areas. Along coastal plains and near mud-dominated rivers, muddy tidal flats tend to prevail. In front of coastal mountain ranges delivering large quantities of sand into a tidal regime, sandier flats and subtidal sand ridges are prominent. An example of the latter situation is the Miocene (Burdigalian) marine molasse in the northern foreland basin of the Alps (Homewood and Allen 1981). It can be assumed that sandy material usually forms narrower tidal flats than finer grained muds, which better withstand erosion and reworking (see below).

The material forming tidal sediments is commonly transported by the flood from deeper water into the tidal flats as suspended load or as bed load. Due to the special creek system which develops in such areas, the incoming water flow is restricted first to the channels or creeks, where it can reach high velocities on the order of 1 m/s (Fig. 3.10b). Later, after the water level has risen, the water spills onto the higher widely extended flats. Consequently, the velocity of the flood current drops significantly, and the suspended particles, flocs or pellets, start to settle. After the tide has attained its maximum height, the thin water layer on top of the mudflats nearly comes to rest for a short time (approximately 1 h).

All particles which have settled to the ground during that time have a chance of staying behind in the mudflats. The rest, including most of the nonflocculated clay-size grains, will flow back seaward with the following *ebb current*. This current in turn will reach high velocities, as soon as the mudflats are drained and the seaward-flowing water becomes confined to the creek system. Therefore the channels are kept clean of fine-grained sediments, and erosion along their banks is very common.

In this way *small slides* are produced which are reworked by the channelized ebb or flood currents, forming *clasts and pebbles of mud*. As a result of the periodic emergence of the mudflats during low tides, the fine-grained sediment at the surface loses water by evaporation and becomes more solid than freshly deposited mud covered permanently by water. For this reason, mud pebbles are fairly resistant and often preserved in the sedimentary record.

Mud Deposition

The above mentioned mud deposition generates a striking contrast between mudflats along protected coasts and beach and foreshore sands along open coasts (Chap. 3.1). On sandy beaches the finer-grained particles are sorted out and carried into deeper water, whereas in tidal flats the fine-grained material is deposited near the coast slightly below or above the mean water line (Fig. 3.10a). This may even occur in medium to high energy tidal environments (waves and currents). This type of mud deposition is favored by the following circumstances and processes:

- Mud flats, marshes, or *coastal sabkhas* are usually found in areas sheltered from the effects of powerful wind-driven waves, i.e., they lie behind barrier islands and within coastal embayments or estuaries (Fig. 3.9b–d).
- The formation of mud flats is favored by gentle offshore slopes and, of course, high concentrations of suspended fine-grained material provided by nearby mud-dominated rivers.
- Mud deposition in tidal flats is enabled by the special distributary system for the incoming and outflowing water masses of the tidal waves, as well as by the flocculation of clay-sized particles or the incorporation of clay and silt particles into faecal pellets.
- After settling, the flocs or pellets stick together to form a cohesive sediment which requires higher current velocities for erosion than sand. In addition, thin mats of microorganisms (Gerdes et al. 1985) and, particularly in supratidal marshes, plants able to withstand the severe environmental conditions often further protect the muddy sediments from erosion.

Because water is lowest on the landward part of the flats, settling flocs and pellets do not need much time to sink to the ground. The slowly starting subsequent ebb current usually carries only part of this material back into deeper water. In this way the highest portions of the mudflats tend to increase in height more quickly than the deeper portions.

The Holocene **mudflats of the North Sea** accumulated at an average sedimentation rate of 1 to 2 mm/a. Over periods of months, however, the mudflats may grow much faster (about 5 to 10 mm/month), but vertical aggradation is frequently interrupted by erosional events (Reineck 1980; Anderson et al. 1981). Long-term rapid upbuilding is limited by tidal range and basin subsidence. In supratidal areas, which are flooded only during spring tides or storms, the rate of deposition decreases considerably.

Tidal marsh sediment along the Delaware river near Philadelphia accumulated at the rate of 0.4 mm/a during the last 2 ka, prior to the modern colonization of this area (Orson et al. 1990).

When muds are deposited under comparatively low-energy conditions, they are little affected by subsequent channeling and thus contain only small portions of cross-bedded sands and sand bars characteristic of the intertidal and subtidal zones.

Specific Sedimentary Structures of Tidal Flats

A special feature of laterally migrating tidal channels is lateral accretion bedding, or *longitudinal cross-bedding* (cf. Fig. 3.12), in contrast to normal, transverse cross-bedding, generated by current ripples and dunes within the channels. Due to the changing direction of flood and ebb currents, these types of small and large-scale cross-bedding often display bidirectional orientation of subsequent sets of laminae, a feature which is referred to as *herringbone stratification* (cf. Fig. 3.12).

Another interesting and diagnostic feature of tidal channel deposits is the large *sand waves* (often 1 to 2 m high), which may indicate not only the diurnal or semi-diurnal tidal cycle, but sometimes also the effects of the fortnightly variation from neap tide to spring tide (neap-spring cycles, Visser 1980; Fig. 3.11), which were found in modern and ancient sediments of different ages. Such sandwaves grow with the dominant current (this often being the ebb current), but are also affected by the weaker opposite current (reworked foresets and migrating smaller ripples).

During slack water at high water stand, clay and silt-sized sediment (flocs and pellets) can settle on top of sandy foresets and particularly in ripple troughs and thus form characteristic *mud drapings,* which partly resist later erosion. In shallow water channels, the accumulation of coarse shells and shell debris, derived mostly from eroded parts of the tidal flats, is common.

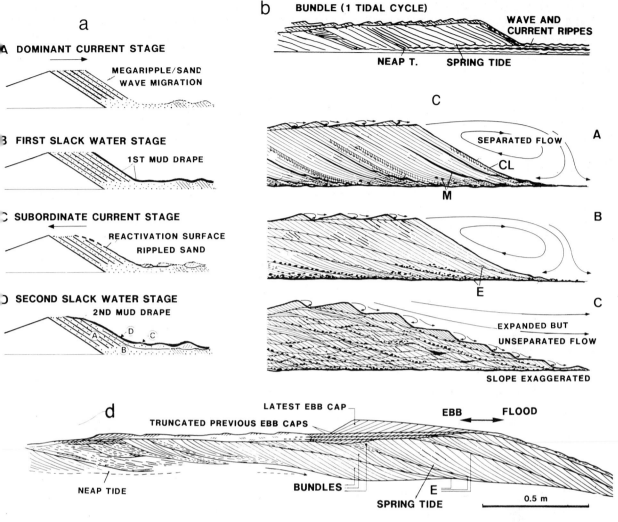

Fig. 3.11a-d. Subtidal to intertidal sand waves as found in estuaries. **a** Formation of a bundle of cross-bed foresets, a reactivation surface, and mud drapes during an ebb-flood tidal cycle. **b** Succession of bundles affected by spring tide and neap tide. (After Homewood and Allen 1981). **c** Different (theoretical) types of sand waves (about 4 m high and 200 m long) in subtidal environment. Subordinate current increases in importance from *A* to *C* (*C* begining of herringbone patterns). *E* erosional surfaces; *M* mud drapes and mud clasts; *CL* cross lamination. (**a** and **c** after Allen 1980 and 1982). **d** Intertidal to subtidal sand waves observed on the Dutch North Sea coast. (Boersma and Terwindt 1981)

Various Types of Tidal Sediments and Climate Control

The sediments of tidal flats proper and particularly those of the supratidal zone are strongly influenced by climatic factors controlling biogenic production, terrigenous sediment input, salinity of coastal waters, etc. Therefore several types of tidal flat sediments cam be distinguished (e.g., Ginsburg 1975):

- *Siliciclastic tidal sediments* (Fig. 3.12).
- *Calcareous tidal sediments* (Fig. 3.13).
- *Stromatolitic tidal sediments* (Fig. 3.14).

The most important characteristics of these depositional systems are summarized in the facies models (Figs. 3.12 through 3.14). A detailed discussion of these various types of tidal sediments with their great variability and numerous special features is not possible here, but the general rules discussed above can be applied to most of them.

Depending on climatic conditions, the mudflats may become vegetated (salt marsh, Figs. 3.9 and 3.10a) or, in tropical to subtropical regions, enable the growth of mangrove with a dense net of rootlets (Fig. 3.13). Such environments are commonly populated by a rich fauna, which is sometimes also found in the fossil record (e.g., Westgate and Gee 1990). In more arid climates, the supratidal zone may be characterized by coastal sabkhas where due to high evaporation some salts are precipitated (Chap. 6.4.1).

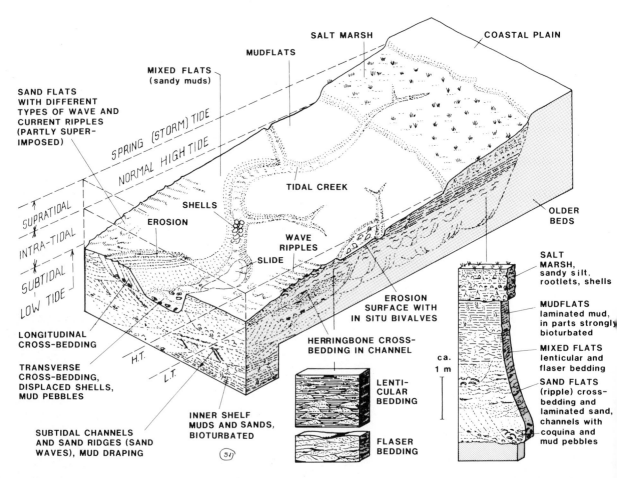

Fig. 3.12. Main features and special sedimentary structures of mesotidal siliciclastic tidal flats. (Based on Van Straaten 1954; Klein 1970; Ginsburg 1975; Reineck 1984, and others). Vertical scale exaggerated approximately 500x

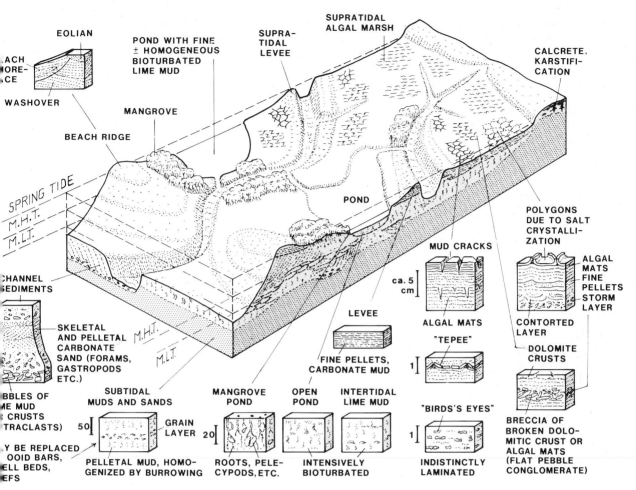

Fig. 3.13. Main features and special sedimentary structures of tidal flats associated with carbonate shelves and platforms in warm, humid climate. A high-energy environment leads to many large channels. (Based on Ginsburg and Hardie 1975; Shinn 1983; Sellwood 1986, and others). Vertical scale of block diagram exaggerated approximately 500x. Note the different scales of sections displaying sedimentary structures

Hypersaline waters favor the growth and preservation of microbial (algal) mats in the intertidal and subtidal zone (Figs. 3.13 and 3.14), because other organisms feeding on them become rare. Such *stromatolitic structures* are frequently found in ancient rocks, particularly in the Proterozoic (Chap. 6.5). They often give rise to the formation of characteristic voids (vugs, fenestrae, tepees, bird's eyes, Fig. 3.14) which are generated by crinkling when the sediments are exposed to air, or by gas due to decomposing organic matter.

Tidal sediments consisting largely of *calcareous sands and muds* can become cemented early and therefore have a particularly high preservation potential. Carbonate platforms (Chap. 12.2) such as the Bahamas are built up in great part as tidal flats. They may develop a sand ridge at their outer margin and ponds within the tidal flats proper, which are filled up by fine-grained lime mud (Fig. 3.13). Idealized sections of various types of tidal flat

sediments are shown in Fig. 3.15. They demonstrate the influence of both the hydraulic regime and climate, but they do not account for sea level changes and large erosional events.

Response of Tidal Flats to Sea Level Changes

As are all nearshore sediments, tidal deposits are strongly affected not only by variations in the influx of allochthonous material, but also by relative sea level changes. Relative *sea level fall* leads to a seaward migration of the tidal complex and usually causes partial erosion of pre-existing tidal sediments (Fig. 3.16a). If such a situation persists for some time, the chances for preservation of tidal deposits, particularly of supratidal and intertidal mudflats, are very limited. In contrast, inter and supratidal deposits can easily follow a *rising sea level* and build up thick sequences, because their sedimentation rate is sufficiently

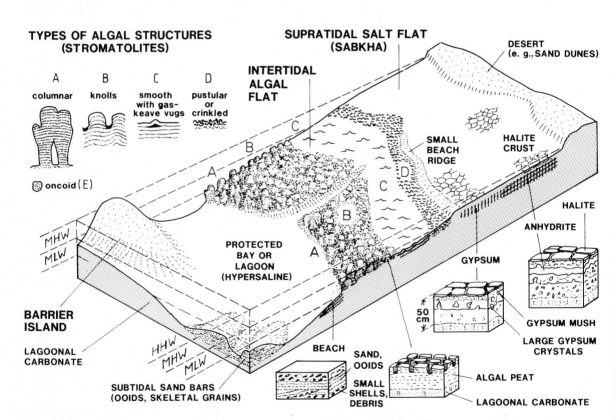

Fig. 3.14. Algal and evaporitic tidal flats in warm arid regions and low-energy environment (few channels!). (Based on Purser 1973; Schwarz et al. 1975; Butler et al. 1982; Schreiber 1986, and others). The intertidal algal flats and coastal sabkhas border protected hypersaline bays and

lagoons. Note the different types of algal structures according to their positions below or above mean sea level. For evaporites in sabkha environments also see Chapter 6.4.1 and Fig. 6.5. Vertical scale exaggerated about 500x

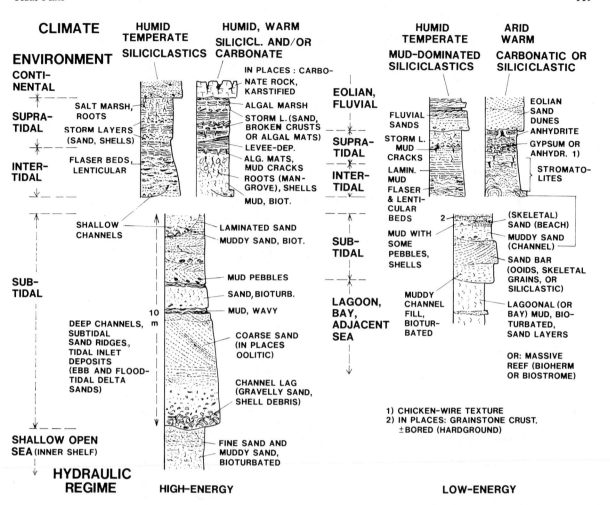

Fig. 3.15. Hypothetical sections to compare tidal deposits formed in areas of different climates and hydraulic conditions. The variability of these sediments increases from the subtidal to the supratidal zone due to climatic influences. Fully oxidized brown-colored primary sediments of the supratidal and partly of the intertidal zones tend to become red during diagenesis. In contrast, the thickness and nature of gray subtidal sediments are chiefly controlled by hydraulic conditions (high-energy open shelf versus low-energy bays and lagoons). The thickness of the different tidal zones is based on the assumption that neither subsidence nor sea level changes occur (see Fig. 3.17). (Compiled after different sources, e.g., Ginsburg 1975; Shinn 1983; Sellwood 1986)

high. In addition, they are fairly capable of resisting wave and current attack when their erodibility is diminished by algal mats and early cementation. If the sea level rise is slow or terminated, mud flats and supratidal sediments may prograde seaward and thus enlarge their areal extent. This is one of the reasons for the surprising thickness and wide areal distribution of tidal sediments in the geologic record. Furthermore, specific tectonic settings, such as long-lasting, slowly subsiding shallow platforms are needed to maintain a tidal flat environment over long time periods.

Remarks to Ancient Tidal Sediments

In spite of their great variety, siliciclastic and calcareous tidal sediments are normally easy to recognize in the ancient record. Their most distinctive features are:

- Sections with alternating subtidal, intertidal, and supratidal sediments.
- Small- and larger scale migrating channel systems filled with reworked sediments (mud pebbles) and shell deposits.
- Current and wave-induced sedimentary structures, partially bipolar, mud drapings, and sand waves indicating diurnal, semi-

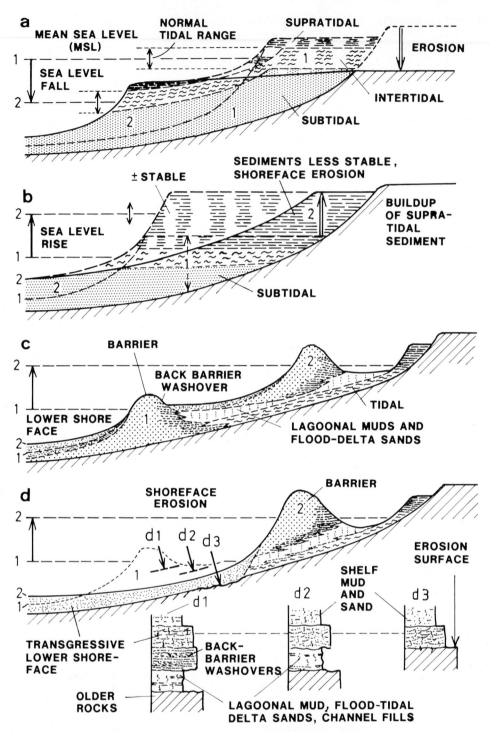

Fig. 3.16a-d. Response of tidal sediments and barrier island complexes to sea-level changes; resulting vertical sequences and lateral facies changes. **a** Falling sea level (Stage *1* to *2*) leading to coastal erosion and preferential loss of older supratidal and intertidal sediments. New regressive tidal sediments prograde or build out. Slow sea level fall may enable the preservation of widely extended subtidal sediments. **b** Rising sea level, transgressive sequence or continuous upbuilding of muddy intertidal and supratidal sediments. **c** Rapidly rising sea level, drowning of former barrier-island complex and buildup of second barrier-lagoonal complex farther inland. **d** Slowly rising sea level causing the barrier-island complex to migrate landward. Shoreface erosion, prograding from stage *d1* to *d3*, removes older sediments and generates vertical sections with differing stratigraphic gaps. (Based on Howard and Reineck 1981; Elliott 1986a, and others). Vertical scales strongly exaggerated

diurnal, and neap-spring tidal cycles.
- Indications of emergence (mud cracks, crinkled microbial mats, etc.).
- Specific epifauna and infauna.
- Characteristic facies associations: coastal barrier sands, lagoonal and estuarine sediment.
- Transitions either into shallow-marine environments or emergence with pedogenesis, plant rootlets, karstification, etc.

For a safe identification of tidal deposits, several of these features should be present.

When the overall fining-upward sequence of low tidal flat, midflat, and high tidal flat deposits is complete, its thickness allows an estimation of the *paleotidal range* (Klein 1971). The preservation of complete tidal sequences is, however, rare, and the lower limit of the tidal range is often difficult to determine (Terwindt 1988).

The sedimentary record provides evidence that tidal action occurred from the Precambrian to the Present (e.g., Klein and Ryer 1978; de Boer et al. 1988). Consequently, the Earth-Moon system must have always operated in a similar way as it does today. In addition, it has been tried to determine the duration of a Proterozoic year by investigating tidal cycles and rhythmites in glacial-influenced ebb-tidal deposits and banded iron formations (Chap. 6.5) in South Australia (Williams 1989a, b). According to this study, the Proterozoic year was about 400 days long; each day lasted approximately 22 h.

3.2.2 Sediments of Barrier-Island Complexes

General Characteristics

On microtidal and mesotidal shorelines (Fig. 3.9b and c), tidal sediments often represent part of a larger facies association which includes the deposits of barrier-islands and lagoons (Davis 1978; Reinson 1984; Elliott 1986a; Boggs 1987; Ward and Ashley 1989). Such environments can be summarized under the term *barrier-island complex.*

Barrier islands are the result of nearshore transport and accumulation of sand-size siliciclastic or carbonate particles along coastlines which are exposed to considerable wave action, but moderate tidal range. Their formation is favored by a low-gradient shoreface-shelf profile and abundant supply of sand-size

sediment. The islands develop from beach ridges, nearshore sand bars, or sand spits prograding parallel to the coastline. They often grow upward with rising sea level, separating an estuary or lagoon from the open sea. The barrier ridge is cut by channels (inlets), allowing tidal waves to enter and exit the lagoon (Fig. 3.17). In this way, a rather complex depositional environment is created, consisting of several subenvironments:

- The subtidal to subaerial barrier-beach complex.
- The subtidal-intertidal ebb-delta associated with shoreface sediments.
- The subtidal-intertidal flood-delta associated with lagoonal and tidal flat sediments (back-barrier zone).
- The (laterally migrating) subtidal-intertidal inlet-channel complex.

Barrier-island systems frequently form long, extended chains along straight coasts (cf. Fig. 2.3). Therefore, their deposits usually represent elongate bodies which parallel the strandline. Three of the above-mentioned subenvironments are characterized by sandy deposits, whereas in the back-barrier zone both sand and muds accumulate. The sediment source may be chiefly terrigenous and/or biogenic (carbonate) and thus magnify the variability of the barrier-island complex as a depositional system. As pointed out above, tidal sediments and fauna in the lagoon strongly reflect the climatic conditions of the neighboring landmass.

Subenvironments of Barrier-Island Complexes

The **barrier-beach complex** consists of, from bottom to top:

- *Shoreface sands* down to a depth of 10 to 20 m below mean sea level, where the normal wave action ceases. These sands generally exhibit decreasing grain size and increasing bioturbation with growing water depth (Fig. 3.17). While the upper shoreface deposits may contain gravelly sands and display multidirectional sedimentary structures (trough cross-beds and low-angle planar cross-beds), the lower ones consist of fine to very fine-grained sand or muddy sand. Here, the primary sedimentary structures, consisting mainly of planar laminated and small-scale cross-bedded material, are frequently obliterated by bioturbation.

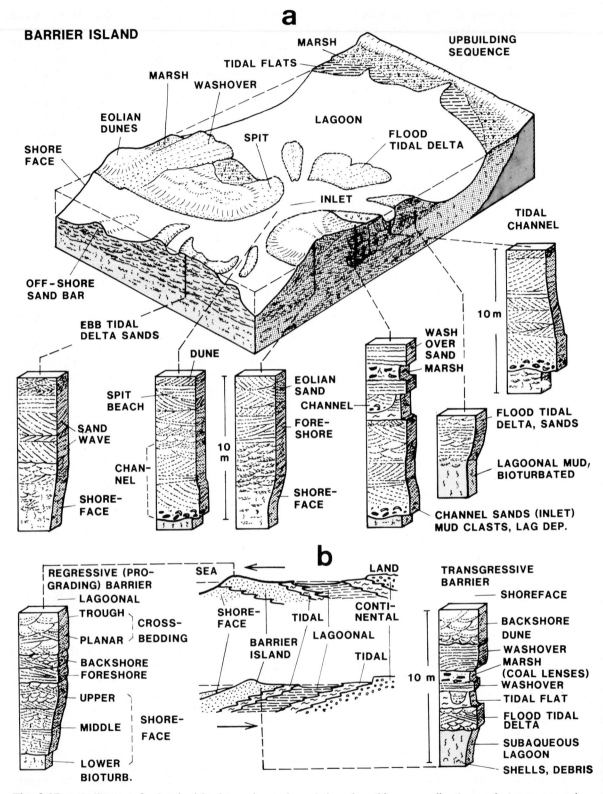

Fig. 3.17. a Sediments of a barrier-island complex and lagoon in microtidal to mesotidal environments. Here chiefly upbuilding of sediments is assumed with laterally shifting inlets. **b** Seaward or landward migrating barrier-island chain and resulting prograding (regressive) or transgressive sedimentary sequences. Note that parts of these sequences may be lost by erosion (cf. Fig. 3.16d). (Based on Reinson 1984; Elliott 1986a, and others)

- *Beach-foreshore sands* displaying parallel to low-angle, seaward-dipping laminations and usually little bioturbation. They are overlain by

- *Backshore eolian dunes* starting with rather small-scale, multi-directional trough and planar cross-beds. If storm surges overrun parts of the islands, they create

- *Washover sands* which extend into the lagoon. These show either subhorizontal planar lamination or, if they reach the lagoon, delta foresets. The washover deposits may make up a significant portion of the total barrier island complex. They reflect the tendency of the islands to migrate landward. They also feed the

- *Back-barrier sand flats* (Fig. 3.17). The sedimentary structures of both eolian dunes and washover sands can be obliterated or modified by plant growth and soil formation.

Ebb-tidal sands accumulate seaward of the barrier and are associated with shoreface sands. In contrast to flood-tidal sands, they are also affected by wind-generated waves and longshore currents. Therefore their sedimentary structures can reflect several processes and current directions. Their faunal content is mainly of open-marine origin, although ebb currents may also transport lagoonal biogenic particles seaward into the ebb-tidal delta.

Flood-tidal sands alternate with lagoonal muds and tidal flat sediments of the back-barrier zone. These delta sands show planar and trough cross-beds which are mainly generated by flood-oriented mega-ripples and sand waves, but a seaward orientation of cross-beds also occurs.

Tidal-channel deposits between the ebb and flood-tidal delta can cover relatively large areas, because the channels and tidal inlets between the barrier-islands frequently migrate parallel to the shoreline. Shifting of inlets involves erosion on one side of the barrier and accretion of sand on the other (spit accretion). Channel fills usually begin at the erosional base with relatively coarse lag deposits. These are overlain by large-scale bidirectional planar and trough cross-beds. On top of this sequence bidirectional small to medium-scale trough or planar cross-beds, and parallel and ripple lamination are observed.

Controls by Climate and Hydraulic Regime

Depending on the climate, lagoonal and tidal flat sediments can vary considerably. If the barrier-island chain is interrupted by many wide inlets (mesotidal hydraulic regime), the lagoonal waters are sufficiently exchanged with the open sea water and therefore are *normal saline*. Under these conditions, lagoonal fauna can develop a highly diverse assemblage, and the fine-grained, often laminated lagoonal muds tend to become thoroughly bioturbated and structureless. However, if there are just a few narrow inlets (microtidal regime), the lagoonal waters become either *brackish* or *hypersaline,* and their fauna is abnormal and of low diversity. Oyster beds, for example, indicate freshwater influence, specific gastropods and seagrass typify hypersaline conditions. Due to storms which produce washovers, the lagoonal environment may change episodically. Simultaneously, shells of normal marine organisms can be swept in.

In humid regions, the lagoons often accumulate organic matter, which later may form coal seams, and their bottom waters can easily become anoxic. *Arid to semi-arid lagoons* tend to collect more carbonate, and their tidal deltas may become oolitic. Marginal portions of the lagoons can periodically dry out and accumulate gypsum (cf. Chap. 6.4.1). Migration of eolian sand over dry lagoons is common. Tidal flat sediments bordering the lagoon may include plant roots, peat, or microbial mats and stromatolites, as well as evaporites formed in coastal sabkhas.

Response of Barrier Island Complexes to Sea Level Changes

Barrier-island complexes are dynamic systems responding to changing sediment input and sea-level changes. Rapid sea-level fall (or emergence of the barrier system by uplift) tends to destroy the sedimentary complex by erosion and is therefore not further discussed here. With constant or rising sea level, the island chain can migrate either landward or seaward and thus bury and preserve some of its older deposits (Fig. 3.17b). If sediment input from the hinterland or alongshore is high, the island chain moves seaward *(prograding or regressive barrier).* In this case, vertical sections show a transition from shallow marine and shoreface facies into the foreshore, backshore, and dune sands of the barrier island. Lagoonal deposits may follow on top of the foreshore and backshore sands.

In contrast, low sediment input favors the *landward transgression of barrier systems*. Then a complete vertical section starts with continental or tidal flat sediments and grades into lagoonal muds, flood-delta sands and channel fills of the inlet system. Finally, washover sands and eolian dunes may follow. Consequently, vertical sections clearly document the regressive or transgressive history of the barrier.

However, in many cases the development of a transgressive barrier system is considerably modified during rising sea level (or submerging coast). If the sea level rises relatively quickly, the pre-existing barrier islands are drowned in-place and form a kind of inner shelf shoal (Fig. 3.16c) as described, for example, from the abandoned western part of the Mississippi delta (Penland et al. 1988). Then, a new barrier may be built up farther landward (Fig. 3.16c). Consequently, the former lagoonal muds and flood-delta deposits are covered directly by shoreface sediments. The drowned sand barriers can later provide interesting reservoirs for hydrocarbons or groundwater.

The second alternative, as mentioned before (Fig. 3.17b), is a slow *landward migration* of the total barrier system with moderately rising sea level. Then, the transgression can be accompanied by significant shoreface erosion or shoreface retreat (Fig. 3.16d). In this case, the sediment eroded from the upper shoreface is redeposited either in deeper water (transgressive lower shoreface) or swept through inlets or storm-generated washovers into the back-barrier region.

Shoreface erosion can affect not only barrier sands, but may also proceed into lagoonal and tidal deposits. As a result, the transgressive beds of the lower shoreface can overlie both back-barrier sands and lagoonal muds (including flood-delta deposits), only lagoonal muds, or directly older bedrock (Fig. 3.16d1, d2, d3).

The bedrock may have been exposed to air prior to the transgression. Deep erosion into pre-existing beds is favored by slow sea level rise and low sediment supply. It often creates planar erosion faces and appears to be rather common in the ancient sedimentary record, whereas the former barrier islands may be completely gone. Only the infillings of deep channels of the former inlet systems have a good chance of being preserved under such conditions. These and other complications have been repeatedly described from modern and ancient examples (e.g., Nummedal et al. 1987; Nummedal and Swift 1987).

Remarks to Ancient Barrier Island Sands and Lagoonal Sediments

Ancient barrier island sands are more difficult to identify than tidal sediments, particularly when their upper portions (foreshore, washover, and eolian sands) are missing. The lower portions of these complexes may be easily confused with the normal shoreface sands of shallow seas or with mouth bar sands of marine deltas (Chaps. 3.4.2 and 3.4.3). In such cases it is necessary to investigate the larger scale facies association of the sand complex under consideration.

Even lagoons are not always readily recognized in the geologic record, because of the variety of associated deposits (Ward and Ashley 1989). Lagoons are shallow features, and modern examples frequently display sedimentation rates on the order of 1 to 2 mm/a. Therefore, they are filled up with sediments in a relatively short time period (Nicols 1989). Thus, lagoonal sediments are volumetrically less significant than many other coastal and shallow-water deposits, particularly deltaic sequences (Boggs 1987). Nevertheless, ancient lagoonal sediments provide valuable information on the position of the paleo-coastline, climate, and biota of the corresponding time period.

3.3 Sediments of Shallow Seas (Including Carbonates)

3.3.1 Predominantly Siliciclastic Sediments

Introduction

The physical processes briefly described in the previous sections (beach and shoreface sediments, tidal deposits, and barrier island complexes) also control, to a large extent, the sediment dispersal and sedimentary structures of shallow-marine environments. Wind-generated waves and currents as well as tidal currents may affect the depositional processes in relatively deep water. The processes are additionally influenced by storm-generated density currents and oceanic currents of different origin (Chap. 5.2).

Epicontinental Seas and Shelf Seas

For several reasons it is useful to distinguish between two types of shallow seas (Fig. 3.18; also see Chap. 1.2):

- *Epeiric* or *epicontinental seas* which develop on continental crust and are bordered on several sides by land areas.
- *Marginal* or *pericontinental seas (shelf seas)* located on top of transitional crust along the boundary between continents and deep oceans. This type corresponds with the present-day shelves, particularly those on passive continental margins.

The depth range of both types of shallow seas is from 10 to 20 m near the coast (apart from the nearshore areas) to roughly 200 m in the case of many modern shelf seas. Some subbasins in epicontinental seas and on the shelves can attain greater depths. Both types of seas are fully marine, and deviations from the mean salt content of the open sea are limited (Chap. 5.2). Of course transitions exist between semi-enclosed epicontinental seas and adjacent seas (Chap. 4), which tend to become brackish or hypersaline depending on the climate of the region.

The physical processes in the two types of shallow seas may show significant differences:

- Continental shelves are normally fully exposed to the hydraulic regime of large neighboring oceans. They may be affected by distant storm centers and long-distant oceanic currents, whereas epicontinental seas are often better protected from these influences (Fig. 3.18). If epicontinental seas are partly enclosed, their areal extent is not sufficiently large for the initiation of long, high-energy waves or significant tidal waves. Although, for example, the North Sea and the Bering Sea are situated on continental crust, they are not typical epicontinental seas in this sense.
- Shelf seas often lose a great part of their sediments (sand and finer grained material) into deeper water by traction currents and gravity mass movements. This occurs less in typical, widely extended epicontinental seas, although even these basins collect thicker sed-

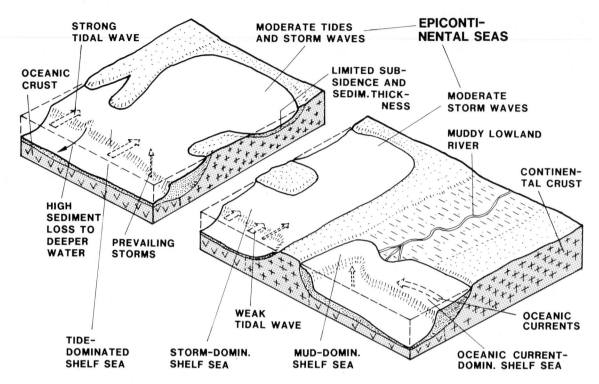

Fig. 3.18. Different types of shallow-marine basins, epicontinental seas and shelf seas, and dominant physical processes operating in these basins. Normal fairweather waves are present in all these environments and not mentioned here

iments in topographic deeps than in shallower regions.

- Because they evolve on thinning transitional crust, shelf seas exhibit stronger subsidence, particularly during their early stages of basin development, and therefore have the potential to accumulate thicker and possibly other sedimentary sequences than epicontinental seas (see Chap. 12.2).

Factors Influencing Shallow Sea Sediments

The sediments of shallow seas may vary considerably for the following reasons:

- Marked differences in the input of terrigenous material from neighboring land areas due to changing climates, presence or absence of large river deltas relevant to the location studied.
- Variations in biogenic production.
- Influence of relative or eustatic sea level fluctuations.
- Reworking of earlier intra-basinal sediments, etc.

Due to all of these influences, it is not possible to describe the various sedimentary facies of shallow seas by only a few simple facies models. In addition, it has been quoted repeatedly (e.g., Johnson and Baldwin 1986) that the modern continental shelves are not yet in equilibrium with their present-day hydraulic regime and sediments sources. These shelves and their sediments still reflect the unusually rapid, high-amplitude Pleistocene regressions and transgressions (cf. Chaps. 7.1 and 7.4). The last transgression (Holocene) terminated only a few thousand years ago! This is a time period in which only about half a meter of shelf sediment can accumulate. The present-day shelf sediments, particularly those on the outer shelf but also those in other areas of low deposition, therefore still reflect conditions of the last and possibly former low sea level stands at depths of more than 100 m. This situation has to be taken into account when modern shelf sediments are studied in order to achieve a better interpretation both of ancient counterparts and, even more so, of the deposits of ancient epicontinental seas.

Prior to the modern concept of sequence stratigraphy (Chap. 7.4), the architecture of shallow-sea sediments was poorly understood.

Older facies models presented elsewhere and partially summarized in this chapter are, however, useful for the interpretation of short-period processes which are not or only little affected by sea level changes. The following simple models, which are deduced theoretically as well as from modern and ancient examples, should be regarded in this sense. They mainly characterize siliciclastic sediments, but also take into account mixed siliciclastic-carbonate deposits.

Facies Models for Shallow Seas

The facies models of this chapter are all based on the assumption that the coastline progrades seaward. They can be used only for short time periods in which the sea level is more or less constant.

In a first step we assume that there is no subsidence. Consequently, the shallow basins are filled up in a comparatively short geologic time span (a few millions of years) if they accumulate sediments at a rate of about 50 to 100 m/Ma. In reality, basins generally subside during sediment accumulation and therefore frequently persist for much longer time periods. This point will be discussed below.

High Terrigenous Input

Low-energy waves, mud input. Figure 3.19a1 demonstrates shallow seas with high terrigenous input, mainly as mud (silt and clay-size particles). Under prevailing low-energy wave conditions, the shoreface sediments are dominated by silty muds overlain by a thin cap layer of beach and foreshore sands. In deeper water, muds of different nature are encountered. One portion of the mud is carried directly from the mouths of rivers as flocs or aggregates to its location of final deposition, whereas the other portion is reworked by moderate storms along the lower shoreface and redeposited as thin mud tempestites in deeper water (Chap. 3.1.2). Such an alternation of autochthonous and allochthonous mud layers is frequently masked or completely obliterated by intensive burrowing on the sea floor. It is mostly overlooked in field exposures. Muddy shorelines and shoreface zones commonly occur downdrift from deltas prograding into low-gradient shallow seas (e.g., Fraser 1989). Well known modern examples are the Louisiana-Texas coast west of the Mississippi delta and the Surinam coast west of the Amazon delta (Wells and Coleman 1981).

Low-energy waves and high input of sand and mud may lead to a situation as shown in Figure 3.19a2. The thickness of shoreface sands increases considerably; it may include offshore sand bars and proximal sandy tempestites (Chap. 3.1.2) in the transition zone to deeper water. Farther offshore, bioturbated mud tempestites alternating with silty and clayey autochthonous host sediments are again a characteristic facies. As in the example in Figure 3.19b, some shell beds may also be present, but biogenic production is generally very diluted by the high input of terrestrial material.

Storm-dominated sea, high input of sand and mud (Fig. 3.19b1). The beach-shelf profile becomes more gentle. The sandy, in places gravelly facies belt, comprising shoreface sands, sands with swaley and hummocky cross stratification (Chap. 3.1.2), and sandy tempestites, expands seaward into deeper water. In such cases, sands represent a great portion of vertical sections through the entire basin fill (e.g., Howard and Reineck 1981). Part of the sands transported by storms into deeper water can be reworked by subsequent storm-induced currents, direct storm wave action, or oceanic currents (see below). Far offshore, mud tempestites should be the dominant bed type.

Tide-dominated seas. As known from present-day tidal- and current-influenced shallow seas, such as the North Sea, the Irish Sea, and the Atlantic shelf of North America, tidal currents may also affect the sedimentary processes at water depths up to 50 m and more (Model b2 in Fig. 3.19). They generate characteristic sand waves, which may show mud draping. In addition, tidal waves form and slowly move flat, large sand ridges tens of kilometers in length, several hundred meters to some kilometers wide, and up to several tens of meters high (Reineck and Singh 1980; Stubblefield and McGrail 1984; Swift 1985; Johnson and Baldwin 1986; Saito 1989b).

A 200 km long and 90 km wide, radiating field of tidal current ridges was observed in the southwestern Yellow Sea at water depths up to 20-30 m (see Fig. 11.2; Liu Zhenxia et al. 1989). The sand ridges consist of fine sand originating from the abandoned deltas of the Yellow River and the ancient Yangtze River and were shaped as a result of the interaction of two tidal wave systems operating in the Yellow Sea.

Bottom currents in between the sand accumulations commonly erode longitudinal furrows which may collect and store coarse lag deposits, if coarse-grained material is present in

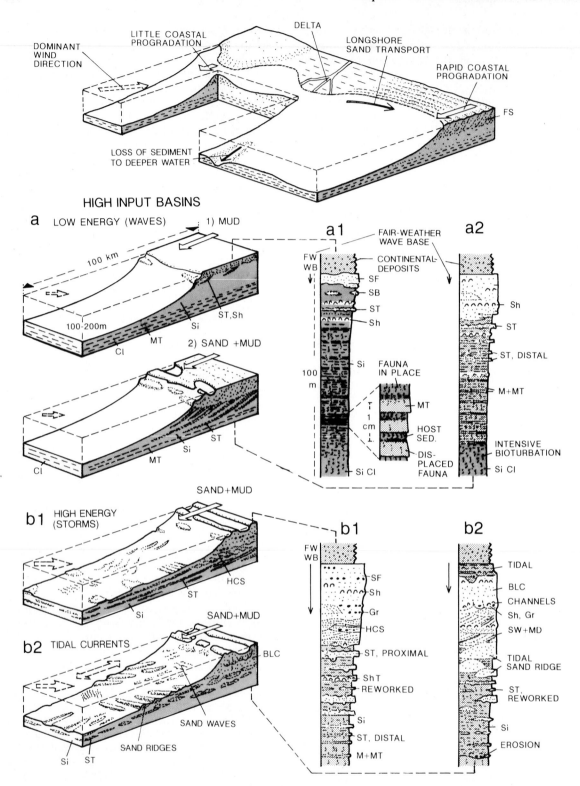

Fig. 3.19. Top Shallow sea with locally low or rapid progradation of shorelines controlled by coastal morphology, position of river delta, and prevailing wind and current directions. **a** and **b** Facies models of rapidly prograding ("regressive") sedimentary sequences of shallow seas. All basins receive high input of terrigenous material, predominantly mud (**a1**) or both sand and mud (**a2, b1,** and **b2**), but differ in their hydraulic regimes. Stable sea level, no subsidence. Legend see Fig. 3.20

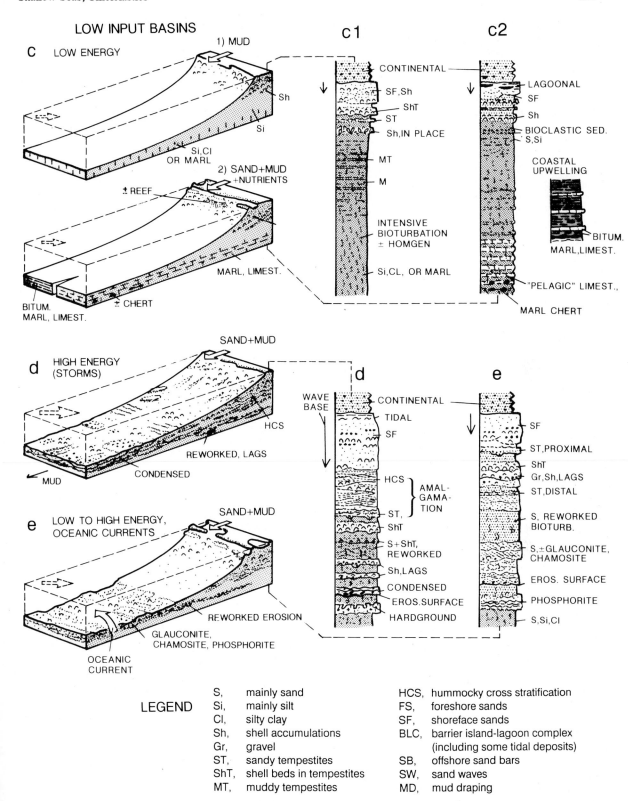

LEGEND

S,	mainly sand	HCS,	hummocky cross stratification
Si,	mainly silt	FS,	foreshore sands
Cl,	silty clay	SF,	shoreface sands
Sh,	shell accumulations	BLC,	barrier island-lagoon complex
Gr,	gravel		(including some tidal deposits)
ST,	sandy tempestites	SB,	offshore sand bars
ShT,	shell beds in tempestites	SW,	sand waves
MT,	muddy tempestites	MD,	mud draping

Fig. 3.20 (Fig. 3.19, continued). **c** through **e** Facies models of slowly prograding ("regressive") sedimentary sequences of shallow seas. All basins receive low input of terrigenous material, predominantly mud (**c1**) or both sand and mud (**c2, d,** and **e**), but differ in their hydraulic regimes and biogenic production. Stable sea level, no subsidence. Coastal upwelling may lead to the deposition of marls and limestones rich in organic carbon and chert (**c2**)

the underlying beds. Different types of ripples and current lineations provide evidence of frequently active bottom currents. Flat ribbons or patches of sand are common features in rather deep water. Thus, vertical sections of such a basin fill show again a very high proportion of sand with a variety of sedimentary structures, some of which are not very well known. They may indicate bidirectional current orientation, although one current direction (either the ebb or the flood current) is usually dominant. The fine-grained river load or muds produced in the basin itself are mainly deposited in basin areas which are protected from tidal and other bottom currents or, in the case of shelf seas, they are transported into deep water.

Storm- and current-dominated seas. The high-energy models (Fig. 3.19b1 and 2) are characterized by widely extended sand bodies. Prominent modern examples are the shelves of the western North Atlantic or the southern North Sea, where large, flat sand-ridges cover large areas. However, these sand-ridges still reflect the last Quaternary rapid sea level fluctuations. It is therefore doubtful whether these sand bodies are analogs of ancient shelf sands.

Low Terrigenous Input

The second series of models shown in Fig. 3.20 represents examples with low input of terrigenous mud and sand. These models are probably realized in nature more frequently than those discussed above. For obvious reasons, such basins also persist for longer time periods than basins receiving large terrigenous influx.

Wave-dominated, mud input. The sedimentary fill of a low-energy, wave-dominated basin with prevailing mud input (Fig. 3.20c1) resembles that in Figure 3.19a1, but the slower siliciclastic sedimentation rate enables higher concentrations of biogenic constituents and more intense bioturbation. For that reason, shell accumulations in the shoreface zone and, subordinately, in offshore tempestites are more pronounceed than in model Fig. 3.19a1. The sediments of the outer shelf may become marly, because they contain relatively high proportions of biogenic carbonate of both benthic and planktonic origin. Mud tempestites tend to become completely mixed with host sediments and thus obliterated.

A further reduction of terrigenous input to an insignificant portion of the total sediment may lead, when low-energy conditions are maintained, to sediments *rich in biogenic components* (Fig. 3.20c2). In this case, a favorable climate and sea water sufficiently rich in nutrients is needed. The latter may be provided by river water or by oceanic currents entering the shelf sea. As a result, the sediments of the inner shelf can again display abundant shell layers and bioclastic sands and silts. Larger shells are commonly crushed to small pieces by predatory organisms. On the outer shelf, the carbonate content may become so high that marls and lime muds are deposited which later form marl-limestone sequences (Chap. 7.2). If the contribution of tiny skeletons of planktonic organisms is high, the limestones resemble deep-water pelagic limestones. They can also contain some chert, but it is usually possible to infer their shallow-water origin from alternations with other sediment types. Bioturbation is again very intense in this environment, thus diminishing the preservation potential of tempestites.

An alternative of this model is an environment with an *oxygen minimum* zone in bottom waters of the outer shelf or in special depressions of epicontinental seas. Such a situation may be caused by upwelling waters rich in nutrients (Chap. 5.3.4) or by restricted water circulation. The resulting sediments are laminated bituminous clays and marls which may contain some limestone beds and chert layers. Bioturbation is limited to certain horizons or completely missing.

Storm- and tide-dominated seas. The model in Fig. 3.20d characterizes high-energy conditions (storms and tides) with low input of sand and mud. Coastal progradation is limited and the slope of the beach-shelf profile becomes rather gentle. In contrast to model b1 (Fig. 3.19), shell beds and finer grained biogenic components may become significant contributors to the total sediment. Repeated reworking due to storm action leads to pronounced amalgamation in the shoreface-offshore transitional zone and to other erosional surfaces with lag deposits and condensed beds in deeper water. Beds already buried and somewhat indurated may become re-exposed to sea water by current action and colonized by faunal assemblages needing firm or hard grounds.

Shelf seas affected by oceanic bottom currents. The last model (Fig. 3.20e) deals with shelf seas which are strongly influenced by oceanic currents entering the shelf and preventing the deposition of fine-grained material in these regions (e.g., Flemming 1980). In the shoreface zone, the depositional processes may resemble those described for models c1 or c2, but at a greater depth strong and repeated reworking including mixing of older and younger sands is common. Gravel may be concentrated in lag deposits, and vertical sections are characterized by many erosional surfaces. If the mean sedimentation rate becomes very low including periods of non-deposition, glauconitic minerals may be formed at the sediment/water interface (Chap. 6.1). Under special conditions, phosphorites can grow and may be concentrated in such an environment. Bottom life may differ from the previous examples, and the intensity of bioturbation tends to change considerably from bed to bed.

Summary

All the facies models described so far (Figs. 3.19 and 3.20) represent so-called "regressive" sequences, although the sea level was kept constant. They are generated solely by a prograding shoreline. Most of these models imply *coarsening-upward sections* and, in view of the thickness of sand bodies, also thickening-upward sequences. Exceptions from these rules are the high-energy models d and e in Figure 3.20. The low-energy models c1 and c2 with an upward-decreasing proportion of biogenic constituents may referred to as sequences with "upward-decreasing biogenics". The low-energy models a1, a2, c1 and c2 mainly represent epicontinental seas. The storm-dominated, tide-dominated, and current-dominated models b1, b2, d and e are realized predominantly in shelf seas. Model e is restricted to shelf seas.

Modifications of the Facies Models

The generalized vertical sequences described above can be modified significantly by a variety of factors not mentioned so far:

1. Under the assumption that there is no subsidence, the thicknesses of the facies sequences shown in Figs. 3.19 and 3.20 correspond to the original water depths of the sedimentary basins. Taking subsidence into account, the thicknesses of the total basin fills, as well as those of single facies types increase.

If subsidence is equal to the sedimentation rate for a certain time period, the facies at each location within the basin remain unchanged. As a result, each particular facies zone, for example the shoreface or the outer shelf sediments, can theoretically reach very great thicknesses.

2. If the basin floor subsides faster than sediment accumulates, the water depth at a certain locality increases. Consequently, the different facies zones migrate landward, and vertical sequences show the opposite tendency as described above, i.e., fining-upward or thinning-upward trends. Similar results are brought about by a rapid sea level rise leading to an increase in water depth. Conversely, tectonic uplift of the basin floor or sea level fall cause a seaward migration of the facies zones and thus a shortening of the vertical sequences. These problems are further discussed in Chapter 7.4 (also see, e.g., Aigner and Dott 1990).

3. Furthermore, the two-dimensional models of Figs. 3.19 and 3.20 do not take into account local control of sedimentary processes resulting from irregular basin morphology, shifting and abandonment of river deltas, input of sediment from different sources, regional and temporal changes of climate, direction of dominant winds and currents, etc. Coevel depositional processes may strongly differ from one location to the other in the same basin (Fig. 3.19, top). At one site, the coastline may retreat landward and at the other prograde seaward at the same time. Such complications should be borne in mind, if the theoretical models are applied to observations in nature.

Large-Scale Shallow-Marine Sand Bodies

The processes of the beach and shoreface zone including the inner and outer shelf region described above may generate widely extended, partially *amalgamated* or *stacked marine sand sheets*. Prerequisites for such a development are effective sediment sources and a relatively fast subsiding basin. In this case, even moderate sea level falls cannot cause emergence and thus lead to subaerial erosion and reworking of the marine sand sheets (cf. Chap. 7.4).

A well exposed and repeatedly studied ancient example of sandy and muddy shelf-slope sediments is the foreland basin of the Western Interior seaway in North America (Shurr 1984; Swift and Rice 1984; Swift et al. 1987; also see Chap. 12.6). The western margin of this seaway was bordered by the emerging Rocky Mountains which shed large quantities of mud and sand into the seaway (Fig. 3.21a and b). Currents generated by the dominant north-

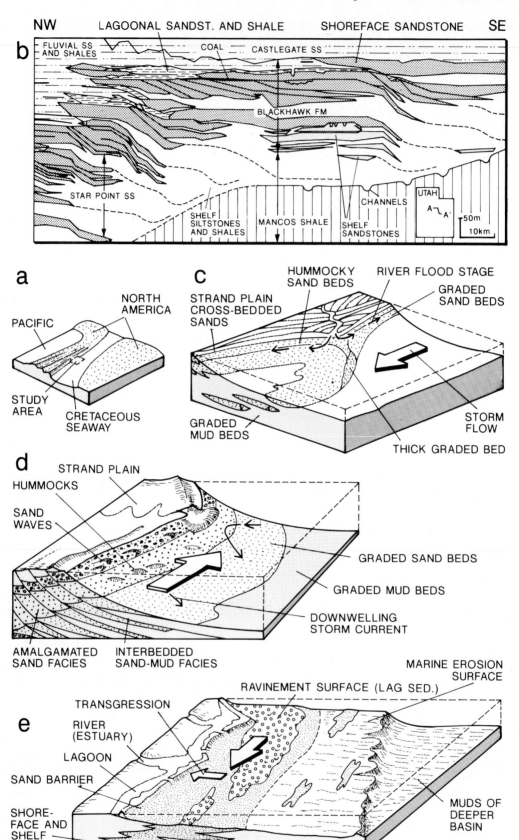

b

NW LAGOONAL SANDST. AND SHALE   SHOREFACE SANDSTONE SE

FLUVIAL SS
AND SHALES   COAL — CASTLEGATE SS

BLACKHAWK FM

STAR POINT SS

SHELF
SILTSTONES   MANCOS SHALE
AND SHALES

CHANNELS

UTAH
A   A'

SHELF
SANDSTONES

50m
10km

a

PACIFIC   NORTH
AMERICA

STUDY
AREA   CRETACEOUS
SEAWAY

c

HUMMOCKY   RIVER FLOOD STAGE
SAND BEDS

STRAND PLAIN   GRADED
CROSS-BEDDED   SAND BEDS
SANDS

STORM
FLOW

GRADED
MUD BEDS   THICK GRADED BED

d

STRAND PLAIN

HUMMOCKS

SAND
WAVES

GRADED SAND BEDS

GRADED MUD BEDS

DOWNWELLING
STORM CURRENT

AMALGAMATED   INTERBEDDED
SAND FACIES   SAND-MUD FACIES

MARINE EROSION
SURFACE

RAVINEMENT SURFACE (LAG SED.)

e

TRANSGRESSION

RIVER
(ESTUARY)

LAGOON

SAND BARRIER

SHORE-
FACE AND
SHELF
SANDS

MUDS OF
DEEPER
BASIN

westerly to northeasterly winds and deflected by Coriolis forces (see Chap. 3.1.2) caused storm tides along the western coastline (Fig. 3.21c and d). The measured opposite paleoflow directions are explained by the special coastal morphology of the study area situated in a large bight where setup of water along the coasts has created geostrophic currents of varying directions (Swift et al. 1987). The shelf sands are either linked to the shoreface sands, or they form large, partially isolated, lenticular sand bodies in between shelf muds. In some areas these elongate sand bodies resemble the sand-ridges and sand-ridge fields described from modern shelf seas (e.g., Tillman and Siemers 1984). Sands entering the basin are distributed either directly from the river delta by longshore transport and storm flow, or from the foreshore zone by downwelling storm currents into deeper water (Fig. 3.21c and d).

The seaward prograding sands consist of several facies units (from top to bottom):

- Cross-bedded sands of the strand plain.
- Relatively coarse grained, cross-stratified sands of the upper shoreface (surf zone) with sand waves.
- Fine grained, amalgamated sands of the middle shoreface with hummocky cross stratification.
- Very fine grained, graded shelf sands (tempestites) alternating with muds.

These storm deposits may rest on a submarine erosion surface characterized by lag deposits and channel fills (Fig. 3.21e and b). Such an erosion surface is not necessarily associated with a relative sea level fall (cf. Chap. 7.4), but may also be caused by waning sediment supply in an unsteadily subsiding basin.

Many other examples of ancient shelf sands may be interpreted in a similar way (e.g., Reineck and Singh 1980; Matthews 1984; Tillman et al. 1985; Knight and McLern 1986; Morton and Nummedal 1989; Aigner and Dott 1990). In any case, the interplay between subsidence, sediment accumulation, and sea level changes is of eminent importance for the thickness and vertical and lateral facies association of such sands.

3.3.2 Carbonate Buildups and Reef-Lagoon Complexes

Introduction

A major part of the present-day and ancient carbonate rocks was, and is formed in very shallow marine water up to a water depth of about 10 to 20 m. Most of this carbonate is produced by organisms (skeletal carbonate) in regions of warm climate in low latitudes, whereas carbonate production is very limited in high-latitude oceans. Here, part of the carbonate may even be dissolved on or near the sea floor.

The physical environment of nearshore and shallow-marine carbonate sediments is in many ways similar to that of sediments dominated by siliciclastic components. The physical processes mentioned in the chapters on beach and shoreface sediments, barrier-island complexes, and shallow seas (Chaps. 3.1, 3.2, and 3.3.1) also operate in carbonate-dominated environments. In spite of these common features, many textbooks and numerous articles treat carbonate rocks more or less separately from siliciclastic rocks for the following reasons:

- Carbonate rocks are strongly influenced by the evolution of carbonate-producing organisms and their different faunal assemblages through the Earth's history. The various aspects of this topic, particularly those of reef structures, are described in many special articles and books (e.g., Wilson 1975, Toomey 1981; Sellwood 1986; Scoffin 1987; Flügel 1989).
- Organisms produce not only a variety of skeletal particles of different size and shape, but also carbonate minerals of differing composition and thermodynamic stability. In addition, biochemically induced growth of carbonate particles as well as purely inorganic precipitation of carbonate minerals from sea and lake waters are common under special environmental conditions. These problems are treated by Bathurst (1975), Purser (1980),

Fig. 3.21a–e. Large-scale distribution and facies association of widely extended shallow-sea sandstones in the Cretaceous North American seaway. a General paleogeographic situation; b Cross section of Upper Cretaceous shale and sandstone formations, Book Cliffs, Utah, displaying shoreface and shelf sandstones prograding step-wise seaward over shelf muds. c and d Models of sand distribution at wave-dominated, prograding deltas. River mouth sand is distributed by littoral currents and storm flow (c), and redistributed by a storm-driven inner-shelf transport system (d). e Coastal retreat, sediment trapping in lagoons, shoreface and shelf erosion as a result of transgression (insufficent sediment supply for prograding). (After Swift et al. 1987)

Füchtbauer (1988), Tucker and Wright (1990), and others.
- Moreover, because of their relatively high solubility in water, carbonate minerals tend to dissolve and reprecipitate much easier and faster than silicate minerals. As a result, loose sediments can become indurated even at or near the sediment-water interface (early diagenesis) and, with passing time, their pore space may be filled with different types and generations of carbonate cements. All these phenomena are described in detail in several texts (e.g., Bathurst 1975; Flügel 1982; Sellwood 1986; Schroeder and Purser 1986; Moore 1989) and can be dealt with here only in a brief summary.

The first prerequisite for the accumulation of carbonate-rich sediments is a low input of siliciclastic material to the site of deposition. Otherwise the predominantly biogenic-produced carbonate is too diluted to form carbonate rocks, even when carbonate production is very high. For this reason, carbonate sediments cannot form at the mouth of large river deltas or along the coasts of mountain ranges, shedding large volumes of weathering products into the sea. If the siliciclastic influx is low, carbonate-rich sediments can accumulate not only in warm but also in temperate waters (see below). However, carbonate production is favored in tropical seas where different groups of organisms secrete thick, large carbonate skeletons and where, in addition, carbonate can also precipitate directly from sea water supersaturated with respect to calcium carbonate. Therefore, thick and widespread carbonate deposits are concentrated nowadays in the tropical and subtropical zones, and we have reason to assume that this was also the case in the geological past.

However, pure or high carbonate deposition alone is not sufficient to explain the special geometry of many carbonate rocks and their relationship to other deposits. Shallow-water carbonates and biogenic carbonate structures have the outstanding ability to build up sedimentary bodies which may rise with relatively steep slopes high above the sea floor. Such structures are referred to as carbonate buildups no matter whether they are mainly produced by organic reefs, or by skeletal sand und mud. Even originally unconsolidated carbonate sediment can form wave-and current-resistant structures if it is lithified by early cementation (see below).

To summarize, the following factors favor the generation of carbonate buildups: Production of biogenic carbonate in excess of the accumulation rate of siliciclastic material, frame-building organisms forming reefs, early cementation of skeletal sands and muds, and sufficient subsidence.

Three types of large-scale depositional environments are important for the buildup of significant carbonate bodies (Read 1982, 1985; Scholle et al. 1983; Scholle et al. 1989; Stanton and Flügel 1989; Tucker et al. 1990):

1. Carbonate ramps.
2. Rimmed carbonate shelves.
3. Isolated carbonate platforms.

Whereas carbonate ramps and shelves are attached to land masses, the typical carbonate platforms are isolated from mainland coasts and therefore ideally meet the requirement of low or absent terrestrial influx.

The best known and frequently described example of this type of carbonate buildup is the Bahama Platform (e.g., Purdy 1963; summaries in Bathurst 1975; Sellwood 1986, and Chap. 12.2.2), where a carbonate massif more than 4 km thick has accumulated since the Cretaceous. Ancient carbonate platforms are of similar extent (e.g., in the Alps, in the southern Apeninnes, and in the Devonian of Canada) but normally do not reach such great thicknesses.

Carbonate ramps may develop without significant contribution from reefs, whereas rimmed carbonate shelves and platforms need reef structures to build up and maintain comparatively steep slopes.

Carbonate Ramps

A carbonate ramp is a gently sloping sea floor (generally less than 1 to 2 degrees) without a marked break in slope and thus deviates from most present-day shelves, which are characterized by a distinct shelf edge (Ahr 1973; Read 1982 and 1985; Wilson and Jordan 1983; Wright 1986).

Well known modern examples of ramp settings are the shallow region of the southern Arabian/Persian Gulf (Purser 1973; Chap. 4.4.3), the deeper shelf and upper slope of West Florida (Mullins et al. 1988), the Libyan coastal zone of the Mediterranean, and the Yucatan shelf.

On a carbonate ramp, the high-energy facies of the wave-dominated nearshore zone gradually pass downslope (1 to 2°) into deeper water and finally into basinal sediments deposited under low-energy conditions (Fig. 3.22). Landward, ramps are characterized by a high-

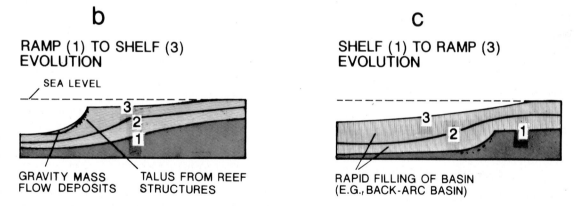

Fig. 3.22. a Idealized carbonate ramp with an inner ramp barrier complex separating a lagoonal-tidal zone from the open marine ramp facies. **b** Evolution of a carbonate ramp to a rimmed carbonate shelf. **c** Evolution of a rimmed shelf to a carbonate ramp (rare). (Based on Read 1982; Aigner 1985; Buxton and Pedley 1989)

energy grainstone belt, whereas rimmed carbonate shelves exhibit such a zone on their
basinward edge. Continuous, large reef structures and sediment gravity flow deposits containing clasts of cemented, shallow-water facies at the foot of the gentle ramp slope are
generally absent.

The nearshore zone of shallow, high-energy
oolitic and peloidal sands or skeletal shoal
complexes separates a protected landward area
with lagoons, coastal and peritidal clastics, and
possibly tidal flats from an open marine,
high-energy foreshore and shallow-water belt
(Fig. 3.22a). The back-barrier area is dominated by low-diversity euryhaline faunal assemblages which tolerate a wide range of salinities. The inner ramp, seaward of the beach
and barrier facies, is an efficient "carbonate
factory", providing additional carbonate for
both the back-barrier region and the outer,
deeper ramp.

Different types of organisms have flourished on the inner
ramp, reflecting the faunal evolution over Earth's history.
In the Cenozoic, coralgal patch reefs, nestling and boring
bivalves, sedentary gastropods, tube-secreting worms, encrusting bryozoa, and large foraminifera have typified this
zone (Buxton and Pedley 1989); in the Triassic crinoidal
ramps were common (Aigner 1985).

In addition to biogenic carbonate production, the formation of oolitic sands may contribute significantly to overall
carbonate production. Hardgrounds appear to be a common
feature on inner ramps. Normal wave action and frequent
storms transport biogenic sands and larger skeletal particles
(calcarenites and calcrudites) along shore and as traction
carpets into somewhat deeper water. Rare large storms produce sandy and muddy tempestites (Chap. 3.1.2) which
extend into the deeper ramp and basin, where they alternate
with lime muds or marls. Here, pelagic sediments, such as
planktonic foraminiferal and nannofossil oozes (cf. Chap.
5.3.1), dominate, but some isolated carbonate buildups
may also occur on the outer ramp. A distinct trend in
ichnofacies is also observable from the inner to the outer
ramp.

It appears that carbonate ramps are common in geologic
periods or environments in which reef builders secreting
large, rigid skeletons are rare. The organic assemblages of
ramps function mainly as grain and mud producers, as well
as trappers and binders to form carbonate banks. Ramps
are possibly more common in temperate climatic zones than
in warmer regions. A modern example of this type is the
shelf of southern Australia (see below).

Carbonate ramps are common on passive (extensional) continental margins and in epicontinental seas. On continental margins, carbonates frequently overlie volcanics, evaporites,
and clastic sediments, representing an early
phase of deposition in a rift basin or a young
ocean basin (cf. Chap. 12.1). Ramps tend to
evolve into rimmed carbonate shelves (Fig.

3.22b) when carbonate production on the developing shelf edge is high, and the production and deposition of carbonate in deeper
water remains limited. The buildup of a rim
with a steepening slope can be promoted by
mud mounds which contain only minor proportions of biogenic framework (see below).
Such structures may be confused with normal
reefs where frame-building organisms play a
great part (Flügel 1989). The transformation
of a rimmed shelf into a ramp is less common
(Fig. 3.22c); it may occur, for example, where
the shelf is drowned and buried by prograding
clastics, or when the basin is filled rapidly
from another side.

Reefs and Other Carbonate Buildups

General Aspects

In terms of large-scale sedimentary bodies,
organic reefs are relatively simple structures
which rise above the sea floor as a result of
high skeletal calcium carbonate production
and low siliciclastic input. In contrast to other
sedimentary bodies, the growth and geometry
of reefs are controlled mainly by organisms,
the reef community, and to a limited extent
by the physical processes of their environment.

The internal structure of reefs is complicated, because they
are composed not only of the solid skeletons of sessil
frame-builders (for example, massive stocks of corals in
the modern seas), but of the skeletons and shells of other,
less resistant benthic organisms attached to the reef. Furthermore, skeletal debris provided by physical breakup or
by boring and rasping organisms (bioerosion) as well as
other sediments can be caught in cavities within the reef
structure and contribute considerably to the buildup of the
reef core (Fig. 3.24b). Encrusting organisms often grow
over the dead parts of the reef surface and aid in stabilizing
the structure. In addition, most reefs are associated with a
mantle of talus on their flanks which, in effect, also forms
part of the reef complex.

It is only the upper part of a reef which is organically
active, because reef-builders take their food and nutrients
from the surface water. Calcareous algae or hermatypic
corals, which live in symbiosis with microscopic algae
(zooxanthellae) need sunlight for photosynthesis. Hence,
the reefs can only grow in shallow water up to a depth of
about 50 to 80 m, but they grow optimally just a few
meters below sea level. Their upward growth terminates
within the tidal zone, where they form a flat surface covered by water during high tide. The growth rate of coral
reefs can be extremely fast in comparison to the sedimentation rate of other biogenic and clastic marine sediments
(Chap. 10.2). Under favorable conditions in the late Quaternary, coral reefs were able to grow upward with the rising sea level as fast as 1 cm per year (or 10 m/ka).

Reef-Building Communities

The modern coral reefs are restricted to warm tropical to subtropical waters of normal salinity and are unable to withstand substantial changes in water chemistry. Furthermore, the main reef-builders cannot tolerate muddy water because they are filter-feeders or micropredators, whose feeding apparatus can be clogged by fine sediment particles. As a result, reefs are unable to survive near the mouth of rivers delivering large quantities of fresh water and mud. It is generally assumed that most ancient reef communities required environmental conditions similar to those of the modern ones.

Nevertheless, several types of reef communities exist, which do not need warm water and much sunlight. Based on studies of modern and Cenozoic examples, many authors distinguish two groups of reef-building communities (e.g., James 1983):

- *Chlorazoan association*, consisting of green algae and hermatypic corals and, in addition, benthic foraminifera, molluscs, bryozoans, calcareous red algae, and barnacles. This association forms the prominent modern reefs in tropical-subtropical waters most of which also contain inorganically grown ooids and pelletal carbonate muds. Apart from normal calcite, aragonite and Mg-calcite are important primary carbonate minerals.

- *Foramol association*, consisting predominantly of benthic foraminifera, molluscs, bryozoans, and partially calcareous red algae and barnacles. These communities can exist in temperate waters and at greater depths than the chlorazoan association, but they grow more slowly and cannot normally develop isolated, high bioherm structures (see below). Another difference is that in primary carbonate mineralogy calcite predominates over aragonite and Mg-calcite, and ooids are absent. At depths below 100 to 200 m, glauconite may be present.

Modern examples of this type of community were found in the western North Sea and in the northeastern Atlantic on the Rockall and Voering Plateau, where very little siliciclastic material is deposited. Widely extended cool water carbonates also occur on the shelf of southern Australia, where an inner shelf association rich in coralline algae, bivalves, and brachiopods passes, without a reef barrier, into an outer shelf association characterized by bryozoans, calcified worm tubes, foraminifera, and sponges (James 1990). Cold-water carbonates were reported from the ancient Permian shelf sea of Tasmania (Rao 1981), where they alternate with tillites and siliciclastic shallow-sea sediments rich in dropstones.

In some periods during Earth's history, other groups of frame-builders played a major role in the construction of reefs, for example stromatolites in the Precambrian, bryozoans, stromatoporids and tabulate corals in the lower Paleozoic, coralline sponges in the Permian, Triassic, and Jurassic, or rudist bivalves in the Cretaceous.

Reef Types

Despite the great variability of the reef-building communities and their internal structures, most of the reef bodies can be classified simply according to their geometry (from flat and small to usually thick and large structures):

- **Flat muddy banks**, one to several km long, rising some decimeters or meters above the surrounding sea floor, partially built by sedentary organisms such as algae, sea grass, branching corals, molluscs and many other groups. Such skeletal-rich banks occur not only in warm coastal seas as, for example, on the Florida shelf, but also in temperate waters.

In the latter case, however, the thick and rapidly growing shells of tropical waters are absent and replaced by representatives of the foramol association. Generally, these banks become more sandy and gravelly (including coarse shell debris and fragments of algal structures, e.g., *Lithothamnium*), and lime mud is scarce or absent. Banks similar to these nearshore structures of the temperate zone are also found in deeper water in environments with little siliciclastic input, especially on top of plateaus at depths between 100 and several hundred meters. Here, of course, algal activity is absent or very limited (red algae), but ahermatypic corals may be present and form patches rising somewhat over the neighboring sea floor. Glauconite is rather common in such temperate environments of low accumulation (e.g., on the Rockall Bank in the northeastern Atlantic). Carbonate banks in shallow and deeper water, under either warm or temperate conditions, can be stabilized by early cementation forming erosion-resistant tops or hardgrounds.

The bank-type reefs of the present-day oceans are considered counterparts of the frequently described *biostromes* in ancient sediments (Fig. 3.23a). By definition, these are also bedded structures, such as shell beds or coral-rich beds exhibiting part of their fauna in situ, but they did not grow fast enough in their central part to form lens-like or mound-like bodies as was the case for *bioherms*. Superposition of biostromes or small bioherms can result in thick reef structures ("stratigraphic" or stratified reefs), but these never developed a substantial primary relief.

In contrast, "ecologic" reefs form prominent topographic, wave-resistant features such as barrier reefs, patch reefs, pinnacle reefs (Fig. 3.23a and d).

- **Reef mounds** are flat, long lenses or steeper conical piles of incomplete reef structures (James 1983), which consist of poorly sorted

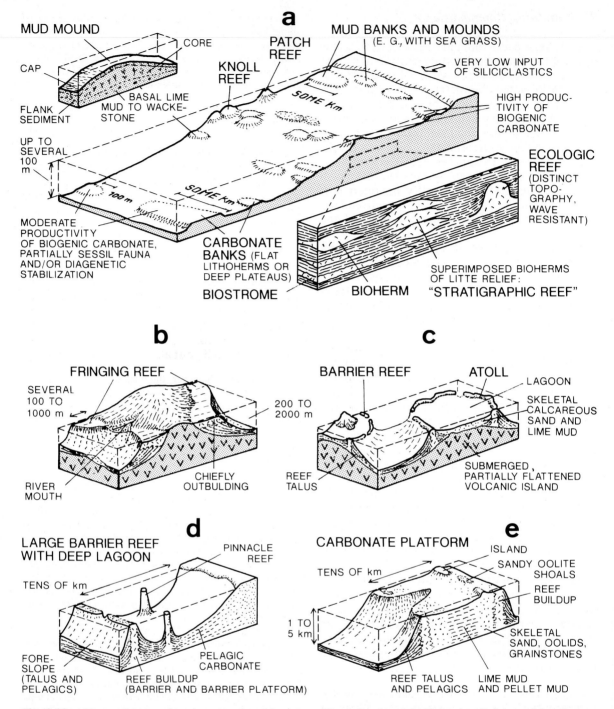

Fig. 3.23. a Nomenclature and environmental position of different reefs and reef-like structures in modern seas and ancient rock sequences, demonstrated on an idealized carbonate shelf. **b** Fringing reefs and c, barrier reefs attached to mainland coasts or islands. **c** Atoll, developed from previous fringing and barrier reefs (**b, c**) on top of submerging volcanic island.

d Large barrier reef with barrier platform and deep lagoon, collecting mainly pelagic coccolith and foraminiferal ooze (chalk) or marly muds. **e** Carbonate platform isolated from mainland and surrounded by deep water. Note the different facies zones of platform rim and shelf lagoon. (Based on James 1983; Sellwood 1986, and others)

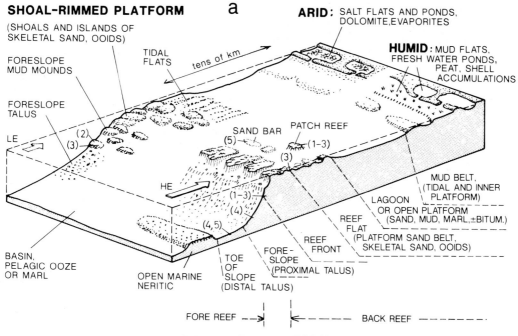

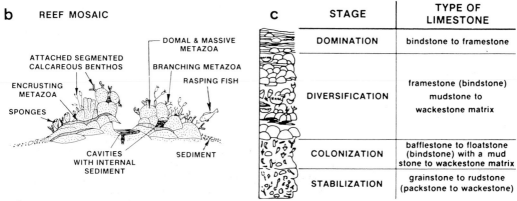

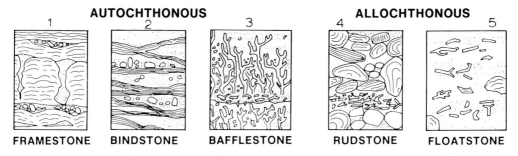

Fig. 3.24. a Topography and facies zones of shoal-rimmed (*upper part* of block diagram) and reef-rimmed carbonate shelves or platforms (*lower half*). Low-energy (*LE*) wave produce a gentler foreslope with mud mounds, sandy shoals, and islands at the shelf break. High-energy (*HE*) conditions can only be tolerated by solid reef structures that have steep foreslopes. The mud belt facies of the inner shelf is influenced significantly by climate. **b** Reef mosaic, showing details of composite reef buildup. **c** Stages of reef growth; high species diversity with domal, massive, lamellar, branching and encrusting reef builders is realized only in the diversification stage. **d** Simple nomenclature of autochthonous and allochthonous reef limestones. (Based on different sources, e.g., Zankl 1971; Wilson 1975; James 1983; Sellwood 1986, modified)

bioclastic lime mud with minor amounts of delicate to dendroid skeletons of sessil organisms growing on top of the muddy matrix (e.g., sponges, algae, bryozoans, branching corals, bivalves).

The cap of such mounds may be formed by a thin layer of encrusting organisms or an early lithified limestone crust (Fig. 3.23a). Larger skeletons of typical frame-builders are generally not found. For that reason, reef mounds cannot grow up to the zone of strong and constant wave turbulence (Fig. 3.24 a). They have lengths on the order of 100 m and occur on the gently dipping slopes of carbonate platforms, in tranquil reef lagoons, shelf seas, and occasionally in deeper basins.

- **Knoll reefs** and **patch reefs** are frame-built, isolated, roughly circular bioherms (Fig. 3.23a). Knoll reefs occur on the upper foreslope of medium-energy shelf or platform margins, where they often form a reef barrier or gently seaward dipping ramp in combination with forereef or interreef debris. Patch reefs usually grow in the shallow outer part of a protected reef lagoon belonging to the back reef zone behind a larger reef barrier. Patch reefs are frequently associated with skeletal and oolithic sand bars. In deep lagoons the patch reefs may develop into isolated conical high *pinnacle reefs* (Fig. 3.23d), surrounded largely by pelagic carbonate.

- **Reef rims, fringing reefs, and barrier reefs**. In high-energy environments the frame-building organisms can produce narrow elongated structures referred to as *reef rims*. They grow up to or close to mean sea level into the zone of greatest wave action. This implies that their primary structure must be sufficiently solid to withstand permanent and occasionally very strong wave attack. As a result, their seaward slope tends to be very steep and their proximal foreslope talus often consists of coarse reef debris and fallen blocks from the active reef zone. Nevertheless, these structures are frequently able to progress not only upward, but also seaward toward their food supply and on top of their foreslope talus (Fig. 3.25a).

If seaward outbuilding prevails and the carbonate body is attached to land masses or islands, the resulting coastal carbonate platforms are called *fringing reefs* (Fig. 3.23b).

Wide structures of this type develop when carbonate production exceeds subsidence and, in addition, the relative sea level remains constant or repeatedly reaches about the same level. Modern fringing reefs are commonly 0.5 to 1 km wide and may include some very shallow lagoons of limited extent. Well-known examples occur, for example,

along the coasts of the Red Sea and many tropical islands where only little terrigenous material is swept into the sea.

Conversely, rapidly rising sea level or strong subsidence often leads to a pronounced vertical buildup of the reef front and thus to *barrier reefs* and atolls separating shallow or sometimes also rather deep and wide lagoons from the open sea (Figs. 3.23c and d). The most prominent recent example of this type is the 2000 km long Great Barrier Reef along the eastern continental margin of Australia. An outstanding fossil example are the Devonian barrier reefs in the Canning basin, western Australia.

The Back Reef Zone and Reef Lagoons.

The salinity and other environmental characteristics of the back reef zone are strongly influenced by water exchange between the open sea and reef lagoons via inlets or deeper channels crossing the barrier reef. If water exchange is sufficient to maintain approximately normal marine conditions, carbonate production in the lagoons remains predominantly biogenic. However, it often lags behind the carbonate buildup along the reef front, with the exception of isolated patch or pinnacle reefs mentioned above, or along inshore banks, for example, of branching corals and algae. Consequently, lagoons associated with barrier reefs tend to become deeper during times of rapid relative sea level rise.

In humid regions with inflow of fresh water, lagoons with limited sea water exchange may develop stratified waters (see Chap. 4.2) and an oxy- gen-deficient bottom water zone. Then lagoonal muds may become bituminous and lack bioturbation. Generally, lagoonal sediments are sensitive to seasonal and longer term climatic changes and frequently record such variations in their sediments.

The inner parts of lagoons, including tidal flats in particular, are strongly affected by climatic change. In humid tropical to subtropical zones, brackish and fresh water ponds with corresponding faunas and rapid plant growth may develop (Fig. 3.24a). In arid regions, hypersaline conditions often promote chemical and biochemical carbonate precipitation, early transformation of calcium carbonate to dolomite, and the generation of evaporitic layers in the supratidal zone (Chaps. 3.2.2 and 6.4.1).

The many *atolls* of the present-day oceans (more than 320 examples in the Indian and Pacific oceans) are ring-like barrier reefs surrounding a shallow lagoon in which a central land mass is lacking (Fig. 3.23c). As already described by Darwin, most of these atolls are associated with submerged, isolated, volcanic seamounts, which previously formed islands. These reef structures can be explained by alternating outbuilding of fringing reefs and upbuilding of barrier reefs related to the rapid Quaternary and slower earlier sea level fluctuations. In addition, a general tendency of the seamounts to subside plays a role. The carbonate buildups of some atolls reach thicknesses in excess of 1000 m and represent time spans of some tens of million years. The central lagoons collect reef debris and sediment produced in the lagoon itself, mainly sand- and silt-size skeletal carbonate.

Composite Structures: Carbonate Buildups

Carbonate shelves and carbonate platforms are composite structures, consisting of the reef front or reef belt, the foreslope proximal and distal reef talus (fore reef), and the back reef, including the reef flats and the lagoon (Figs. 3.24a and 3.25a). In the case of carbonate shelves, the coastal mud belt and tidal flats form an additional part of the total structure (cf. Chap. 3.2.2).

The lateral sediment distribution on carbonate platforms, as for example observed on the Great Bahama Bank, may include (from land to sea):
- Calcareous mud (biofacies adjusted to normal to hypersaline conditions).
- Pellet mud.
- Oolite shoals and oolitic grapestone.
- Reefs and rocky bottom along the platform edge.

These different facies zones can migrate landward or seaward with time in response to relative sea level fluctuations, changes in carbonate production or in the hydraulic regime. Thus, some of the facies zones can become more, others less important as contributors to the total carbonate buildup. Nevertheless, both the isolated and land-attached platforms may grow more or less continually, regardless of the specific carbonate production of frame-building organisms, the various reef types discussed above, and the different facies zones of the reef. Mechanical reef destruction and *bioerosion* by endolithic organisms causing a downward erosion on the order of 0.3 m/ka (Stearly and Ekdale 1989) lead to a redistribution of carbonate on platforms and their slopes.

In some shallow-water environments, carbonate may also precipitate inorganically from supersaturated sea water, as observed in the Persian Gulf and on the Great Bahama Bank (e.g., Shinn et al. 1989). These so-called "whitings" are characterized by milky water containing about 10 mg/l fine grained calcium carbonate, mostly aragonite.

Types of Reef Limestones

Apart from lagoonal and tidal sediments, the different facies zones of a reef complex are represented by various types of reef lime-

stones (Fig. 3.24a through e). These can be subdivided into two groups (James 1983):

1. Autochthonous reef limestones displaying in-place fossils, which support the framework of the reef. We can distinguish:
- Framestones, consisting largely of frame-building skeletons.
- Bindstones, showing in-place tabular or lamellar fossils, which encrust or bind the sediment together during deposition.
- Bafflestones, exhibiting in-place stalked fossils, which trap sediment by baffling.
2. Allochthonous reef limestones, the material of which was not produced at the site of final deposition but delivered from nearby or more distal sources:
- Rudstones, clast-supported limestones, containing a high proportion of coarse-grained particles.
- Floatstones, containing more than 10 % particles with diameters greater than 2 mm, which, in contrast to the rudstones, are matrix-supported.

The occurrence of these limestone types within a reef complex is indicated by numbers in Figure 3.24a.

Response of Carbonate Buildups to Sea Level Changes

Vertical Buildup and "Drowning"

Because carbonate shelves and platforms are sensible gauges of sea level changes, their growth is in turn strongly affected by relative sea level changes (Kendall and Schlager 1981; cf. Chap. 7.4).

A relatively simple structure is the pure vertical buildup of an isolated carbonate platform with sea level rise (or subsidence of the basin floor; Fig. 3.23e). If lateral shifting of the facies zones is limited, vertical sequences of such platform carbonates can theoretically show very thick successions of the same facies, for example superimposed knoll reefs or bioherms, fore-slope reef talus, lagoonal muds, etc. (Fig. 3.23e). At the same time, interreef basinal muds or even deep-sea pelagic sediments can accumulate close to the carbonate platform. If the platform is subdivided by large and deep channels or troughs, as for example the Bahama Platform by the Tongue of the Ocean, shallow water carbonates can be bordered by carbonate turbidites and other deposits transported by gravitiy mass movements.

a PROGRADING REEF (CARBONATE PLATFORM) SEQUENCE

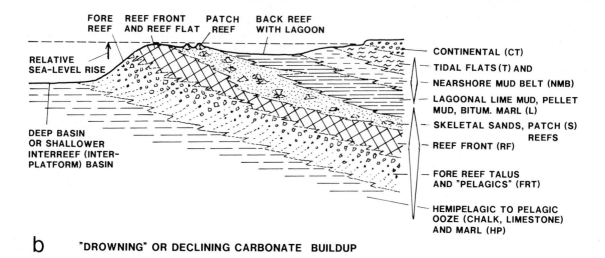

FORE REEF · REEF FRONT AND REEF FLAT · PATCH REEF · BACK REEF WITH LAGOON

RELATIVE SEA-LEVEL RISE

DEEP BASIN OR SHALLOWER INTERREEF (INTER-PLATFORM) BASIN

CONTINENTAL (CT)

TIDAL FLATS (T) AND NEARSHORE MUD BELT (NMB)

LAGOONAL LIME MUD, PELLET MUD, BITUM. MARL (L)

SKELETAL SANDS, PATCH (S) REEFS

REEF FRONT (RF)

FORE REEF TALUS AND "PELAGICS" (FRT)

HEMIPELAGIC TO PELAGIC OOZE (CHALK, LIMESTONE) AND MARL (HP)

b "DROWNING" OR DECLINING CARBONATE BUILDUP

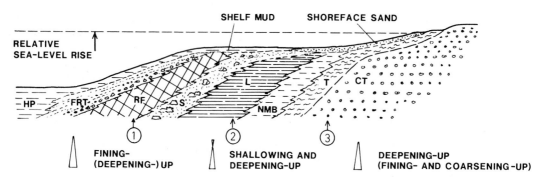

RELATIVE SEA-LEVEL RISE

SHELF MUD   SHOREFACE SAND

HP   FRT   RF   S   L   NMB   T   CT

① FINING-(DEEPENING-) UP

② SHALLOWING AND DEEPENING-UP

③ DEEPENING-UP (FINING- AND COARSENING-UP)

c EMERGING CARBONATE BUILDUP

C1 FALLING SEA-LEVEL

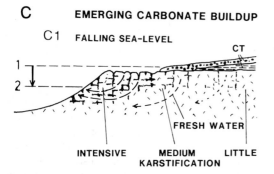

CT

1
2

FRESH WATER

INTENSIVE   MEDIUM   LITTLE
KARSTIFICATION

C2 FALLING AND RISING SEA-LEVEL

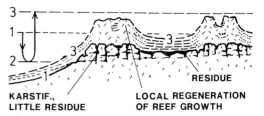

3
1
2

3

RESIDUE

KARSTIF., LITTLE RESIDUE

LOCAL REGENERATION OF REEF GROWTH

Fig. 3.25. a Complete, idealized sequence of a prograding carbonate buildup (reef front). Note the limited thickness of each facies (in contrast to Fig. 3.23e), the changing proportions of different facies zones during the carbonate buildup, as well as repeated coarsening (shallowing) upward and fining upward. **b** "Drowning" or declining reef, due to rapid sea level rise or, more likely, to deteriorated conditions for reef growth. Landward migration and subsequent termination or replacement of different facies zones of the carbonate buildup. For abbrevations see (a).

Note differences in the vertical sections 1 to 3. **c** Permanently (**c1**) or temporarily (**c2**) emerging and karstified carbonate buildups. **c1** Different intensities of karstification in relation to climate, duration of subaerial exposure, etc., or to nature and thickness of overlying beds. **c2** Reef growth or general carbonate buildup starts with preferred initiation on topographic highs of previously karstified land surface. (Consequences for carbonate diagenesis are shown in Fig. 3.26)

In many cases, however, the relatioship between relative sea level rise and the growth potential of carbonate buildups leads to more complicate structures. The reasons for these complications are that (1) sea level changes occurred frequently and at differing rates over the Earth's history (Chap. 7) and (2) the growth rates of carbonate buildups varied with the type of reef builders involved and changing environments. One can distinguish several situations (Kendall and Schlager 1981):

- *Start-up phase.* Carbonate accumulation lags behind the rapidly rising sea.
- *Catch-up phase* (after initial drowning). Carbonate accumulation exceeds the rate of coeval sea level rise. Then two cases are possible: (a) The fast-growing rim and patches of the interior platform are able to catch up with sea level rise and survive, whereas the platform interior is transformed into a deep lagoon (Figs. 3.23c and d and 3.25c2). (b) The total shelf or platform builds to sea level, but may include shallow lagoons (Fig. 3.24a).
- *Keep-up phase.* Carbonate accumulation closely matches (or exceeds) the rate of sea level rise. The top of the buildup remains flat and closely to sea level; deep lagoons cannot develop. An excess in carbonate production may cause a seaward prograding of the flat-topped platform (Fig. 3.25a); the platform is frequently capped by evaporites.
- *Drowning phase ("give-up").* Sea level rise exceeds carbonate growth potential; Shallow-water carbonates are overlain by hardgrounds and/or sediments of deeper environments (Fig. 3.25b).

Apart from rapid sea level rise, reef growth and the production of biogenic carbonate can also be hampered, or completely subdued, by other changes in the environmental conditions such as salinity, supply of food and nutrients, muddy waters, oxygen deficiency of sea water, etc. As a result of the extremely fast sea level rises during short intervals of the Quaternary, part of the pre-existing reefs really drowned. However, in pre-Quaternary times, when sea level fluctuations had smaller amplitudes and probably showed a slower rate of change, drowning of active carbonate buildups appears to be less likely. It is assumed that such sedimentary bodies as a whole can grow upward as fast as 200 to 1000 m/Ma (0.02 to 0.1 cm/a, Hallock and Schlager 1986), which is much less than the optimum growth rate of single reef structures (up to about 1 cm/a; also see Section 10.2). Nevertheless,

even the much smaller growth rates of carbonate buildups are probably sufficient to compensate for most of the relative sea level rises of Phanerozoic times.

On the other hand, reef growth and carbonate buildups ended more or less abruptly and were replaced by other deposits several times during nonglacial periods, for example during the Mid-Cretaceous (Hallock and Schlager 1986). The cause of such global crises in the growth of carbonate platforms is not yet clear, but there is a striking coincidence of world-wide "drowning" and oceanic anoxia (Schlager and Philip 1990).

Pro- and Retrograding Carbonate Buildups

Figure 3.25b shows a model of landward migrating carbonate facies zones and their termination under a transgressive sea. The vertical sections of such a sedimentary body vary according to their location far off or close to the paleo-coastline. Under these transgressive conditions, fining and deepening-upward sequences prevail (Locations 1 to 3 in Fig. 3.25b).

Due to the fact that reef fronts are able to grow seaward on top of their talus and subsiding basin floor, prograding carbonate buildups are common in the geological record (Fig. 3.25a). In this case, the different facies zones reach only limited thicknesses in vertical sequences. Typical successions below the reef front show coarsening- and shallowing-upward trends. Farther landward, where lagoonal and tidal sediments cover the reef core, the vertical sections on top of the reef core exhibit a fining- and deepening-upward trend toward the central lagoonal deposits, thereafter some coarsening.

A classic example of a large, prograding carbonate buildup is the Permian reef complex of the Guadalupe Mountains region in Texas and New Mexico (Newell et al. 1953; Ward et al. 1986), which is also known for its rich oil and gas fields. Hydrocarbons have accumulated mainly at the contact of the lagoonal sediments (here dolomites) with the coastal evaporites, and in the basinal channel-fill clastic carbonates. The cavernous pore space of the marginal reef facies is commonly filled with water.

Emergence and Continued Growth of Carbonate Buildups

Because both isolated and land-attached carbonate buildups usually grow close to mean sea level, even minor relative sea level falls lead to mechanical abrasion or to an emergence of the carbonate platform. If the platform surface remains uncovered by continental deposits, dissolution of carbonate, starting at the surface and penetrating into deeper parts of the carbonate body, can bring about *karstification* and *soil formation* (Fig. 3.25c1). These phenomena are absent or less pronounced, if the carbonate surface is only briefly exposed to weathering before it is covered by poorly permeable material or additional carbonate-bearing layers.

Sea level rise subsequent to emergence can reactivate local reef growth or extensive carbonate buildup on top of a karstified, irregular surface (Fig. 3.25c2). Thereafter, reef growth starts again on topographic highs, and inter-reef lagoons or basins tend to become deeper than the pre-existing ones, providing the reefs can keep pace with the rising sea level. For this reason and also due to a succession of sea level changes with the long-term tendency to a slow net rise, a previously flat-topped large platform may become subdivided into a series of ecological superimposed reefs and small inter-reef basins. Exposure of carbonate buildups to the air and karstification are also important factors controlling carbonate diagenesis (see below).

"Pelagic" Carbonate Platforms

Drowned or subsided carbonate platforms below the depth favorable for reef growth may be referred to as *"pelagic" carbonate platforms* (Franke and Walliser 1983; see Chaps. 5.3.7 and 12.2.2). In an intermediate stage between shallow-water and pelagic carbonate deposition, where the platforms are still exposed to storm and current action, the sequence overlying a carbonate buildup may become discontinuous and interrupted by intervals of omission and erosion.

Such condensed carbonate sections are common in the geologic record and were described from many regions, for example from the Jurassic and Cretaceous (e.g., Bernoulli and Jenkyns 1974; Fürsich 1979; Bergner et al. 1982; Jenkyns 1986) and from late Devonian and early Carboniferous strata in various European countries and North Africa (Tucker 1974; Wendt and Aigner 1985; Wendt 1988).

These platform carbonates are frequently characterized by irregular bedding and a nodular appearance, red color, ferromanganese nodules, corrosion surfaces and hardgrounds with a sessil fauna, and accumulations of special fossils, for example cephalopods and crinoids.

Average sedimentation rates on such platforms are very low (0.1 to 1 m/Ma), while adjacent basins are filled at rates between 20 and 100 m/Ma. Laterally, such condensed horizons may show transitions to clastic shelf sediments or to brecciated, gravity-deformed, and resedimented slope sediments. A clear distinction between this type of "pelagic" platform and the deep marginal and oceanic plateaus discussed in Chapter 5.3.7 appears to be difficult.

Diagenesis of Reefs and Carbonate Buildups

General Processes of Carbonate Diagenesis

As previously mentioned, reefs and other carbonate sediments often undergo marked changes immediately after their deposition (early pre-burial diagenesis) and also later on when they are buried under younger sediments (late or burial diagenesis). The most important processes involved are:

- Dissolution of pre-existing carbonate minerals creating void space, for example, in the unsaturated vadose zone of emerged platforms (Fig. 3.26b).
- Cementation, i.e., growth of new carbonate crystals in the void space, thus leading to lithification.
- Replacement of single crystals or mosaics of fine crystals by other, often larger carbonate minerals (neomorphism).

Numerous articles have been written on this topic, and several textbooks discuss the various aspects of this problem in detail (e.g., Bathurst 1975; Purser and Schroeder 1986; Sellwood 1986; Füchtbauer 1988; Moore 1988; Scholle et al. 1989). The following brief summary is based partly on these texts, but physico-chemical considerations are omitted here (also see Chapter 13).

Rapid and intensive diagenetic transformation of most of the primary material of a carbonate buildup is favored by:

- Relatively good solubility of skeletal carbonate and thermodynamic instability of some carbonate minerals (aragonite and Mg-calcite) under changing physico-chemical conditions.
- High porosity and permeability of many reef structures, reef talus, and skeletal sands.
- "Internal draining systems" within the carbonate buildup, i.e., zones of better permeability acting as conduits for pore fluids flowing laterally and upward, or for meteoric water with an initially downward flow.

- Frequent emergence of carbonate platforms resulting from both major and minor sea level falls, which lead to significant alterations in the pore water chemistry.

- Carbonate diagenesis is also affected by the hydraulic regime of the water body overlying the sediment. High energy conditions may promote rapid cementation through *sea water pumping* (see, e.g., summary in Tucker and Wright 1990).

These criteria explain the great importance of allochemical diagenesis, which is characteristic for these rock types and is especially demonstrated by dolomitization. The latter process requires the introduction of substantial amounts of magnesium from outside. It must be transported by large volumes of pore-water circulating through the rocks (cf. Chap. 13.2).

Cement Sequences

Carbonate rocks exhibit different types of cementation:

1. *Early marine cements* (fibrous or micritic Mg-calcite and aragonite), form very close to the contact with sea-water at the reef front, foreslope, and on top of a carbonate buildup (Fig. 3.26a). These cements contribute considerably to the mechanical resistivity of reef structures and some sandy shoals against wave attack.

2. *Early meteoric cements* (sparitic or micritic low-Mg-calcite) grow during periods of emergence in zones of fresh-water influence, mainly in the phreatic zone (Fig. 3.26b).

3. *Late calcite cements* are represented by coarse sparry crystals, which frequently show an increasing ferrous iron content (deep-burial diagenesis, Fig. 3.26b).

In a closed system (isochemical diagenesis), cements filling void space have to be procured by a simultaneous dissolution of metastable carbonate minerals (Land 1967; Bathurst 1975). Under increasing overburden load, this process is promoted by *pressure dissolution* at the grain contacts (Chap. 13.3). As a result, the carbonate sediment becomes more solid and the total structure undergoes compaction. Reef cores are subject to less compaction than their surrounding finer grained sediments. Therefore, their relief is enhanced and, in addition, they tend to drain the expelled pore-water from their neighboring sediments (Fig. 3.26b). Neomorphism usually occurs simultaneously with the filling of pore space. Skeletal aragonite and fine crystalline mosaics of

other carbonate minerals are replaced by coarser calcite. Magnesium, released by dissolution of high-Mg-calcite can form some dolomite.

Dolomitization

Significant dolomitization of carbonate buildups is, however, only possible in open systems, if large quantities of Mg are transported by circulating pore waters, particularly sea water, to the site of diagenesis (see Chap. 13.3). For reasons mentioned above, reefs are especially susceptible to early and late dolomitization. Under arid conditions, lagoonal water may become hypersaline and attain a salt concentration sufficiently large for the precipitation of gypsum (Chap. 6.4.1). If calcium sulfate is precipitated, the Mg/Ca ratio of the brine far exceeds that of normal sea water (Mg/Ca = 5.2) and consequently initiates dolomitization of skeletal carbonate. The relatively dense lagoonal brine tends to sink into the underlying rocks and to flow back toward the sea (sabkha model with evaporative reflux, Fig. 3.26c). In addition, skeletal opaline silica can be dissolved and later reprecipitated as chert in the mixing zone between brine and normal pore-water.

In a humid climate, lagoonal water may become brackish or fresh and, due to its lower density, form a lens-shaped water body reaching into the subsurface (Fig. 3.26b). Where fresh (or brackish) waters and marine pore waters mix, the solution becomes supersaturated with respect to dolomite. Hence, even these conditions may cause replacement of calcium carbonate by dolomite (early dolomitization, "Dorag model"). Similarly, fresh groundwater from emerged platforms, or from the continent can flow seaward and partially flush out the primary marine pore water. The resulting mixing zone between meteoric and sea water may persist for a long time span and thus bring about substantial dolomitization. However, this mixing-zone model has been questioned recently as a viable explanation for massive dolomitization (Hardie 1987).

The "Coorong model" derived from present-day lagoons and coastal plain lakes of Southern Australia can explain lagoonal and lacustrine dolomites (Fig. 3.26 d).

According to this model, Mg-rich groundwaters flow seaward through an extensive aquifer and mix with sea water seepage through a seaward prograding belt of coastal sand dunes, composed mainly of skeletal carbonate (von der Borch and Lock 1979; Warren 1990). Whereas the outer lagoon maintains its connection to the open sea, older lagoons farther inland are separated from direct sea water

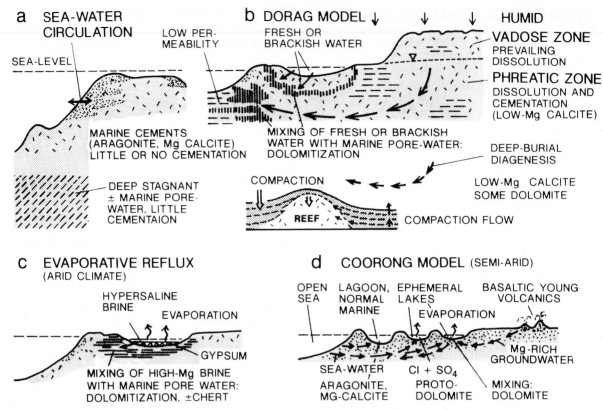

Fig. 3.26a-d. Diagenesis of reefs and carbonate buildups.
a Early marine cementation. **b** Dissolution and cementation
in the zones influenced by meteoric water as well as dolomitization in the mixing zone between fresh/brackish water
and marine pore water. Compaction flow and deep-burial
diagenesis (calcite and some dolomite). **c** Dolomitization by
evaporative reflux of lagoonal brine (after precipitation of

gypsum). **d** Lacustrine primary dolomitic muds on top of
marine and lagoonal skeletal carbonates, caused by Mg-rich
groundwater/sea-water mixing (Coorong model). (Based on
Blatt et al. 1980; Sellwood 1986; Schroeder and Purser
1987; Von der Borch and Lock 1979; Warren 1990, and
others)

influence and presently form very shallow ephemeral lakes.
Intensive evaporation during the summer leads to highly
concentrated, Mg-rich brines, which enable the true primary precipitation of dolomite and the deposition of dolomitic
mud. Dolomite is associated predominantly with Mg-calcite
and, in the more arid settings, with magnesite (or rarely:
hydromagnesite). Dolomite makes up about 10 % of the
carbonate minerals in the surface sediments across the
coastal plain. Highly soluble salts of chlorides and sulfates
are flushed back into the sea during the winter when the
lakes are flooded. The lacustrine dolomitic muds consist to
a great part of pellets and are mostly laminated. They rest
on top of skeletal marine carbonates and organic-rich early
Holocene lacustrine muds and show extensive mud cracks
at their surface.

The Coorong model should not be used to
explain widespread dolomitization in ancient
supratidal and shelf carbonates (Warren 1990).
    Dolomitization largely destroys the primary
organic reef structures and physical sedimentary structures. It usually leads to a "secondary" porosity, which can be of great economic

importance (Chap. 13.3), but other diagenetic
processes may modify this trend. Long-lasting
burial diagenesis generally tends to close the
pore space completely.
    The nature and duration of carbonate dia-
genesis in relation to deposition largely control
the great economic significance of carbonate
buildups as reservoirs for hydrocarbons,
groundwater, and host rocks for mineral deposits, particularly Pb/Zn ores (Chap. 13.5).
Reservoir potential and quality of reef carbonates as material for building and industrial
purposes vary significantly with regard to reef
types (Flügel 1989). Organic reefs and reef
mounds tend to become exposed to subaerial
weathering and karstification during their
early history and therefore commonly provide
abundant pore space, while mud mounds
grown at a greater depth below sea level seem
to be of minor importance as reservoirs and
host rocks.

3.4 Sediments of Marine Delta Complexes

3.4.1 Types of Marine Deltas

Introduction

Modern, large marine deltas commonly represent complex depositional systems, composed of sediments of greatly differing characteristics and environments (Fig. 3.27c). Their facies range from fluvial plains over shallow lakes, lagoons, tidal flats, estuaries, beach and shorefaces, to subaqueous delta fronts, delta platforms, and prodelta slopes. In addition, the offshore shelf may be affected significantly by deltaic sedimentation. Many deep-sea fans are associated with large marine deltas shedding huge volumes of terrestrial material into the sea. For these reasons, the depositional environments of marine deltas are discussed here with reference to the less complex subenvironments described earlier (particularly in Chaps. 3.1 and 3.2).

In spite of the variety in depositional environments, the different facies types of lateral or vertical deltaic sequences represent a characteristic, diagnostically significant facies association. The following brief summary is based on special publications and books on marine deltas (e.g., Scruton 1960; Allen 1965; Morgan and Shaver 1970; Wright and Coleman 1973; Weimer 1976; Galloway and Hobday 1983; Elliott 1986b; Fraser 1989; and others), but cannot cover all variations and details observed in recent and ancient case studies.

Prograding Fan Deltas

Marine deltas develop where rivers enter the sea and cause a seaward prograding of the coastline due to their high sediment load. In the case of a narrow coastal plain in front of a high mountain range, such a progradation can take place rather uniformly along a wide coast-parallel zone and generate a series of coalescing fan deltas (Fig. 3.27a). Coarse material transported during river floods reaches the coast and may form comparatively wave-resistant blocky and gravelly beach ridges. Lagoonal muds can accumulate behind such ridges. The foreshore is characterized by sorted gravel and sand; the shoreface profile tends to be relatively steep and may show indications of delta foresets similar to those of lake deltas (Chap. 2.5). Most of the sand transported into shallow water depth, however, is reworked repeatedly by storm waves and currents and therefore loses its primary deltaic sedimentary structures.

Larger Marine Deltas (Overview)

The greater and more typical marine deltas are built by rivers which drain large areas and reach the sea after a long passage through alluvial plains. Hence their sediment load is finer grained and consists predominantly of silt and clay with minor proportions of sand. Gravel is often absent or transported only in very small quantities. The shape of these deltas is mainly controlled by the rate of sedi-

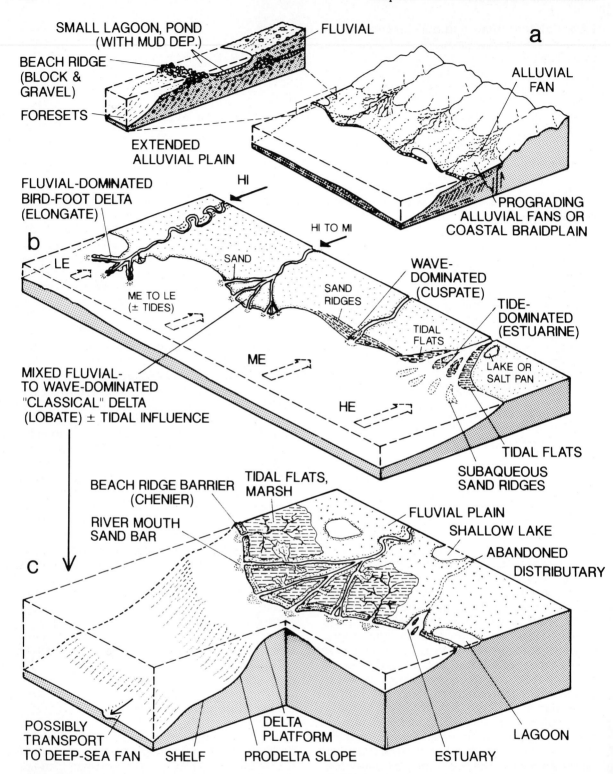

Fig. 3.27. a Marine fan delta caused by seaward prograding alluvial fans or braidplain. Note coarse-grained beach ridges (blocks and gravel) protecting small lagoons or ponds from wave and current action. Older beach ridges and lagoonal silts and muds may be overridden by fluvial deposits.

b Different types of large marine deltas controlled by waves and tidal currents (*LE* low energy; *ME* medium energy; *HE* high energy conditions) as well as sediment input (*HI*, high input; *MI*, medium input). **c** Different subenvironments of a large, lobate, wave- and tide-influenced delta system (similar to the Niger delta)

ment input and the hydraulic regime of the sediment-receiving basin. Strong waves and tides hamper or prevent the outbuilding of fluvial sediments into the sea, whereas low-energy conditions favor this process. Using these criteria, one can distinguish several types of marine deltas (Fig. 3.27b):

- *Fluvial-dominated birdfoot delta, elongate:* Due to high sediment input and low counteraction by the sea forces (low-energy conditions), the single main distributaries of the delta prograde seaward separately and create a coastline similar to the shape of a birdfoot. A prominent example of this type is the modern Mississippi delta.

- *Mixed fluvial- to wave-dominated, "classical" delta, lobate.* Higher wave energy, which may be accompanied by moderate tides, prevents the separate outbildung of fluvial systems. Instead, all distributaries advance more or less uniformly and thus generate the characteristic protruding lobate or triangular coastline which the Greek compared with their capital letter "delta". Many modern rivers build this type of delta, for example the Danube into the Black Sea, the Nile and Ebro into the Mediterranean (e.g., Sestini 1989). In all these cases, the influence of tides can be neglected. However, the deltas of the Niger and Orinoco entering the Atlantic Ocean also belong to this type, although they are moderately affected by tides.

- *Wave-dominated deltas, cuspate:* Strong wave action prevents the local outbuilding of a delta front. Temporarily deposited river load is reworked and transported from the river mouth along the high-energy coast or into deeper water. Splitting of the main river into several distributaries and avulsion (shifting) of channels are less frequent than in birdfoot deltas. Overall, wave-dominated deltas advance more slowly over a broader front. Modern examples include the deltas of the Rhone (Mediterranean), the Brazos entering the Gulf of Mexico and the Sao Francisco and other South American rivers discharging into the South Atlantic.

- *Tide-dominated delta, estuarine.* Strong tides and tidal currents migrate, as discussed in Chapter 3.2, some distance up the river and therefore widen the river mouth to form an estuary. In combination with wave action, the river load is swept out into the shallow sea or transported longshore into areas of lower energy conditions, where it can settle in tidal flats or deeper water. The main river may split into several distributaries with islands in bet-

ween, and part of the sand load forms separate subaqueous sand ridges in front of the river mouth. The coastline does not protrude seaward and the total delta complex progrades comparatively slowly. Present-day examples of this delta type are the Ganges-Brahmaputra, the Colorado, and the Rhine.

3.4.2 Sedimentary Processes and Facies of Various Delta Types

Fluvial-Dominated and Lobate Deltas

Distributary Channels and Levees

The processes taking place at the mouth of an individual distributary channel of the elongate to lobate delta type are schematically shown in Fig. 3.28. On both sides of the channel, flat natural levees are built up during times of peak flood. In places where the levees are not high enough, the flood spills over into the adjacent marshland and generates sandy *crevasse splays*. The levees consist of fine sand and silt (cf. Chap. 2.2.3) and can also be traced as *subaqueous sand bars* from the river mouth some distance into the sea. Because large delta areas usually subside but are flooded discontinuously, their subaerial levees display irregular vertical aggradation. Plant cover and soils may develop in times of non-deposition. The walls of channels cutting such levees, marsh deposits, or bay muds can be very steep, but often the channels migrate laterally, eroding one of their banks and aggrading on the other. Channels may also be filled with sand bars or mud.

Interdistributary Delta Plain

The interdistributary space between the rapidly prograding subaerial levees of a birdfoot delta or behind the uniformly advancing beach ridge barrier of a lobate delta is occupied by *marshland, swamps, and fresh water lakes*. In a tidal regime with outlets to the sea, tidal flats develop behind the beach ridge barrier. The sediments of these areas are predominantly siliciclastic and strongly influenced by the climate of the region as described in Chapter 3.2. Under humid conditions, the interdistributary, slowly subsiding swamps offer ideal prerequisites for the accumulation and preservation of peat and allochthonous plant debris. Many coal deposits

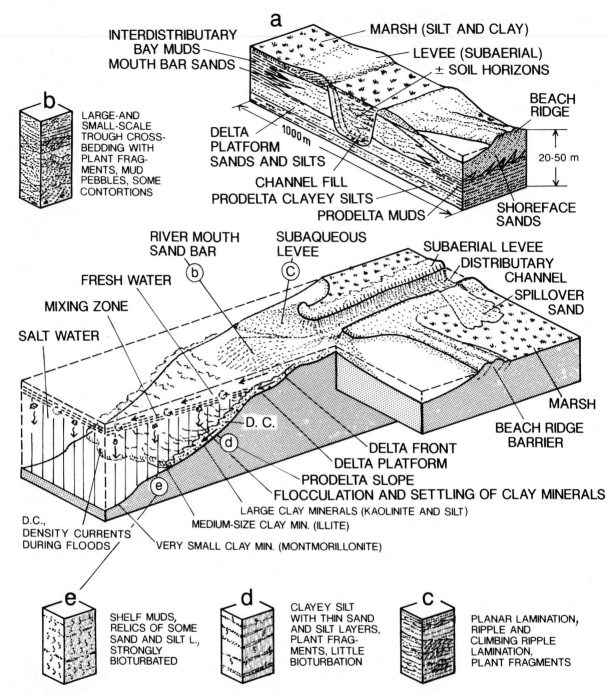

Fig. 3.28a-e. Model of a distributary channel (bird-foot or lobate delta type) and depositional processes at its mouth.

a Cross section through channel with adjacent sediments. **b** through **e** characteristic sedimentary structures of some subenvironments

were formed in such an environment, and hydrocarbons found in deltaic sediments also partially derive from source rocks of this type. Shallow, large lakes developing in the upper part of the delta plain may be filled by rapidly prograding lacustrine deltas fed by fine-grained material of the main river as observed on the Mississippi delta plain (Tye and Coleman 1989). These deltaic sediments represent short basin-filling episodes and differ from the common, *Gilbert-type* lacustrine deltaic sequence (Chap. 2.5.1) in displaying extensive distributory mouth-bar sands and parallel-laminated prodelta muds instead of coarser grained, steeply inclined prodelta foresets.

River Mouth and Delta Front

The bedload of the distributary channels is deposited directly in front of the subaerial delta, commonly at water depths in the range of 5 to 30 m. The sediment load of large river systems draining extensive alluvial plains (apart from highlands far from the coast) is commonly fine grained. Consequnetly, the *river mouth bars* are composed mainly of fine sand displaying large- and small-scale trough cross-bedding of rather consistent current direction (Fig. 3.28 b). At a lobate delta, river mouth bars, subaqueous levees, and foreshore sands of the beach ridge barrier may combine to form a more or less continuous sand sheet *(delta front sands)* of locally varying thickness on the order of 5 to 30 m.

The *beach ridge barrier* is fed by sands transported alongshore from the river mouths. Separately advancing channels of the birdfoot delta type, however, generate isolated elongate sand bodies, the so-called *bar finger sands* (Fig. 3.28b).

Typical features of a rapidly prograding delta front are growth faults, slope gullies, and mud diapirs originating from differential compaction and failure of underconsolidated sediments. They were intensively studied in the Mississippi delta and Niger delta regions (see below), but are also known from ancient examples (e.g., Galloway and Hobday 1983; Prior and Coleman 1984; Martinsen 1989; Pulham 1989; Doust and Omatsola 1990; also see Chaps. 5.4.1 and 13.2).

Delta Platform and Prodelta Slope

The river mouth and its accompanying shoreface sands prograde as the *delta front*. During times of peak flood, however, the river water with its suspended load may become denser than sea water. Then it flows basinward as undercurrent and drops its sandy and silty bedload in deeper water. Seaward of the delta front a kind of *delta platform* is observed at shallow water depths (10 to 30 m). This feature results from the combined action of rapid deposition and wave action during the constructive phase, but may also be shaped during a subsequent destructive phase of the delta

development (see below). The *prodelta slope* farther offshore is generally very gently inclined (less than one up to a few degrees) toward the floor of a shelf sea or a deeper oceanic basin.

A particularly great part of the silt- and clay-size material of the river load is deposited on the prodelta slope. The transport and settling behavior of sand and silt particles is little affected at the transition from fresh water to sea water. The suspended clay minerals, however, tend to flocculate and form aggregates as soon as they pass, under conditions of normal river discharge, from the less dense, overlying fresh-water wedge into the zone of mixed brackish water or normal sea water (Fig. 3.28). The clay aggregates sink much faster than isolated clay particles and thus significantly contribute to the prodelta slope sediments.

Since comparatively large clay minerals such as kaolinite produce denser aggregates and settle earlier than do very small ones (e.g., montmorillonite), there may be a kind of lateral fractionation of these minerals. Kaolinite tends to accumulate near the river mouth, while illite and montmorillonite are normally transported further into deeper water.

Sedimentary Structures

Characteristic sedimentary structures of all these environments are shown in Fig. 3.28b through e, as well as in the sections especially dealing with these subenvironments (Chaps. 2.5, 3.1 through 3.3). Of particular interest are the *prodelta sediments*, because they usually reach considerable thicknesses and have a good chance of preservation, even if the upper part of the deltaic sequence is eroded. Prodelta sediments typically consist of clayey silt or silty clay (hemipelagic terrigenous material) with thin (mm to cm) intercalations of fine sand or coarse silt reflecting episodes of minor and major river floods. Plant fragments may be abundant, but the percentage of autochthonous marine biogenic components is generally low. Due to a high sedimentation rate (on the order of a few mm to tens of cm per year), burrowing by bottom dwelling organisms is sparse. Therefore, primary laminations within the clayey silts, although indistinct and irregular, are frequently preserved.

Growth Faults and "Depobelts"

The prograding prodelta sediments of major rivers form huge *clastic wedges* which may reach thicknesses in excess of 10 km and overlie oceanic crust (e.g., in the Niger delta region, Doust and Omatsola 1990). Such sediment wedges are strongly affected by synsedimentary and postsedimentary normal faults. In these cases the prograding delta complex can be divided into a number of major, growth fault-bounded sedimentary units or "depobelts". Within each depobelt, the sediment buildup may occur stepwise or cyclic in response to continued deformation and sea level changes. Even the gentle foot of the prodelta slope may undergo deformation characterized by imbricated "toe thrusts".

The shelf sediments in front of a large delta also tend to be dominated by fine-grained siliciclastic material delivered by the river. With decreasing sedimentation rates bioturbation becomes a significant process masking or obliterating all kinds of bedding.

Wave-Dominated Deltas

These deltas are characterized by beach ridge/foreshore sands advancing with the active delta lobe in a wide coastal-parallel front basinward (Fig. 3.29a). In relation to the wave energy and wave base, these coastal barrier sands form sheets of considerable thickness (up to tens of meters). Due to intensive reworking, the coastal barrier sands tend to be better sorted than the distributary channel and mouth bar sands of the fluvial-dominated deltas. They coarsen upward as a result of the prograding shoreface and often contain abundant shell debris which is produced by shallow water organisms and added during the processes of coastal sand transport (Fig. 3.29a and c). Sideward shifting and return of the active delta lobe to the subsiding initial site may generate a second or several coastal barrier sand sheets on top of the first one (see below). Behind the active coastal beach ridge, the space between older barrier sands and other parts of the interdistributary area may be occupied by mangrove swamps, marshes, or lakes. The suspended river load of wave-dominated deltas is usually removed from the river mouths and widely distributed in the adjacent basins. As a result, the prodelta facies of fluvial-dominated deltas is poorly developed or more or less missing.

Tide-Dominated Deltas

In contrast to the other delta types, tide-dominated deltas lack an advancing subaerial delta front, but accumulate most of the river sediment in front of their wide, estuarine mouth in shallow water or in the adjacent deeper sea (Fig. 3.29b). Thus, an extensive subaqueous delta platform is produced, on which at least part of the bedload and suspended river load comes to rest. Unlike the fluvial-dominated deltas, the sands are not dropped as localized mouth bar sands, but repeatedly reworked and redistributed by tidal currents mainly directed perpendicular to the general coastline. Hence, they form elongate subparallel sand ridges between the estuarine distributaries of the river mouth, as well as on the deeper delta platform. The sideward migrating distributaries are filled with tidal current-influenced sands, exhibiting bidirectional cross-bedding and ripple lamination, in addition to estuarine muds with mixed fauna (fresh water to normal saline).

Although the estuaries of a tide-dominated delta give little evidence of progradation, the total complex of such a delta also slowly advances seaward. The resulting facies association, however, resembles that of a normal tide-dominated coastline (some distance away from an entering river, see Chap. 3.2) rather than the associations found in the other delta types. A typical vertical facies succession for such a constructional phase includes (from top to bottom):

- Alluvial plain sediments.
- Supratidal marsh or, depending on the climate, evaporitic beds of a coastal sabkha or other supratidal deposits.
- Extensive intertidal sediments.
- Point bar and tidal-influenced sands of distributary channels, estuarine muds with variable fauna.
- Tidal-influenced sediments of the delta platform with subtidal sand ridges and sand waves.
- Shelf sediments, still more or less influenced by tidal currents.

Figures 3.29d and e display characteristic vertical successions of an estuarine channel fill and a sand ridge on a delta platform. Coastal barrier sands play a subordinate role in this type of delta. These deltaic sediments and their facies architecture are not discussed further here, because they are more related to subtidal, tidal, and estuarine deposits described in Chapters 3.2.1 and 4.2.

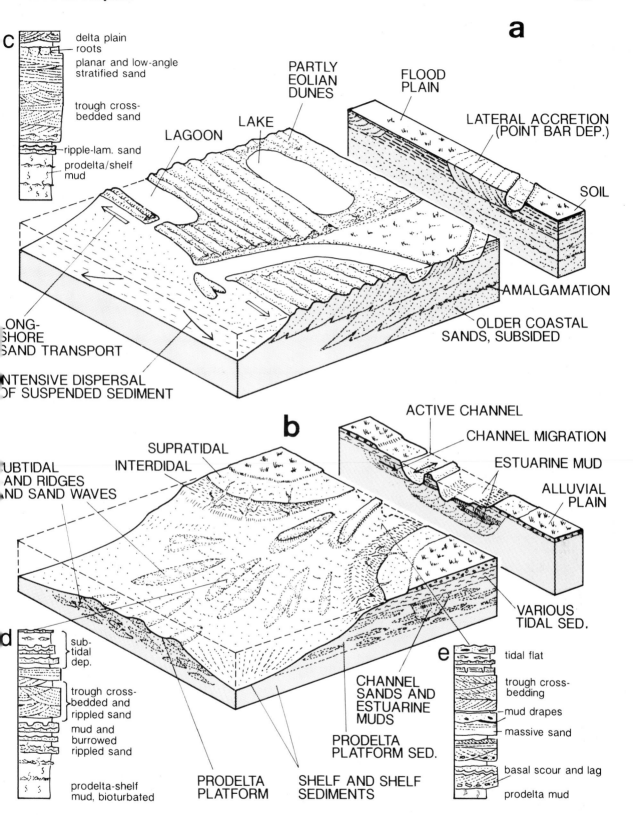

Fig. 3.29a,b. Conceptual models of **a** wave-dominated and **b** tide-dominated delta. **c** Vertical section of an advancing foreshore/coastal barrier sand ridge. **d** and **e** Vertical sections of estuarine channel fill and tidal sand ridge on prodelta platform of tide-dominated delta. (**c** through **e** based on Galloway and Hobday 1983).

3.4.3 Facies Architecture, Constructional and Destructional Phases

Facies Architecture of Lobate Deltas

Constructional Phase

The final facies architecture of a delta complex is controlled not only by its constructional phase during outbuilding, but also by destructional periods of reworking and redistribution of sediments delivered by rivers. The lobate and wave-dominated delta types are less susceptible to extensive destruction than the birdfoot delta. In the latter case, the far seaward protruding distributaries can be reworked easily by waves and currents, especially when they are cut off from their sediment source by crevasse splays which evolve into new distributaries. Figure 3.30a shows the pure constructional facies association of a lobate delta consisting of five main facies groups (from bottom to top):

- Prodelta muds (with some intercalated silt and sand layers) prograding seaward.
- River mouth and coastal barrier sands (delta front sands) prograding seaward similar to the prodelta muds.
- Point bars and infillings of distributary channels (predominantly sand) and levee sands and silts (possibly including soils).
- Sediments of the interdistributary delta plain (bay or lagoonal muds, possibly sediments of tidal flats, marshes, lakes), locally overlain or interrupted by crevasse splay silts. Delta plain deposits are characterized by relatively slow upbuilding (aggradation) on top of the subsiding delta complex.
- Fluvial sediments of the prograding alluvial plain consisting of point bars, channel fills, and finer grained deposits of the flood plain (cf. Chap. 2.2.3).

The elevation of the actual delta front deposits is always adjusted to the sea level, but older equivalents of the deposits are affected by sediment compaction and subsidence of the crust under the increasing sediment load. Therefore, the delta front sediments buried below the prograding subaerial delta plain dip slightly landward. The different types of sediment accretion discussed here, i.e., progradation of prodelta and delta front deposits generate an unusual pattern of the isochrones within the total delta complex (Fig. 3.30a) and contrast to the prevailing vertical aggradation of delta plain sediments.

Destructional Phase

Destruction of delta lobes commonly takes place where main distributaries have prograded too far into the sea and lose their sediment supply by channel diversion. Such a development is mostly initiated by breaches in the levees, crevasse splays, and crevasse channels that find a shorter and steeper course into the sea. Another cause of coastal retreat is subsidence and/or rising sea level (also see below). Destruction is brought about by wave action and wave-induced currents (Chap. 5.2) which predominantly rework and redistribute the river mouth and delta front sands, the coastal sand barrier close to the distributaries, and part of the interdistributary sediments behind the sand barrier. This process may produce a chain of retreating beach ridges and barrier islands (Fig. 3.30b) which protect flooded portions of the delta plain from further erosion. As a result, widely extended shallow bay muds and salt marsh deposits of limited thickness can accumulate.

In front of the retreating coastal sands, a *veneer of reworked, relatively coarse foreshore and shoreface sands* may rest on truncated mouth bar sands or directly on prodelta muds. These sands frequently contain shell concentrations of mixed origin, part of which are derived from eroded bay and lagoon deposits. Fine grained material reworked from the interdistributary areas is swept into deeper water by currents. Such a destructional phase continues until a new equilibrium is established between the forces of the sea and the capability of the accumulated sediments to resist erosion. A new delta lobe may prograde and incorporate the retreated coastal sands into the subaerial delta plain of a subsequent constructional phase.

Vertical Facies Successions

Vertical sections in delta sediments are very variable, particularly in cases, when the delta was subjected to destructional periods (Fig. 3.30c through g). A characteristic feature is coarsening-upward of prodelta silts and clays, mouth bar sands, and shore face/beach ridge sands. The opposite tendency is observed in vertical sections through the infillings of distributary channels and minor channels associated with crevasse splays. On or near the channel floor, mud clasts, logs, and sometimes reworked nodules of ironstone, formed during early diagenesis, are common. An irregular

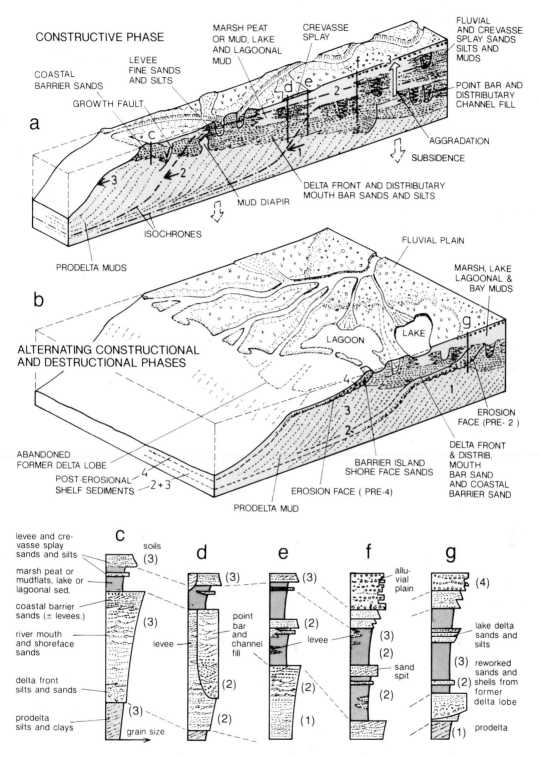

Fig. 3.30a-g. Facies association of a classical lobate delta, generalized, with landward dipping coastal and fluvial deposits due to continuous subsidence (similar to the Niger delta, see, e.g., Allen 1970; Doust and Omatsola 1990). **a** Constructive phase with permanent outbuilding (progradation) of prodelta muds and aggradation of sediments of the delta plain. Note the unusual position of isochrones.

b Discontinuous progradation, interrupted by periods of partial delta destruction (erosional faces *1* and *2*), which may remove river mouth bar and subaqueous levee sands and silts, as well as part of the primarily overlying sediments of shallow bays, lagoons, marshland, and lakes. **c** through **g** Vertical sections (locations shown in **a** and **b**), see text for explanation

coarsening-upward trend is also common in interdistributary sequences, i.e., from lagoonal muds to tidal flats, from salt marsh and crevasse splay silts to sands, or from lake bottomsets to lake delta foresets and topsets of the alluvial delta plain. Such coarsening-upward sections may occur repeatedly and are often described as autocyclic deltaic sequences (Chap. 7.1).

The most important sedimentary structures of these subunits are described in the previous chapters. Normally graded sand and silt layers may occur in the prodelta environment as a result of density currents (Fig. 3.28), as well as in crevasse splay sheet sands and silts due to waning flow conditions after peak flood. Delta abandonment and coastal retreat lead to truncated vertical sequences (Fig. 3.30g). Usually the upper part of a prograding constructional sequence is more or less eroded and, after deposition of a thin veneer of reworked material, overlain by sediments of a younger constructional phase.

Lateral Facies Transitions

The facies in the prodelta region gradually change laterally from prodelta silts and clays to shelf sediments (Fig. 3.28d and e; cf. Chap. 3.3). In shallower water and particularly within the delta plain, the facies strongly vary laterally and may display the following transitions:

- From lower part of point bar sands and infillings of main channels to mouth bar sands, subaqueous levee sands and silts, and shoreface sands.
- From infillings of channels, levee and crevasse splay sands and silts to sediments of bays, lagoons, tidal flats, salt marshes, or salt pans, depending on the climate.
- Successions of lakes and swamps including their minor deltas may pass laterally into overbank deposits of the advancing alluvial river plain.

One has to bear in mind that such lateral facies transitions occur not only within sediments of the same age, but also between older and younger units of the complicated facies architecture of a marine delta (see isochrones in Fig. 3.30a).

The facies also change from the upper to the lower delta plain. In the upper plain, channel sands play a main role and the interdistributary area is largely occupied by levee and crevasse splay deposits. Truncation due to delta lobe abandonment is insignificant. However, relatively thick bay-fill and lagoonal sequences prevail in the lower delta plain, particalurly in the case of birdfoot deltas (see below). The seaward portions of these deposits are exposed to delta destruction.

Facies Architecture of Birdfoot Deltas

The facies association of a birdfoot delta is in many ways similar to that described in Figure 3.30, but some important differences exist. During constructional phases the distributaries build, on top of prodelta silty clays or clayey silts, elongate isolated bodies of delta front and mouth bar sands out into the sea (Fig. 3.31). The area in between is occupied by bays accumulating muds with marine fossils. If such elongate delta lobes are abandoned, part of their delta front sands retreat and form sand barriers which convert the former interdistributary marine bays into lagoons of varying salinity. Later, marsh and/or lake deposits may follow, before the prograding fluvial plain, its crevasse splays, and overbank fine-grained muds take over. Therefore it is common for marine prodelta sediments or bay muds to be directly overlain by lagoonal, intertidal, marsh, or lake sediments in the interdistributary area of a birdfoot delta.

The interdistributary environment is a particularly favorable site for the accumulation and preservation of organic matter and thus for the generation of coal seams. Outcrops in this area may show similar sequences over fairly large distances and therefore give little evidence of distributary channels and their specific facies association (also see below). In total, the vertical and lateral facies successions display the same variations as the classical lobate delta. However, they include a larger percentage of marine bay sediments.

Delta-Lobe Switching and Response to Sea Level Changes

"Shallow-" and "Deep-Water" Delta Prograding

In both types, the classical lobate and the birdfoot deltas, a delta lobe advancing over an abandoned part of the former delta can prograde faster than a lobe which is built out into deeper water. Such a *shallow-water delta* produces a thin delta front sequence and tends to

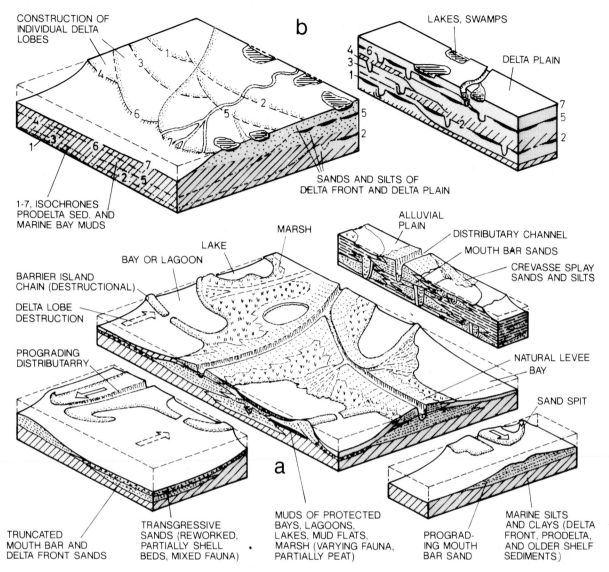

CONSTRUCTION OF
INDIVIDUAL DELTA
LOBES

b

LAKES, SWAMPS

DELTA PLAIN

1-7, ISOCHRONES
PRODELTA SED. AND
MARINE BAY MUDS

SANDS AND SILTS OF
DELTA FRONT AND DELTA PLAIN

ALLUVIAL
PLAIN

DISTRIBUTARY CHANNEL

MOUTH BAR SANDS

CREVASSE SPLAY
SANDS AND SILTS

MARSH

LAKE

BAY OR LAGOON

BARRIER ISLAND
CHAIN (DESTRUCTIONAL)

DELTA LOBE
DESTRUCTION

PROGRADING
DISTRIBUTARRY

NATURAL LEVEE

BAY

SAND SPIT

a

TRUNCATED
MOUTH BAR AND
DELTA FRONT SANDS

TRANSGRESSIVE
SANDS (REWORKED,
PARTIALLY SHELL
BEDS, MIXED FAUNA)

MUDS OF PROTECTED
BAYS, LAGOONS,
LAKES, MUD FLATS,
MARSH (VARYING FAUNA,
PARTIALLY PEAT)

PROGRAD-
ING MOUTH
BAR SAND

MARINE SILTS
AND CLAYS (DELTA
FRONT, PRODELTA,
AND OLDER SHELF
SEDIMENTS)

Fig. 3.31. a Facies association of a bird-foot delta (Mississippi-type), generalized, with two seaward advancing, isolated distributaries (constructional phase) and two abandoned distributaries (delta destruction). Note that delta front and mouth bar sands represent elongate linear sand bodies (in contrast to the sand sheet in Fig. 3.29a); hence, interdistributary sediments of the delta plain may rest directly on prodelta deposits or on marine bay muds. **b** Large-scale facies architecture of a birdfoot delta with imbrication of delta lobes partially representing "shallow-water" and "deep-water" depositional conditions. See text for further explanation

form continuous delta front sheet sands in contrast to *deep-water deltas*, which deposit thicker, coarsening-upward delta front sequences as well as thick prodelta sediments. In the latter case, the distributaries tend to become widely spaced and therefore generate laterally discontinuous mouth bar sands (birdfoot delta). Figure 3.31b demonstrates the large-scale facies architecture of a predominantly fluvial-dominated delta composed of a series of delta lobes similar to those known from the Holocene Mississippi delta (Frazier 1967). The upper (landward) part of these individual delta lobes usually exhibits the characteristics of shallow-water deltas, whereas the lower part represents a deep-water delta and therefore a thick delta front sequence. As a result, the generalized and simplified sections perpendicular and parallel to the coastline (Fig. 3.31b) show combined delta front and delta plain deposits of rather varying thicknesses.

In spite of the obvious *imbrication of delta lobes,* the overall, large-scale delta complex commonly renders the impression of a cyclic sequence in which layers rich in organic matter (coal seams) are particularly arresting. More thorough investigations, however, frequently reveal that the coal seams switch laterally to deeper or higher positions in the total sequence, and the same is true of all the other facies types.

The Scale of Marine Deltas and Depositional Cycles

The problems associated with bed correlation and cyclic sequences, mentioned above and described in Chapters 7.2 and 7.3 in more detail, underline the necessity to discuss the scale of delta facies models.

The delta plains of present-day large marine deltas cover areas of thousands of square kilometers. The delta front of the Holocene Mississippi delta is approximately 400 km wide and protrudes 100 to 150 km into the Gulf of Mexico. Similarly, the delta lobe of the Niger is about 400 km wide and has prograded seaward 100 to 200 km. Several other recent delta plains reach about the same size, and we can assume that ancient delta plains were of the same magnitude.

On the other hand, the thoroughly investigated facies associations of deltaic coal fields usually cover smaller areas and therefore can barely reveal the complete architecture of geological bodies as large as those of entire delta complexes shown in Fig. 3.31b. Instead, the local studies usually give rise to the impression that individual beds can be traced over relatively long distances and that cyclicity is rather regular and possibly of regional or even global importance. This is not true in those cases in which the vertical sequences are caused solely by delta lobe switching. However, some deltaic coal cyclothems appear to be controlled by eustatic sea level changes originating from processes outside of the deltaic depositional area (cf. Chaps. 7.4 and 14.5).

Response of Marine Deltas to Sea Level Changes

Such a situation is demonstrated in Fig. 3.32 for the classical lobate delta type. A rapid rise in sea level leads to extensive coastal retreat on top of the submerging former delta plain. If the sea level rise does not exceed a few tens of meters and then persists for some time, the delta front can prograde again as a shallow-water delta (Fig. 3.32a). As soon as the position of the former delta front is reached, further outbuilding takes place in deeper water and therefore will proceed more slowly (Fig. 3.32b, deep-water delta). Falling sea level promotes coastal advance in conjunction with the prograding delta front to a deeper level. The former delta plain emerges above sea level by tens of meters and will therefore be dissected and partially eroded by the main river and its tributaries. The eroded material contributes significantly to the further outbuilding of the lowstand delta. Part of the emerged deltaic terrace may also be removed by large slumps and mass flows (Chap. 5.4.1). A subsequent sea level rise to its initial level tends to restore the former coastline and to generate a transgressive sequence again (Fig. 3.32c, similar to that shown in a).

The vertical facies successions resulting from such sea level changes vary from locality to locality within the delta complex (Fig. 3.32d through f). Interdistributory sites landward or seaward of the initial coastline show alternations between the following facies groups (listed from land to sea):

- Lower delta plain association (lagoonal, tidal, marsh, crevasse splay deposits) alternating with the alluvial plain association (flood plain, point bar, channel fill, lake deposits, etc.; Fig. 3.32f).
- Lower delta plain association alternating with shoreface and delta front deposits (delta front association; Fig. 3.32e).
- Shoreface/delta front alternating with the inner shelf/prodelta association (Fig. 3.32d).

Most of these different facies successions show variations between two principal groups and are restricted to certain parts of the delta complex. Some of them may, however, occur and follow each other in the same area and thus display greater facies variations. Other complications may arise from additional switching of delta lobes. Point bar deposits, channel fills, and their associated levee sediments may become stacked due to alternating periods of downcutting and refilling of the distributaries (Fig. 3.32a).

When the sea level fluctuations reach amplitudes on the order of 50 to 100 m, prodelta sediments and inner shelf associations play a greater part in the facies successions than described in the model of Fig. 3.32. Then, the sea may transgress beyond the former delta

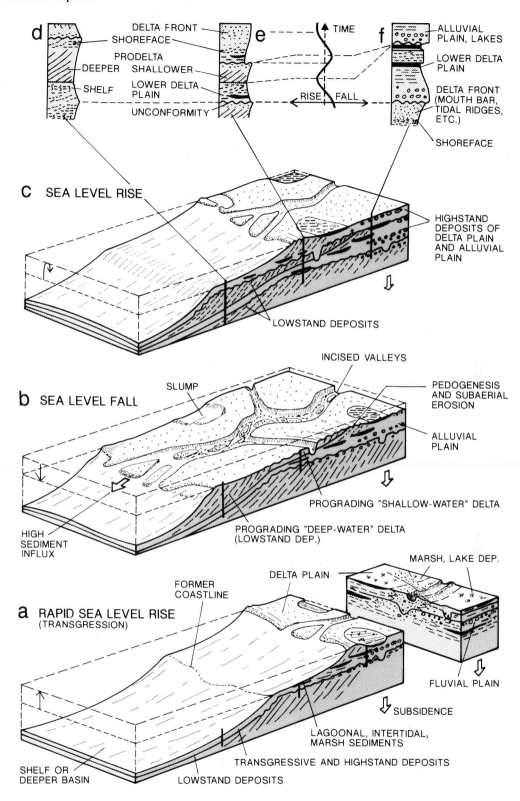

Fig. 3.32a–f. Effects of sea level changes on the depositional environments of a classical, lobate delta. **a** and **c** Rapid sea level rise causes coastal retreat, followed by a slowly prograding coastline during highstand (shallow-water delta overriding former delta plain). **b** Prograding shallow and subsequent deep-water delta during sea level fall and lowstand. Sections **d** through **f** demonstrate mainly interdistributary facies successions. See text for further explanation

plain and prograde a long way upriver and thus completely change the configuration of the depositional area. Such scenarios are, however, rarely discussed in studies on the effects of sea level fluctuations on deltaic sediments. In any case, the flat lowlands of a delta plain are the prototype of an environment, where even small sea level oscillations can affect the depositional processes of large areas.

4 Sediments of Adjacent Seas and Estuaries

4.1 Introduction

Adjacent seas are shallow or deep water-filled basins of limited extent which are connected with the world oceans by a gateway or strait. Such basins may develop on both continental and oceanic crust, as well as in diverse tectonic settings. Prominent present-day examples are the Red Sea, representing an ocean in its initial stage, the Persian Gulf, a foreland basin, the Black Sea, a kind of remnant basin, and the Gulf of California, which originated from strike-slip movements along a plate boundary. Wide river estuaries, although usually covering only limited areas, also show some features characteristic of adjacent basin.

The nature of adjacent seas is largely determined by the mechanism controlling the connection with the neighboring ocean. Depending on the dimension of this opening and the climate, such basins may have a more or less normal salinity, or they may tend to become brackish (hyposaline) or more saline (hypersaline) than average sea water (Seibold 1970; Grasshoff 1975). Because such systems react very sensitively to small changes in the entrance area, the sediments of adjacent basins may vary significantly with time, definitely more so than normal marine deposits.

Some adjacent seas represent immature ocean basins (e.g., the Red Sea) which evolve into wider ocean basins. Then the sediments of the adjacent sea stage may become buried deeply under younger, fully marine slope sediments or continental margins sediments (Thiede 1978; cf. Chap. 12.2). Land exposures, when the area was later uplifted, or drillholes in such sequences can reveal the early climatic-oceanic history of such a basin development. To understand the highly variable facies of adjacent basin sediments, we should briefly consider some general rules concerning the water circulation of such basins (also see, e.g., Seibold 1970; Seibold and Berger 1982; Pickard and Emery 1982). Some rules for oceanic water circulation are discussed in Chapter 5.2.

4.2 Water Circulation and Sediments

The general rules for thermohaline circulation in the oceans (Chap. 5.2) also apply for circulation in estuaries and adjacent basins. In addition, the circulation is largely controlled by the geometry of the connection between the open sea and the basin, i.e., by the width and depth of the opening, frequently characterized by a sill. Furthermore, river inflow and the climate of the neighboring land are important factors.

Estuaries

The wide mouth of a river, i.e., an estuary, may be regarded as a comparatively small, specific, frequently occurring type of an adjacent sea. Such a semi-closed elongate basin has a relatively wide connection to the open sea and on the landward end a substantial inflow of fresh water (Figs. 3.8c and 4.1). The river water entering the estuary mixes to some extent with the denser sea water, but

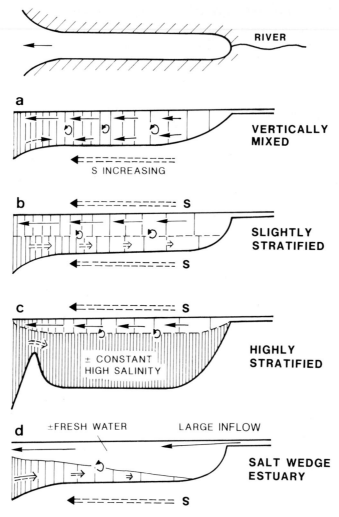

Fig. 4.1a-d. Estuary in plan view (*at top*) and models of estuarine water circulation in cross section (a through d). *Arrows* marked with *S* and *narrowing hatching* indicate increasing salinity

eventually flows out in the form of a surface current into the open sea. The inflowing sea water enters the estuary as a bottom current and later mixes with outflowing fresh water. This type of water exchange is referred to as *estuarine circulation*. It is interesting to note that an increase in river flow also causes a somewhat higher inflow of sea water. However, the salinity of the estuarine water is reduced under such conditions. Estuaries display either vertically mixed or stratified waters. The latter type is controlled by the markedly differing salt contents of the outflowing and inflowing waters, in contrast to the open sea where (smaller) density gradients are generated primarily by differences in water temperature.

The models in Fig. 4.1 show the four principal cases identified in nature. In a shallow estuary, the entering river water and sea water may become completely mixed and therefore *unstratified* (Fig. 4.1b), but salinity increases from the river inlet toward the open sea. If mixing of sea water and fresh water is less effective, which occurs particularly in deep estuaries with a sill, the water body displays a more or less pronounced and stable *density stratification* (Fig. 4.1c and d). Very stable conditions are typical for deep fjords, as for example along the Norwegian coast. A large river inflow can create a fresh water layer over the total length of the estuary, which is underlain by a wedge-like salt-water body. As a result of some mixing of surface and bottom water, the outflowing surface current transports more water into the sea than the entering river into the estuary. Thus, the outflow at the mouth of the estuary can be ten or more times the volume of the river flow.

The sedimentation rate of terrigenous material tends to be generally high in estuaries, reflecting the high suspended load of many rivers. It is, however, difficult to develop a simple facies model for estuaries. An individual estuary may pass from one circulation system into another as a result of strongly variable river discharge. It may be affected by tides and tidal currents (Chap. 3.2.1), causing coastal erosion, reworking and redeposition of sediments in both landward and seaward directions. Supported by a high, fluvial nutrient supply, organic productivity in estuaries is commonly high, but the number of species is restricted in brackish water subjected to frequent changes in salinity. Plant debris delivered by the river and marine fauna entering the estuary often become mixed in the estuarine sediment. Sediment reworking and redeposition are common in shallow estuaries, whereas organic matter and fossil remains have a good chance of being preserved in deeper estuaries due to rapid sediment accumulation. Where estuarine water masses are stratified (Fig. 4.1c), the oxygen content of bottom water may drop and hamper benthic life.

"Humid Adjacent Seas" with Estuarine Water Circulation

In larger adjacent seas, evaporation from the water surface of the basin becomes an important factor (Fig. 4.2). In humid climates, precipitation, P, and river discharge, R, into the basin are greater than water loss by evaporation, E. This situation is similar to that observed in estuaries. Therefore the term *estuarine circulation* is also used for such "humid basins".

Present-day examples of this basin type are the Baltic Sea and the Black Sea (Figs. 4.2a

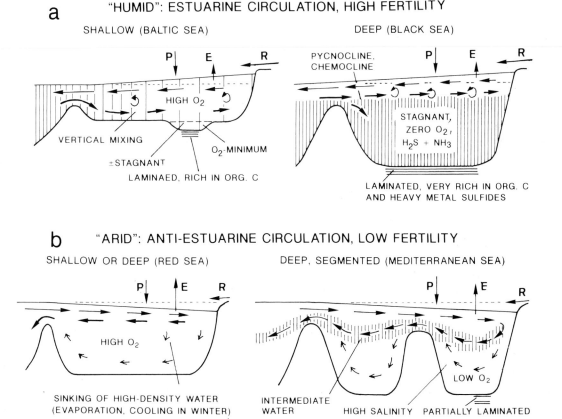

Fig. 4.2a,b. Circulation models for adjacent seas with shallow, narrow entrances. See text for further explanation

and 4.4). They are characterized by a narrow, shallow entrance, allowing the denser sea water to enter the basin as subsurface inflow, while the less saline water from the basin flows as surface current out into the sea. The Gulf of California also displays estuarine circulation. However, in contrast to the previous examples, its opening to the Pacific is wide and deep and therefore readily allows subsurface inflow of cold and comparatively dense intermediate ocean water (Fig. 4.5, see Chap. 4.3 for further details).

Baltic Sea. This fairly shallow basin (average depth about 100 m) corresponds more or less to the vertically mixed estuary model (Fig. 4.1b). Its water mass is largely unstratified and was well oxygenated in pre-industrial times, apart from some deeps with water depths up to 460 m. The salinity of the Baltic Sea (usually less than 10 to 15 ‰) increases from the landward side toward the opening(s) to the normal saline North Sea. Consequently, stagnant stratified waters only developed in some marginal bays and deep depressions in the central part of the basin. However, the redoxcline rose during the last 20 years due to man made "eutrophication" (cf. Chap. 2.5.1) with the result that about one third of the Baltic Sea bottom ($\geq$ 75 m of water depth) is now covered by laminated sediments (Jonsson et al. 1990). Nutrients for aquatic life are delivered by rivers from the neighboring land areas as well as by erosion of sediments in the coastal zones. Therefore, production of organic matter can be high, and some of it may be preserved in the sediments, particularly in the deeps mentioned above where benthic life is reduced or missing due to oxygen deficiency (for more details see, e.g., Hinz et al. 1971; Seibold et al. 1971; Kögler and Larsen 1979). Similar depositional conditions are also realized in many deep fjords along formerly glaciated coasts where freshwater enters the sea.

Black Sea. The modern, approximately 2000 m deep Black Sea is an example of estuarine circulation with distinctly stratified water masses (Figs. 4.1d, 4.2a, and 4.4). Its connection with the Mediterranean Sea via the Bosporus strait is narrow and only about 40 m deep. Because the Black Sea also belongs to the "humid" basin type, its outflow at the Bosporus is stronger than subsurface inflow. Nevertheless, the limited inflow of sea water has filled the major part of the modern deep basin with saline water. Thus, the bottom and intermediate water mass of the Black Sea has a salinity of about 22‰ and is therefore much denser than the surface water. Mixing between these two water masses is hampered by the pronounced pycnocline, which also acts as a stable chemocline separating well oxygenated, nutrient-rich surface water from stagnant, oxygen-deficient deeper water. The surface water exhibits phytoplankton productivity, including calcareous nannofossils which accumulate on the sea floor. The anoxic deeper water contains H_2S and does not allow any benthic life apart from anaerobic micro-organisms. For this reason, the bottom sediments are well laminated and fairly rich in organic matter and heavy metal sulfides (see Chap. 4.3 for further details and the sedimentary history of this basin).

"Arid Adjacent Seas" with Anti-Estuarine Water Circulation

Adjacent "arid basins" are characterized by an excess in evaporation, E, over the sum of precipitation, P, and river runoff, R. As a result, the water surface of these basins is slightly inclined towards the land (Fig. 4.2b), particularly if the connection to the open sea is limited. In contrast to the "humid basin", sea water flows as a warm surface current into the arid basin *(anti-estuarine circulation)* and tends to become increasingly saline due to continued high evaporation. Examples of this basin type are the Red Sea, the Persian Gulf, and the Mediterranean Sea.

Red Sea. At the northern end of the Red Sea, far away from the opening to the Gulf of Aden, the salinity becomes 42.5 ‰, and the water temperature reaches 30 °C in summer. In winter the surface water begins to sink as a result of the combined effect of increased salinity and falling temperature (about 20 °C). Thus, the deeper water is forced to leave the basin as subsurface outflow, which may form an intermediate water layer in the Gulf of Aden. In the Red Sea, the deep water is surprisingly warm ($\geq$ 20 °C) up to a depth of more than 2000 m and has a salinity of $\geq$ 40 ‰. Apart from some special deep brine pools (Degens and Ross 1969), water stratification is absent, and the oxygen content is high throughout the basin. However, planktonic and benthic biogenic productivity in this basin are only moderate, because river discharge and nutrient influx are very low. On the other hand, these conditions are favorable for the growth of coral reefs, mainly fringing reefs which border the coasts on both sides of the Red Sea. The Holocene deep-sea sediments of this basin consist mainly of biogenic carbonate; laminated beds with high organic carbon contents are absent. At the northern end of the Red Sea, carbonate ooids are formed in shallow water. The sedimentary history of this basin is briefly described in Chapter 4.3.

Persian Gulf. In contrast to the Red Sea, the Persian Gulf represents an example of a shallow adjacent sea with a somewhat wider opening to the ocean (cf. Fig. 4.6). The mean depth of the gulf is only 25 m; water temperatures vary between 15 °C in the southeastern part near the entrance and about 35 °C in the shallow western and southwestern regions. In the latter areas, the salinity increases (40 to 50 ‰) and may reach values up to 100 ‰ in the tidal zone. The tidal range is mostly between 0.5 and 1 m. In spite of the inflow of a large river, the Persian Gulf shows the characteristics of an "arid basin". It is well oxygenated and provides a limited supply of nutrients for surface productivity. Biogenic carbonate is the most important constituent of the sediments; it also contains plankton influenced or carried by surface currents from the Indian Ocean. On the particularly arid southwestern side of the basin, calcareous oolites are common, and aragonite is precipitated in the hypersaline intertidal zone as cement, thin crusts, or grain coatings (ooids). Some reefs grow along the coast and on topographic highs within the shallow southwestern part of the gulf. In the supratidal zone behind the reefs and barrier islands, coastal sabkhas occur (Chap. 6.4). Some information on the history of this basin is given in Chapters 4.3 and 14.3.

"Arid basins", or lagoons with a small opening to the ocean, also tend to develop highly concentrated brines and to precipitate marine salts (Chap. 6.4). The present entrances of the Red Sea (about 25 km wide and 140 m deep) and Mediterranean Sea are too wide and deep to permit significant brine concentrations, but during the late Miocene thick salt deposits were generated in these basins (Chap. 6.4.1). As in the Black Sea, the Red Sea and the Persian Gulf experienced drastic changes during their history, leading to pronounced variations in their sedimentary records (Chap. 4.3).

The Composite Basin of the Mediterranean Sea

The Mediterranean Sea is divided into a western and an eastern deep basin by a sill between Sicily and the North African coast (Fig. 4.2b). The eastern basin in particular is characterized by a large excess in evaporation over precipitation, leading to relatively high water temperatures and salinities. Three different water masses can be distinguished in this type of basin: Surface water, intermediate water, and deep bottom water, which interact in several ways. Intermediate and bottom waters are formed in winter, when cooling leads to an increase in density. The intermediate water flows back over the mid-basin sill and out as an undercurrent through the Strait of Gibraltar into the Atlantic Ocean. The deep water is somewhat cooler, and it circulates more slowly than the other water masses. It mixes to some extent with intermediate water and flows out into the Atlantic.

Today, the oxygen content of the deep water is fairly high in both basins, but during the early Holocene and during several intervals within the Quaternary, the bottom water in the eastern basin, including the Adriatic Sea, was more stagnant and repeatedly depleted of oxygen. During the last 0.5 Ma, up to 12 discrete, basin-wide, organic-rich sapropel layers formed in the eastern Mediterranean, which may be associated with periods of high, monsoonal precipitation in eastern Africa and the general climatic changes on the nearby land masses (Rossignol-Strick 1985; Ten Haven et al. 1987; Murat and Got 1987). As a result, fresh water flooding by the Nile River may have caused the influx of land-derived organic matter and have provided nutrients for high marine organic productivity. In addition, the increased freshwater discharge favored the formation of stratified waters and thus prevented convective overturn and ventilation of bottom water. Whether or not a current reversal in the Strait of Gibraltar occurred during the transition from the last glacial phase to the Holocene, as well as whether additional reversals resulted from earlier Pleistocene climatic changes, is controversial (Diester-Haass 1973; Huang and Stanley 1974).

All the larger and smaller basins of the Mediterranean Sea contain similar fills (Bouma 1990), showing the following sequence:

- Pliocene to Quaternary predominantly clastic sediments.
- Miocene evaporites.
- Older deep-water sediments.

The present-day basinal sediments of the Mediterranean are largely hemipelagic muds with low to medium carbonate contents produced mainly by foraminifera and coccoliths. In addition, several mid-sized fans in front of major rivers (e.g., Rhone, Ebro, and Nile Rivers) and many smaller ones are observed. In some regions, various volcaniclastic sediments play a role. Sedimentation rates are on the order of 10 to 30 cm/ka, but locally may be considerably higher, particularly where mud flows and mud turbidites play a role (for further details see Stanley 1972, 1977; Emelyanov 1972; Wezel 1980; Stanley and Wezel 1985; Van Hinte et al. 1987).

Basins with Horizontally Separated Inflow and Outflow

Some marginal ocean basins have a horizontally separated inflow and outflow (Fig. 4.3 b; Pickard and Emery 1982). Due to the gain or loss of heat, salinity, and fertility in the adjacent sea, the characteristics of the outflowing water may deviate considerably from those of the inflow. Consequently, organic production and sediments may differ on the inflowing and outflowing sides of the marginal sea. Present-day examples of this basin type are the Caribbean Sea (the American counterpart of the Mediterranean) and larger, marginal parts of the world oceans, such as the Norwegian-Greenland Sea.

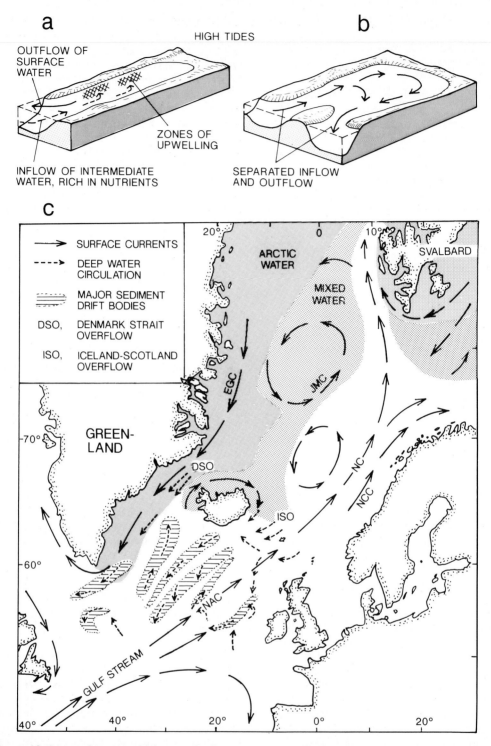

a

HIGH TIDES

OUTFLOW OF
SURFACE
WATER

ZONES OF
UPWELLING

INFLOW OF INTERMEDIATE
WATER, RICH IN NUTRIENTS

b

SEPARATED INFLOW
AND OUTFLOW

c

→ SURFACE CURRENTS

- - -> DEEP WATER
CIRCULATION

≈≈≈≈ MAJOR SEDIMENT
DRIFT BODIES

DSO, DENMARK STRAIT
OVERFLOW

ISO, ICELAND-SCOTLAND
OVERFLOW

ARCTIC
WATER

MIXED
WATER

SVALBARD

GREEN-
LAND

DSO

ISO

EGC

JMC

NC

NCC

NAC

GULF STREAM

70°

60°

40°

40°

20°

0°

20°

20°

0°

10°

Fig. 4.3. a and **b** Contrast between widely opened adjacent
basin with surface outflow and intermediate water inflow
(e.g., Gulf of California, cf. Fig. 4.5), and basin with
horizontally separated inflow and outflow of surface water
(e.g., Caribbean Sea). **c** Norwegian-Greenland Sea with
simplified surface current pattern and bottom currents in
the neighborhood of the Greenland-Iceland-Scotland Ridge.
Note special sites of bottom water overflow and associated
huge sediment drift bodies to the south of the ridge. Asymmetric sediment distribution is briefly discussed in text.
NAC North Atlantic Current; *NC* Norwegian Current; *NCC*
Norwegian Coastal Current; *EGC* East Greenland Current;
JMC Jan Mayen Current; *WSC* West Spitsbergen Current.
(Based on Stow and Holbrook 1984; Thiede et al. 1989;
Bohrmann et al. 1990)

The Caribbean Sea

The surface water circulation of the warm Caribbean Sea is characterized in the southeast by the inflow of a portion of the trade wind-driven South Equatorial Current through the Lesser Antilles Islands, and in the northwest by outflow of Caribbean water through the Straits of Florida (Gordon 1966). Surface water temperatures are high throughout the year (25 to 28 °C), and salinities are slightly above average (34 to 36.5 ‰). Such conditions, in conjunction with limited nutrient supply, enable optimal coral reef growth. The shape and extent of carbonate shelves, associated with the Caribbean islands and coral reefs, are among other factors influenced by the wind pattern and surface current system (Adey and Burke 1977; Geister 1983). The shelf off west Florida and the platforms of Florida and Campeche represent outstanding examples of a carbonate province.

In contrast to the Red Sea, the water masses of the Caribbean are distinctly stratified due to the wide connection to the Atlantic. Consequently, the Caribbean deep water has a much lower temperature and is less saline than the surface water (Gordon 1966). The bottom waters of the deep Colombian and Venezuelan basins are well oxygenated, apart from a small area in the southeast (Cariaco Trench). The deep-sea sediments of this region, however, exhibit no clear relationship with the separated inflow and outflow of surface currents. This may result from the topographic and tectonic complexity of the Caribbean Sea and its transition into the Gulf of Mexico (Nairn and Stehli 1975). Some of the basins, such as the western part of the Gulf of Mexico and the Colombian basin in the western Caribbean, receive large amounts of terrigenous material delivered by major rivers, as well as volcaniclastics from different sources within the area. These sediments build delta cones and deep-sea fans, including turbidites, and dilute the biogenic sediment production in various ways. The Neogene pelagic sediments are predominantly nannoplankton-foraminiferal oozes, reflecting the high temperature and limited nutrient content of the surface waters.

The Norwegian-Greenland Sea

This Sea represents the northern North Atlantic and forms a subpolar deep-sea basin to the north of the Greenland-Iceland-Faeroe-Scotland Ridge (Fig. 4.3c). The depositional environment of this basin is markedly asymmetric as a result of basin morphology and water circulation (Thiede et al. 1986). On the eastern side, the relatively warm surface water of the Norwegian Current (4 to 13 °C, salinity about 35 ‰), originating from the northern branch of the Gulf Stream, flows northward and transports nutrient-poor water masses and heat into the basin. The sediments below this current tend to be relatively rich in biogenic carbonate, although the sediments of the Norwegian-Greenland Sea are predominantly derived from terrestrial sources. By contrast, the outflowing East Greenland Current is cold (- 1.5 to + 2° C) and less saline (31 to 34 ‰), due to waters entering from the Arctic Ocean. It is richer in nutrients than the Norwegian Current and enables a relatively high production of siliceous phytoplankton (diatoms) during the warmer season and thus the deposition of sediments rich in biogenic silica (Bohrmann 1988). In addition, the East Greenland Current carries ice-rafted materials southward. At present, the deep waters of the entire basin have high oxygen contents.

The sediments of the Norwegian-Greenland Sea reflect the Neogene to Quaternary changes in climate and current patterns (Bohrmann 1988; Bohrmann et al. 1990). In the lower to middle Miocene, biosiliceous sediments relatively rich in warm-water species and marine organic matter prevailed. High fertility may have been caused by river-supplied nutrients (not by upwelling). With an increasingly cooling climate in the upper Miocene and Pliocene, the Norwegian Current and import of pelagic carbonate became stronger. Simultaneously, bottom water from the Norwegian Greenland Sea began to flow over the Greenland-Scotland Ridge into the North Atlantic Ocean and caused the formation of very large, elongate sediment ridges (or "sediment drifts", Fig. 4.3c; Stow and Holbrook 1984; Swift 1984).

In the western part, the Greenland Sea, mainly biosiliceous sediments were deposited, while the eastern basin, the Norwegian Sea, was characterized by both calcareous and siliceous sediment components. Ocean circulation varied from stronger to more reduced inflow of Atlantic surface water. As a result, the sediment composition fluctuated from carbonate-rich to opal-rich sections. Intensified bottom water circulation is also evident from several hiatuses found in drill cores. A significant contribution of ice-rafted material and pronounced, glacial-induced cyclicity in the sediments began not earlier than 1 to 1.2

Ma B.P. (Henrich 1990). During cold periods, the southward prograding polar front, sea ice, and lowered sea level led to a marked decrease in water exchange with the Atlantic, reduced oxygenation of the deep water masses, and sediments characterized by higher proportions of organic matter and biogenic silica.

Estuarine and Anti-Estuarine Large Ocean Basins

Finally, the terms "estuarine and anti-estuarine circulation" can also be applied to large ocean basins (Berger 1970, 1976), which may also exchange their water masses. For example, the present-day Atlantic Ocean exhibits deep-water outflow into the Pacific and thus anti-estuarine circulation. Consequenctly, the Atlantic tends to accumulate more carbonate and less silica than the Pacific, which in turn exhibits higher fertility, especially for biogenic silica, and increased carbonate dissolution. Generally, oceanic basins with estuarine circulation tend to develop stagnant bottom waters and to accumulate anaerobic sediments. Another consequence of slow inter-oceanic water exchange is the fact that, for example, Pleistocene carbonate-rich layers in deep-sea sediments are not necessarily coeval in the Atlantic and Pacific Oceans (Grötsch et al. 1991).

4.3 Sedimentary History of Some Modern Adjacent Seas

The Black Sea

The young sedimentary history of the Black Sea is particularly interesting (Fig. 4.4), but still controversial (Degens and Ross 1974; Hsü 1978; Steininger and Rögl 1984). Apart from climatic variations, there are two reasons for the special development of this basin: (1) The Black Sea is not connected with the world oceans, but with another marginal sea, the Mediterranean, and is therefore strongly affected by the history of the latter basin; and (2) the shallow, narrow opening of the Bosporus has probably existed only for short time intervals, i.e., at present, in the Late Pleistocene, and in the Miocene. During these periods, the depositional environment and sediments of the Black Sea were influenced by inflowing marine waters (Fig. 4.4e, Sections A and C).

Miocene-Pliocene Development

In the late Miocene, the Black Sea and Caspian Sea were euxinic basins with marine fauna. During the subsequent "salinity crisis" in the latest Miocene (Chapter 6.4), the eastern Mediterranean Sea was reduced to some small relic basins (Steininger and Rögl 1984) and, simultaneously, the Black Sea (as well as the Caspian Sea) were closed off from the ocean and formed lakes, which is indicated by the presence of endemic fauna and shallow-water carbonates (Fig. 4.4e).

In the long interval between the late Miocene and late Pleistocene (Fig. 4.4, Section B), the Black Sea was a fresh to slightly brackish water lake with predominantly carbonate deposition. This period is characterized by pronounced siderite formation and climatically controlled alternation of marly and calcareous beds accumulated under changing redox conditions (Fig. 4.4). During the Late Pleistocene, the depositional environment fluctuated several times from lake to adjacent sea conditions with marine influence, because the glacial sea level in the Mediterranean repeatedly rose above or dropped below the sill depth of the Bosporus. The organic matter of the last lake phase is prdominantly of terrestrial origin (Calvert and Fontugne 1987).

The Holocene Black Sea

The modern situation of the Black Sea began about 7000 years ago, when the Holocene transgression had reached the sill depth of the Bosporus and enabled the spillover of sea water into the former fresh to brackish water lake (Degens and Ross 1974; Degens and Stoffers 1980; Stanley and Blanpied 1980; Glenn and Arthur 1985). It took another 4000 to 6000 years until the present-day environment was established. During this interval, saline, anoxic bottom waters expanded and finally filled the major part of the basin. This development can be deduced from hydrographic reasoning, as well as from dated sediment cores taken from the abyssal plain and basin slope. Surface water productivity was high during this transitional interval and led to the accumulation of laminated sapropel rich in organic matter (up to about 15 % organic carbon) and a carbonate content on the order of 10 % (Calvert and Fontugne 1987). A typical deep-water sequence is shown in Fig. 4.4d, but intercalations of mud flow deposits are omitted, which may occur between the uppermost two units.

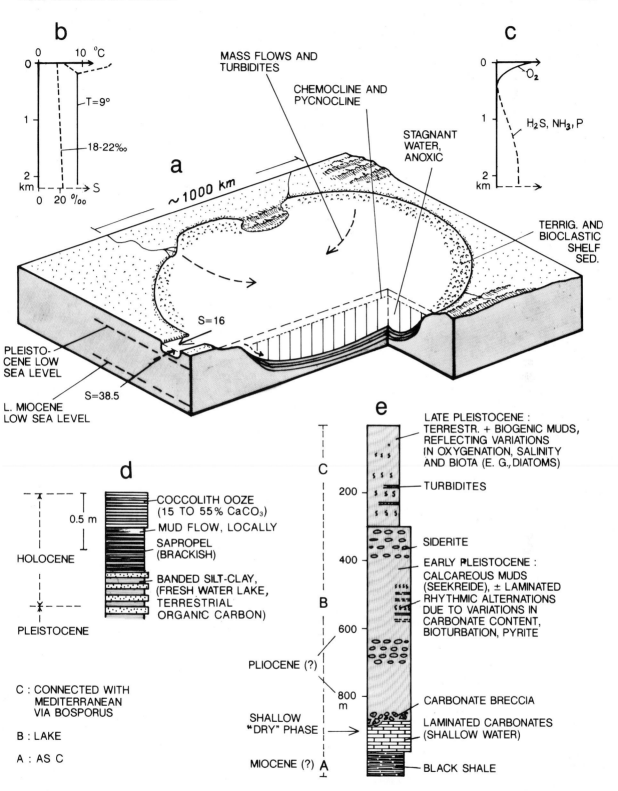

Fig. 4.4a-e. Model of a "humid", Black Sea-type basin with surface outflow of low salinity and more saline subsurface inflow. **a** Basin configuration and chemocline. **b** and **c** Relationship between salinity *(S)*, temperature *(T)*, and chemical parameters with depth below water surface.

d Pleistocene/Holocene sediments of the basin center. **e** Neogene sediments drilled at DSDP Site 380 in the Black Sea (simplified sequence; age determinations are uncertain). (Based on Degens and Ross 1974; Hsü 1978; Degens and Stoffers 1980; Calvert and Fontugne 1987)

The last one to two thousand years in Black Sea evolution are characterized by a shallow, rather stable chemocline, separating oxygenated, fairly fertile surface water from a thick deep-water mass which is anoxic and contains considerable quantities of H_2S, NH_3, and P (Fig. 4.4b and c; also see Grasshoff 1975). Most of the hydrogen sulfide is produced by bacterial sulfate reduction within the water column. After a peak in organic carbon deposition about 5000 years ago, the rate of organic carbon accumulation has tended to slow, because the rising chemocline/pycnocline has reduced the recycling of nutrients and thus surface water fertility. The uppermost, varved coccolith ooze on the abyssal plain commonly contains between 1 and 4 % of organic carbon and between 15 and 55 % of carbonate. The accumulation rate of organic carbon is low in comparison to that of modern, highly productive coastal upwelling zones (Chap. 5.3.4).

Consequently, the formerly widely held opinion that the Black Sea provides a particularly effective depositional environment for hydrocarbon generation, should be significantly modified. Such a favorable situation only existed during relatively short periods, as for example during the last 7000 years (also see Chap. 10.3.3).

Gulf of Aden and Gulf of California

Both the Gulf of Aden and the Gulf of California are rather narrow, deep oceanic basins affected by strike-slip movement (Cochran 1981; Moore and Buffington 1968; Curray, Moore et al. 1982a). Their spreading axes exhibit several offsets and thus generate a number of deep subbasins. Both gulfs have wide, deep openings to the world oceans, which allow the exchange of intermediate and bottom waters without the restraints of a barrier. For this reason, the salinity in these basins is normal marine.

The Gulf of Aden

The hemipelagic Neogene sediments of the Gulf of Aden consist predominantly of nanno oozes, with varying proportions of land-derived material and calcareous, shallow-water biogenic components. The planktonic component does not deviate markedly from that of the Indian Ocean (pelagic nanno oozes and minor proportions of oozes rich in foraminifera and radiolaria, Cronan et al. 1974).

In contrast to the Red Sea (see below), surface salinities and temperatures during the last glacial period did not differ very much from present values, but plankton productivity may have been lower. Thus, the "glacial" sediments in this region show lower carbonate contents, lower organic carbon values, and a reduced abundance of aragonitic pteropods than the present-day sediments. While the strait of Bab al Mandab is kept practically free of sediment by the outflowing bottom currents, the sedimentation rates in the Gulf of Aden beyond the influence of these bottom currents are unusually high (100 to 170 cm/ka for wet sediment, containing about 50 % carbonate; Einsele and Werner 1972).

The Gulf of California

The narrow, deep Gulf of California shows some special features, which are summarized in a simplified model (Fig. 4.5). After a proto-gulf stage on continental crust, characterized by a shallow-marine basin, the modern Gulf of California began to grow 4 Ma B.P. as a result of rifting and strike-slip movements along transform faults. Presently, the gulf is in an early drifting stage in which new oceanic crust is generated in several specific spreading troughs separated by comparatively deep sills (Fig. 4.5, cf. Chap. 12.8; Curray, Moore et al. 1982a). The long, rather deep gulf is surrounded by mountainous regions and exhibits rugged, complicated sea floor topography (van Andel 1964; Einsele and Niemitz 1982), which favors the entrapment of sediments within the gulf. Because of its wide, deep opening, the tidal range in the gulf is high at its shallowing, narrowing northern end. The climate ranges from arid in the northern region to moderately humid in the south. In spite of the high relief and several rivers entering the basin, the hemipelagic deep-water sediments and redeposited gravity mass flows (mud turbidites) are predominantly fine-grained.

The most striking hydraulic feature of this basin is the subsurface inflow of very large volumes of cold, nutrient-rich, intermediate ocean water and the outflow of warm surface water (Roden 1964; Calvert 1966). Seasonally changing offshore winds cause upwelling of nutrient-rich deeper water masses along the eastern or western margin of the gulf. These generate plankton blooms, with diatoms being the most abundant group with minor proportions of radiolaria, calcareous nannofossils and foraminifera. Most of the silica necessary for

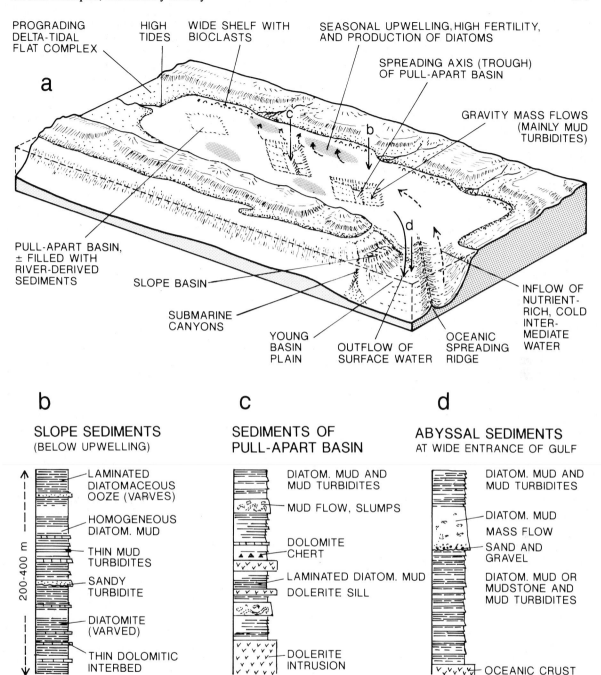

Fig. 4.5. a Gulf of California-type marine strike-slip basin, with several deep spreading centers and wide, deep opening to the ocean. Inflow of intermdiate, cold, nutrient-rich water, causing seasonal upwelling and high plankton pro
ductivity along both coasts of the gulf. **b** through **d** Simplified sediment sections at different sites within the gulf (locations marked in **a**). (Based on van Andel 1964; Calvert 1966; Curray, Moore et al. 1982a)

diatom skeletons is derived from inflowing ocean water. Partial mineralization of the plankton leads to an oxygen minimum zone at intermediate depths along the slopes, as well as to annual varves, consisting mainly of diatoms (seasonal high plankton production) and fine-grained detrital material swept into the gulf by seasonal rains.

The exceptionally high primary producivity of these upwelling systems is also indicated by the enrichment of trace elements, such as P, S, Mo, Se, and Cd, in comparison to average shales (Brumsack 1986a). Calcareous skeletons are partially dissolved and normally constitute only a few percent of the sediment. During diagenesis, thin, interbedded dolomites can form.

The sedimentation rate on the slopes below the zones of coastal upwelling is very high (about 400 to 600 m/Ma of porous sediment, corresponding to approximately 100 to 150 m/Ma of dry, solid rock).

A large proportion of the slope sediments is redeposited and accumulates as mud turbidites in the deepest troughs of the subbasins. The mean sedimentation rate of the 4 Ma old modern gulf is about 50 m/Ma (dry, solid material). This value does not represent the present sedimentation rate, but is calculated from the total sediment volume, taking into account the growth of the gulf (cf. Chap. 11.3: extensional basin filling). Biogenic silica has contributed about 15 % of the total sediment volume. Locally, the sedimentation rates have varied from practically zero on bathymetric highs to 1000 m/Ma in some of the spreading troughs (Einsele and Niemitz 1982).

The *organic matter* in the basin and slope sediments is predominantly of marine origin, derived mainly from diatomaceous protoplasm. Its concentration is highest (2 to 4 % organic carbon) in the sediments accumulated below the oxygen minimum zones along the basin slopes. Sediments in the deep basins contain less organic carbon (mostly 1 to 2 %), because the bottom water characteristics fluctuate between anaerobic and slightly aerobic. Part of the organic matter transported by mud turbidites into the deep subbasins, however, is land-derived. Varved sections are richer in organic carbon than homogeneous layers. On the average, the sediments of the

Gulf of California contain about 2 % of organic carbon and are therefore potentially good sources of oil and gas (Chap. 14.1). Despite high heat flow, the organic matter is thermally immature (see Chap. 14.2), except for sediments affected by the intrusion of basaltic sills and hydrothermal processes in the actively spreading oceanic subbasin centers. Here, thermogenic hydrocarbons occur, and diatomaceous oozes and terrigenous turbidites may be thermally altered to an epidote-zoisite facies.

Quaternary climatic changes and sea level fluctuations obviously exerted only a limited influence on the sediments of the Gulf of California. Drilling cores display thick alternations between laminated and homogeneous, bioturbated sediments. It is possible that the latter sections correspond with low sea level stands.

The Red Sea

The Red Sea is a young, narrow ocean basin in transitional from rifting to drifting mode (Fig. 4.6a, cf. Chaps. 12.1 and 12.2). The northern part of this basin is still in the stage of rifting, which generated a wide "main trough" on top of thinned, intensively faulted continental crust (Cochran 1983a; Le Pichon and Cochran 1988). This period of extension and dike injection lasted for about 20 Ma (since the early Miocene, Fig. 4.6b) and affected an area which is 100 to 160 km wide. The structural basin evolution is characterized by an early stage of tilted blocks caused by substantial extension, which was followed by the formation of horsts and grabens, indicating rapid subsidence (Montenat et al. 1988). After evaporite deposition (see below), extension continued, accompanied by halokinetic structures. In the middle and more southern part of the Red Sea, the main trough is bisected by a deep axial trough, formed by sea floor spreading during the last 4 Ma. To the north and south of this region, the axial trough is discontinuous and represented by a series of deeps, alternating with shallower intertrough zones.

Fig. 4.6. a Sedimentary history of Red Sea-type, "arid" basin in a transitional stage from rifting (main trough) to drifting (axial trough). Early rifting is accompanied by coastal and normal to hypersaline, shallow-marine sediments, which are followed by thick sabkha salts and deeper water evaporites. Post-evaporite sequences in shallow and deep water environments reflect normal to slightly hyper- saline conditions. **b** Simplified cross section of northern Red Sea. (After Montenat et al. 1988). **c** Idealized section of drill hole in axial trough (based on Stoffers and Ross 1974). **d** Glacial/interglacial variations in water circulation of the Red Sea and resulting pelagic sediments. (Based on Locke and Thunell 1988)

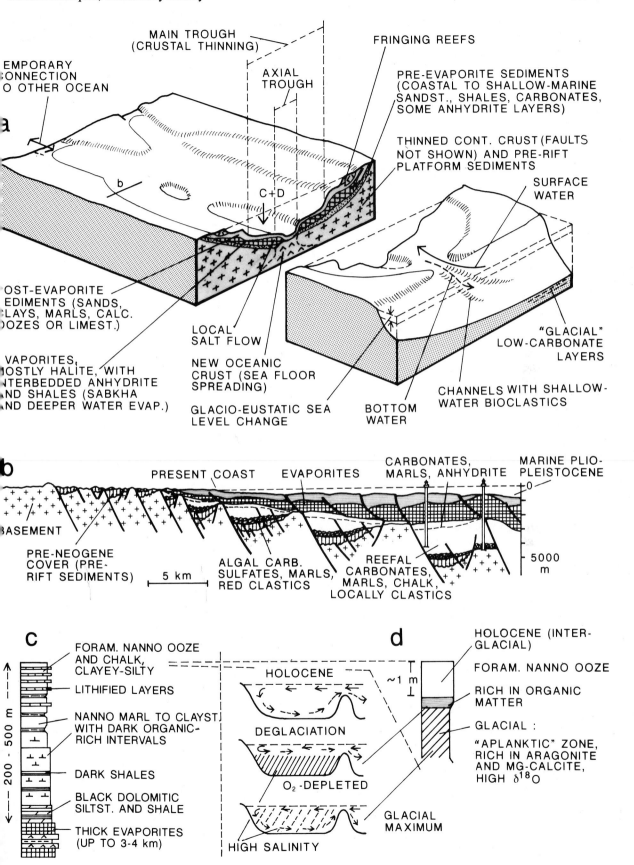

a

TEMPORARY
CONNECTION
TO OTHER OCEAN

MAIN TROUGH
(CRUSTAL THINNING)

AXIAL
TROUGH

FRINGING REEFS

PRE-EVAPORITE SEDIMENTS
(COASTAL TO SHALLOW-MARINE
SANDST., SHALES, CARBONATES,
SOME ANHYDRITE LAYERS)

THINNED CONT. CRUST (FAULTS
NOT SHOWN) AND PRE-RIFT
PLATFORM SEDIMENTS

SURFACE
WATER

C+D

POST-EVAPORITE
SEDIMENTS (SANDS,
CLAYS, MARLS, CALC.
OOZES OR LIMEST.)

EVAPORITES,
MOSTLY HALITE, WITH
INTERBEDDED ANHYDRITE
AND SHALES (SABKHA
AND DEEPER WATER EVAP.)

LOCAL
SALT FLOW

NEW OCEANIC
CRUST (SEA FLOOR
SPREADING)

GLACIO-EUSTATIC SEA
LEVEL CHANGE

BOTTOM
WATER

CHANNELS WITH SHALLOW-
WATER BIOCLASTICS

"GLACIAL"
LOW-CARBONATE
LAYERS

b

PRESENT COAST   EVAPORITES

CARBONATES,
MARLS, ANHYDRITE

MARINE PLIO-
PLEISTOCENE

BASEMENT

PRE-NEOGENE
COVER (PRE-
RIFT SEDIMENTS)

5 km

ALGAL CARB.
SULFATES, MARLS,
RED CLASTICS

REEFAL
CARBONATES,
MARLS, CHALK,
LOCALLY CLASTICS

5000
m

c

FORAM. NANNO OOZE
AND CHALK,
CLAYEY-SILTY

LITHIFIED LAYERS

NANNO MARL TO CLAYST.
WITH DARK ORGANIC-
RICH INTERVALS

DARK SHALES

BLACK DOLOMITIC
SILTST. AND SHALE

THICK EVAPORITES
(UP TO 3-4 km)

200 - 500 m

d

HOLOCENE

DEGLACIATION

O_2-DEPLETED

GLACIAL
MAXIMUM

HIGH SALINITY

HOLOCENE (INTER-
GLACIAL)

FORAM. NANNO OOZE

RICH IN ORGANIC
MATTER

GLACIAL :

"APLANKTIC" ZONE,
RICH IN ARAGONITE
AND MG-CALCITE,
HIGH $\delta^{18}O$

~1 m

Rift-Stage Sediments with Evaporites

During early rifting, continental, coastal, and shallow-marine sediments of mostly normal salinity were deposited (Miller and Barakat 1988; Montenat et al. 1988). In addition, some evaporites and dolomites formed during short intervals.

The subsequent period (middle to late Miocene) is characterized by thick evaporites (also see Chap. 6.4.1), which accumulated partially in shallow-water and coastal sabkha environments (Stoffers and Kühn 1974; Miller and Barakat 1988), but probably also in the deeper water of the rapidly subsiding basin. The evaporites reach thicknesses of more than 1000 m in the northern part and up to 3000 or 4000 m in the central and more southern parts of the basin. All these evaporites probably formed prior to the generation of the axial trough, i.e., on thinned continental crust, although evaporites are also found in the axial trough. This indicates subsequent salt flow from the flanks of the main trough into the axial trough (Girdler and Whitmarsh 1974). In the central Red Sea (DSDP Site 227), the drilled upper section of the evaporites contains, besides halite, several intervals of anhydrite, dark shales, and dolomites, which clearly indicate that evaporite deposition occurred discontinuously. During its "salinity crisis", the Red Sea was probably connected with the Mediterranean Sea and closed at its southeastern end. Later, the strait of Bab al Mandab opened and connected the Red Sea with the Gulf of Aden and the Indian Ocean.

Post-Evaporite Sediments

The post-evaporite Pliocene and Quaternary sediments recovered during the Deep Sea Drilling Project from the central graben of the Red Sea also record variable depositional conditions (Fig. 4.6c; Stoffers and Ross 1974). The first strata on top of the evaporites are dark grey, dolomitic silty claystones and early diagenetic dolomites. Intervals with high organic carbon and pyrite contents point to periods of restricted water circulation in the basin (cf. Fig. 4.6d). Further development is characterized by hemipelagic nanno oozes, marls and marly claystones, which again exhibit dark, organic-rich interbeds. Locally, volcanic ashes may be intercalated. The youngest sediments consist of foraminiferal nanno ooze and chalk, which are more or less diluted by clay and silt from terrestrial sources. This occurred particularly during periods in which the climate was wetter than today.

In the main trough and axial trough, the sedimentation rate of post-evaporite deposits was generally 30 to 60 m/Ma, but much higher and lower values occurred locally. In the axial trough of the southern Red Sea, young wet sediment accumulated at a rate of about 25 cm/ka (250 m/Ma, Einsele and Werner 1972).

A characteristic feature of the young, unconsolidated (or only slightly consolidated) Red Sea sediments is the occurrence of many lithified layers, containing high proportions of chemically precipitated aragonite and Mg-calcite. These layers obviously formed during times of glacial sea level lowstand and increased salinity (Reiss and Hottinger 1984). Further evidence for this assumption is the faunal and isotopic composition of Holocene and late Pleistocene sediments in the central and southern Red Sea (Reiss et al. 1980; Locke and Thunell 1988; Thunell et al. 1988). Glacial surface and bottom water salinities were significantly higher than at present, possibly by more than 10 ‰. Water exchange between the Gulf of Aden and the Red Sea was markedly reduced and the bottom waters were low in oxygen. At the same time the climate was extremely arid.

During such periods, fertility in the photic zone rose in the northern Red Sea (Gulf of Aqaba), whereas planktonic carbonate production was limited in the southern regions. The resulting "aplanktic" layers also exhibit high $\delta^{18}O$ values (Fig. 4.6d). During deglaciation, i.e., about 11 000 to 8000 years B.P., the sea level began to rise, but the bottom water in the Red Sea was still hypersaline and rather stagnant beneath a surface water zone of lower salinity and increasing exchange with the open sea. As a result, the sediments representing the transition from the last glacial maximum (about 18 000 years B.P.) to the Holocene are relatively rich in organic matter.

The Persian Gulf

Since the Upper Cretaceous, the Persian Gulf has been a shrunken foreland basin in front of the Zagros fold belt (Fig. 4.7), and its deepest portion (80 to 100 m) is a structural feature running parallel to this mountain range in the northeast (Purser 1973, Murris 1985). The sediments in front of the fold belt are mainly siliciclastic sandstones, silts, and shales, but farther to the southwest an open-marine basin

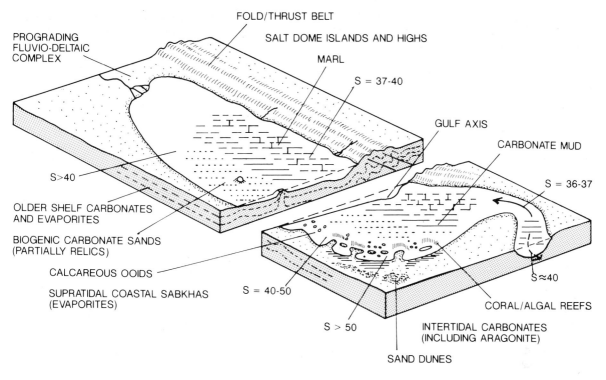

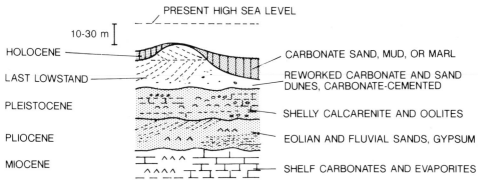

Fig. 4.7. Shallow, "arid" adjacent sea (Persian Gulf-type), tectonic setting, salinity (S in ‰), and distribution of most important sedimentary facies. (Based on Seibold 1970; Sarnthein 1972; Purser 1973; Stoffers and Ross 1979; and others). Miocene to Holocene sediment sequence mainly characterizes shallow regions of gulf opposite to the mountain range. (After Kassler 1973)

has been maintained which in turn has been bordered by a shelf region characterized by carbonate and evaporite deposition. In the central basin, marls and calcareous shales have accumulated, because carbonate production has been diluted by suspended river load carried into the basin from the east and northwest. In the northwest, the Tigris-Euphrates River system has built a fluvio-deltaic complex out into the gulf.

Oligocene and Miocene sandstones and limestones, which are sealed by evaporites, are

the principal reservoirs of several giant oil fields. They store hydrocarbons derived from older source rocks (Chap. 14.3). The present configuration and hydrographic situation of the gulf is unfavorable for the accumulation of significant amounts of organic matter.

The *late Pliocene and Quaternary* history of the gulf was strongly affected by the glacial regressions and transgressions (Stoffers and Ross 1979). During low sea level stands, the basin became nearly completely dry. Emerged shallow-marine skeletal carbonates, tidal carbonates, and oolites were cemented and form-ed lithified layers, or they were reworked. Fluvial sediments and eolian sand dunes migrated into and across the basin (Fig. 4.7b). Consequently, the Holocene marine sediments rest on terrestrial deposits in large areas of the gulf (Kassler 1973).

The *Holocene* marine deposition began at about 12 000 years B.P. with carbonate sands and light muds rich in aragonite needles and pellets (carbonate content ≥ 70 %). Under the present conditions, silty marls (carbonate content about 55 %) accumulate in wide areas of the gulf. They are rich in biogenic consti-tuents and contain abundant high magnesian calcite. Thicknesses and sedimentation rates of the Holocene carbonate sands and calcareous muds vary according to the irregular pre-Holocene topography. Average sedimentation rates are on the order of 10 to 20 cm/ka; in depressions, sediments accumulated at a rate up to about 300 cm/ka, whereas topographic highs may be almost free of Holocene sedi-ments.

5 Oceanic Sediments

5.1 General Aspects

Introduction

The ocean basins are the ultimate sink for all
the material transported by rivers or blown by
winds into the sea. In addition, they produce
large quantitities of autochthonous biogenic
material. In their areal extent, the sediments
of the present-day deep ocean basins surpass
all the other sedimentary environments by far.
Even measured by volume and for a certain
time slice, deep-sea sediments have presumably predominated over other sediment types,
for example shelf deposits, for the last millions of years. However, in the ancient record,
a relatively large proportion of former deep-
sea sediments is missing. These materials were
subducted and partially transformed into
metamorphic rocks at convergent plate margins, incorporated into orogenic belts, and
eroded as a result of uplift. Nevertheless, even
in ancient rock sequences on the continents,
nonmetamorphic deep-sea sediments play a
great part, and their identification and interpretation are an important objective in basin
studies and paleogeographic reconstructions.

As a result of intensive research during the
last few decades, including the Deep Sea Drilling Project, our knowledge about deep-sea
sediments in the present and ancient oceans
has increased enormously. Much has been
written on the distribution of marine sediments in space and time in relation to the
changing configuration of the ocean basins,
varying climate, and paleoceanographic conditions (e.g., Kuenen 1950; Emery 1960; Shepard
and Dill 1966; Shepard 1973; Schopf 1980;
Berger 1981; Emiliani 1981; Kennett 1982;
Seibold and Berger 1982; Pickering et al. 1986;
etc.). This brief summary emphasizes some
general principles for the generation and distribution of deep-sea sediments and includes
an overview of the most important processes
controlling oceanic circulation (Chap. 5.2).

Without an elementary knowledge of these processes, many sedimentary features and their regional variations cannot be properly understood. Special case studies from both present-day and ancient ocean basins are mentioned in the references.

Sediment Sources

Marine sediments are derived from several sources. These sources contribute, on a global scale, approximately the following proportions to the *present-day* total sediment accumulation in the oceans, including shelf environments (Gorsline 1984):

1. *Allochthonous sediments* from the continents (also see Chaps. 2 and 9):
   - Suspended load of rivers (≥ 90 %).
   - Eolian dust (≤ 1%).
   - Volcaniclastic material (≤ 1 %).
2. *Autochthonous biogenic sediments* (around 10 %).

These values may considerably underestimate the importance of both eolian dust and fine grained volcanic ash, which are easily overlooked in marine muds (cf. Chaps. 2.3 and 2.4). If we exclude the enormous sediment accumulation in shelf seas, which catch the mayor part of the incoming river load, and only consider deep-sea environments, the minor contributors can reach much higher percentages than those listed above. In areas far away from the continents, the allochthonous portion of pelagic sediments probably consists mainly of wind-borne material and oceanic volcanic input. Chemical precipitation of salts in deep oceanic basins does not occur today, though it was significant during some periods in the past in deep marginal basins (Chap. 6.4).

Transport Mechanisms and Sediment Types

In addition to their source, deep-sea sediments also can be classified in terms of transport mechanisms and deposition:

- Slow pelagic settling (Chap. 5.3.1).
- Redeposition of shallow-water and slope sediments by gravity mass movements (Chap. 5.4).
- Redeposition of deep-sea sediments subsequent to winnowing and current-reworking (Chap. 5.5).

Sediments primarily formed by *slow pelagic settling* are subdivided into two groups:

1. *Hemipelagic sediments,* deposited in the neighborhood of continents, which contain a large proportion of terrigenous silt and clay. At least 25 % of their grain size fraction ≥ 5 µm is terrigenous and volcanogenic in origin, or it is derived from shallow-marine sediments (Berger 1974, Jenkyns 1986). The median grain size of hemipelagic sediments is ≥ 5 µm. This group may be further subdivided into:

   - Terrigenous muds and mudstones ($CaCO_3$ content ≤ 30 %).
   - Volcanogenic muds (predominantly volcanic ash, $CaCO_3$ content ≤ 30 %).
   - Calcareous muds and marlstones ($CaCO_3$ content ≥ 30 %).

2. *Pelagic sediments,* which are usually deposited at greater distances from the continents, and contain less than 25 % terrigenous material of the fraction ≥ 5 µm. Their median grain size is ≤ 5 µm, apart from authigenic minerals and skeletons of microfossils. This group is subdivided into:

   - Pelagic (silty) clays and claystones containing ≤ 30 % $CaCO_3$ and biogenic SiO_2 (calcareous clays or siliceous clays and claystones).
   - Calcareous oozes, marls and marlstones, chalk and pelagic limestones, $CaCO_3$ ≥ 30 %.
   - Siliceous oozes, silicified claystones, porcellanite, diatomites, radiolarites, chert, SiO_2 ≥ 30 %.

Within these subgroups, more special types of sediments can be defined. *Black shales* containing relatively high amounts of organic matter may be either rich or poor in carbonate or silt-size terrigenous material (cf. Chap. 10.3.3). Therefore they cannot be clearly attached to one of the groups or subgroups listed above. Hemipelagic and pelagic sediments are frequently redeposited and partially mixed with shallow-water material by gravity mass movements, and they may be winnowed and reworked by deep bottom currents (Chaps. 5.4 and 5.5).

Specific Behavior of Muddy Sediments

Erosion, transport, and deposition of fine-grained sediments do not follow the comparatively simple rules found for cohesionless materials such as sand and gravel. Clay and fine silt-size particles stick together due to van der Waals forces or electrostatic attractions which are large relative to the small weight of the single particles. In addition, they may be bonded by organic secretions of bacteria, algae, diatoms, and other micro-organisms living on top of or within the sediment (McCave 1984). Fecal pellets, consisting mainly of inorganic, fine-grained particles bound together by organic compounds, are abundant on widely extended areas of the sea floor. Furthermore, the same assemblage of particles may vary in water content or degree of compaction. The combined result of all these processes is described and measured as *cohesion*. Consequently, the same general type of mud may display quite different values of cohesion and thus require lower or higher critical current stresses for its erosion, which are difficult to predict. It is, however, evident that many marine muds need higher current velocities to be eroded than clean coarse silts or sands, even if their grain fabric is disturbed by burrowing organisms or other discontinuities. Erosion usually does not occur particle by particle, but in the form of variously sized aggregates.

Similarly, transport and deposition of fine-grained sediment depends to a large extent on the size and settling velocity of *aggregates* (Gibbs 1985; Kineke and Sternberg 1989). Small particles and aggregates colliding in the water column can adhere to one another and form larger aggregates. Fully flocculated suspensions settle much faster than their individual particles do in a disaggregated state. In most marine environments, such aggregates have diameters greater than 1 mm, but under higher fluid shear close to the sea bed they break up into sizes ≤ 100 μm unless they have strong biological binding as, for example, fecal pellets. In this broken up state the aggregates are deposited on smooth or rough beds, where some sorting between larger, more stable aggregates and smaller aggregates may take place.

5.2 Water Circulation in the Oceans

Salinity and Density of Ocean Water

Circulation of water masses in the oceans is the result of an interplay between the atmosphere and the oceans. Both atmospheric and oceanic circulation are driven by the energy provided by the sun's radiation. The nature of these two circulation systems is, however, quite different. Atmospheric circulation, which affects a medium of very low density, is extremely fast and complex, and is not restricted to certain basins with strict boundaries like the oceans. Circulation in the oceans is much slower, which is chiefly caused by the higher density of water in comparison to air. Thus, much more energy is needed to move thick water masses than to drive the atmosphere.

Besides absolute density, *variations in density* also play an important part in controlling circulation systems (e.g., Pickard and Emery 1982). The density variations of sea water are controlled by *salinity* (S), water temperature (T), and to a minor extent by pressure. The average salinity of present-day ocean water is about 35 g of salts per kg of sea water (for exact determinations all carbonate has to be converted to oxide, while bromine and iodine are replaced by chlorine). Salinity is usually written as, for example, S = 35 ‰ or 35 ppt (parts per thousand). Whereas the total concentration of dissolved salts, i.e., the salinity, varies from place to place and at different depths, the ratios of the main components of sea water are found to be almost constant in all parts of the oceans (see below). This fact is explained by the permanent mixing of water masses not only within certain oceanic basins, but also by water exchange from one basin to another. On the other hand, this homogenization of sea water is counteracted by processes continually diluting or concentrating the salt content of sea water in specific areas, for example by entering fresh water from the continents, by forming or melting sea ice, or by high evaporation in warm, shallow seas.

In the large oceans, salinity only varies between about 33 and 37 ‰ (Fig. 5.1 a and b). Therefore its influence on the density of sea water is generally lower than that of *temperature variations* (Fig. 5.2a). The surface temperatures of present-day ocean waters vary between about -2 and +30 °C (Fig. 5.1a and c). In low latitude regions the difference in temperature of surface water and deep water (>1000 m depth) is greater than 20 °C. Provided salinity and pressure are kept constant, a substantial increase in temperature, ΔT, leads to a significant decrease in the density of sea water (e.g., for ΔT = 30 °C and S = 35 ‰, the density changes from about 1.028 to 1.022 g/cm³, Fig. 5.2a).

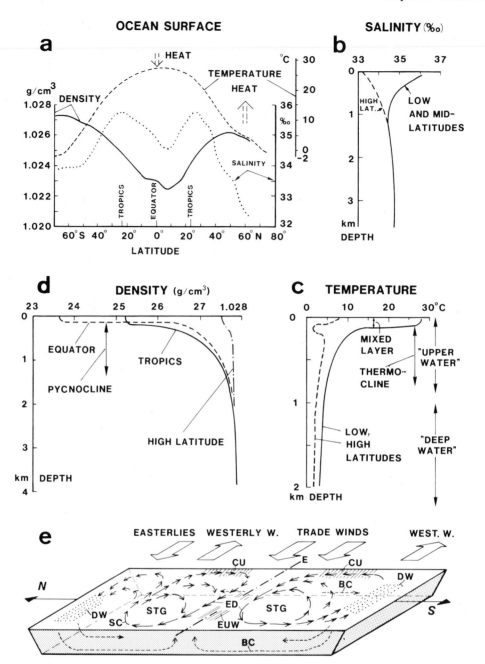

Fig. 5.1. a Variation in surface temperature, salinity, and average density relative to latitude. **b** Salinity-depth profiles in the Pacific. **c** Temperature-depth profiles, thermocline, and zones of "upper water" and "deep water". **d** Typical density-depth profiles for low and high latitudes. All data in a-d from present-day oceans. (After Pickard and Emery 1982). **e** Rectangular ocean with idealized, simplified circulation system as a result of wind forces and thermohaline effects (based on different sources). Note the distinct asymmetry of the gyres (*STG* subtropical gyre; *SPG* subpolar gyre), the intensified surface *(SC)* and bottom *(BC)* currents along the western margin of the basin (longer arrows), and the zones of coastal upwelling *(CUW)* along the eastern margin. *ED* equatorial divergence associated with equatorial upwelling *(EUW); STC* and *PC* subtropical and polar convergence giving rise to downwelling *(DW)*

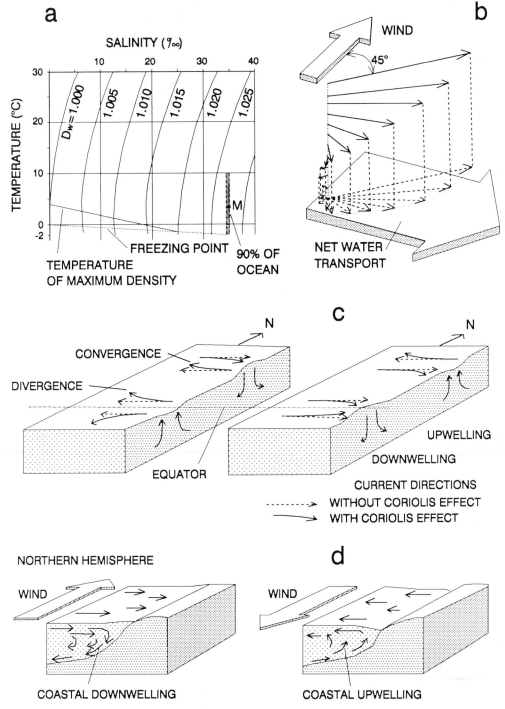

a SALINITY (‰)

TEMPERATURE (°C)

$D_w = 1.000$ 1.005 1.010 1.015 1.020 1.025

FREEZING POINT

TEMPERATURE OF MAXIMUM DENSITY

90% OF OCEAN

M

b WIND

45°

NET WATER TRANSPORT

c N   N

CONVERGENCE

DIVERGENCE

EQUATOR

UPWELLING

DOWNWELLING

CURRENT DIRECTIONS

------> WITHOUT CORIOLIS EFFECT

———> WITH CORIOLIS EFFECT

NORTHERN HEMISPHERE

WIND

COASTAL DOWNWELLING

d WIND

COASTAL UPWELLING

Fig. 5.2. a Density of water *(Dw)* as a function of salinity and temperature (at atmospheric pressure). Note changing temperatures of maximum density and freezing point. *M* mean value for present world ocean; *hatched field* 90 % of ocean. (After Pickard & Emery 1982). **b** Deflection of wind-driven currents due to the Coriolis effect in the nor thern hemisphere (Eckman spiral). (Stowe 1979). **c** Oceanic convergence and divergence causing downwelling and upwelling as a result of the Coriolis forces in the northern hemisphere. (After Pickard and Emery 1982). **d** Coastal downwelling and upwelling caused by winds along the coast, northern hemisphere. (After Stowe 1979)

Although the effect of pressure changes on the density of sea water may be relatively high, for example between surface waters and deep ocean waters, its influence on water circulation is usually neglected. The reason for this is that water masses at the same depths are subject to the same pressure. Thus no horizontal density gradient due to pressure is generated which may initiate or enhance lateral flow. Therefore, the pressure component of sea-water density will not be further discussed here. The effects of salinity and temperature on the density of sea water are generally depicted in Fig. 5.2a. At high temperatures, the density change with ΔT becomes more pronounced than at lower temperatures. The change in density with ΔS is about the same at all temperatures. It should, however, be noted again that about 90 % of the ocean water covers only a very narrow range of salinities; the temperature range also remains rather limited if surface waters at low latitudes are excluded.

Furthermore, it is of interest that both the temperature of maximum density as well as the freezing point of sea water drop with increasing salinity. Low saline water sinks before freezing, because its maximum density is already reached at temperatures above the freezing point. Consequently, the total water column is overturned until it all reaches the temperature of maximum density. On further cooling, the surface water becomes lighter and finally freezes from the surface downward. At salinities above 25 ‰, the maximum density is reached at approximately -2 °C at the freezing point. Thus the overturning of the water column continues up to this temperature. As a result freezing is delayed.

Density Stratification and the Pycnocline

Everywhere in the ocean, the most dense water tends to sink to the bottom and the least dense water comes up to the surface. Particularly in small, closed basins and basins with restricted water circulation, but also in equatorial and tropical regions of the present-day oceans, this process often leads to a rather stable *density-stratified water mass*. Characteristically, water density does not increase uniformly with depth. Usually there is a relatively shallow, wave-affected water layer of lower density underlain by a thick denser water mass. The transitional layer between these zones is called the *pycnocline;* it may frequently also act as a *thermocline* and/or

halocline (boundary between water masses of different salinity). Here, density increases rapidly with depth (Fig. 5.1d). As long as the surface water remains less dense than the deep water body, for example in warm climates, and the bottom water cannot be replaced, the density stratification and the pycnocline persist for long time periods.

This can be observed in many lakes in warm, low-latitude regions (Chap. 2.5), in the Black Sea, or in the Eastern Mediterranean. In all these cases, the surface water may be characterized by turbulence and circulation, but the bottom waters are little affected by atmospheric forces.

The *pycnocline* acts as an effective barrier to water movement, either downward or upward. It is often associated with a thermocline, separating an upper warmer water mass from a deeper cooler one (Fig. 5.1c), and a *chemocline* indicating a major change in water chemistry. The stability of such an internal boundary layer is, however, essentially based on the rapid change in water density with depth.

In mid and high-latitude regions, where surface waters undergo cooling during winter time or are permanently cool, density stratification tends to be less stable or absent (Fig. 5.1d). It can be only maintained if the surface water has a considerably lower salinity than the bottom water and thus always remains less dense. Such special conditions are found, for example, in some Norwegian fjords or in parts of the Baltic Sea.

Oceanic Water Circulation

Circulation in the oceans is essentially driven by two mechanisms: Variations in water density and winds. Since density is controlled by water temperature and salinity, the first mechanism is called *thermohaline circulation*. It enables deep water circulation. The second component is *wind-driven near-surface circulation*.

Thermohaline circulation

Thermohaline circulation is essentially controlled by the temperature gradient between the poles and the equator. The horizontal and vertical density distribution in the modern oceans (Fig. 5.1a and d) deviates from the ideal, horizontally stratified water masses consisting of two layers of constant temperature and salinity indicated above. The density of surface water changes in a north-south section through the present-day oceans as a function of water temperature and salinity (Fig. 5.1). Although salinity decreases from low latitudes toward the poles, density is greatest at high latitudes due to a lower temperature and, in places, by the ejecting of salt when ice freezes. Furthermore, density-depth profiles (as well as salinity and temperature) near the equator and tropical seas considerably differ from those at high latitudes (Fig. 5.1b, c, and d). Although both locations show the expected increase in density with depth, low latitudes are characterized by a big jump in density creating a two-layer ocean, whereas at high latitudes, the vertical density gradient is small.

Ocean water displaying such a density distribution cannot be in a stable condition; it circulates. The relatively dense surface water at high latitudes sinks and is replaced by surface water from lower latitudes. The sinking dense water approaches the sea floor, and then flows as bottom currents towards the equator and then somewhere upwards. Such thermohaline circulation is kept in operation as long as the nonequilibrium conditions can be maintained. This is accomplished by continuous heat transfer from the warmer atmosphere to the low-latitude surface waters, or by heat loss from the oceans to the cooler atmosphere at high latitudes. Therefore, small ocean basins and long, narrow, east-west extended ocean regions which belong entirely to the same climatic zone, are but little affected by thermohaline circulation. If there was a period of more or less uniform climate on earth, even large oceans extended widely from north to south could hardly develop pronounced thermohaline circulation.

Wind-Driven Circulation

Wind blowing over the surface of a standing water body causes both waves and motion of the surface water. In the open ocean, waves transport energy but not water masses (see Chap. 3.1). Hence in this case, we must deal only with the second consequence of atmospheric circulation: *wind-generated currents*. On a gross scale, wind-driven currents reflect the global wind pattern and only move surface waters. However, surface currents are modified by the ocean basin boundaries and by *Coriolis forces*. The combined effect of these factors is shown in Fig. 5.1e for a hypothetical rectangular ocean resembling the present-day Atlantic Ocean. Trade wind belts to

the north and south of the equator give rise to a north equatorial and south equatorial current flowing westward. Near the coast they turn either to the north or to the south until they reach the west wind belts in the mid-latitude regions. Then the surface waters flow eastward, until they are divided into two branches, a northeast current and a southeast current.

In the northern hemisphere, the southeast current turns south along the eastern continental margin and finally feeds into the north equatorial current, thus forming a large gyre with clockwise water circulation. In the southern hemisphere, similarly a counter-clockwise gyre is completed. Those parts of the eastward-directed mid-latitude currents which turn poleward can again form gyres, a counter-clockwise gyre in the northern hemisphere and a clockwise gyre in the southern hemisphere. Another conspicuous phenomenon, related to the rotation of the Earth, is that the currents become narrower and faster on the west side of an ocean, where the water is piled up somewhat before it flows back in other directions, both in the northern and southern hemisphere. Between the westward-flowing equatorial currents, an eastward-directed counter-current may develop.

This idealized current pattern in a rectangular ocean is, of course, substantially modified by deviating atmospheric circulation and, in particular, by the irregular topography of a single or several connected ocean basins. The present South Atlantic, for example, is openly connected with the other large oceans. Therefore, water masses flowing southward feed into an Antarctic circumpolar current system, where large volumes of water sink due to cooling and initiate thermohaline bottom circulation flowing northward (Antarctic bottom water). Wind-driven surface currents as well as currents in deeper water are also affected by *Coriolis forces*. Consequently, surface current directions deviate from those of the wind (Fig. 5.2b). In the northern hemisphere, the Coriolis force deflects wind-driven currents to the right, in the southern hemisphere to the left. The clockwise gyre north of the equator and the counter-clockwise gyre south of the equator reflect this mechanism (Fig. 5.1e). At the surface of a deep water body, the angle between the directions of wind and current may be up to 45° (Fig. 5.2b), in shallow water less. Below the surface, each layer of water is dragged along by the layer above it. Hence, due to the Coriolis force, each sublayer flows to the right of the overlying layer. Since the Coriolis deflection increases with slower current speeds, the slower moving deeper waters turn even more to the right than the surface currents. The resulting distribution of current directions at different depths is called the *Eckman spiral*. The direction of net water transport over all depths is 90° to the right of the wind direction. In shallow waters the deflection of net water transport is less than 90°.

Combined Effects of Thermohaline and Wind-Driven Circulation

The phenomenon of the Eckman spiral clearly demonstrates that wind-driven surface currents and thermohaline circulation prevailing in deeper waters cannot act independently. Ocean currents are the result of the combined effects of both factors. Wind primarily causes horizontal motion, whereas the thermohaline effects primarily generate motion with a vertical component. Combined horizontal surface currents in places can accelerate vertical motion (downwelling and upwelling) and thus intensify deep bottom currents. On the other hand, there are large areas in the ocean with little exchange between the upper water layer (300 to 1000 m deep including the thermocline) and deep water, whereas regions with substantial exchange between these two water masses are limited.

In the open ocean, currents generated by easterly winds near the equator tend, due to the Coriolis effect, to diverge to the north or to the south of the equator (Fig. 5.2c). The resulting *equatorial divergence* of surface waters leads to *upwelling* of deeper water along this zone (Fig. 5.1e). The same phenomenon is created at the boundary of oppositely directed currents if, for example, in the northern hemisphere the more northerly current flows westward (Fig. 5.2c).

In contrast, converging surface currents tend to pile up water and thus generate a downward motion *(downwelling)*. The most prominent example in the present-day oceans is the subtropical convergence and the Antarctic polar front in the southern hemisphere, where eastward-directed currents are forced to converge. Here, as well as in northern polar regions, downwelling is intensified by the cooling of surface waters. Downward moving water reaches, according to its density, intermediate depths or the sea bottom and feeds into relatively slow subsurface or bottom currents. These currents are also subjected to the Coriolis effect and thus tend to turn to the right in the northern hemisphere and to the left in the southern hemisphere.

Of particular interest to sedimentologists are *bottom currents,* because they may be able, at least locally, to transport sedimentary particles and even to rework silty to fine-sandy sediments. Equator-directed bottom currents derived from downwelling, high latitude waters travel preferentially on the western boundary of an ocean, for example at the base of the continental margin (Fig. 5.1e). They more or less follow the contours of the ocean basin and are therefore referred to as *contour currents.* Such currents may also occur on the eastern margin, but they usually flow poleward.

Upwelling and downwelling also occur in coastal waters. Onshore surface currents tend to stack up water along the coast, which will consequently begin to sink and replace deeper water *(coastal downwelling)*. Offshore currents generate *coastal upwelling* of intermediate or deep water. Due to the effect of the Eckman spiral, these processes are initiated or intensified by winds blowing parallel to the coast line, for example in front of a coastal mountain range (Fig. 5.2d). These rules allow one to predict from the wind direction and the position in the northern or southern hemisphere whether or not coastal upwelling or downwelling exists.

If an ocean extends into *polar regions* and is partly covered by sea ice during a glacial period, its circulation pattern will be modified. The high-latitude area of downwelling, high-density water will move toward the equator, but the comparatively stable water stratification at low latitudes will persist. The formation of large quantities of sea ice may lead to an increase in ocean salinity by 1 to 2 ‰ and thus stabilize the deep water mass. In addition, temporarily melting ice can locally dilute surface water and further promote density stratification. As a result of both processes, thermohaline circulation may slow somewhat and transport of oxygen into deep water may be reduced.

Current Speeds, Transport of Water Masses and Nutrients

Surface current speeds can locally reach values as high as 150 to 250 cm/s (Gulf-Stream, Kuroshio off Japan), but are often considerably lower. The present-day Antarctic circumpolar current has a speed of 4 to 15 cm/s. Deep bottom currents can locally develop about the same velocity (contour currents up to 15 to 20 cm/s), but are generally much slower (1 to 3 cm/s).

Because oceanic currents affect *water masses* with very large cross sections (in terms of both width and depth) they transport enormous volumes of water (e.g., the Gulf Stream and Kuroshio from 30 to $150 \cdot 10^6$ m^3/s, the Antarctic circumpolar current $110 \cdot 10^6$ m^3/s, and the northward flowing Antarctic bottom water in the South Atlantic $18 \cdot 10^6$ m^3/s). Hence, oceanic circulation also transfers huge quantities of *heat* from equatorial zones into high latitudes and, vice versa, provides for cooling in low latitudes regions. In this way, the oceans also exert a very large influence on the climate and life of neighboring continents. Surface currents carry fine-grained, suspended particles and living organisms, including their larvae, over long distances and thus distribute life and sediments. Currents, particularly those upwelling from intermediate and deep water, transfer *oxygen* and *nutrients* released by the mineralizsation of organic matter or dissolution of biogenic skeletal particles (see following sections) into coastal or equatorial regions. There, in the euphotic zone to depths of several meters to several tens of meters penetrated by sun light, the nutrients enable high phytoplankton productivity and strong activity by its consumers (Chap. 10.3.2).

Paleoceanography

All observations discussed so far have pertained to modern oceans under the present climatic conditions, with a strong temperature gradient from the poles to the equator and thermohaline circulation which is driven by the formation of cold bottom water in the polar regions. This and the present-day configuration of the ocean basins has led to a circulation pattern which prevents the existence of large bodies of stagnant bottom water, apart from adjacent basins and very limited coastal sea areas (Chap. 4).

However, we still live in a "glacial world" and have to take into account that the modern oceans are poor analogs of the ancient oceans. Over the Earth's history, the configuration of the ocean basins has undergone profound changes and has significantly affected the climatic zones of the globe. The above mentioned physical rules controlling oceanic circulation may have led to a circulation pattern quite different from the present one. Many workers believe, for example, that the Cretaceous oceans developed rather stagnant, warm, and more saline bottom water. This is ascribed to a warmer, more equable climate, which generated denser water in arid shallow water regions and, during high sea level stands, in widely extended marginal seas. This water flowed into the deep basins and formed more or less stagnant bottom water (e.g., Thierstein 1979; Berger 1981; Hay 1987; several papers in Brooks and Fleet 1987). Thus, most of the oceanic circulation at this time may have been restricted to surface and intermediate waters. Some of the consequences of such a pattern are discussed below and in Chapters 10.3.3 and 7.5.

We are just beginning to understand such global changes in paleoceanography. In understanding the most important features of present-day oceanic currents, one can proceed then to develop conceptions about the circulation patterns of ancient ocean basins. Physical oceanographers are trying to quantify and simulate all these processes, in addition to many other phenomena not mentioned here.

First attempts have already demonstrated that it may also become possible, through a joint effort of oceanographers and marine sedimentologists, to simulate the current patterns of ancient oceans and thus significantly improve our understanding of ancient marine sediments (also see, e.g., Witzke 1987; Berger et al. 1989; Barron and Peterson 1991).

5.3 Hemipelagic and Pelagic Deep Sea Sediments

5.3.1 Overview

The Terrigenous Component

The terrigenous component of hemipelagic and pelagic sediments is delivered as the suspended load of rivers and wind-borne dust derived either from arid to semiarid regions or from volcanic eruptions (cf. Chaps. 2.3.4, 2.4, and 9.4). During floods, highly concentrated suspensions near the river bed can directly reach the deep sea as underflows, but less concentrated suspensions may float for a short time as a fresh water wedge on top of sea water. As already mentioned in Chapter 3.4, clay suspended in fresh water is generally much less aggregated than in sea water. Thus it can be widely distributed (Fig. 5.3) before it comes in contact with more saline water promoting aggregation. Once flocculated, the settling velocity of the aggregates increases to values of several meters up to several tens of meters per day (McCave 1984). Thus their

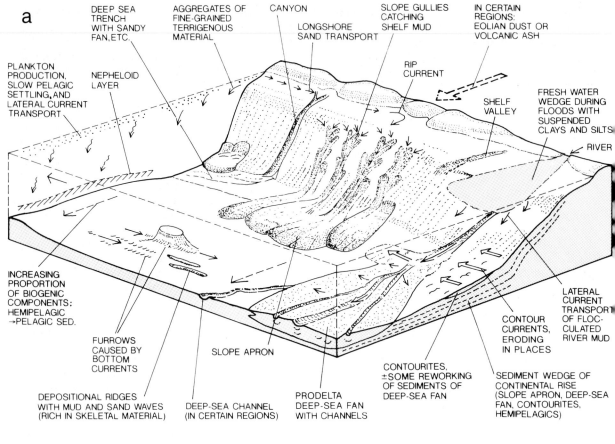

a

PLANKTON
PRODUCTION,
SLOW PELAGIC
SETTLING, AND
LATERAL CURRENT
TRANSPORT

DEEP SEA
TRENCH
WITH SANDY
FAN, ETC.

NEPHELOID
LAYER

AGGREGATES OF
FINE-GRAINED
TERRIGENOUS
MATERIAL

CANYON

LONGSHORE
SAND TRANSPORT

SLOPE GULLIES
CATCHING
SHELF MUD

RIP
CURRENT

IN CERTAIN
REGIONS:
EOLIAN DUST OR
VOLCANIC ASH

SHELF
VALLEY

FRESH WATER
WEDGE DURING
FLOODS WITH
SUSPENDED
CLAYS AND SILTS

RIVER

INCREASING
PROPORTION
OF BIOGENIC
COMPONENTS:
HEMIPELAGIC
→PELAGIC SED.

FURROWS
CAUSED BY
BOTTOM
CURRENTS

SLOPE APRON

DEPOSITIONAL RIDGES
WITH MUD AND SAND WAVES
(RICH IN SKELETAL MATERIAL)

DEEP-SEA CHANNEL
(IN CERTAIN REGIONS)

PRODELTA
DEEP-SEA FAN
WITH CHANNELS

CONTOURITES,
±SOME REWORKING
OF SEDIMENTS OF
DEEP-SEA FAN

CONTOUR
CURRENTS,
ERODING
IN PLACES

LATERAL
CURRENT
TRANSPORT
OF FLOC-
CULATED
RIVER MUD

SEDIMENT WEDGE OF
CONTINENTAL RISE
(SLOPE APRON, DEEP-SEA
FAN, CONTOURITES,
HEMIPELAGICS)

b

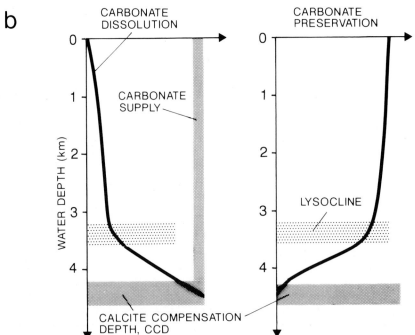

CARBONATE
DISSOLUTION

CARBONATE
PRESERVATION

WATER DEPTH (km)

CARBONATE
SUPPLY

LYSOCLINE

CALCITE COMPENSATION
DEPTH, CCD

Fig. 5.3. a Principles of sediment distribution by surface and bottom currents in deep-sea environments and the accumulation of hemipelagic and pelagic sediments. **b** Supply, dissolution, and preservation of carbonate as a function of water depth and the positions of the lysocline and calcite compensation depth, *CCD*. Note that lysocline and CCD may rise or fall with time and from basin to basin. See text for further explanation

additional lateral transport is a function of water depth and the velocity and depth range of the ocean currents.

This can be demonstrated by a very simple example, which does not take into account effects of large-scale turbulences: If a surface current has a mean velocity of 0.2 m/s (= ca. 17 km/d) to a depth of up to 200 m, an aggregate settling from sea level with a velocity of 10 m/day can reach a point 340 km away from its original position.

However, there are also other means for transporting fine-grained particles or aggregates into deeper waters. Silts and clays are sorted out of river input by wave action and move offshore via *nepheloid plumes* associated with bottom currents (Gorsline 1984). Depending on the stratification of the water masses and the densitiy of the suspensions, such turbid plumes may migrate near the surface, at mid-depth, or they may reach the deep sea where they drop most of their suspended material. The finest-grained part of the suspension can, however, be suspensed for a long time by deep bottom currents and form the so-called ne-pheloid layer in many parts of the modern oceans (Fig. 5.3). The thickness of this layer is on the order of 100 to 1000 m, but the concentration of suspended matter is very low.

Although fine-grained terrigenous particles are distributed in a number of different ways (including gravity mass flows as described below) over relatively long distances, it is also obvious from the current patterns discussed earlier that large areas in the oceans probably receive very little terrigenous mud. Most of this material probably stems from dustfall from the stratosphere, where fine-grained particles are rapidly distributed around the globe (cf. Chap. 4.3.5).

The Biogenic Component

The biogenic component of the hemipelagic and pelagic sediments is produced either in surface waters (plankton) or on the sea floor (benthos). In the deep sea, the latter contribution to the sediment mass is of minor importance, although it may be a significant indicator of ancient water depth. Primary planktonic production varies significantly in different parts of the oceans (Chap. 10.3.2); in addition, a great part or, in certain regions, virtually all of the organic and skeletal remains is mineralized or dissolved on its way to the sea floor. Thus, the ultimate contribution of biogenic components to the deep-sea sediments of an individual basin may differ great-

ly in space and time. For example, the following situations can be distinguished:

1. The biogenic autochthonous component is strongly diluted by allochthonous terrigenous material resulting in hemipelagic silty clays or clayey silts, a relatively high sedimentation rate (greater than ca. 5 cm/ka), and low biogenic carbonate and/or opaline silica contents. Many fine-grained sediments on gentle continental slopes belong to this category, but they may also contain redeposited material (see below).
2. The biogenic component is moderately diluted by terrigenous material, creating hemipelagic calcareous and/or siliceous silty clays or clayey silts, and a medium sedimentation rate (often 2 to 4 cm/ka).
3. The biogenic component is only slightly diluted by terrigenous material, resulting in pelagic calcareous or siliceous oozes, chalks, limestones, diatomites, radiolarites etc., usually with a sedimentation rate of ≤2 cm/ka. These sediments are particularly sensitive to relatively small environmental changes and therefore frequently produce rhythmic and cyclic sequences (Chaps. 7.1 and 7.2).
4. Partial or entire dissolution of the biogenic component and a very low influx of terrigenous matter may produce pelagic clay (usually reddish brown, known as red deep-sea clay) and a very low sedimentation rate (≤1 cm/ka).

Under euxinic conditions and in poorly oxygenated areas (Chap. 10.3), the sediments of groups (1) to (3) can be transformed into "black shales" of rather differing composition. Group (4) is frequently associated with the occurrence of manganese nodules (see below).

5.3.2 Deep-Sea Carbonates and Carbonate Dissolution

Biogenic Carbonate Production

Planktonic carbonate is produced mainly by coccolithophores (nannoplankton), foraminifera, and free-swimming molluscs over wide areas of the tropical, subtropical and temperate zones of the modern oceans. It is less abundant at high latitudes where cool temperatures cause the living conditions for these organisms to deteriorate and, in addition, promote their dissolution in carbonate-undersaturated waters. Total oceanic carbonate production is a function of the evolution of life, i.e., the efficiency of organisms to extract calcium carbonate from sea water, the availability of calcium delivered by rivers from the continents, or other (minor) sources, and the recycling of calcium and nutrients (particularly nitrogen and phosphorus) by the oceanic circulation system. In addition, plankton growth depends on the presence of special organic compounds, such as vitamins. These "micronutrients" are mainly produced by bacteria and required by many phytoplankton species. A summary on these and other growth factors is given by Grant Gross (1980); rates of carbonate production are discussed in Chapter 10.2.

Over the Earth's history there have been times with low and high global marine production and preservation of carbonate. A peak in shallow-marine carbonate production occurred in the middle to late Paleozoic; an additional era of high carbonate production began around 100 Ma ago which has persisted into present times. Since that time the fixation of calcium carbonate as calcite by marine plankton has become an important factor.

Carbonate Dissolution

The modern oceans and their living communities, including those in shallow-water environments (Chap. 3.3.2), extract considerably more calcium carbonate per unit time than can be replaced by river influx of calcium hydrogen carbonate and by smaller contributors, for example hydrothermal sources along mid-oceanic ridges. However, the deficiency in biogenic calcium carbonate consumption is balanced by the oceanic circulating system which recycles part of the extracted material. Calcium carbonate is dissolved in the deep sea by cold bottom currents coming from high-latitude regions (in the present oceans by Antarctic and Arctic bottom waters). Prior to their downwelling, these water masses flow poleward as surface currents. On their way to high latitudes, they become depleted in calcium carbonate but rich in free CO_2 due to near-surface productivity, cooling, and uptake of CO_2 from the atmosphere.

Later, after they have sunk into intermediate depths or to the sea floor, they can take up CO_2 from the decomposition of organic matter on their way toward the equator (cf. Fig. 5.1e). Thus, they become *undersaturated with respect to calcium carbonate* and therefore dissolve carbonate shells settling through deep waters or already resting on the sea floor. Aragonite shells and thin skeletons of calcite are more readily dissolved than robust calcitic shells. Carbonate wrapped into fecal pellets has a better chance of being preserved than do free floating shells.

From mapping the sea floor and testing its surface sediments, one can find two boundaries (Figs. 5.3b and 5.4a):

1. A contour zone on the sea floor, where dissolution of less stable shells (consisting of aragonite) starts to reduce significantly the normal carbonate content of the sediment found at higher elevations. This boundary is referred to as the *lysocline*.
2. The second, deeper boundary, is known as the *calcite compensation depth*, or *CCD*. Here, the planktonic calcium carbonate supply rate is balanced by the dissolution rate.

Below this depth, the sediment should be free of calcium carbonate. In the present-day oceans the CCD is observed generally at depths between 3.5 and 5.5 km (Seibold and Berger 1982). It is usually depressed in low-latitude regions and below zones of equatorial divergence (upwelling) due to the high production of calcium carbonate, and it is raised in high-latitude areas with low carbonate production. A higher CCD is also found along western continental margins, where coastal upwelling is widespread. In this case, dissolution of carbonate near or at the sea floor may be enhanced by decomposition of organic matter, which releases CO_2 provided by the highly fertile photic zone (see below).

The Dynamic Equilibrium Between Production and Dissolution of Carbonate

For a limited geologic time period which is not affected by major climatic or other global changes, the position of the lysocline and CCD can be expected to remain constant. These boundaries then signify that a kind of steady-state condition is being maintained between (biogenic) precipitation of calcium carbonate, retrieval of calcium carbonate from the ocean by dissolution, and supply of calcium carbonate from outside sources. Figure 5.4e shows such a situation for the sediments of a mid-oceanic spreading center. With continuing accretion of new oceanic crust, the ridge crest moves away from its former position (from location 1 to 2) and the cooling crust subsides. As a result, the depositional area sinks below

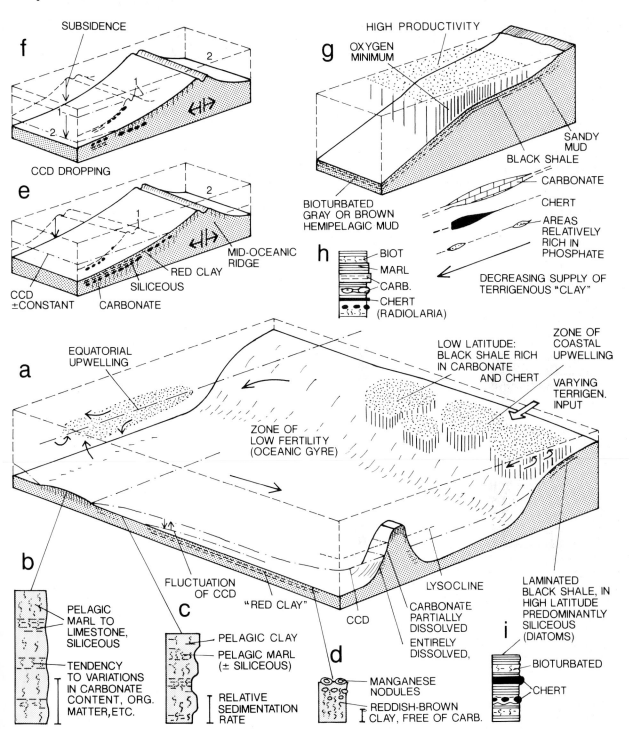

Fig. 5.4. a Zones of high and low fertility, calcite compensation depth *(CCD)* and lysocline, and corresponding distribution of pelagic sediments (**b, c, d**) in a hypothetical ocean segment from the equator to polar regions (diatomaceous oozes of polar and subpolar zones omitted). Note that a fluctuating CCD can cause carbonate-bearing beds to alternate with carbonate-free interbeds (**c**). **e** and **f** Succes-

sions of calcareous, siliceous and clayey sediments on top of oceanic crust adjacent to mid-oceanic ridge, as controlled by a more or less constant (**e**) or lowering CCD (**f**). **g, h, i** Onshore-offshore (**g**) and latitudinal variations (**h** and **i**) in the composition and bed-sequences of black shales accumulated under coastal upwelling. For further explanation see text

the CCD, and the calcareous sediments sitting directly on pillow lava or basaltic flows are overlain by light gray or red siliceous sediments and later by red clay.

In the history of many ocean basins, however, the CCD (including the lysocline) did not maintain a constant position, particularly when long time periods are considered. Biogenic carbonate production in shallow waters and carbonate dissolution in deep waters may be considered as a system being in a dynamic equilibrium. The steady-state model mentioned above implies a raising or lowering of the CCD, if one or several of the parameters controlling the system are changed. Decreasing oceanic circulation, for example due to a more balanced climate between the poles and the equator, can lead to a drop in the overall fertility of the oceans and thus to lower carbonate production. This in turn will reduce the need for carbonate dissolution in deep waters to keep the carbonate budget balanced and consequently cause a drop of the CCD. A similar result can be achieved by an increase in the input of hydrogen carbonate from the continents or by dissolution of emerged carbonate buildups during low sea level stands.

In contrast, intensified oceanic circulation and mineralization of organic matter can enhance fertility and thereby raise the CCD. The same effect may be generated by increased carbonate production on shelves and in reefal structures.

From information gathered in the Deep Sea Drilling Project, the CCD has moved up and down several times (by as much as 2 to 3 km) in the last 100 Ma (e.g., Hay 1987). Sometimes these variations were synchronous in the three major ocean basins, sometimes the CCD behaved quite differently in each. The CCD was very low around the Cretaceous/Tertiary boundary in the Pacific Ocean, while it stayed relatively high in the Atlantic and Indian Oceans. A CCD markedly higher than at present was found for the early Tertiary, with the result that far fewer pelagic deep-sea carbonates were preserved than during other periods.

In the transitional zones between full carbonate preservation above the lysocline, carbonate-bearing sediments between the lysocline and the CCD, and siliceous red clay (see below), a fluctuating CCD can generate an alternating succession of these two bed types (Fig. 5.4b and c, cf. Chap. 7.2). In addition, the carbonate system of the oceans is closely linked with the carbon dioxide content of the atmosphere. Warm periods in the Earth's history appear to have been associated with high CO_2 partial pressures in the atmosphere,

which in turn promoted carbonate dissolution in the oceans (e.g., Berger 1985). This topic is currently being investigated by many workers from various disciplines.

Adjacent seas with narrow openings to the world oceans and therefore limited exchange of water masses (Chap. 4, Fig. 4.2) are characterized by their own carbonate budget and a CCD which may deviate significantly from those of the major ocean basins. Such smaller basins are controlled mainly by the climate and river input of neighboring continents. A high influx of calcium hydrogen carbonate and limited carbonate production, as frequently observed in humid, mid-latitude zones, cause the CCD to drop to or below the sea bed. Dissolution of calcium carbonate can then only take place within the sediment as a result of organic matter decomposition (e.g., in the Baltic Sea, cf. Chap. 4.2). Similarly, adjacent basins surrounded by hot, arid regions will not necessarily develop a high CCD, because their carbonate production is usually kept low due to a shortage of nutrients, if their water exchange with the open ocean is limited. Thus, a CCD appears to be absent in most of these basins, i.e., it lies below their basin floors (e.g., in the Red Sea).

The Chalk Problem

Young, soft, calcareous muds are commonly described as calcareous oozes. Hemipelagic to pelagic oozes are transformed into chalk as a result of incomplete, moderate diagenesis (e.g., Neugebauer 1974; also see Chap. 13.1). The term *chalk* refers to a variety of ancient, white to light grey calcareous marine sediments, which are still porous, friable, and composed predominantly of tiny, calcitic skeletal remains, such as coccolithophorides. Apart from pure chalk, there are also chalks containing sands, larger shells, glauconite, pyrite, phosphorite (see sections below), and considerable proportions of clay (marly chalk) and organic matter (bituminous chalk).

Although chalk consists predominantly of planktonic skeletal remains, it was formed not only in the deep sea above the calcite compensation depth, but also in shallow seas, particularly in the upper Cretaceous. The presence of abundant sighted ostracods in chalk suggests that the late Cretaceous sea floor in Alabama and Mississippi was within the photic zone, i.e., not deeper than 65 to 90 m (Puckett 1991). Most of the typical, widespread chalk deposits in Europe, North America, in the

Middle East, and Australia were generated during this period (Hattin 1986b, 1989; Hancock 1989). The formation of chalk on shelves and in epicontinental seas was favored by a high sea level stand and low influx of terrigenous material, high production of calcitic nannoplankton, and scarcity of aragonitic skeletal remains which commonly promote lithification and cementation. Wide and stable shelves, as well as flooded continental regions were exposed to wind patterns and current systems providing nutrients for planktonic productivity.

In the examples mentioned above, the pure chalk facies may grade landward into marly chalk and mudstone or, in high-energy environments, into shelly chalk and skeletal limestone. Basinward, the chalk tends to become laminated and richer in organic matter. On slopes and in special morphological depressions, part of the material forming chalk may be derived from shallower water and be redeposited by mass flows and turbidity currents (Chap. 5.4). On submarine highs where continuous chalk deposition is prevented by wave and current actions, or hampered by low sea level stands, the chalk sequences are frequently condensed and contain hardgrounds and omission surfaces. Many chalk sequences contain chert and show rhythmic bedding in the Milankovitch frequency band (Chap. 7.2).

5.3.3 Red Clay and Manganese Nodules

Red Deep-Sea Clay

The water masses of large oceanic gyres (Fig. 5.1e) have little exchange with nutrient-rich coastal waters or intermediate waters welling up in equatorial regions. Therefore the planktonic productivity of the central ocean basins is low and the CCD tends to rise. These basins receive little terrigenous input, and their bottom waters are well oxygenated. Under these conditions, the so-called *red clay* is deposited (Fig. 5.4a and d). In a fresh state, this extremely slowly accumulated material has a reddish brown color, because it contains finely dispersed iron oxyhydroxides as pigment. It has lost nearly all its organic matter by oxidation as a result of long-term exposure to oxygenated bottom water. The clay mineral assemblage of red clays is controlled mainly by their source areas and therefore varies in space and time. Illite, smectite, kaolinite, and chlorite are major constituents in wide regions. Eolian dust is assumed to contribute a

large portion of such red clays, and even some cosmic dust (spherules of nickel-iron, minerals of chondrites) has been discovered (Bryant and Bennett 1988).

In the transition zone between highly and poorly fertile surface waters, biogenic carbonate may already dissolve in the water column, whereas part of the siliceous remains may reach the sea bottom. In this case, red clays contain considerable proportions of radiolaria (predominantly living in warm equatorial waters) or diatoms (mainly from colder waters).

Manganese Nodules

In large areas of the present oceans, particularly in the central Pacific, the red clay is covered by *manganese nodules* which may also contain various amounts of iron and relatively high quantities of nickel, cobalt, copper, molybdenum, and other trace metals (e.g., Seibold 1978; Calvert and Piper 1984; Halbach et al. 1988). The nodules are concentrated on the sea bed and mostly become sparse some tens of cm below the sediment-water interface. For this reason it has been postulated that the nodules dissolve under reducing conditions at some depth below the sea floor and that Mn migrates by diffusion to the interface and precipitates at the contact with sea water. If the chemical environment in interstitial waters remains oxidizing, the nodules are preserved. It has been also noted that bacteria play a role in the fixation of manganese.

Most nodules are flat at their base and show concentric growth structures. Their growth rates vary from relatively high values (several mm/ka) to very low rates (several mm/Ma). The low values are even lower than the sedimentation rate of pelagic red clay (0.5 to about 5 mm/ka, Andreyev and Kulikov 1987; Halbach et al. 1988). No relationship was found between the growth rate of nodules and their host sediments. The finding that most manganese nodules have a higher age than the surrounding surface sediment leads to the conclusion that they "migrate" upward with the sea bed. This may be accomplished either by continuous dissolution and reprecipitation, by burrowing organisms lifting the nodules, or by bottom currents, which are occasionally strong enough to move and turn over porous nodules. Bottom currents also prevent continuous sediment accumulation and thus are responsible for the very low average sedimentation rates characteristic of deep-sea sediments rich in manganese nodules.

The source of manganese, iron, and the other metals is a matter of debate. They may derive from the metal content of organic compounds, which were produced in surface waters, but were decomposed at, or near the sea bottom. Other sources may be submarine exhalations and normal bottom currents containing the metals in very low concentrations, either in solution or adsorbed to suspended particles. In addition, the composition of the nodules is influenced by diagenesis, particularly their variable Mn/Fe ratios.

In recent years, these widely distributed nodules and particularly their contents in Co, Ni, and Cu have aroused considerable interest, and tests were run by several groups for underwater mining of these enormous ore deposits. These activities, including our present knowledge on the formation and geochemistry of manganese nodules, were recently summarized in a special volume (Halbach et al. 1988). In ancient sediments, manganese nodules formed in the deep sea are relatively rare (Cronan et al. 1991). If they occur on top of paleo-seamounts or in former basinal settings, they resemble lithologically, mineralogically, and geochemically the modern ferromanganese deposits. In many cases, however, they have either been subducted with their accompanying sediments and transformed by metamorphism, or they have been dissolved as mentioned above.

5.3.4 Sediments in Zones of Upwelling

General Aspects

In zones of coastal and equatorial upwelling, where nutrient-rich waters come to the surface, the productivity of phytoplankton (e.g., coccoliths and diatoms) and zooplankton (e.g., foraminifers) is significantly enhanced (Fig. 5.4a). The availability of abundant food attracts larger organisms (e.g., fish) which feed on the micro-organims. Although most of the primary production is used up and mineralized in the water column, the overall sedimentation rate of biogenic material increases considerably in zones of upwelling as compared to areas of normal fertility. Therefore, deep-sea sediments associated with equatorial upwelling usually still contain biogenic carbonate and silica and more organic matter than their neighboring sediments. This means that the CCD and possibly also the lysocline drop below the sea floor (Fig. 5.4a).

Coastal Upwelling

Preservation of organic matter, biogenic silica, and carbonate is particularly significant in zones of coastal upwelling where the nutrient supply from intermediate waters is concentrated in rather narrow zones on the upper slope and outer shelf along continental margins (Fig. 5.1e). The modern oceans display four major areas of year-round coastal upwelling as part of the eastern boundary currents in the Atlantic and Pacific Oceans (Thiede and Suess 1983): off northwest and southwest Africa, western North America, and northwestern South America. There are also areas of seasonal upwelling associated with prevailing wind regimes, for example in the northern Indian Ocean, off northwest Australia, and in the Gulf of California.

The lower part of the water column below coastal upwelling is often characterized by a pronounced *oxygen minimum zone*. Oxygen demand for total mineralization of the large quantities of organic matter produced in surface waters is higher than oxygen supply in the relatively short (shallow) water column. In addition, the particulate organic matter needs less time to sink through the water column and reach the sea floor than in the deep ocean. Thus, oxygen transfer from outside the zone of coastal upwelling cannot compensate for rapid oxygen consumption below the area of high biogenic productivity. Consequently, benthic life at the sea bottom is reduced or entirely absent, the sediment can become laminated, and a large part of the organic matter reaching the sea floor is preserved. The centers of upwelling are, however, not fixed to a certain location, but may move along the coast and seaward. Similarly, the intensity of upwelling varies with time, because it is controlled by fluctuations in the wind and oceanic current patterns. For example, rapid oxygen depletion by a sudden plankton bloom can cause a mass mortality of fish and other organisms (Brongersma-Sanders 1957).

Sediments Below Coastal Upwelling

For reasons mentioned above, *black shales* originating from coastal upwelling usually display frequent changes from anoxic to low-oxygen conditions, as can be seen by the appearance or disappearance of a particular ichnofauna (Savrda and Bottjer 1987, 1989). Similarly, the occurrence of molluscan shell beds in black shales indicates an interval, in which the upper boundary of the oxygen minimum zone dropped to the sea floor (Schneider and Wefer 1990). Due to a high productivity of diatoms and/or radiolaria, sediments from upwelling zones are often rich in *siliceous layers* or chert nodules. Diatoms prevail in cool, high-latitude waters where the production and preservation of carbonate are low. In warm, lower latitude regions, the carbonate content in upwelling sediments can become rather high and attain those values typical of marls, and sometimes even those of marly limestones. During diagenesis, dolomite may form, a process which is favored by the decomposition of organic matter (Lippmann 1973; Suess et al. 1987).

The fauna preserved in upwelling sediments is characterized by a limited number of species and indicates a cooler environment than in adjacent regions. The organic matter in such sediments usually reaches a few percent, but sometimes also up to 10 % by dry weight. It is entirely or predominantly of marine planktonic and bacterial origin (Summerhayes 1983; ten Haven et al. 1990), which is an important indicator for the type of black shale. The sediments in zones of upwelling may also contain abundant fish debris (bone beds) and a relatively high amount of phosphorus due to the accumulation of abundant fecal pellets and coprolites. Several *trace metals*, particularly U, Mo, Cd, Zn, and Ni occur in considerably higher concentrations than in normal deep-sea sediments (e.g., Baturin 1983; Brumsack 1986a and b). The sedimentation rates in coastal upwelling zones are to a large extent controlled by terrigenous influx.

In the Peru upwelling region, the mean sedimentation rates of late Pleistocene sediments (drilled at water depths of about 250 and 450 m) are about 7 and 17 cm/ka, respectively (Wefer et al. 1990); older sediments accumulated more rapidly.

Coastal upwelling was certainly also an important process in ancient marine environments. The formation of many black shales is now ascribed to this mechanism not only in the Cenozoic and Mesozoic, but also in older times (Parrish et al. 1983; Parrish 1987; Pedersen and Calvert 1990). Most of these black shales occur along west-facing coasts, in settings similar to the present-day major centers of upwelling. In the completely different paleogeographic situation in the Paleozoic, however, it can be assumed that coastal upwelling also operated along east-west trending coastlines.

5.3.5 Siliceous Sediments

Biogenic Production of Opaline Silica

Skeletons of opaline silica are produced in the modern oceans by several groups of organisms (diatoms, radiolaria, silicoflagellates, and sponges) in large quantities. By number, diatoms are the most abundant siliceous organisms; in tropical regions they make up between about 30 and 75 % of the total amount of opal-concentrating organisms, in Antarctic waters 99 % (Lisitzin 1972; Blueford 1989). The most abundant group by weight, however, are the radiolarians. On the average, 1 mg of silica corresponds to 2000 radiolarian skeletons, but 100 000 diatom cells and 250 000 dinoflagellates equal that weight (thus yielding a weight ratio of 250:5:1). For that reason, the radiolarians today reach a weight percentage of about 60 to more than 95 % of the silica production in tropical waters, but ≤1 % in the Antarctic. In addition, the radiolaria have a much higher potential for the preservation in marine sediments than the other two groups (see below). Nevertheless, it is assumed that the importance of the radiolarians in the formation of siliceous sediments was even greater in pre-Cenozoic times.

As with carbonates, the organisms extract much more silica from ocean waters than is added by rivers and volcanic activity. Consequently, the sea water near the surface is extremely *undersaturated* (it normally contains about 1 mg/l SiO_2), particularly with respect to opaline silica. Therefore most of the tiny tests of microorganisms (solubility around 110 mg/l; in contrast to quartz, about 5 mg/l) are already dissolved within the upper water column before they reach deeper waters, which have higher silicon concentrations, and finally the sea floor. But even here a great part of the remaining tests is still dissolved within the uppermost sediment layer (Hein and Obradovic 1989). For average conditions in the present oceans, it is assumed that perhaps 1 % of the originally produced biogenic silica is stored in the sediment. For that reason, there are large areas on the sea floor where biogenic silica is completely absent or can be neglected as an important constituent of deep-sea sediments.

Biosiliceous Sediments

Preservation of Opaline Silica

The chances for preservation of siliceous skeletal remains is enhanced if they settle comparatively fast through the water column. This
is true for relatively robust skeletons, such as
those of radiolaria, and of course in shallower
waters, for example in zones of coastal upwelling. Particularly favorable for their preservation is the incorporation of microorganisms into fecal pellets, which protect the
skeletons from being dissolved. In addition,
submarine volcanic activity, releasing substantial amounts of silica into the ocean, may
locally reduce the dissolution of opaline silica.
The most effective way to increase the preservation of this material is, however, high productivity in zones of upwelling. One has to
bear in mind that the input of silica alone
(e.g., by volcanism) into surface waters cannot
enhance fertility unless other nutrients such as
nitrogen and phosphorus are also available.

In order to have high concentrations of opaline silica in sediments, the preserved part of
the primary production must not be diluted by
terrigenous material or carbonate. Consequently we can expect highly siliceous or biosiliceous sediments only in areas with (1) high
primary productivity where (2), in addition,
the influx of river-borne material is very low,
and (3) the carbonate production also is so low
or the lysocline and CCD are so high that most
of the carbonate is dissolved in the water
column or within the sediment.

Areal Distribution of Biosiliceous Sediments

In the present oceans such conditions mainly
occur at high latitudes far away from the
coasts. Here, diatomaceous oozes are deposited
in two wide, circumpolar belts, particularly
around the Antarctic and in the northern
Pacific. Sediments very rich in diatoms and
poor in carbonate are also found in several
regions of present-day coastal upwelling, for
example off southwest Africa, Peru, Baja
California, and in the Gulf of California.
During the Miocene, great parts of the coastal
waters around the Pacific (e.g., along North
America) were characterized by rapid deposition of diatomaceous oozes (Monterey Formation).

Another, less well developed belt displaying
biosiliceous sediments is observed in zones of
equatorial upwelling. Here the sediments are
dominated by radiolarians (see above) and
contain variable amounts of biogenic carbonate. At the transition to red clay below the
CCD, carbonate is dissolved, which may lead
to a relative enrichment of biogenic silica. The
widespread radiolarites in the Triassic-Liassic
and Middle und Upper Jurassic Tethyan
Ocean probably were also associated with an
equatorial zone of high fertility (Jenkyns and
Winterer 1982; Hein and Parrish 1987; Hein
and Obradovic 1989). In addition, the CCD
must have been very high, because radiolarites
were found not only in deep basins, but also
on subsiding carbonate platforms and outer
shelves. A simplified model of such a scenario
is shown in Fig. 5.4f. Since prolific planktonic
carbonate production tending to raise the CCD
is not known prior to the Late Jurassic, the
high CCD was probably caused by large,
widely extended carbonate platforms bordering the northern and southern rim of this
east-west-oriented seaway. They extracted
and fixed most of the carbonate available in
the Tethyan Ocean.

In zones of coastal upwelling with bottom
waters low in oxygen, Neogene and Recent
siliceous oozes are frequently laminated and
rich in organic matter. They form thick, dark
layers and may contain intercalations of calcareous beds, carbonate nodules and phosphorite (see below). Siliceous oozes deposited in
deeper water, for example on mid-oceanic
ridges and in ponded basins on oceanic crust,
tend to become fully oxidized and therefore
red-colored (Fig. 5.5). Bioturbation in such
layers is sparse or absent, because there is not
enough organic matter left to live on.

Bedded Cherts

The silica of lithified biosiliceous sediments is
frequently concentrated in chert bands and
layers of chert nodules. Some of these features
are of primary origin, other due to diagenesis.
In contrast to layers deposited from pelagic
settling, many rhythmically bedded radiolarites also represent bioclastic, fine-grained turbidites deposited by low-density turbidity
currents. Their individual beds display a sharp
lower boundary and some grading (either in
grain size or in composition, see Chap. 5.4.2).
They are derived from the slopes of submarine
highs (ridges, hills, deep plateaus; Fig. 5.5)
which receive little terrigenous material (e.g.,
Barrett 1982). If they accumulate below the
CCD, the siliceous turbidites are devoid of
carbonate. These types of red or greenish

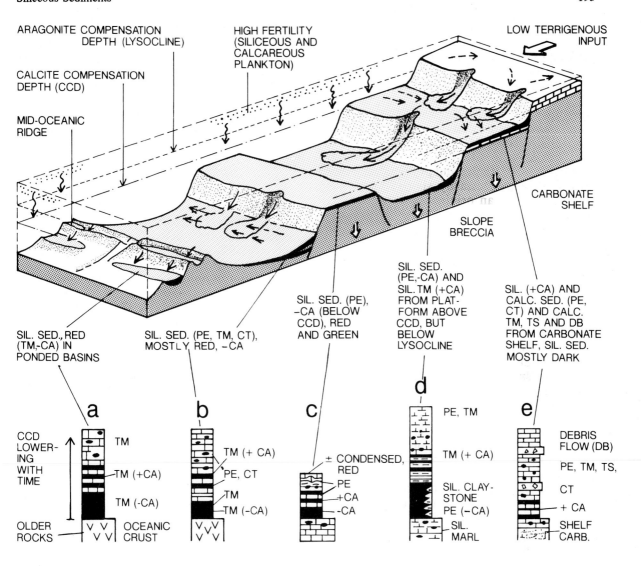

Fig. 5.5a–e. Siliceous sediments, e.g., radiolarites, in a relatively narrow oceanic basin with regionally high fertility, bordered by a wide, block-faulted continental margin (e.g., a former uniform carbonate platform). Note the different types of radiolarites (deposited by pelagic settling, bioclastic mud turbidites, or current sorted, with or without carbonate) in relation to the submarine topography and the elevation of the *CCD* and lysocline. **a** through **e** Vertical changes in lithology due to lowering *CCD* (raising *CCD* from an initial low position causes reverse sequences). (Based on several sources, e.g., Bernoulli and Lemoine 1980; Hein and Obradovitch 1989; Vecsei et al. 1989; Ruiz-Ortiz et al. 1989)

colored, banded chert layers may alternate with more or less siliceous red or greenish mudstones and are usually interpreted as deep-water deposits.

Carbonate-bearing biosiliceous sediments and cherty pelagic limestones indicate a depositional depth above the CCD or between the aragonite and calcite compensation depth (Fig. 5.5 and Chap. 5.3.2). Successions of purely pelagic, biosiliceous sediments overlain by siliceous carbonates and pure carbonates do not, however, necessarily signify shallowing sequences, but can originate from a drop in the CCD as well (Fig. 5.5a through e). Similarly, pelagic limestones followed by chert layers are not always the result of subsidence and consequent deepening of the basin in question.

In particular, *radiolarites* in ancient rocks are frequently ungraded and show well developed laminations and low-angle cross-bedding. Because it is unlikely that beds with these sedimentary structures have been deposited from suspension currents, they are assumed to originate from normal traction currents at the sea bottom, which sorted and concentrated hydrodynamically equivalent particles (e.g., Barrett 1982; Gursky 1988; Ruiz-Ortiz et al. 1989; Vecsei et al. 1989). Due to subsequent strong compaction of the highly porous primary beds, the initially steeper angles of the cross-laminae have been modified. Such beds may have formed at different water depths and may thus contain carbonate and siliciclastic material (Fig. 5.5).

Silica Diagenesis

Because of the relatively good solubility of opaline silica, which is enhanced with increasing temperature at growing burial depths, deep-sea siliceous sediments are susceptible to marked diagenetic modifications. Near the sea floor and up to several hundred meters sub-bottom, siliceous oozes normally remain more or less unchanged and maintain a very high porosity (75 to 90 %). Below this depth range, which depends significantly on the temperature gradient and other factors, the original skeletal material (opal-A) is dissolved and reprecipitated as cristobalite and tridymite (opal-CT). Sediments rich in this phase are still porous and are referred to as *porcellanite*. The first skeletons to be destroyed are the delicate tests of diatoms, which are followed by the more robust radiolarians and sponge needles. Later, with increasing temperature

and burial depth, these SiO_2 modifications are in turn transformed into *quartz (chalcedony)*.

These processes have been described by many authors in detail (e.g., Lancelot 1973; Calvert 1974; Kastner 1981; Pisciotto 1981; Isaacs et al 1983; Thein and von Rad 1987; Füchtbauer 1988; Hein and Obradovic 1989; Ruiz-Ortiz et al. 1989).

Coeval with these mineral transformations, the porosity of the siliceous sediments is drastically reduced in two steps (Fig. 5.6a, cf. Chap. 13), because (1) the intratest porosity of the microfossils is diminished, and (2) the normal pore space between the grains is filled with quartz cement. Layers particularly rich in primary SiO_2 and/or layers more porous than others may be the loci of preferential reprecipitation of quartz (Tada 1991). Thus, they are transformed into chert beds of secondarily enhanced SiO_2 concentration (Fig. 5.6a). Migration of SiO_2 and growth of chert bands and nodules cease when the sediments become impermeable as a result of cementation.

In Mesozoic-Cenozoic clayey siliceous oozes below the present sea floor, the transformation of opal-A into opal-CT and quartz took place about 50 Ma after their deposition, provided the sedimentation rates and thermal gradients were about average (Thein and von Rad 1987). However, the silica in calcareous oozes, diatomaceous oozes, and oozes associated with reactive volcanic material can be transformed into porcellanite and quartz after shorter time periods, particularly if the thermal gradient is high.

Part of the silica released from microfossils may also be utilized to form clay minerals (zeolites) if sufficient Al_2O_3, alkali and alkaline earth ions are available in the pore water. Thus the formation of chert bands or nodules is hindered or completely prevented if the opal content in the primary sediment was limited. In the case of silica-poor calcareous oozes which cannot provide enough additional ions for the generation of zeolites, the silica is diagenetically concentrated in chert nodules (Fig. 5.6b).

Their precipitation appears to start rather early in diagenesis, because their cores are frequently richer in SiO_2 than their outer rims. More features of silica and carbonate diagenesis are shown in Fig. 5.6c and d. Precipitation of both phases can be more or less coeval, that is, chert may form the center of a carbonate nodule, or pre-existing carbonate concretions may influence the formation and shape of chert. In this respect it may be important whether or not carbonate was originally present as unstable aragonite or high-Mg calcite in some quantity. In clayey and marly sediments, gravitational compaction continues after the formation of concretions, in special cases deforming

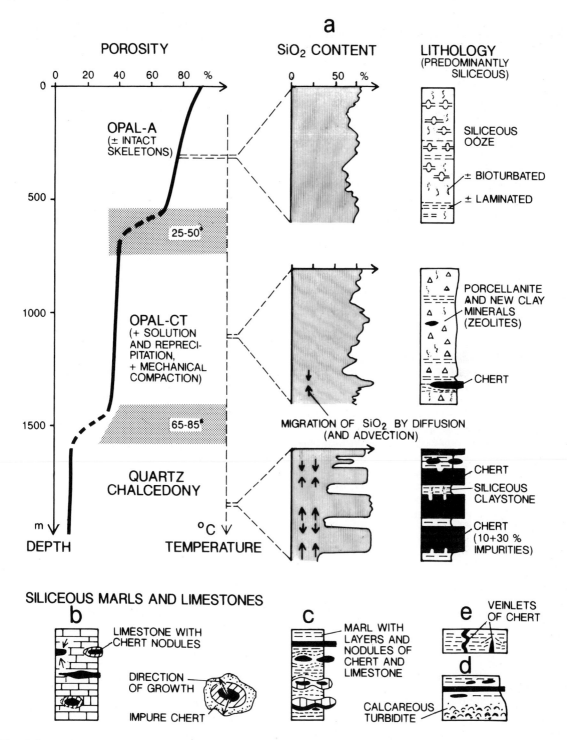

Fig. 5.6. a Processes and results of silica diagenesis in pelagic, siliceous oozes (based on several sources, see text). Note that banded cherts may be partially generated by a kind of rhythmic unmixing, which accentuates primary minor differences in texture or composition of beds. **b** through **d** Diagenetically formed bands and concretions of chert or carbonate in limestones and marls. **e** Veinlets of chert or laminated marl deformed subsequent to quartz precipitation. (After Lancelot 1973)

chert "veinlets" (Fig. 5.6c and d). Siliceous turbidites and calcareous turbidites rich in opaline silica are affected by diagenesis in a way similar to that described for pelagic sediments (Fig. 5.6e, see, e.g., Elorza and Bustillo 1989).

5.3.6 Marine Phosphorite Deposits

General Aspects

Phosphorite deposits have attracted the interest of geologists for a long time not only because of their economic potential, but also because they are significant indicators of a certain depositional environment. Numerous papers on this topic were collected and summarized by Bentor (1980), Slansky (1980), Kolodny (1981), Baturin (1982), Burnett and Riggs (1990), and Notholt and Jarvis (1990).

Phosphorite studies deal with a variety of problems, for example physicochemical considerations for their precipitation, trace elements including uranium and rare earths, radiometric dating, budget and recycling of phosphorus, paleoceanographic reconstructions for their depositional environment, the role of diagenesis and replacement of carbonate, effects of repeated reworking and weathering, economic evaluations, etc. Some of these problems are still not very well understood. We can touch here only a few general points.

Since Neogene times and probably in older periods, phosphorite preferentially formed atop submarine highs, along the edges of continental shelves where mainly pelagic sediments accumulated (also see below), and in shallow seas. Water depth appears to be less important for the accumulation of phosphorite or its parent sediment than a specific topographic situation or low sedimentation rate and reworking of sediment. Phosphorites are frequently observed in the following sequences (also see Föllmi et al. 1991):

- In condensed beds on top of drowning shallow-water carbonate platforms, often followed by pelagic sediments.
- At the top of shallowing, coarsening upward sequences, overlain by sediments of deeper water.
- In condensed intervals, rich in organic matter, topped by siliceous sediments indicating an increase in coastal upwelling.
- In transgressive-regressive sequences in rather shallow marine environments with low terrigenous input, where organic-rich sediments (with phosphorus) deposited during the transgression are partially or entirely reworked

(relative enrichment of phosphorite) during the subsequent regression (e.g., Glenn and Arthur 1990).

All these situations have in common a *changing depositional environment with phases of condensation and reworking. Although* many sediments deposited in coastal upwelling zones are also characterized by high phosphorus contents, enhanced upwelling is probably not required for phosphogenesis (McArthur et al. 1988). Some modern phosphorites, for example on the shelves off eastern Australia and on the western margin of South Africa, occur in environments where upwelling is weak and the bottom sediments are not rich in organic matter. In regions of strong present-day upwelling, for example off Peru and Namibia, significant phosphorite concentrations in fact tend to occur in the transitional zone between fully developed laminated black shales, rich in organic matter and chert, and adjacent, better oxygenated sediments seaward or landward (Figs. 5.4g and 5.7a). With rising or falling sea level, however, such a zone of phosphorite formation may migrate upslope or downslope and thus create a broader deposit along a shelf-slope or bank-ridge setting.

The phosphate phase of the common phosphorites consists of the mineral *francolite,* i.e., *carbonate fluorapatite.* This mineral contains variable amounts of trace elements, particularly uranium and rare earth elements, which are not further discussed here.

Phophate Precipitation

For some time it was believed that, under certain conditions, such as high phosphorus content in upwelling sea water, decreasing partial pressure of CO_2, and increasing temperature, francolite could be chemically precipitated directly from sea water. This view is now rejected by most of the experts on phosphorites (see, e.g., Kolodny 1981). Instead, it is assumed that phosphate is precipitated within the uppermost layer of organic-rich sediments during early diagenesis.

Buried organic matter contains considerable amounts of phosphorus (often around 1 %), which is released by microbial decomposition at the sea floor or several cm to tens of cm below (e.g., Baturin 1982; Froelich et al. 1988; Glenn 1990). In addition, the phosphorus in bones (e.g., fish), fecal pellets, coprolites, and adsorbed to iron-hydroxides (inorganic P sources, see below) may be dissolved and thus

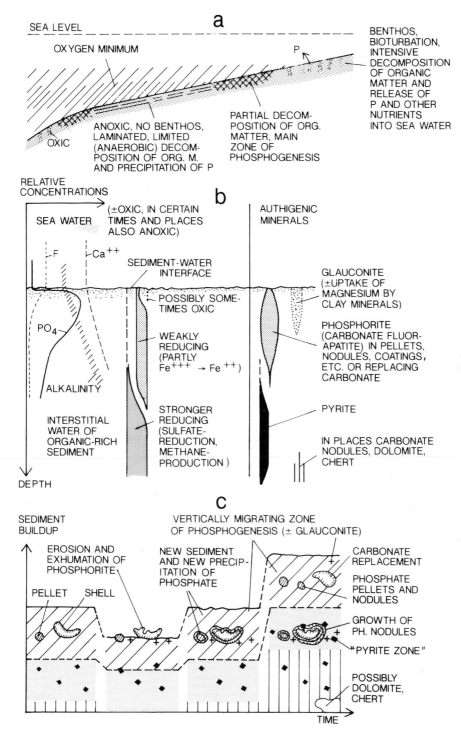

Fig. 5.7. a Schematic cross section of continental margin or submarine plateau, showing main zones of phosphogenesis in relation to oxygen minimum and intensity of decomposition of organic matter. (After Slansky 1980). **b** Chemical environment within uppermost sediment (not to scale), which enables microbial release of P from organic matter (and/or bone remains) and precipitation of carbonate fluorapatite and/or replacement of carbonate by phosphate. The weakly reducing milieu which can release ferroan iron also is favorable for the formation of glauconite. Pyrite and dolomite as well as chert may be formed at greater depths below the sediment-water interface. (Based on several sources, e.g., Ouwehand 1986; McArthur et al. 1988; Froelich et al. 1988; Glenn and Arthur 1988). **c** Time-sediment buildup demonstrating vertical migration of the zones of phosphogenesis, pyrite precipitation, and possibly dolomite and chert formation. For further explanations see text

increase the phosphorus concentration in interstitial waters to values much higher than in sea water.

Phosphorus is also provided by the suspended load of rivers, particularly those draining regions with lateritic soils (Glenn and Arthur 1990). In this case, phosphorus is predominantly adsorbed to iron hydroxides and released later, when the river load has been deposited in a shallow sea where the sediments are subject to weakly reducing conditions below the sediment/water interface.

Recently, some workers have pointed out the significance of microbial mats for the uptake of phosphorus and the precipitation of phosphate (O'Brien et al. 1981; Lucas and Prévot 1984), which can also take place in lagoonal environments (Soudry and Lewy 1988).

The uppermost sediment zone with phosphorus release appears to be characterized by a weakly reducing, mildly anaerobic environment (Fig. 5.7b). Hence, nitrate and some ferric iron can be reduced, thus providing ferroan iron for the generation of glauconite.

Sulfate reduction and the precipitation of iron sulfide commonly do not yet take place at this shallow burial depth. These processes require strongly reducing conditions up to the sediment-water interface, which only occur in a very weak current regime. In this case, pyrite may be precipitated simultaneously with phosphate.

Reprecipitation of phosphorus takes place within the sediment or directly below the sediment-water interface (Fig. 5.7b). Due to the concentration gradient between the pore water and sea water, phosphorus tends to migrate upward by diffusion, and fluoride from sea water (with its higher concentration) downward into the sediment. Poorly soluble carbonate fluorapatite will precipitate preferentially where

1. Phosphate nuclei are already available (e.g., in fecal pellets or coprolites).
2. Calcium concentration in the porewater is also high, and/or
3. Calcium carbonate is present.

The phosphate forms ooid-like pellets, nodules, and crusts, or coatings on other materials, particularly on the surface of carbonates and calcareous shells, which can be more or less replaced by phosphate in the course of time.

Secondary Enrichment of Phosphorite

Because the phosphorus content in the organic matter of marine sediments, including those in zones of high fertiliy, is limited, a sediment layer with a certain thickness can produce only a small quantitity of phosphate. Such a situation can be observed in some present-day and ancient sediments which were deposited more or less continuously in zones of upwelling. They contain thin layers of phosphate particles or phosphatized lenses and laminae consisting of fecal pellets, microbial filaments, or foraminifera (Garrison et al. 1987; Föllmi et al. 1991).

Significant concentrations of phosphorite can only be generated by repeated alternation of sediment accumulation, in situ phosphogenesis, winnowing and reworking (Fig. 5.8b and c, also see, e.g., Baturin 1982; von Rad and Kudrass 1984). Sea level changes can favor such a process as mentioned above. In consequence, the net sedimentation rate over a considerable time period approaches zero or becomes even negative, i.e., erosion prevails over deposition. Simplified examples representing such a situation on a large scale are shown in Figure 5.9. Phosphorite is concentrated on submarine highs, on shelf breaks, or in shallow, channel-like depressions, where high wave or current energy leads to maximum erosion and long-term nondeposition. The original parent sediment for primary phosphate precipitation may be completely dissolved and reworked at the site of maximum phosphorite concentration (Fig. 5.8c and 5.9b).

For these reasons, the phosphorite in one individual deposit may show a considerable age range, as identified with the aid of included fossils or radiometric age determinations (e.g., von Rad and Kudrass 1984; Thomson et al. 1984; McArthur et al. 1988).

Furthermore, there have been certain time periods of enhanced phosphorite formation in the Earth's history, for example in the late Proterozoic and Cambrian, Permian, late Cretaceous/early Tertiary, and Miocene (Slansky 1980; Valeton 1988), but phosphogenesis also takes place at some locations in the modern oceans, for example in the coastal upwelling zones off Baja California, Namibia, and Peru. However, as previously mentioned, upwelling is not the only prerequisite for phosphogenesis.

Weathering processes on land in emerged formations can further enhance the concentration of phosphorus which originated from

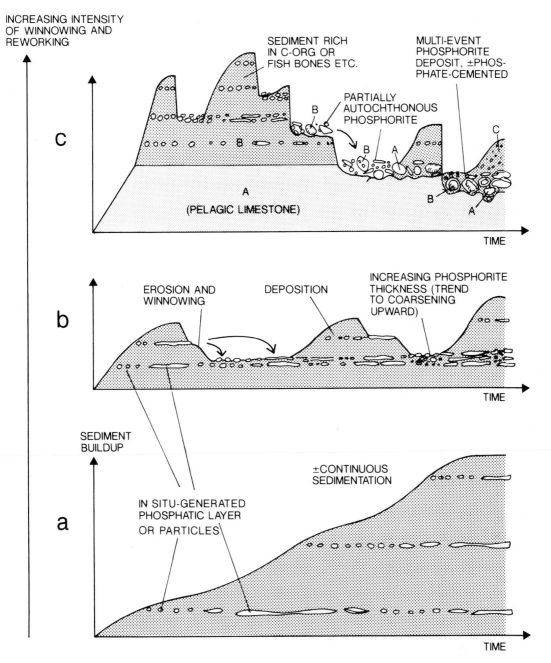

Fig. 5.8a-c. Time-sediment buildup diagrams demonstrating different types of phosphorite deposits in relation to winnowing and reworking. **a** More or less continuous accumulation of sediment rich in organic matter with thin layers of early diagenetic phosphate particles or phosphatic layers. **b** Discontinuous accumulation and winnowing of phosphate-bearing sediment generating increasingly thick, more or less authochthonous layers of phosphorite. Net sedimentation rate approaches zero. **c** Different episodes of phosphorite concentration by winnowing, reworking, and possible redeposition (allochthonous phosphate) of pre-exixting phosphate formed within the sediment. Note that only layer *B* was rich in organic matter or bone fragments delivering P; however, layer *B* may be completely gone except for phosphatized pebbles preserved with pebbles from layer *A* in the final, partly phosphate- or carbonate-cemented phosphorite deposit. Net erosion exceeds net sedimentation. (Based on different sources, e.g., von Rad and Kudrass 1984; Föllmi et al. 1991, greatly modified)

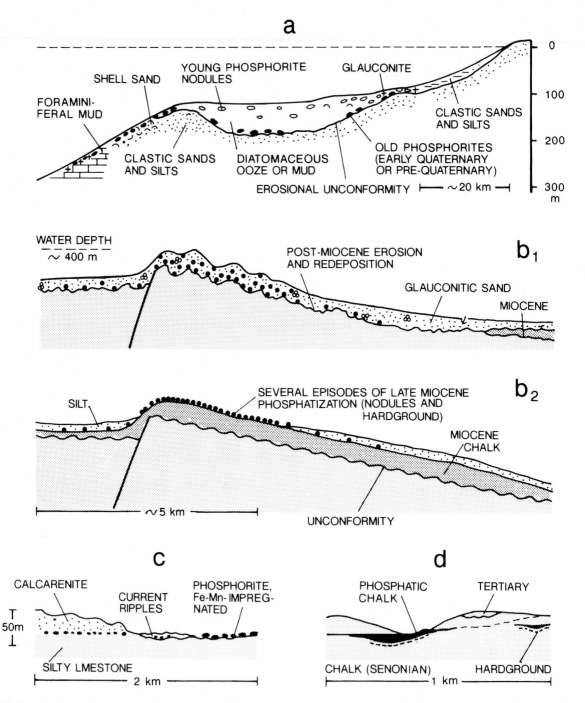

Fig. 5.9. a Schematic cross-section of continental margin demonstrating zones with different sedimentation rates, condensation, and phosphorite ages, similar to the present-day situation off Namibia and Peru. (After Baturin 1982). **b** Phosphorites on the Chatham Rise to the east of New Zealand, in two stages of development (**b1** and **b2**), concentrated at the ridge crest by pronounced erosion and reworking. Note that the primary phosphate-bearing strata (Late Miocene) can only be observed at the ridge flanks. (After von Rad and Kudrass 1984). **c** Model of phosphorite lag deposits of the inner Blake Plateau, derived from Miocene strata and exhumed by the Gulf Stream. (After Manheim et al. 1980). **d** Senonian phosphorites of the Anglo-Paris Basin, concentrated on top of hardgrounds in current-eroded channels and troughs. (After Jarvis 1980)

marine deposits, for example, by preferential dissolution of the accompanying calcium carbonate (also see Valeton 1988).

Co-Existing Authigenic Mineral Phases

The weakly reducing environment enabling phosphogenesis to occur, particularly in combination with repeated winnowing and reworking, is also favorable for the formation and relative enrichment of *glauconitic minerals* (Chap. 6.1). Therefore, this mineral is frequently associated with phosphorite deposits and may occasionally occur in large quantities, as, for example, in the upper Cretaceous of Egypt (Glenn and Arthur 1990). Pyrite may also be present in phosphorites, but is frequently precipitated subsequent to phosphogenesis at somewhat greater burial depths, when sulfate reduction has already started (Fig. 5.7b and c). The occurrence of chert, carbonate nodules and dolomite, as well as calcite cement in conjunction with phosphorite, is assumed to be a result of later diagenesis.

5.3.7 Sediments on Marginal and Oceanic Plateaus, Ridges, and Seamounts

By applying the general rules described in the previous chapters, most of the sedimentary processes observed on deep submarine plateaus, on isolated seamounts, and on ridges and bank tops of block-faulted, topographically structured shelf areas can be easily understood. Their deep position below sea level is frequently the result of long persisting subsidence which, in contrast to shelf seas, could not be compensated for by sediment accumulation. Some simplified examples for such a situation are shown in Fig. 5.10a. In all these cases, sedimentation on older rocks started with shallow-marine deposits, but was later followed by hemipelagic and pelagic material. In this chapter we are dealing only with this second depositional phase.

Deep Plateaus and Seamounts (Weak Currents)

Deep, isolated oceanic or marginal plateaus (Fig. 5.10a) which are little affected by surface and intermediate currents can preserve a fairly complete record of pelagic to hemipelagic sedimentation. Elevated above the CCD,

they tend to be carbonate-dominated, for example, the Rockall Plateau in the North Atlantic (Zimmerman et al. 1984). During certain times and particularly at high latitudes, platform sediments may also become rich in biogenic silica. Sea floor spreading and associated plate tectonic motions may shift their position in the course of time, for example from higher to lower latitudes or from regions of weak to strong oceanic currents, which can lead to unconformities. In such a way their sedimentary facies can gradually change and, in some cases, include periods of equatorial or coastal upwelling with the formation of phosphorite, chert, and glauconite. The sediments on deep plateaus also frequently record minor climatic oscillations, which generate rhythmic or cyclic calcareous ooze (limestone)-marl successions (Diester-Haass 1991) or, in subpolar regions, alternations between layers rich in biogenic carbonate or silica as reported from the Meteor Rise in the South Atlantic (Shipboard Scientific Party 1988a).

In the present-day oceans, such phenomena can be likewise observed at the tops of old, *sunken seamounts* which lie at considerable depths below sea level (more than 100 to 200 m). They are often covered by pelagic material consisting of nannofossils, foraminifers and/or siliceous remains and the corresponding lithified sediments. If they are situated below zones of equatorial upwelling or have passed through such a zone due to plate tectonic motion, they can also be rich in phosphorites.

The likelihood that submarine plateaus and seamounts show a kind of steady-state, undisturbed depositional environment for a very long geological time span is low. Ongoing subsidence of these structures, in conjunction with changes in climate and morphology of the ocean basin involved, may modify or completely alter the regional oceanic current pattern and thus strongly influence terrigenous and biogenic sedimentation at these sites.

Current-Influenced Plateaus

If a fairly deep marginal plateau is subjected to the influence of powerful, deep-reaching surface currents, deposition of pelagic and some benthic carbonate as described above is drastically curtailed or completely prevented (Fig. 5.10a, example for tropical-subtropical waters). The resulting sequence is markedly *condensed* and displays *hardgrounds* with sessile benthos (sponges, serpulids, sometimes deep-water corals, signs of boring organisms etc.). In zones of particularly strong current

a

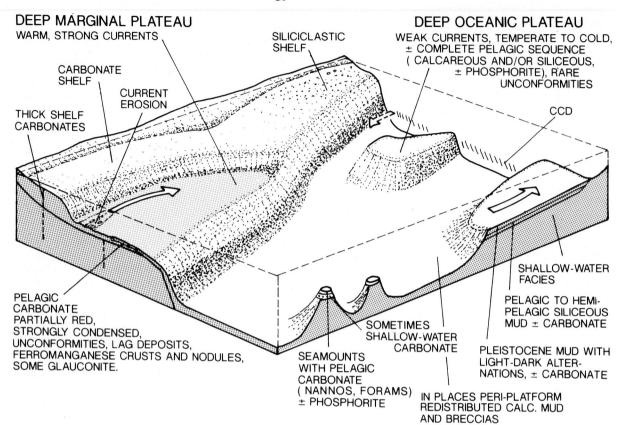

DEEP MARGINAL PLATEAU
WARM, STRONG CURRENTS

DEEP OCEANIC PLATEAU
WEAK CURRENTS, TEMPERATE TO COLD,
± COMPLETE PELAGIC SEQUENCE
(CALCAREOUS AND/OR SILICEOUS,
± PHOSPHORITE), RARE
UNCONFORMITIES

CARBONATE
SHELF

CURRENT
EROSION

SILICICLASTIC
SHELF

CCD

THICK SHELF
CARBONATES

PELAGIC
CARBONATE
PARTIALLY RED,
STRONGLY CONDENSED,
UNCONFORMITIES, LAG DEPOSITS,
FERROMANGANESE CRUSTS AND NODULES,
SOME GLAUCONITE.

SOMETIMES
SHALLOW-WATER
CARBONATE

SEAMOUNTS
WITH PELAGIC
CARBONATE
(NANNOS, FORAMS)
± PHOSPHORITE

IN PLACES PERI-PLATFORM
REDISTRIBUTED CALC. MUD
AND BRECCIAS

SHALLOW-WATER
FACIES

PELAGIC TO HEMI-
PELAGIC SILICEOUS
MUD ± CARBONATE

PLEISTOCENE MUD WITH
LIGHT-DARK ALTER-
NATIONS, ± CARBONATE

b

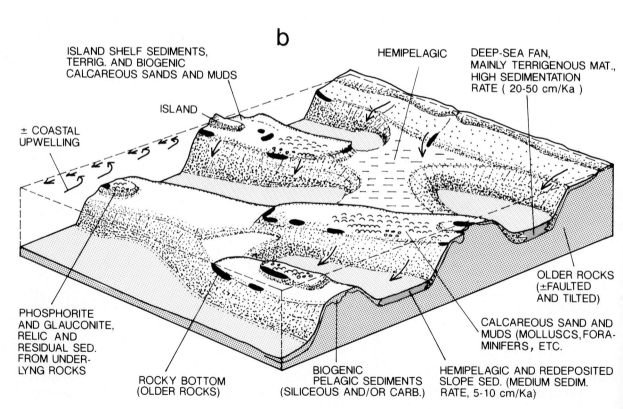

ISLAND SHELF SEDIMENTS,
TERRIG. AND BIOGENIC
CALCAREOUS SANDS AND MUDS

HEMIPELAGIC

DEEP-SEA FAN,
MAINLY TERRIGENOUS MAT.,
HIGH SEDIMENTATION
RATE (20-50 cm/Ka)

ISLAND

± COASTAL
UPWELLING

OLDER ROCKS
(±FAULTED
AND TILTED)

PHOSPHORITE
AND GLAUCONITE,
RELIC AND
RESIDUAL SED.
FROM UNDER-
LYNG ROCKS

ROCKY BOTTOM
(OLDER ROCKS)

BIOGENIC
PELAGIC SEDIMENTS
(SILICEOUS AND/OR CARB.)

HEMIPELAGIC AND REDEPOSITED
SLOPE SED. (MEDIUM SEDIM.
RATE, 5-10 cm/Ka)

CALCAREOUS SAND AND
MUDS (MOLLUSCS, FORA-
MINIFERS, ETC.

action, for example near the shelf slope, *erosional unconformities and lag sediments* are common. At least some of these carbonate sediments will become completely oxidized and later red-colored.

Fossil examples are the widespread occurrences of condensed nodular red limestones of the Ammonitico Rosso type in the Jurassic Tethys (Bernoulli and Jenkyns 1974; also see Chap. 3.3.2).

During times of virtual nondeposition, nodules and crusts of ferromanganese may develop. Other periods allowing some deposition of clay and the preservation of some organic matter may enable the formation of glauconite.

The 800 to 1000 m deep **Blake Plateau** off the eastern Florida shelf belongs to this type of deep plateau (Sheridan and Enos 1979). Its older sediments represent shallow-water carbonates, but at the beginning of the Paleocene deep water environments were established. In addition, the sediments deposited since that time also contain considerable quantities of phosphorite and some chert. The phosphorite occurs in the form of phosphatized carbonate debris, pellets, pebbles and conglomeratic layers, as well as continuous pavements (Manheim et al. 1980, also see Section 5.3.6). It is a lag deposit derived from Miocene strata which accumulated in an oceanographic regime quite different from the present situation under the Gulf Stream.

Another example for a subsiding, sediment-starved, Jurassic-Cretaceous continental margin at low latitudes is the **Mazagan Plateau** off Morocco (Bernoulli and Kälin 1984; Jansa et al. 1984; Winterer and Hinz 1984). Here, the constructional phase of carbonate buildup ended in the early Cretaceous. Hence, greatly condensed, reworked, middle to upper Jurassic reddish limestones are overlain by hemipelagic Tertiary beds.

In cooler, temperate and subpolar waters with less current influence but greater influx of terrigenous material, the pelagic to hemipelagic sediments of a marginal plateau display mixtures of terrigenous mud, and siliceous and calcareous ooze. Depending on changing climate and oceanic circulation, one of these three main components may dominate over the others.

The example in Fig. 5.10a is based on drilling results from the approximately 1000 m deep **Voering Plateau** off the Norwegian coast (Eldholm et al. 1987). Here, pelagic sedimentation began in the Miocene, after the plateau had subsided from a position near the sea level down to a considerable water depth. Since that time, the composition of the 200- to 300 m-thick sediment cover varied between dominantly siliceous, mixed siliceous and calcareous, to dominantly siliciclastic deposits. Carbonate-dominated layers rarely occurred. With the onset of glaciation in the northern hemisphere in the Late Pliocene, the plateau sediments changed. They were replaced by alternating dark and light mud layers. The dark ones are poor in carbonate and contain ice-rafted debris. They represent glacial periods, while the light layers are richer in carbonate and sand deposited during interglacials. Only two groups of benthic organisms, sponges and foraminifers, are abundant throughout the entire sequence, which is more or less bioturbated. The mean sedimentation rate in the Neogene, including the Pleistocene, was around 2 cm/ka and thus considerably higher than for the Blake Plateau. Nevertheless, reworking and hiatuses indicate the frequent activity of bottom currents. In certain time intervals, the biogenic carbonate settling on the sea floor was entirely dissolved within the sediment.

Slopes and Basins Close to Submarine Plateaus

The slopes and basins adjacent to submarine plateaus may be considerably affected by sediments shed from the plateaus (also see Chaps. 12.2.2 and 12.2.3). This occurs, as long as the plateaus are within the reach of oceanic currents strong enough to move sediment. The sediments of slope aprons and peri-platform oozes reflect to some extent the environmental conditions of the submarine plateaus. They have been described from the present oceans and ancient examples (e.g., Mullins and Cook 1986; Schlager and Camber 1986; Schlanger and Premoli Silva 1986; Eberli 1989). On shallow carbonate platforms, aprons and peri-platform sediments are markedly controlled by sea level variations, because carbonate production and thus the sediment supply into deeper water can be high only during sea level highstands. In contrast, deeper plateaus may increase carbonate production during sea level lowstands.

Fig. 5.10. a Sediments on deep submarine plateaus under different environmental conditions, for example temperature, currents, nutrient supply. (Based on several sources, e.g., Sheridan and Enos 1979; Stow 1986; Eldholm et al. 1987). **b** Sediments of complex, segmented shelf (Californian borderland-type) with basins and ridges of various depths below sea level; structural elements of wrench tectonics omitted (see, e.g., Howell et al. 1980; Teng and Gorsline 1989). Note that only the youngest sediments are shown, which may still be influenced by lower sea level stands and temporal variations in the intensity of coastal upwelling. (Based mainly on Emery 1960; Garrison et al. 1987). Older sediments may be strongly affected by alternating up and down movement of crustal blocks, causing unconformities, pinch outs of certain layers, and slump aprons of varying size and position along the foot of the slopes. For further explanations see text

Submarine Ridge and Basin Topography

Some continental margins display a series of topographic basins and ridges with markedly diversified sediments.

Such settings may originate from listric faulting along divergent plate boundaries (provided that sediment accumulation is too low to smooth out the submarine relief), from strike-slip motions, or from the accretion of terranes.

A very simplified, generalized view of one of the most prominent modern examples, the Californian continental borderland (Ingersoll and Ernst 1987), is presented in Figure 5.10b. The narrow elongate basins and ridges of such a borderland have various depths below sea level.

Basin Sediments

Most sediments are delivered to the basins via gravity mass movements (Chap. 5.4). The inner, landward basins are fed primarily by submarine canyons collecting sand and mud from the continent and inner shelf; they are filled mainly by deep-sea fans and reach sedimentation rates on the order of 20 to 50 cm/ka (Gorsline and Douglas 1987; Teng and Gorsline 1989). The outer basins are cut off from direct terrestrial sediment supply. They accumulate pelagic to hemipelagic sediments derived from their slopes and produced in the water column; minor amounts of material may also come from island shelves. As a consequence the sedimentation rate drops to moderate values (5 to 10 cm/ka).

Sediments on Submarine Ridges and Bank Tops

Deposition ontop of these highs is controlled by several factors, the most important ones being distance from river mouths and other siliciclastic sediment sources, climate and oceanic current patterns, water characteristics, and local topographic features. In the example of Figure 5.10b, the climate is semi-arid, sediment supply by rivers is fairly high, and the region is subjected to moderate coastal upwelling. Under these conditions one can distinguish between several zones (Fig. 5.10b, mainly based on Emery 1960; Garrison et al. 1987):

- Rocky bottoms with some sessil organisms, practically devoid of sediment. This situation is observed on hills rising above the level of island shelves and large, flat bank tops, at the outer edges of island shelves and large bank tops, on the upper parts of steep basin slopes, and on the walls of submarine canyons.
- Mainland continental shelves and island shelves. Terrigenous, and mixed terrigenous and biogenic sand and mud, decreasing grain size and increasing carbonate content with depth (for further details see Chap. 3.3).
- Large bank tops at medium depths (about 100 to 300 m). Biogenic sands and muds, relic sediment and residual sediment from underlying rocks, abundant phosphorite and glauconitic sand. Depressions on the flat ridges are filled with sediment.
- Large bank tops at greater depths (≥ about 300 m), far away from the coastline. Thick cover of biogenic pelagic sediment, in this case mainly siliceous. Layers of small nodules of phosphorites and other authigenic sediments are, if present in certain horizons, diluted and buried.

The phosphorite may occur in small grains, nodules, and slabs, or it may form thin coatings on rocky grounds and gravel. It is usually covered by a film of manganese oxide. Frequently, older nodules are cemented together by younger phosphorite to form a nodular conglomerate. The phosphorite is often associated with glauconitic foraminiferal sand (banktop glaucophosphorites), which is found either within or surrounding the nodules. In the Californian borderland, the age of the phosphorite dredged from the surface of large banks is Late Miocene to Quaternary. Hence the net sedimentation rate at these locations was extremely low during the last 10 to 15 Ma. It is, however, not clear whether part of this phosphorite was precipitated directly from upwelling sea water or whether all of it is a product of diagenesis, which would mean that the host sediment was more or less eroded.

5.4 Gravity Mass Flow Deposits and Turbidites

5.4.1 Gravity Mass Flows

Introduction

In the present-day oceans, as well as in many ancient sedimentary basins, the deposits of gravity mass movements form a great portion of the total sediment body (e.g., Kelts and Arthur 1981; Stow and Piper 1984; Thornton 1984; Stanley 1985). Such deposits are particularly important in deep basins nearby continents, and they often result from laminar mass flows and turbulent suspension currents. Widespread, thick flysch sequences in the Alps and other orogenic belts consist predominantly of redeposited sediments. Since around 1950, numerous articles and some special books have been published on turbidites and other gravity mass flow deposits (e.g., Walker 1978, 1984a,c; Stanley and Kelling 1978; Saxov and Nieuwenhuis 1982; Schwarz 1982; Prior and Coleman 1984; Mutti and Ricci Lucchi 1978; Mutti et al. 1984; Mutti and Normark 1987). The flow behavior of the different types of gravity mass movememts is treated for example in Middleton and Hampton (1976), Blatt et al. (1980), Stow (1980), Allen (1982), Lowe (1982), Komar (1970, 1985), Postma (1986), and summarized by Stow (1986).

In this chapter, the most characteristic features and depositional environments of these mass flow deposits in the marine realm are desribed, using a few, simplified conceptual models. The general characteristics discussed for deep-sea environments can also be applied to lake sediments (Chapter 2.5), but large-scale phenomena are commonly missing.

Nomenclature

Up to now, there is no uniform, consistent nomenclature for all the bed-scale phenomena and internal sedimentary structures resulting from gravity mass flows and turbidity currents of various magnitude and grain size distribution. Bouma (1962) proposed a simple nomenclature to describe the succession of internal sedimentary structures in sandy turbidites (T_a through T_e), which has generally been used for many years. This nomenclature is well suited for medium-grained sandy beds, but it suits neither distinctly proximal types of relatively coarse-grained turbidites nor pure mud turbidites (see below) or mass flow deposits. Therefore, additional terms and symbols were introduced (Lowe 1982; Walker 1984a, c; summary in Stow 1986). However, these different nomenclatures are difficult to combine into one broad, consistent system. For that reason, new, purely descriptive symbols are proposed (Fig. 5.14), which are used in all the other figures of this chapter.

For example, sandy turbidites are discriminated from mud turbidites by using the symbols *TS* and *TM*, respectively; *ig* signifies inversed grading, *lm* laminated mud, etc. All types of redeposited beds can be described with these symbols. Carbonate turbidites may be referred to as *CTS* and *CTM*. This system can be expanded or modified as needed. For the more large-scale phenomena of deep-sea fans, additional terms are briefly mentioned in Chapter 5.4.2.

Types of Gravity Mass Movements

The most important gravity mass movements found in both ocean and lake basins are summarized in Fig. 5.11. They can be subdivided into several groups:

- **Mass movements of lithified, jointed rocks**: Rockfall along coastal cliffs or steep submarine slopes and fault scarps (Fig. 5.11a). The transport distance of such fallen rocks is very limited, unless there is a possibility of a composite mass movememt, as indicated in Fig. 5.11i.

- **Creep, sliding and slumping** of semi-solid to soft sediments (Fig. 5.11c through e) on slopes of various angles (as little as a few degrees). Movement takes place if the shear stress exceeds the shear strength of the sediment at some depth below the sedimentary surface, which is usually tested by stability analysis (Fig. 5.11b). The *shear stress* increases with the slope angle and depth below the sea floor. Deep below the surface of a gentle slope, the shear stress can be as high as on a steep slope at shallow depth. For this reason, there is a tendency for thick mass movements to develop on gentle slopes, whereas thin ones are characteristic for steep slopes. The shear stress can, in addition, be significantly enhanced by earthquakes and, in shallow water, by the effect of storm waves on the sea bed. Furthermore, the sudden loading by an approaching slide or slump often generates secondary failure planes and propagation of the mass movement. Similarly, a drop in sea level or tectonic uplift can lead to an additional loading of emerging sediments which when losing buoyancy may trigger slope failure.

The *shear strength* of sediments usually increases with depth below the sea floor, but is reduced by high pore pressure in underconsolidated sediments. Such a situation frequently occurs in areas where fine-grained sediments are deposited rapidly, for example in front of deltas. A further reduction in shear strength is caused by the rather common development of biogenic gas in the uppermost tens of meters in slope sediments rich in organic matter. Finally, the release of methane from crystallized gas hydrates due to increasing water temperature (resulting from climatic changes or shifting of current systems) may locally diminish the shear strength of sediments and thus cause mass movements.

In creep and slides, the sediment masses do not change their mechanical state, i.e., they move as a kind of rigid plug without sig-

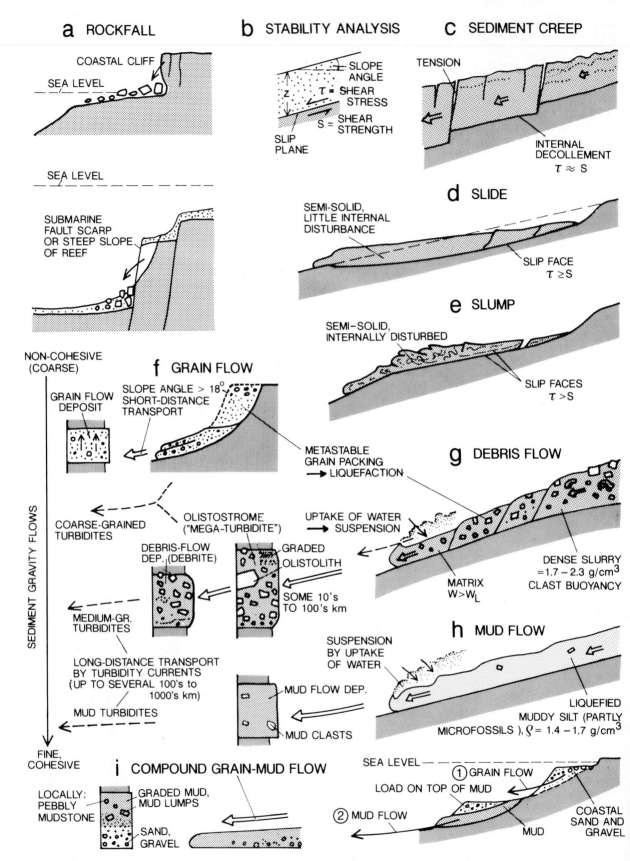

a ROCKFALL

COASTAL CLIFF

SEA LEVEL

SEA LEVEL

SUBMARINE
FAULT SCARP
OR STEEP SLOPE
OF REEF

b STABILITY ANALYSIS

SLOPE
ANGLE

z τ = SHEAR
STRESS

S = SHEAR
STRENGTH

SLIP
PLANE

c SEDIMENT CREEP

TENSION

INTERNAL
DECOLLEMENT
$\tau \approx S$

d SLIDE

SEMI-SOLID,
LITTLE INTERNAL
DISTURBANCE

SLIP FACE
$\tau \geq S$

e SLUMP

SEMI-SOLID,
INTERNALLY DISTURBED

SLIP FACES
$\tau > S$

NON-COHESIVE
(COARSE)

f GRAIN FLOW

GRAIN FLOW
DEPOSIT

SLOPE ANGLE > 18°
SHORT-DISTANCE
TRANSPORT

METASTABLE
GRAIN PACKING
→ LIQUEFACTION

g DEBRIS FLOW

UPTAKE OF WATER
→ SUSPENSION

COARSE-GRAINED
TURBIDITES

OLISTOSTROME
("MEGA-TURBIDITE")

DEBRIS-FLOW
DEP. (DEBRITE)

GRADED
OLISTOLITH

SOME 10's
TO 100's km

MEDIUM-GR.
TURBIDITES

MATRIX
W>W$_L$

DENSE SLURRY
=1.7 – 2.3 g/cm³
CLAST BUOYANCY

h MUD FLOW

SUSPENSION
BY UPTAKE
OF WATER

LONG-DISTANCE TRANSPORT
BY TURBIDITY CURRENTS
(UP TO SEVERAL 100's TO
1000's km)

MUD TURBIDITES

MUD FLOW DEP.

MUD CLASTS

LIQUEFIED
MUDDY SILT (PARTLY
MICROFOSSILS), ϱ = 1.4 – 1.7 g/cm³

SEDIMENT GRAVITY FLOWS

FINE,
COHESIVE

i COMPOUND GRAIN-MUD FLOW

SEA LEVEL

① GRAIN FLOW

LOAD ON TOP OF MUD

② MUD FLOW

LOCALLY:
PEBBLY
MUDSTONE

GRADED MUD,
MUD LUMPS

SAND,
GRAVEL

COASTAL
SAND AND
GRAVEL

MUD

nificant internal disturbance. The distance of transport is very short. Slumps show considerable internal disturbance, slump folding, and frequently several slip faces (Fig. 5.11e). They often evolve into debris or mud flows.

- **Sediment gravity flows.** Flow may occur in several different ways:

1. *Viscoplastic flows* with internal shear planes and virtually no movement at the base of the flow. Such gravity movements usually need a rather high slope angle (5 to 10°) and reach limited thicknesses.
2. *Slide-debris flows,* or *slideflows* move as a more or less rigid plug over a basal shear zone, where the water pressure is in excess of hydrostatic pressure and thus reduces the shear strength of this material, which may even be liquefied. Such flows can reach great thicknesses and occur on very gentle slopes (0.1 to 1°). In addition, they can develop erosional features in the form of broad channels.
3. *Liquefied mass flows.* As a result of their high in-situ water content and meta-stable grain packing, sediment masses on subaqueous slopes can be frequently transformed by earthquake shocks into fluids of high density and viscosity. Partial or entire remolding of the sediment creates a small surplus of pore water which cannot immediately escape. As a result, the shear strength of the material drops drastically and approaches zero without the uptake of additional water (Einsele 1989). The liquefied masses start to flow downslope, even on very gentle slopes ($\leq 0.5°$), and become further remolded and disorganized. Similarly, already moving slump masses may be converted into slow, plastic debris flows or mud flows. Typical examples of this type of liquefaction are non-cohesive or low-cohesive sands and silts, but it appears that many sediments rich in diatoms, nannofossils and other micro-organisms are also susceptible to this process.
4. *Grain flows* consisting of pure sand are characterized by their frictional strength. To overcome friction between the grains, a kind of dispersive pressure must develop. This can only be achieved on fairly steep slopes (18 to 37°), as at the head of submarine canyons and on some prodelta slopes. Grain flows require an environment with ready supply of sand, and they usually travel only short distances. The deposits of individual grain flow processes are thin (several centimeters) and may show reverse grading (see, e.g., Shepard and Dill 1966; Lowe 1976; Dingler and Anima 1989).

Because many sediment gravity flows evolve from laminar to fully turbulent systems (flow transformation, see also Lowe 1979; Postma 1986), an exact correlation of natural flows to idealized flows (fluidized flow, liquefied flow, grain flow, mud flow or cohesive debris flow) is often difficult. In the oceans, the most important flow types are mud flows and cohesive debris flows (see below). They contain varying proportions of mud, which provide them with cohesive matrix strength supporting larger particles. When their excess pore water dissipates, the flow masses come to rest.

By uptaking additional water from the overlying water body, individual gravity flows or parts of them can evolve into masses of lower density and viscosity and, if there is a long, sufficiently steep gradient, finally generate turbulent suspension currents of high velocity (turbidity currents).

- **Compound mass movements** result from the sudden loading of slope sediments by other masses, e.g., slides, slumps, or grain flow deposits originating from higher slope areas (Fig. 5.11i). In this way, the underlying sediment is transformed into a loaded, undrained condition with reduced shear strength.

Moore DG et al. (1982) reported an interesting case from the southern Gulf of California, where coastal sand and gravel were first transported as a grain flow and debris flow via a steep submarine canyon, down to a mid-fan and lower slope region (around 2500 m deep). There, the superposition of a great load of coarse clastics onto siliceous silty clays (porosity 70 to 80 %) triggered a second, considerably larger mass movement on a much gentler slope (about 1.5°), which carried the coarse material down to a 3000 m deep marginal basin plain (slope angle 0.1°).

Fig. 5.11a-i. Summary of submarine mass movements and gravity mass flows (based on many sources, e.g., Middleton and Hampton 1976, Walker 1978, Moore DG et al. 1982, Prior and Coleman 1984; Stow 1986, Einsele 1989). Rockfalls (**a**), slides (**d**), and slumps (**e**) occur if the shear stress exceeds the shear strength of the rock or sediment at some depth (**b** stability analysis). **g** and **h** Debris and mud flows usually originate from slides and slumps by liquefaction of the primary, metastable grain packing (in situ water content, w, greater than liquid limit, w_l). Turbidity currents evolve by uptaking of additional water. **i** A composite, two-step grain-mud flow mechanism can explain long-distance transport of pebbles and gravel into the deep sea

The mass spread over an area of approximately 300 km^2, forming a sheet several tens to about 100 m thick. Due to the remolding and differential settling of the flow mass particles, the coarse material now forms the base of the mass flow deposit, 70 km away from its source area.

From this example one can see that such compound mass movements provide a mechanism for transporting coarse gravel over long distances and on very gentle slopes into the deep sea.

Occurrence, Volume, and Transport Distance of Gravity Mass Movements

Preferential Sites of Mass Movements

Figure 5.12 summarizes the most important tectonic and environmental settings for the occurrence of large gravity mass movements. Most of theses settings provide both high influx of sediment and strong relief.

Marine deltas exhibit a variety of mass movements, including mud diapirs and large-scale creep generating growth faults (Fig. 5.12a). Very common are slides and mud flows on extremely gentle slopes in shallow water as described for example for the Mississippi delta (Prior and Coleman 1984); even more widespread are the *huge deep-sea fans* in front of the deltas of major rivers. They are fed directly by suspension currents from the river, or by slumps, mud flows, and turbidity currents originating at the prodelta slope. Well documented examples include the 1929 Grand Banks slump and turbidity current off the Laurentian Channel of North America (Piper and Shor 1988; Hughes Clarke et al. 1990) and the 1979 mass-wasting at the prodelta slope of the Var river in the northwestern Mediterranean (Auffret et al. 1988). The deep-sea fans of major modern rivers (e.g., Bengal, Indus, Mississippi, Amazon, Congo fan) are 250 to 2500 km long and are cut by channels 5 to 25 km wide (Stow 1981; Mutti and Normark 1987).

Similarly, *submarine canyons* collecting shallow-water sediments from the foreshore zone, or directly from river input, can generate large deep-sea fans of many tens to several hundreds of kilometers in length. These fans tend to have high sand contents and even may incorporate some gravel, in contrast to many slope aprons which are predominantly fed by fine-grained, muddy slope sediments. Rapid subsidence and active faulting during rifting and early drifting (right-hand side of Fig. 5.12a) can create an environment particularly favorable for extensive gravity mass movements along the *shelf break* of young basins (Bourrouilh 1987; Eberli 1987). Due to "shelf break erosion", the debris flows may carry soft and lithified material of various age.

Subduction-related depositional environments provide several possibilities for different types of gravity mass movements. *Deep-sea trenches* adjacent to a main continent (not shown in Fig. 5.12b), such as the modern Peru-Chile trench, frequently receive high quantities of river material, and therefore can be rapidly filled up with both sandy and muddy gravity flow deposits (Thornburg and Kulm 1987). Trenches far away from significant terrestrial sediment sources (Fig. 5.12b) are fed by slumps and debris flows originating from an accretionary wedge as well as young, autochthonous slope sediments. For that reason, they often contain material of varying age and nature (polymict clast composition), sometimes including ophiolites and metamorphic rocks. The sedimentary fill of *forearc basins*, and to some extent that of *backarc basins* (not shown in Fig. 5.12b), is usually characterized by a high proportion of volcaniclastic material, provided by a volcanic arc and transported by gravity mass movements into the basin.

Redeposited gravel-sized rock fragments and skeletal material (rudites), sand-sized shell fragments (arenites), and finer-grained lutites (including pellets) play a great part in deep-sea carbonate depositional environments (Fig. 5.12c). The main sources of these materials are *carbonate shelves* and *isolated carbonate platforms*. Due to early differential lithification and the extremely steep slope of many reef buildups, rock falls including large boulders are fairly common. Oversteepened slopes may also lead to *collapse events*, generating large-scale slumps and debris flows. Turbidity currents transport coarse and fine-grained shallow-water carbonate farther basinward (see e.g. Remane 1960; Wilson 1975; Cook and Enos 1977; Scholle et al. 1983a; McIlreath and James 1984; Eberli 1987).

Large mass movements in particular tend, as mentioned above, to evolve from slides or slumps into debris and mud flows which in turn may lose part of their mass to turbidity currents (see below).

a

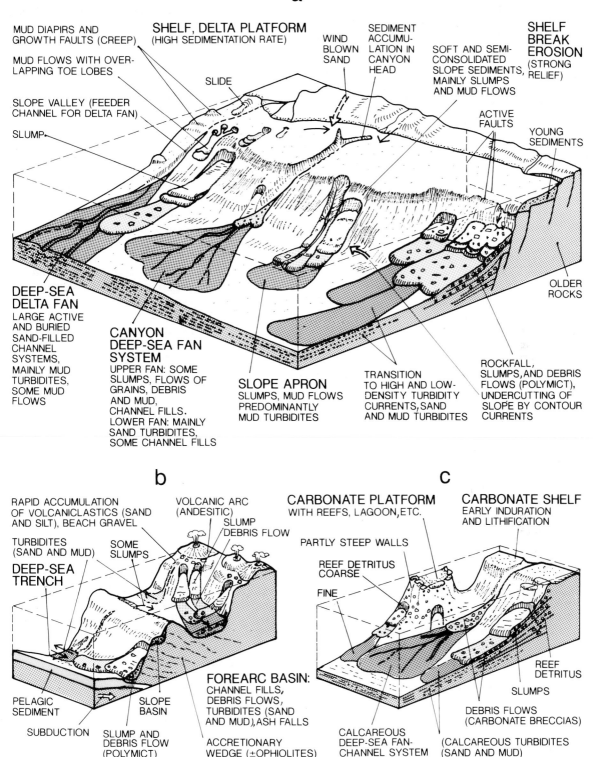

MUD DIAPIRS AND
GROWTH FAULTS (CREEP)

MUD FLOWS WITH OVER-
LAPPING TOE LOBES

SLOPE VALLEY (FEEDER
CHANNEL FOR DELTA FAN)

SLUMP

SHELF, DELTA PLATFORM
(HIGH SEDIMENTATION RATE)

SLIDE

WIND
BLOWN
SAND

SEDIMENT
ACCUMU-
LATION IN
CANYON
HEAD

SOFT AND SEMI-
CONSOLIDATED
SLOPE SEDIMENTS,
MAINLY SLUMPS
AND MUD FLOWS

SHELF
BREAK
EROSION
(STRONG
RELIEF)

ACTIVE
FAULTS

YOUNG
SEDIMENTS

OLDER
ROCKS

DEEP-SEA
DELTA FAN
LARGE ACTIVE
AND BURIED
SAND-FILLED
CHANNEL
SYSTEMS,
MAINLY MUD
TURBIDITES,
SOME MUD
FLOWS

CANYON
DEEP-SEA FAN
SYSTEM
UPPER FAN: SOME
SLUMPS, FLOWS OF
GRAINS, DEBRIS
AND MUD,
CHANNEL FILLS.
LOWER FAN: MAINLY
SAND TURBIDITES,
SOME CHANNEL FILLS

SLOPE APRON
SLUMPS, MUD FLOWS
PREDOMINANTLY
MUD TURBIDITES

TRANSITION
TO HIGH AND LOW-
DENSITY TURBIDITY
CURRENTS, SAND
AND MUD TURBIDITES

ROCKFALL,
SLUMPS, AND DEBRIS
FLOWS (POLYMICT),
UNDERCUTTING OF
SLOPE BY CONTOUR
CURRENTS

b

RAPID ACCUMULATION
OF VOLCANICLASTICS (SAND
AND SILT), BEACH GRAVEL

TURBIDITES
(SAND AND MUD)

DEEP-SEA
TRENCH

SOME
SLUMPS

VOLCANIC ARC
(ANDESITIC)

SLUMP
/DEBRIS FLOW

PELAGIC
SEDIMENT

SUBDUCTION

SLOPE
BASIN

SLUMP AND
DEBRIS FLOW
(POLYMICT)

ACCRETIONARY
WEDGE (±OPHIOLITES)

FOREARC BASIN:
CHANNEL FILLS,
DEBRIS FLOWS,
TURBIDITES (SAND
AND MUD), ASH FALLS

c

CARBONATE PLATFORM
WITH REEFS, LAGOON, ETC.

PARTLY STEEP WALLS

REEF DETRITUS
COARSE

FINE

CARBONATE SHELF
EARLY INDURATION
AND LITHIFICATION

REEF
DETRITUS

SLUMPS

DEBRIS FLOWS
(CARBONATE BRECCIAS)

CALCAREOUS
DEEP-SEA FAN-
CHANNEL SYSTEM

(CALCAREOUS TURBIDITES
(SAND AND MUD)

Fig. 5.12a-c. Summary of the most important depositional environments, where large gravity mass movements and turbidite sedimentation take place. **a** Passive continental margins, including an early rifting-drifting stage with active faults. **b** Convergent margin with forearc basin. **c** Reposition associated with carbonate shelf and platform. (Based on different sources, e.g., Stow 1986)

Volume of Dislocated Sediments

In the present-day oceans, a number of extremely large slides have been reported, the volume of which reaches 1000 to 20 000 km³ (summary in Schwarz 1982); many mass movements range between 0.001 and several 100 km³. The same orders of magnitude are characteristic of the volume of large mud flows and turbidity currents which evolved from slides and slumps. Of course, there are also numerous smaller displacements of sediments with volumes between 10^3 and 10^6 m³, which are usually not further described in reports on the modern sea floor. However, this is the category in which we may notice local, large mass movements or thin, more widespread turbidite beds in normal field exposures.

Travel Distances

Debris flows and mud flows can travel distances of several 100 up to 1000 km, as observed in the present-day oceans (Akou 1984; Simm and Kidd 1984); turbidite flows may redeposit material as far as several 1000 km away from its primary location. In ancient rocks (Eastern Alps, Apeninnes, Pyrenees) it is possible to trace specific marker beds across 100 to 170 km (e.g., Hesse 1974; Ricci Lucchi and Valmori 1980; Mutti et al. 1984). These observations clearly indicate that gravity flow deposits have a high potential for being widely distributed in relatively large and deep basins.

Deposits of Debris Flows and Mud Flows

The principal features of debris flows and mud flows are summarized in Fig. 5.13. They reflect the flow process prior to deposition (Lowe 1982), which can be characterized as a more or less laminar, cohesive flow of a comparatively dense, sediment-fluid mixture of plastic behavior. The sediments of these mass flows are supported by the cohesive matrix and their clasts, at least partially, by *matrix buoyancy*. Blocks float on the debris as a result of small density differences between the blocks and the debris, plus the cohesive strength of the clay-water slurry (Rodine and Johnson 1976). Debris flows come to rest as the applied shear stress drops below the shear strength of the moving material. The flows "freeze", which is accomplished either by *cohesive freezing* or, in the case of a cohesionless sandy matrix, by *frictional freezing*, or by both processes.

Debris flow deposits (debrites, DF) and *olistostromes* (very thick, extensive debrites) consist of a medium to fine-grained matrix and a varying proportion of *matrix-supported clasts* (Fig. 5.13a). The typical debrite is rich in clasts of different sizes; the clasts may be derived from older sediments and rocks within the basin (intraclasts), or from sources outside the basin (extraclasts). In olistostromes, single clasts or blocks can reach the size of a house and more. They are called *olistoliths* and resemble blocks or intact rock bodies that are frequently observed in tectonic mélange zones. Internally, most debrites lack any bedding phenomena or imbrication of clasts. Except at the top of the debrite, in situ traces of burrowing organisms are completely missing. In some examples, elongate clasts are aligned horizontally, indicating the direction of flow.

The base of debrites may be scoured or, when particularly coarse-grained material on steep slopes is involved, show plug-flow channels (Cas and Landis 1987); the basal sediments often display a thin sheared zone. The lowermost part of a debrite is frequently inversely graded, due to prograding frictional freezing. The higher portion may exhibit indistinct normal grading; the top of the bed is either sharp or grades into an overlying turbidite, thus forming a compound debrite-turbidite couplet (DF-TS/TM, see below). In places, the top of a debrite may be current-winnowed and therefore transformed to a clast-supported layer. In some cases, a debrite or mud flow deposit is directly overlain by a subsequent debrite or an overlapping lobe of the same mudflow (also see Fig. 5.12a). The two layers are then combined by amalgamation.

Calcareous debrites resulting from large-scale slope collapse often form sheet-like megabreccia beds, which are primarily clast-supported and contain little fine-grained matrix material (Mullins and Cook 1986). Brecciation of semi-lithified platform carbonates may also result from tectonic deformation, for example during rifting, and thus generate "internal breccias" which can be incorporated into debris flows (Füchtbauer and Richter 1983).

Mud flow deposits have much in common with debrites Fig. 5.13b), and there is no sharp boundary between these two end-members of the same group. They have a muddy matrix with a high silt (or micro-fossil) content and contain only a small amount of clasts, mostly intraclasts which are frequently deformed by

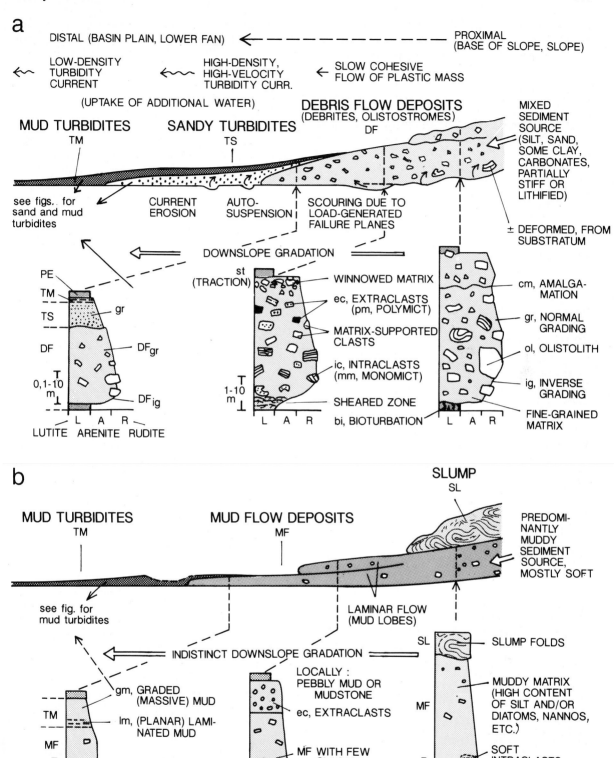

Fig. 5.13a,b. Conceptual models for proximal and distal debris flows (**a**) and mud flows (**b**), finally evolving into high and low-density turbidity currents. (Based on different sources, e.g., Middleton and Hampton 1976; Lowe 1982; Stow 1986; Bourrouilh 1987; Souquet et al. 1987). See text for further explanation

the preceding slumping and flow process. Locally, the admixture of gravel or other coarse material from submarine canyons leads to pebbly mud or mudstone.

Due to their structureless, homogenized matrix and relation to earthquakes, volcanic eruptions, and tsunamis, the terms "unifites" (Feldhausen et al. 1981), "homogenites" (Cita and Ricci Lucchi 1984) and, for thick beds of frequently compound origin, "megaturbidites" or "seismoturbidites" (Mutti et al. 1984) are used by some workers to characterize gravity flow deposits of exceptionally large volume and areal extent.

Elmore et al. (1979) described a modern megaturbidite, 500 km long, more than 100 km wide, and up to 4 m thick, of upper Pleistocene age from the Hatteras abyssal plain in the western Atlantic. This bed consists predominantly of fluvially derived sand and shelf mud with a large proportion of mollusc shell fragments. In proximal regions, the poorly sorted lower part of the bed ($\geq$20 % mud) may have been deposited as a sandy debris flow, whereas its upper part and more distal portions reflect deposition from a turbidity current. Another compound, but carbonate-bearing, debrite-turbidite was observed in the Exuma Sound, Bahamas (Crevello and Schlager 1980). This bed is 2 to 3 m thick and covers an area of more than 6000 km^2. Hicke (1984) reported a Holocene example from the Ionian abyssal plain in the Mediterranean. Here, a 12 m thick homogenized mud layer containing around 50 % carbonate, partially from intermediate and possibly even from shallow waters, covers an area of 1100 km^2 in about 4000 m of water. Locally, the layer has a sandy base composed of shell fragments.

Couplets of debrites (mud flow deposits) with sandy and muddy turbidites, indicated in Fig. 5.13 by the symbols DF-TS-TM or MF-TM (see Fig. 5.14), were also described from several ancient sedimentary sequences (e.g., Stanley 1982; Mutti et al. 1984; Bourrouilh 1987; Souquet et al. 1987).

5.4.2 Turbidites and Deep-Sea Fan Associations

General Characteristics of Turbidites

Sediments deposited from suspension currents show a variety of distinctive features which vary depending on the magnitude and velocity of turbidity currents, the material in and the distance from the source area, the morphology of the basin, as well as other factors. In fact, turbidity currents were postulated as a mechanism for producing graded, sheet-like sand beds (sandy turbidites) from having studied rhythmic bedding and their internal structures in ancient rock sequences (Kuenen and Migliorini 1950). This fascinating concept can explain all the different observations in the field with one group of simple processes:

- The episodic transport of large volumes of shallow-water sands, including their fauna, into the deep sea and their distribution over wide areas.
- Marked linear erosion or nondeposition in submarine canyons (turbidity current feeder channels) and widely extended erosion of the uppermost centimeters of the pre-existing, soft, deep-sea sediments on deep-sea fans and basin plains.
- The generation of certain bed types, as well as vertical and lateral successions of sedimentary structures, including traces of bottom dwelling organisms.

Reports of breakage of submarine telegraph cables on continental slopes testify to the great power of turbidity currents. This occurred, for example, as a result of the famous Grand Banks earthquake 1929 off Newfoundland (Piper and Normark 1982; Piper and Shor 1988), as well as in front of submarine canyons at the mouth of some large rivers. The Grand Banks turbidity current eroded deeply and entrained sediment from the sea floor over a distance of at least 200 km (Hughes Clarke et al. 1990). However, due to their infrequent occurrence in relation to human life spans, such large-scale turbidity currents could never be directly observed in operation. Direct measurements were performed on low density undercurrents in lakes or water reservoirs, as well as in some small-scale flume experiments.

Suspension or turbidity currents result either directly from suspended sediments delivered by rivers in flood state, or from unstable submarine sediment accumulations which have failed. In the marine realm, the density of suspensions caused by river floods is, however, usually not high enough to produce *density currents* in the sea (density of sea water 1.027 g/cm^3). Therefore, the *failure of large sediment accumulations* in shallow waters is probably the most important prerequisite for turbidity currents. Such accumulations are common on prodelta slopes, or they are accomplished by longshore currents (Chap. 3.1) carrying material into the head of submarine canyons. Rip currents and storm action generally transport sediment into deeper water and to the shelf edge.

Turbidity currents evolve from high-density gravity mass flows such as slumps, grain flows or mud flows (Fig. 5.15a) by uptaking addi-

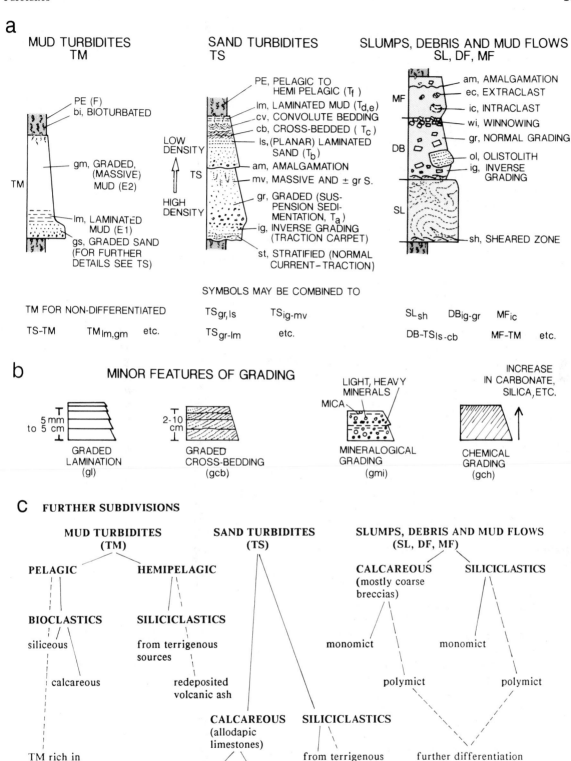

Fig. 5.14. a Descriptive terms and symbols for the internal structures of redeposited sediments (association of slumps, debris and mud flows, sand and mud turbidites); Bouma divisions *in parentheses*. Many other characteristics such as sole marks and trace fossils (lebensspuren) are omitted. **b** Minor features and different types of grading. **c** Further subdivisions of the main groups; additional terms and symbols may be introduced as needed

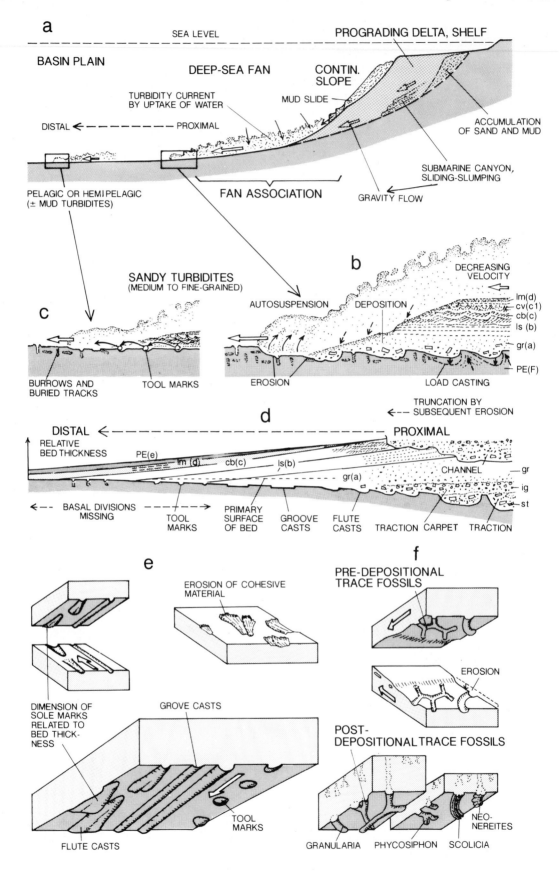

tional sea water (or lake water). As a result, the sliding or flowing masses become less dense and therefore turn from a slow cohesive flow into a more rapid, *turbulent density flow*. The sediment particles lose their mutual contact and go into suspension. They are kept in this state by the upward component of fluid turbulence. The driving force of the turbidity current is primarily a function of the difference in density of the suspension and the overlying water body (in contrast to river flow under air!), the submarine relief, i.e., the angle and length of slope, and the thickness of the suspension current. Thus, high-velocity suspension currents can be achieved only when at least two of these factors become comparatively great.

This is usually true for submarine slopes along continental margins and deep, narrow ocean basins. Although these slopes usually have angles of only a few degrees, they are much steeper than the gradients of large, subaerial rivers (0.1 to 0.01°). In addition, submarine mass movements also can produce turbidity currents of considerable density (up to 1.10 to 1.17 g/cm^3) and thickness (several hundred meters, see e.g. Piper and Shor 1988). In combination, these factors can cause high-velocity currents (up to 10 to 20 m/sec).

High- and Low-Density Turbidity Currents

We distinguish between high- and low-density turbidity currents. *High-density turbidity currents* reach high velocities and can carry relatively coarse-grained sand, pebbles, and intraclasts. Within the confines of submarine channels, large-scale turbidity currents have the competence to transport gravel (up to at least 10 cm in diameter) as bed load, and thus may generate lenses of conglomerate at the foot of prodelta slopes (Komar 1969; Piper and Shor 1988). In addition, they have the capacity to erode cohesive muds on extensive areas of the sea floor. The eroded material feeds the suspension with new sediment (a kind of feedback system), which replaces coarser material

settling out of suspension in the slackening body and tail of the current (Fig. 5.15b). In this way, and by the maintenance of turbulence by gravitational forces (auto-suspension), the current is kept in motion and can travel over long distances.

Low-density turbidity currents flow slowly and can therefore keep only silt and clay-sized material in suspension, or larger aggregates consisting of particles of these sizes. Their erosional capacity is very low or nonexistent, but weak turbulence maintains such suspension currents for relatively long periods of time. They can attain considerable thicknesses and distribute their suspended load as a thin bed over wide areas. It can be assumed that low-density, muddy turbidity currents often are the final stage of sand-bearing suspension currents, which have lost their coarser grain size fraction (Chough 1984; Stanley 1985).

Proximal and Distal Turbidites

With the aid of internal sedimentary structures in the turbidite beds and different sole marks (Fig. 5.15e and f), one can distinguish between *proximal* (near-source) and *distal turbidites* (Fig. 5.15d). However, this concept should be used with caution, particularly in deep-sea fan associations (see below; further discussion in Macdonald 1986; Stow 1986). Proximal turbidites are often relatively coarse-grained and thick-bedded, and their tops may be truncated by a subsequent turbidity current (amalgamation). Downslope, the turbidites successively tend to lose their basal divisions; they become thinner and finer-grained. In many turbidite sequences, a striking correlation between the dimensions of sole marks and the thicknesses of the corresponding sandy beds can be noticed. For example, large flute casts or groove casts are often associated with particularly thick sandstone layers (Fig. 5.15 e and f). This and the following observations strongly support the turbidity current hypothesis explaining all features of a turbidite by one single sedimentological event.

Fig. 5.15a-f. Conceptual model for the generation of proximal and distal sandy turbidites and their internal structures. **a** Transition from slope failures (slumps, debris and mud flows, grain flows) on continental slopes and canyon heads to turbidity currents. **b** "Classic", complete turbidite (more or less proximal) showing the total succession of Bouma divisions *(in parentheses)* with modified descriptive symbols (Fig. 5.14a), as well as autosuspension due to turbulence. **c** More distal turbidite, no erosion, basal Bouma divisions missing. **d** Idealized proximal-distal development of turbidite bed; proximal channel fills may show sedimentary structures due to traction by normal currents *(st)*, as well as *traction carpets (ig* inversely graded). **e** Different sizes of flute and groove casts in relation to bed thicknesses. **f** Pre-depositional and post-depositional trace fossils. (After Seilacher 1962; Kern 1980)

Pre- and Post-Event Trace Fossils

As Seilacher (1962) has pointed out, we can distinguish between *pre-event trace fossil associations* living in the pelagic to hemipelagic mud interval, and post-event *assemblages* which recolonize a freshly deposited turbidite sand layer. The burrows of the first group are exhumed by the erosive force of the turbidity current and filled up with sand (lebensspuren on the sole of the sand bed, Fig. 5.15g). The recolonizing assemblage has to dig down into the sandy layer from a new, higher level Fig. 5.15h). If the turbidite is thin, some of the burrowing organisms may reach its base and feed on the background sediments. If the sandy layer is thick, only its top sections can be burrowed, and only those assemblages able to live on sand can persist.

Coarse-Grained Turbidites

High-density turbidity currents carrying pebbles and clasts as bed load and finer-grained material in suspension may show an initial stage of *traction sedimentation* (coarse-grained conglomeratic sand with plane lamination, cross-bedding, and internal scour). This is then overlain by thin, horizontal layers displaying inverse grading. This division represents *traction carpet deposits* resulting from mixed frictional freezing and suspension sedimentation (Lowe 1982). The uppermost division of such a proximal turbidite is generated by rapid settling from suspension *(suspension sedimentation)*, and may be either structureless or normally graded. Typical features are water-escape structures (pillar and dish structures). This division more or less corresponds with the lowermost division of the "classic" sand turbidite (Bouma division T_a; here referred to as TS_{gr}, see Fig. 5.14).

Medium-Grained Sandy Turbidites (Siliciclastics and Carbonate)

Sedimentary Structures of Individual Beds

The divisions of a "classic", medium-sized sand turbidite deposited from suspension currents of moderate density can be interpreted in a similar way as above. The divisions consisting of plane laminated and cross-bedded sand (Figs. 5.14 and 5.15, ls and cb; Bouma division T_b and T_c) reflect traction structures, while the division showing laminated mud, lm

(Bouma T_d), may be explained as mixed traction/suspension sedimentation. Finally, the following structureless and indistinctly graded mud interval, gm (Bouma T_e), originates solely from suspension sedimentation. There is also an important group of fine-grained mud turbidites consisting only of the muddy divisions, lm and gm (Bouma divisions T_d and T_e), which are described further below.

Upward-decreasing grain size (normal grading) can be observed not only within division gr (Bouma T_a) and generally from bottom to top in a turbidite bed, but often within the parallel laminated and small-scale cross-bedded divisions, ls to lm (Bouma T_b to T_e, Fig. 5.14b). In these cases, the thicknesses of individual laminae or cross-bedding sets decrease from bottom to top; this is known as *graded lamination,* or *graded cross bedding*.

It is, however, important to note that the presence and degree of grading in a turbidite bed are also controlled by the availability of a range of grain sizes. Well sorted material in the source area does not allow the formation of distinctly graded beds.

An extreme example of this type, indicated in Fig. 5.12a, is nongraded, well-sorted, yellowish sand turbidites in the eastern Atlantic, which are derived from Saharian desert sands. These were blown by offshore winds during glacial low sea level stands onto the shelfbreak, and carried by slumps and turbidity currents into the deep sea (Sarnthein and Diester-Haass 1977).

Bed Sets and Sequences

Vertical sequences in predominantly sandy turbidites do not show only one bed type, for example thick and partially amalgamated, relatively coarse-grained, so-called proximal turbidites, but usually also display thinner and finer-grained beds of a more distal appearance. Nonetheless, a certain trend from sequence to sequence is often apparent (Fig. 5.16a). Such vertical variations in bed types are common in channelized deep-sea fans (see below), where thick-bedded channel fills differ greatly from levee deposits and overbank sediments.

If turbidity currents reach abyssal areas below the carbonate compensation depth (CCD), they can drop calcareous beds alternating with pelagic clays free of carbonate. This fact has often been used for paleoceanographic reconstructions (e.g., Berger and von Rad 1972; Hesse 1975; Hesse and Butt 1976). Furthermore, sandy and muddy turbidity currents can carry organic matter from terrestrial

a   PREDOMINANTLY SILICICLASTIC SANDY TURBIDITES

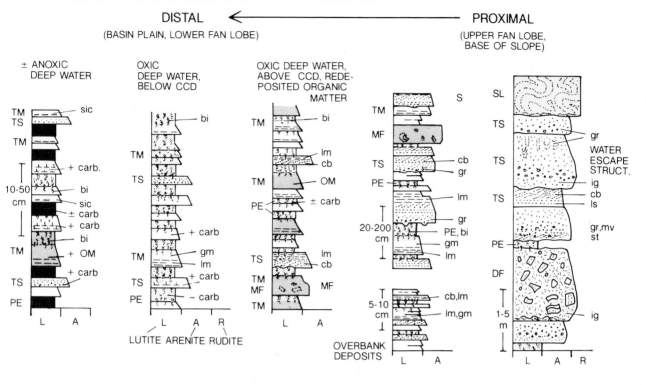

b   GRADED CALCARENITES (ALLODAPIC LIMESTONES)
(AND ASSOCIATED CARBONATE SEDIMENTS OF SLOPES)

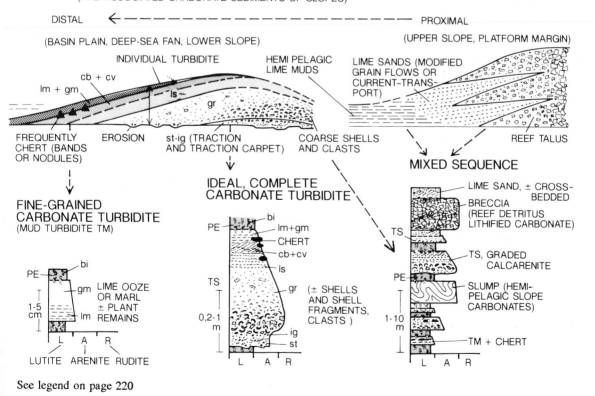

See legend on page 220

sources and low-oxygenated slope areas into well-oxygenated deep basins, where it may be preserved as allochthonous organic material (Fig. 5.16 a). Contrarily, autochthonous black shales can alternate with turbidites containing little organic material and displaced fauna characteristic of an oligotrophic environment.

Carbonate Turbidites (Allodapic Limestones)

So far we have not considered the *composition* of the principal material forming sandy turbidites. Essentially there are two groups of sand-sized materials: siliciclastic sands and carbonate sands, which may, of course, occur in all kinds of mixtures. In the case of quartz and feldspar being the main constituents, one observes that grains of equal diameter behave hydraulically in a similar way. If settled from suspension, a poorly sorted sand with a "wide" unimodal grain size distribution will build up a regularly graded bed, without distinct "jumps" in grain sizes, or enrichments of certain minerals in particular divisions of the bed.

In contrast, carbonate sands consisting chiefly or partly of skeletal material produced on carbonate platforms may behave quite differently. Therefore, turbidite sands and muds rich in carbonate have often been treated separately from siliciclastic turbidites, and called *allodapic limestones* (Meischner 1964). In this case, grains with diameters much greater than 2 mm may behave hydraulically like quartz sand, or sand-sized microfossil shells may be transported and settle in a manner similar to compact silt grains. As a result, the graded carbonate layer deposited from a turbidity current can show distinct jumps in its vertical internal succession of grain sizes and/or sediment composition.

For example, the graded division, gr (Bouma division T_a), can be replaced by a comparatively coarse shell bed, and instead of a carbonate silt division, lm (T_d), a chert layer may be present, which was formed diagenetically from hydraulically equivalent siliceous sponge needles and radiolarians (Fig. 5.16b).

Due to the uptake of eroded deep-sea mud, the amount of autochthonous fauna (nekton, plankton) often increases toward the top of a turbidite bed.

Redeposited carbonate sediments are often coarser than their siliciclastic counterparts. The primary source of sand-sized and larger carbonate grains are the margins of carbonate shelves and platforms (Figs. 5.12c and 5.16b). Abundant shell material, reef detritus, and early lithification of different types of carbonates provide various coarse-grained materials. Hence, allodapic limestones may alternate with carbonate breccias and sands derived directly from platform margins. The maximum thickness of an individual bed is not commonly attained in the neighborhood of the source area, but at some distance downcurrent (Eder et al. 1983); it then decreases distally.

Mud Turbidites

General Aspects

Overlooked for a long time was the fact that a great part of the fine-grained muddy sediments in a deep basin or submarine fan may also be transported by low-density turbidity currents (Stow and Piper 1984; Piper and Stow 1991), instead of being distributed and accumulated by normal, more or less steady pelagic settling.

Mud turbidites may be regarded either as an end-member of gravity mass flows of mixed granulometry, or they may be derived from muddy sediment sources such as prodelta slopes and other fine-grained slope sediments. In addition to gravity movements, large river floods or muddy sediments stirred up by storms in shallow seas can contribute to the formation of mud turbidites. The sediments of several modern submarine fans (e.g., Nile cone, Indus fan, see below) on the continental rises of present-day large oceans, and in smaller oceanic basins (e.g., Black Sea, Gulf of California), consist to a large degree of mud turbidites. Because these beds are deposited from low-density suspension currents, they

---

Fig. 5.16. a Vertical successions of pelagic or hemipelagic beds and event deposits (gravity mass flows, sandy and muddy turbidites) in more proximal or distal regions and in different chemical environments *(CCD* carbonate compensation depth). Note the different vertical scales. Thin-bedded overbank deposits may occur in mid-fan regions, as well as in more distal areas (cf. Fig. 5.18). Organic matter

(OM) is either autochthonous or allochthonous, i.e., laterally transported into the basin by mud turbidites. **b** Proximal and distal carbonate turbidites more or less associated with platform margins and slope sediments; various vertical scales. (After Meischner 1964; Scholle et al. 1983a; McIlreath and James 1984)

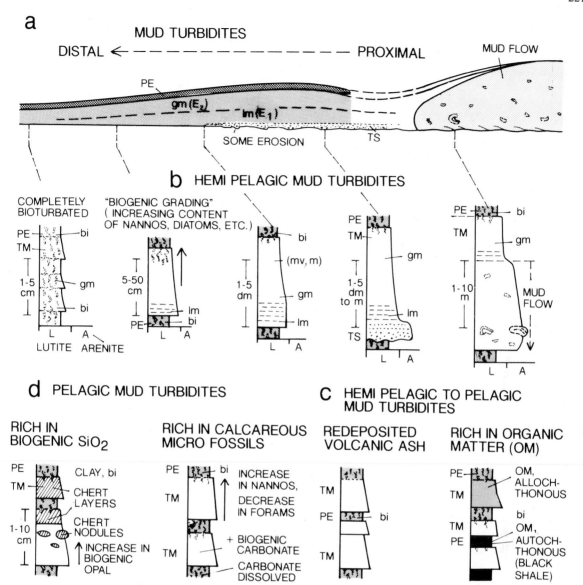

Fig. 5.17a-d. Different groups of mud turbidites. **a** Proximal-distal trend. **b** and **c** Hemipelagic, predominantly siliciclastic (**b**), volcaniclastic, or rich in organic matter (**c**). **d** Pelagic mud turbidites, rich in either carbonate or opaline silica

usually are thin; but rather thick, proximal mud turbidites also occur, for example in the Gulf of California (Einsele and Kelts 1982), probably representing a transitional stage from mud flows to muddy suspension currents (hyperconcentrated flow). The hydraulics of muddy suspensions and their behavior during sedimentation is described by Piper and Stow (1991).

Types of Mud Turbidites and Their Internal Structures

Similar to sandy turbidites, mud turbidites can be subdivided into proximal and distal types (Fig. 5.17). Taking into account the source and composition of their material, Kelts and Arthur (1981) distinguish between *hemipelagic, pelagic* (fine-grained bioclastics), and *volcaniclastic (ash) mud turbidites*. The most characteristic features of mud turbidites are a sharp basal contact to the underlying bed, internal positive grading, and bioturbation at their tops, if the environment at the sea bottom was oxygenic. Grading may become very indistinct, but in modern examples, the bulk densities of such turbidites were still found to decrease towards the tops of such beds. Therefore, it appears often sufficient to distinguish between a laminated division, lm (according to Piper E1, in conjunction with Bouma's divisions a to e, or A to E), and a structureless, indistinctly graded division, gm (E2).

Proximal mud turbidites may contain a thin sand layer at their base, which displays some of the Bouma divisions for sand turbidites. *Distal mud turbidites* of the hemipelagic group, and many mud turbidites of the pelagic group often become so thin that they are completely reworked by burrowing organisms. When this happens, their nature as mud turbidites is obscured, and they can only be recognized if their material and/or fauna differ substantially from the pelagic or hemipelagic background sediment (host sediment, PE).

Because biogenic components are often a significant source material for marine muds, many pelagic and hemipelagic mud turbidites contain hydraulically sorted, slightly altered microfossil assemblages (Brunner and Ledbetter 1987). Because foraminifera, calcareous nannofossils, diatoms or radiolarians behave hydraulically differently, mud turbidites may also show *chemical grading*. Consequently, the carbonate content of an individual sandy or muddy turbidite can increase or decrease from bottom to top; likewise, biogenic silica, later forming chert layers, may be concentrated in an upper division of the turbidite.

Frequently, both biogenic carbonate and silica are transported by turbidity currents into deep basins below the carbonate compensation depth (CCD), where the background sediments are poor in or free of carbonate or opaline silica. In this way, a succession of layers alternatingly with and without carbonate or biogenic silica *("banded" sequences)* can be produced. Similarly, mud turbidites containing high amounts of organic matter from their source area (e.g., slope sediments under regions of upwelling) can alternate with deep sea sediment poor in organic carbon (Fig. 5.17c). In this case, the organic matter is rapidly buried, and therefore at least partly preserved from decomposition in oxygenated environments. Examples of this mechanism are known from the present-day oceans, e.g., from the Cape Verde Rise or the Biscaya abyssal plain in the eastern Atlantic (Kelts and Arthur 1981, Degens et al. 1986). In contrast,

"normal" black shale sediments slowly deposited by vertical settling from the overlying water body in anoxic deep water can be repeatedly interrupted and modified by interbedded, muddy turbidites poorer in organic matter, as in the eastern Mediterranean (Stanley 1986).

Deep-Sea Fan Associations

Large-Scale Phenomena

The "classical" hypothesis, which explains lateral facies changes within individual turbidite beds in terms of proximal-distal trends, was developed for turbidites which originate from a highly efficient sediment source and are deposited on the plain of elongate basins *(type I turbidite deposits*, according to Mutti and Normark 1987). This type may include the fringe of the lower, unchannelized part of deep-sea fans or detached fan lobes (Fig. 5.18, D), where turbidity currents are not affected by submarine channels and their levees. This situation is also common at the foot of continental slopes, where relatively low sediment input by slope failures creates a nonchannelized slope apron (Fig. 5.12a). The predominance of such a "line source" is also characteristic of calcareous deep-sea sands and muds which were produced on carbonate shelves and platforms and later transported into deeper water.

However, as mentioned earlier and as indicated in Fig. 5.16 a, the sedimentary facies of many deep-sea fans fed by sediment input from large rivers, i.e., from an efficient point source, are strongly affected by a *channelized sediment distribution system* (Fig. 5.18 A through E; *type II turbidite deposits* of Mutti and Normark 1987). Several thoroughly studied modern examples have shown that submarine canyons do not end at the foot of the slope, but mostly continue into depositional fans and basin plains (e.g., Barnes and Normark 1985; Mutti and Normark 1987). In addition, the long-term input of large volumes of river sands generates migrating, sinuous *channel-levee complexes* with crevasse splays similar to those of subaerial meandering rivers (Nelson and Maldonado 1988; Damuth et al. 1988). The channels of large submarine fans are many kilometers wide and have gradients of the order of 1 %. In the upper and middle fan region, their levees may rise above the surrounding sea floor by tens and, in extreme cases, more than 100 m. Large slumps and

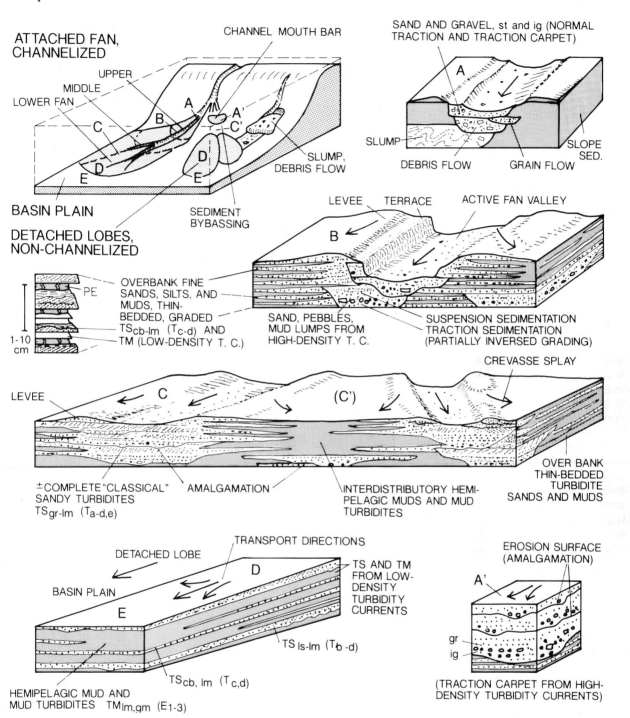

Fig. 5.18. Model of the facies association of a deep-sea fan. (Based on several sources, e.g., Mutti and Ricci Lucchi 1978, Nelson et al. 1978, Walker 1978, Shanmugam and Moiola 1985, and author's own observations).

Note the difference between channelized, attached fan and detached, nonchannelized fans which show a more regular turbidite sequence. For explanation of symbols see Fig. 5.14

debris flows may bury parts of the pre-existing channel-levee system from time to time and thus cause the formation of a new distributory system. In addition, smaller mass flows down the backside of the levees have been observed (Amazon fan, Damuth et al. 1988).

Sediment Distribution System

From the study of both modern and ancient sediments, several models for sediment distribution in such deep-sea fans have been developed (e.g., Mutti 1977; Mutti and Ricci Lucchi 1978; Nelson et al. 1978; Walker 1984a, c; Shanmugan and Moiola 1985; Mutti and Normark 1987). All of them show a great variety of bed types in the channel system, its levees and overbank regions, and the lobes of the upper to lower fan region.

Moving down the feeder channel, one may encounter more or less *channelized deposits* of cohesive mass flows, coarse grained sands and conglomeratic sandstones. The latter two are generated either by normal current traction transport or by mixed frictional freezing and suspension sedimentation *(traction carpet* displaying inverse grading, ig, Fig. 5.18, A). In addition, normally graded beds preferentially showing divisions gr and ls (Bouma $T_{a,b}$), as well as indications of vertical water escape (dish and pipe structures), are common. These channel fills cut into or pass laterally into thick or thin-bedded turbidites of the more classical types on the levees and fan lobes (Fig. 5.18, B and C).

In the transition zone between relatively high-gradient channelized flow and unconfined flow on a gentler gradient at the beginning of a depositional lobe *(channel-lobe transition),* turbidity currents may suddenly change their flow regime from rapid to more tranquil flow (Mutti and Normark 1987). Such a *hydraulic jump* is accompanied by increased turbulence and enlargement and dilution of the suspension flow. The sea bed is frequently marked by large-scale scour features, mud clasts, and rapid deposition of sand and coarser material. Mud-dominated currents tend to deposit the bulk of their suspended load downstream of the hydraulic jump. In the zone of channel-lobe transition, they usually cause less scouring and leave only some cross-stratified sands.

Whereas thin, relatively dense, fast turbidity currents tend to flow basinward within the confines of the channels and their levees, low-velocity, less dense thicker suspension currents

build up *levees* and drop their fine-grained load in *interchannel areas (type III turbidite deposits).* Within these deposits, sands are almost entirely restricted to the infills of minor channels. The direction of flow may be deflected from the main channel by Coriolis forces. Thus, originating in the upper fan area, the turbidity currents form thin-bedded *overbank deposits* (Fig. 5.18, B and C), showing predominantly the upper divisions of turbidite sands, cb, lm, and gm (Bouma T_{c-e}). Correlation of such beds becomes difficult even over short distances, since levee erosion can take place locally. Fine-grained intercalations between these sand layers for the most part represent redeposited material: mud turbidites. Downslope, on the fringe of the lower fan and on the basin plain, large-scale turbidity currents usually deposit extensive sand and mud layers of the more distal type (Fig. 5.18, D). Only very large, rare debris and mudflows and their subsequent turbulent flows spread their load over large areas of the total fan and basin plain.

According to a comparative study of sand layers in present-day ocean basins (Pilkey et al. 1980), the *percentage of sand layers* in the total sediment volume decreases distally, as also observed in ancient flysch sequences. The thickest layers were found in basins which have large drainage areas. Single sand beds could be traced over distances as great as 500 km (Hatteras abyssal plain in the western Atlantic). A more detailed description of the characteristics of deep-sea fans, the adjacent basin plains, and slope aprons is given by Stow (1986) and Mutti and Normark (1987).

Active and Inactive Phases of Fan Deposition

An ideally prograding slope-fan association with a more or less fixed channel system produces an upward increase in *proximality,* i.e., a coarsening, thickening sequence. However, as a result of shifting channels and changing sites of mass movement on slopes, vertical sections of deep-sea fan sediments can show a succession of beds of apparently widely differing proximality (Fig. 5.18, B and C, and Fig. 5.16a). Normal, current-transported material may alternate with debrites and mud flow deposits, sandy and muddy turbidites, and hemi-pelagic or pelagic sediments. *Switching of fan lobes* in conjunction with migrating channel systems (Fig. 5.19a) can generate both fining (and/or thinning) upward as well as coarsening (and/or thickening) up-

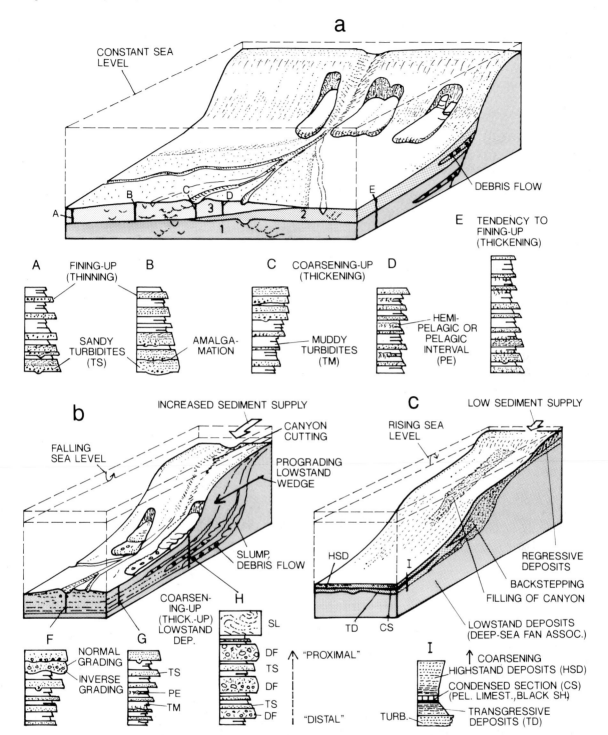

Fig. 5.19a-c. Facies association and development of deep-sea fan. **a** Constant sea level and steady position of continental slope, permanently high sediment input via submarine canyon; switching fan lobes *(1, 2, 3)* and migrating fan valleys. Note that both fining (and/or thinning) upward *(A and B)* and coarsening (and/or thickening) upward sequences *(C and D)* occur; the overall tendency is fining upward *(E)*. **b** Relative sea level fall favors rapid prograding and upbuilding of fan due to high sediment input. Most of the sections tend to coarsen (and/or thicken) upward *(F, G, H)*. **c** Relative sea level rise and resulting large reduction in sediment supply may terminate fan growth. Channel fills and turbidites are replaced by fine-grained transgressive deposits *(TD)*, a thin, condensed section *(CS* black shales, pelagic ooze, etc.), and highstand deposits *(HSD)* of increasing thickness (again, coarsening upward). (Partially based on Walker 1978; Bally 1987, greatly modified)

ward sequences. Under constant sea level and a persisting slope, or a landward retreat of the slope, vertical aggradation of fan sediments causes a slight tendency for fining-upward sequences to develop (Fig. 5.19, E). Similarly, substantial basin subsidence may lead to an aggradational fan system with insignificant vertical but marked lateral facies change (Macdonald 1986).

In response to the amount of sediment input by their feeder system, deep-sea fans undergo *phases of rapid progradation* and/or upbuilding, or *periods of inactivity,* including some reworking. The former case is usually correlated with a lowering of sea level, when the gradients of rivers entering the sea are steepened, and former coastal and shallow water sediments are eroded and swept into deeper waters. As a result of slope and fan progradation, one may get coarsening (and/or thickening) upward stratigraphic sequences (Fig. 5.19b, sections F through G). In contrast, sea level rise usually leads to reduced sediment supply from terrestrial sources and may bring about the filling of submarine valleys and canyons. Consequently, gravity mass movements usually come to an end during such periods, and fan deposits may be draped by normal hemipelagic to pelagic sediments (Fig. 5.19c, section I). The transition from the transgressive to the highstand phase may be characterized by particularly reduced sediment accumulation. The resulting condensed section is often represented by pelagic oozes, limestones, or thin black shales. For further consequences caused by sea level variations see Chap. 7.4.

Example: the Mississippi Deep-Sea Fan

The Late Pliocene-Pleistocene Mississippi fan in the Gulf of Mexico is one of the most thoroughly studied modern deep-sea fans. It represents an example of a mud-dominated fan fed by a major river whose delta migrated from southwest to northeast with time (Figs. 5.20a and 5.21a). Similarly, the channel system of the Indus deep-sea fan, the second largest submarine fan in the world covering an area of $1.2 \cdot 10^6$ km^2, migrated from west to east (Mchargue and Webb 1986; Kolla and Coumes 1987). Both canyon cutting and fan deposition were strongly affected by eustatic sea level changes of large amplitude and high frequency.

In the case of the Mississippi fan, at present, only the youngest feeder canyon is preserved, which was cut more than 500 m deep into older shelf and slope sediments during the late Pleistocene (prior to 30 ka; Goodwin and Prior 1989). Older canyons are completely filled with sediments and incorporated into the prograding slope, which also includes localized slope fans. Since 19 ka B.P., even the youngest canyon was partially filled with the material of a prograding delta and deposits of gravity mass movements, which occurred along the originally steep canyon flanks (Fig. 5.20c, d, and e). Thus, the canyon widened by retrogressive slope failure and mass movement processes resulting in arcuate re-entrants which contain slumped debris and residual knolls (Fig. 5.20c). Erosional unconformities within the channel fill indicate two to three episodes of deposition and subsequent erosion (Fig. 5.20e). The canyon was rapidly filled with prodelta sediments of a lowstand delta, which alternate with mud and debris flow deposits (Fig. 5.20d). About 7.5 ka B.P. the delta moved northward away from the canyon head. Since that time, hemipelagic muds drape the channel fill.

The Mississippi fan sediments are up to 4 km thick and can be subdivided into 13 to 17 depositional sequences (Figs. 5.20b and 5.21c), reflecting the same number of sea level oscillations (Feeley et al. 1990; Weimer 1990). Most of the fan sediments accumulated during the last 2.4 Ma at rates up to more than 1000 m/Ma (for more details see Wetzel and Kohl 1986). The location of the older, buried channel systems in the upper and middle fan region was investigated using multifold seismic data (Weimer 1990).

The result is a complicated pattern of superposed, mostly branching channels (Fig. 5.21b) which generally young from west to east. The individual sequences consist of channel-levee systems, which may be branched, and associated overbank deposits (Fig. 5.21c). Several sequences are bounded by erosional unconformities truncating the upper portion of the pre-existing fan surface. Deposits of slides and gravity mass flows frequently form the base of a new depositional sequence. They are interpreted to have formed during a lowering sea level, whereas the channel-levee systems were deposited when the sea level was near its lowest position. With rising sea level, sediment transport into the deep sea decreased, and during high sea level stand a thin layer of hemipelagic sediment accumulated (condensed section, see above and Chap. 7.4).

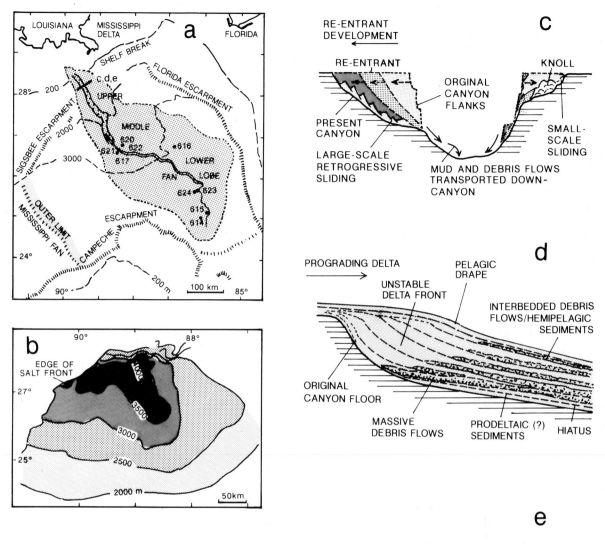

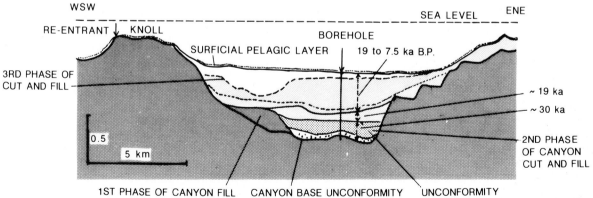

Fig. 5.20. a Extension of the late Pliocene-Pleistocene Mississippi deep-sea fan in the northern Gulf of Mexico. Numbers refer to DSDP drilling sites; *shaded area* is shown in Fig. 5.21b and c in more detail. (After Wetzel and Kohl 1986). **b** Isopach map of upper and middle fan sediments shown in (**a**). (Weimer 1990). **c** Schematic transverse section of late Pleistocene Mississippi canyon on upper slope, indicating widening of canyon by gravity mass movements and erosion subsequent to incision. **d** and **e** Longitudinal and transverse section of canyon, respectively, showing stepwise filling interrupted by phases of erosion which caused unconformities. See text for further explanation. (After Goodwin and Prior 1989)

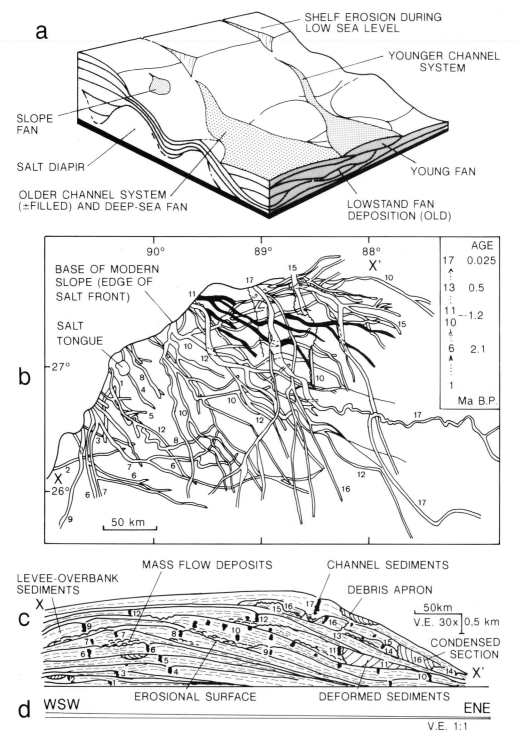

Fig. 5.21. a Very simplified model of shifting, Mississippi-type deltaic canyon-deep-sea fan system. Canyon cutting and fan building occur mainly during lowering sea level. Note smaller slope fans and other complications caused by salt diapirism. (Feeley et al. 1990). **b, c,** and **d** Superimposed channel-levee systems of 17 depositional sequences of late Pliocene-Pleistocene age, revealed by seismic investigations and several drill holes in the upper and middle portions of the Mississippi fan (cf. Fig. 5.20a). Channels are numbered from *1* to *17* (oldest to youngest) and partially dated (Ma before present). **c** Cross section through proximal fan (location shown in **b**). Note eastward migration of depositional sequences with time; strong vertical exaggeration as compared to true vertical scale (**d**). (After Weimer 1990)

Deep-Sea Channels

A continental margin/deep-sea fan system characterized by a submarine canyon and channelized fan lobes may continue into a deep-sea channel which finally ends in an abyssal plain. Such deep-sea channels were observed in the present oceans, where they reach lengths of 2000 to more than 3000 km and widths of a half to several kilometers (Carter 1988; Hesse 1989a). The most prominent example is the Northwest Atlantic mid-ocean channel of the Labrador Sea, which represents a submarine, river-like drainage system with several tributaries. The main deep-sea channels have levees and a low longitudinal gradient similar to large subaerial rivers. The infillings of these channels consists of sandy to gravelly fining-upward sequences far away from any land source. On the levees, parallel laminated, thin mud turbidites were observed (Hesse and Chough 1980; Hesse 1989a). The channel system may end an a kind of submarine braidplain.

Current Directions and Paleocurrent Patterns

Current directions in turbidity currents can easily be derived from sole marks (Fig. 5.15e), internal structures such as cross-bedding, clast and grain orientation etc., and current ripples (see, e.g., Collinson and Thompson 1982).

In several early studies on ancient turbidite sequences, one of the most striking phenomena, besides regular rhythmic bedding, was that the directions of turbidity currents were surprisingly constant over large areas. Later it became evident that these examples predominantly represent basin plain and lower fan environments in elongate basins (type I deposits), which are fed by one main sediment source. In more proximal fan associations as well as in basins supplied with sediment from varying major sources, the paleocurrent patterns become less regular and sometimes rather complex. In cases where the fan lobes have room to switch (Fig. 5.18), the sediments are dispersed radially over the course of time.

Small-scale turbidity currents flowing within the levees of a meandering, distributing channel system are forced to change their flow directions significantly. Larger, thicker currents spill over the levees, are deflected from the main current, and may even cause some levee erosion (Piper and Normark 1983). Slackening currents tend to be further reflected by Coriolis forces. Finally, one should take into account that in certain areas the current patterns of turbidites can be overprinted by contour currents (Stow 1986). Slow, low-density turbidity currents may even turn into contour currents (Hill 1984).

Paleoslope orientations can be inferred from slide scars, slump folds, and sometimes from the imbrication of clasts in debrites and mud flows. For reliable measurements, good, large exposures are needed.

Sedimentation Rates and Frequency of Turbidite Events

Deep-sea fans and adjacent basin plains are areas of high *sedimentation rates* (cf. Chap. 10.2), particularly during relatively low sea level stands or in times of tectonic activity creating increasing relief. In modern fan environments, including Miocene to Pleistocene deposits, average sedimentation rates between 100 and 1000 m/Ma are common, but near the sediment source and in over-supplied basins (Mutti et al. 1984), higher values also occur. The giant Bengal fan, Indus fan, Amazon fan, Mississippi fan, and some other elongate, present-day deep-sea fans represent wedge-shaped sediment bodies which reach maximum thicknesses on the order of 5 to 10 km which have been built up in a time span of a few Ma to 20 Ma (Curray and Moore 1974; Bouma et al. 1985; Bouma et al. 1986; Kolla and Coumes 1987; Damuth et al. 1988). The same applies to many thick, ancient flysch sequences.

The *frequency* of cohesive mass flows and turbidite events is related to the rate of sediment accumulation in their source area, though it also varies greatly among the different fan and basin plain environments (Fig. 5.22). In addition, frequent earthquakes, volcanic eruptions, or rapid uplift in the source area (Klein 1985a) may cause a relatively short recurrence time for redepositional events. As a result, only thin beds of limited areal extension may be generated. There appears to be an inverse (logarithmic) relation between bed thicknesses and frequency in ancient turbidites (Piper and Normark 1983).

In the Sea of Japan, representing a modern backarc basin, one thin-bedded mud turbidite (with an average thickness of 6 mm) has been deposited at time intervals of 50 years (Chough 1984); larger turbidite events on the Toyama submarine fan occurred in the Pleistocene with periodicities between about 3000 and 8000 years, in the late Miocene to Pliocene at time intervals of about 60 000 years (Klein 1985a). Somewhat longer recurrence times have been reported from the Coral Sea basin (Klein 1985a). On the

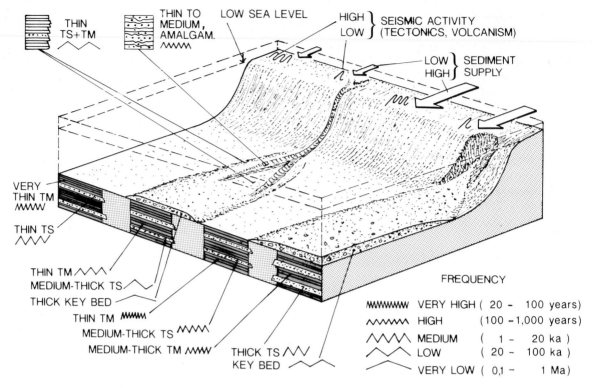

Fig. 5.22. Frequency of turbidite events and bed thicknesses in relation to sediment supply, seismic activity, and location within a deep-sea fan/basin plain facies association

California continental borderland, the recurrence time of gravity mass movements is one event every 200 to 500 years (Malouta et al. 1981). Where river floods generate thin, fin-grained turbidites, as for example in the Santa Barbara basin, recurrence intervals of 50 to 120 years are common (Thornton 1984). From the Mediterranean, Rupke and Stanley (1974) report one mud turbidite, on the average, every 300 to 400 years for the last 20 000 years. In the central and southern part of the Gulf of California, relatively thick (often several dm) Quaternary mud turbidites have been generated with a frequency between 2000 and 10 000 years (Einsele and Kelts 1982); a similar order of magnitude is typical of many sand turbidites with small to medium thicknesses in lower fan and basin plain environments.

In submarine channels and overbank deposits in middle fan environments, Piper and Normark (1983) assume frequencies of one event per 10 to 100 years and 100 to 1000 years, respectively. This signifies that, according to a rough estimation, one out of ten turbidity currents is thick enough to spill over channel levees and deposit its suspended load. Similarly, the frequence of preserved turbidites appears to be less in distal fan regions than in channels and overbank settings (Klein 1985a). However, most of the smaller turbidite events

do not leave much sediment in the higher parts of the channel system, where erosion and amalgamation are common.

The *longest recurrence* times are to be expected for thick mud flows and turbidites (megaturbidites) which form extensive sheets on submarine fans and basin plains (Fig. 5.22). In ancient rocks, such key beds, often rich in redeposited carbonate due to high carbonate production in shallow waters, occur once approximately every 50 000 years to 1 Ma, depending on the amount of sediment supply (Mutti et al. 1984). They may have been triggered by extremely strong earthquakes with very long recurrence times. In a deep-sea drillhole at the foot of the slope of Baja California, only one of these event deposits was found in a mud turbidite sequence representing 3 to 4 Ma (Moore et al. 1982). This, and some other even larger mass movements observed in the present-day oceans, occurred during the last glacial sea level lowstand (16 to 17 ka ago). A famous example is the compound slide/mudflow/turbidity current in the Canary basin (Embley, in Saxov and Nieuwenhuis 1982), where a 10 to 20 m thick mud flow deposit covers an area of 30 000 km^2, and the subsequent turbidity current travelled over 1000 km.

Although in this and other cases earthquakes are quoted as the triggering mechanism, the recurrence time for such "mega-events" is probably controlled primarily by the period of high amplitude sea level changes, as well as by the availability of large volumes of unconsolidated or only partially lithified sediments at the heads of submarine canyons, along the shelfbreak, and on the upper slope. Tectonic activity alone as the cause of large mass movements appears to be an important factor only under certain conditions, for example during the rifting stage of a rapidly subsiding basin or in subduction-related environments.

Concluding Remarks

All gravity mass movements, particularly debris and mud flows, as well as high and low-density turbidity currents, can be regarded as a family of related processes and sediments, including their occurrence and different bed types in the deep sea. As demonstrated in some conceptual models, the more distal types frequently evolve from the proximal ones. Sandy and muddy turbidites show a great variety of bedding phenomena. Proximal carbonate turbidites often tend to be coarser grained than their siliciclastic counterparts; more distal ones may display some special features such as chert nodules. Sandy and muddy turbidites alternate with pelagic or hemipelagic sediments deposited above or below the CCD, in well oxygenated or euxinic bottom waters.

Mud turbidites in particular may contain considerable amounts of allochthonous organic matter as well as datable fauna from their source area. One subgroup of mud turbidites consists predominantly of skeletal carbonate or silica. Textural grading may be complemented or replaced by chemical and mineralogical grading.

The facies patterns of sandy and muddy turbidite beds in deep-sea fan environments, including cohesive mass flows and beds deposited by current traction, differs considerably from the regular bedding and consistent transport directions in lower, detached fan lobes and basin plains. Paleocurrent directions in such systems vary a great deal, and individual beds cannot be traced over long distances. Coarsening and fining-upward sequences may be controlled by several processes.

Submarine fans and basin plains receiving materials from gravity mass movements are regions of relatively high sedimentation rates (100 to 1000 m/Ma and more). The recurrence interval of mass flow and turbidite events varies greatly, from relatively frequent, thin mud turbidites (50 to several 100 years) to thick, extensive key beds ("mega-turbidites", 50 000 to more than 1 Ma). It also varies within a submarine fan association characterized by amalgamated channel-levee complexes and thin-bedded overbank deposits. In addition, the availability of large volumes of unconsolidated sediments and relatively low sea levels are considered a primary control on the abundance and magnitude of gravity mass movements and their deposits. Frequent seismic activity tends to cause more, but smaller mass flows than those found in quieter zones with rare, but larger events.

5.5 Erosion and Reworking of Deep-Sea Sediments

Erosional Features and Depositional Ridges

Recent investigations of the sea floor, aided by drilling, underwater photography, current measurements, and bio- and magnetostratigraphy have increasingly revealed that deep-sea sediments are subjected to winnowing and reworking by bottom currents, as indicated in Fig. 5.3, in many regions (e.g., Heezen and Hollister 1971; Kennett 1982; Tucholke and Embley 1984; Sarnthein and Mienert 1986; Stein et al. 1986). In the North Atlantic, for example, large *erosional furrows* (several meters deep, 100 to 1000 m long) were discovered, which also display smaller-scale sand dunes and current ripples. Of the same size or even larger are *depositional ridges,* which are tens of kilometers wide and have been traced over hundreds of kilometers. They are partly mantled with large flat mud waves a few tens of meters high. Both furrows and ridges form parallel to the bottom currents. The depositional structures are built of mud, silt, and sand sorted out from older sediment. At least part of the coarser fraction consists of skeletal material.

Contour Currents and Contourites

Due to Coriolis forces, equator-directed bottom currents are deflected to the western margin of oceanic basins in both the northern and southern hemispheres. They follow the contours of the continental slope and rise (Chap. 5.2) or the foot of slopes of oceanic plateaus and ridges. Conversely, poleward flowing bottom currents converge along the eastern boundaries of such basins. These so-called *contour currents* evidently sometimes reach velocities (probably ≥30 cm/s) sufficiently strong enough to erode and transport fine-grained sediments. Therefore, the erosional and depositional features mentioned above are particularly well developed along the paths of such contour currents.

In times of strong climatic contrast between the poles and the equator, contour currents are able to undercut continental slopes and trigger very large submarine slides and slumps in certain regions (Sheridan 1981; von Rad and Wissmann 1982). Not only pelagic and hemipelagic sediments are affected by such contour currents or other tractive bottom current processes, but sands and finer-grained materials of deep-sea fans and slope aprons can also be partially reworked and redeposited as *contourites*, or just as layers displaying tractive transport (Stow 1986; Stanley 1988a). For example, the basal, graded subdivision of an individual turbidite (cf. Chap. 5.4.2) may be preserved, while the higher portions are replaced by structures characteristic of normal tractive transport. Such phenomena can be only found if the structures of such current-reworked sediments are not destroyed by contemporaneous or subsequent bioturbation. For that reason, contourites are often masked or difficult to recognize.

On mature passive continental margins, the sediments at the foot of the slope, i.e., on the *continental rise,* form thick wedges which thin seaward (Figs. 5.3 and 12.15). These sediments may include gravity mass flows and channel sediments deposited as slope aprons or deep-sea fans, materials from different provenances reworked and redeposited by contour currents, and normal hemipelagic sediments. The discrimination of fine-grained turbidites, contourites, and pure hemipelagic material in such settings is often difficult, particularly in bioturbated sediments.

Stratigraphic Gaps in the Deep-Sea Record

Seismic records, numerous gravity cores, and hundreds of drill holes from the Deep Sea Drilling Project have shown that erosional features are not restricted to the present sea floor (e.g., Van Andel et al. 1977; Moore et al. 1978; Thiede 1981; Ehrmann and Thiede 1985; Burckle and Abrams 1987). Unconformities and stratigraphic gaps of both short and long duration are ubiquitous in the deep ocean and occur throughout the history of these large basins. In the South Atlantic about one third of the aggregated time represented by all drill sites is occupied by stratigraphic gaps. Generally, more than half of the record is missing in the Paleogene, while in the Neogene between one tenth and one half are missing (Berger 1981). Hiatuses have formed even in periods in which thermohaline circulation is believed to have been sluggish. There is hardly a single location in the present-day oceans where a really complete stratigraphic section can be found. However, because of the limits of biostratigraphic resolution, only long hiatuses can usually be clearly identified.

MacLeod and Keller (1991) have stressed the importance of hiatuses as a significant factor for the far-reaching global events at the Cretaceous/Tertiary (K/T) boundary. They postulate that mass extinction, abrupt shifts in carbon isotope abundances, and specific trace element concentrations in deep-sea sediments are related to long intervals of nondeposition. The stratigraphic hiatuses resulted from a marked sea level rise during the latest Maastrichtian and earliest Danien. This caused the locus of sediment deposition to migrate landward across the continental shelf, where stratigraphic sections are more or less complete over the same time interval (cf. Chap. 7.4). In any case, the effects of incomplete or extremely condensed stratigraphic deep-sea records should be taken into account when rare events such as bolide impacts and their consequences for the evolution of marine organisms are discussed.

6 Special Depositional Environments and Sediments

6.1 Green Marine Clays

Introduction

Sediments greenish in color are fairly common in the present-day oceans at various water depths. They are also frequently observed in ancient sediments, but in these cases the minerals causing the green color may differ from those found in young sediments due to diagenesis. The correct identification, crystallographic and geochemical characterization of fine-grained green minerals is difficult and can be done only in the laboratory using special methods in clay mineralogy. For simple rock descriptions in the field, barely more than the greenish color can be noted, which in turn may be modified by other sediment components such as organic matter and fine-grained particles of different composition.

For these reasons and the unavailability in the past of special investigative techniques, much uncertainty existed about the nature of greenish marine sediments. Most geologists and sedimentologists used the mineral names glauconite, chamosite, and chlorite for green particles, which were often poorly defined. In the last decade, however, much new information has become available in this field (Odin and Matter 1981; Van Houten and Purucker 1984; Odin 1988; Odin and Morton 1988). Some important aspects of our present knowledge are briefly summarized.

Remarks to the Nature of Green Particles

Green particles in the oceans are authigenic aluminum silicates (clay minerals, mica) rich in iron and magnesium. In addition, they may contain various amounts of potassium and other cations. While magnesium and potassium are present in sea water in relatively high quantities, the concentration of iron in oxygenated waters is extremely low. Therefore, normal sea water cannot be the *source of iron*.

Iron is delivered to the oceans by two principal sources (Fig. 6.1): (1) by rivers from the continents in the form of detrital particles and colloids of iron oxyhydrates, and (2) by volcanic activity within the oceans, such as is common at mid-oceanic ridges, island arcs, and other volcanic islands. In the second case, iron can be leached from basaltic rocks either on the sea floor or from deeper sections of the oceanic crust by sea water circulating through basaltic rocks and generating hydrothermal systems.

The green minerals contain either ferrous iron or iron in both the ferrous and ferric states. Thus, their formation is associated either with overall slightly reducing conditions, or with reducing microenvironments within the sediment near the sea floor. The authigenic growth of green pigment is referred to as *verdissement* (Odin 1988) and includes *glauconitization, verdinization,* and *chloritization*.

Greenish Sediments in the Modern Oceans

In the present-day oceans, several zones of greenish or brown, iron-rich sediment facies can be distinguished from shallow to deep water (Fig. 6.1).

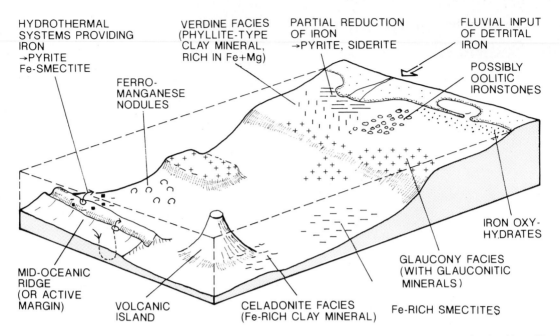

Fig. 6.1. Generalized model demonstrating source and distribution of iron-rich marine sediments, mostly green clays. Note that iron is provided either by fluvial input or by volcanic activity, including submarine leaching of basaltic rocks. See text for explanation of the various facies types. (After Odin 1988)

- *Early diagenetic pyrite and siderite* occurs in sediments in front of rivers entering the sea. The iron is derived from detrital iron minerals which are deposited and frequently redistributed in such areas. In conjunction with simultaneously delivered organic matter, reducing conditions are established within the sediment. Iron is released from terrigenous material and made available as easily soluble ferrous iron to form pyrite and siderite.

- *Verdine facies.* In tropical to subtropical warm regions, this facies zone follows the delta-front zone on the inner shelf (at water depths of 20 to 60 m). The verdine facies occurs in areas of fairly rapid sediment accumulation and is characterized by various marine green clays of the phyllite group. Thus, this facies differs from the glaucony facies (see below) in both the depositional environment and mineral composition. Similar to the glauconitic minerals, these green clays are found as pigment in microtest chambers, fecal pellets, or in pores between other sediment particles. The green clays are predominantly ferric and rich in magnesium. In the present oceans this facies is known from shelf sediments off the mouths of the rivers Amazon, Orinoco, Niger, Senegal, Congo, and other tropical rivers (Odin et al. 1988a). Green marine shales of considerable thickness are also common in the ancient record, for example in the Lower Cretaceous.

- *Iron ooides.* If delta-front and inner shelf sediments rich in iron are frequently reworked and redeposited in areas some distance away from the zone of rapid sedimentation, the ferrous iron compounds may become partially oxidized, transformed to ooids and concentrated as oolitic iron deposits (see Chap. 6.2).

- *Brown iron oxyhydrates.* Detrital iron minerals transported along the coast in shallow water are chemically and/or biochemically altered, particularly so in tropical waters. The iron released is precipitated as brown iron oxyhydrate (limonite, goethite).

- *Glauconitic minerals* commonly form in deeper water along the shelf edge and upper slope or on the tops of submarine highs, i.e., outside the influence of deltaic or various other types of continuous sedimentation. Glauconitic minerals constitute a wide spectrum of minerals from K-poor glauconitic smectites to K-rich micas and are known only from marine environments. Marine sediments characterized by the green pigment of these minerals may be referred to as *glaucony facies* (Odin 1988).

Glauconitization takes place in rather cold waters and is a long-lasting process in environments of slow, discontinuous deposition.

The sediment should contain clay minerals and mica and develop microenvironments with slightly reducing conditions. As a result of the interaction between sea water and such a favorable substrate, the initial minerals are slowly altered to green grains or thin films on the surfaces of larger particles or hardgrounds. Glauconitic minerals may fill small cavities, for example the tests of microfossils, or partially replace fecal pellets and bioclasts. After deep burial, glauconitic minerals are subject to modifications, particularly in their chemical composition.

- *Iron-rich smectite*. In the deep sea, iron is frequently concentrated in either iron-rich smectites or ferromanganese nodules (Chap. 5.3.3). At least some of the iron and manganese required for these sediments is delivered by volcanic activity (known as juvenile iron).

- *Celadonite* or *celadonitic minerals* are typical clay minerals rich in ferric iron and potassium, which have a higher silica content than glauconitic minerals. The iron is derived from oceanic basalts. It is incorporated into clay minerals at the contact with sea water around volcanic islands and along mid-oceanic ridges (Fig. 6.1). Celadonitic minerals fill vesicles, veins, and other small voids in porous lava flows. Celadonite-bearing rocks may be regarded as an additional, independent green facies characteristic of marine environments.

- *Pyrite* and *iron-rich smectite* also occur at active mid-oceanic ridges, where sea water intrudes into basaltic rocks and returns as hot submarine springs to the ocean. The water in such hydrothermal systems is frequently depleted in oxygen and can leach considerable quantities of ferrous iron and other metals (Mn, Cu, Zn, Co, Cd, Pb, etc.) on its way through hot basaltic rocks. The vents of these hydrothermal fluids are known as "black smokers", because they discharge water containing tiny, suspended particles of metal sulfides. These are distributed over the sea floor in the neighborhood of the vents. In this way, sediments rich in pyrite are formed which may pass laterally into sediments characterized by *iron-rich smectites*.

Other portions of the metal sulfides derived from mid-oceanic hydrothermal systems may be precipitated as massive ore deposits on and below the sea floor (called "stockwork mineralization", see e.g. Oberhänsli and Stoffers 1988).

In contrast to earlier views, it is now thought that these various green minerals are not the result of a slow transformation of pre-existing clay minerals and mica, but represent newly formed mineral phases. Each of these mineral groups (verdine, glauconitic minerals, celadonitic minerals) characterizes a specific marine depositional environment. The verdine facies found in shallow seas off the coasts of subtropical regions may also be widespread in ancient rocks and typify rather thick sequences. The glaucony facies, on the other hand, develops on various initial substrates and occurs both at considerable depths (about 50 to 500 m) and across many latitudes. Due to its relation to very low or interrupted sedimentation, the typical glaucony facies is usually restricted to thin strata or special layers in both siliciclastic and carbonate rocks.

6.2 Oolitic Ironstones

Oolitic ironstones are known throughout the Phanerozoic, but seldom from Neogene sediments. Due to the lack of modern analogs, and stirred by substantial economic interest, much has been written and speculated about these perticular deposits. In the last decades, however, mining of these ore deposits has declined, because their iron content is considerably lower than that of the large Precambrian banded iron deposits (Chap. 6.5.2) which are now exploited in several continents.

Depositional Environment

The genesis of oolitic ironstones is still controversial and obviously not fully understood (e.g., Gygi 1981; Maynard 1983; Odin et al. 1988; Schneider and Walter 1988; Young and Taylor 1989). Most of these deposits contain marine fossils and exhibit sedimentary structures and bedding sequences characteristic of shallow-marine environments. However, iron ooids (ooliths) were also found in lakes, for example in Lake Chad, North Africa (Lemoalle and Dupont 1976). The formation of ooids, i.e., sand size grains consisting of a smaller nucleus surrounded by concentric layers of goethite and/or chamosite, requires repeated reworking under turbulent hydrodynamic conditions (Fig. 6.2). This is indicated by the shrinkage, fractioning, and overgrowth of individual ooids. Such an environment should be permanently oxidizing, but the presence of ferrous iron in chamosite, which is a diagenetic chlorite mineral, indicates that the ooids

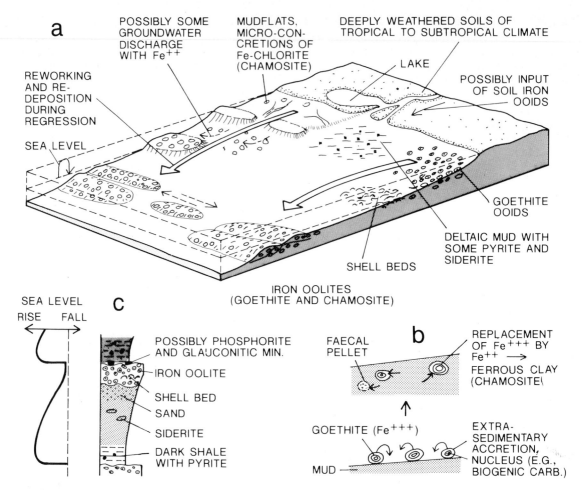

Fig. 6.2. a Model depicting most common hypothesis on iron supply by suspended river load from deeply weathered soils or groundwater rich in ferrous iron. Repeated winnowing, reworking, and final deposition of iron oolites below the sea level of the lowermost stage of transgressive-regressive cycles. **b** Two-stage formation of ooids at the sea bottom (oxidizing conditions, limonite-goethite coatings including some organic matter and silica) and under mud cover (reducing conditions, replacement by ferran clay minerals). Note that fecal pellets can also become enriched in ferrous iron. **c** Idealized transgressive-regressive sedimentary cycle with iron oolites as "roof beds". (Based on several sources, e.g., Maynard 1983; Bayer et al. 1985; Van Houten and Purucker 1984; Odin et al. 1988; Bayer 1989a)

also developed partially within the sediment under reducing conditions.

An alternative hypothesis is given by Dahanayake and Krumbein (1985), who assume that iron ooids form under coatings of microbial films. Fecal pellets and algal-produced oncoid particles may also be transformed into iron-rich ooid size particles due to their organic matter content favoring a reducing microenvironment. In any case, after burial, the primary iron-rich sediment and its constituents are more or less modified by diagenetic overprint, including the precipitation of some pyrite and/or siderite. Carbonate ooids may be replaced by iron compounds (diagenetic ferruginization, see, e.g., Kimberley 1979; Maynard 1983).

Provenance of Iron

A further problem is the **sources of iron** in thick ironstones. As with oolitic carbonates, an environment is needed where input or deposition of terrigenous silicates is low or virtually absent. However, unlike the formation of carbonate ooids, which are precipitated from sea (or lake) water supersaturated with respect to calcium carbonate, the iron required for the ironstones has to be delivered by rivers draining deeply weathered, vegetated soils. Another mechanism supplying iron may be oxygen-free groundwater rich in ferrous iron, which discharges directly into the sea (Fig. 6.2). The latter process is probably not sufficient to produce iron concentrations on the order required for larger ore bodies.

In the case of river input, regardless of whether iron is supplied in the form of detrital iron-bearing minerals or as mixed iron oxyhydrate-organic matter colloids, this process is accompanied by the transport of considerable quantitities of other clastic materials into the sea. These cause high sedimentation rates near the river mouths and thus dilution of any kind of chemical or biogenic sediment. Consequently, comparatively pure iron oolites of substantial thickness can only form at some distance away from deltaic areas. This also applies to those iron oolites which may consist predominantly of eroded ooids formed in lateritic soils, as assumed by Siehl and Thein (1978) for the large, shallow-marine Jurassic iron deposits of the minette ores in Lorraine, eastern France. The nuclei of the ooids in many oolitic ironstones, however, originate from marine environments and therefore indicate that the ooids were produced in shallow seas (Maynard 1983).

Effects of Reworking and Sea Level Changes

Some of the above mentioned problems can be solved by assuming that iron oolites are the product of *repeated reworking, winnowing,* and *sorting* of sediments containing disseminated ooids. During the first stage, detrital, iron-bearing minerals may accumulate in marine deltaic regions, and iron may be precipitated as pyrite, siderite, green particles, and limonite as desribed in the preceding chapter (Fig. 6.1). After abandonment or lateral migration of the delta away from the site of subsequent iron concentration, the bulk of the fine-grained sediment is reworked, and the various pre-existing iron compounds are partially oxidized, dissolved, and replaced by iron ooids, possibly with the aid of microorganisms. Alternating exposure to aerated waters and reducing conditions after shallow burial below mud may lead to oolith growth under changing redox conditions (Fig. 6.2b). During repeated transgressions and regressions, the sand size ooids are sorted out and concentrated in sand bars in the foreshore zone or on submarine swells. This zone alternatively migrates landward and seaward with rising and falling sea level (Hallam and Bradshaw 1979; Bayer et al. 1985; Bayer 1989a). If the sea level becomes particularly low, the ooids move far basinward and tend to become stable and buried under the subsequent finer grained transgressive sediments.

In fact, many iron oolites represent the *roof beds of transgressive-regressive sedimentary cycles,* beginning with dark pyritic shales (peak of transgression), followed by normal to greenish shales with some siderite concretions (Fig. 6.2c), then predominantly fine grained quartz sand, and culminating in ironstone deposition (Hallam and Bradshaw 1979; Bayer et al. 1985). The latter frequently show distinct cross bedding which may be bipolar, thus indicating subtidal environments. The iron oolites often alternate with shell beds and other lag deposits and are sometimes overlain by thin layers containing phosphorite nodules and glauconitic minerals. Oolitic ironstones are frequently excellent marker beds (Dreesen 1989).

Summary

In summary, oolitic ironstones occur predominantly in clastic sequences and represent a kind of condensed lag sediment. Most of the iron appears to have been supplied as clastic particles by rivers draining deeply weathered lateritic soils, where the iron was already pre-concentrated and the quartz content low. The ooids and other iron-rich particles probably formed in coastal waters under repeatedly changing environmental conditions, partially at the sea bottom and partially within the sediment. Transgressive-regressive cycles led to winnowing, repeated migration, and final deposition of oolitic sand bodies in the form of offshore bars or more widely extended sand sheets during the deepest sea level stands. In tidal-dominated regimes, the iron oolites mostly represent sediments of the subtidal to lower intertidal zones.

Most Phanerozoic ironstones formed during long periods of high, but short-term fluctuating sea level (Ordovician to Devonian and Jurassic to Paleogene) in epicontinental seas (Van Houten and Arthur 1989). During these times, the climate was warm and humid, terrigenous sediment influx into the sea was low, and deeper water masses tended to become poorly oxygenated. Thus, black shale deposition, coeval with, or shortly before or after iron oolite formation was common.

6.3 Red Beds

General Aspects

Red beds include claystones, sandstones, arkoses, radiolarites, some limestones, and occasionally other rock types such as tephra layers. The striking reddish color of red beds, in contrast to the drab gray color of most other ancient sedimentary rocks, has always roused the interest of both lay persons and professional geologists alike. It was frequently assumed that red beds indicated ancient arid environments, because they are often associated with evaporites. Indeed, this type of red bed is particularly common and occurs in regions of low paleolatitude. In the Permian and Triassic, thick, widespread red beds accumulated on all continents. However, red sands or soils are rare and not typical in present-day deserts. In addition, red beds were identified in a variety of other depositional environments, including marine sediments. In principle, they can occur even in glacial sequences. These findings led to much controversy over the origin of red beds. Summaries on the current state of knowledge were published by Glennie (1970), Van Houten (1973), Turner (1980) and in several textbooks on sedimentology and stratigraphy (e.g., Dunbar and Rodgers 1957; Leeder 1982; Füchtbauer 1988).

Most authors assume that the *staining pigment* of red beds is very fine-grained, uniformly dispersed *hematite* (Fe_2O_3), whereas hematite concentrated in larger crystals or at certain spots does not cause red color. Torrent and Schwertmann (1987) have pointed out that the red color is produced by the special optical behavior of tiny hematite clusters; the color of synthetic hematite-clay mixtures varies with the grain size of the hematite crystals. The hydrous ferric iron oxides, such as *goethite* and the less well defined *limonite*, are brown to ocher in color. Less than 1 % of hematite is sufficient to generate the red staining of clays and silts. The total iron content of claystones, however, is commonly two to three times higher than that needed by the staining pigment, because clay minerals and other mineral phases also contain ferric and ferrous iron. In alternating red and green claystones, the iron content of red beds is, if at all, generally only

insignificantly higher than that of green beds. The *red color of sandstones* is caused by grain coatings containing hematite; their hematite content may be even less than in claystones. Reddish colored top layers of sandstones may also be caused by mechanical infiltration of detrital clay (Turner 1980).

The presence of limonite or hematite indicates that the sediment is in an *oxidizing state*. This is generally better achieved and maintained in continental environments above the groundwater table than in aquatic systems. Therefore, red beds are common in continental deposits, although they also occur in special lacustrine and marine sediments. During the early history of the Earth, when the atmosphere was poor in or free of oxygen (Chap. 6.5.1), the formation of red beds was limited or impossible.

Oxidizing conditions within a sediment can be maintained for a long time when little or no organic matter or reducing pore waters are available to reduce the ferric to ferroan iron. Then, iron cannot be incorporated into iron-bearing carbonates, clay minerals, or form iron sulfide (pyrite). Furthermore, ferric iron cannot be carried away by circulating pore water like dissolved ferrous iron. In contrast, in buried sediments containing sufficient organic matter to reduce all the pre-existing limonite and hematite, the brown or red pigment cannot survive. Hence, they become drab gray in color. Secondary reddening can occur only if the excess organic matter is destroyed by oxidizing pore water or some other mechanism.

Having these basic rules in mind, the occurrence of red beds in various depositional environments is easy to understand.

Primary, Allochthonous Red Beds

Red, *lateritic soils* develop preferentially in tropical regions with seasonally wet and dry climates (wet-dry tropics, see e.g., Valeton 1983). These soils contain hematite in addition to hydrated iron oxide and other stable minerals such as quartz and clay (cf. Chap. 9.1). Such soils can be eroded and carried by rivers or wind into regions with differing climates (Fig. 6.3a). In rare cases, red dust and red-stained sand may reach deserts and form primarily red sand dunes and clay dunes (cf. Chap. 2.3; Millot 1964; Solle 1966). The reddening of sand dunes in Australia, however, is only observed in old eolian sands where originally brown-coated sand grains have become red by aging (Wopfner and Twidale 1988, see below and Chap. 2.3.2).

More frequently, *river-transported red soils* accumulate in semi-arid and arid lowlands and may alternate with carbonates and evaporites deposited in playa lakes. If the redeposited soils contain sufficient organic matter, hematite and ferric iron hydroxides are slowly reduced. As a result, the bed becomes drab gray or greenish in color. Alternations of red and green beds may be caused by minor differences in primary organic matter content. Primarily red silts and sands, or marly limestones, which have low hematite and ferric iron hydroxides contents, require less organic matter for decoloring than do silty clays and clayey silts with relatively high ferric iron contents. For that reason, sandstones, limestones, and evaporites, intercalated into red claystones frequently show drab gray colors.

Redeposited lateritic soils can also occur on river floodplains in semi-arid regions, where sparse vegetation cannot provide sufficient organic matter for ferric iron reduction. Ancient examples of this type of red bed frequently contain in situ pedogenic limestone nodules (caliche, cf. Chap. 2.2.3), indicating a relatively dry paleoclimate. Similarly, rapid deposition of carbon-poor prodelta sediments in lakes and in the sea may allow the maintainance of the primary red color in parts of the prodelta sequences. Red prodelta deposits are known, for example, from the marine deltas of the Orinoco, Amazon, and Yangtze Rivers, i.e., rivers draining tropical and subtropical regions. In shallow seas with limited inputs of lateritic soils, the red pigment mostly vanishes at depths of a few centimeters to decimeters below the sediment-water interface, as observed in the Adriatic Sea (Hinze and Meischner 1968) or in the western part of the Gulf of Aden (pers. observ.).

Furthermore, red soils are carried into lakes and swamps in humid regions where they are usually reduced and form gray deposits alternating with organic-rich layers such as peat. Occasionally, however, a thick red bed may be preserved within a sequence with coal seams, as for example in some Tertiary coals in Europe. Another instructive example is the barren red Carboniferous coal measures of Britain, which occur adjacent to the productive coal measures of the same age (Glennie 1970; Turner 1980). Conclusive evidence for the primary nature of these red beds is, however, not available. They may also result from originally brown material or reflect post-depositional oxidation processes (see below).

In Situ (Secondary) Formation of Red Beds

Many red beds do not result from redeposited lateritic soils, but reflect in situ processes and may therefore be referred to as *secondary red beds*. In this case, the source area and its climate are of minor importance. The principal factor controlling the subsequent sediment color is a *well oxygenated depositional environment* in which the organic matter, either autochthonous or allochthonous, is mineralized completely or to a very high degree. Then the potential of the relic organic matter to reduce all the ferric iron compounds present in the sediments is too low (Fig. 6.3).

The precursors of such red beds are either *brown soils* or *brown sediments*, because ferric iron hydroxides left behind after the consumption of organic matter are commonly brown or ocher in color at or near the sediment surface. With increasing burial depth, temperature, and age, these iron compounds lose their water and are transformed into hematite. Hence, one can expect that present-day brown sediments, found below the zone of active aeration and bottom life, alter after burial and become red. The formation and preservation of brown colored sediments are favored by:

- High iron content in sediment (e.g., siliciclastic material rich in biotite, pyroxene, amphibole, olivine, and iron-bearing clay minerals).
- Low production and rapid mineralization of organic matter in the depositional area (e.g., fluvial plain with sparse vegetation in semi-arid climate).
- Slow or intermittent sedimentation, allowing organic matter consumption in the zone of active bottom life (soil zone or subaquatic benthic epifauna and infauna).

a   PRIMARY ALLOCHTHONOUS RED BEDS

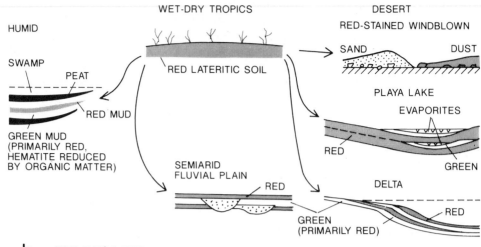

b   RED BEDS FORMED IN SITU

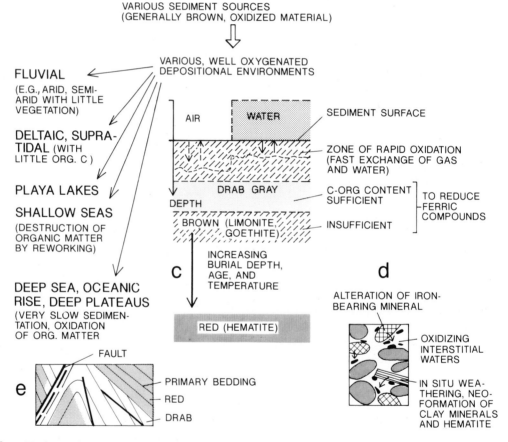

Fig. 6.3a-e. Various origins of red beds. **a** Allochthonous primary red beds formed by eroded, redeposited lateritic soils. **b** Secondary red beds resulting from brown, well oxygenated sediments which do not contain enough organic matter for the complete reduction of ferric iron in limonite and goethite below the zone of near-surface rapid oxidation. **c** Slow transformation of brown limonite and goethite into hematite (red pigment) with increasing burial depth, temperature, and age. **d** In situ alteration of iron-bearing minerals (e.g. pyroxene, biotite) by interstitial water with high redox potential; release of ferric iron and neo-formation of hematite within sediment. (After Walker 1967, 1976). **e** Secondary discoloring of red beds by circulation of waters with low redox potential along faults. (After Franke and Paul 1980)

- Redox and pH conditions in interstitial water favoring the formation of iron oxide after burial, if subsequent reduction is absent (Walker 1967, 1976, 1979).

Depositional Environments of Brown and Red Beds

Examples of depositional environments with a high potential of forming brown beds can be easily deduced from these general rules. They include both continental and marine systems (Fig. 6.3b). The fluvial red mudstones of many formations can be interpreted as originally brown floodplain sediments derived from muddy suspension load or aggregated river bed load (Rust and Nanson 1989; Chap. 2.2.3).

Environments with relatively rapid sedimentation such as river floodplains, prodelta deposits in lakes and in the sea, or supratidal sediments can maintain an oxidized state only when their primary organic carbon content is low. In contrast, slowly deposited pelagic sediments on submarine plateaus and in the deep sea may originally contain relatively high amounts of organic carbon, but benthic life in oxygenated waters has sufficient time to destroy the organic matter before it is finally buried under younger sediments. *Red deep-sea clay* accumulating below the calcite compensation depth (Chap. 5.3.2) is representative of this group of red beds, which does not depend on a specific clay mineral source.

Red deep-sea sediments are restricted to basinal sites and mid-oceanic ridges without major clastic influx (Franke and Paul 1980), although they may contain biogenic silica and carbonate. Red clays may alternate with gray turbidites which are deposited rapidly and obtain their organic material from source areas with higher sedimentation rates (Faupl and Sauer 1978).

Carbonates poor in iron only require a little preserved organic carbon to be held in a reduced state. Therefore, most ancient carbonates are light gray in color. Red limestones indicate a particularly well oxidized environment, low sedimentation rate, and in some cases possibly emergence above sea level. In addition, some red limestones, sandstones, and to a lesser degree claystones with irregular color boundaries across primary bedding planes may result from postdepositional processes, such as circulating groundwaters with a high redox potential. Similarly, reducing pore waters can destroy red staining, particularly along zones of high fracture permeability (Fig. 6.3e). Transgressions of the sea over brown or red continental sediments normally leave behind gray deposits, but can also affect the underlying sediments by providing circulating, reducing pore waters. Thus, the uppermost portions and locally deeper zones of these sediments may be discolored.

Finally, it is important to note that iron is also released from unstable, iron-bearing minerals during diagenesis (Walker 1967, 1976, 1979; Füchtbauer 1988). Oxidizing interstitial waters lead to *in situ weathering* of these particles and to the formation of new clay minerals, as well as ferric iron hydroxide. The latter slowly alters to hematite and thus causes red staining as mentioned earlier. This process commonly takes place in sediments which were originally brown and did not contain sufficient organic matter to maintain reduced conditions in the subsurface.

6.4 Marine Evaporites

(Continued under the heading of 6.5)

6.4.1 Models for Evaporite Deposition

Introduction

Marine sea water-derived salt deposits are more uniform and, in several respects, easier to interpret than continental evaporites (Chap. 2.5). Their large representatives, however, impose some difficult problems (see below). Salt deposits are known from the Precambrium and throughout Phanerozoic time, but there were some major epochs of halogenesis, for example in the early Cambrian, the Devonian, the Permian, and the late Triassic to early Jurassic (e.g., Zharkov 1981). Due to their importance in geology and for the economy, salt deposits have received much attention from earth scientists of different specific disciplines; thus there are numerous publications summarizing our knowledge in this field (e.g., Lotze 1957; Richter-Bernburg 1968; Braitsch 1971; Kirkland and Evans 1973; Nissenbaum 1980; Kendall 1983; Sonnenfeld 1984; Dronkert 1985; Peryt 1987b; Müller 1988; Peryt 1987b; Schreiber 1988a and b; Sonnenfeld and Perthuisot 1989; Warren 1989). Because marine evaporites originate from sea water of more or less constant composition (at least since the Cambrian), their mode of precipitation can be predicted rather well. There is, however, only a limited number of cases in which the sequence and amount of marine salts can be explained just by evaporation of a certain volume of sea water.

Normal sea water has a density of 1.025 g/cm^3 and contains about 35 g per liter dissolved constituents. These mainly consist of the following (water-free) salts (in percent by weight of the total salt content):

78 % NaCl (halite),
18 % potash salts, i.e., chlorides and sulfates of K and Mg (e.g., sylvite KCl, carnallite $MgCl_2 \cdot KCl \cdot 6H_2O$, kainite $MgSO_4 \cdot KCl \cdot 3H_2O$, kieserite $MgSO_4 \cdot H_2O$, etc.),
3.5 % $CaSO_4$ (gypsum and anhydrite),
0.3 % carbonates, and some minor constituents such as bromides, fluorides, borates.

Common Models for Evaporite Deposition

The following models for the depositional environment of evaporites were developed from recent examples and the study of ancient salt deposits. All of them have in common that the depositional basin lies in an arid to semi-arid zone and loses more water by evaporation than it receives by precipitation and inflow of river water. Consequently, ancient salt deposits are important paleoclimatic markers, indicating arid, normally low-latitude zones as opposed to tropical or temperate and cold high-latitude regions.

As a result of an excess in evaporation, the water level in a basin tends to decrease and salt concentration and density of the water

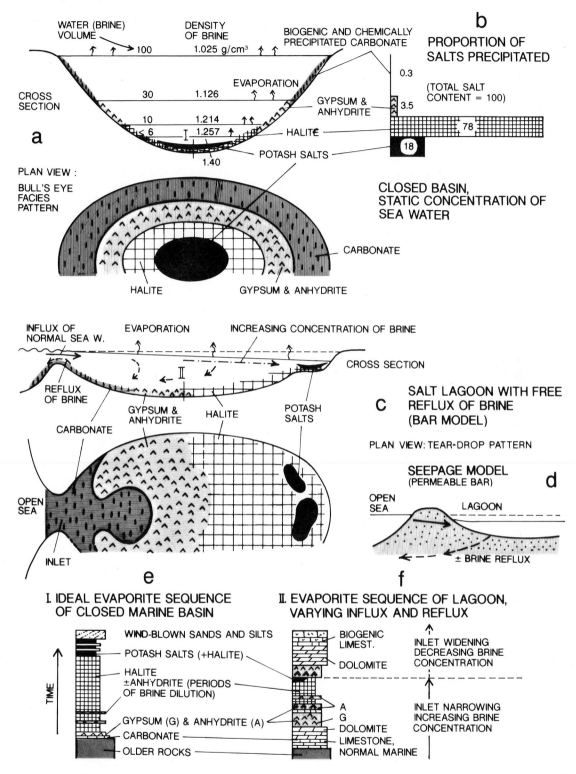

Fig. 6.4a-c. Models of evaporite deposition. **a** Closed basin with "static" concentration of sea water and "bull's eye" facies pattern. **b** Proportions of different evaporites precipitated under conditions of **a**. **c** Silled salt lagoon, open system with brine reflux ("dynamic" system of evaporite deposition), frequently causing a "tear drop" facies pattern. **d** Lagoon separated from the open sea by per-meable bar (sand dunes, reef etc.) allowing slow influx of sea water and reflux of brine. Situation similar to **c** with tendency to high brine concentration and therefore predo-minance of halite. **e** Idealized evaporite sequence of closed basin (**a**). **f** Idealized example of evaporite sequence of silled lagoon (**c**) with reversed development (prograding dilution) in upper half. For further explanations see text

increase. We distinguish several basic models of evaporite deposition:

1. *Closed Basin*

This model is based on the following conditions: the initial salt content is equal to that of normal sea water; the full basin is assumed to hold 100 volume units of water; there is no inflow of sea water or outflow of brine (Fig. 6.4a).

During the *first stage* of development, the water volume is reduced to about 30 volume units and its density is raised to 1.126 g/cm³. Throughout this period a small quantity (about 0.3 % of the total salt content) of biogenic carbonate and later possibly chemically precipitated aragonite are deposited. Upon the extraction of Ca, the Mg/Ca ratio of the developing hypersaline water increases and thus favors the early diagenetic transformation of aragonite and calcite to dolomite.

In the *second stage*, the brine volume is reduced from about 30 units to 10 units (density 1.214 g/cm³) by evaporative drawdown, and gypsum is precipitated (3.5 % of the total salt content). The formation of anhydrite at normal field temperatures ($\leq$25 °C) is possible only in the last part of this period, when the concentration of the brine has nearly reached saturation for halite.

During the *third stage* of development, when the brine volume drops below 10 units, the major part of rock salt (halite) is deposited. Mostly, the crystals start to grow at the air-water interface before they settle through the water column. Precipitation of some anhydrite may continue, particularly in winter time or during periods of temporary brine dilution caused by ephemeral fresh water inflow.

The *final stage* of this "static brine concentration" begins with a volume of about 6 units (density 1.257 g/cm³) and is characterized, in addition to continuing halite precipitation, by the deposition of chlorides and sulfates of K and Mg (18 % of the total salt content of the original sea water). Halite makes up about 78 % of all salts precipitated (cf. Fig. 6.4b).

The areal distribution of the different salts reflects the evaporative drawdown (Fig. 6.4a). The comparatively poorly soluble carbonates and calcium sulfates occupy the margin of the basin, whereas the most soluble salts are found in the center of the basin. Such a facies distribution is referred to as a "bull's eye" pattern.

The dry salt pan may be overlain by wind-blown sand and silt. An ideal vertical section of such a development is shown in Figure 6.4e.

This model is seldom verified in nature in its pure form. Additional influx of sea water may augment the volume of salts precipitated, but cannot change the contribution of the different evaporites to the total salt body. However, the salt content of a closed basin with initial normal sea water composition is often also supplemented and modified by inflowing surface and ground water of differing composition. Thus, the basin is slowly transformed into a continental salt lake or playa, where additional salts may gain importance (Chap. 2.5).

2 (2a). *Salt Lagoon or Shallow Barred Basin*

This basin type is connected with the open sea by a small opening (Fig. 6.4c). Lowering of the water level by intensive evaporation leads to *influx of surface water* from the open sea. Salt concentration and density of this water increase toward the landward side of the lagoon and causes the surface water to sink and flow back seaward as underflow and over the sill into the sea (*brine reflux*, cf. Chap. 4, arid adjacent seas). Consequently, salts are transported continuously into and out of the lagoon, but since the influx of salts is greater than their reflux, part of the salts can be precipitated in the lagoon. Such a situation can be described as a "dynamic system of evaporite deposition".

The *type of salts* precipitated depends on the salt concentration which is reached in different parts of the lagoon. If the entrance to the lagoon is relatively large, the water exchange between open sea and lagoon is little hampered, and the concentration of the lagoonal water remains low. With decreasing area of inflow, the salt concentration in the lagoon increases. Thus, a certain opening in conjunction with the surface area of the basin may, for example, cause a brine concentration which leads to the precipitation of gypsum in large parts of the lagoon, but not yet to the crystallization of halite. In this case, the ratio of the areas of the evaporite basin and its inlet is often between 10^7 and 10^8 (Lucia 1972). Under these conditions, NaCl is returned by brine reflux to the open sea before it reaches saturation. Such a situation can be maintained for a considerable time period and thus enables the deposition of a thick gypsum or anhydrite layer.

Consequently, the lagoon may experience different stages of development. It may start with the preferential deposition of carbonate followed by gypsum and anhydrite. Then the halite stage is reached and finally also potash salts (chlorides and sulfates of K and Mg) can be precipitated. However, in contrast to the closed basin model, the thickness or volume of these different evaporites is no longer a function of normal sea water composition; it rather depends on how long a certain stage of evaporite deposition is maintained.

Figure 6.4f shows a vertical section of such a *lagoonal evaporite sequence*, in which the carbonate and calcium sulfate stages lasted much longer than the halite stage and therefore generated comparatively thick layers. In this example, the stage of potash salt precipitation was realized for a short period, but in many cases in nature neither this nor the halite stage were ever reached.

Deepening or widening of the inlet or shallowing of the lagoon as a result of salt accumulation may again cause increased water circulation in the lagoon and thus dilution of the brine. Consequently, the development of the depositional system can be reversed and stages of high salt concentration are followed by the precipitation of lower soluble evaporites (upper part of section in Fig. 6.4f).

Particularly in shallow lagoons, the concentration of the brine may also change laterally and become higher in a landward direction. As a result, different evaporites can form simultaneously as shown in Fig. 6.4c, i.e., carbonates precipitate near the entrance of the lagoon, while rock salt and possibly K and Mg salts are deposited at its landward end. In plan view, such a facies distribution is called a "tear drop pattern".

A famous recent example described in many textbooks (see above) is the Kara Bogas Gol (lagoon) on the eastern side of the Caspian Sea.

It covers an area of approximately 20 000 km^2 but has a maximum depth of only 8 m. Although the Caspian Sea has a salinity of only 1,3 % and its salt composition differs from that of normal sea water, the lagoon inspired Ochsenius in 1877 (see, e.g., Sonnenfeld 1984; Müller 1988) to propose the depositional model of a silled lagoon ("Barren-Theorie") and also to apply this model to marine evaporites. Carbonate and gypsum are precipitated near the entrance of the Kara Bogas lagoon; glauberite (Na_2SO_4.-$CaSO_4$) and halite follow landward.

The sill between the open sea and the lagoon or shallow adjacent sea may be caused by tectonic movements, but is also generated by

sedimentary processes such as reef structures, nearshore sand bars, or barrier islands (Chap. 3.2). In the ancient record, shelf carbonates and reefs of different nature (e.g., algal structures) in front of a reef lagoon are frequently associated with evaporites, because warm climate favors both carbonate production including reef growth and high evaporation.

Seepage basin. The lagoon may also be separated from the open sea by a permeable subaerial bar or ridge (Fig. 6.4d). Similar to the process in the silled basin, evaporative drawdown of the water level in the lagoon then causes sea water to flow underground into the lagoon and denser brine to flow back into the ocean. Due to the restricted inflow and reflux, this system creates a transitional situation between a closed and a silled basin. In contrast to the silled basin model, the ratios of the different salts precipitated tend to deviate less from those given by the composition of normal sea water. Consequently, the predominant salt in a seepage lagoon will often be halite. A well studied example of this type of basin is the MacLeod evaporite basin in western Australia (Logan 1987), where carbonates, gypsum, and halite were precipitated in the Holocene.

The brine reflux in both the subaerially separated and the barred basin may cease when the permeability of the underground ridge becomes too poor or the water level in the lagoon drops too deep to enable reflux, respectively. The latter case is already verified for the Kara Bogas Gol (see above). In this manner, a kind of closed system is established which, in contrast to model (1), is fed for a certain period of time by inflowing sea water (also see below).

2b. *The Synsedimentary Subsidence Model*

This model is a modification of the barred basin model (2) (Fig. 6.6a). It was proposed to explain the formation of salt deposits thicker than the limited depth of the model 2 basins (e.g. Sonnenfeld 1984). Normal slow tectonic subsidence is enhanced by the isostatic effect of the rapidly accumulating salt deposits on top of (usually) continental crust (see Chap. 8.1). In addition, differential subsidence within the basin may cause lateral variations in the evaporite thickness. If the basin is filled up, the residual brine precipitating K and Mg salts may be collected in ponds above the areas of maximum subsidence.

Except for rapidly subsiding rift zones and strike-slip basins (Chap. 1.2), the great thicknesses of salt deposits frequently observed cannot be interpreted adaquately by this mechanism. If the basin is filled up rapidly by one evaporative phase, the time for the crust to react to the applied load is too short to cause substantial additional subsidence. Thick salt deposits can be deposited only if a basin experiences several evaporative cycles interrupted by long intervals of starved sedimentation or non-deposition. During these intervals, continued tectonic subsidence and the isostatic effect of the former salt load can create sufficient space for further significant salt accumulation (see below).

3. *Coastal Sabkhas*

Coastal sabkhas represent supratidal flats which are rarely flooded during spring tides (Chap. 3.2). They usually develop behind barrier island chains and more or less hypersaline lagoons (Fig. 6.5a). They are built up by the accumulation of marine sediment transported onshore, or by eolian and fluvial material. The surface of such sabkhas is controlled by the groundwater table, which normally gently rises from the sea level in a landward direction. Continental groundwater from the land may flow seaward and mix with sea-derived groundwater. Where unconsolidated sandy and silty sediment above the capillary fringe of the groundwater becomes dry, it may be removed by wind as long as the sabkha is not overridden by large volumes of dune sand or fluvial deposits.

Salt precipitation is accomplished by evaporitive *pumping of groundwater* up to the sabkha surface, where the concentration of the brine exceeds saturation for calcium sulfate and finally also for halite. The minerals precipitated in the different zones of the intertidal and supratidal area are shown in Fig. 6.5b). The precipitation of gypsum begins in the lower sabkha within the algal mats. *Primary anhydrite* can be formed in the higher outer part of the sabkha, where the brine concentration approaches saturation for halite. Halite usually is found as crusts at the surface, but it is often re-dissolved during floods which also deliver new brine into the system. Locally, more halite and even potash salts may accumulate in small ponds excavated by earlier floods.

It is assumed that a kind of brine reflux may take place underground, possibly along older buried valley or channel fills. Provided the concentration of the brine becomes sufficiently high in the pore system, gypsum or anhydrite are also generated below the groundwater table, particularly in the upper supratidal zone (Fig. 6.5b, c, and d). Primary gypsum may be overgrown later by anhydrite or transformed into anhydrite, which often shows characteristic contorted beds, small diapiric structures, or a "chicken-wire" pattern (Fig. 6.5 d). Under the influence of increasing Mg/Ca ratios, primary aragonite and calcite are partially or entirely converted to *dolomite*.

The host sediment of the mostly nodular and lenticular sabkha evaporites is either carbonate sand and mud, siliciclastic, or a mixture of both components. Because of the lack or oxidation of organic matter in the middle and upper supratidal zone, the sediment tends to be brown and later, after more pronounced diagenesis, red in color.

The upper limit of one *cycle of sabkha evaporites* is controlled by the capillary fringe of the groundwater table. When this fringe can no longer reach the surface and is overlain by dry sandy material, the evaporation of groundwater practically ceases. Consequently, the sabkha evaporites are usually overlain either by windblown sand or fluvial deposits (Fig. 6.5 d). The lower boundary depends on the concentration of brine reached below the water table. Thus, assuming constant sea level and neglecting subsidence, the thickness of such a cycle is approximately one to a few meters, and its areal extent is also limited. However, under conditions of changing sea level and subsidence, sabkha evaporites can reach greater thicknesses and form cyclic sequences (Fig. 6.5e). Channels, storm erosion, and redeposition of sabkha sediments in the intertidal or subtidal zone can complicate such a simple cyclicity model.

As a result of relative sea level changes, sabkha evaporites often affect larger areas than observed in the many present-day examples (see, e.g., Purser 1985). Due to frequent sea level changes in the past, this type of evaporites is very common in the ancient record. In relation to the local situation, there are many modifications of the conceptual model shown in Fig. 6.5. In the context of this chapter, the most important alternative, not shown in Fig. 6.5, is the transition from a salt lagoon to a sabkha. In this case, lagoonal evaporites formed adjacent to coastal sabkhas may eventually be overlain by sabkha evaporites. This development is frequently observed in ancient sediments.

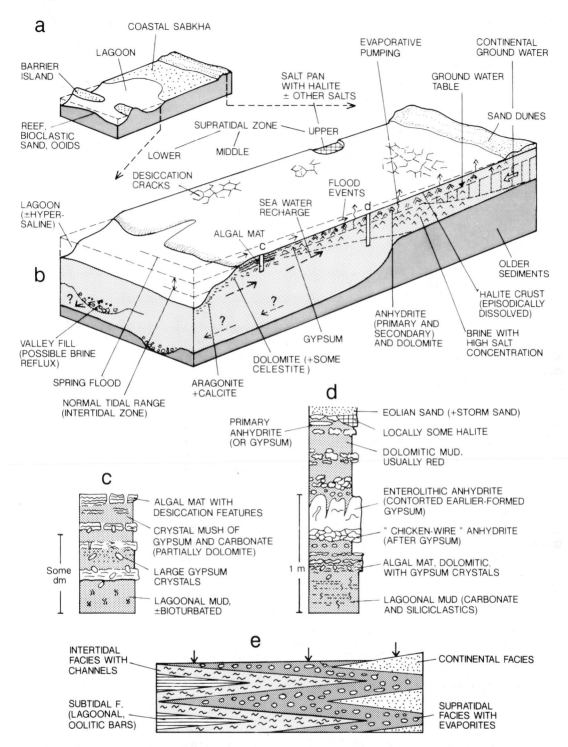

Fig. 6.5a-d. Model of coastal sabkha (based on studies of the Trucial Coast, Abu Dhabi) of the Arabo-Persian Gulf and other areas, see, e.g., Purser 1985). **a** General setting; note that the lagoonal water is already hypersaline. In the case of a narrower opening to the sea, salts may precipitated in the lagoon and underly the sabkha evaporites. **b** Coastal sabkha with evaporative pumping of inflowing sea water (underground or epidodic by floods) and continental groundwater, lateral change in evaporite precipitation above and partially also below the groundwater table. **c** and **d** Vertical sections of lower and higher supratidal zones, location shown in **b**. **e** Scheme of landward and seaward migrating facies belt including sabkha evaporites. Note the different vertical successions indicated by *arrows*. (After Shearman 1978; Shinn 1983, and others).

4. *The Deep Basin-Deep Water Model*

This model (Fig. 6.6b) is another attempt to explain a rapidly deposited very thick evaporite sequence without synsedimentary subsidence (Schmalz 1969). The principal points of this model are briefly summarized as follows.

Stage 1. Due to a narrowing gateway to the open sea, the former deep-reaching water circulation with normal marine deposition of siliciclastic and biogenic sediments in the deep adjacent basin (on thinned continental or on oceanic crust) is replaced by a circulation system restricted to the upper water layer (Fig. 6.6b). In this stage, the lower water body has already reached a higher density (1.07 to 1.08 g/cm^3) than the surface water, where the production of phytoplankton continues. The stagnant deep water favors the development of euxinic conditions and the deposition of bituminous marls or limestones (Stinkkalke).

Stage 2. In the succeeding stage, caused by further narrowing or shallowing of the sill, the surface brine increases in density and starts to sink on the landward side of the basin. At the beginning of this process, gypsum and halite precipitate at the water surface and are redissolved when they sink into less concentrated deeper water.

Stage 3. The subsequent development of the basin is controversial. According to Schmalz (1969), halite, gypsum, and some carbonates are deposited rapidly and more or less simultaneously, generating a lateral facies succession as shown in Fig. 6.6c. Once the basin is partially or nearly filled up, residual brines occupy smaller ponds on top of halite. In contrast, Kendall (1988) points out that the brine concentration of a deep, silled basin must be more or less laterally homogeneous, because differences in water density can hardly be maintained in such a basin. Therefore, the deep-water basin appears to be incapable of precipitating different mineral facies simultaneously.

The **Miocene evaporites below the Red Sea** rest on thinned continental and, probably due to postdepositional salt flow, locally also on young oceanic crust (see Chap. 4.3 and Fig. 4.7). They were explored by seismic investigations and scientific and commercial deep sea drilling. Not taking salt diapirs into account, the evaporites reach a thickness of up to 3 to 4 km and a width of 100 km (Kinsman 1975b). Some authors believe that these evaporites represent a prominent example of deep basin-deep water evaporites, but others argue that at least in the upper section and in marginal regions of the evaporite basin, shallow-water and sabkha conditions were prevailing (Stoffers and Kühn 1974, also see below and Chap. 4.3). In any case, the connection to the open ocean, most likely to the north, must have been much narrower than the present-day Straits of Bab el Mandeb in the southeast. The present inlet permits the exchange of very large water volumes and therefore controls a water circulation which allows slightly hypersaline conditions only in the northernmost part of this basin.

5. *The Deep Basin-Shallow Water Model*

Several salt deposits, which most likely were formed in deep basins, clearly show evidence for shallow water and even sabkha environments. These occurrences are explained by two related processes: (1) evaporative drawdown of the water table after the basinal brine had already reached a relatively high concentration, and (2) closing of the barrier to the ocean. Seepage below the barrier or episodic spill-overs may deliver sea water, besides some fresh water, into the shrinking basin. Saturation for the different evaporites is established in a similar way as described for model 1. In order to initiate halite precipitation, the water level must fall deep below the original niveau, and the final stages of evaporite precipitation occur under playa and sabkha conditions. The areal distribution of the resulting salt deposits therefore resembles a bull's eye pattern (Fig. 6.4a). Most workers believe that the Late Miocene evaporites below the Mediterranean Sea were formed in such a way (see below).

Complex Models for Evaporite Deposition (Giant Salt Deposits)

General Aspects

Many ancient evaporite deposits of relatively small to medium dimension can be both interpreted with the models described above and compared to present-day small evaporite basins. However, as already mentioned earlier, these basic models in their pure form cannot explain all the phenomena observed in ancient regions of salt deposition. Especially the origin of the so-called "saline giants" (Hsü 1972) has been and still is the object of lively debates (Jauzein 1984; Sonnenfeld and Perthuisot 1989; Bussot 1990).

One of the reasons for this uncertainty is the lack of present-day examples for such enormously thick (up to several thousand meters) and widely extended evaporites (several thousand km long and hundreds of km wide) as known from Precambrian to Miocene times.

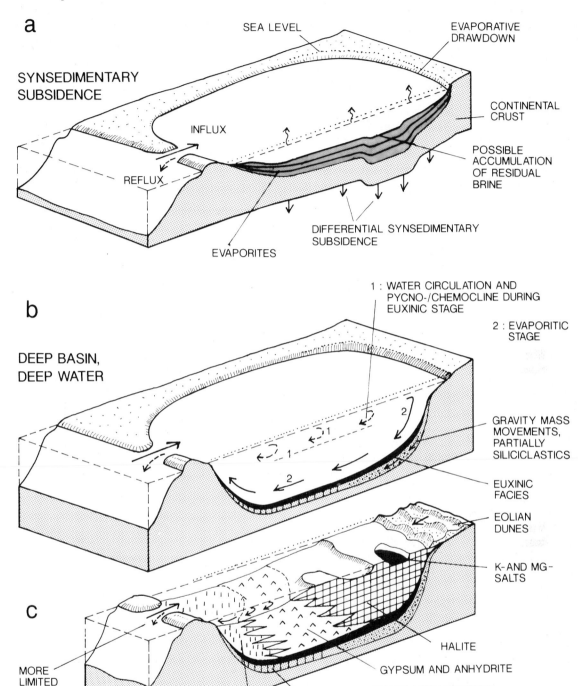

Fig. 6.6. a Synsedimentary subsidence model with brine reflux explaining fairly thick salt deposits precipitated in shallow barred basin (see Fig. 6.4b). **b** and **c** Deep basin, deep water model with brine reflux. (After Schmalz 1969). Normal marine conditions are followed by *1* an euxinic stage with high plankton productivity, but stagnant deep water body. *2* Increasing brine concentration leads to deep circulation and later to the precipitation of gypsum, anhydrite, and halite in the center and landward part of the basin. The basin is rapidly filled up with salts; K and Mg salts may be formed in special ponds; the dry salt pan is overridden by eolian sand

There is no real large or deep salt basin on the present Earth, probably because we are living in an interglacial period, in which the temperature gradients between the equator and the poles are unusually high, and the continents are widely distributed over the surface of the globe. Therefore, climatic zones and tectonic settings favorable for the generation of large evaporite basins are limited.

The *complex evaporite basin model* is a combination of several simple models discussed above and takes into account substantial *relative sea level changes*, which occurred throughout the Phanerozoic and probably even in the earlier history of the Earth (Chap. 7.5).

The model is largely based on studies on the classical Zechstein basin in Central and Northwestern Europe (Füchtbauer and Peryt 1980; Richter-Bernburg 1985; Paul 1987; Peryt 1987a), and particularly on the review by Smith (1980). Earlier discussions concerning this example are discribed by Hsü (1972).

Basin Development

The development of such a basin is briefly summarized in the following simplified way (Fig. 6.7):

- *Desert drainage basin.* A continuously subsiding large inland basin on continental (or transitional) crust is separated from the open sea by a land barrier and receives, due to arid climate and low relief in its drainage area, little sediment. If such a situation persists for a long time period, the floor of the desert drainage basin drops substantially below mean sea level (several hundred meters), even under conditions of slow subsidence.
- *Flooding by the sea.* During a phase of particularly high sea level, the entire basin is flooded and an *inflow-reflux system* established (Fig. 6.7a; model 2). As long as the water exchange between the open ocean and the inland sea is sufficiently great, normal marine sediments and fauna (including pelagic carbonate and in places redeposited material) are distributed over the entire basin. In marginal, shallow parts of the basin, *shelf carbonates* and even lagoonal and sabkha evaporites may be deposited. Shelf edges and submarine ridges and highs are preferential sites of *algal reef growth* in warm waters.
- *Onset of sea level fall.* In the course of time and forced by slowly falling sea level in the neighboring ocean, the salt concentration of the inland sea begins to increase and may first lead to stagnant, but only slightly hypersaline

bottom waters in parts of the basin (cf. Model 4, Fig. 6.6b). During such a period, *black, bituminous shales* may be deposited.

A famous example of this type of sediment is the "Kupferschiefer", a laminated marl rich in trace metals, particularly lead, zinc, and copper. It accumulated at the beginning of the first Zechstein evaporite cycle (Paul 1982).

The lateral extent and upper limit of the oxygen-depleted bottom water can be inferred from bioturbated sediments on the basin margin and on submarine highs, respectively. Later, besides continued carbonate accumulation in shallow waters, *gypsum* is precipitated in parts of the shelves, on slopes (often in the form of thick, massive beds), and in the deeper parts of the basin (frequently as thin varves). Warmer and more highly concentrated marginal waters tend to accelerate salt precipitation and cementation and thus produce thicker evaporites in the shallow parts of the basin than in the deeper central basin.
- *Continued falling sea level* further reduces water exchange between the ocean and the adjacent basin, until the connection is completely interrupted (Fig. 6.7b). The immigration of normal marine fauna is terminated. The water level in the *closed basin* is drawn down by evaporation. Entering rivers start to incise valleys into the emerged shelf and slope areas. Marginal carbonates are subjected to karstification. Earlier formed lagoonal and sabkha evaporites are partially leached and their salts transported by surface runoff and groundwater into the *shrinking inland sea*. During this phase, the major part of halite is rapidly deposited still in rather deep water; it may alternate with thin layers of anhydrite indicating minor climatic variations including annual varves (see below). The evaporites may reach considerable thicknesses in the basin centers, because (1) they now cover only parts of the total basin, and (2) the brine was already highly concentrated prior to evaporative drawdown. Marginal areas of the inland sea again exhibit lagoonal and sabkha evaporites. The main basin may be subdivided into *several subbasins* (Fig. 6.7b) with differing salt concentrations and salt deposition. In relation to the hydrologic regime of the drainage area, the basin may finally fall dry except for some playa lakes containing highly concentrated brine including potash salts.
- The *dry, huge salt flats* are more or less covered by thin continental beds and some playas. During this period, subsidence driven by the additional salt load continues and may

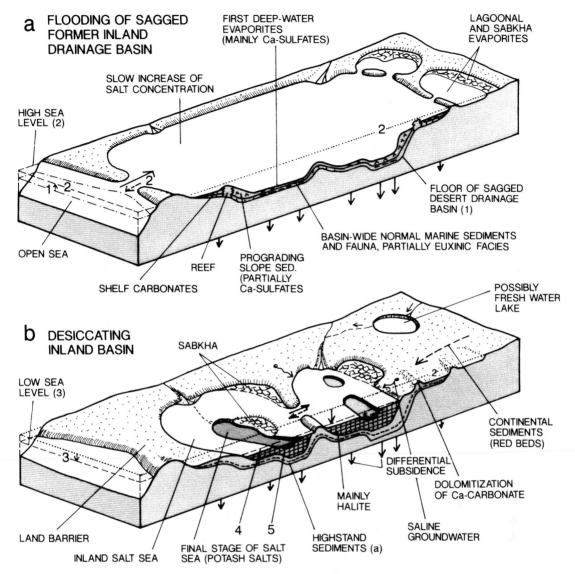

a FLOODING OF SAGGED FORMER INLAND DRAINAGE BASIN

FIRST DEEP-WATER EVAPORITES (MAINLY Ca-SULFATES)

LAGOONAL AND SABKHA EVAPORITES

SLOW INCREASE OF SALT CONCENTRATION

HIGH SEA LEVEL (2)

2

OPEN SEA

FLOOR OF SAGGED DESERT DRAINAGE BASIN (1)

BASIN-WIDE NORMAL MARINE SEDIMENTS AND FAUNA, PARTIALLY EUXINIC FACIES

REEF

PROGRADING SLOPE SED. (PARTIALLY Ca-SULFATES

SHELF CARBONATES

b DESICCATING INLAND BASIN

SABKHA

POSSIBLY FRESH WATER LAKE

LOW SEA LEVEL (3)

2

CONTINENTAL SEDIMENTS (RED BEDS)

3

DIFFERENTIAL SUBSIDENCE

DOLOMITIZATION OF Ca-CARBONATE

MAINLY HALITE

SALINE GROUNDWATER

LAND BARRIER

INLAND SALT SEA

4   5

FINAL STAGE OF SALT SEA (POTASH SALTS)

HIGHSTAND SEDIMENTS (a)

Fig. 6.7a,b. Model of complex large evaporite basin on continental or transitional crust (Zechstein-type basin), demonstrating one large depositional cycle in two main stages of development: **a** Particularly high-rising sea level *(2)* causes flooding of long-persistent former inland drainage basin (during stage *1* of sea level), which has subsided up to several hundred meters below mean sea level. After flooding sediments with normal marine fauna accumulate which are basin-wide correlatable; then salinity increases and evaporite deposition can begin.

b Lowering sea level (stage *3)* restores land barrier and creates large hypersaline inland sea; evaporative drawdown of water level *(4)* leads to rapid "deep-water" salt precipitation (mainly halite). Finally large areas fall dry and potash salts may accumulate in residual playa lakes *(5)*, before continental deposits take over. Note that during all stages of development shallow-water lagoonal and sabkha evaporites may be precipitated along the margin of the basin. For further explanations see text

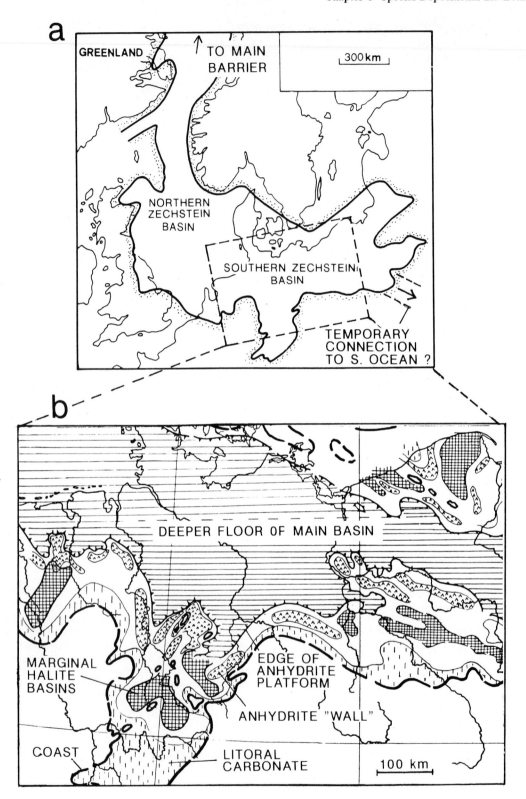

Fig. 6.8. a Configuration of Zechstein Basin in central and northern Europe with main barrier to the north. (After Peryt 1987a, slightly modified after Ziegler 1988). **b** Southern central part of Zechstein Basin demonstrating the irregular topography and evaporite distribution, with thick halite deposition, in marginal parts of the basin during the first evaporite cycle, Z1. (After Richter-Bernburg 1985, cf. Fig. 6.10g)

restore the situation to the one at the beginning of this development, until a pronounced high sea level rise causes a further flooding of the inland depression and initiates a new cycle of evaporite deposition.

Examples

The **European Zechstein basin** experienced four to five large depositional cycles. Their total thickness varies considerably from some tens of meters in the marginal zones to more than 2000 m (maximum thickness 3500 m including Rotliegend evaporites) in special troughs. The thicknesses of the salt deposits of a single large cycle often reach several hundred meters, but also vary considerably from marginal to central parts of the basin (Fig. 6.8), thus indicating both pronounced differences in synsedimentary subsidence as well as variations in sedimentation rate (see below). Local differential subsidence may originate from strike-slip movements (Ziegler 1989). The large depositional cycles are superimposed by minor cycles (Richter-Bernburg 1985, Langbein 1987).

The **Late Miocene (Messinian) evaporites below the Mediterranean** were detected and explored by deep-sea drilling and compared with uplifted Messinian sections on land. They represent two evaporite cycles and generally reach thicknesses of several hundred meters, in places 1000 to 2000 m, which were deposited in a very short time interval. They cover an area of about 2.5×10^6 km$_2$ and lie, at least partially, on top of oceanic crust which was created prior to the "salinity crisis" (Hsü et al. 1977; Dronkert 1985; Busson 1990). The salts are underlain and covered by hemipelagic deep-water sediments; rivers entering the Mediterranean have incised deep valleys and canyons to adjust to the drop of water level. For these and other reasons, most workers believe that these evaporites originate from a closure of the basin in the region of the present-day Straits of Gibraltar, leading to the drying of a formerly water-filled basin of about 1500 m depth. Hence, the salts must have precipitated in shallow water in a desiccated deep basin (previous section, model 4). A drastic change of climate was not necessary for this development. Actually, even lagoonal and sabkha evaporites in marginal zones have been found (cf. Fig. 6.7), whereas potash salts were discovered in the basin centers. There are many interesting details such as high bromium contents in the lower salt unit indicating evaporation of sea water (see below), or isotopic studies giving evidence for the occurrence of brackish environments in the upper unit due to the influx of river water. A critical point is the salt budget which requires either a high brine concentration prior to the closure of the opening to the west, or repeated surface inflow, or exceptionally intensive seepage of ocean water into the Mediterranean.

Other Ancient Salt Deposits

Large marine salt deposits are known from the late Precambrian throughout the Phanerozoic, but a peak of evaporite deposition was obviously reached in the Permian to Early Jurassic, when the supercontinent of Pangaea was assembled and started to break up (cf. Chap. 7.5). The rifting and early drifting stage of proto-oceans provided a number of restricted, rapidly subsiding basins favorable for the deposition of thick evaporite sequences. For this reason and due to widely extended arid climate, the large salt occurrences below the present passive continental margins and slopes (e.g., around the Atlantic ocean) were formed. Other important evaporites were deposited in marginal and epicontental basins on continental or transitional crust. Examples are listed and described in the literature mentioned earlier.

6.4.2 Sequences, Sedimentary Structures, and Sedimentation Rates of Evaporites

Characteristic Sequences with Salt Deposits

As already demonstrated above (Fig. 6.4), progressing or decreasing brine concentrations cause characteristic successions of salt deposits including their accompanying sediments such as marine bituminous shales, carbonates, and terrestrial beds.

In addition to the large depositional cycles mentioned above, many evaporite sequences show a greater number of *minor cycles*, which may originate from sea level changes with periods within the Milankovitch band (e.g., 40 ka; Chap. 7.1).

A prominent example of such evaporite cycles is the Pennsylvanian of the Paradox Basin in the United States, where thick salt deposits, consisting of 29 to 40 cycles, are flanked laterally by thinner carbonates (Kendall 1988). During low sea level still allowing influx of sea water but preventing brine reflux into the open sea, carbonates on a shelf barrier probably formed coevally with the precipitation of evaporites in a restricted deeper basin (Fig. 6.9a). Conversely, high sea level led to brine reflux and decrease in the salinity of the basin. When the bottom water became stagnant during high sea level, black shales were deposited (Fig. 6.9b, cf. Fig. 6.6b).

Generally or at least in some cases, the longest residence time and hence the *highest brine concentration* in a basin occurs during sea level rise (Fig. 6.9c), as long as the brine does not flow back into the open sea. Thus, in contrast to the general opinion, well soluble salts may be precipitated preferentially during slowly rising sea level, whereas less soluble evaporites, followed by terrestrial deposits, tend to

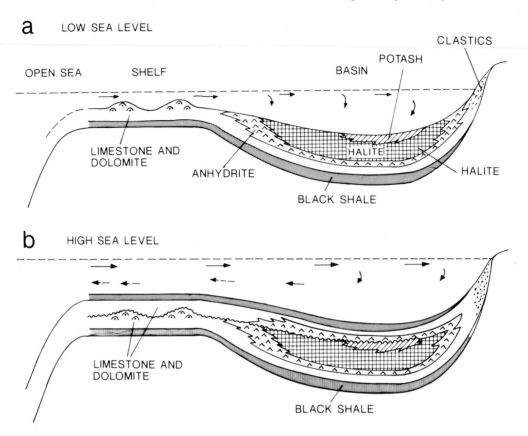

Fig. 6.9a,b. Model for the origin of minor cyclic evaporite sequences caused by sea level changes in a shelf/restricted basin setting. (After Hite 1970, in Kendall 1988, modified). **a** No brine reflux during low and intermediate sea level; algal reef growth on the shelf barrier occurs coevally

with the deposition of halite and other salts in the basin. **b** Brine reflux during high sea level leads to decreasing salinity in the basin and to a circulation favorable for the accumulation of black shales

accumulate in times of sea level fall (Kendall 1988).

Relative sea level changes, regardless of whether they are caused by eustacy or by tectonic movements, also generate characteristic *transitions from siliciclastic deposits to evaporites and carbonates.* The carbonates normally represent the high sea level stand. Transitions between two or three of these sediment types occur frequently in the geologic record; they have been recently desribed, for example, from the Persian Gulf (see Chap. 14.3), the northern part of the Red Sea, and Tunisia (Purser et al. 1987; Ben Ismail and M'Rabet 1990).

Sedimentary Structures

Sedimentary structures of evaporites are either of primary origin or predominantly generated by diagenetic processes (see below). One of the most striking phenomenon in subaquatic deposited evaporites is *primary parallel lamination* (Fig. 6.10a) which frequently occurs in all types of salt rocks (gypsum and anhydrite, halite, potash salts) as well as in carbonates associated with them (Richter-Bernburg 1960). The laminated evaporitic facies is most regular and best preserved in relatively deep parts of the basin, but it is also observed

---

Fig. 6.10. a Thickness of annual varves in halite, anhydrite, and carbonate. **b**, Anhydrite section in cycle 1 (Z1, Goslar) of Zechstein showing repeated facies change probably due to (climatically controlled?) brine concentration, see **c**. **c** Decreasing brine concentration and diminishing early cementation of evaporites with depth causes a vertical facies change in sulfates. **d** and **e** Generation of shelves in

evaporite basin due to partial early cementation and subsequent differential compaction. **f** Collapse breccia caused by early cementation of top layers and subsequent differential compaction. **g** Lateral facies change in Zechstein Basin during three cycles. (**a, b,** and **g** based on Richter-Bernburg 1985; **c** through **f** after Langbein 1987)

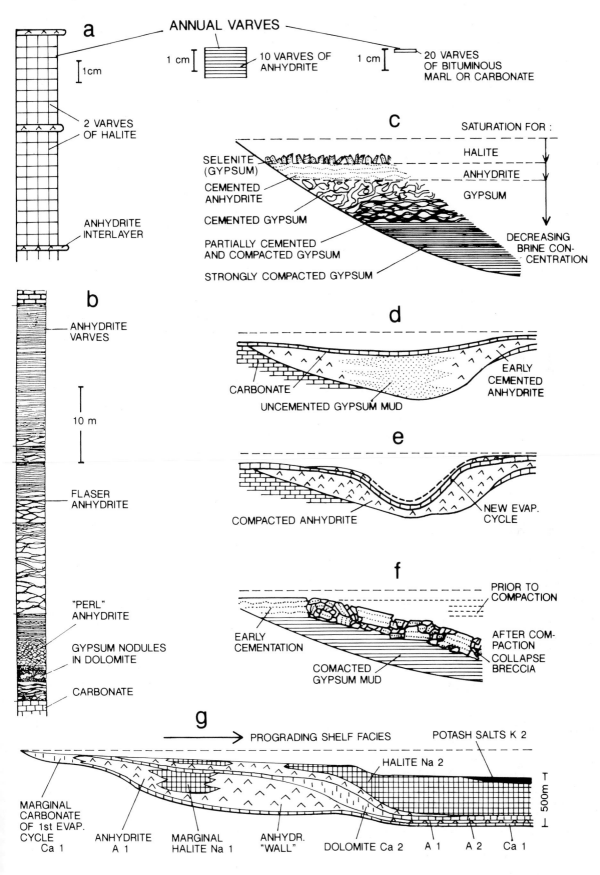

a   ANNUAL VARVES

1 cm
10 VARVES OF ANHYDRITE

1 cm
20 VARVES OF BITUMINOUS MARL OR CARBONATE

1cm

2 VARVES OF HALITE

ANHYDRITE INTERLAYER

c   SATURATION FOR :

SELENITE (GYPSUM)
CEMENTED ANHYDRITE
CEMENTED GYPSUM
PARTIALLY CEMENTED AND COMPACTED GYPSUM
STRONGLY COMPACTED GYPSUM

HALITE
ANHYDRITE
GYPSUM

DECREASING BRINE CON-CENTRATION

b

ANHYDRITE VARVES

10 m

FLASER ANHYDRITE

"PERL" ANHYDRITE

GYPSUM NODULES IN DOLOMITE

CARBONATE

d

CARBONATE
UNCEMENTED GYPSUM MUD
EARLY CEMENTED ANHYDRITE

e

COMPACTED ANHYDRITE
NEW EVAP. CYCLE

f   PRIOR TO COMPACTION

EARLY CEMENTATION
COMACTED GYPSUM MUD
AFTER COM-PACTION
COLLAPSE BRECCIA

g   PROGRADING SHELF FACIES
POTASH SALTS K 2
HALITE Na 2

500m

MARGINAL CARBONATE OF 1st EVAP. CYCLE Ca 1
ANHYDRITE A 1
MARGINAL HALITE Na 1
ANHYDR. "WALL"
DOLOMITE Ca 2   A 1   A 2   Ca 1

in marginal shallow-water areas. Here the laminae of the same evaporite facies tend to become thicker than in the basin center. This observation is explained by higher brine concentration, faster salt precipitation, and intensive early cementation in warmer, shallower parts of the basin. The thickness of the laminae *(varves)* generally depends on the rock type. In the Zechstein evaporites, but also in other marine evaporites, the following characteristic (rounded) values and variations in thickness are frequently observed:

- Carbonates (Stinkkalke) 0.05 mm (0.04 - 0.1 mm).
- Gypsum and anhydrite   0.5   mm (0.2 - 30   mm).
- Halite   50   mm (20 - 150 mm).

The mean ratio of these varve thicknesses is approximately 1:10:1000 and reflects the solubility of these components as well as their rate of precipitation when a certain height of water evaporates. If, for example, the brine has just reached saturation for $CaSO_4$, the evaporation of 1 m of water (annual water loss of NaCl-saturated brines in warm arid regions, Sonnenfeld and Perthuisot 1989) can precipitate about 1 mm anhydrite. In the case of a NaCl-saturated brine, the same rate of evaporation generates a halite layer with a thickness in the order of 15 cm. If the brine is somewhat undersaturated with respect to these salts, the same annual evaporation renders thinner salt laminae. The laminae are always separated by much thinner layers of a less soluble salt, for example halite by anhydrite or anhydrite by an extremely thin carbonate film. All these observations indicate that the laminae are annual varves and originate from the seasonal climatic change.

Furthermore, it was pointed out that evaporite varves may show the 11 year solar cycle (Richter-Bernburg 1960). In the Zechstein basin, anhydrite varves have been correlated over distances of up to 300 km, but in vertical sections, the varved units are frequently interrupted by diagenetically overprinted units (Fig. 6.10b) or by erosional surfaces. Therefore it remains problematic to determine the time span for the deposition of a thick evaporite sequence with the aid of varve counts.

Evaporites of comparatively deep basins also contain *redeposited carbonates* and salts from shallow water in the form of mass flows and turbidites (Chapter 5.4). These features are, however, frequently difficult to identify and to distinguish from diagenetic structures (see below).

Sedimentation Rates

The sedimentation rates for the different evaporites can be determined in undisturbed, varved evaporite sections. Using the above data, one obtains the following orders of magnitude:

- Carbonates   5 cm/ka
- Gypsum and anhydrite   50 cm/ka
- Halite   5000 cm/ka

Of course, these values vary considerably within a certain evaporite sequence or from one salt deposit to another. Whereas the sedimentation rate for carbonates compares well with normal marine shelf carbonates (Chap. 3.3.2), it is obvious that the deposition of marine salts, especially that of halite and the potash salts, can be extremely fast in comparison to other marine sediments. These high rates are, however, only valid for basins where the whole water column has reached saturation for the corresponding salt to be precipitated. Prior to this situation or as a result of episodic brine dilution, the sedimentation rates can be considerably lower.

Nevertheless, the assumption that a rather deep basin is rapidly filled up with halite appears to be well justified by the observations on varves. If, for example, a 1000 m deep, NaCl-saturated basin is cut off from sea water influx, drying of this basin can produce a halite layer of about 150 m thickness throughout the entire basin. If halite precipitation occurs only in parts of the shrinking basin, the halite deposit can become much thicker. Such a rapid deposition of halite is possible only in a pre-existing deep basin, because subsidence cannot proceed with the same high rate. Thus, the frequently observed rapidity of evaporite deposition supports the deep basin model discussed above for "saline giants". However, the deposition of thick units of gypsum and anhydrite requires an open brine reflux system, because the solubility of $CaSO_4$ is too low to allow an adequate storage of this component in the brine of a closed basin. In the case of cyclic evaporite sequences, most of the geologic time comprises the intervals of non-evaporite deposition. Sabkha evaporites, mainly consisting of anhydrite and carbonate, show vertical growth rates of the order of 1 m/ka and horizontal progradation of 1 km/ka (Schreiber and Hsü 1980).

6.4.3 Diagenesis and Geochemical Characteristics of Evaporites

Diagenesis

After burial, the primary salt deposits, particularly those precipitated under a standing water body, are strongly affected by mechanical and physicochemical-mineralogical diagenesis. In order to interpret the origin of ancient evaporites, this substantial diagenetic overprint has to be taken into account.

In an early, more or less euxinic stage of a salt basin development (Chap. 6.4.1), relatively small quantities of gypsum settling together with organic matter can be removed by *microbial sulfate reduction*. Thus, the beginning of the sulfate precipitation may be obliterated or, as observed in some cases, halite directly overlies bituminous marls or carbonates.

In subaqueous environments, the salt minerals and their primary deposits are basically formed in three different ways:

- Crystal growth at the water surface or at the pycnocline (halocline) between lower mineralized surface water and the deeper brine; settling of the crystals without any further change.
- Crystal growth predominantly at the sediment-water interface.
- Crystal growth within the sediment (displacive crystals).

Under the latter two conditions, the size of the crystals tends to be large and the porosity of the fresh salt deposits is limited. Finer grained crystals settling through the water column may, however, create a rather porous sediment. *Mechanical compaction* of such evaporites is substantial and can therefore bring about great differences in thickness between the above mentioned types of salt deposits. Compaction moves both pore water and water of crystallization out of the evaporites. The updip or upward migrating, highly concentrated fluids can affect the surrounding sediments and, for example, cause *dolomitization* of calcarous rocks or, in some cases, also *calcitization* of already dolomitized strata. Frequently, the precipitation of salts from migrating brines leads to cementation and thus to plugging of porous rocks.

Early Cementation and Differential Compaction

A further characteristic feature observed in evaporites is the occurrence of *early cementation* by gypsum, anhydrite, halite, and other salt minerals (Fig. 6.10c through f). Primary sedimentary features are best preserved under conditions of early complete cementation (e.g. banded and massive anhydrite beds). If, however, early cementation is incomplete, nodular and flaser anhydrite result. Differential compaction between early cemented marginal and non-cemented central evaporites (undergoing relatively late and strong compaction) may accentuate differences in bathymetry of the depositional basin and therefore mislead reconstructions of the original basin topography and the subsidence history of the basin (Fig. 6.10d and e).

The buildup of thick marginal evaporites as a kind of shelf brings about a lateral prograding of salinar facies as indicated in Fig. 6.10g (cf. Fig. 6.8). In addition, the compaction of early, but only partially cemented calcium sulfate deposits frequently causes the formation of *collapse breccias* (Fig. 6.10f, Langbein 1987). These should not be confused with sedimentary breccias originating from the reworking of salt crusts on tidal flats, or with mass flows (olistostromes, Chap. 5.4) in deep water-filled basins. Similarly, intensively deformed anhydrite or gypsum layers exhibiting *micro-folds* are usually not caused by sliding, but reflect a kind of liquefaction. This results from the release of water in conjunction with the transformation of gypsum to anhydrite in poorly permeable strata at some depth below the basin floor (see below).

Subsurface Dissolution

Stratigraphic correlations and hence also paleogeographic interpretations of ancient salt deposits are often complicated by subsurface dissolution resulting from circulating groundwater or more or less stagnant formation water (Kendall 1988). It may eliminate easily soluble evaporite intervals, preferentially in marginal zones of the basin (Fig. 6.11). Thus, an originally onlapping sequence can be transformed into an apparently offlapping stratigraphic pattern with an erosional unconformity. Such a sequence may be incorrectly interpreted as an evaporite system according to the bull's eye model (Chap. 6.4.1). Salts leached from marginal zones of the basin (salted-out halite and

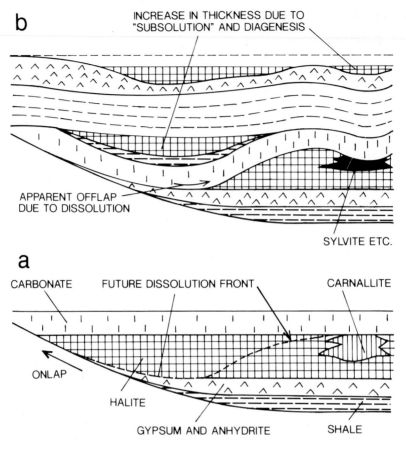

Fig. 6.11a,b. Effects of subsurface solution of halite and conversion of water-bearing salt minerals (e.g., carnallite) into water-free salts (e.g., sylvite) on stratigraphic rela tionship and thickness of subsequent evaporite deposits. (After Kendall 1988)

potash salts) are frequently reprecipitated in more central parts of the basin. Highly concentrated NaCl brines can cause substantial changes of the primary mineral composition of the potash salts.

Phase Changes of Salt Minerals

Already a moderate increase in temperature (and pressure) induces phase changes of several salt minerals. Primary gypsum is converted to anhydrite at temperatures between 18 and 56 °C; carnallite loses its crystalline water at 85 °C, polyhalite and kieserite at higher temperatures. In all cases, water is released and may, if it cannot readily escape, build up excess pore pressure and promote rock deformation, for example small diapiric structures in sulfate deposits. If *volume loss* in the subsurface by subsolution and/or phase changes occurs concurrently with precipitation of over-

lying salts, the basin floor becomes depressed. As a result, thickened salt deposits are formed locally or in limited parts of the basin (Fig. 6.11b).

These brief remarks can only indicate the particularly great importance of diagenetic processes on sedimentary structures and mineral composition of evaporites. Detailed descriptions of these phenomena including illustrations and a special nomenclature for the manifold structures, particularly those in calcium sulfate rocks, are given in the references mentioned above, e.g., by Richter-Bernburg (1985), Langbein (1987), Schreiber (1988a). It has to be borne in mind that many of the diagenetically generated or overprinted structures can be further modified by salt diapirism (see below) and subrosion. Anhydrite is again transformed into gypsum near the land surface and may cause rock swelling, if not enough sulfate is removed by circulating water in solution.

Trace Elements, Stable Isotopes, and Accessories in Marine Evaporites

Much work has been done on the geochemistry of evaporites (see, e.g., Sonnenfeld 1984; Pierre 1988), but only a few points can be mentioned in this text. Of particular interest are the elements strontium, barium, bromine, fluorine, and boron. *Strontium* is a minor constituent of aragonite, gypsum and anhydrite. It is released in conjunction with recrystallization of these minerals and can form an own mineral, celestite $SrSO_4$, which is less soluble than gypsum and therefore can be precipitated in small quantities before saturation for the calcium sulfate minerals is reached. Celestite is found along the margin or on subaqueous highs of some evaporite basins (Müller 1988). It also occurs when gypsum is leached by subrosion. Similarly, *barium* may be precipitated as barite, $BaSO_4$, prior to gypsum and anhydrite. Both celestite and barite hence tend to occur in clays, marls and carbonates in the lower part of an evaporite sequence. *Fluorine* can be precipatated as fluorite, CaF_2, and is sometimes found in dolomites.

Bromine and *boron* are enriched in highly concentrated brines. Since bromine substitutes to some extent for chlorine, the bromine content of halite, sylvite and other salt minerals increases with the brine concentration. Therefore, the bromine content is used to determine the stage of evaporation as well as to correlate salts deposited in the same basin. Boron is predominantly found in residual brines.

Many evaporites also contain or are interbedded with siliciclastic material, mostly silt and clay, which are swept by either currents or wind into the basin. Alkaline brines often dissolve substantial amounts of silica which is later reprecipitated in form of euhedral quartz crystals or, in the neighborhood of the evaporites, as *chert*.

Stable isotopes provide specific information on the changing conditions of deposition and diagenesis of evaporites (Pierre 1988). They indicate the origin of brines precipitating salt minerals, and the concentration, residence time, and recycling of brines in the depositional system. For example, the isotopes Br and Sr may serve to separate marine from nonmarine evaporites. Similarly, heavier oxygen isotopes in fluid inclusions within the salt point to a marine environment. The proportion of the sulfur isotope ^{34}S in sulfate minerals has changed through earth history. With the aid of this isotope it is therefore possible to approximately determine the age of evaporites. This method is particularly useful if salts of different ages are involved in diapirism (see below).

6.4.4 Salt Structures

Halokinesis

Many thick, widely extended, originally more or less uniform salt deposits of the subsurface are transformed by halokinesis into very irregular and often isolated salt structures (Fig. 6.12). These structures may result from nontectonic and tectonic processes (Jenyon 1986). Here, mainly the nontectonic origin of such structures is briefly discussed.

Salt domes and other salt structures are commonly explained as the result of buoyancy (Trusheim 1960; Ramberg 1981). Where lower-density salt is buried under higher density strata, the salt tends to flow upward through the overlying sediment *(buoyancy halokinesis)*. In this case, the burial depth of the salt should be at least 900 to 1200 m; otherwise the overlying rocks do not reach the high density required for this process (see below). However, apart from buoyancy, overburden anomalies *(differential loading halokinesis)* may cause salt sinks and salt dome growth (Jackson and Talbot 1986; Kehle 1988). This mechanism does not depend on the density of the overlying sedimentary rocks and therefore accounts for the initiation of salt structures at shallow burial depth (as little as 100 m) as observed in several salt provinces. The combined effect of both mechanisms can be summarized as follows.

Mechanics of Salt Flow

It is assumed that rock salt behaves like a Newtonian or *viscoelastic fluid*. In both cases, the fluid starts to flow if there is a hydraulic gradient within the fluid. If rock salt represents a viscoelastic fluid, flow of salt takes place only if the differential stress exceeds the yield point of the salt body. A hydraulic gradient is defined by a difference in hydraulic head between two points within the salt layer (Fig. 6.13a). The hydraulic head of a certain point within the fluid is the sum of its gravity potential, G, at its elevation, z, above an arbitrary datum line, and the fluid pressure head, p, at this point, i.e., the height of fluid column in a manometer adjusted at this point.

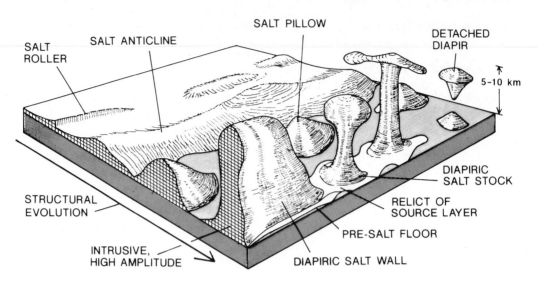

Fig. 6.12. Main types of salt structures and their structural evolution, from left to right. (After Jackson and Talbot 1986)

Salt flow always occurs in the direction of the maximum hydraulic gradient, i.e., from areas of higher to lower hydraulic heads. A tilted salt layer can flow upward, if the decrease in pressure, Δp (e.g., $p_B - p_A$ in Fig. 6.13a), is greater than the increase in gravitational potential, ΔG, defined by the difference in elevation between two points and the density of the fluid (e.g., $\Delta G = z_B \cdot D_S - z_A \cdot D_S$). This occurs when the average rock density, D_R, is greater than that of the salt, D_S; otherwise the salt flows donwhill (from C to B in Fig. 6.13a). Hence, total overburden load plays no direct role in this model; solely the gradients of pressure and gravitational potential are important and, in addition, the yield point of the viscoelastic salt body must be overcome.

If a horizontal salt layer is buried under horizontal and laterally uniform younger sediments, the hydraulic gradient within the salt equals the gradient of the gravity potential, because the overburden weight is the same everywhere. In this case no salt flow occurs. However, if the salt surface is locally elevated (elevation z in Fig. 6.13b), a gradient in the hydraulic head of the salt layer is built up. Salt flows toward the irregularity, if the density of the overlying rocks in the range of z exceeds that of salt, and it flows away from the irregularity for rocks of lower densities. Similarly, a local increase in overburden pressure (differential loading), for example by a prograding delta lobe, can cause a significant hydraulic gradient in the salt layer and thus force the salt to flow away from the delta

lobe. Many other sedimentary load anomalies are common, for example reefs, sandy shoals, desert dunes etc., and can trigger salt flow in the subsurface.

Evolution of Salt Structures

The evolution and shape of some idealized *salt structures* is shown in Figure 6.14. In nature, these structures are highly variable dependent on their position within a basin, the original thickness of salt deposits, and the history of subsidence and sediment accumulation (Trusheim 1960; Jackson and Talbot 1986; Kehle 1988). Salt domes are typically one to several km in diameter and have steeply dipping or even overhanging sides that may extend several km downward. They commonly exhibit the following stages in their development:

1. Initiation and salt pillow formation.
2. Erosional truncation of overlying rocks.
3. Extrusion of salt domes or pillars through erosional holes in the sedimentary cover (piercement salt domes, diapirs).
4. Collapse and burial.

Much salt is lost during the stage of extrusion through erosion and dissolution (Chap. 9.2). Once the salt layer next to the dome is completely evacuated, the upward movement of salt ceases, and the salt structure reaches a stable configuration, regardless of whether or not the dome is buried under an increasing

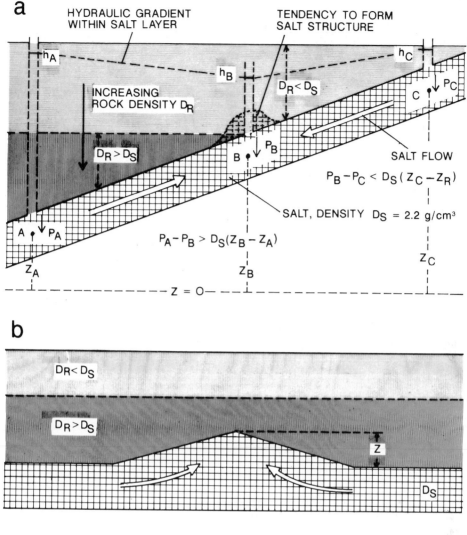

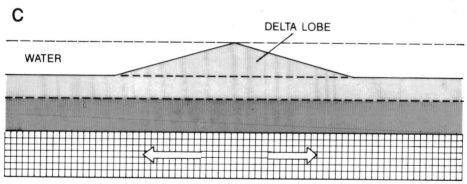

Fig. 6.13a-c. Flow of salt in the subsurface in relation to the hydraulic gradient within the salt layer. (After Kehle 1988). **a** Downhill and uphill flow of salt due to the lower or higher density D_R of sedimentary rocks overlying the tilted salt layer with density D_S. At point B the converging salt flow tends to create a salt structure. **b** Positive irregu-larity on surface of salt layer initiates salt flow into the irregularity, if rock density D_R within depth zone z exceeds D_S of salt; with $D_R \leq D_S$ salt flows away from the irregu-larity. **c** Differential loading of horizontal salt layer, e.g., by prograding delta lobe, causes salt to flow away from the area of maximum loading

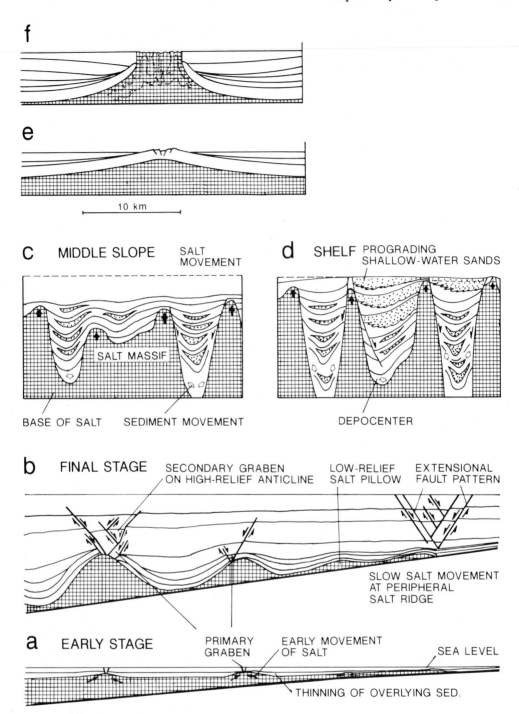

Fig. 6.14a-f. Different types of salt structures. **a** and **b** Evolution of salt pillows and salt anticlines from originally basinward thickening wedge of evaporites. (After Kehle 1988). **c** and **d** Development of depotroughs over thick salt deposits on a continental slope/shelf setting. Differential movement of salt and younger sediment is initiated on lower slope and accentuated by increasing sediment accu-mulation due to prograding shelf edge. Sands may form reservoirs for hydrocarbons. (After Jackson and Galloway 1984, in Kehle 1988). **e** and **f** Two stages of the growth history of salt domes: **e** Salt pillow formation and begin-ning of erosional truncation, **f** piercement of overlying rocks, and extrusion. Final collapse and burial is not shown. (After Kehle 1988)

sediment cover. The internal structure of many salt domes is characterized by tight folds and highly deformed salt bodies of different nature (Fig. 6.14f). Their associated rocks show systems of normal faults and graben structures over the tops of the domes (Fig. 6.14b) and frequently upturned strata along the flanks.

Part of these phenomena, as well as the structures of subsiding depotroughs (Fig. 6.14c and d), are favorable for the migration and accumulation of hydrocarbons (Chap. 14). Due to their plastic behavior at shallow burial depths, evaporites act as seal rocks for upward migrating oil and gas.

6.5 Nonactualistic (Precambrian) Depositional Environments

6.5.1 The Evolution of the Atmosphere, Hydrosphere, and Climate

Introduction

This brief chapter is added to remind the reader that most of our knowledge on depositional systems summarized in this book is deduced from observations in modern environments. It is based on the frequently quoted principle: " The Present is the key to the Past". Furthermore, it is widely accepted that the depositional environments did not change fundamentally during the Phanerozoic. This view is derived from the fossil record and from the results of various geochemical investigations. Even long-term, profound climatic variations from an icehouse to a greenhouse state of the Earth (Chap. 7.5) can be explained by minor changes in the composition of the atmosphere and the volume and salinity of the oceans. A major step in the evolution of sedimentary environments occurred in the late Paleozoic when the continents were colonized by plants. Prior to this stage of plant evolu-

tion, all continents were more or less barren of higher life and represented huge deserts. In contrast to the present-day situation, a large part of these deserts must have received considerable amounts of rain and should therefore have undergone rapid mechanical and chemical erosion. Nevertheless, we can understand these processes from observations in modern, mountainous regions with sufficient precipitation, but sparse vegetation.

Serious problems arise for the interpretion of the *Precambrian*, that is the Archean and Proterozoic spanning the times from about 4600 to 2500 Ma and 2500 to 590 Ma before present, respectively (Windley 1984). Since weathering of rocks and sedimentary processes were always closely related to the properties and circulation of the atmosphere and the oceans, as well as to the evolution of life, we have to consider the question whether and in which way these factors deviated in pre-Phanerozoic times from the younger Earth's history.

Early Life

The earliest life forms are microfossils representing possibly cell walls and having shapes and sizes like bacteria. They occur in dark cherts and shales as old as about 3500 Ma (e.g., Schopf 1983; Schopf and Packer 1987). They were capable of either utilizing inorganic compounds for synthesis of organic molecules (autotroph bacteria, e.g., methane bacteria), or of living on organic molecules generated by abiotic chemical synthesis (fermentation bacteria). Stromatolites, i.e., microbially precipitated layered structures, have been preserved in several metamorphic series (greenstone belts) as old as 2600 to 3500 Ma (Fig. 6.15). Their structure was most probably generated by mats of *blue-green algae* (cyanobacteria). These cyanobacteria may therefore have started to produce oxygen as early as about 3500 Ma ago. Results from stable carbon

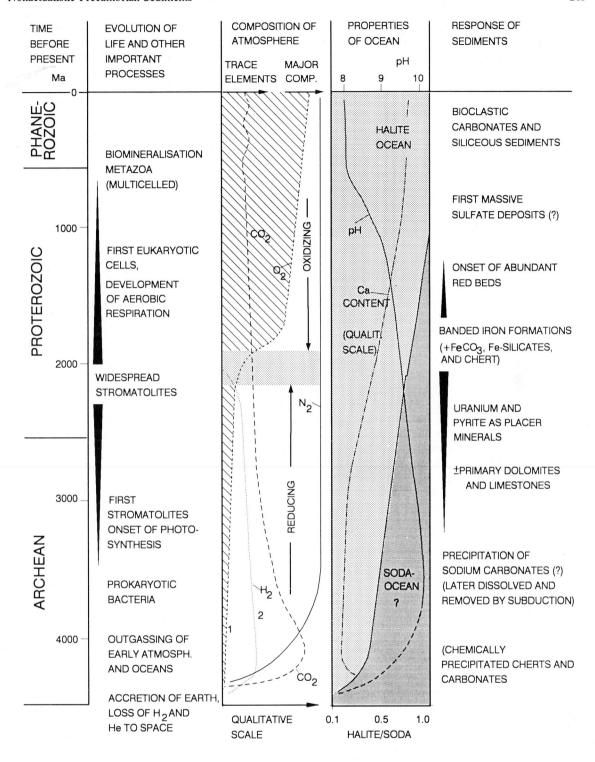

Fig. 6.15. Some characteristics of the (hypothetical) Precambrian atmosphere, ocean, and sediments. (Overview, based on several sources, e.g., Holland 1984; Windley 1984; Kempe and Degens 1985; Veizer 1988). See text for further explanation

isotopes studies on limestones and dolomites (Schidlowski et al. 1975; Schidlowski 1987) indicate that substantial amounts of carbon, accumulated by autotrophic organisms, underwent biologically mediated fractionation. This led to a depletion of light carbon in carbonates as common in younger carbonates. Stromatolite growth was restricted, however, to shallow subaqueous environments.

With the beginning of the *Proterozoic*, micro-organisms including bacteria, algae, fungi, and possibly lichen-like plants became more abundant and diversified, but the more complex, multicellular metazoa obviously did not yet exist. In the period between 2300 and 600 Ma, the *stromatolites* were the most common, widespread fossil structures, which built thick sequences of limestones, dolomites, and cherts in the tidal and subtidal zone. The amplitude of single structures reached many meters (cf. Chap. 3.2.1). It appears that since that time the importance of cyanobacteria mats gradually declined.

The evolution of *multicellular metazoa* did not start until after about 1,800 Ma. It is assumed that they required an atmosphere containing oxygen (see below) and an ozone screen in the upper atmosphere to protect them from ultra-violet radiation. Early types of metazoa were not capable of building shells or skeletons. They include jelly fish, worms, sponges, and soft corals and are summarized under the term "Ediacaran fauna", which was first found in the Flinders Ranges in southern Australia (e.g., Conway Morris 1990). The main evolution of the metazoa occurred in the Phanerozoic; the formation of hard skeletal parts by "biomineralization" started at the Precambrian/Cambrian boundary.

Atmosphere and Climate

The Early Atmosphere

Although the atmosphere represents a very small mass in relation to the total mass of the Earth, it exerts a dominant influence on the shaping of the landscape, the existence and evolution of life, weathering of rocks, transport and deposition of sediments. For this reason, an atmosphere differing from the present-day situation must have had a profound impact an all these aspects.

The evolution of the atmosphere was closely related to the formation of the Earth's crust and hydrosphere. Most experts hold the opinion that the primordial atmosphere of the Earth resembled that of Jupiter, which has sufficient mass to retain light molecules and noble gases. In the case of the Earth, however, the light, volatile components, such as hydrogen and helium, were early lost to space and replaced by an atmosphere mainly caused by outgassing of the Earth's mantle (Salop 1983; Holland 1984; Budyko et al. 1987). The accretion of planetesimal material, meteoric impacts, and enhanced radioactive decay raised the temperature on Earth to such a degree that the primarily more or less homogeneous mass melted (e.g., Wetherill 1990). This led to a differentiation of the accreted mass into core, mantle, early crust, and a secondary atmosphere derived from the volatiles of the magma and/or late phases of accretion.

This atmosphere attained a very high density comparable to that of Venus. Its main components were water vapor, carbon dioxide, and nitrogen. Minor constituents included hydrogen, methane, and ammonia. Light volatiles were continuously lost to space; acids (HCl, HF) and H_2S exhaled from volcanoes were washed out by precipitation. Subsequent rapid heat loss to space led to cooling of the crust and condensation of water. Evidence for a first hydrosphere on Earth comes from water-laid sediments ranging back as far as about 4000 Ma.

Oxygen in the Early Atmosphere

Whether or not oxygen was present in small quantities in the early atmosphere is controversial (e.g., Schidlowski et al. 1975; Clemmey and Badham 1982; Holland 1984; Budyko et al. 1987; Walker and Drever 1988). A limited amount of oxygen produced by photolysis, i.e., by dissociation of water vapor as a result of ultra-violet radiation in the upper atmosphere, was rapidly used up by oxidation of volcanic gases, ferrous iron dissolved in sea water (see below), and weathering of rock-forming minerals. It was not before the plants had started *photosynthesis* by using water, carbon dioxide and solar energy to generate organic compounds and release molecular oxygen that larger volumes of oxygen could be produced than were permanently consumed. However, an excess in free oxygen was only achieved by the storage of organic matter in sediments. From then on, the oxygen content of the atmosphere slowly increased (Fig. 6.15). Rock weathering and the formation of carbonates were very efficient in binding carbon dioxide during this early stage of atmospheric evolution and thus did not allow the maintainance of high carbon dioxide pressure for long (Kempe and Degens 1985). The transition from an oxygen-free or oxygen-poor to an oxygen-bearing atmosphere occurred between 2500 and 1500 Ma B.P., but temporal variations in the contents of oxygen and carbon dioxide cannot be excluded during this stage (Budyko et al. 1987).

Climate

With the establishment of a hydrosphere, the climate on Earth probably became rather stable (Henderson-Sellers and Henderson-Sellers 1989; Kasting 1989). On the one hand, the young Sun had a lower luminosity than today, but on the other hand, the early oceans probably covered a much higher proportion of the Earth's surface, generating a thalassocratic epoch. Thus, the Earth sent back a smaller fraction of the received solar energy to space than at present. The effect of the following slow increase in solar luminosity on the surface temperature of the Earth was probably counteracted by a decrease in atmospheric carbon dioxide (and possibly other gases) diminishing the "supergreenhouse" effect (cf. Chap. 7.5). Thus, a reasonable temperate climate and the persistence of the oceans were maintained for at least the last 3500 Ma.

Evolution of the Hydrosphere

The water of the hydrosphere is derived from the differentiation and outgassing of magma and late phases of meteorite accretion (e.g., chondrites and comets containing ice). It was estimated that the Earth's mantle contains approximately three times the water mass present in the modern oceans (Salop 1983). The initial high partial pressure of CO_2 in the atmosphere must also have raised significantly the CO_2 content of sea water. The salt content and the nature of the early ocean are less clear. It was probably hot and rich in both carbonic acid and stronger acids and therefore capable to rapidly extract alkali and earth alkali ions, iron, and silica from volcanic rocks. For this reason, the various acids became neutralized, and the pH of ocean water was presumably higher than assumed by some authors (at least pH 6 according to Walker and Drever 1988). The total salt concentration of the early ocean may have been higher than that of the modern ocean. Sodium delivered by rocks and

chlorine provided as hydrogen chloride by volcanic exhalations were important constituents. Most of the other major ions, such as calcium and magnesium, tended to form less soluble mineral phases which precipitated. Potassium released from primary rock minerals was largely used to form clay minerals.

It was recently pointed out that the early ocean might have been a "soda ocean" in analogy to modern soda lakes, which occur in volcanic regions (Kempe and Degens 1985; Degens 1989; Kempe et al. 1989). Since volcanoes commonly deliver more CO_2 than HCl, not only contained the early atmosphere more CO_2, but also the initial ocean may have had higher concentrations in HCO_3^- and CO_3^{2-} than in chloride. This, in turn, may have caused a situation similar to that in modern soda lakes, which are highly alkaline and contain little calcium and magnesium due to the precipitation of these ions as carbonates. If this is correct, the early ocean water must have been rich in sodium carbonate and have had a high pH, ranging between 9 and 11 (Fig. 6.15; for kinetic and mass balance considerations related to this problem see Kempe and Degens 1985). Such a sea water, particularly if it was hotter than the present ocean, could dissolve large amounts of silica.

As long as only little free oxygen was available, the conditions in sea water were reducing and allowed, in addition, the solution of considerable amounts of ferrous iron (approximately 1000 times the mass of ferric iron dissolved in the modern ocean). Thus, the early ocean may have been a *"soda ocean" rich in silica and ferrous iron*, but relatively poor in calcium and magnesium, and devoid of sulfate, because there was not sufficient oxygen for the oxidation of hydrogen sulfide and sulfur. Later, as a result of the growing continents, the time span necessary for the recycling of marine sediments including carbonates and organic carbon increased. At the same time, sodium-rich pore waters were incorporated into subduction complexes where sodium was used to form sodium feldspars (albite and plagioclas) in the growing granodioritic continental crust. Thus, the soda-dominated ocean was gradually transformed into a *halite-dominated ocean* (Fig. 6.15).

6.5.2 Precambrian Sediments

Archean and Proterozoic rocks are known from all continents and represent a major part of the so-called *shields* or *cratons*. The most common rock associations are old, high-grade gneiss and low-grade greenstone belt complexes. They include banded ironstone formations and various siliciclastic rocks, for example turbidite sequences. Here, some special sedimentary rocks are briefly discussed.

The above mentioned views on the early evolution of the atmosphere and hydrosphere are derived from the sparse Precambrian fossil record and rocks, which are commonly strongly affected by repeated periods of metamorphism and tectonism. In spite of these difficulties, Precambrian rocks including sedimentary sequences have been studied intensively in various aspects and have provided significant results. Chemical sediments in particular have shed some light on the nature of the early atmosphere and hydrosphere.

Banded Quartzites

Some of the oldest sedimentary sequences in the Archean ($\geq$3500 Ma) are characterized by metamorphic quartzite-amphibolite associations. The bedded quartzites can reach 1000 m in thickness, are recrystallized, and do not show any internal structures or clastic textures, such as the contours of primary grains (Salop 1983). These rocks can be interpreted as mineralogically and texturally mature sandstones, as known from younger, repeatedly recycled quartz deposits, although they appear to contain some clastic interbeds with zircon and sillimanite grains. Therefore, these quartzitic rocks are thought to be chemically precipitated as chert layers in a similar way as observed in present-day highly alkaline soda lakes of the Magadi type (Chap. 2.5). Such an explanation supports the above mentioned "soda ocean" theory, allowing the dissolution of high amounts of silica, which precipitate when water evaporates, for example in marginal shallow seas, or as soon as the pH is lowered and/or the water temperature drops. Without advocating the soda ocean theory, Drever et al. (1988) assume that the silica concentration of the early ocean was about 20 times greater than that of the modern oceans (6 ppm) which are depleted in silica by the abundant growth of diatoms and radiolaria.

Dolomites

The dominance of dolomites over limestones in Proterozoic and some older sequences may have been caused by a high Mg/Ca ratio

(Tucker 1982; Budyko et al. 1987) and by a possibly high pH of the early ocean. These rocks precipitated either directly from ocean water, which may have been a soda-rich ocean, or formed during early diagenesis (see Chap. 13.3). In addition, all carbonates are comparatively rich in iron, manganese, and silica, and they may alternate with chert layers. Furthermore, some authors also assume that strontium-rich aragonite precipitated from Proterozoic ocean water in both peritidal and open marine subtidal environments (Grotzinger 1986; Peryt et al. 1990). In any case, both dolomites and limestones were generated by chemical or biochemical precipitation. Biogenic skeletal carbonate was not available prior to the Phanerozoic (Fig. 6.15). Periods with a decrease in the partial pressure of CO_2 and an increase in pH may have favored the precipitation of carbonates from ocean water.

Banded Ironstone Formations

A third, frequently discussed phenomenon of the Precambrian is the occurrence of thick and widespread, economically important banded ironstone formations (e.g., Cloud 1973; Trendall and Morris 1983; Holland 1984; Zhu et al. 1988). Most of these deposits have an age between 2000 and 3000 Ma with a maximum around 2000 Ma, but some are also older or younger. In their typical facies, they consist of alternating thin layers of chert and red, iron-rich beds composed of hematite, magnetite, siderite and iron silicates. In addition to the signature of the depositional environment, the mineralogical composition of these rocks is also affected by metamorphism and subsequent weathering processes (e.g., Trendall and Morris 1983; Weggen and Valeton 1990).

The coexistence of chert, iron carbonate, iron silicates, and iron oxyhydrates (as precursors of hematite and magnetite) reflect precipitation from an alkaline, alternatively reducing and oxidizing solution rich in dissolved silica and ferrous iron. The thin-bedded strata of Proterozoic ironstones probably formed in widely extended shallow seas, partially in the tidal zone. The surface waters of the oceans were possibly somewhat more oxygenated than deeper water masses (Drever et al. 1988; Veizer 1988). In addition, algal blooms may have produced oxygen periodically for the rapid oxidation and precipitation of iron present in sea water. Accompanying clastic beds may contain pyrite. Some occurrences of iron-rich beds are associated with deposits of deeper water, mafic tuffs and subaqueous volcanic exhalations which may have delivered dissolved iron and silica (Simonson 1985; Breitkopf 1988). Archean representatives of banded iron-formations are more difficult to explain because of the scarcity of all types of microfossils; some of them are thought to be of volcanic origin.

Placer Deposits and Paleosols

Another indicator of an oxygen-poor atmosphere are the placer deposits of *uraninite* (U_3O_8) in fluvial sandstones and conglomerates. Such deposits are older than about 2200 Ma and occur, for example, in the Witwatersrand in South Africa and in the Elliot Lake region in Canada (e.g., Pretorius 1981; Holland 1984). In oxidizing conditions, this mineral is unstable and cannot survive exposition to weathering and fluvial transport for long. Pyrite in such placer deposits has also been quoted as evidence for a reducing atmosphere, but since this mineral can easily form during diagenesis, it appears less useful.

The evidence of both uraninite and pyrite including some other arguments for a reducing early atmosphere discussed here are not generally accepted (e.g. Clemmey and Badham 1982). Many experts currently hold the view that the early atmosphere was not completely devoid of oxygen. This opinion is supported by the occurrence of *Precambrian paleosols* of various ages, which show all criteria of soil formation (Grandstaff et al. 1986; Farrow and Mossman 1988; Zbinden et al. 1988). Some red beds, weathering crusts at the surface of pillow-basalts, and the occurrence of thin sulfate beds indicate the presence of oxygen in the Archean and early Proterozoic atmosphere.

Stromatolites, Phosphorites

The occurrences of thick and widespread calcareous and dolomitic stromatolites reflect an environment, where minor local effects, such as the uptake of CO_2 by blue-green algae from sea water supersaturated with repect to calcium carbonate and dolomite, caused biochemical precipitation of carbonates. Examples of this mechanism have been described from modern alkaline lakes, which have low contents of calcium (Kempe and Kazmierczak 1990). A 4500 m thick upper Precambrian sequence in the Peking area consists predominantly of carbonates with stromatolites and some siliciclastic beds. These deposits accumulated in a long-persisting subtidal, intertidal, and supratidal environment and thus reflect very stable tectonic and environmental conditions (Song and Gao 1985).

Due to its low calcium content, a soda-rich ocean may also have contained relatively high concentrations of phosphate, as observed in present soda lakes. *Phosphorites* in economic concentrations appear, however, not before the late Proterozoic (Veizer 1988). In contrast to younger occurrences, these and most Cambrian phosphorites are commonly nonpelletal, i.e., a direct relationship to biogenic activity cannot be established. Thus, chemical precipitation of phosphorite at the sediment-water interface, which is also discussed for some Phanerozoic phosphorite deposits (e.g., Sheldon 1981), cannot be excluded. A rise in calcium content of the ocean and minor changes in the physico-chemical properties of sea water may have led to supersaturation with respect to calcium phosphate. Diagenetically formed phosphorites caused by the release of phosphorous from decomposition of stromatolitic algal tissue (cf. Chap. 5.3.6) may have acted as nuclei for precipitation of phosphorite from sea water or interstitial water. On emerging phosphate-bearing stromatolites phosphorite may also have formed crusts ("phoscrete") similar to calcrete as observed in Cambrian carbonates in Australia (Southgate 1986).

Phosphorites were described from Precambrian sequences in several regions (e.g., Salop 1983). In China, widespread, partially thick phosphorite beds in association with cherts, dolomites, limestones, carbonaceous shales, and stromatolites are known from the early and late Proterozoic (e.g., Sang and You 1988). They were deposited on stable platforms which received little terrigenous input.

Red Beds

Red beds (cf. Chap. 6.3) only became abundant after the period of banded iron formation. This may be related to the fact that the early Earth was dominated by oceans and the area of emerged continents was limited. In addition, a great part of the red beds formed on the continents may have been eroded later. Nevertheless, many workers have assumed that the appearance of extensive red beds coincided with a substantial increase in the oxygen content of the atmosphere (Fig. 6.15).

Evoparites

The question whether or not significant *evaporite deposits* accumulated in the Archean is difficult to answer, because subsurface dissolution, diagenesis, and metamorphism may have obscured and obliterated most of them. Nevertheless, relics of gypsum crystals and sulfate nodules have been found in a chert-barite unit as old as 3500 Ma in Western Australia (Buick and Dunlop 1990). However, thick sulfate layers could only precipitate after sufficient oxygen was available for the oxidation of hydrogen sulfide and sulfur. Such sulfate deposits appear in the second half of the early Proterozoic, and the majority of gypsum and anhydrite is younger than 800 Ma (Budyko et al. 1987; Veizer 1988). If the soda ocean concept is applied, the late appearance of thick, massive gypsum in the upper Proterozoic may also result from the depletion of the earlier ocean in calcium (Kempe and Degens 1985). In the following period of the late Proterozoic and early Cambrian, thick evaporites with high proportions of halite, potassium and magnesium salts, and dolomites accumulated. In addition, the Ca/Mg ratio in carbonates tended to increase and their contents in manganese and silica to decrease. Organic matter and pyrite are present in Precambrian sedimentary rocks, but their contribution to the composition of sediments preserved in the geologic record becomes significantly greater in the Phanerozoic.

Clastic Deposits

Apart from these special sediment types and environments, many Precambrian *siliciclastic* and *volcaniclastic deposits* closely resemble their Phanerozoic counterparts. Hence, facies models derived from modern environments are more and more used for the interpretation of Precambrian depositional systems.

7 Depositional Rhythms and Cyclic Sequences

7.1 General Aspects

Individual Beds, Rhythmic Bedding, and Sedimentary Cycles

Many sedimentary sections exhibit a kind of rhythmicity due to regularly alternating beds traceable over long distances, or a repetition of larger units which are referred to as depositional cycles. Rhythmic and cyclic sequences occur world-wide on various scales in presumably every environmental and stratigraphic system. They have attracted the interest of earth scientists for a long time, but their origin is still a matter of debate in many cases. In fact, they may be caused by a variety of processes and special feedback systems which are difficult to resolve in detail. Cyclic phenomena in particular are becoming increasingly important in stratigraphic correlations and possibly in the dating of sedimentary successions. Special volumes have summarized our current knowledge and concepts in this field (e.g., Weller 1964; Duff et al. 1967; Schwarzacher 1975; Einsele and Seilacher 1982; Einsele et al. 1991), and many publications on facies analysis deal with this subject. In the following discussion we proceed from small units observed in field exposures, to large sequences which are studied in basin cross-sections and compiled from the sections of several basin fills.

The basic sedimentological unit described in sequence analysis is the laterally traceable, relatively uniform *bed*. *Bed thicknesses* vary from a few centimeters to several meters, but commonly are some 5 to 40 cm thick. Their lateral extent ranges from a few meters to thousands of kilometers. Beds are separated from each other by thinner or thicker, usually weaker intercalations, i.e., *interbeds*, of differing composition or structure (for more detailed definitions see, e.g., Campbell 1967; Reineck and Singh 1980; Collinson and Thompson 1982). The bed and overlying interbed form a *bedding couplet*. When the interbed becomes very thin, as found in bed-dominated alternations, it may essentially represent a *bedding plane*. Bedsets or *bundles* represent several bedding couplets, separated by thicker interbeds (Fig. 7.1g). Bedsets without interbeds are formed by the amalgamation of event layers (see below). Depositional rhythms and cycles are defined as follows (Fig. 7.1):

- **Rhythmic bedding or rhythmic sequences** consist of two alternating bed types (succession AB, AB, etc), i.e., beds and interbeds, or a succession of bedding couplets. This group is subdivided into two different categories:

1. Bedding variations result from abrupt changes in sedimentation due to *depositional events* or episodes at random to quasi-periodic time intervals (forming stochastic, episodic, and *discyclic bedding;* Fig. 7.1 a and b). The most prominent examples of this group are tempestites and turbidites, flooding episodes in alluvial plains, and repeated volcanic ashfalls into basins receiving normal background sedimentation.

2. Bedding variations are caused by *slow, gradual repeated changes* in deposition. This type may also be called *cyclic bedding,* or when strictly periodic, *periodic bedding* (Fig. 7.1c and d). Typical examples are unlithified chalk-marl sequences and pelagic limestone-marl alternations.

- **Sedimentary cycles and cyclic sequences** are composed of at least three different beds and interbeds which form a repeated succession (e.g., ABC, ABC, etc.; Fig. 7.1e and f).

The term *cyclothem* (e.g., Weller 1964) is widely used in North America in a purely descriptive way, especially for coal-bearing sequences, particularly those of Pennsylvanian age (Heckel 1986; Klein 1989; Riegel 1991). This term describes a *basic cycle,* i.e., a package of lithologies representing the smallest cyclic unit of a sequence. The succession of beds and interbeds in a cyclothem may be symmetric, semisymmetric, or completely asymmetric (Fig. 7.1 e and f). In seismic stratigraphy (Chap. 7.4) this term is replaced by *depositional sequence,* which is defined by its lower and upper boundaries as well as by lowstand, transgressive and highstand deposits, or, if it is a smaller unit, by the term *parasequence* (van Wagoner et al. 1987; Vail 1987; Vail et al. 1991). In this book, primarily the terms depositional sequence or deposition-

al cycle and, for a succession of such cycles, cyclic sequence are used.

Autogenetic and Allogenetic Processes

Genetically, one can distinguish between two groups of mechanisms leading to rhythmic and cyclic sequences. These mechanisms were originally defined by Beerbower (1964) to describe cyclic styles in fluvial systems:

Autocyclic sequences (or "autogenetic sequences", Dott 1988) are controlled primarily by processes taking place in the sedimentary basin itself. Their beds usually show only limited stratigraphic continuity. Prominent examples in this category are storm-generated tempestites, turbidite sequences caused by mass flows, the migration and superposition of channel and lobe systems in fluvial environments as well as in the deep sea, and successions of coal seams produced by switching lobes of a subsiding delta plain.

Allocyclic sequences (or "allogenetic sequences") are caused mainly by variations external to the considered sedimentary system (e.g., the basin), such as climatic changes, tectonic movements in the source area, and global sea level variations. Such processes tend to generate cyclic phenomena of a greater lateral continuity and time period than autocyclic processes. Most characteristically, allocyclic processes may operate simultaneously in different basins in a similar manner. Thus it should be possible to correlate part of the allocyclic sequences over long distances from one basin to another, and perhaps even from continent to continent.

However, it is often not possible to distinguish sharply between autocyclic and allocyclic processes. For example, regional tectonics may affect both the drainage area outside the depositonal basin as well as tectonic structures within the basin and thus initiate the influx or redeposition of coarse-grained sediments. The occurrence and frequency of essentially autocyclic mass flows and turbidites is commonly also controlled by allocyclic eustatic sea level changes. Climate-induced rises in lake level will affect outbuilding and upbuilding of delta lobes. In fact, there are many depositional sequences displaying the results of both allocyclic and autocyclic phenomena. Nevertheless, the terms *autogenetic* and *allogenetic* still appear useful when applied in the sense that the processes are dominated by one of these two processes.

RHYTHMIC BEDDING

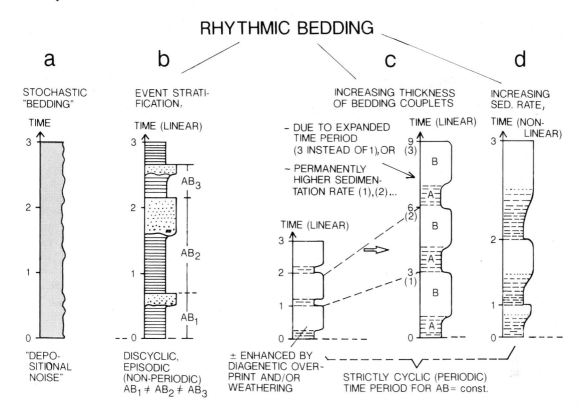

a
STOCHASTIC "BEDDING"

TIME

"DEPO-
SITIONAL
NOISE"

b
EVENT STRATI-
FICATION,

TIME (LINEAR)

AB_3

AB_2

AB_1

DISCYCLIC,
EPISODIC
(NON-PERIODIC)
$AB_1 \neq AB_2 \neq AB_3$

c
INCREASING THICKNESS
OF BEDDING COUPLETS

TIME (LINEAR)

- DUE TO EXPANDED
  TIME PERIOD
  (3 INSTEAD OF 1), OR

- PERMANENTLY
  HIGHER SEDIMEN-
  TATION RATE (1), (2) ...

TIME (LINEAR)

± ENHANCED BY
DIAGENETIC OVER-
PRINT AND/OR
WEATHERING

STRICTLY CYCLIC (PERIODIC)
TIME PERIOD FOR AB = const.

d
INCREASING
SED. RATE,

TIME (NON-
LINEAR)

NORMAL FIELD-SCALE CYCLES

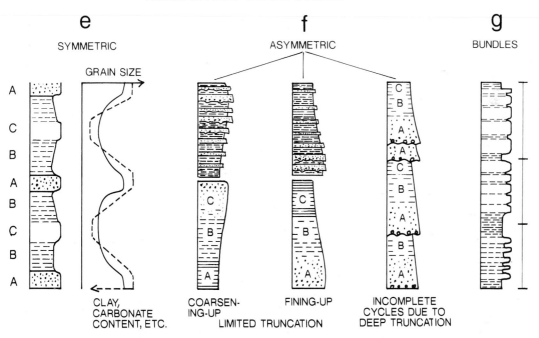

e
SYMMETRIC

GRAIN SIZE

CLAY,
CARBONATE
CONTENT, ETC.

f
ASYMMETRIC

COARSEN-
ING-UP

FINING-UP

LIMITED TRUNCATION

INCOMPLETE
CYCLES DUE TO
DEEP TRUNCATION

g
BUNDLES

Fig. 7.1a-g. Different types of rhythmic (**a** through **d**) and cyclic (**e** through **g**) sedimentary successions and definitions of some terms. Alternations between beds and interbeds and the time periods of cyclic sequences may be either strictly cyclic (periodic), quasi-cyclic (**c** and **d**), or discyc- lic (nonperiodic; **a** and **b**). The thicknesses of correspond- ing bed types can change from cycle to cycle, due to va- riations either in the time period (**c**) of succeeding cycles or in the sedimentation rates of (**d**). See text for explanation. (Einsele et al. 1991)

Scales of Rhythmic and Cyclic Phenomena

A simple, primarily descriptive classification of rhythmic and cyclic phenomena is shown in Fig. 7.2, which is based mainly on the thicknesses of beds and larger sedimentary cycles. Here, rhythmic and cyclic sediments are subdivided into four groups:

- Varve-scale laminations.
- Bed-scale rhythms and cycles.
- Field-scale sedimentary cycles (including the third and fourth order cycles and parasequences of the nomenclature proposed by Haq et al. 1987; Vail et al. 1991; Table 7.1).
- Various orders of macro-scale cyclic sequences (supercycles and megacycles).

Except for the macro-scale sequences, this descriptive classification can be easily applied in the field and in well logs without any information on genesis, sedimentation rates, and associated time spans.

Varve-scale laminations or bed-scale alternations are well-known features, but none are the product of only one specific process in a certain environment, and each may represent quite different time periods. Field-scale sequences represent the typical outcrop cycle, several meters to tens of meters thick. Many of these field-scale marine cycles are interpreted today as representing global and relative sea level changes on the order of 100 ka to several Ma. However, in lacustrine environments, where overall sedimentation rates are usually much higher than in the deeper sea, cycles of the same thickness are commonly controlled by short-term climatic fluctuations on the order of 20 to 100 ka. Climatic variations with time periods on this order are also believed to be the cause of many rhythmic, marine marl-limestone sequences. Vail et al. (1977, 1991) and Haq et al. (1987) proposed subdividing depositional sequences into six orders of cycles, defined primarily in terms of time periods (Table 7.1).

These groupings define the time period over which the process is active or will not change markedly in character, or they signify the duration of a cycle or type of event (if the term "event" is used in a broader sense than in the following).

Macro-Scale Cycles

Cycles of the first and second order normally cannot be seen in single field exposures. They comprise successions of considerable thickness

Table 7.1. Orders of time periods, origins, and consequences of various cyclic phenomena. (After Vail et al. 1991, modified)

Order	Duration (Ma)	Tectonic origin, consequences, distribution in space	Eustacy (global)
1	≥ 50	Formation of sedimentary basins (regional)	Major continental flooding epochs
2	3 – 50	Specific phases in the evolution of sedimentary basins: Major transgressive-regressive facies cycles (regional)	Sequence cycle
3	0.5 - 3	Folding, magmatism, diapirism: Disturbances of general transgressive-regressive trends (local)	Sequence with systems tracts
4	0.1 - 0.5		Parasequences
5	0.02- 0.1		Milankovitch cycles
6	≤ 0.02		

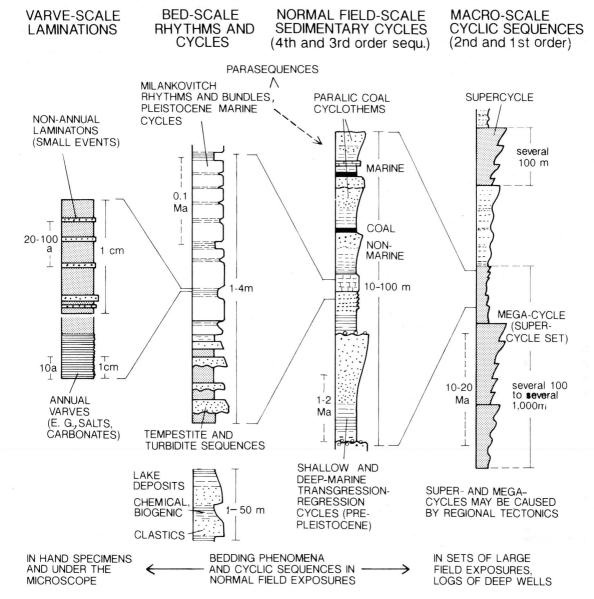

Fig. 7.2. Descriptive terms for rhythmic and cyclic sedimentary sequences of various scales, time periods, and different origins in the field. Orders of cycles *in paren-* *theses* after Vail et al. (1977) and Haq et al. (1987). The term "mega-cycle" is used here for "supercycle set" in Haq et al. (1987)

(100 m up to several km) and represent long time periods (usually between 10 Ma and more than 100 Ma). They are commonly caused by long-term tectonic processes generating sedimentary basins and controlling their specific phases of evolution. Hence, these cycles are primarily associated with crustal extension, lithospheric flexure, thermal contraction, ocean spreading, and subduction, but also with uplift in the basin drainage area and other tectonic processes (Chaps. 1 and 12). They are characterized, for example, by long persisting onlap of coastal sediments on continental margins (Watts et al. 1982; Sheridan 1987). Although tectonic processes operate on a regional scale, they may cause global major, first order continental flooding epochs or second order transgressive-regressive facies cycles, when regional processes change the total volume of the oceans (Vail et al. 1991). By contrast, depositional sequence cycles of higher orders are thought to be caused by changes in ocean water volume (glacio-eustatic cycles).

Field-Scale Cycles (Third to Fifth Order)

Conspicuous cycles in the field are third order sequences with a time range of 0.5 to 3 or 4 Ma. Fourth order cycles represent eustatic variations around the larger eccentricity cycles of the Milankovitch band (E 3 and E 4, Chap. 7.2). The so-called *parasequences* of the Vail nomenclature fall into this category. Furthermore, some of the coal cycles seem to have this periodicity. Fifth order sea level variations represent the generally shorter periodicities of the *Milankovitch frequency band*, such as glacio-eustatic cycles (0.02 to 0.1 Ma). Variations on these orders of magnitude include field-scale and bed-scale phenomena, as expressed for instance in peritidal carbonate cycles (Strasser 1990). Sea level changes with higher frequencies than the Milankovitch variations (with periods ≤20 ka) are denoted as sixth order cycles (Table 7.1).

The expression of all these cyclic phenomena in sedimentary sequences is significantly modified locally by various depositional environments. It is the aim of sequence stratigraphy to divide the stratigraphic record into genetically related units, to date these units and their sequence boundaries, and to investigate the processes which have caused them.

Periodic, Quasi-Periodic, and Discyclic Sequences

Categorizing cyclic phenomena into various orders and using the term sedimentary cycle may foster the opinion that cycles within a sequence represent equivalent time periods. This is not necessarily true, and in fact most workers use the term cycle not in this sense, but only as a convenient way to describe repeated successions of certain lithologies and facies types. In rhythmic sequences caused predominantly by autogenetic processes (e.g., turbidite and tempestite sequences), it is clear that they are the result of frequently but irregularly recurring (discyclic) sedimentological events. Smaller irregularities in the mode and rate of deposition are sometimes referred to as *depositional noise*, resulting in purely stochastic sequences (Fig. 7.1a). By contrast, some types of cyclic bedding and certain cyclic sequences may reflect allogenetically dominated processes with a regular time period (strictly cyclic or periodic sequences; Fig. 7.1c and d).

It is, however, very difficult to prove whether or not a given cyclic sequence is really caused by a mechanism with a constant time period. It may even be difficult to determine the average cycle duration from a series of subsequent cycles if the beginning and end of the cyclic sequence is not precisely known, as recently pointed out for Pennsylvanian coal cycles in North America (Klein 1990). Numerical techniques, such as time series analyses, are important testing tools for the problem of cycle duration (e.g., Schwarzacher 1975; Weedon 1991). The most prominent examples of periodicity affecting depositional patterns are the Earth's orbital cycles of precession, obliquity, and eccentricity with periods of 21, 41, and about 100 and 400 ka (i.e., Milankovitch cycles, see, e.g., Berger et al 1984; Fischer 1991; Einsele and Ricken 1991). If, as demonstrated for the Pleistocene oxygen isotope curve, a depositional system is repeatedly affected by phenomena with constant or quasi-constant recurrence times, the resulting sediments may be termed *periodites* (Einsele 1982c). However, even in this case, the thicknesses of single layers A and B may vary due to changing sediment composition or fluctuating sedimentation rates (Fig. 7.1c and d).

Symmetry and Asymmetry of Sedimentary Cycles

Sedimentary cycles may be either symmetric or asymmetric (e.g., coarsening-upward or fining-upward sequences), as well as complete or incomplete (Fig. 7.1e and f); the latter commonly results from low subsidence values. The so-called "punctuated aggradational cycles" (Goodwin and Anderson 1985) represent truncated, shallowing-upward sequences deposited in shallow water environments under a fluctuating sea level. The superposition of two or several periodically recurring processes can generate a succession of bundles (Fig. 7.1g; Schwarzacher and Fischer 1982). A more detailed description of various types of sea level-related asymmetric sedimentary cycles is given by Einsele and Bayer (1991).

Cycle Hierarchy

Sedimentary cycles are caused by various processes, including regional and global tectonics, sea level fluctuations, changes in climate, and variations in the water volume of lakes. They are affected by variations in biogenic sediment production and modified by local factors. In addition, biological processes such as the appearance and mass production of new organisms in specific environments, or the mass extinction of pre-existing groups, may lead to significant, repeated biological events which leave behind a signature in the sedimentary record.

When cyclic phenomena of different frequencies and natures are superposed, the resulting sedimentary record may show a complex multi-cyclic pattern. This reflects not only the added effects of two or more cyclic processes, but the different processes also exert some influence on each other. The term *cycle hierarchy* includes such an interplay between various factors controlling the actual stratigraphic record (for more details and examples see Chaps. 7.3 and 7.6).

Biologic Response to Sedimentological Events and Cycles

One of the best means for discriminating between subaquatic cyclic beds and event beds is the study of benthic organisms and their burrows preserved in beds and interbeds. The reactions of benthic marine and lacustrine fauna, or of plants in a fluvial environment, to *gradual*, cyclic fluctuations versus *short episodic events* are fundamentally different. Long-term environmental changes on the order of Milankovitch cycles will be accommodated by shifts in the faunal and floral spectrum and species dominance. Physical events, in contrast, will be experienced as catastrophes, wiping out the existing bottom life (Brett and Seilacher 1991; Seilacher 1991). Re-establishment of the original bottom community can begin only after the depositional event. In the case of tempestites and turbidites, the post-event community may differ from the background fauna, if the new substrate on the sea floor provides different conditions for epifauna and infauna (e.g., Kidwell 1991; Sepkoski et al. 1991).

Such *taphonomic feedback systems* also operate in regressive and transgressive situations in shallow seas. *During regressions*, the muddy fraction is winnowed and transported into deeper water, leaving behind a shell layer which can be settled by a new epifauna. Reduced siliciclastic influx *during transgressions* may allow the establishment of sessile epifaunal organisms (such as some species of corals and oysters), who live in quiet waters near or below the storm wave base and achieve stability through their large, massive skeletons. *Rare storm events* may rework such outsized

bioclasts and mix them with diagenetically formed concretions (see below). Thus, over significant periods of time, the production of *outsized bioclasts* and *concretional diaclasts* in combination with episodic burial and winnowing may eventually generate thick shell beds or mounds even in regions where overall fine-grained substrate conditions persisted (Brett and Seilacher 1991).

Epibenthic shell beds occur throughout the Phanerozoic record. They have been frequently used as marker beds, because they occur over considerable distances and commonly coincide with biozonal boundaries. They may have been formed during either regressions or transgressions. The latter situation applies if the coarse particles are produced in place by biological processes and are enhanced by diagenesis, and if the sedimentary structures reflect short turbulence events rather than the background regime.

Diagenetic Overprint

Primary bedding features, including sedimentary structures, bedding planes, bedding rhythms and larger sedimentary cycles, can be significantly modified by diagenetic overprints, particularly in carbonates and siliceous sediments (Chaps. 13.3, 3.3.2, and 5.3.5; also see Ricken 1986; Ricken and Eder 1991; Tada 1991). Even minor variations in primary composition and pore space are sufficient to cause significant diagenetic modifications. As a result, primary structures and bedding phenomena may be enhanced, modified, or in some cases obliterated.

Selective Cementation and Pressure Dissolution

In rhythmic sequences, selective cementation and pressure dissolution are the most important processes enhancing compositional variations. At burial depths of several hundred meters, carbonate or silica starts to dissolve from the potential interbed and to reprecipitate as pore cement in either the overlying or underlying bed. Unstable carbonate minerals or, in the case of silica, increased temperatures, promote these processes. As a result, the beds become lithified and resist further mechanical compaction, while the interbeds continue to compact and lose dissolved matter by diffusion. Thus, the original thicknesses of the interbeds are substantially reduced, while the thicknesses of the cemented beds are essentially retained. In addition, the original differences in composition between beds and interbeds are amplified. *Solution seams* and *stylolites* are sites where intensive dissolution has taken place. In sequences with high carbonate contents, bedding eventually becomes brick-like, consisting of limestone layers of similar thickness alternating with thin, highly compacted, carbonate-depleted marl beds (compare Fig. 13.18). In skeletal carbonates, early diagenetic differential dissolution and cementation may lead to selective lithification of specific layers.

Hardgrounds, Concretions, and Nodular Limestones

Other common features generated by diagenetic overprint are hardgrounds, concretion layers, and the cementation of beds adjacent to the event bed (e.g., underbeds). Hardgrounds are formed by early cementation at the sediment surface in various calcareous environments when the sea floor is subject to winnowing and erosion, as on carbonate platforms, shelves, and continental slopes. Cemented hardgrounds with erosional surfaces may occur repeatedly in a sequence and thus indicate a kind of rhythmicity or cyclicity.

Concretions of carbonate, silica, and phosphate generally form early in diagenesis near the sea floor, but they also grow at greater depth below the sediment-water interface (Raiswell 1987). Concretions are frequently concentrated in layers which deviated in their original composition and/or texture from the underlying and overlying sediments. They may have had higher carbonate, biogenic silica, or organic carbon contents, or relatively high porosities and somewhat differing grain size distributions. For this reason, most concretion horizons reflect subtle primary depositional variations and are found to be more or less parallel to the primary bedding.

Bedded, nodular limestones are generated by early concretionary cementation of bioturbated carbonate muds and by pressure dissolution during the later stages of diagenesis. *Flint layers* in chalk may develop independently from the primary bedding rhythm (Ricken and Eder 1991). Cementation of graded beds can also affect a thin layer below the event bed, thus forming an *underbed*, which apparently enhances the thickness of the event bed.

Effects of Weathering

Further alterations of rhythmic and cyclic sequences are caused when sections are exposed to weathering. In the case of limestone-marl alternations, strata with carbonate contents ≤ 65 to 85 % are commonly subject to rapid physical disintegration, while layers with higher carbonate contents resist. Hence, rhythmic bedding may become more conspicuous in field exposures than in drill cores, but weathering may also obscure the existence of bedding couplets if the carbonate contents in both the bed and interbed are below or above the "weathering boundary".

7.2 Special Features and Examples of Rhythmic Bedding

Cyclic Versus Discyclic (Event) Bedding

Cyclic Bedding

Some characteristics of the entirely different processes generating rhythmic sequences are shown in Fig. 7.3. Cyclic bedding is caused by slow, gradual variations in primary sediment composition and associated variations in sedimentation rates (Fig. 7.3a). These lead to a vertical sediment buildup that changes smoothly with time and is characteristic for various sediment types, particularly for calcareous and siliceous deposits. However, a small variation in the primary composition, texture, or fabric may be sufficient to promote a secondary differentiation into beds and interbeds accentuated by diagenetic overprint, as mentioned above. When not interrupted by depositional events, the intensity of burrow mottling is continuous in succeeding beds, although slow variations may occur in number and type of bottom dwelling species as well as planktonic organisms (see below). Anoxic intervals may further modify the alternating beds and their biological record.

The *recurrence intervals* of the various cyclic deposits can theoretically be strictly periodic, quasi-periodic, or nonperiodic. The 11 year sun spot cycles found in varved marine salt deposits and lake sediments (Richter-Bernburg 1960; Glenn and Kelts 1991) represent a strict periodicity. Milankovitch cycles (Chap. 7.1), however, hold some complications. Whereas, for example, the individual orbital

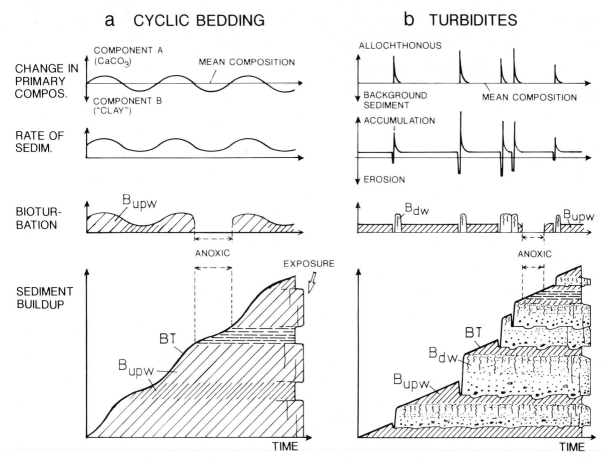

a   CYCLIC BEDDING

b   TURBIDITES

Fig. 7.3a,b. Vertical sediment buildup time curves (*BT* curves) to characterize different types of rhythmic bedding. **a** Cyclic or periodic bedding due to slow gradual changes in accumulation and composition of sediments. **b** Event bedding caused by repeated erosional and depositional episodes of different magnitudes at random time intervals. Both **a** and **b** may also show anoxic intervals. B_{upw} zone of bioturbation migrating slowly upward; B_{dw} new community of bottom dwelling organisms burrowing downward. (After Einsele 1982c)

parameters seem to have constant amplitudes and time periods, their complicated combined effects on climate and sediments tend to create a periodicity with considerably varying amplitude and changing importance of the various time intervals (Einsele and Ricken 1991; Fischer 1991). The response of a depositional system may lag behind the orbital signal. Cyclic sequences caused by relatively long-term, third order sea level changes (Chap. 7.4) also show varying time periods within a certain order of magnitude. It is the superposition of several mechanisms creating differing time periods and intensities which often renders it difficult to discover from the response of sediments whether there is a true periodic process in operation above the depositional background noise. For these reasons, many allocyclic phenomena in sediments appear to be quasi-periodic, and some even nonperiodic, rather than strictly periodic.

Discyclic Bedding

By contrast, alternations due to *episodic phenomena* (e.g., tempestites, turbidites, river floods, and sequences recording volcanic episodes) show a very irregular sediment build-up-time curve (Fig. 7.3b). Slow, but more or less continuous vertical accumulation of fine-grained background sediment is interrupted irregularly by erosional and immediately succeeding *depositional episodes*. In the case of turbidite sequences (cf. Chap. 5.4.2), background deposition is discontinuously interrupted by laterally transported, allochthonous turbidite sediment. Therefore, the rates of sedimentation also change abruptly. The bio-turbated surface layer of the background sediment, occupied by bottom dwelling organisms which normally slowly migrate upward with the sediment surface, is episodically truncated and replaced by an *event deposit*. Recolonization of the event layer is achieved by specially adapted fauna burrowing from the new surface downward.

As in cyclic sequences, intervals of anoxic conditions may occur. Whereas turbidites always contain material from distant sources, often mixed with sediments taken up underway, the tempestite material is either entirely autochthonous or derived from nearby sources (Chap. 3.1.2). In many other respects, as well, tempestites resemble turbidites.

Most event beds show a kind of *amalgamation* and *cannibalism,* i.e., large events wipe out the record of earlier, smaller ones including part of their host sediment. As a result, one event bed may directly follow another or several other ones, the tops of which were truncated (Fig. 7.1f and 7.3b). Generally, the thicker the event bed and the deeper it has eroded, the longer is the recurrence interval of the event. The sedimentological episode is normally a matter of hours or days. The areal extent of individual event beds is commonly limited to parts of a particular sedimentary basin. However, there are also periods in the ancient record in which allogenetic processes, such as global sea level falls, have triggered the generation of discyclic rhythmic bedding (see below).

Combinations of Both Cyclic and Discyclic Bedding

Some limestone-marl sequences (see below) and sediments from oxygen-deficient environments, such as black shales, phosphorite-bearing strata, and varved siliceous oozes, may show both gradual cyclic variations and episodic bedding types. Cyclic phenomena are commonly ascribed to variations in climate causing changes in the oceanic current systems and thus in the distribution of nutrients and oxygen in sea water. Event beds are frequently represented by various turbidites including muddy and bioclastic types; storm-related winnowing of mud may produce bioclast-rich lags. Long-term rhythms or cycles are often superposed by short-term fluctuations, generating thin, varve-scale (bio)laminations, or they may contain benthic and bioturbated intervals due to water mass mixing and turn-

over (Gerdes et al. 1991; Savrda et al. 1991; Oschmann 1991).

Examples of Marine Cyclic Bedding

The most prominent examples of cyclic bedding in the marine environment are:

- Hemipelagic to pelagic limestone (or chalk)-marl alternations.
- Black shale-carbonate rhythms (redox cycles).
- Pelagic banded cherts.

These types of rhythmic bedding are mentioned in Chapter 5.3.7 and described in detail in Einsele et al. (1991). Here, some important points are briefly summarized.

Pelagic Limestone-Marl Alternations

Pelagic limestone-marl alternations have been formed since the upper Jurassic, after substantial planktonic carbonate production began in deep and shallower marine environments which were little affected by waves and bottom currents. In earlier periods, particularly in the Paleozoic, fine-grained carbonate deposition was restricted to the upper slopes, shelves, and epicontinental seas, where either well-bedded or nodular limestones were formed.

In order to produce pelagic carbonates, the rate of planktonic carbonate deposition must have been three to four times higher than terrigenous silt and clay input (Seibold 1952; Einsele and Ricken 1991). On the shelf and in epicontinental basins, rhythmic limestone-marl sequences of some thickness needed special requirements, such as the coincidence of subsidence and sea level rise, which may have restricted their occurrence to relatively short time spans of 1 Ma to a few Ma.

Generally viewed, three basic processes can produce limestone-marl alternations (Einsele and Ricken 1991; Arthur and Dean 1991):

1. Variations in carbonate production.
2. Periodic carbonate dissolution.
3. Periodic terrigenous dilution.

Each of these types represents an end member of alternations with simultaneous oscillations of several inputs. Cycles formed by varying carbonate production and dissolution show rhythms with thick limestone beds and thin marly interbeds, whereas terrigenous dilution is indicated by the opposite trend. Limestone-marl successions can be formed in the following environments (Fig. 7.4):

- *Outer shelves, lower slopes of carbonate ramps, central portions of epicontinental seas, marginal plateaus.* Rhythmic bedding is caused predominantly by periodically increased terrigenous influx *(dilution cycles)* in relation to variations in climate, vegetation, and/or sea level. These types of limestone-marl rhythms may contain intercalations of distal carbonate and siliclastic tempestites. With increasing terrigenous influx, limestone-marl successions are replaced by claystone-marl sequences,

displaying thicker bedding couplets, but thin marl beds. Shallowing of the sea favors reworking, omission, channeling, and finally interfingering and replacement of the bedded pelagic limestones by bioclastic carbonate arenites and a reef association.

- *Marginal deep seas, deep-sea plateaus, and environments around isolated carbonate platforms* appear to provide the most suitable environments for the development of limestone-marl rhythms. Carbonate accumulates at water depths above the lysocline and is affected by dissolution only under the presence of substantial amounts of organic matter (e.g. Diester-Haass 1991). Thus, variations in carbonate content reflect *changes in productivity*

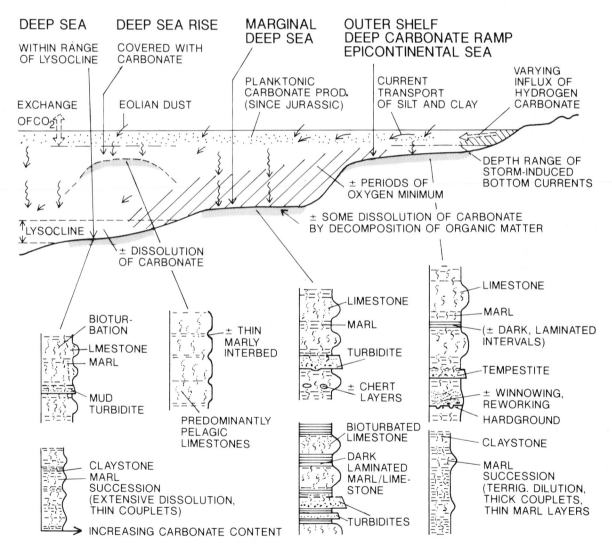

Fig. 7.4. Overview of depositional environments and processes generating different types of pelagic to hemipelagic limestone-marl, claystone-marl, or calcareous redox rhythms. See text for discussion

rather than terrigenous dilution. Productivity of planktonic carbonate ooze may change according to oceanic circulation and nutrient supply. Under oxygen deficient conditions, redox cycles with dark, laminated interbeds are formed (see below). Below upwelling zones carbonate-bearing sequences may also show rhythmic chert accumulations. If the supply of clay is very low, the formation of marly interbeds ceases.

- Deep-sea environments between the lysocline and the CCD. Here, variations in *carbonate dissolution,* due to vertical oscillations of the lysocline, are the main control on the development of limestone-marl rhythms. As a result, sequences with large variations in $CaCO_3$ are formed (e.g., Farell and Prell 1987). In deepsea regions near the shelves and continents, periods of rising and high sea level appear to favor the accumulation of sediments rich in carbonate, while during falling sea level carbonate deposition tends to become more diluted or replaced by siliciclastic materials, including turbidites.

The *periodicities of limestone-marl bedding couplets* described in the literature are in the Milankovitch frequency band (around 20, 40, 100, and 400 ka). This is the case for various types of carbonate rhythms in the Quaternary (e.g., Berger et al. 1984; Grötsch et al. 1991; Diester-Haass 1991) as well as for pre-Quaternary examples which developed during times of little to no continental ice (e.g., Hattin 1986a; De Visser et al. 1989; Einsele and Ricken 1991). Consequently, substantial sea level oscillations cannot be the primary factor causing these limestone-marl alternations. It is therefore assumed that relatively weak orbital signals, i.e., small changes in seasonality, are enhanced by a kind of *atmospheric-oceanic feedback system.* This leads to shifting climatic belts, changes in oceanic thermohaline circulation, and modifications within the global carbon cycle, including atmospheric CO_2 and carbonate deposition. Thus, both planktonic carbonate productivity and carbonate dissolution (namely a rising or falling CCD) can be affected. For example, a 13% increase in atmospheric CO_2 may cause a CCD rise of about 1 km (Bender 1984).

In addition, an increase in carbonate production on the shelves can lead to a rise of the CCD and lysocline in the deep sea and vice versa. High river input of dissolved carbonate due to a dense, widely extended vegetation cover on the continents, on the other hand, can cause a CCD lowering and thus further carbonate preservation in the deep sea. At the same time, dense vegetation tends to reduce mechanical erosion and the supply of terrestrial silt and clay to the oceans, which in turn prevents strong dilution of carbonate deposition. In conclusion, such a balanced system is very sensitive to minor changes in single processes. The *sensitivity of the oceanic circulation system* was probably even greater during periods of more uniform global climate than it is today under the conditions of a significant temperature gradient between the poles and the equator (see below).

Black Shale-Carbonate Rhythms (Redox Cycles)

Bedding rhythms with *black shales as interbeds* are known throughout the Phanerozoic (Wetzel 1991), but they are particularly common in the Jurassic and Cretaceous record. Bed-scale successions of this type occur, for example, in the middle Cretaceous of the Atlantic and Tethyan Oceans (Weissert et al. 1979; Pratt 1984; Cotillon 1985; Dean and Arthur 1986; Herbert et al. 1986; Ogg et al. 1987; Arthur and Dean 1991; de Boer 1991; Sageman et al. 1991). Hemleben and Swinburne (1991) describe Mesozoic *plattenkalk facies* of the Solnhofen type consisting of pelagic carbonate beds with very thin marly intercalations.

In most cases, the pelagic to hemipelagic bedding couplets of *clay-carbonate redox cycles* consist of light carbonate beds and dark, organic-rich interbeds with lower carbonate contents. The black shale intervals are frequently thinly laminated, but their laminae represent small sedimentological events, or minor, short-term climatic variations rather than annual varves (Cotillon 1991). The cyclic sequences commonly exhibit repeated transitions from aerobic to anaerobic conditions (Savrda et al. 1991; Oschmann 1991). Similar to limestone-marl rhythms, black shale-carbonate rhythms are deposited in environments above the CCD or the lysocline (Fig. 7.4).

The sensitive oceanic circulation system mentioned above also controlled carbonate redox cycles. Periods with better oxygenated waters usually produced lighter colored, bioturbated beds richer in carbonate than periods characterized by oxygen deficiency, resulting in dark, laminated interbeds. Frequent alternations between these two bed types were favored by bottom water in which oxygen demand for remineralization of organic matter was either slightly lower or higher than oxygen supply by oceanic circulation. The organic matter in the dark layers may have been derived from both marine productivity and terrestrial sources.

During times of reduced temperature differences between the poles and the equator and the generally *warmer climate* than at present, such as in the Cretaceous (e.g., Barron and Washington 1985; Sloan and Barron 1990), thermohaline oceanic circulation tended to slow. Instead of cold polar deep-water masses, warm dense saline water, originating from flooded low-latitude shelves and adjacent seas, may have become the major bottom water source (Arthur et al. 1987). Since the solubility of oxygen decreases with increasing temperature and salinity, the sinking, more saline, warm surface waters soon became oxygen depleted. This, in turn, hampered the mineralization of organic matter in intermediate and deep waters, as well as the release and recycling of nutrients controlling surface productivity. These processes may have caused short-term, *expanded, intensified oxygen-minimum zones* at intermediate water depths and thus the widespread deposition of black shale layers, even in epicontinental sea settings. Subtle changes in climate and oceanic circulation were sufficient to periodically restore more oxygenated conditions with higher planktonic carbonate production.

Rhythmic Bedding in Siliceous Sediments

Alternations between marine, biogenic siliceous and non-siliceous beds and their origin were already mentioned in Chap. 5.3.5 (Fig. 5.6). Their rhythmic bedding is generated by both cyclic and discyclic (event-related) depositional processes. In addition, calcareous beds may contain layers of chert nodules which indicate primary alternations between beds poor and rich in opaline silica. Rhythmic bedding in siliceous sediments was described from many regions, particularly in Mesozoic sediments of the Tethys Ocean (Hein and Obradovic 1988) and in Tertiary sequences around the Pacific (Iijima et al. 1985), i.e., regions of high fertility in siliceous plankton. Rhythmicity may be expressed in *annual varves* or in *banded ribbon radiolarites* (e.g., Jenkyns and Winterer 1982; Jenkyns 1986), which are modified and enhanced by silica diagenesis (Chap. 5.3.5). The most important facts about such alternations were summarized recently by Decker (1991). In this section, cyclic phenomena in the range of the Milankovitch frequency band are discussed.

Similar to limestone-marl alternations, cyclic sequences in siliceous sediments result from (1) variations in siliceous *plankton productivity,* (2) periodic *dilution of siliceous sediments* by terrigenous material, and (3) variations in *dissolution of opaline silica* in the water column and at the sea floor. However, productivity cycles appear to be more effective than dilution cycles in generating rhythmically interbedded chert layers. Thus, climate-induced global changes in oceanic circulation and the recycling of nutrients through upwelling may predominantly control the generation of beds rich in opaline silica. At the same time, carbonate production should be low, or the position of the lysocline or CCD should be well above the sea floor in order to prevent strong dilution of opaline silica by carbonate. The occurrence of banded chert is not restricted to deep water; it rather indicates an environment of high siliceous plankton production and very limited terrigenous influx.

Peritidal-Lagoonal Carbonate Cycles

Cyclic carbonate sequences formed in peritidal environments, i.e., in the shallow subtidal, intertidal, and subtidal-lagoonal zones, commonly display thicker individual depositional cycles and are more complex than deep-water limestone-marl or black shale-carbonate alternations. They may be caused either by autocyclic (intra-basin sedimentary processes) or allocyclic processes (sea level changes; Strasser 1988 and 1991).

Peritidal carbonate cycles have been described from many regions since the pioneering study by Fischer (1964) in the Triassic Dachstein limestone in the northern calcareous Alps. Similar carbonate cycles are known from other parts of the widely extended Triassic carbonate platforms of the western Tethys (e.g., northern and southern Alps, inner Carpathians, and Hellenids; e.g., Bechstädt 1975; Goldhammer et al. 1987), but also from the Paleozoic in Europe and North America (e.g., Koerschner and Read 1989).

Peritidal-lagoonal carbonate sequences commonly reach thicknesses of several hundreds of meters. Individual cycles are typically one to several meters thick and may be symmetric or asymmetric; they can be correlated laterally over distances of more than 10 km (Haas 1990). Their periodicities mostly range between 20 and 50 ka and thus fall within the

Milankovitch frequency band. Hence, the average rate of carbonate deposition (around 5 to 20 cm/ka) is much greater than in the deep sea (see above and cf. Chap. 10.2). One can distinguish two end member types:

1. Transgressive Lofer cycles or Loferites represent *deepening-upward sequences* with subsequent erosional unconformities (Fig. 7.5b) consisting of (from top to bottom):

- Erosional unconformity
  (emergence, karstification, paleosol)

- Subtidal calcarenite bed with diverse marine fauna
- Intertidal and subtidal algal mats
- Supratidal argillaceous bed
- Erosional unconformity
  (emergence, karstification, paleosol)

2. Regressive, shallowing-upward sequences (Fig. 7.5b; Strasser 1991):

- Erosional unconformity

- In places: soils, calcrete, beach ridges, eolian deposits, etc.
- Supratidal algal marsh, freshwater to hypersaline ponds, sabkha
- Intertidal carbonate mud and sand in channels
- Subtidal carbonate muds and sands, stromatolites, reefs

Shallowing-upward sequences displaying little erosion and no emergence can be formed in slowly subsiding areas without sea level changes (Fig. 7.5d). Carbonate sediment produced in shallow-marine environments is transported toward the tidal flats and builds sediment up to sea level (Phase 1). With decreasing sediment supply, the sediment surface maintains its position near sea level for some time (Phase 2), but then drops due to continued subsidence (Phase 3). As soon as sufficient sediment becomes available after a certain lag time, sediment buildup starts again and forms the base of the next cycle.

By contrast, cyclic *sequences showing emergence, karstification, pedogenesis, and erosion* can be interpreted only with the aid of relative sea level changes (Fig. 7.5be). The amplitude of such oscillations may be very small in comparison to Pleistocene sea level changes, but they must allow subaerial erosion, karsti-

fication, soil formation under vegetation, or the existence of a coastal sabkha. Relatively deep erosion may remove the regressive part of a cycle and thus lead to an asymmetric, deepening-upward type, as in the Lofer cycle.

A simple facies model for symmetric and a-symmetric peritidal-lagoonal cycles is shown in Figure 7.5f. On the landward side of the lagoon, cyclic sediments tend to be intensively dolomitized as a result of evaporative pumping (Chap. 6.4.1) and truncated by erosion. Many shallow-water carbonate cycles seem to be delicately balanced between high-frequency eustatic variations and autocyclic processes related to laterally migrating tidal flats.

There is increasing evidence that such high-frequency sea level variations occurred throughout the Phanerozoic (e.g., Einsele et al. 1991). Small sea level variations with amplitudes of 1 to 10 m may result from storage and release of continental water, affecting lake and ground water reservoirs, as well as possibly some mountain and polar ice.

Examples of Discyclic Event Bedding

The most prominent examples of discyclic, event-related rhythmic bedding are:

- All types of *turbidites*, sandy and muddy (siliciclastic, calcareous, and siliceous) and associated minor *gravity mass flow deposits* (Chap. 5.4.2).
- Siliciclastic and calcareous *tempestites* (Chap. 3.1.2).
- *Flood event deposits* on alluvial plains (inundites; Chaps. 2.2.1 and 2.2.3).
- *Tephra layers*, particularly widely dispersed ashfalls (Chap. 2.4).

These types of event beds do not always form conspicuous rhythmic bedding. Frequently, one event bed may follow an earlier one without an intercalated interbed representing the more or less autochthonous background sedimentation. A sequence of fine-grained, thin mud turbidites, for example, is commonly not referred to as rhythmic bedding, but rather as massive siltstone or claystone (Kelts and Arthur 1981). Other event beds such as thick mud flows, debris flows, and olistostromes, or thin silt layers originating from dust storms, tsunami- and bolide-related deposits, and sediments reflecting earthquake shocks without being displaced (seismites) are either too rare, too thick, or too thin, and not sufficiently extensive to generate rhythmic bedding (also see Clifton 1988).

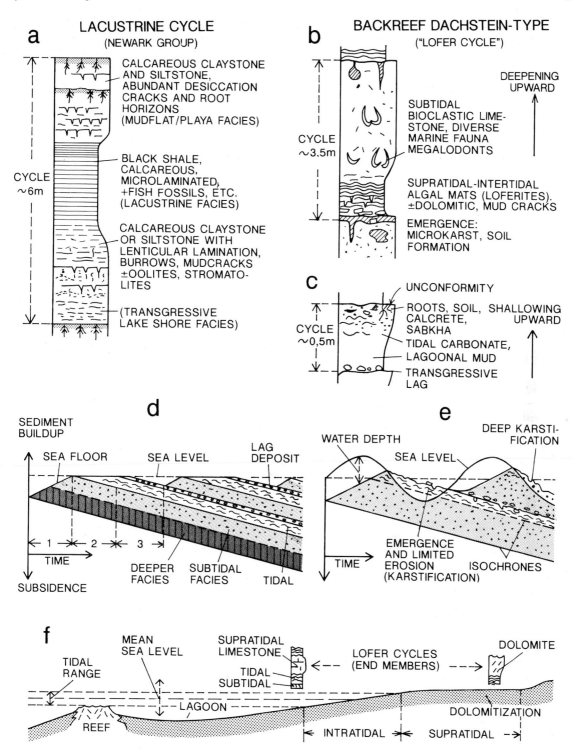

Fig. 7.5a-f. Idealized sedimentary cycle in the lacustrine Triassic Newark Group, North America (**a** Fischer 1991) and peritidal-lagoonal carbonate cycles from the Triassic in the Alps and southern Europe (**b**) and Purbeckian in the Swiss-French Jura Mountains (**c** after Haas 1991; Strasser 1991). Lofer cycles in the Dachstein limestone (**b**) display transgressive, deepening-upward sequences truncated by emergence, karstification, and pedogenesis, while Purbeckian cycles (**c**) are shallowing-upward. **d** and **e** Models to explain peritidal cycles by subsidence alone (autocyclic, **d**) or by sea level oscillations (allocyclic, **e**; based on Strasser 1991; Koerschner and Read 1989). **f** Facies model for the generation of symmetric and asymmetric peritidal carbonate cycles, with or without dolomite. (After Haas 1991)

The following considerations on the recurrence intervals of rhythmic phenomena take into account only the common types of macroscale rhythmic bedding.

Recurrence Time and Dating of Cyclic Bedding Couplets

Although the recurrence time of cyclic and episodic phenomena in sedimentary sequences spans a very wide scale, many common features in marine environments such as tempestites, turbidites, limestone-marl rhythms and banded chert layers occur in about the same, narrower range of frequencies, i.e., between 1000 and several 100 000 years. This implies that the thicknesses of individual bedding couplets are also often in the same range, regardless of whether they are generated by sedimentological events or gradual cyclic changes. Since both types of bedding rhythms can occur in the same sequence, distinguishing between cyclic and discyclic bedding and determining the number of cycles in a particular section may be problematic in such cases. However, this is only true for environments where the sedimentation rates are of the same order of magnitude, as for example in wide areas of the oceans. Most lake sediments accumulate much faster, with the result that cycle periods similar to those in the sea generate thick packages of sediment, but event deposits maintain their dimensions or tend to decrease in size.

A special problem is the identification of the various *periodicities of the Milankovitch frequency band*. The Earth's axial precession (P) has a mean period of 21 ka, the change in axial obliquitiy (O) spans 41 ka, and the orbital eccentricity (E) varies in several cycles: E1 at 95 ka and E2 at 125 ka form the ca. 100 ka E cycle (e.g., Herbert and D'Hondt 1990; Fischer 1991); E3 has a period of 413 ka, while E4 is ca. 1300 ka. If only one or two of these frequencies, for example P and O, or O and E, are recorded in the sediments, it is normally difficult to decide which of the different periods they represent. A better determination is possible when P, E, and possibly E3 occur in the same sequence and form a bundle, or when O, E, and E3 are present. Then, in combination with other methods for stratigraphic dating, (e.g., radiometric dating and magnetostratigraphy) very precise stratigraphic resolution can be achieved (e.g., Schwarzacher 1987; Kauffman et al. 1991).

7.3 Depositional Cycles in Lakes, Fluvial and Deltaic Systems

7.3.1 Cyclic Sequences in Lakes

Lakes represent particularly sensitive depositional environments which respond immediately and markedly to various kinds of change in their drainage area (Chapter 2.5). When lake sediments accumulate rapidly under stratified water conditions, they may allow very accurate time resolution, including seasonal variations. Hence, lake sediments can record a wide range of cyclic and episodic phenomena in various climatic zones. Their potential for registering short-term cycles and events has been recently summarized by Glenn and Kelts (1991). Here, cyclic sequences related to climatic changes within the Milankovitch frequency band are considered.

Effects of Changes in Climate

Salt Lakes

Pleistocene and Holocene variations in climate have commonly left a very distinct record in many lake sediments. Present-day hypersaline lakes and playas in regions of arid to semiarid climates experienced cooler, wetter conditions during glacial periods or during the transitions from glacials to interglacials. Closed lake systems were transformed into open systems and vice versa (Chap. 2.5), with the result that deposition of carbonates and evaporites was largely replaced by siliciclastic material. The thicknesses of such Pleistocene *clastic-evaporite cycles* depend on the sedimentation rate in the lake. One cycle may reach tens to hundreds of meters in thickness.

An outstanding example of a continuous, annual climatic record has been reported from varved Permian evaporites of the Delaware Basin in New Mexico (Anderson 1982). The measured sequence consists of calcite-laminated anhydrite and anhydrite-laminated halite caused by seasonal changes in temperature and evaporation (compare Section 6.4.2). It represents a time span of 260 ka, but can be subdivided into cycles of various lengths (about 100 ka, 20 ka, 2.7 ka, and 200 years). The 2.7 ka period may indicate episodes of basin freshening similar to the climatic changes known from the Holocene.

Post-Glacial "Humid" Lakes

In more humid climates, the post-glacial sediment successions known from many glacier-shaped lakes along the margins of mountain ranges frequently display a *half-cycle* consisting of clastic sediments followed by carbonates (seekreide) and organic-rich muds.

Sediment-Starved Lakes

In the past, relatively sediment-starved, long-persisting lake systems accumulated less sediment over time periods of the same order as the Pleistocene climatic variations (Milankovitch cycles). Hence, their Milankovitch cycles are thinner and occur more frequently than in depositional systems with high sediment influx.

Some of the Triassic rift basins associated with the breakup of Pangea (Chap. 12.3), for example the Newark rift in the eastern United States, existed over a period of 30 to 40 Ma (Smoot 1991). Their sediment fills include various lake deposits of both perennial and ephemeral lake systems (cf. Chap, 2.5). Dry phases may be characterized, for example, by playa mud flats, whereas wet phases led to laminated calcareous black shales (Fig. 7.5a; van Houten 1964).

In the so-called Newark Supergroup, the mean thickness of the basic cycle is 5 to 6 m; in the basin center, the cycles may contain some evaporites. In such depositional settings, all signals of the Milankovitch frequency band (P, O, E, E3 and E4, cf. Chap. 7.2) could be identified (Olsen 1986; Fischer 1991).

Cyclic deposits from a very large playa lake also occur in the upper Triassic (Keuper) of Europe (cf. Chap. 12.3). Here, mostly red shales alternate with thin, dolomitic limestones and form bedding couplets 1 m thick and less. Locally, fluvial and deltaic sands or gypsum and anhydrite are intercalated.

Some generalized types of lake sequences which may occur cyclically are discussed in Chapter 2.5. Normally, one cannot expect that many of these cycles are actually realized in an individual lake system, because the life times of lakes are limited. Either a lake is filled up and overlain by fluvial deposits, or the lake floor and its surroundings subside below sea level and are transformed into a marine basin (cf. models in Chap. 12.1).

7.3.2 Sediment Successions in Fluvial and Deltaic Systems

Autogenetic Sediment Successions

Alluvial fans and rivers represent systems transporting material from more elevated regions into lowlands and the sea, but many of them do not leave behind any fossil record. Fluvial deposits of some thickness, and thus sequences displaying some sort of cyclicity, can be only expected in areas of substantial subsidence which are continuously filled. Under such conditions, even *alluvial fans* can show sequences of fluvial material alternating with deposits of sheet floods and debris flows (Chap. 2.2.2), and thus form a kind of discyclic succession, with irregularly spaced thin and thick event beds. Similarly, *fan deltas* into lakes and shallow seas can exhibit special flood events triggering subaquatic gravity mass flows.

Repeated sequences of larger areal extent can be generated in *braided and meandering river systems* (Chap. 2.2.3). As a result of lateral channel migration, relatively coarse-grained channel fills can form wide, elongated sand sheets. Particularly in braided systems, older channel fills may be partially reworked and their material incorporated into subsequent channel fills, which ultimately produces an architecture of vertically and laterally stacked channels (Fig. 2.9). Idealized channel sections then consist of an *asymmetric sequence* with a fining-upward trend, which is commonly terminated by the erosional face of the subsequent channel. Meandering systems show channel fills with a better developed fining-upward character and less erosional truncation, because flood deposits predominate over channel sediments. Abandoned channels (ox bow lakes) may be filled with mud or organic detritus (Fig. 2.16). *Pedogenesis* under various climates (soil horizons with roots, formation of calcrete or silcrete) characterizes long intervals between large flood events and sediment accumulation affecting the total flood plain.

In this way, autocyclic or autogenetic sequences are generated, consisting of channel fills, overlying flood plain deposits, and swamps which are partially truncated by the subsequent channel cutting process. Many *coal seam successions* were generated in the purely fluvial environment (Fig. 2.18) of a subsiding basin. Coal seams, and soil horizons in particular, can be traced laterally over distances of many kilometers, but they may be interrupted locally by channel sands.

Records of major river floods show repetitions of 10 to 150 years. Measured in terms of a human life span or the written historical record, these floods are large and rare, but not gigantic enough to be transmitted to the rock record (Dott 1988; Clifton 1988). The recurrence interval of large flood episodes which can be preserved in the geological record probably ranges from hundreds to thousands of years.

Allogenetic Sequences

Autocyclic or autogenetic processes related solely to the dynamics of the depositional system itself may be superposed by other processes of a more regional or global nature. As shown in Fig. 2.19a, tectonic uplift in the hinterland of the fluvial basin commonly causes a longer-term coarsening-upward trend in the irregular autocyclicity of an alluvial fan and its accompanying river deposits. Subsidence or the wearing down of highlands in the drainage area produces the opposite trend (Fig. 2.19b). *Climatic variations* on the order of Milankovitch frequencies can bring about a cyclic change from the relatively coarse-grained, channel fill-dominated deposits of a braided river system to the finer grained, flood plain-dominated sediments of a meandering or anastamosing river system.

Flood plains in lowlands near the sea may also be affected by eustatic sea level variations and thus be overprinted by a global, possibly periodic to quasi-periodic signal. During sea level highstands, *marine ingressions* cover large areas and leave behind shallow-marine strata with marine fossils. After the following regression, beach, lagoonal, tidal and tidal marsh, swamp and fluvial sediments again take over. Many of the Paleozoic coal cyclothems and younger transgressive-regressive sequences with coal are explained in this way (e.g., Ryer 1983; Riegel 1991).

Similarly, *cyclic phenomena in deltas* entering the sea may originate from both autogenetic and allogenetic processes. The irregular switching of delta lobes, related to the degree of sediment accumulation and the geometry of the delta complex, may lead to alternations between sediments of the delta plain and marine incursions (Chap. 3.4). Interdistributary areas cut off from sediment supply subside below sea level and are flooded either by fresh water or the sea (Fig. 3.32). When deltaic progradation returns to this area, marsh, lagoonal, and shallow-marine layers are again covered by sediments of the fluvial delta plain. In this way, a purely autocyclic sequence may be established with alternating fluvial, lagoonal, lacustrine, and marine sediments. Holocene cycles in the Mississippi delta plain span a time period of approximately 1000 to 1500 years (Tye and Kosters 1986). However, as in the case of low-lying river plains, these autocyclic processes can be overprinted and modified by global sea level oscillations.

One of the means for discriminating between allogenetic and autogenetic processes is correlation. While global and regional phenomena can be traced over long distances and from basin to basin, beds and sedimentary cycles related to autocyclic processes pinch out in a relatively short distance.

7.4 Sea Level Changes and Sequence Stratigraphy

7.4.1 General Principles and Terms

Sea level changes were already mentioned in previous chapters as a mechanism controlling vertical and lateral facies transitions (e.g., Chaps. 3.2 and 3.3), as well as a factor influencing subsidence (Chapter 8). Minor variations in sea level were discussed in Chapters 7.2 and 7.4 on peritidal carbonate cycles and fluvial-deltaic cyclic sequences. In this chapter we shall deal mainly with *third order sea level* changes (duration 0.5 to 3 or 4 Ma, Table 7.1), which have recently been described from many regions throughout the Phanerozoic (e.g., Hallam 1984 and 1988; Haq et al. 1987, Bond et al. 1988) and even in the Proterozoic (e.g., Christie-Blick et al. 1988; Southgate 1989).

Modern *sequence stratigraphy* has been developed for cyclic successions produced by this type of mechanism, but it may also be used for cycles of lower and higher orders and combinations of sea level oscillations of different orders. The well-known Quaternary sea level variations caused by the buildup and melting of enormous ice sheets had *amplitudes* on the order of 100 m, but their time period was much shorter (largely fifth order, see Table 7.1) than that of the third order cycles. The amplitudes of the pre-Quaternary, third order sea level oscillations, on the other hand, are poorly known. Estimations range mostly from ten to a few tens of meters.

It should be emphasized that there is still great uncertainty not only as to the time period and amplitude (e.g., Burton et al. 1987), but particularly as to the cause of pre-Quaternary sea level fluctuations (e.g., Donovan and Jones 1979; Matthews 1984b; Watts and Thorne 1984; Cloetingh 1986; Hubbard 1988; Ziegler 1988).

At present, it does not yet appear to be sufficiently proven whether or not the third order sea level oscillations are really synchronous on a global scale. The exact and reliable dating of relatively short-term processes and their correlation from continent to continent is still difficult. Another reason for this uncertainty is that eustatic oscillations are frequently superposed by regional and local tectonic processes (cf. Chaps. 7.1 and 7.6). In addition, local variations in sediment supply and the location of the sections investigated within a particular basin play a significant role in the recognition and dating of sedimentary cycles (e.g., McGhee and Bayer 1985).

The purpose of this chapter is to show the response of sediment accumulation to relative sea level changes, and to introduce the concept of sequence stratigraphy which can explain the large-scale structure of depositional systems much better than earlier approaches. Our knowledge of sea level fluctuations through geologic time is increasing rapidly, and discussions on the global nature and synchroneity of these oscillations are still continuing, but these topics are beyond the aims of this text.

Relative Sea Level Change and Sediment Accumulation

For the description and understanding of stratigraphic sequences, one should be familiar with some general principles and special, recently introduced terminology (Posamentier and Vail 1988; Van Wagoner et al. 1988; also see Chap. 7.4.3).

The old terms *transgression* and *regression* of the sea are not used unanimously and may sometimes even be misleading. Figure 7.6 demonstrates that during rising sea level the coast line does not always retreat landward (transgressive sea), but can also prograde seaward, provided that the influx of sediment is high at this locality. Therefore, the terms transgression and regression refer here only to relatively rising or falling sea level and do not necessarily include landward or basinward shifting of the coast line.

Sedimentary basins are areas of long-term subsidence and sediment accumulation. If we define *relative sea level change* as the change in water depth at a certain location in the basin, then we mean that it is controlled by the interplay of the rates of subsidence, sediment accumulation, and rise or fall of sea level (e.g., Pitman and Golovchenko 1983, 1988; Vail et al. 1984). The interrelationship between sea level change and subsidence is demonstrated in Fig. 7.7 for three different rates of subsidence. It is assumed that sea level oscillates regularly like a sine curve and subsidence is constant over a long period. If there is no se-

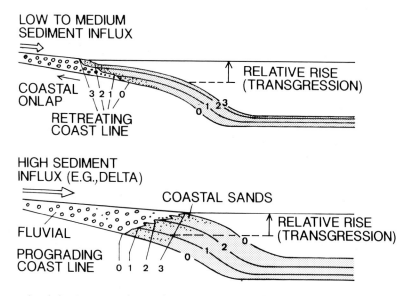

Fig. 7.6. Relative sea level rise (transgression) and retreat or progradation of coast line due to differing sediment influx

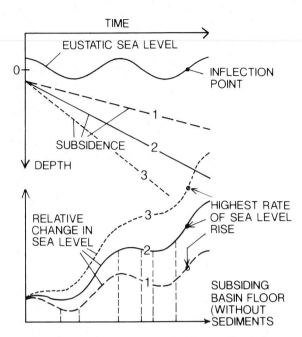

Fig. 7.7. Relationship between absolute sea level change *(above)*, different rates of subsidence, and relative sea level with time. For further explanation see text

diment accumulation on top of the basin floor, then the relative change in sea level through time is described by a curve as shown in the lower part of Fig. 7.7.

A relative sea level fall takes place only when the rate of sea level fall is higher than that of subsidence (Curve 1). If subsidence is fast (Curve 3), the sea level rises continuously, but with changing rates. The varying space available between sea level and the subsiding basin floor can be called the *"sediment accommodation potential"*. this is reduced by sediment accumulation, which ultimately controls the water depth in the basin. Furthermore, it should be noted that the highest rate of sea level rise coincides with the so-called *inflection point* of the sea level curve (Figs. 7.7 and 7.12). This point indicates the time of most rapid coastal onlap associated with a minimum sediment supply to the somewhat deeper parts of the basin (Fig. 7.9c and f).

In addition to the interplay between absolute sea level change and subsidence, the water depth in the basin is further modified by *sediment accumulation* on top of the basin floor (Fig. 7.8). Constant sediment supply to a cer-

tain location within the basin may, over a longer period, just compensate for relative sea level change (Fig. 7.8a, Curve 2) and thus lead periodically to a filling up or deepening. The same sediment influx causes a stepwise deepening of the basin, if the subsidence rate is greater than the sedimentation rate (Curve 3). When the sediment supply is constant as before, the effect of sea level variations on the depositional system is most pronounced during times of slow subsidence (Curve 1). Here, when relative sea level falls, some of the arriving sediment cannot be deposited and therefore will bypass shallow-water regions. Even more important is the fact that sediment accumulated earlier will be eroded and swept into deeper water. In the end, the basin tends to be filled up in this region, but sediment accumulation is repeatedly interrupted by erosional periods.

Finally, Fig. 7.8b demonstrates a case in which, over a long length of time, relative sea level change is compensated for by sediment accumulation, but *sediment supply also varies*. Influx is high during falling sea level, but low during the transgressive phase. Such an assumption agrees with our observations in nature. The model shows that even under conditions of long-term sediment/subsidence compensation there are periods with excess sediment supply or erosion. Consequently, the sedimentation rates at a given location and at other locations farther basinward vary markedly. Times of rapid accumulation are interrupted by periods of *omission and erosion*, which in turn are followed during the transgressive phase by relatively long-lasting sediment-starved conditions before the next sedimentary cycle begins. Intervals of very low sediment supply lead to condensed sections, e.g., black shales, which are discussed further below.

Figure 7.8 also shows vertical sediment sequences which originate from the described conditions. A significant difference exists between the normally measured sediment thickness and a *chronostratigraphic sequence* which has a linear time scale. Only the latter method can clearly demonstrate stratigraphic gaps and show whether a certain layer is deposited rapidly or slowly, as for example the condensed layer in Figure 7.8b. Similarly, the water depth curve plotted along the linear time scale deviates from the sediment thickness curve (Fig. 7.8b).

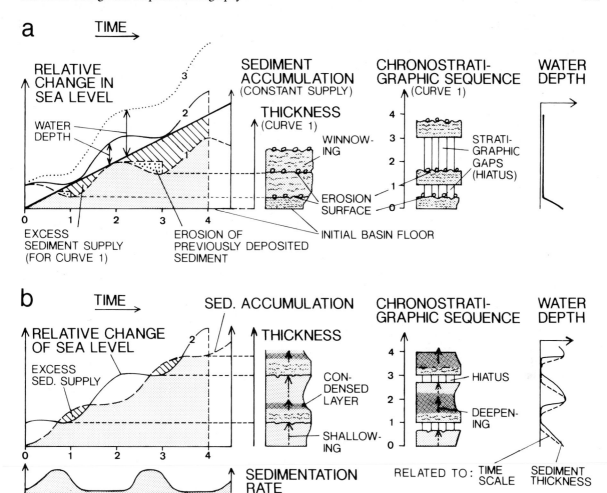

Fig. 7.8a,b. Relationship between relative sea level change (controlled by absolute sea level fluctuation and subsidence, cf. Fig. 7.7) and sediment accumulation versus time. Resulting sedimentary sequences and changing water depth are plotted either versus thickness or versus time.

a Constant sediment supply, with vertical sequences corresponding to Curve *1*. **b** Changing sediment supply, with vertical sequences originating from sea level Curve *2*. For further explanation see text

7.4.2 Changes in Sea Level and Storm Wave Base in Shallow Basins

Models for Basins with Differing Subsidence

The simplistic models in the previous section allowed sediment aggradation up to sea level. In marine basins, however, this is rarely and only locally realized, because waves and currents prevent sediment from settling in a wide foreshore zone. Especially storm waves and their resulting bottom currents (Chap. 3.1.2) remove and transport sediment either to the coast or into deeper water. Hence, the depth range of storm waves and their associated geostrophic currents largely control the water

depth which is kept sediment-free in the shallow parts of marine basins. For reasons of simplicity, this depth is referred to in the following as the *wave base*.

Model Parameters

The wave base is taken into account in the models in Fig. 7.9, which may represent two common settings:

- Shelf basin. Storm wave base 50 m deep, intermediate subsidence rate of 50 m/Ma.
- Epicontinental sea. Storm wave base only 20 m deep, subsidence rate 10 m/Ma.

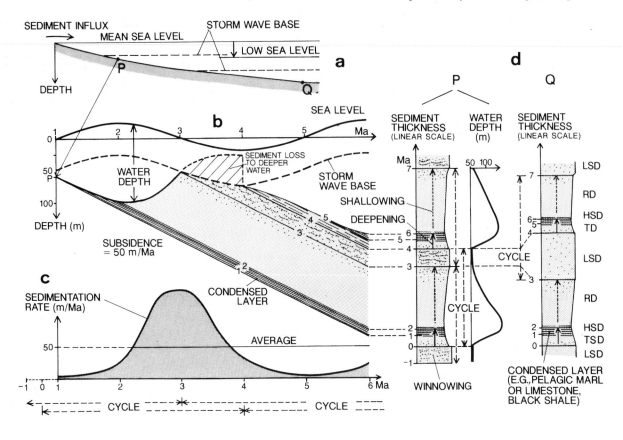

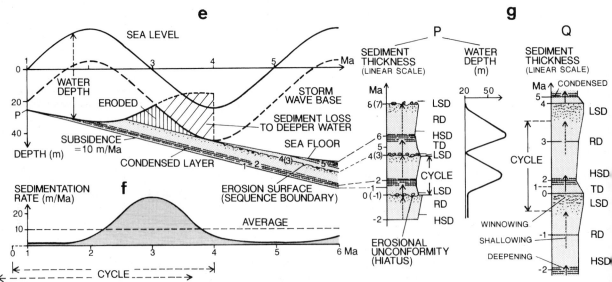

Fig. 7.9a-g. Idealized models showing the relationships between sea level change, as well as its corresponding storm wave base, and sediment accumulation for single locations (**a** points P and Q) in shallow marine basins with constant subsidence. **b, c, d** Rate of sea level fall less than subsidence. **e, f, g** Rate of sea level fall greater than sub-sidence. Note that vertical scales in **e** and **g** differ from **b** and **d**; sections in **d** and **g** display sediment thicknesses versus non-linear time scales. For abbrevations in **d** and **g** see Fig. 7.12a. See text for further explanation. (Einsele 1985; after Baum and Vail 1987; Vail 1987).

The amplitude of sea level change and the duration of one regression-transgression cycle is assumed to be equal in both cases (50 m and 4 Ma, respectively).

Under these conditions, the rate of sea level fall is either lower (Fig. 7.9b) or higher (Fig. 7.9e) than the rate of subsidence at certain time intervals. The sediment supply (Fig. 7.9c and f) varies, as in the models of Fig. 7.8. Then, sedimentation at points P and Q proceeds in the way shown in Figs. 7.9b and e. Point P is situated slightly below the storm wave base during mean sea level, point Q is below the storm wave base of low sea level (Fig. 7.9a).

The Shelf Basin Model

In the shelf model (Fig. 7.9b), the sediment accommodation potential at point P is reduced during the regressive phase by a falling storm wave base. As a result, some of the sediment is carried into deeper water, and the sediment left behind may show signs of winnowing. Erosion of previously deposited material is not required, but it may take place locally in this depth zone by cutting shallow submarine channels. The transgressive phase is characterized by rapid deepening, because it is combined with low sediment influx from terrigenous sources. The sedimentation rate tends to be lowest near the inflection point of the transgressive phase and may therefore generate *condensed layers* consisting largely of pelagic material or, in certain environments, of black shales or thin beds rich in authigenic minerals, such as phosphorite or glauconite (Loutit et al. 1988). Sediment lost at location P during the regressive phase can be accumulated at location Q in deeper water. Consequently, vertical sections at P and Q differ markedly from each other (Fig. 7.9d). At both locations, the water depth versus sediment thickness curves show "rapid" transgression and "slow" regression (only plotted for P), although both phases have the same duration in the models.

The Epicontinental Sea Model

In Figure 7.9e the rapidly falling storm wave base (faster than subsidence) leads at location P to an *erosional unconformity* during the regressive phase. If the previously deposited sediments are cohesive (clays, silty clays, marls) and contain layers which are already diagenetically indurated (e.g. hardgrounds, limestone or siderite concretions, limestone layers), submarine erosion often ends at such horizons (Fig. 7.10e and f). As a result, lags of the eroded material may be left behind (e.g., reworked fossils and concretions, iron ooids, clasts of semi-solid mudstone, etc.; Fig. 7.10d). Sessile fauna which need hard substrates may establish a new community, including rock-boring species. Storm wave erosion may proceed stepwise from shallower to deeper indurated layers and locally cut channels (Fig. 7.10e and g). In all other aspects, the situation in Fig. 7.9e is similar to that in Fig. 7.9b, including an asymmetric water depth curve (Fig. 7.9g).

Vertical Sequences and Their Interpretation

In order to deduce sea level changes from the resulting depositional sequences, transgressive and regressive sections as well as sequence boundaries must be identified. Whether this can be done easily or not depends on the method used and on the location or area considered within a basin fill. Seismic stratigraphy along a cross section from the margin to a more central part of the basin is a powerful tool in solving this task (see Chap. 7.4.3 and, e.g., Payton 1977; Vail et al. 1984; Van Wagoner et al. 1987), if the sequence boundaries and sections between these boundaries can be dated. Field exposures or drill cores, on the other hand, allow a thorough sedimentological description and sampling of the sections for biostratigraphic and geochemical investigations, but are commonly less useful for the establishment of a continuous, two-dimensional cross-section through the basin. This is necessary for the identification of low-angle, large-scale types of coastal offlap or onlap and other important features (see below). Thus, a combination of both methods should be applied whenever possible (e.g., Baum and Vail 1987).

In the case of Figure 7.9e representing field exposures or drill holes, it is obvious that erosional unconformities should be used as *sequence boundaries* (Type 1 after Vail 1987, see below). Then the sediment package between two of these unconformities can, as a first approximation, be referred to as a *transgressive/regressive sedimentary cycle* consisting of several subunits, an erosional phase (represented by a lag deposit), a deepening-upward section, and a shallowing-upward section.

EMERGENCE   SUBMARINE EROSION

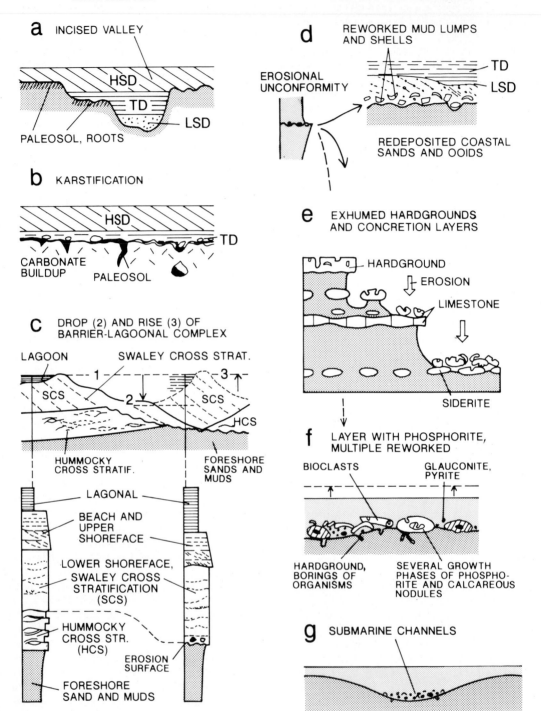

Fig. 7.10. a-f Various types of subaerial and submarine erosional unconformities related to sea level changes. See text for explanation.

(After **a** Weimer 1986; **c** Plint 1988; **d** Bayer et al. 1985 and Hallam 1988; **e** Einsele 1985; **f** Brandt 1985)

In the other examples in Fig. 7.9, erosional unconformities are missing and the sequence boundaries are therefore less distinct (Type 2 boundaries, if the sea level fall is slower than subsidence), or the section was always deposited below the storm wave base (locations Q). Here, the transition from the shallowing unit (regressive deposits RD) to the lowstand deposits (LSD) is usually taken as the sequence boundary.

Other terms and symbols of the concepts of sequence stratigraphy listed in Fig. 7.9 are explained below (also see Fig. 7.12). Furthermore, it is important to note that the sequence boundaries in general do not coincide with the lowstands (or highstands) of the eustatic sea level curve, but may migrate up and down in time according to their position near to or far away from the edge of the basin (also see Fig. 7.11).

Cross-Sections Through Epicontinental Basins

In the models described above only certain locations in a shallow sedimentary basin were evaluated. In the following model (Fig. 7.11), the response of sediments to sea level changes is demonstrated in a cross-section through the broad marginal zone of a basin which receives little influx of terrigenous material. This is a common situation for epicontinental seas. Basin subsidence is assumed to increase from the hinge line near the coast toward the center of the basin, and long-term sediment accumulation (below the storm wave base of low sea level) approximately equals subsidence. The duration and amplitude of sea level oscillations may vary, as shown in the chronostratigraphic sequence of Fig. 7.11.

During times of transgression and coastal onlap, a new sedimentary cycle can begin with a transgressive lag deposit on top of older rocks or sitting on an earlier submarine erosional surface. Pre-existing channels are filled with transgressive sediments (Fig. 7.10a). The coastal onlap of the transgressive phase varies with the changing amplitude of sea level oscillations. If the time period of the transgressive-regressive cycles is long enough (e.g., on the order of 1 Ma), a considerable part of the coastal onlap sediments is removed on the landward side by mechanical and chemical denudation (Fig. 7.10a and b; also see Chap. 9.2) during the subsequent regressive phase, unless it is protected by overlying continental sediments. However, this is rarely the case under the conditions outlined above.

In addition to subaerial weathering and soil formation, part of the transgressive sequence is eroded by the storm wave base during the subsequent regressive phase (Figs. 7.10c and 7.11a). Eroded sands are redeposited in the form of tempestites and prograding barrier islands (Chaps. 3.1.2 and 3.2.2). Vertical sections from this region may show sharp-based shoreface sequences with erosional gaps (Fig. 7.10c; Plint 1988). Calcareous or iron ooids formed in shallow water during high sea level stand may be transported seaward and come to rest on *submarine erosional surfaces* (Fig. 7.10d). The basinward extension of these (Type 1) unconformities also varies, according to the differing amplitudes of sea level change, but in deeper water the erosional surfaces are well preserved under the cover of subsequent transgressive sediments. However, the depth range of erosion and the intensity of reworking decrease basinward (Fig. 7.11b) until the erosional surfaces merge into sediments which are no longer affected by storm wave erosion. Consequently, the erosional surfaces are in a strict sense *diachronous,* and the same unconformity can sit on a transgressive sequence (landward) or on the subsequent regressive unit (basinward).

Beyond the depth range of the storm wave base, the chronostratigraphic sequence should be devoid of significant stratigraphic gaps if deeper, erosive bottom currents can be excluded. Here, the sequence often includes thin horizons of *maximum sediment starvation,* i.e., the condensed horizons mentioned earlier. They are associated with the inflection point of sea level rise, when the rate of sea level rise has reached its maximum and little terrigenous sediment is swept into deeper water.

Figure 7.11c shows the same cross-section as in (b), with the addition of sediment thicknesses. Here, the erosional unconformities of the successive T–R cycles merge landward and may form composite beds. Unconformities associated with the regressive phase are frequently overlain directly by transgressive lags, or they are mixed with such lags.

Finally, it is important to note that the *number of cycles* which can be identified in vertical profiles differs along the cross-section. In the deepest part of the basin shown in Figure 7.11c, it may be difficult to recognize cyclicity when sandy regressive deposits are missing and the condensed section is poorly developed. Similarly, the true number of cycles cannot be found at the very edge of the basin. Even somewhat farther seaward, the number of readily visible cycles varies due to

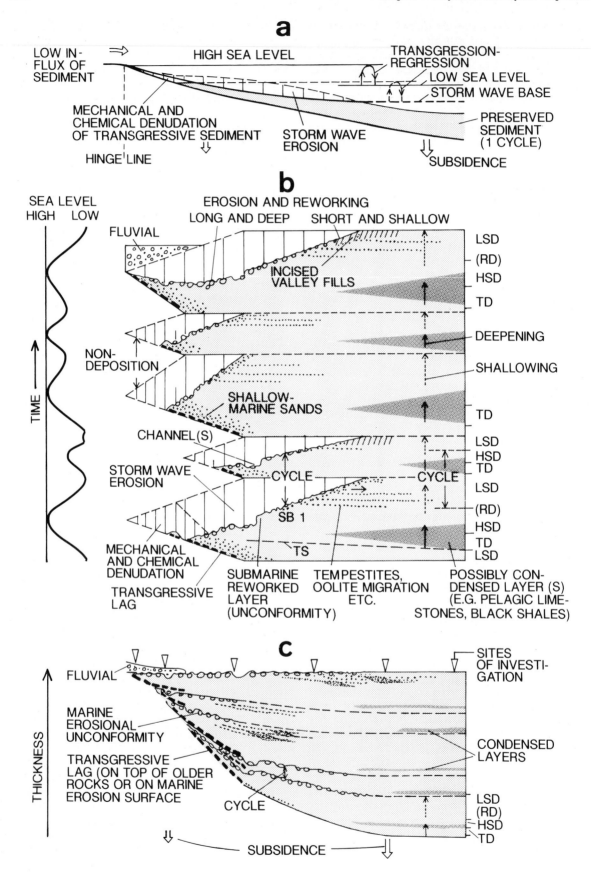

the differing basinward extension of the unconformities. Such local differences are one of the major reasons why the number and correlation of T-R cycles for a certain region and time period often remain controversial.

Many regional studies in various shallow-marine environments have recently contributed to our knowledge in this field (see, e.g., Aigner and Dott 1990).

7.4.3 Sequence Stratigraphy in Continental Margin Settings

General Aspects

The Concept of Sequence Stratigraphy

Relative sea level changes affect not only coastal areas and shallow seas, but also have far-reaching consequences for the sediments of deeper basins. This was first recognized by a working group of Exxon Production Research Company under P. Vail, which evaluated seismic records and well logs from present-day and ancient continental margins in many parts of the world (Payton 1977; Schlee 1984; Vail et al. 1984). This method became known under the term *seismic stratigraphy*. Based on this work, the general concept of *sequence stratigraphy* was developed (Baum and Vail 1987; Vail 1987; Van Wagoner et al. 1987; Posamentier and Vail 1988). This concept is particularly suitable for continental margin settings with their shelf break, slope and deeper basin, but it can also be applied to basins with a ramp margin, as shown in the previous section. Sequence stratigraphy is the study of rock relationships within a chronostratigraphic framework of repetitive, genetically related strata.

The Sequence and its Boundaries

In order to define and apply this concept, several special terms have been introduced. Although some of these terms may still be subject to change, most of them are listed and explained to some extent in Fig. 7.12. The fundamental unit of sequence stratigraphy is the *sequence,* which is bounded by unconformities and their correlative conformities, which replace the unconformities basinward (Fig. 7.12b). A sequence results, as demonstrated previously, from one transgression-regression cycle. An *unconformity* is a surface along which there is evidence of subaerial and/or submarine erosion, indicating a significant hiatus. Such a regional surface is called a *Type 1 sequence boundary* (SB 1). It is associated with a basinward shift of the facies and coastal onlap of overlying strata. It is formed in areas where the rate of eustatic sea level fall exceeds the rate of basin subsidence. Sequences can also be bounded by regionally extended *Type 2 sequence boundaries* (SB 2), which lack both significant erosion and a basinward shift in facies. They occur in areas where the rate of eustatic sea level fall is less than the rate of basin subsidence. If subsidence increases basinward, a Type 1 boundary may therefore pass into a Type 2. Type 2 boundaries are commonly less distinct than Type 1 boundaries, but they mark the transition from sea level highstand to lowstand (see below) and thus can usually be identified.

Systems Tracts

A sequence can be subdivided into so-called *systems tracts,* as is shown in Fig. 7.9d and g and Fig. 7.11b dealing with shallow basins, although thus far these terms have been only rarely applied to these types of depositional setting. The term systems tracts, however, is used here only as a heading for the subunits. For the designation of individual subunits the simpler and more common term deposits is used (see, e.g., Baum and Vail 1987). For example, "highstand systems tracts" is referred to here as "highstand deposits". The boundaries of the system tracts for the most part do not coincide with the peaks or lows of the eustatic sea level curve (e.g., the highstand deposits, or HSD, Fig. 7.12). Each sequence consists of three different systems tracts (from bottom to top):

Fig. 7.11a-c. Cross-section through shallow basin and response of its sediments to relative sea level changes of differing time periods and amplitudes (ramp setting without shelf break, differential subsidence, little influx of terrigenous sediment). **a** Margin of basin with oscillating sea level and associated storm wave base.

b Chronostratigraphic section: for abbreviations for systems tracts see Fig. 7.12. **c** Cross section showing sediment thicknesses; note difficulties in recognizing and correlating cycles between different vertical profiles marked by *arrowheads*. For further explanation see text. (After Einsele 1985)

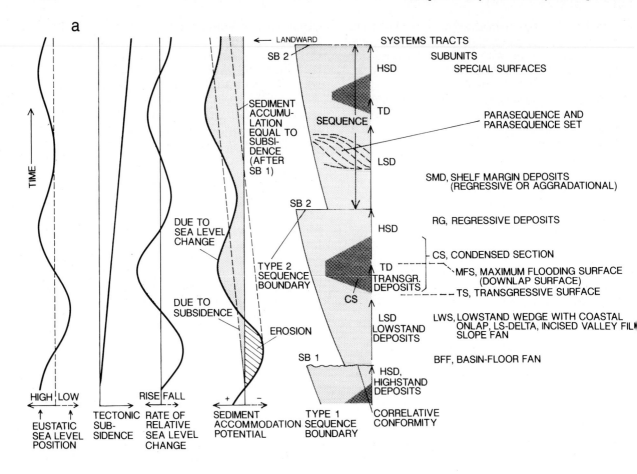

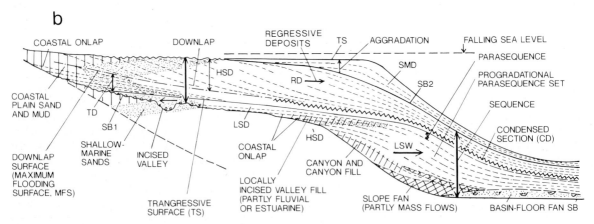

Fig. 7.12. a Relationship between eustatic sea level curve, subsidence, rate of sea level change, and sediment accommodation potential versus time, as well as associated sediment sequences with coastal onlap and terms of sequence stratigraphy. **b** Cross-section through sediments of continental margin with shelf, shelf break, slope, and deep basin. Positions of transgressive-regressive sequences in shallow and deep water, their boundaries *(SB)* and subdivision into system tracts *(LSD, TD, HSD)* and additional subunits. For explanation see text. (Mainly after Baum and Vail 1987; Haq et al. 1987; van Wagoner et al. 1987; Christy-Blick et al. 1988)

- **Lowstand deposits** (LSD) typically lie on top of a Type 1 sequence boundary (SB1). They develop during the phase of rapid eustatic sea level fall up to the early part of the rise. In a basin with a distinct shelf break, the lowstand deposits can be subdivided into three separate units, a basin-floor fan, a slope fan, and a lowstand wedge. The *basin-floor fan* is usually associated with the erosion of canyons into the slope and the *incision of fluvial valleys* into the shelf, particularly in its inner portion where subsidence is usually slow. The base of the submarine fan is the lower sequence boundary. The *slope fan* is characterized by turbidite and debris flow deposition. It may be coeval with the basin-floor fan or the early lowstand wedge. The *lowstand wedge* (LSW) progrades basinward, with *downlapping surfaces* onto the basin-floor fan or the slope fan. In front of a river mouth, a *lowstand delta* may be built out into the deeper sea. In the late stages of the development of a lowstand wedge, the sediments onlap onto the underlying sequence boundary, while the shoreline may still migrate seaward or become more or less stable. The reason for this behavior is the fact that now most of the incoming sediment is deposited in shallow water (cf. Fig. 7.6) and no longer in the deeper, prograding lowstand wedge. Simultaneously, incised valleys on the shelf are filled with fluvial, estuarine, or shallow-marine sediments. The top of the lowstand wedge is a marine flooding surface called a *transgressive surface* (TS), which marks rapid landward retreat of the shoreline.

If lowstand deposits lie on a Type 2 sequence boundary, they are called **shelf margin deposits** (Fig. 7.12b). In shallow basins without a shelf break (i.e., with ramp margins) the lowstand deposits only form a thin wedge. During relative sea level fall, the shoreline steps rapidly basinward, enabling stream incision and sediment bypassing. When the sea level starts to rise slowly, the incised valleys are filled and a thin lowstand wedge is built out seaward (cf. Fig. 7.11).

- **Transgressive deposits** (TD) begin with the *transgressive surface* (TS) and form during rapid sea level rise. Landward, they *onlap* onto the lower sequence boundary, but seaward they *downlap* onto the transgressive surface. The top of the transgressive deposits is the so-called *surface of maximum flooding* (MFS). This surface marks the change from a landward-onlapping and upbuilding sedimentary unit to a basinward-prograding wedge with

downlap onto it. It is therefore also referred to as a *downlap surface*.

At some distance basinward from the shoreline, the time of maximum flooding also signifies a time of *maximum sediment starvation*. The resulting facies consists of thin hemipelagic or pelagic beds deposited at very low sedimentation rates. Such a *condensed section* (SC) occurs largely within the transgressive and distal highstand deposits (Loutit et al. 1988).

- **Highstand deposits** (HSD) represent the late part of a eustatic sea level rise, its stillstand, and the early part of its fall. Highstand deposits are widespread on the shelf or in the marginal areas of shallow basins. A *lower aggradational unit* is succeeded by a *seaward progradational unit with downlap* onto the top of transgressive deposits, the maximum flooding surface (downlap surface). The subsequent rapid sea level fall generates a Type 1 or Type 2 sequence boundary on top of the highstand deposits.

Sequences and Parasequences in Various Depositional Environments

Sequences and their systems tracts may be subdivided into *parasequences* and *parasequence sets* (Fig. 7.12). A parasequence is defined as a "relatively conformable succession of genetically related beds or bedsets bounded by marine flooding surfaces and their correlative surfaces" (e.g., Bally 1987; Van Wagoner et al. 1988). They are progradational and their beds shoal upward. Parasequence sets are successions of genetically related parasequences which form distinctive stacking patterns. A parasequence may consist of smaller units representing time periods on the order of the Milankovitch frequency band (see above and Chap. 7.6).

The subdivision of sedimentary deposits into sequences, systems tracts, and their subunits provides a powerful methodology for basin analysis. The identification, detailed mapping, and correlation of sequence boundaries and other characteristic surfaces enables a better understanding and interpretation of large-scale vertical and lateral facies relationships in a depositional system.

Although all sequences, originating from relative sea level changes, show principally the same succession of systems tracts, their lithologic and stratigraphic expressions vary enormously from the nearshore zone to deep water

(Fig. 7.12). In the coastal zone, sequence boundaries are represented by erosional unconformities, and the sequence is frequently composed of fluvial and shallow-marine sands. In deep water, the same transgressive-regressive cycle may be bounded by the base of a submarine fan and, at its top, by an inconspicuous, conformable surface within a hemipelalgic unit with an increasing sedimentation rate. Here, the most distinctive horizon is a condensed section.

In the deep-water sequence, pelagic organisms and occasional volcanic ash layers are used to date the cycle. In shallow water, relics of benthonic life are commonly predominant, and transgressive or regressive erosional lag deposits mostly contain mixed faunas from reworked sediments of various ages. Therefore, an exact dating of sequence boundaries is frequently problematic (see, e.g., Baum and Vail 1987).

Models for Siliciclastic, Calcareous, and Deltaic Environments

Continental Margins Dominated by Siliciclastic Sediments

In order to demonstrate the deposition of sequences in a passive continental margin setting, the model in Fig. 7.13 shows several stages of development. Similar to the model in Fig. 7.11, it is assumed that the subsidence rate increases from the hinge line toward the basin. The influx of terrigenous material is moderate and the sediments are predominantly siliciclastic. This model demonstrates that the depositional history of a transgressive-regressive cycle can also be subdivided into two steps. During the transgressive and highstand phase, most of the incoming sediment is trapped in shallow water and on the shelf, while little material is transported into deeper water. In contrast, during falling sea level and lowstand (lowstand deposits) the locus of major sediment accumulation is the slope and continental rise, where relatively high sedimentation rates are reached. With an increasing rate of subsidence, however, the lowstand deposits may also be concentrated on the outer shelf and upper slope as shelf margin deposits (SMD) on top of a Type 2 sequence boundary (SB 2). In this case, only a short hiatus between HSD and SMD is present on the landward side.

Carbonate-Dominated Continental Margins

Figure 7.14 displays a situation similar to that of Fig. 7.13, but terrigeneous input is low and thus carbonate sedimentation prevails. The sedimentary dynamics of carbonate systems differs considerably from siliciclastic systems, however, particularly with changing sea level (e.g., Sarg 1988; also see Chap. 12.2). Carbonate production on shelves and carbonate platforms commonly provides much more sediment volume per unit area and time than planktonic organisms for pelagic sediments.

On a *flooded shelf*, abundant benthic carbonate production generates fairly pure carbonates behind rapidly growing reefs near the prograding shelf break. During the late highstand or at the *beginning of the lowstand phase*, when the inner platform is infilled and the prograding highstand deposits have reached the former shelf break, reef debris and reworked shelf carbonate tend to be transported to the foot of the slope or basin floor by mass flows or turbidity currents. If, in addition, a terrigenous sediment source is present, mixed carbonate-siliciclastic sediments accumulate in the basin (e.g., Yose and Heller 1989).

The steep slopes of carbonate platforms favor instability and slope collapse during falling sea level and thus the formation of large debris flows. If part of the carbonate was already indurated or semi-lithified prior to failure, the resulting mass flow deposits have the appearance of sedimentary megabreccias containing gravel-size and even larger clasts with little matrix.

During a *lowstand*, however, at least part of the shelf is emerged and benthic carbonate production on the shelf as a whole decreases markedly. Then, two different situations must be distinguished:

- In the absence of terrigenous sediment sources, relatively thick, possibly cyclic pelagic limestones and marls may accumulate in deep water, as the calcite compensation depth (or CCD; see Chap. 5.3.2) is depressed during low sea level.
- In systems with terrigenous influx, streams entering the sea may cut valleys into the shelf, develop steeper gradients, and gain in ability to transport terrigenous material. Hence, the composition of slope and basin-floor sediments may change significantly from predominantly carbonates to deposits rich in siliciclastic material, i.e., silty clays and turbidite sands, or mixtures of both carbonates and siliciclastic sediments.

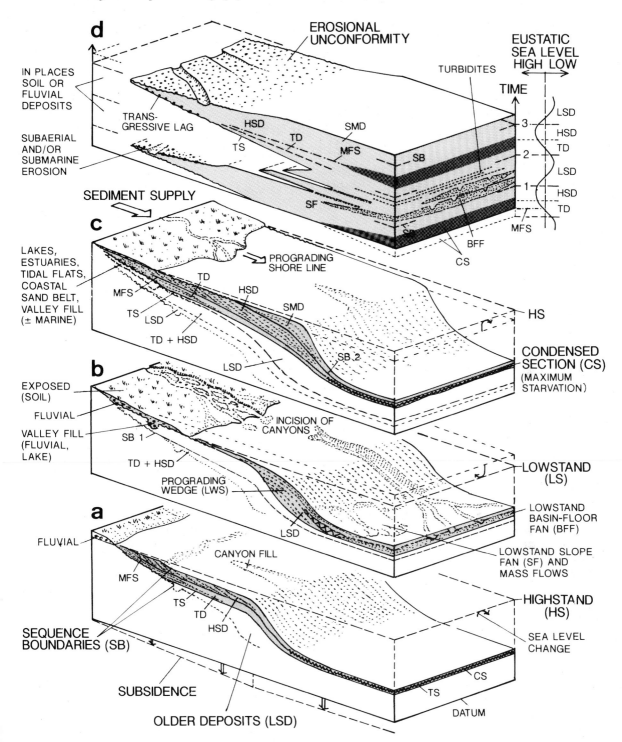

Fig. 7.13a-d. Model of predominantly siliciclastic sequence and systems tracts in continental margin-deep sea setting, controlled by sea level change and differential subsidence. **a** Transgressive *(TD)* and highstand deposits *(HSD)*. **b** Subsequent lowstand deposits *(LSD)*. **c** Subsequent trans-gressive, highstand, and shelf margin deposits *(SMD)*. **d** Chronostratigraphic succession from **a** through **c** and approximate correlation of systems tracts with eustatic sea level curve. For abbrevations see Fig. 7.12. (After Haq et al. 1987; Baum and Vail 1987; Posamantier and Vail 1988)

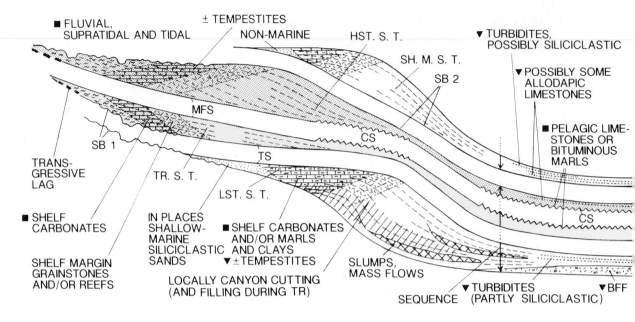

Fig. 7.14. Model of carbonate shelf with transition to deep sea, affected by sea level changes and differential subsidence. The different systems tracts are separated from each other along the sequence boundaries *(SB)*, the transgressive surface *(TS)* and the maximum flooding surface *(MSF)*. For further explanation see text and Fig. 7.12. (After Vail and Sangree 1988; Sarg 1988)

During subsequent transgressive and early highstand phases, the input of terrigenous material into deeper water is greatly reduced. As a result, the condensed section in deeper water is frequently represented by marls and pelagic oozes (limestones), which may contain high proportions of organic carbon. At water depths below the CCD, thin clay beds with some authigenic minerals or black shales may develop. A rapid rise in sea level may thereby lead, along with other factors, to a "drowning" of the carbonate buildup along the platform margin, while closer to land the carbonate accumulation continues (Vail et al. 1991).

A further alternative is discussed by Robaszynski et al. (1990) in their study on a 800 m thick Turonian limestone-marl section in central Tunisia. These beds accumulated in a basin close to a carbonate shelf and consist of three main components: terrigenous muds, detrital platform-derived carbonate, and pelagic carbonate. Changes in sea level led to the following modifications (Table 7.2).
In this case, the terrigenous mud component dominated during high sea level stands, when sediment supply and progradation overtook the slowing eustatic rise. Then vertical pelagic

carbonate flux was diluted by lateral mud input. Lowstands are characterized by reworked skeletal material (wackestones, packstones, and grainstones) derived from the carbonate shelf, whereas the transgressive deposits exhibit an increasing proportion of pelagic calcareous nannofossils and foraminifers. Stratigraphic boundaries rich in ammonites appear to be flooding surfaces in transgressive system tracts.

Marine Deltas

In a deltaic depositional system, not only are the delta plain and delta front facies affected by sea level changes (Chap. 3.4), but also the deeper water sediments in front of the delta. Figure 7.15 presents the scenario of a fluvially dominated delta with a generally high sediment input. Sediment influx varies, however, between highstand and lowstand conditions. During highstand the stream gradient is low, resulting in a meandering system depositing large amounts of sediment onto a coastal alluvial plain. Therefore, during the transgressive and highstand phase, the delta front pro-

Table 7.2. Proportions of various components in basin sediments

	Terrigeneous mud	Detrital carbonate	Pelagic carbonate
Lowstand (shelf margin wedge, shelf partially exposed)	High/low	High	Low (dilution)
Highstand	High	Low	Low (dilution)
Transgressive	Low	Low	High
Lowstand (shelf partially exposed)	High/low	High	Low (dilution)

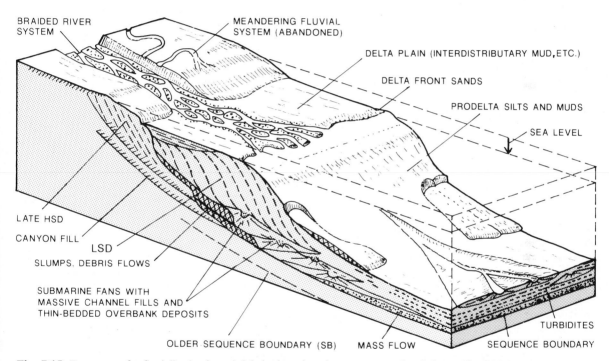

Fig. 7.15. Response of a fluvially dominated deltaic depositonal system to sea level change from highstand to low stand. For abbrevations see Fig. 7.12. (After Vail and Sangree 1988; Vail et al. 1991)

grades relatively slowly. In contrast, lowering sea level increases the stream gradient and favors the development of a braided system, cutting a wide channel into the previously accumulated coastal plain and transporting large quantities of terrigenous material basinward. This process leads to rapid progradation of a lowstand deltaic wedge, including various types of gravity mass movements (Chap. 5.4).

Such deposits have a high potential of being presevered and are therefore described from a number of regions, for example from the Natashquan river entering the Gulf of Lawrence (Long et al. 1989) or from ancient deltas on the North American Atlantic continental margin (Poag et al. 1990).

At the foot of the prodelta slope, debris flows and slumps, sometimes containing blocks of indurated sediments, may be abundant. Far-

ther basinward, finer grained and more extensive mass flows are common which are overlain by deep-sea fans fed from submarine
valleys cut into the prodeltaic sedimentary
body. The landward shift of such a succession
of channelized fans reflects the change from
falling to rising sea level and the onset of submarine valley fill on the slope. The fills of
slope valleys and fan channels often consist of
more or less massive sands, possibly including
some gravel; the overbank deposits of the fans
are made up of thin-bedded muds and turbidite sands, but may also contain larger
sheets of turbidites originating directly from
prodelta slope instabilities.

Models and examples of these and other
depositional environments are concentrated in
specialized volumes (Payton 1977; Schlee 1984;
Wilgus et al. 1988) and currently published in
several journals.

7.5 Long-Term Cyclic Phenomena in the Earth's History

Introduction

Long-term trends and possibly cyclic developments in Earth's history have been of primary
interest since geological and stratigraphic research began. This topic concerns various
specialized fields in the earth sciences, such as
tectonic evolution and orogenesis, volcanism
and magmatism, climatic trends, greenhouse
and icehouse effects, the development of the
atmosphere and the oceans, black shale events,
rock cycling as a result of subduction and
mantle convection, geochemical cycles, etc.,
and last but not least the history of life. Here,
some major factors controlling trends and
long-term cycles in sedimentary sequences and
depositional systems are briefly discussed.
These are:

- Plate tectonic evolution and basin formation.
- Long-term climatic trends.
- Special trends in the deposition of sedi-
  ments.

Plate Tectonic Evolution and Continental Flooding Cycles

Plate Tectonic Megacycles

The most important results of global plate
tectonic evolution through Earth's history are
the aggradation of individual continents, separated by ocean basins, into super-continents
and their subsequent breakup to form new
continents which drift away from each other.
Continental aggradation occurred in the late
Proterozoic and Permian, and breakups in the
early Cambrian and early Jurassic (Fig. 7.16).
Possibly, there was also a third epoch in the
early Proterozoic (between about 1800 and
2000 Ma) in which continental aggradation
and breakup took place (Hoffmann 1988,
1989). The time period from one breakup to
the next may be referred to as a plate tectonic
megacycle or a first order cycle (Chap. 7.1).
Russian authors assume 12 orogenic-tectonic
cycles for the entire Precambrian (Salop 1983),
which have about the same length as the
younger megacycles mentioned above, but the
older cycles are very poorly known.

The youngest two or three megacycles in
the plate tectonic evolution had profound
consequences in several respects. For one
thing, the volume of the ocean basins underwent changes. During times of slow ocean
spreading and continental aggradation associated with subduction, collision, and mountain
building, mid-oceanic ridges became smaller
and lower and the average depth of the ocean
basins deeper (Pitman 1978; Worsley et al.
1984; Heller and Angevine 1985). Hence, the
global sea level fell if it is assumed that the
volume of ocean water is constant. By contrast, rapid spreading rates generated high,
broad oceanic ridges and thus reduced the
ocean basin volume. Consequently, the sea
level rose and partially flooded neighboring
continental lowlands. In addition, other factors
such as submarine magmatism and changing
sediment influx from the continents may have
modified these general trends.

Continental Flooding-Regression Cycles and Their Consequences

The rifting and breakup of the Pan-African
supercontinent in the latest Proterozoic and
early Cambrian (e.g., Porada 1989) and subsequent rapid ocean spreading led to a major era
of flooding, with sea level highstand and
widely extended marine sediments on conti-

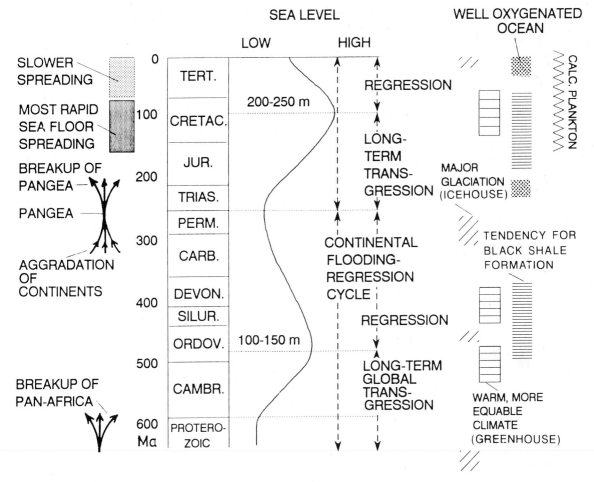

Fig. 7.16. Long-term, first order cycles in Earth's history associated with aggradation and dispersion of continents. Note coincidence of low sea level with major glaciations (icehouse state) and well oxygenated ocean waters, or high sea level with warmer, more equable climate (greenhouse state) and tendency for black shale deposition. (After Fischer 1982; Vail et al. 1991)

nental crust in the early Ordovician. The peak in sea level may have been 100 to 150 m above the present level (Bond et al. 1988). It was followed by a long-term, slow sea level fall which reached its deepest level with the completion of Pangea at the end of the Permian.

The late Permian and Triassic represent an epoch of radical change from falling to rising sea level. The overall climate changed from an icehouse state with polar ice caps and large latitudinal temperature gradients to a greenhouse state (Veevers 1988, 1990). In the Triassic, large volumes of sediment did not reach the oceans, but were deposited as nonmarine strata in incipient rift zones (Chapters 12.1 through 12.3) and in lowlands adjacent to the sea. The base level was low and the sediment supply exceeded the space being created by slow relative rise in sea level, resulting in

major regression (Vail et al. 1991). The tendency for arid climates favored the deposition of evaporites, red beds, and eolian sediments.

In the oceans, relatively narrow, low-latitude zones were characterized by a high supersaturation with respect to calcium carbonate and a high Mg/Ca ratio (Mackenzie 1990). As a result, carbonate platforms were dolomitized to a great extent (e.g., in the Alps), and aragonite was the dominant mineral forming pore-lining cements and ooids in the so-called "aragonite ocean" (Sandberg 1985; Opdyke and Wilkinson 1990). Several major groups of organisms became extinct, but others persisted with modifications; reef-builders developed new types of framework.

During the Jurassic and early Cretaceous, the sediment supply could not keep up with the new space created by an accelerated sea level rise. Thus, predominantly marine sediments encroached onto the continental margins of the drifting continents, originating from the breakup of the Pangea supercontinent. Ocean spreading was fast in the early Cretaceous and accompanied in the Pacific by the extrusion of large volumes of basaltic magma (Schlanger et al. 1981; Arthur et al. 1985). As a result, global sea level reached an absolute high represented by several peaks in the period from the Turonian up to the Campanian (Kauffman 1983; Sahagian 1987; Hancock 1989). It is assumed that the sea stood 200 to 250 m above its present level. Large areas of the continents were flooded and covered by shallow-marine sediments. After this highstand, a long-term, slow sea level fall was initiated by decreasing ocean spreading which has lasted into the present time. This trend, as in earlier times, has been superposed by many shorter-period oscillations. As a consequence of the overall conditions, the upper Cretaceous and Cenozoic marine sediments were restricted to the continental margins and deep ocean basins.

The overall influx of terrigenous material into the world ocean was reduced during high sea level (Mackenzie 1990). Nannofossil carbonates accumulated in widely extended areas on shelves and in epicontinental seas. The Mg/Ca and Sr/Ca ratios in sea water were relatively low. The dominant abiotic carbonate mineral in enlarged, low-latitude ocean regions (so-called "calcite oceans") became calcite; reef growth was acchieved by a few species of reef builders; evaporites were rare, and glaciers were, if at all present, restricted to high mountain ranges on the continents.

The Greenhouse State of the Earth

The association of fast ocean spreading with increased volcanic activity and CO_2 in the atmosphere is probably a major factor in the establishment of a greenhouse state during times of high sea level. In addition, the enlarged ocean surface with its low albedo was able to absorb more solar energy than the reduced water surface during times of sea level lowstand. This mechanism contributed to the development of a warmer, more equable, humid climate with reduced temperature gradients between the poles and the equator. Deep ocean water reached temperatures similar to those of surface waters. Thermohaline circulation was fundamentally different from the present situation, which does not differ

profoundly from the icehouse state represented by three major glaciation periods (Fig. 7.16).

Warmer oceans with slower circulation resulted in poorly oxygenated waters and thus promoted the deposition of black shales (Schlanger and Cita 1982; cf. Chaps. 5.2 and 10.3.3), either in stagnant basins or in enlarged zones of coastal upwelling and oceanic divergence (Parrish and Curtis 1982; Parrish 1987). At the same time, the vegetation cover on the continents expanded into high latitude regions and may have contributed to the enrichment of organic matter in marine sediments (e.g., Simoneit and Stuermer 1982; Stein et al. 1989).

This development was accompanied by a decrease in the $\delta^{13}C$ values of carbonate rocks (Arthur et al. 1988), reflecting the transfer of sedimentary organic carbon to the oxidized inorganic carbon in limestones. At the same time, sulfur was transferred from evaporitic sulfate rocks to the reduced sulfur reservoir of pyrite in sediments, resulting in an increase of $\delta^{34}S$ values of evaporitic sulfate minerals (Mackenzie 1990). In addition, oolitic ironstones and phosphorites appear to have been formed in relatively large quantities (Van Houten and Arthur 1989).

"Bio-Events" in the Evolution of Life?

Whether or not and in which way these long-term continental flooding and regression cycles have influenced the evolution and distribution of life in the sea and on the continents is an open question. Many authors postulate special "bio-events", including the punctuated evolution of species, dispersal events, and mass extinctions, which have controlled the distribution, survival, and recovery of life (e.g., Nitecki 1981; Fischer 1982; Sepkoski 1989; Kauffman and Walliser 1990). Other authors are more cautious and hold the opinion that our data base is too limited and incomplete for the establishment of a periodicity in the evolution of life (e.g., Hoffmann 1989). In any case, major bio-events cannot be correlated directly with the first order flooding-regression cycles. The occurrence of abundant calcareous plankton since the upper Jurassic, for example, appears to be independent of sea level change and other related factors.

Some of the bio-events may have been associated with rapid changes in the depositional environment, such as the famous "Kellwasser crisis" in the upper Devonian (Frasnian-Famennian, see, e.g., Schindler 1990; Buggish

1991). This event is characterized by mass extinction of pelagic marine organisms in tropical regions and a widespread black shale horizon. The sudden deterioration in living conditions occurred during a long-term trend towards a cooler climate, i.e., during the transition from a greenhouse to an icehouse state of the Earth (see above). In addition, the bio-event was associated with the narrowing of the Paleotethys Ocean.

The Pleistocene period with its rapid climatic change is the most understood example for the influence of environmental stress on life. Glaciation at high latitudes caused not only the migration of plants and animals into lower latitude regions, but also led to the extinction of various families and species. It appears, however, that most of the bio-events in the Phanerozoic cannot be explained by one single mechanism (Raup and Jablonski 1986). On the other hand, one cannot entirely reject the possibility that there is some relationship between the physical processes outlined above and the evolution and extinction of species.

7.6 Superposition of Cycles of Various Orders and Differing Origins

General Aspects

The previously mentioned, long-term plate tectonic megacycles (first order cycles in Table 7.1) in the Phanerozoic had durations of about 250 and 350 Ma. These periods correspond approximately to the life times of major sedimentary basins, such as passive continental margin basins or passive margins evolving into foreland basins (cf. Chaps. 8.5 and 12.6.3). The evolution of such basins may be comprised of several tectonically controlled phases, for example two or three phases of extension (passive margin settings, e.g., Hynes 1990) or the transition from an extensional to a compressional regime (foreland basin). As a result, the subsidence rates change with time, and the subsidence-time curves show transitions from concave to convex-upward forms (Fig. 7.17b; Vail et al. 1991). These *subsidence rate changes* lead to regional relative rises and falls in sea level (Fig. 7.17c) and thus to major transgressive-regressive facies cycles. Such second order, tectonically controlled cycles typically last several tens of Ma (Table 7.1) and are superposed on the long-term (1st or-

der) continental flooding cycles (Fig. 7.17a; Fischer 1982; Vail et al. 1991). They may then be superimposed in turn by third order cycles and possibly another one or two shorter term eustatic cycles (Fig. 7.17).

The Signature of Long and Short-Term Sea Level Oscillations

Transgressive-regressive sequences on all scales are controlled by the nature of eustatic variations, the degree of sea floor subsidence, and the rate of deposition. Erosion of previously deposited sediments can only occur when the rate of sea level fall is greater or equal to the rate of sea floor subsidence (Pitman and Golovchenko 1983 and 1988; Vail et al. 1984). In this case, shallow-marine sediments may undergo subaerial erosion, or storm-induced currents can remove older marine deposits and produce Type 1 sequence boundaries. However, when the rate of maximum sea level fall is lower than the rate of subsidence, sediment accumulation at a specific site is continuous, and no distinct sequence boundary is generated.

It is important to note that the maximum rate of sea level fall is much greater for higher frequency oscillations than for lower ones (Fig. 7.17d). Consequently, high-frequency eustatic variations have a greater potential for causing erosion and, thus, sequence boundaries. This holds true even when a long-term sea level rise, which normally cannot produce submarine unconformities, is superposed with shorter oscillations (Fig. 7.17d). The erosion potential of these oscillations is restricted, however, to shallow-water environments. Particularly peritidal carbonates, which undergo early cementation and lithification, have been found to record and preserve such high-frequency eustatic variations (Chap. 7.2).

Thus, short-term cycles producing a great number of unconformities are particularly evident in field exposures of limited extent, while longer-term sea level changes are better expressed in large-dimensional cross sections and seismic records.

Milankovitch Cycles Within Third Order Sea Level Variations

The Milankovitch climatic cycles discussed in Chapter 7.1 have time periods which are 10 to 100 times shorter than the third order eustatic cycles (Table 7.1). They represent the parase-

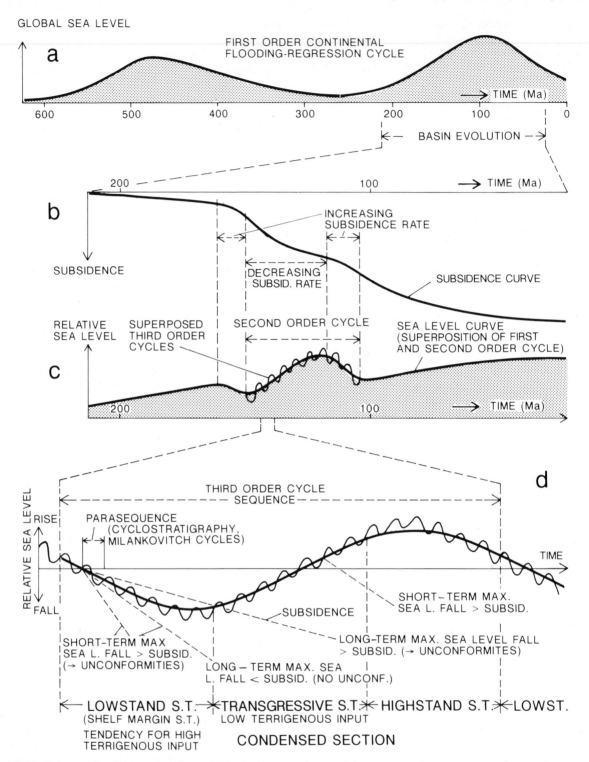

Fig. 7.17a-d. Superposition of eustatic sea level variations with different time periods and amplitudes and tectonically controlled regional subsidence. **a** Global first order cycle. **b** Regional subsidence time curve of multiphase extensional basin. **c** Corresponding second order transgressive-regressive cycles caused by **a** and **b**. **d** Third order eustatic sea level variation and associated systems tracts *S.T.* (After Van Wagoner et al. 1987), superposed by short-term Milankovitch-type oscillations (fourth to fifth order cycles, Table 7.1). Note that short-term oscillations display higher rates of maximum sea level fall and thus are more capable of producing erosional unconformities in coastal environments than longer term sea level changes. See text for explanation

quences, simple sequences, or even shorter units in the Vail nomenclature. If the orbital signals of the Milankovitch cycles are transformed into climatic variations which cause minor oscillations in sea level, then third order sea level variations are superposed by a great number of *high-frequency* but, in nonglacial times, *low-amplitude eustatic oscillations* (Fig. 7.17d).

The sedimentological expression of these high-frequency (fourth and fifth order) eustatic variations depends on several factors, such as their temporal position within the third order cycle and the depositional environment. In *shelf and tidal flat carbonates,* superposed high-frequency oscillations permit the deposition of thicker carbonate cycles with higher proportions of subtidal sediments during slowly rising sea levels than during slowly falling sea levels (Koerschner and Read 1989). If the rate of long-term sea level fall is equal to subsidence, only intertidal and supratidal sediments can accumulate, interrupted by unconformities.

In *siliciclastic shallow-water environments,* the potential to generate and preserve high-frequency cycles is less favorable as compared to peritidal carbonate cycles which undergo early lithification. *Coal cycles* (cyclothems) containing marine beds may represent the siliciclastic counterparts of the shallow-water carbonate cycles.

Marl-limestone rhythms, black shales, and rhythmic sequences containing various types of event beds are associated with different phases of the third order transgressive-regressive cycle. Periodic limestone-marl successions and black shale-limestone alternations, deposited on the outer shelf or in the deeper sea, form preferentially during the transition from the transgressive to the highstand phase (condensed section; Fig. 7.17d). *Episodic event beds* such as mass flows, tempestites, and turbidites, however, seem to be associated with the prograding phases of the lowstand and highstand systems tracts.

Concluding Remarks to Depositional Rhythms and Cyclic Sequences

The superposition of cyclic and episodic phenomena of various natures and scales produces a kind of *sequential ordering of beds,* which is found in almost all stratigraphic successions. The identification of such an ordering and the unraveling of the various mechanisms and time periods creating a particular

sequence are tasks of modern sedimentology and stratigraphy. One of the main problems in pursuing this is the fact that global and regional causal factors are frequently modified and obscured by local processes in the specific depositional environment. This is the reason why we observe such a great variety of sedimentary sequences in the geologic record.

On the other hand, some of the concepts of sequence stratigraphy need further refinement and should be tested in more types of environments. Some of the terms and their definitions, for example the use of erosional surfaces and their lateral correlatives as sequence boundaries, are not generally accepted (e.g., Galloway 1989). Isostatic compensation, sediment compaction, and sedimentary dynamics have to be taken into account to evaluate more accurately the geometry of stratigraphic sequences (Steckler et al. 1990). Reliable *stratigraphic correlation* of sediments and inferred sea level curves from different basins far away from each other is still a major problem (e.g., Ziegler 1988; Hancock 1989).

Much uncertainty still exists concerning the *frequency and amplitude of pre-Quaternary sea level changes.* The greatest frequency of transgression/regression cycles for a given time span appears to occur in sedimentary sections which allow high stratigraphic resolution. In the Turonian of central Tunisia, for example, the time period of third order cycles was found to be less than 1 Ma (Robaszynski et al. 1990). Furthermore, the signature of high-frequency sea level changes appears to be best preserved in *low-energy, slowly subsiding basins* with sufficient sediment input to generate a fast, seaward prograding depositional system (Fulthorpe 1990).

Further studies on ice volume-induced and thus sea level-related excursions of oxygen isotopes may provide some of the means to get more information for resolving these problems (Matthews 1984b; Burton et al. 1987; Boer 1991).

The occurrence of both cyclic and episodic phenomena in a certain region can be used to refine stratigraphic correlations. Biostratigraphic boundaries or datable volcanic ash layers provide the chronostratigraphic framework. Between these boundaries, short-term rhythms or cycles and the succession of various event beds, including biological and geochemical marker horizons, permit the subdivision of large stratigraphic zones into relatively short subunits. These can provide a detailed succession of relative time lines within a basin fill. Under favorable conditions, many of these phenomena can be traced over large distances within an epeiric basin or shelf basin and sometimes even between several of these basins. This method is referred to as *"high resolution cyclic and event stratigraphy"* (Kauffman 1988; Kauffman et al. 1991).

In the pelagic realm, where Milankovitch-type variations are commonly much better preserved than in shallow seas, it may be possible in the future to develop detailed *cyclostratigraphy*.

Part III
Subsidence, Denudation, Flux Rates and Sediment Budget

8 Subsidence

8.1 General Mechanisms Controlling Subsidence

Isostasy

Substantial sediment accumulation and the formation of sedimentary basins result from crustal subsidence. At least in the beginning of basin formation, tectonic subsidence must predate sedimentation, whereas later, subsidence may also be actively driven by an increasing sediment load (total subsidence). In this chapter, the different mechanisms leading to crustal subsidence as well as some models quantifying the development of subsidence versus time, i.e., subsidence history curves, are briefly introduced. A more comprehensive treatment of this topic is given by Allen and Allen (1990).

In most cases, tectonic subsidence of the land surface or sea floor is controlled by the principle of isostasy, thermal contraction of the lithosphere, and/or flexural loading (Fig. 8.1). According to the present view of *isostasy*, the elevation of the top of the crust is a function of the thicknesses and densities of several layers (sea water, sediments, solid crust consisting of igneous and metamorphic rocks, and solid upper mantle or mantle lithosphere, Fig. 8.1) resting on the viscous asthenosphere (mantle asthenosphere). Within the asthenosphere, *horizontal surfaces of constant pressure* can be assumed, which implies that the mass per unit area of the overlying rock column is everywhere the same. Hence one can write:

$$\rho_w h_w + \rho_s h_s + \rho_c h_c + \rho_m h_m + \rho_a h_a = \text{constant} \qquad (8.1)$$

where the layers (with thickness h) are sea water, w, sediment, s, crust, c, mantle lithosphere, m, and mantle asthenosphere, a. Density values, ρ, of these layers are listed in Fig. 8.1. The mass of the atmosphere per unit area is neglected in Eq. (8.1).

The base of the lithosphere is assumed to be a temperature boundary (approximately 1350 °C). Therefore, the hotter mantle asthenosphere has a somewhat lower density than the overlying mantle lithosphere. Thickening of the crust at the expense of the mantle lithosphere causes uplift of the land surface or sea bottom, whereas thickening of the mantle-lithosphere at the expense of the crust is followed by subsidence. For that reason, thick continental crust usually rises above sea level, whereas the top of much thinner oceanic crust lies several km below. Newly formed oceanic

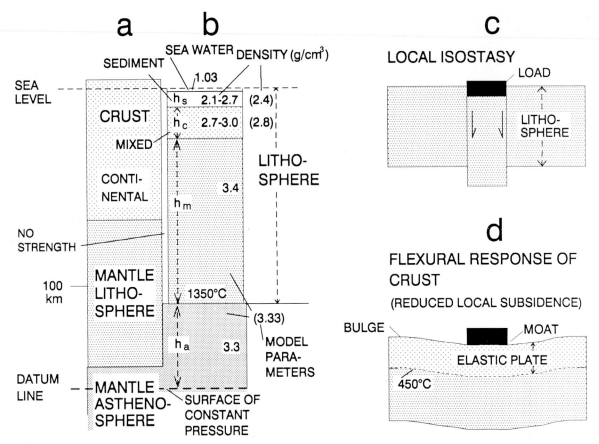

Fig. 8.1a-d. Principle of isostasy demonstrated by two crustal columns (**a** high plateau; **b** continental shelf) consisting of layers of different thickness and density on top of a surface of constant pressure. Values in *parenthesis* are

used in the models (see text). **c** Local isostatic response of lithosphere to additional load. **d** Reduced but more widely extended subsidence due to flexural response of rigid crust

lithosphere is encountered at an average depth of about 2.5 km below sea level. This is also approximately the tectonic subsidence of a basin resulting from the partial replacement of thinning continental crust by upwelling denser upper-mantle material (Fig. 8.1b). Old oceanic crust on top of cooling mantle lithosphere may subside more than 6 km below sea level.

In order to quantify the mechanism of isostasy, two relationships have to be considered:

1. The thicknesses of the different layers on top of a given datum in the asthenosphere may change, but the mass of the total rock column per unit area will remain constant. Hence the sum of all changes in mass above the datum line is zero:

$$\Delta(\rho_w h_w) + \Delta(\rho_s h_s) + \Delta(\rho_c h_c) + \Delta(\rho_m h_m) + \Delta(\rho_a h_a) = 0 \qquad (8.2)$$

2. The changes in thickness of all layers above the datum line, including the thickness change, Δh_g, of the atmosphere, is zero.

$$\Delta h_g + \Delta h_w + \Delta h_s + \Delta h_c + \Delta h_m + \Delta h_a = 0$$

hence, the subsidence, S_i, of the land surface or sea floor (below the base of sediments) in relation to a fixed sea level or other datum line is

$$S_i = \Delta h_g + \Delta h_w + \Delta h_s \qquad (8.3)$$

The law of isostasy was first proposed by Airy; it is only valid for cases in which each crustal column behaves completely independently from its neighboring columns and does not support any adjacent loads. In addition, it as assumed that equilibrium conditions are established, which is not the case shortly after a rapid change in the load resting on the datum line within the asthenosphere. The type of isostacy under equilibrium conditions may be described as *local isostatic compensation,* in contrast to the flexural response of an elastic crust to a surface load (Fig. 8.1c and d).

Initial Subsidence of Water-Filled Basins (Without Thermal Effects)

In simplified cases, subsidence, S_i, due to local isostatic compensation, neglecting any thermal effects, can be calculated under the following assumptions: The densities of the different layers remain constant and the thicknesses of the layers (except that of sea water) are known or will be given as variables. Mantle lithosphere and mantle asthenosphere are combined into the thickness h_{ma} with the density ρ_{ma} (= 3.33 g/cm^3). Prior to subsidence, the surface of the crust is at sea level and covered neither by water nor sediment. For these conditions, one can simplify Eq. (8.2) to

$$\rho_w \, \Delta h_w + \rho_c \, \Delta h_c + \rho_{ma} \, \Delta h_{ma} = 0 \qquad (8.4)$$

and Eq. (8.3) to

$$\Delta h_w + \Delta h_c + \Delta h_{ma} = 0 \; ,$$

where $\quad S_i = \Delta h_w$ (Fixed sea level) $\qquad (8.5)$

After equating (8.4) and (8.5), one obtains

$$\Delta h_w = \Delta h_c \, \frac{\rho_c - \rho_{ma}}{\rho_{ma} - \rho_w} \qquad (8.6)$$

Inserting the density values listed above and in Fig. 8.1 (e.g., $\rho_w = 1.03$ g/cm^3 for sea water), we obtain

$$\Delta h_w = -\,0.23 \, \Delta h_c \, , \; \text{or} \; \Delta h_c = -\,4.34 \, \Delta h_w \; .$$

If under these simplified conditions the crustal thickness h_c is reduced, for example, by 15 km (around half of its normal thickness), the top of the crust subsides from sea level to a depth of 3.5 km and creates space for a water body of the same depth. Similarly, the water depth of a basin can increase by 1 km, if the thickness of the mantle is increased by 4.34 km at the expense of the crust.

A more general expression for the initial subsidence of a basin filled with water up to the (fixed) sea level, derived from Eqs. (8.2) and (8.3), is (e.g., Suppe 1985):

$$S_i = \frac{(\rho_c - \rho_a)h_c + (\rho_m - \rho_a)h_m}{\rho_w - \rho_a} \left(1 - \frac{1}{\text{ß}}\right), \quad (8.7)$$

where ß = *stretching factor,* which describes the amount of stretching of a certain segment of the lithosphere. For z = original thickness and b = original width (Fig. 8.2), the new width after extension is increased to bß and the thickness reduced to z/ß. This formula, however, still does not take into account any sediment load or thermal effects which are usually associated with crustal thinning and the formation of a topographic low.

Inserting the values from the previous example (ß = 2, $\rho_m = \rho_a = 3.33$ g/cm^3, $h_c = 30$ km, and $h_m = 100$ km) into Eq. 8.7, the result is $S_i \approx 3.5$ km. With $\rho_m = 3.4$ g/cm^3, $\rho_a = 3.3$ g/cm^3 (Fig. 8.1), and the other parameters from this example remaining unchanged, the initial isostatic subsidence for the water-filled basin is $S_i \approx 5.5$ km.

These examples demonstrate that the results of such calculations strongly depend on the density values assumed for the mantle lithosphere and mantle asthenosphere, as well as on the thickness of the crust and lithosphere prior to stretching. Some authors have pointed out that stretching of a crust thinner than 18 km will generate uplift, whereas stretching of thicker crust causes subsidence. However, this boundary condition also varies considerably with the densities and thicknesses of the crust and mantle lithosphere (Jarvis 1984).

Initial Subsidence of Sediment-Filled Basins (Without Thermal Effects)

The *sediment load* of a partially or entirely filled basin causes additional subsidence due to isostatic compensation. This effect can be easily calculated if the former water-filled basin (water depth, h_w) is completely filled with sediments up to sea level (sediment thickness, h_s, and average density of sediments, for example, $\rho_s = 2.4$ g/cm^3). Then the elevation above the isostatic compensation depth remains the same as before, and from Eq. (8.3) we can assume:

$$h_s - h_w + \Delta h_a = 0 \; . \qquad (8.8)$$

Similarly, the mass above the compensation level does not change (Eq. 8.2), hence we obtain

$$\rho_s h_s - \rho_w h_w + \rho_a \Delta h_a = 0 \; . \qquad (8.9)$$

Equating (8.8) and (8.9) gives

$$h_s = \frac{(\rho_w - \rho_a)}{(\rho_s - \rho_a)} \, h_w \, , \qquad (8.10)$$

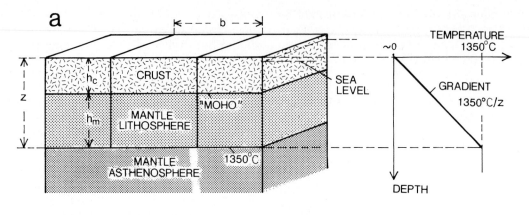

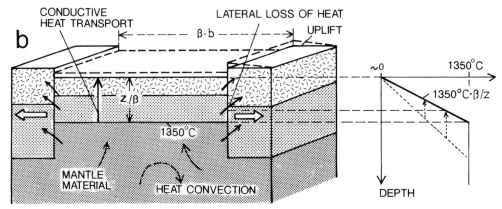

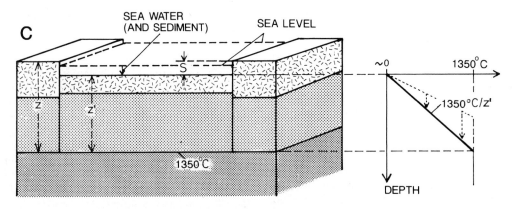

Fig. 8.2a-c. Finite-length extensional rift basin model. **a** Prior to rifting. **b** Initial subsidence due to isostatic adjustment at the end of short rifting event, buildup of high geothermal gradient. **c** Thermal subsidence due to slowly cooling lithosphere, geothermal gradient approximately reduced to original state. See text for explanation. (After McKenzie 1978)

where h_s is the maximum sediment thickness up to sea level.

With an average sediment density of $\rho_s = 2.15$ g/cm^3, a sediment thickness twice the initial water depth can develop (*amplification factor* 2); with $\rho_s = 2.55$ g/cm^3 the amplification factor is about 3.

In our example of a water-filled basin (based on Eq. 8.7) we obtained $h_w \approx 3.5$ km. Using the density values mentioned above and listed in Fig. 8.1, from Eq. (8.10) one gets $h_s \approx 8.7$ km. This signifies that a sediment-filled segment of this model basin will subside 2.5 times more than a purely water-filled segment, due to isostatic response of the crust. In comparison to an air-filled depression below sea level, the amplification factor for a corresponding sediment-filled basin becomes even greater (in our example approximately $1.46 \times 2.5 \approx 3.6$, also see Jarvis 1984).

Thermal Uplift and Subsidence

If a column of the lithosphere is heated above the temperature of the surrounding rocks, its mass remains unchanged, but its volume increases and may cause uplift at the land surface or sea bottom. Upwelling hot mantle material, for example at oceanic spreading centers, generates oceanic ridges which later, due to cooling of the oceanic crust and thickening of the underlying mantle lithosphere, subside considerably with age by several km. Similarly, upwelling hot mantle material below a thinning continental crust will first tend to raise the surface of the thinned crust, if this thermal expansion is not exceeded by the effects of isostasy, and then cause subsidence due to cooling. This type of subsidence is referred to as thermal subsidence, S_t. Thermal subsidence and uplift, apart from changes in temperature and geometrical factors, are controlled by the coefficient of thermal expansion (or contraction), $\alpha = 3.4 \times 10^{-5}$/°C, and the thermal diffusivity of the lithosphere, $k = 8 \times 10^{-7}$ m^2/s.

If for example the temperature of a 50 km high rock column is raised or lowered uniformly by 300 °C, its height will increase or decrease due to thermal expansion or contraction by ca. 0.5 km. This is, of course, an unrealistic example, which only demonstrates the order of magnitude of such thermal effects. In nature we observe a thermal gradient from the base of the mantle lithosphere (1350 °C, Fig. 8.2) to the land surface or sea water (slightly above 0 °C). With changing lithosphere thickness, this gradient and thus the temperatures at different depths vary, but the combined effect of this process causes approximately the same amount of expansion or contraction as that calculated for a rock column which is subjected to the same change in mean temperature as the lithosphere.

As a result of mantle upwelling, a steep temperature gradient is established. Later, this gradient is slowly reduced by heat loss to the atmosphere, until the original gradient will be restored (Fig. 8.2b and c). This process is described by the thermal diffusivity of the lithosphere. Cooling of an abnormally heated lithosphere takes tens of million years, and therefore thermal subsidence is a long-term, slowly decaying process. The mathematical treatment of thermal expansion and contraction of the lithosphere is beyond the purpose of this text (see, e.g., Royden et al. 1980; Watts and Thorne 1984; Sawyer 1988), but the following basin modeling does include thermal effects.

Flexural Response of the Crust to Loading

It is a well-known fact from geological observation that the lithosphere adjacent to local or regional loads (water, ice, sediment accumulation, isolated seamounts on top of oceanic crust) or unloading (lake dessication, ice melting, rock denudation) reacts by downwarping or uplift, respectively (Fig. 8.1 c and d). The best studied examples are continental ice caps, which cause not only isostatic compensation below the ice, but also some subsidence in areas adjacent to the ice load. Later, after ice melting, these regions begin to rise again in conjunction with the formerly ice-loaded crust. This phenomenon may affect a zone 50 to 300 km wide adjacent to the loaded and unloaded crust.

Several authors have made an attempt to simulate the flexural response of the crust to a given load (e.g., Watts et al. 1982; Watts and Thorne 1984). They treat the lithosphere either as an elastic or visco-elastic plate underlain by the viscous mantle. In the case of *pure elastic behavior*, the mechanics of lithospheric flexure can be compared with the bending of sheets or beams having a certain flexural rigidity. This flexural rigidity, D, in turn is a function of the elastic constants of the rock material, i.e., Young's modulus, E, and Poisson's ratio, σ, as well as thickness, h_e, of the elastic lithosphere. Then we obtain

$$D = \frac{E\,h_e^3}{12\,(1 - \sigma^2)} \qquad . \tag{8.11}$$

Equation (8.11) clearly shows that the elastic thickness, h_e, is the most important factor controlling the rigidity or flexural strength of the lithosphere.

As has been derived from observations in nature, the rigidity, D, of the lithosphere may vary by four orders of magnitude (Fig. 8.3). It is primarily a function of the plate age at the time of loading. Young oceanic crust at or near a mid-oceanic ridge has the lowest rigidity, while old continental crust, for example, below the fill of foreland basins or Precambrian shields under an ice load, show very high values. The base of the elastic plates is defined by the depth of the 450 °C isotherm (cf. Fig. 8.15). Thus, plate thickness, h_e, is relatively thin during or shortly after a major thermal event, but increases substantially with age after this event. The mechanical properties of both oceanic and continental lithospheres appear to be similar (Karner et al. 1983).

Furthermore, the vertical displacement of the bending crust from its original elevation is affected by the densities of the underlying viscous mantle material and, if present, the overlying water body or sediments (see, e.g., Bott 1982; Watts et al. 1982). The resultant of these two effects is an upward pressure or an additional buoyant resistance of the asthenosphere to the bending.

The mathematical treatment of these combined mechanisms for locations at various horizontal distances away from the load is rather complicate. In addition, the load on the plate can be distributed nonuniformly. Therefore, only some results of such modeling can be discussed here.

In the case of a relatively small load of limited areal extent (Fig. 8.4a), an elastic lithosphere can support this surface load by distributing it onto a larger area. Hence, subsidence is less than that of a locally compensated load, and bending of the lithosphere is not very pronounced. For greater loads with an extent wider than the thickness of the lithosphere (Fig. 8.4b) and long loading times, at least the center of the load tends to approximate isostatic equilibrium and the flexural response of the adjacent crust can become very important. The time necessary to reach such an equilibrium depends on the rheological properties of the lithosphere (see below). The initial response of the lithosphere to an instantaneously applied load is a rapidly downwarped flexural moat around the load. In general, the vertical deflection decreases in a sinusoidal manner away from the load. The mass of material displaced from beneath the downwarped basin must approximately equal the mass of the applied load. Beyond the zone of decaying subsidence, an upwarped peripheral bulge is usually developed (Fig. 8.4). Where old, thick, elastic lithosphere is being flexed, the width of the flexure is much broader than that of a young or recently heated thin plate, provided the load and its geometry are about the same.

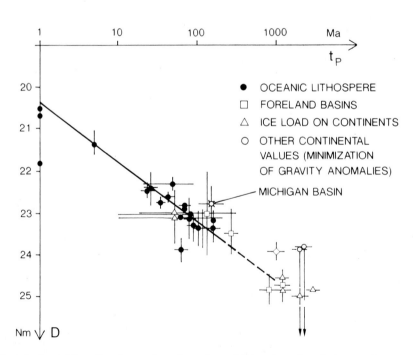

Fig. 8.3. Effective elastic rigidity, *D*, of oceanic and continental lithosphere in relation to plate age, t_P, at the time of loading after a major thermal event. (After Karner et al. 1983)

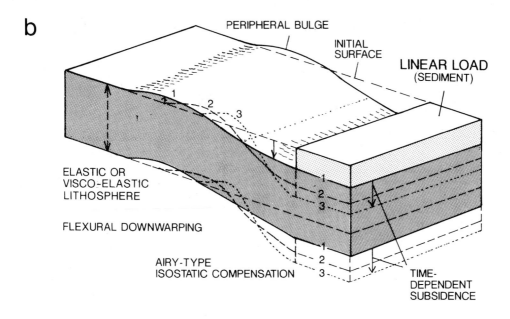

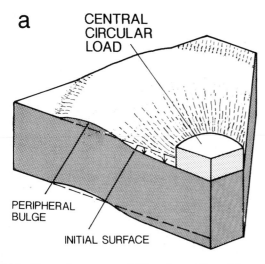

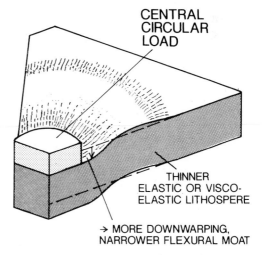

Fig. 8.4a,b. Flexural response of lithosphere adjacent to (a) local small or (b) large, wide linear load, not to scale. **a** A load on top of relatively young, thin lithosphere creates a deeper, narrower flexural moat than a load on thick, old lithosphere. **b** Under a long-persisting load, rapid initial subsidence *(1)* may be followed by further slow subsidence *(2)* until the load is ultimately compensated by local isostasy *(3)*. Simultaneously, the peripheral bulge migrates toward the load. (Partially based on Quinlan and Beaumont 1984).

Aside from such general rules, the results of modelling the lithospheric flexure often remain uncertain and even problematic. The flexural rigidity of the lithosphere appears to be dependent not only on plate age at the time of loading, as mentioned above, but also on the duration of loading (e.g., Bott 1982). An initially high rigidity may decrease with time following loading. Short-period loading (on the order of 10 000 years) generates less bending of the lithosphere than loading times of the order of 100 Ma. This phenomenon may be related to thinning of the elastic plate with age due to a rising 450 °C isotherm under the long-persisting load, or it may be ascribed to a visco-elastic lithosphere and a softening of the crust with time.

If the load remains in place for a long time period, deformation of the lithosphere changes in a manner as shown in Fig. 8.4b. After a first stage of rapid downwarping, lithospheric material at depth under the load relaxes stress, thereby creating a deeper central depression in conjunction with a narrower basin. The peripheral bulge is progressively uplifted and migrates toward the load. Ultimately, relaxation will evolve toward a state of local isostatic equilibrium.

Such a time-dependent development is primarily controlled by the mechanical state of the lithosphere. Although some workers strongly believe that models based on elastic condition plates can fairly well predict the processes observed in nature, others favor a uniform visco-elastic (Maxwell) model of the lithosphere. A third possibility is the assumption of a Maxwell layer with temperature-dependent viscosity overlying a low-viscosity fluid (e.g., Quinlan and Beaumont 1984). Few of the models have taken into account additional local heat sources associated with special magmatic or volcanic events during the history of a continental margin or other flexure-influenced basins.

Summary of Factors Controlling Subsidence

In summary, subsidence of the floor of a sedimentary basin (or uplift of the land surface) is controlled by the following principal factors:

- Thinning (or thickening) of the lithosphere due to horizontal extension (or compression, underplating).
- Upwelling of mantle material in response to crustal thinning.
- Increase (or decrease) in lithospheric density, for example due to
  1. cooling (or heating),
  2. pervasive dike intrusion or other injections of magma,
  3. phase changes in the crust such as melting and magma crystallization as well as transitions of minerals of relatively low density into minerals of higher density (or vice versa).
- Isostatic subsidence (or uplift) in response to sediment loading (or erosion).
- Flexural loading.
- Subcrustal magma convection.

The subsidence models of Chapter 8.3 have been worked out for relatively simple cases and deal primarily with the effects of isostasy, including sediment loading and thermal subsidence as well as flexural downwarping. The results obtained for simplified model basins can be compared with subsidence curves observed in nature for characteristic basin types of similar tectonic setting.

8.2 Methods to Determine Subsidence of Sedimentary Basins

Introduction

The *subsidence history* of sedimentary basins may be described in different ways. Structural geologists and geophysicists are primarily interested in that part of subsidence which is controlled by crustal processes (tectonic or thermo-tectonic subsidence), whereas sedimentologists and stratigraphers deal with the progressive burial history of sediments (total or cumulative subsidence) and often want to interpret subsidence curves in terms of sedimentation rates, paleo-water depth, and sea level changes.

Proceeding from endogenetic, geodynamic to exogenetic processes, the subsidence history at a certain location within a basin is controlled by the following mechanisms (Fig. 8.5):

- *Thermo-tectonic subsidence* due to processes within the crust and the mantle lithosphere (extension, thermally induced changes in thickness and density, see below). The resulting, essentially exponential subsidence curve (A) for an air-filled basin represents the minimum amount of subsidence which is not affected by any other mechanism (such as filling by water or sediment, see below).
- *Subsidence caused by water load.* When the tectonically driven, subsiding basin is filled up to sea level with water, the amount of subsidence versus time increases considerably (curve B in Fig. 8.5).
- *Subsidence caused by sediment load.* If a tectonically subsiding basin is permanently filled with sediments up to a fixed sea level, subsidence in relation to a water-filled basin is magnified by a factor about 2.5 (see above), but the resulting subsidence curve (C) is still exponential and smooth.

The regular subsidence curve for a sediment-filled basin can be modified by several minor factors:

1. At certain times, the *basin is not completely filled with sediments*. Hence, the total load on top of the basin floor is reduced according to the paleo-water depth, and subsidence decreases (Fig. 8.5, Curve D). Occasional erosion of sediment leads to additional unloading with similar effects (Curve E).

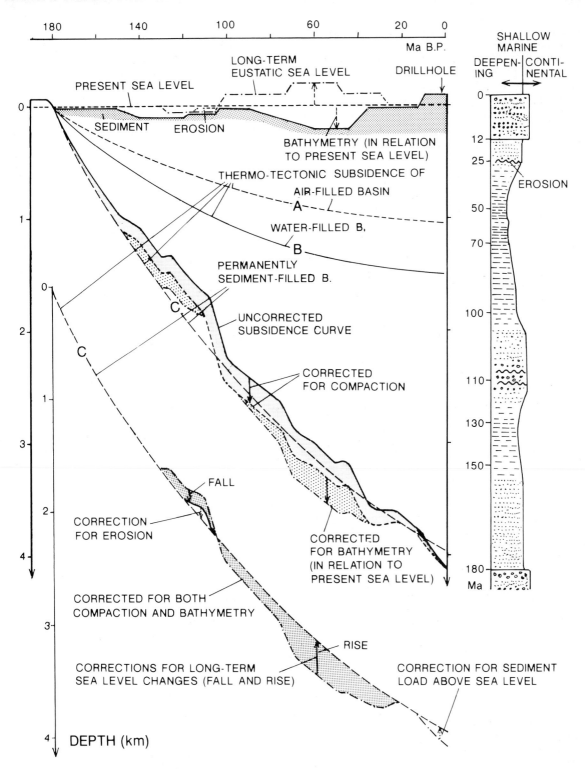

Fig. 8.5. Synthetic, semi-quantitative example showing drilled vertical sediment succession and derived, uncorrected subsidence curve at a given location within a basin. Purely thermo-tectonic subsidence in an air-filled basin (Curve *A)* is magnified by a water load (*B,* related to present sea level) and the load of a permanently sediment-filled basin *(C)* up to the present sea level. Curve *C* is modified by sediment compaction, a reduced sediment load (increasing paleo-water depth or bathymetry), and eustatic sea level changes. In addition, erosion can diminish subsidence, or a sediment load above sea level can enhance subsidence

2. In contrast, *eustatic sea level rise* brings about a greater water load and sometimes more sediment load, thus causing increased subsidence (Curve F). In periods of sea level fall, subsidence is again reduced. These effects apply to long-term eustatic sea level fluctuations; short-term variations are not considered here.

3. *Compaction* reduces the initial thickness of the sediments, particularly in the lower part of the section where the overburden load becomes high. Consequently, subsidence curves derived from the present thicknesses of special stratigraphic units (Curve G) do not show a true subsidence history. The older units yield subsidence rates too low and the younger units rates too high in comparison with true subsidence.

The combined effect of the last three factors (change in water depth, sea level change, and compaction) may cause marked excursions from the regular curve (C), but these factors can also more or less counterbalance each other. The second case occurs when increasing paleo-water depth coincides with a high sea level stand. In summary, the uncorrected subsidence curve (G in Fig. 8.5), which is found from present sediment thicknesses of known stratigraphic age, represents the cumulative result of the effects of several factors. It is usually a rather irregular curve comprising a systematic error due to sediment compaction.

An evaluation of the role of all factors controlling subsidence is possible when the sedimentary fill of a basin is largely preserved and nature and thickness of the underlying crust can be determined. Sediment thicknesses deposited at known water depths during certain time intervals allow to derive the subsidence history of the basin at chosen locations.

Techniques to Determine Subsidence

In order to determine the subsidence of a chosen location in a sedimentary basin as well as to discriminate between the various mechanisms mentioned above, special techniques must be used. If the tectonically driven part of subsidence is the objective, this technique is referred to as "backstripping" (Watts and Steckler 1979; Gradstein et al. 1985; Hegarty et al. 1988; Steckler et al. 1988). The purpose of backstripping is to calculate and remove the effects of compaction, sediment loading, changing paleo-bathymetry, and sea level variations. If backstripping is combined with

the search for information on the development of the depositional system, such as data on sedimentation rates, paleo-water depth or eustatic sea level changes, often the term "geohistory analysis" (van Hinte 1978) or burial history is used (Guidish et al. 1985).

Sediment Decompaction

The first step in backstripping is to reconstruct the original sediment thicknesses, h_{s1}, of the growing sedimentary fill from the basin floor up to dated stratigraphic boundaries in particular exposures or well logs. Provided a sedimentary column with present thickness, h_{sp}, and mean porosity, n_p, had an original mean porosity, n_1, then its *original thickness*, h_{s1}, was

$$h_{s1} = \frac{1 - n_p}{1 - n_1} h_{sp} \qquad (8.12)$$

The initial mean porosities for particular lithologic units and depths below the sedimentary surface can be taken from well logs or published porosity-depth curves (e.g., Hegarty et al. 1988; cf. Chap. 13.2). A graphical method approximating decompaction in combination with Eq. (8.12) for sections consisting of different lithologic units is shown in Fig. 8.6. This procedure has to be repeated for all time steps and units considered. The resulting curve corrected for decompaction demonstrates that without this correction subsidence is underestimated for the first phase and overestimated for the last phase of basin development.

Effects of Paleo-Bathymetry and Sea Level Changes

The second step in backstripping is the removal of the effects of paleo-bathymetry and sea level changes. The simplest way to account for changes in sediment and water loading is the assumption of local, Airy-type isostatic adjustment. Figure 8.7 demonstrates some cases which can be easily calculated under the following premises: The thermo-tectonic subsidence of a particular location within a completely sediment-filled basin proceeds in the way shown from (a) to (b). This "normal" development (Curve AB in Fig. 8.7b') may be modified by several processes:

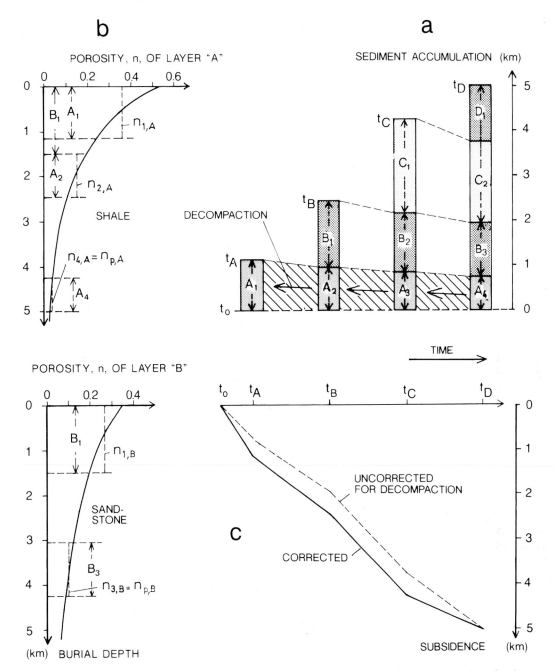

Fig. 8.6a-c. Graphical method for decompacting present sediment thicknesses of units A_4, B_3, C_2 to their original thicknesses (A_1, B_1, C_1). **a** Growth of sedimentary column with time, t, and progressive compaction of buried units. **b** Two examples of porosity-depth curves for different sediments demonstrating decompaction of units A and B (A_4 to A_1 or A_2, respectively, and B_3 to B_1) using Eq. (8.12) in text. **c** Resulting corrected subsidence curve in comparison to uncorrected curve; no changes in paleo-water depth and eustatic sea level

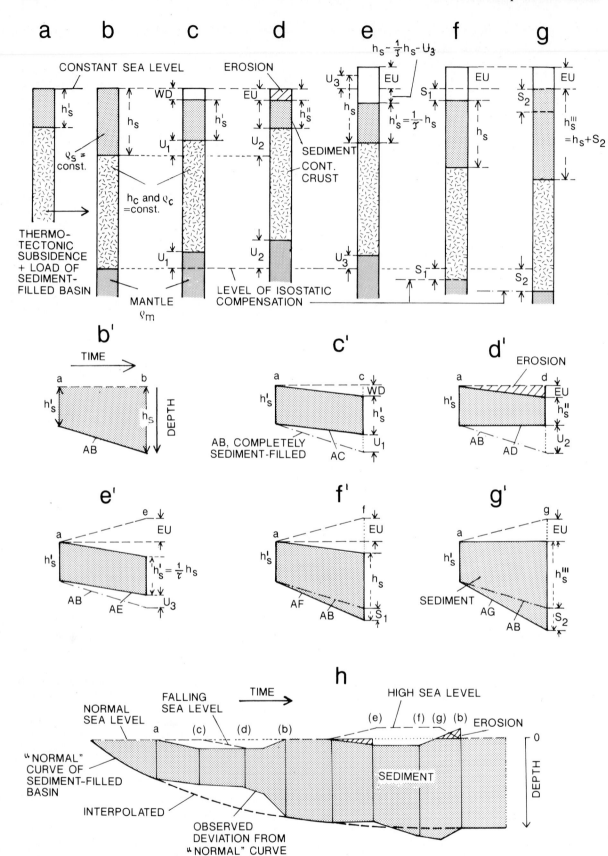

1. **Reduced or ending sediment deposition** and thus deepening of the basin (increase in paleo-water depth, WD). Using Eqs. (8.1), and (8.3) or (8.5) from Chapter 8.1, we can write (cf. Fig. 8.7b and c)

$$h_s' = h_s - WD - U_1 , \qquad (8.13)$$

where U = uplift in relation to normally subsiding basin floor, and

$$h_s'p_s = h_s p_s - U_1 p_m - WD p_w . \qquad (8.14)$$

Inserting of Eq. (8.13) into Eq. (8.14) yields

$$U_1 = WD \cdot \frac{\rho_s - \rho_w}{\rho_m - \rho_s} \qquad (8.15)$$

With $\rho_s = 2.4$ g/cm^3, $\rho_m = 3.33$ g/cm^3 (as in the example of Chap. 8.1), and no further sediment accumulation with time (h' as in a) relative uplift is $U_1 \approx 1.5$ WD. Hence, the rate of total subsidence is reduced (Curve AC in Fig. 8.7c') in comparison to a basin which is permanently kept sediment-filled (Curve AB). For more accurate calculations, it must be taken into account that the average sediment density, ρ_s, grows with increasing thickness of the sedimentary column.

2. **Eustatic sea level fall** (-EU) prevents sediment accumulation and may even lead to erosion (Fig. 8.7d) in a shallow-marine environment. As in the previous example, we compare columns (b) and (d) and get

$$h_s'' = h_s - U_2 - EU \qquad (8.16)$$

and $U_2 \rho_m + h_s'' \rho_s = h_s \rho_s .$ (8.17)

Equation (8.16) inserted into (8.17) yields

$$U_2 = EU \cdot \frac{\rho_s}{\rho_m - \rho_s} \qquad (8.18)$$

With the same values as those used above one obtains for uplift in relation to the normally subsiding basin floor U_2 = 2.58 EU. This signifies that a falling sea level in association with submarine erosion can produce a very significant negative deviation (Fig. 8.7d', relative uplift U_2 and curve AD) from the "normal" subsidence curve (Curve AB).

3. **Eustatic sea level rise and stagnant sedimentation** (Fig. 8.7e, i.e., the sediment thickness remains the same as in a) may still cause uplift, U_3, of the basin floor in relation to the normally subsiding basin. If the amplification factor for the growing sediment thickness from (a) to (b) is τ, then $h_s = \tau h_s'$ or $h_s' = 1/\tau \, h_s$. By comparing columns (b) and (e) on top of the isostatic compensation depth and again using Eqs. (8.1) and (8.3), we find

$$U_3 = (1 - \frac{1}{\tau}) \, h_s \frac{\rho_s - \rho_w}{\rho_m - \rho_w} - EU \, \frac{\rho_w}{\rho_m - \rho_w} \qquad (8.19)$$

In our example, with $\tau \approx 1.7$ and other values as above, we obtain $U_3 = 0.24 \, h_s - 0.45$ EU. Due to stagnant sedimentation, we get less subsidence in this case (Fig. 8.7e', Curve AE) than in the normally sediment-filled basin without sea level rise (Curve AB).

4. **Eustatic sea level rise** is accompanied by **continuous sedimentation** (Fig. 8.7f) as from (a) to (b). Comparing columns (b) and (f) and applying Eqs. (8.1), and (8.3) or (8.5), we obtain the additional subsidence, S_1, superposed on the general subsidence trend:

$$S_1 = EU \frac{\rho_w}{\rho_m - \rho_w} \qquad (8.20)$$

For $\rho_m = 3.33$ g/cm^3 we obtain $S_1 = 0.45$ EU. In this example, however, the water depth is not only increased by the eustatic rise, EU, but by EU + S_1 = 1.45 EU.

Fig. 8.7a-h. Deviations (U and S) from the "normal", thermo-tectonically-driven subsidence curve (*AB* in **b'**) of basin permanently filled with sediment up to (constant) sea level (transition from stage **a** to **b**, also see **h**). Sediment thickness increases from h_s' to h_s. Deviations are caused by changes in paleo-water depth (WD) and eustatic sea level (EU). **b** through **g** Crustal parameters (thickness and density) as well as average sediment density are assumed to remain constant, whereas sediment thicknesses, h_s, WD and EU vary. For local, Airy-type isostatic compensation, the excursions from the "normal" curve (*AB* in **b'**) are shown in **c'** through **g'** and further explained in text. **c** and **c'** Subsequent to **a**, sedimentation ceases and basin deepens to WD: reduced subsidence (Curve *AC*). **d** and **d'** Eustatic sea level fall prevents sediment deposition and even leads to some erosion: greatly reduced subsidence (Curve *AD*). **e** and **e'** Eustatic sea level rise and from **a** on stagnant sedimentation: reduced subsidence (Curve *AE*). **f** and **f** Eustatic sea level rise and sediment accumulation as from **a** to **b**: increased subsidence (Curve *AF)*, but also increase in WD (Eu + S_1). **g** and **g'** Sediment-fill up to "normal" sea level: strongly increased subsidence (Curve *AG*), WD $\geq$ EU. **h** Long-term subsidence curve composed of intervals with sediment-filled basin (**a, b**) at normal sea level and intervals as shown in **c** through **g**. "Normal" curve can be approximated by interpolation. Deviations from "normal" curve may be used to calculate paleo-water depth and eustatic sea level changes. All illustrations not to scale

The additional subsidence S_1, caused by sea level rise, is the so-called hydro-isostatic effect or *hydro-isostasy*. The resulting subsidence (Curve AF) is steeper than the "normal" curve AB (Fig. 8.7f').

5. **Eustatic sea level rise** is associated with **increased sediment accumulation** ($h_s''' = h_s + S_2$) up to the pre-existing sea level (Fig. 8.7g). By comparing columns (b) and (g) we obtain

$$S_2 = EU \ \frac{\rho_w}{\rho_m - \rho_s} . \qquad (8.21)$$

By inserting the values used above, one obtains $S_2 = 1.11$ EU, i.e., subsidence increases significantly (curve AG in Fig. 8.7g').

In summary, these simple examples have yielded the following deviations from the "normal" subsidence curve of a completely sediment-filled basin at constant sea level (where U = relative uplift, decreasing the rate of subsidence; and S = relative subsidence, increasing the rate of subsidence):

(c) Ending sedimentation,
increasing water depth, WD: $U_1 = 1.50$ WD
(d) Ending sedimentation,
sea level fall, and erosion: $U_2 = 2.58$ EU
(e) Ending sedimentation
and sea level rise: $U_3 = 0.24\ h_s - 0.45$ EU
(f) Sea level rise,
continuous sedimentation: $S_1 = 0.45$ EU
(g) Sea level rise,
increasing sedimentation: $S_2 = 1.11$ EU

From these data one can see the order of magnitude by which the subsidence curve of a particular basin may deviate from a smooth, exponentially decaying curve, which is usually expected for purely endogenetic processes such as thermo-tectonic subsidence. Conversely, one can deduce the effects of paleo-water depth, varying sedimentation rates including erosion, and eustatic sea level changes from these deviations, if the "normal" subsidence curve can be inferred from the uncorrected curve (Fig. 8.7h).

This method can be used when the subsidence history comprises intervals for which the sea level can be assumed to have been at average level or constantly high or low. In addition, the basin investigated must have been filled up to sea level with sediments during these intervals. Then the curve for a permanently sediment-filled basin is found by interpolation, and the amount of deviation for a particular time span can be quantified. With the aid of sedimentation rates derived from stratigraphic dating and sedimentological-paleontological criteria for determining paleowater depth, an attempt can be made to discriminate between the different factors causing such deviations. Equations (8.15), (8.18), and (8.19) through (8.21) may enable a rough quantitative evaluation of these factors.

Thermo-Tectonic Subsidence

The purely thermo-tectonic part, T, of total subsidence can be calculated from the uncorrected subsidence with the equation (Steckler et al. 1988):

$$T = h_{s1} \ \frac{\rho_m - \rho_s}{\rho_m - \rho_w} + WD + \frac{\rho_m}{\rho_m - \rho_w} EU \qquad (8.22)$$

where h_{s1} = decompacted sediment thickness; other symbols as above.

In many cases, for example on continental margins (see below), the response of the lithosphere to surface loads is not that of Airy-type isostatic compensation. Due to the strength of the upper lithosphere, the region around a load will also deform by flexural downwarping (Chap. 8.1). Hence, subsidence directly beneath the load will decrease. Consequently, all the results based on the equations in this chapter will yield maximum values for the amount of deviation from a given regular curve as well as minimum values for purely thermo-tectonic subsidence (Eq. 8.22).

In contrast to backstripping, "forward modeling" is a technique which first takes into account crustal parameters (extension, thermal heating and cooling, flexural behavior) in order to predict thermo-tectonic subsidence. Then additional loads (water depth, sediment thickness, eustatic sea level changes) are introduced which amplify and modify purely tectonic subsidence. Some results of this method are presented in the following chapter.

8.3 Modeling of Rift Basins and Observed Subsidence Curves

Rift Models

The McKenzie Extension Model

Many sedimentary basins or part of them originate from continental rifting. These elongate, more or less symmetrical basins have a simple geometry, and the processes controlling their subsidence are relatively well known. Thus, several authors have made an attempt to simulate the subsidence history of such basins by means of a one- or two-dimensional model (e.g., McKenzie 1978; Cochran 1983b; Steckler et al. 1988). A one-dimensional model considers the uplift or subsidence of crustal columns or blocks which do not influence each other, neither mechanically nor by heat transfer, whereas two-dimensional models take into account such phenomena. The models are based on data shown in Fig. 8.1 and the following additional assumptions:

- A relatively *short phase of horizontal stretching* (extensional model for a finite-length rifting event) and crustal thinning are accompanied by passive *upwelling of upper mantle material* (Fig. 8.2b). This advection of heat above the original upper boundary of the asthenosphere (having a temperature of approximately 1350 °C) produces thermal expansion and thus uplift. The amount of isostatic adjustment of the thinned crust is greater than thermal uplift. Therefore, the combined effect of both factors is rapid initial subsidence during rifting. If the primary crustal thickness was 40 km, in a one-dimensional model this initial subsidence accounts for about 40 % of the total subsidence, including subsequent thermal subsidence (e.g., solid curves and curve B in Fig. 8.8).

- The temperature within the uppermost mantle is assumed to be kept more or less constant by convection, but the solid lithosphere acts as a heat conductor to the atmosphere. Rapid extension of the lithosphere and upwelling of upper-mantle material cause an *increase in the geothermal gradient* between

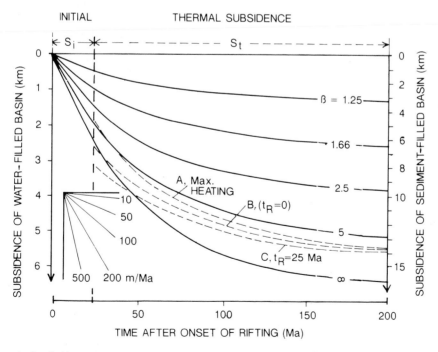

Fig. 8.8. Theoretical subsidence curves *(solid lines)* for different values of stretching factor, ß, and finite rifting time, t_R, of 25 Ma followed by thermal subsidence (one-dimensional model proposed by McKenzie 1978, after Sawer et al. 1982, modified). *Dotted curves* are calculated for ß = 4 and describe: *A* subsidence curve equivalent to mid-oceanic ridge; *B* thermal subsidence after instantaneous rifting, and *C* thermal subsidence after t_R = 25 Ma. Steckler et al. 1988). Note difference in subsidence of water- and sediment-filled basin (amplification factor 2.5, see text)

the mantle top and the crustal surface (sea bottom) and thus, as already mentioned, an increase in the average lithospheric temperature. This initial phase is followed by a long period of *thermal subsidence* as the lithosphere cools and thickens to its original, pre-extension equilibrium thickness (Fig. 8.2c). The graphic expression of this subsidence is that of a decaying exponential curve (Figs. 8.8 and 2.10). Under the premises mentioned above, thermal subsidence may account for about 60 % of the total subsidence (Fig. 8.8, except curves A and C). In the absence of other factors (e.g., flexural effects), the asymptotic level to which the basin floor subsides is controlled by crustal thinning, expressed by the stretching factor ß (Fig. 8.8), as well as by the amplitude of the exponential curve as a function of lithospheric heating. The more thinning and heating the crust undergoes, the greater is the total subsidence of the basin floor.

Finite Rifting Time and Lateral Heat Transfer

The rather simplistic "McKenzie extension model" can frequently describe the first-order features of rifts and other sedimentary basins, but the detailed subsidence history of particular basins derived from field data (Chaps. 8.3 and 12.1), possibly including periods of stillstand or even uplift, often deviates from simulations based on this model. It clearly neglects a number of additional factors (Chap. 8.1) which may considerably influence or modify the subsidence curves.

One of the assumptions in the McKenzie model is that extension has occurred instantaneously. Later calculations have shown that the error due to this assumption is small as long as rifting time does not exceed 20 Ma. However, the rifting stage of a basin may last longer (20 to 50 Ma and more; see, e.g., Ziegler 1988). Furthermore, the fact that there is also *lateral heat conduction* across the horizontal temperature gradient set up by lithospheric thinning should also be evaluated (Fig. 8.2b). The model proposed by Cochran (1983b) takes into account these two additional factors.

Example: Model parameters as in Chap. 8.1 and Fig. 8.1; also, $h_c = 31,2$ km, thickness of lithosphere $h_c + h_m = 125$ km, $\rho_m = \rho_a = 3.33$ g/cm^3, temperature at base of lithosphere $T_o = 1333$ °C, thermal conductivity 0.0075 cal/(deg· cm· s). The lithospheric thermal time constant is 62.7 Ma.

Using these parameters, old, 125 km thick, continental lithosphere at sea level is in isostatic balance with oceanic

lithosphere containing a 5 km thick young crust 2.5 km below sea level, i.e., a ridge crest, later subsiding toward a final depth of 6.4 km.

The following results are of interest:

- The history and mode of subsidence of the model basin (Fig. 8.9) are strongly influenced by the rifting time and by lateral heat transfer into the flanks of the basin. *Instantaneous rifting* without lateral heat loss (one-dimensional model, Fig. 8.9a) in the water-filled basin leads to equal amounts of initial (syn-rift) and thermal (post-rift) subsidence (in total 3.4 km). The basin floor (points B and C) subsides uniformly, and there is no change in elevation of the shoulder of the basin (point A).
- In contrast, *finite rifting times* (5, 20, 50 Ma) and *lateral heat transfer* bring about an increase in syn-rift subsidence (Fig. 8.9b through d). The longer the rifting time, the greater will be the proportion of syn-rift subsidence in the total subsidence.
- During early stages of the subsidence history, lateral heat transfer causes the margin of the graben (point B) to subside faster than its center (point C). It further raises the graben shoulder (point A). Toward the end of thermal subsidence, however, points B and C in the graben have reached about the same final depth, and the graben shoulder, due to thermal contraction following early expansion, approaches again its original elevation.

The subsidence curves (versus time) of these examples are shown in Fig. 8.10. They are calculated for the same stretching factor (ß = 2), but take into account neither any sediment fill nor the possible effect of flexural response of neighboring crustal segments. An increase in the width of the hinge zone (e.g., from 20 km to 40 km, compare Fig. 8.9) has a large effect on the subsidence pattern near this zone, but little influence on the basin center.

Sediment Fill and Crustal Flexure

In the cross-sections in Fig. 8.11, the additional effects of sediment fill and flexural response of the adjacent crust are evaluated for the same model basin as discussed above. It is assumed that the basin will be filled up to sea level with sediment all the time. During the syn-rift phase, the sediment fill is compensated by local isostasy due to active faulting during that stage. Later, the sediment load may be compensated in a regional manner through lithospheric flexure as further discussed below. (Flexural rigidity was defined by the depth of the 450 °C isotherm, Fig. 8.1d,

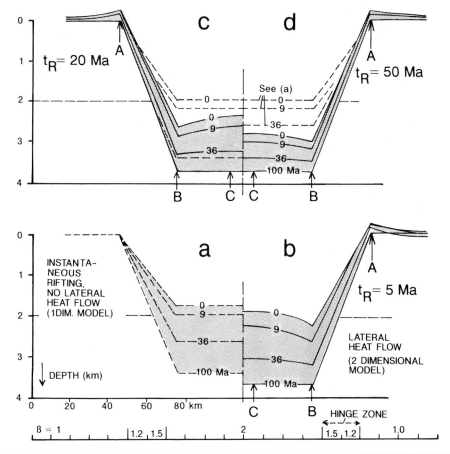

Fig. 8.9a-d. Subsidence in cross section of model basin showing effects of a particular finite rifting time; development of subsidence (0, 9, 36, and 100 Ma after end of rifting), and lateral heat flow (*solid lines,* compare Fig. 8.2 b). *Dashed lines* indicate development of basin due to instantaneous rifting without lateral heat loss. Basin water-filled, width 80 km; stretching parameter of basin floor ß = 2, at the hinge zone less. (After Cochran 1983b)

and thus varies in time and space across the basin).

This model variation demonstrates that lithospheric flexure is most pronounced if the phase of rifting and local isostatic adjustment is short. In this case, post-rift subsidence and thus total subsidence in the center of the sediment-filled basin are reduced. This means that the amplification factor to convert the depth of a water-filled basin into that of a sediment-filled basin becomes smaller than found from Eq. (8.10). This effect can also be seen from Fig. 8.10c, where the subsidence curves of the water- and sediment-filled model basin are shown for different rifting times.

On the other hand, a short rifting phase in conjunction with a strong flexural response of adjacent crustal segments causes wide areas outside the rift basin to subside and collect sediments. Hence, the rift shoulders also tend to subside considerably in post-rift times after a short syn-rift period of uplift (Fig. 8.10a and b). However, the size of the basin increases only slightly with time. With increasing rifting time, expansion of the basin out of the rifting region is delayed, and thus the width of the basin decreases (Fig. 8.11). However, the progressive onlap onto the unrifted area becomes more pronounced.

The transition from local to *regional isostatic compensation* at the end of the rifting phase is accompanied by reduced tectonic subsidence within the rift basin and may even lead to a short period of slight uplift. During that time, the rift shoulders are also still being uplifted by lateral heat flow from the rift into the hinge zone. Hence a slight unconformity between syn-rift and post-rift sediments may develop (not shown in Fig. 8.11), a phenomenon which is also known from examples in na-

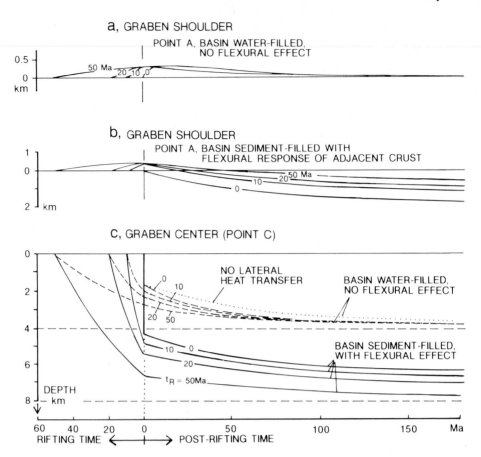

Fig. 8.10a-c. Subsidence curves for three locations *A*, *B*, and *C* of water-filled basin without flexural effects (Fig. 8.9) and sediment-filled basin including flexural effects (Fig. 8.11) for different rifting times t_R. (After Cochran 1983b). Stretching factor in rift graben for all curves ß = 2. Note that amount of syn-rift subsidence increases with growing rifting time at the expense of post-rift subsidence

ture. Pronounced "breakup unconformities" in ancient sedimentary basins are, however, associated with the transition from rifting to an early stage of drifting and accretion of new oceanic crust (Chap. 12.2), which is not considered in this model basin.

The simulated subsidence curves in Fig. 8.10 are modified somewhat further when the stretching factor, ß, does not increase linearly throughout rifting, as previously assumed, or when the lithosphere is thinned by an amount greater than that required by crustal extension. The latter situation may lead to some uplift during the initial phase of crustal thinning, but otherwise these variations in the model parameters affect the subsidence curves very slightly. Similarly, greater compaction of deeply buried sediments in comparison to younger deposits, a factor which is not taken into account in the simulations, does not significantly change the results.

Modifications of the Rift Model and Other Factors Controlling Subsidence

Several workers have refined or modified the subsidence models described above by introducing additional factors, for example, the *injection of melt* segregated from the mantle into the thinning continental crust (Keen et al. 1983). However, this mechanism does not significantly affect the results gained by the simpler models.

Another alternative supported by geophysical data from passive continental margins (see below) is the *nonuniform or depth-dependent extension model* (Royden and Keen 1980; Chenet et al. 1983). Whereas the uniform extension model is based on the assumption that the entire lithosphere is subjected to the same stretching factor, ß, the nonuniform model takes into account differences in the amount of stretching between the upper and lower portions of the lithosphere.

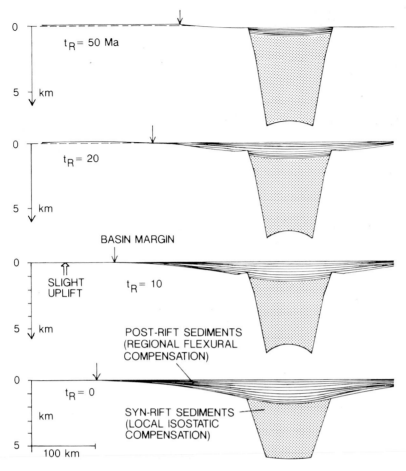

Fig. 8.11. Cross sections of sediment-filled and flexurally influenced model basin with lateral heat loss (same model parameters as for Fig. 8.9b through d) after post-rifting thermal subsidence time of 100 Ma, but varying rifting times, t_R. Syn-rift sediments are assumed to be compensated by local isostasy, but post-rift sediments cause regional (flexural) response of the crust. (After Cochran 1983b)

Observations in nature reveal that the upper crust in rift zones may show a series of tilted blocks separated by listric faults (Fig. 8.15a and b). The stretching factor derived from this geometry is usually much smaller than that needed to explain the substantial thinning of the underlying, ductile lower crust. The additional attenuation of the lower crust is assumed to result from mass loss from the lithosphere to the mantle asthenosphere. Consequently, in this model tectonic subsidence is caused by two main mechanisms acting simultaneously: (1) extension by stretching the entire lithosphere, and (2) additional thinning of the lower crust by thermal processes.

Some authors favor the *phase transition from gabbro* (average density 3.0 g/cm³) *to the denser eclogite* (average density 3.5 g/cm³) in the lower lithosphere as a principal mechanism of subsidence (summarized, e.g., in Wyllie 1971; Artyushkov and Sobolev 1983). Dependent upon depth within the lithosphere (pressure),

this phase transition is accomplished at temperatures of 700 to 800 °C and accelerated by the presence of small quantities of water. Anomalous heat transfer from the deeper mantle can raise this phase boundary and thus reduce lithospheric thickness without changing the mass of a given lithospheric column on top of the isostatic compensation depth.

If, for example, the phase boundary migrates 10 km upward and consequently a 10 km thick column of gabbro is transformed into eclogite, then the thickness of the lithosphere is reduced by approximately 1.5 km (without taking into account a possible additional sediment load or the simultaneous thermal expansion of the lithosphere). Because this mechanism does not require horizontal extension of the lithosphere, it is quoted as the main factor for the formation of certain inland seas, such as the Black Sea, the Caspian Sea, or the Michigan Basin in North America.

Phase transitions due to deep crustal metamorphism may explain unloaded subsidence (or uplift) up to the order of 2 km. Such transitions may take place over a very long time period (order of 100 Ma) when small quantities of water are not available in the lithosphere to speed the process. But it is doubtful whether phase transitions alone can cause the strong and rapid subsidence observed in many rift settings. Therefore it appears that deep crustal metamorphism is not the dominant mechanism generating subsidence during rifting.

Finally, it should be mentioned that all the rift models discussed do not take into account the differential subsidence known from half-graben structures associated with "simple shear" (Chap. 12.2.1). These features appear to be very common in nature and frequently show an asymmetric basin fill as a result of stronger subsidence on the side of the basin with the master fault.

Subsidence of Rift Basins Inferred from Stratigraphic Sections

Young Rift Systems

One of the classic examples of a young tectonic graben is the **Upper Rhine valley** in central Europe (cf. Chap. 12.1). Continental rifting started in the Middle Eocene and has continued into Present time. However, pure extension of the lithosphere is superimposed by a left-lateral strike-slip movement. Partly for this reason, subsidence within the graben varies considerably by a factor of approximately 3. Figure 8.12b shows curves of maximum total subsidence in three parts of the graben. Typically, the highest rates of subsidence (approximately 100 to 300 m/Ma) occurred during the first 10 to 20 Ma after the onset of rifting and therefore primarily represent local isostatic compensation. Subsidence has ended in the southern and middle part of the graben, while farther north, subsidence still proceeds at a slow rate. In fact, in the southern graben there is now a small amount of uplift, accompanied by a relatively strong uplift of the graben shoulders.

The **Dead Sea**, which is also affected by strike-slip motion, has experienced subsidence rates of the same magnitude (up to 200 to 500 m/Ma, cf. Chap. 12.1). Other young continental rift zones, such as the western branch of the East African rift system which started to evolve 7 to 12 Ma ago, exhibit maximum average subsidence rates of 500 to 1000 m/Ma (based on data summarized by Ebinger 1989). These high values are valid for the master fault in asymmetric half-graben structures, and they differ markedly from basin to basin in the segmented rift zone.

Older Continental Rift Zones

Older continental rift zones frequently display a large degree of total subsidence over the long term, although the rate of subsidence may be relatively low. Three wells from the **Paris basin** demonstrate that central parts of the basin, i.e., the former rift basin, began to subside earlier than locations outside of the rift in the flexurally controlled portion of the basin (Fig. 8.13b). The **North Sea basin** has a long and complicated history of subsidence (Ziegler 1988), causing deposition of Permian to Mesozoic sediments up to several kilometers in thickness.

A generalized curve of total subsidence in the sediment-filled Viking graben, in the northern North Sea, shows three distinct phases of increased, fault-related subsidence during the Triassic, late Jurassic, and early Tertiary (Fig. 8.13b, from Beach et al. 1987). Most wells drilled in the central North Sea, representing a failed rift zone, also indicate accelerated post-Middle Cretaceous subsidence (around 20 m/Ma), the general trend of which is shown in Fig. 8.13b. Whether this subsidence was entirely due to continental stretching (e.g., Sclater and Christie 1980; Sclater and Shorey 1989; Beach et al. 1987) or caused mainly by mineral phase transition in the lower crust (Artyushkov and Baer 1989) still appears to be an open question (also see Chap. 12.1).

Continental sag basins show long-persisting subsidence at very low rates (average 10 to 20 m/Ma, compare Chap. 12.4).

8.4 Passive Continental Margins

Concepts for Modeling Passive Margin Subsidence

Passive continental margins evolve from a rift basin into a wider oceanic basin due to the accretion of new oceanic crust. They commonly develop on both margins of the growing ocean and therefore include one half of the former rift basin. Continuing tectonic subsidence and sediment accumulation usually cause a landward and seaward extension of the depositional area (cf. Fig. 8.16). Under these

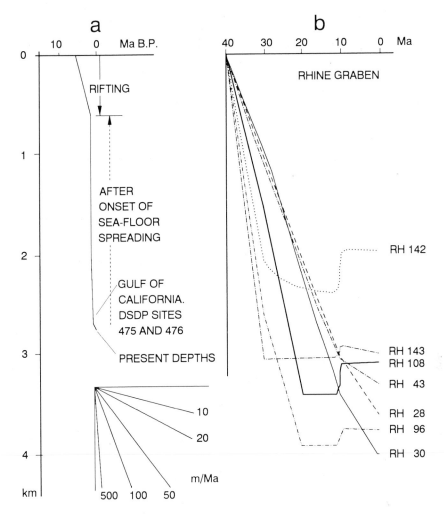

Fig. 8.12a,b. Subsidence of young passive continental margin at southern end of Baja California (Curray et al. 1982b, cf. Figs. 4.5 and 12.37) and decompacted curves of maximum subsidence for Rhine graben, southwest Germany (Roll 1979, cf. Fig. 12.2). In contrast to northern part (*Rh 28, 30, 47*), subsidence ceases earlier in middle (*Rh 96* and *108*) and southern part (*Rh 142* and *143*) of the Rhine graben, and is followed by slight uplift. *Insert* subsidence rates in m/Ma

conditions, flexural response of the adjacent continental and oceanic lithosphere becomes a primary factor in passive margin subsidence in addition to local, Airy-type isostatic compensation.

As already discussed in the previous section, some recent simulations of rift basin development use a two-layer model of the lithosphere. In addition, many seismic investigations have revealed that there is a discontinuity between the upper and lower continental crust, commonly at a depth of about 10 to 15 km (Fig. 8.15a and b). This discontinuity is explained as the *transition from the brittle to the ductile state of the crust.* Extension of the brittle material leads to the formation of listric faults and tilted blocks, whereas the ductile lower crust is deformed by creep (Fig. 8.15b). Although the manner of stretching is different, it is nevertheless a uniform extension of the crust, i.e., the thicknesses of both layers are reduced by the same stretching factor (*one-layer model*).

This amount of stretching is not sufficient, however, to predict the actual amount and history of subsidence observed on many passive continental margins. Therefore it has been proposed that an additional thinning of the lower part of the lithosphere takes place (Fig. 8.15a and b), which is caused by heat transfer from the mantle to the lithosphere but does not involve extension (Chenet et al. 1983).

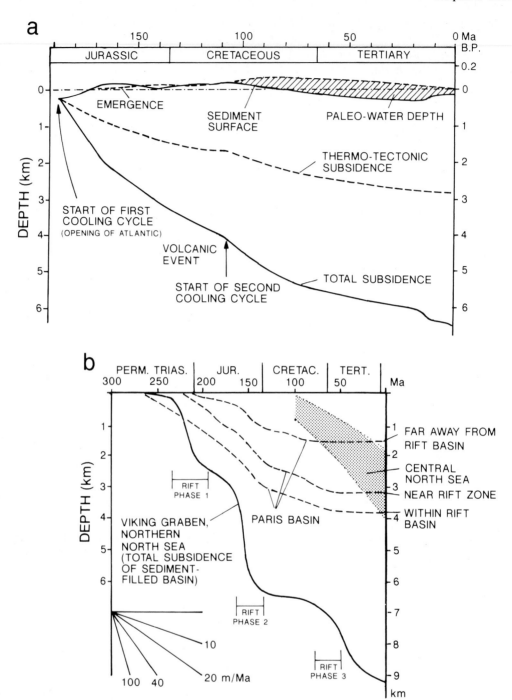

Fig. 8.13. a Total and thermo-tectonic Jurassic to Neogene subsidence at well C.O.S.T. B-2 on U.S. Atlantic continental margin, as affected by long-term sea level changes, variations in bathymetry (including periods of emergence), and volcanism (initiation of second cooling cycle). (After Greenlee et al. 1985). **b** Total subsidence determined from wells in central and marginal parts of Paris basin (Cochran 1983b) and general trend of subsidence in North Sea basin (curve NN from Beach et al. 1987; trend CN from Sclater and Christie 1980). See text for explanation

Hence, the apparent stretching factor, ß, of the lower layer of the crust becomes greater than that of the upper layer (*two-layer model*). This thermal effect may also lead to an upward shift of the 450 °C isotherm and thus to a thinning of the effective elastic thickness of the lithosphere adjacent to the rift basin (Fig. 8.15 c). This in turn will reduce the flexural rigidity of the lithosphere and therefore promote downward bending and the formation of a sediment wedge adjacent to the rift basin.

The following examples of passive continental margins take into account these processes in the development of the rift basin, as well as thermal thinning of the effective elastic thickness of the adjacent lithosphere. First, some subsidence curves for single sites are described, before we proceed to differential subsidence along transects across continental margins.

Examples of Simulated and Observed Subsidence

Young Passive Margin, Gulf of California

The rifting phase of the very young passive continental margin at the southern tip of Baja California (cf. Chaps. 12.1, 4.3, and 12.8, Figs. 9.13, 4.5 and 12.37) is characterized by a subsidence rate (around 200 m/Ma) of the same magnitude as the early rifting phase in the Rhine graben (Fig. 8.12) and in other young continental rifts. This value was found in two drillholes (DSDP Sites 475 and 476, Curray et al. 1982b) on the relatively steep continental slope 6 to 10 km away from the outer edge of the subsided continental crust. With the onset of ocean spreading in the southern Gulf of California 3 to 4 Ma ago, subsidence at these sites accelerated up to at least 1000 m/Ma for a very short interval, before slowing down again.

Subsidence and Burial History at Single Sites in the Northwestern Atlantic

Real subsidence curves for single locations within a sedimentary basin have been worked out for many regions. They are usually based on deep drillholes with well studied stratigraphic sections which include paleoenvironmental interpretations (cf. Chap. 8.2).

The results of such a study from the **Grand Banks off Newfoundland**, i.e., on the north-western Atlantic continental margin, are shown in Fig. 8.14 (Agterberg and Gradstein 1988, for location see Fig. 12.7).

Well Hibernia O-35 was drilled on the edge of a graben which has been subsiding since the early Jurassic; Puffin B-90 is near the southern transform margin of the Grand Banks; and Gabriel C-60 lies farther east close to the edge of subsided continental crust. The latter two wells were affected by rifting later than Hibernia (Fig. 8.14c). All three sites are characterized by a lower Cretaceous clastic wedge (sedimentation rate 10-25 cm/ka) with a major sand influx around 120 Ma (Fig. 8.14a). By contrast, sediment accumulation was markedly reduced between about 80 to 40 Ma B.P., as a result of high sea level and a hinterland of low relief.

Figure 8.14b depicts the *total subsidence and burial history* of individual stratigraphic units at the Hibernia site, taking into account changing paleo-water depths and increasing compaction with burial. Short intervals in the curves characterized by steep gradients reflect time periods with high sedimentation rates. Thermo-tectonic subsidence differs considerably between the three sites mentioned above; the Gabriel well on the outer margin exhibits greater subsidence than the other two sites.

All three wells, however, show less subsidence than would be expexted for purely oceanic crust or continental crust with 60 % dike injection. The principal reason for this difference is the fact that subsidence in the theoretical curves starts at a depth of about 2 or 2.5 km below sea level, while the observed curves begin near sea level.

A second example from the **shelf off New Jersey** (well C.O.S.T. B-2, Greenlee et al. 1988) displays the effects of changing bathymetry, long-term sea level fluctuation, and an early Cretaceous volcanic episode on total and thermo-tectonic subsidence (Fig. 8.13a). In fact, the long-term sea level curve was partially inferred from the observed total subsidence and from the other parameters controlling subsidence. In this example, the combined effect of volcanism and rising sea level has caused a significant positive excursion of the actual subsidence curves from "normal behavior". Neglecting these processes, the reduced sediment accumulation during the late Cretaceous and Paleogene would have led to decreasing subsidence rates.

Passive Continental Margin Transect, Northern Biscay

A particularly interesting example for studying the tectonic subsidence of a passive conti-

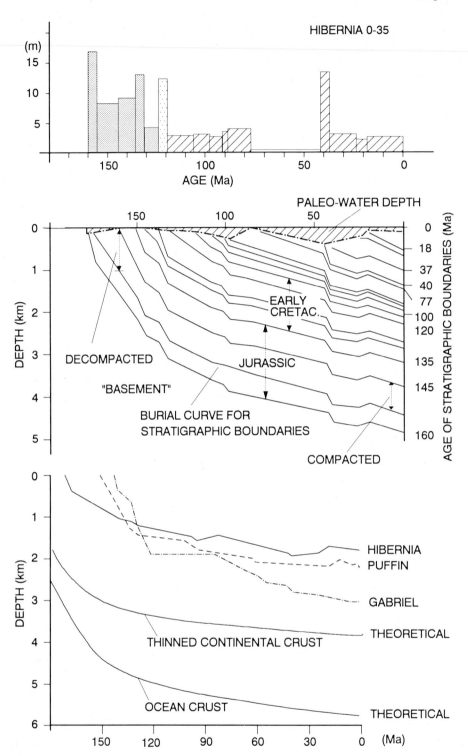

Fig. 8.14a-c. Example of subsidence and burial history from Grand Banks continental margin, northwestern Atlantic. **a** Sedimentation rates at well *Hibernia O-35*. **b** Variations in water depths, compaction and burial histories of stratigraphic units (total subsidence) at well *Hibernia O-35*.

c Inferred thermo-tectonic subsidence of three Grand Banks wells as compared to theoretical curves for thinned continental crust with 60 % dike injections and pure oceanic crust. (Agterberg and Gradstein 1988)

nental margin transect is the **northern Biscay (Goban Spur)**, because here the sediment cover is relatively thin (cf. Chap. 12.1). In this sediment-starved area, active rifting took place during the lower Cretaceous in a pre-existing shallow-marine basin without any indications of doming. The thicknesses of the upper and lower crust were reduced in a way similar to that shown in Fig. 8.15a and b.

At the end of rifting in Aptian time (120 Ma B.P.), a submarine trough 2.5 km deep had developed, although the upper crust was extended by only 10 to 15 % (stretching factor ß = 1.1 to 1.15). The topography of the sea floor during that time can be reconstructed using information from drill holes, dredging and seismic data. In a transect through this passive margin, post-rift subsidence, i.e., from late Aptian to Present time, increased seaward until it reached a similar amount as the oceanic crust at the continental-oceanic crust boundary (Fig. 8.15c).

If the thinning of the total crust (due to both extension and thermal attenuation) is taken into account (using the thinning ratio, ß, in the same way as the stretching factor in the uniform extension model), the relationship shown in Fig. 8.15d can fairly well predict the observed data. A second possibility is the evaluation of different stretching factors for the upper and lower layers of the lithosphere. Most of the post-rift subsidence (up to 4000 m) can be attributed to cooling of the lithosphere, which was previously thinned during the rifting phase.

Atlantic Passive Margin Transect, North America

One of the best studied examples of flexural response of the crust to sediment loading is the Atlantic passive margin of North America. In order to apply the method of "forward modeling" (Chap. 8.2) for determining subsidence and sediment accumulation, the assumptions shown in Fig. 8.16a and b were made (Steckler et al. 1988). Seaward of the hinge zone, the crust and lithosphere are thinned by similar amounts, but landward solely the lower part of the lithosphere is affected by thermal contraction.

Continental breakup and rifting started during the Triassic and ended during the middle Jurassic. The resulting subsidence and thickness of syn-rift sediments (Fig. 8.17 a) are primarily controlled by local isostasy as described above. Directly seaward of a 30 to 50 km wide hinge zone, which marks the transition from relatively undeformed thick conti-

nental crust to highly thinned and heated crust of the rift basin, the syn-rift sediments apparently reach their greatest thickness (cf. Fig. 8.6). With the onset of drifting, local subsidence in the former rift basin slowed and was followed by a broad downsagging of the evolving continental shelf and slope including the neighboring continent.

As a result of flexural downwarping, a landward thinning wedge of coastal plain sediments was developed, which began in the late Jurassic and extended during the lower Cretaceous more than 300 km inland. The fact that Jurassic sediments are nearly absent at the base of this sediment wedge (observed data shown in Fig. 8.17b) strongly indicates additional heating of the lithosphere adjacent to the rift basin. During the late Cretaceous, most of the flexure-induced part of the shelf basin was filled up. Near the hinge zone the thickness of this sediment wedge is about 4 km.

Passive Margin Transect, Southern Australia

A third example (Fig. 8.17c) displays results of a similar "forward model" for the passive margin of southern Australia. The *syn-rift phase* is characterized by rapid tectonic subsidence of about 50 m/Ma (even higher values for total subsidence!). The maximum thickness of syn-rift sediments (more than 5 km), comprising 25 to 30 Ma (during the early Cretaceous), is again found directly seaward of the hinge zone, although here the thinning of the lithosphere (ß = 3.5 to 4.2) is less than farther seaward (ß = 6.2).

The high stretching factor of ß = 6.2 at the continent-ocean boundary signifies that the initial thickness of the continental crust (31.2 km) is thinned to average oceanic crust thickness (5.0 km).

The syn-rift sediments also reach great thicknesses on the 100 km wide hinge zone (ß = 1.2 to 1.6). In the subsequent *post-rift or drifting phase* of 100 Ma, subsidence is controlled by cooling and contraction of extended lithosphere, as well as by the flexural response of the lithosphere landward of the hinge zone. The thickness and extent (200 km) of the landward sediment wedge is less than in the previous example, probably due to the wide hinge zone.

Some simulated and observed tectonic subsidence curves are compared in Fig. 8.18. They have been derived from drill holes on the south Australian continental margin and

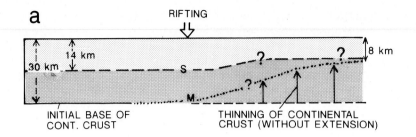

a

RIFTING

14 km

30 km

S

?   ?

M   ?

8 km

INITIAL BASE OF
CONT. CRUST

THINNING OF CONTINENTAL
CRUST (WITHOUT EXTENSION)

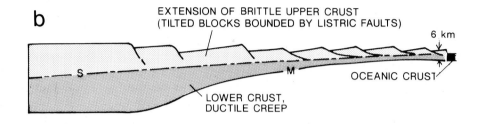

b

EXTENSION OF BRITTLE UPPER CRUST
(TILTED BLOCKS BOUNDED BY LISTRIC FAULTS)

6 km

S

M

OCEANIC CRUST

LOWER CRUST,
DUCTILE CREEP

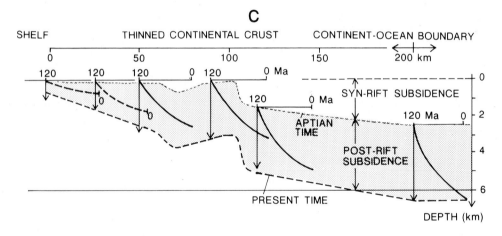

c

SHELF   THINNED CONTINENTAL CRUST   CONTINENT-OCEAN BOUNDARY

0   50   100   150   200 km

120   120   120   0   120   0 Ma

0

120   0 Ma

SYN-RIFT SUBSIDENCE

0

0

APTIAN
TIME

120 Ma   0 Ma

2

POST-RIFT
SUBSIDENCE

4

PRESENT TIME

6

DEPTH (km)

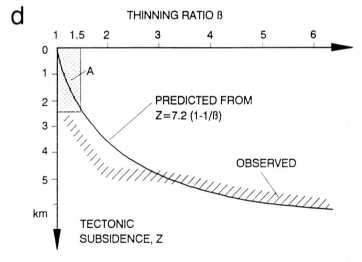

d

THINNING RATIO ß

1 1.5   2   3   4   5   6

0

1

A

2

PREDICTED FROM
Z=7.2 (1-1/ß)

3

4

OBSERVED

5

km

TECTONIC
SUBSIDENCE, Z

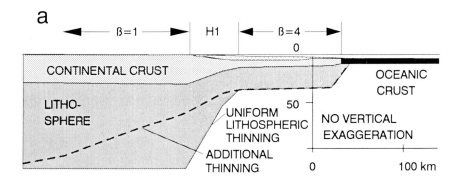

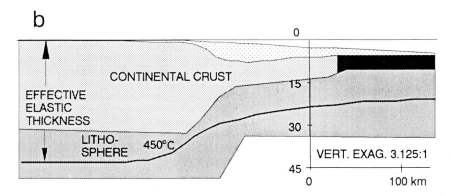

Fig. 8.16a,b. Conceptual models (cross-sections) of U.S. Atlantic passive continental margin showing extension and thinning of continental crust and lithosphere as a result of rifting and drifting with accretion of new oceanic crust. a Additional thinning of the lithosphere landward of the hinge zone (besides uniform thinning seaward and within the hinge zone, ß stretching factor). b Thickness of effective elastic plate defined by 450 °C isotherm. See text for explanation. (After Steckler et al. 1988)

Fig. 8.15a-d. Conceptual model of passive continental margin showing combined effect of a thermal thinning and b stretching of the crust, generating tilted fault blocks in the brittle upper crust and ductile creep in the lower crust. S discontinuity of seismic velocity at the transition from brittle to ductile material; M Mohorovicic discontinuity. (After Chenet et al. 1983). c Theoretical curves of post-rift subsidence versus time for different locations on a transect through sediment-starved northern Biscay margin similar to b. (After Montadert et al. 1979). d Relationship between thinning ratio (corresponding to stretching factor, ß, calculated for total crust regardless of whether it is extended or thermally thinned, hence analogous to uniform stretching model) and subsidence of sediment-starved passive margin, northern Biscay. Hatched observed subsidence of crust backstripped from sediment load since rifting. A maximum subsidence computed from extension of upper crust. (After Chenet et al. 1983, based on Le Pichot and Sibuet 1981)

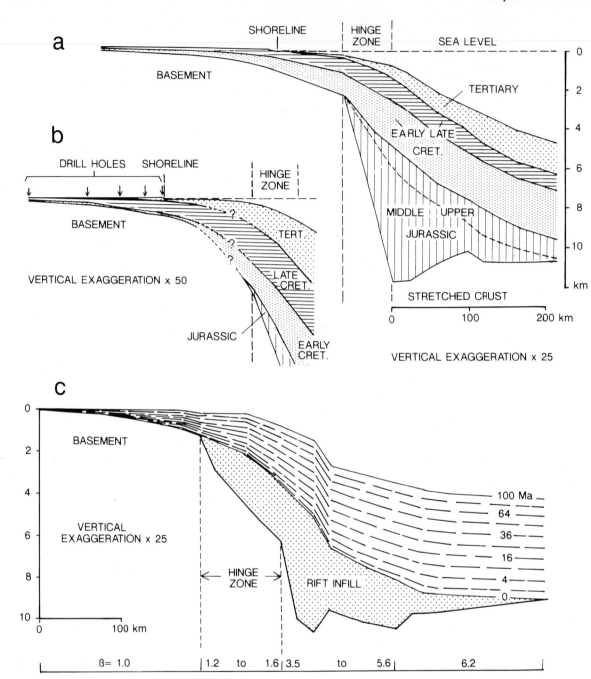

Fig. 8.17a-c. Synthetic stratigraphic cross sections of passive continental margins using flexural loading model. **a** Coastal plain and shelf region of New Jersey, two-layer model with thinning of elastic lithosphere landward of hinge zone (cf. Fig. 12.7; after Steckler et al. 1988). Note pinching out of Jurassic sediments beneath coastal plain and onlap of early Cretaceous onto basement.

b Observed New Jersey section based on drill holes. **c** Southern Australian margin, one-layer model, stretching factor (ß = 6.2) seaward of the hinge zone larger than in **a**, where ß = 4. Onlap of post-rift sediments onto basement where no stretching occurred (ß = 1); both subsidence and sedimentation rates decrease versus time. (After Hegarty et al. 1988)

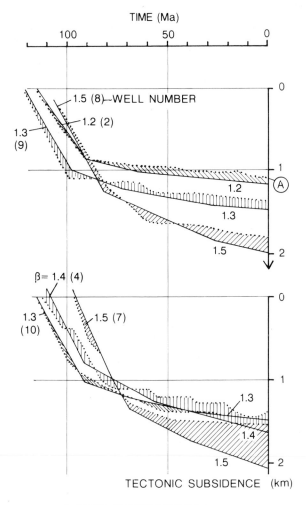

TIME (Ma)

1.5 (8)—WELL NUMBER

1.2 (2)

1.3 (9)

1.2 (A)

1.3

1.5

β = 1.4 (4)

1.3 (10)

1.5 (7)

1.3

1.4

1.5

TECTONIC SUBSIDENCE (km)

A, TOTAL SUBSIDENCE 2.6 km

Fig. 8.18. Comparison of model results *(solid lines)*, using varying stretching factors, ß, and rifting times between 25 and 30 Ma, with observed tectonic (not total) subsidence *(dotted lines)* for wells drilled on the passive continental margin of southern Australia. Total subsidence due to sediment loading is about 2.5 times tectonic subsidence. (After Hegarty et al. 1988)

corrected for sediment loading. The good fit of many synthetic curves with the real curves is taken as evidence that the model assumptions are adequate.

The subsidence history can be subdivided into two phases, a rifting stage with rapid tectonic subsidence lasting around 25 to 30 Ma, and a post-rift margin development (drift phase) with distinctly slower rates of subsidence as a consequence of cooling and contraction of the lithosphere.

The isochrones for post-rift development (Fig. 8.17a and c) display decreasing sediment

accumulation versus time, which is a result of the assumption that the average water depths in the cross section have remained the same all the time (this is also assumed for the model in Fig. 8.17a). In reality, the sedimentation rates may vary considerably and therefore bring about changes in water depths as well as in the location of the shelf edge which often does not coincide with major structural boundaries of the continental margin. Minor modifications of these models are caused by sea level changes which may generate erosional unconformities (Chap. 7.2).

*Concluding Remarks
on Passive Continental Margins*

All these examples of passive continental margins show a relatively good agreement between simulated and observed subsidence curves. In many cases, however, subsidence histories are more complex. Rifting and crustal extension may occur in several distinct phases and thus cause subsidence curves with repeatedly steepening gradients. Such a situation was found, for example, for the Mesozoic shelves of the alpine Tethys (Swiss Plateau and Jura Mountains), which experienced major subsidence phases in the Triassic, early Jurassic, middle Jurassic, and early Cretaceous (Wildi et al. 1989), but only part of these phases were active at one location.

The sedimentary sequences of subsiding passive continental margins are discussed in Chapter 12.2.

8.5 Subsidence of Basins Related to Tectonic Loading, Subduction, and Strike-Slip Motion

Foreland Basins

General Aspects and Problems

The primary factor producing foreland basins in front of overthrusting mountain belts is *flexure of the lithosphere.* Foreland basins develop on continental crust loaded by the vertical weight of an approaching overthrust belt as well as by the lateral force exerted by plate compression (cf. Chap. 12.6). Although the quantification of overthrust loads appears to be very difficult, some authors (e.g., Watts et al. 1982; Quinlan and Beaumont 1984) have made an attempt to model such a situation for different time steps. There are, however, examples (e.g., the Apennines and Carpathians) where the mountain belt overthrust load appears to be insufficient to create the existing foreland basins. In these cases, a deep basin already existed on top of the foreland plate prior to overthrusting. Thus the approaching load did not produce a marked topographic high and its effect on the underlying plate was limited.

Another serious problem, as already discussed in Chapter 8.4 and summarized by Allen et al. (1986), is the *rheological state of the crust* being flexed. Neither the model assumption of a purely elastic nor of a visco-elastic plate appears to be adequate to sufficiently explain the observed phenomena. The elastic plate model fails to account for the observed rapid relaxation of the lithosphere to a loading event, and the visco-elastic model implies the unrealistic prediction that flexural stresses vanish to zero over geological time. Therefore, additional assumptions are being proposed for the modeling of foreland basins. A thermo-rheological model including time-dependent effects may be better suited to explain both short-term relaxation and long-term finite rigidity of the lithosphere. Rapid subsidence and sedimentation in active foreland basins produce thermal anomalies within the sediments and the lithosphere which later decay. Consequently, the mechanical state of the lithosphere changes considerably through time.

In this context, it is also important to note that foreland basins, at least their internal parts, are often superimposed on a passive continental margin, i.e., on lithosphere which was already mechanically extended and had thermally subsided. As overthrusting continues, the orogenic belt progressively loads more rigid lithosphere on the landward side of the former margin. *Variations in flexural rigidity* along the strike of mountain belts may lead to unpredictable lateral changes in the subsidence history of foreland basins. This is, for example, postulated for the northern margin of the Indian plate which was exposed to the load of the overthrusting High Himalayas (Lyon-Caen and Molnar 1985).

Finally, in some cases the behavior of the foreland plate may be strongly affected by low-angle subduction from the side opposite to the approaching overthrust belt. As a consequence of the doubled thickness of the lithosphere, the foreland plate undergoes significant uplift and may therefore greatly influence the geometry and subsidence history of the foreland basin. Such a mechanism (*sublithospheric loading*) is postulated for the uplift of the Colorado Plateau in North America and the rapid subsidence of the adjacent basins during the Upper Cretaceous (Cross 1986).

For all these reasons, modeling the development of foreland basins probably remains a difficult task and is therefore not further treated here. A qualitative understanding of the processes involved is very useful, however, in studies of the development and subsidence history of such basins.

Examples of Subsidence Curves

The examples in Fig. 8.19 demonstrate some general trends found in several foreland basins of various ages. During the *pre-molasse stage*, total subsidence frequently resembles that of passive continental margins characterized by exponential decay. The rates of subsidence depend on the stage of evolution of this passive margin as well as on the location of the site investigated. Sites close to the former ocean or remnant basin experience higher subsidence than landward sites (Fig. 8.19b, example of Venetian basin). With the *approaching overthrust belt* (cf. Fig. 1.2), subsidence accelerates and may operate at very high rates in the subalpine region (total subsidence 100 to 300 m/Ma) for a short time interval.

For the Swiss Alps foreland, Homewood et al. (1986) inferred a thrust dip progagation of 7 mm/a and a convergence rate decreasing from around 8 to 2 mm/a during basin evolution (also see Chap. 12.6.3).

As soon as *plate collision* is complete and convergence comes to an end, the downgoing plate is not further subject to increasing loads. Consequently, subsidence slows down quickly. In the northern foreland basin of the Alps, rapid subsidence is followed, after a very short time interval, by uplift (Fig. 8.19b). This may be explained by *crustal rebound* due to isostatic adjustment of the thickened crust, by removal of the negative thermal anomaly built up in the lithosphere under overthrusting (Kominz and Bond 1986), as well as by *erosional unloading* of the emerged, high-relief frontal overthrust and fold belt.

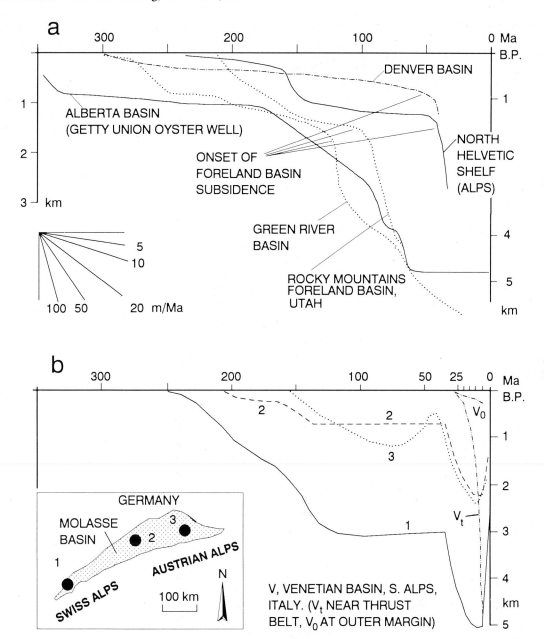

Fig. 8.19a,b. Subsidence histories of various foreland basins. All long-term curves show a pre-molasse phase, commonly characterized by slow to medium subsidence rates, and a molasse phase of rapid subsidence with the onset of foreland basin evolution. **a** Examples of the Rocky Mountains foreland in North America and the northern Helvetic zone of the Alps (Funk 1985). Note that Denver and Green River curves (Kominz and Bond 1986) indicate tectonic subsidence, while the other curves represent total subsidence (Alberta curve corrected for compaction; Utah curve from Cross 1986). **b** Examples of total, uncorrected subsidence curves from northern Alpine molasse basin (After Lemcke 1974) and Venetian basin, southern Alps (Massari et al. 1986). Note strong rebound, particularly in western portion of molasse basin in front of Swiss Alps

Subduction-Related Arc-Trench Systems

As in foreland basins, flexure of the lithosphere in conjunction with subduction is an important mechanism for generating *deep-sea trenches*. In this case, however, the crust is oceanic and much thinner than continental crust. The subducting oceanic plate is underthrust beneath an accretionary prism and an island arc or continental margin (cf. Chap. 12.5). The vertical load of these overriding bodies varies with the dip angle of subduction and the landward increasing thickness of the accretionary prism and magmatic arc. In addition, lateral forces exerted by convergence and the subduction rate play some role, at least in parts of the arc-trench system.

As described in Chapter 12.5 (Fig. 12.26), the seaward and central portions of an accretionary prism, as well as the magmatic arc, tend to gain in elevation with time, while the forearc and backarc regions subside. The depth of the trench floored by the subducting oceanic plate may be comparatively stable through time due to the establishment of steady-state conditions, i.e., a kind of balance between incoming and accreting sediment (Chap. 12.5.2).

Subsidence of the *forearc region* is controlled by the nature of the underlying crust, which may be oceanic or intermediate. Seaward parts of the basin fill may rest on deformed sediments of the accretionary wedge (composite forearc basins, Chap. 12.5.3). The subsidence of *backarc regions* varies in relation to the lack or occurrence of backarc spreading and overthrusting of the magmatic arc. Consequently, both the total amount and rate of subsidence are very variable in forearc and backarc basins, but they may reach high values during their evolution. It is evident from these points that the subsidence histories are problematic to simulate for such tectonic settings .

Pull-Apart Basins

Strike-slip movements commonly operate at high rates, but commonly persist only for limited time periods (Chap. 12.8). Particularly in transtensional systems, they cause significant *crustal extension* or the opening of a crustal "gap" which is filled by upwelling magma. Consequently, the early stage of basin formation is accompanied by rapid subsidence which, due to lateral heat loss, may be even higher than that of rift basins. In this case, subsidence can be modeled in a way similar to that described for rift basins (see above and Fig. 12.35).

However, many pull-apart basins show a complex subsidence history with more than one pulse of rapid subsidence. In addition, subsidence may be followed by a period of uplift. Such complications have been described from the onshore Tertiary basins in California, such as the Los Angeles, Sacramento, Salton Trough, Ventura, and other basins (Dickinson et al. 1987; Mayer 1987). Changes in subsidence trends are probably caused by the transition from transtensional to transpressional deformation (see Chap. 12.8), but other processes such as volcanism associated with wrench faulting may also play a part.

Conclusions

As summarized by Dickinson and Yarborough (1976) and evident from the discussion in this chapter, the various basin types display a specific trend in their subsidence history. Rift basins, pull-apart basins, passive continental margin and oceanic basins show a concave-up shape of their subsidence curves indicating exponential decay. By contrast, orogenic basins subject to tectonic loading (remnant and foreland basins, many backarc basins) are characterized by convex-up sections in their subsidence curves. In the later stages of evolution, however, subsidence of these basins slows and is frequently followed by uplift. Failed rifts and aulacogens exhibit composite, S-shaped subsidence curves with an early concave-up and later convex-up section. In addition, the (total) subsidence curves of many sedimentary basins are modified significantly by local tectonic processes, variations in paleobathymetry, volcanic events, and other factors mentioned in this chapter.

9 Denudation: Solute Transport and Flux Rates of Terrigenous Material

9.1 Weathering and Soils

Introduction

Most of the sediment deposited in sedimentary basins is derived from land areas exposed to subaerial weathering. To some extent, this is also true for biogenic and chemical sediments, insofar as their production depends on the dissolved river load delivered into the sedimentary basins. The land-derived, solid material is usually referred to as the terrigenous, allochthonous, clastic component of a sedimentary deposit.

Due to weathering and erosion, the land surface is slowly lowered. The average amount of this lowering in a certain region over a given time span is defined as the *denudation rate*. Denudation involves the transport of bedrock, soil, and vegetation mass into continental basins or into the oceans. This mass transfer is brought about by a variety of physical, chemical, and biological processes. Rainfall, surface runoff, and river transport are the most important transport agents, but groundwater circulation, ice and wind transport also play a significant part in many regions. Mass transfer to a marine basin may take place discontinuously due to intermediate storage as soils, continental sediments, and lake deposits.

Disintegration of rocks due to stress release, as well as thermal expansion and contraction is apparent even at a depth of several tens of meters below the land surface (Fig. 9.1b). Newly formed joints near the surface and increasing pore space favor the circulation of water and thus accelerate *mechanical and chemical weathering*. The zone of most pronounced weathering is near the surface in the pedosphere, where biological activity also promotes the alteration of rocks and minerals. In general, large, thermodynamically unstable minerals are transformed into smaller, stable components such as clay minerals ("mineralo-

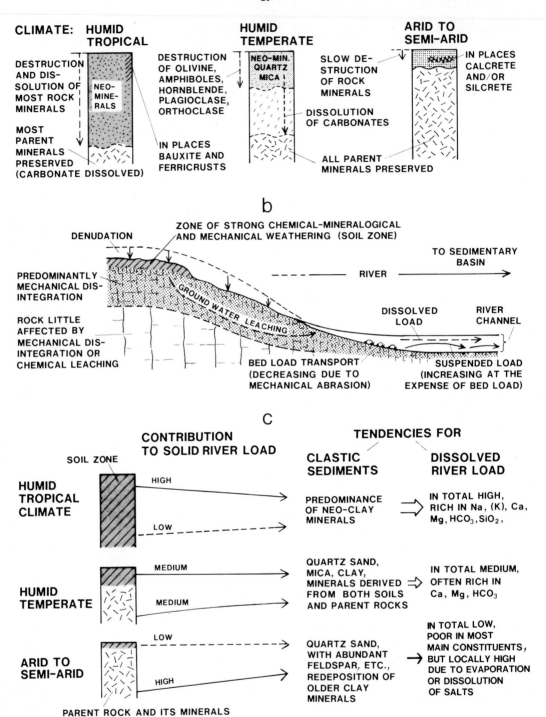

Fig. 9.1a-c. Qualitative relationship between climate, weathering, and delivery of clastic sediments as well as dissolved materials into sedimentary basins. **a** Intensity and depth of destruction (or preservation) of parent rock minerals and formation of clay-type neo-minerals (including oxides and hydroxides of Al and Fe). **b** Different zones of weathering, resulting denudation, and river transport of particulate and dissolved load. **c** General trends in climatically-controlled production of clastic sediments and the input of dissolved constituents into sedimentary basins. Note that the temperate and especially the arid to semi-arid regions can deliver substantial amounts of unchanged parent rock material and nonaltered minerals (including clay minerals)

gical" weathering). At the same time, the parent minerals lose part of their solid matter by chemical reactions and dissolution ("chemical" weathering). Thus, the entire alteration process should really be referred to as "chemical-mineralogical" weathering, but it is frequently called *chemical weathering*. Several minerals are completely dissolved, such as carbonate minerals.

Whereas in arid climates rocks are broken down mainly by mechanical processes, chemical-mineralogical and biological weathering is dominant in wet climates. The products of chemical weathering are best developed in lowland areas or on flat plateaus, where the altered material, the regolith or saprolite, cannot readily be removed from its parent rock. Hence, deeply weathered profiles of completely altered parent rock are common in tropical lowlands (Fig. 9.1a).

Regions of high topographic relief in conjunction with substantial rainfall are most important in delivering particulate matter into rivers. As Carson and Kirkby (1972) and others have pointed out, erosion on hillslopes is controlled by the weathering rate and transport processes which remove the disintegrated, loose material at the surface. Erosion may be "weathering-limited", if the capacity of the transport process exceeds the rate at which erodible material is generated by weathering. As a result, the bedrock is exposed and the slopes tend to steepen, thus further promoting mechanical denudation. In contrast, erosion is "transport-limited", if more material can be supplied by weathering than can be removed by transport processes. In this case, thick soil profiles develop, and transport of materials in solution predominates.

Weathering, Soils, and Composition of Terrigenous Sediments (Overview)

Climate Control

In *humid tropical and subtropical regions*, carbonate is completely removed in solution from the regolith. The newly formed clay minerals (neo-minerals), for example kaolinite, reflect the climate of their drainage area rather than the mineralogical composition of the source rocks. As a result of the pedogenic formation and enrichment of iron as Fe_2O_3 (hematite), many of these soils become red (laterites; cf. Chap. 6.3). If the common clay minerals, as aluminum silicates (e.g., kaolinite), also become unstable and their silica

content is dissolved and carried away by running water, hydroxides and oxides of aluminum and iron are left behind as residue. In some places, these *ferrallitic* end products (soils called *pedalfers*) accumulated as bauxite deposits, or they form thick, hard lateritic crusts (*ferricrusts*, Fig. 9.1a). Detailed descriptions of these processes are given in several text books (e.g., Ollier 1969; Birkeland 1984; Chorley et al. 1984; Weaver 1989; Retallack 1990).

In more *temperate humid zones*, illite is a characteristic weathering product (cf. Chap. 9.3), whereas in high-latitude regions, as well as in arid zones, soil generating processes and hence the formation of neo-minerals are slow. Such regions can provide even relatively unstable minerals as sediment components for basin fill, such as minerals of the pyroxene and hornblende groups, different types of feldspar, chlorite, etc. Consequently, the terrigenous material derived from such source areas is controlled primarily by source rocks in the drainage area.

In *semi-arid zones*, where little water is available to remove calcium in solution, even calcium carbonate can be reprecipitated in the soil zone. It often forms nodules close to the surface (*calcrete* or *caliche*, Fig. 9.1a). Such soils are called *pedocals*. Similarly, dissolved silica can reprecipitate to form siliceous crusts and near-surface *silcrete* (see, e.g., Twidale and Milnes 1983; Thiry and Milnes 1991).

Rock-Type Control

The nature of the parent rocks exposed to weathering is also of considerable interest when weathering processes and sedimentary rocks from various geological epochs are compared. In the early history of the Earth, the land masses were entirely or predominantly composed of igneous rocks and subject to *first cycle weathering and deposition*. Later, as a result of successive cycles of weathering, sedimentation, and mountain building, sedimentary and metamorphic rocks became increasingly important on the land surface. Thus, the composition of the weathering products, and hence that of the sediments also changed through geologic time (cf. Chap. 6.5).

First cycle sedimentary rocks may already contain a high proportion of stable minerals. If they are exposed to weathering again, either the proportion of stable minerals in the new sediment will be enhanced (increased *mineralogical maturity* in warm and wet climates), or the second cycle deposit will strongly reflect the composition of the parent sedimentary rocks. Claystones and mudstones, for example, can be easily and rapidly broken down to their fine-grained fundamental components by mechanical processes (chiefly by repeated drying and wetting), if they are not protected by a vegetated soil cover. Thus, particularly in arid and semi-arid zones, the nature of the clay particles delivered into a sedimentary basin may be controlled by the pre-existing exposed rock types rather than by the climate of the drainage area (cf. Chap. 9.3).

Rock type is, of course, also important for all the coarser grained materials provided by mechanical disintegration in the drainage area. In areas of high relief, marked soil formation is prevented by rapid erosion (weathering-limited conditions). In this case, slightly altered parent rocks are carried away by rivers as gravel and even boulders into lower regions. But during this transport, particularly the gravel and to some extent the sand grains become mechanically rounded and tend to decrease in size along the river's course. In this way, the parent rocks add additional material to the fine-grained clastics, mainly to the silt fraction.

Clastic Sediments and Dissolved River Load

The relationship between the climate of the source area and the weathering products of the bed rock (in this case granite), providing both clastic sediments and dissolved river load for a sedimentary basin, is summarized in Fig. 9.1c. Tropical regions tend to deliver predominantly neo-minerals and a small portion of the most stable minerals from pre-existing bed rocks as terrigenous sediment components. Temperate, humid regions supply both neo-minerals and stable pre-existing minerals, whereas arid regions may deliver comparatively unstable minerals. This simplified model may be modified by grain sorting during sediment transport. Fine-grained suspended river load moves much faster and usually over longer distances than the coarser, sandy, gravelly bed load (Fig. 9.1b) and can therefore be separated from the bed load.

The *dissolved river loads* in the three climatic zones shown in Fig. 9.1c also differ markedly both in the concentrations of various chemical species and in their freight rates. Due to intensive precipitation and chemical weathering, more solid matter per unit area and time is dissolved in the tropics than in

other zones. As a result, denudation by dissolution is often much greater than that caused by mechanical processes. Even in temperate humid zones, chemical denudation frequently exceeds mechanical denudation, whereas in arid regions, mechanical processes are more important than dissolution. These aspects will be discussed quantitatively in the following chapter.

9.2 Chemical and Mechanical Denudation Rates from River Loads

9.2.1 Chemical Denudation Rates

General Aspects

Chemical denudation is the lowering of the land surface by total or partial dissolution of soils and rocks in water at or below the surface. The dissolved constituents are carried away by running surface water or groundwater. The chemical denudation rate can be measured quantitatively by analyzing the *dissolved species in the runoff* of small and large river systems. Such measurements should be carried out at short intervals for several years in order to evaluate seasonal or short-term changes in water chemistry. In addition, the volume of discharged water, as well as the areal extent of the river catchments must be known. Moreover, it should be ascertained that no substantial quantity of groundwater is lost by subsurface flow from the area studied.

An alternative method of determining the chemical denudation rate is to quantify *chemical residues in soil profiles*, i.e., the degree of dissolution which the parent rock has undergone since it was exposed to weathering. This method can only be applied under favorable conditions, for example, when during the last glaciation a fresh rock surface was formed by ice action and subsequent mechanical denudation can be ruled out. Thus far, only sparse data have been obtained by this method (e.g., Colman and Dethier 1986; April et al. 1986; also see Chap. 9.5.1). The following discussion is therefore based mainly on the "river discharge method".

The *chemistry of natural surface waters and groundwaters* is the result of interaction between infiltrating rainwater and soils and rocks. Short-term overland flow during storm events, as well as shallow-circulating near-surface runoff or interflow, are not in contact with the soil particles and the rocks sufficiently long to reach a state of equilibrium between the concentrations of dissolved species in the water and the solid phases. As a result, the concentrations of most of the dissolved constituents are lower than they could be under equilibrium conditions. On the other hand, more deeply circulating groundwater usually has a long enough residence time to reach equilibrium conditions with most rock components, before it emerges again at the surface and contributes to the runoff of rivers.

Because the groundwater component and long-term interflow make up 60 to 80 % of the total discharge of many river systems, we can assume that the chemistry of river water during low and intermediate runoff (between peak floods) is often close to equilibrium with the most abundant rock minerals present in the drainage area. In other words, the composition of river water reflects the major soil and rock types present in the catchment area rather than other environmental conditions (e.g., climate and relief), but the presence or absence of vegetation cover may also play a significant role, particularly in carbonate karst terrains (see below). However, streams solely draining areas of high relief, polar or subpolar regions, and ephemeral streams resulting from storm events in arid zones, show rather low concentrations of dissolved species (apart from highly soluble salts in limited areas).

Apart from rock type, the chemical denudation rate is of course also largely controlled by the *effective rainfall* (creating runoff), i.e., precipitation minus evapotranspiration. Thus, chemical denudation can be substantial even in regions of cold climate and/or areas of marked nonequilibrium between water-soil-rock interactions, if the effective precipitation is high. For these reasons, differences in chemical denudation rates in large regions (up to and including continents) are not as great as those for mechanical denudation rates (see below).

Chemical denudation in a particular region may also vary considerably with the *nature and thickness of the soil profile*. According to several authors (e.g., Carson and Kirkby 1972; Stallard 1985), there is presumably an optimum soil thickness which maximizes the rate of chemical rock weathering. For less than optimum soil thickness, the soil cannot retain the water supplied by precipitation long enough to interact sufficiently with the altered or fresh subsurface bedrock. Hence, chemical denudation decreases. If soil thicknesses exceed the optimum, more water is stored in the soil and evapotranspiration can reach its maximum. As a result, less water is available for the weathering of bedrock and output of dissolved material. In addition, the formation of thick, impermeable soil layers can induce a further reduction in the weathering rate. Thus, the denudation rate for a certain area may decrease through time, for example after 10 to 100 ka of soil formation, even if the climatic conditions remain constant.

Rock-Specific Phenomena, Groundwater Chemistry

Dissolution of Salt Rocks, Sulfate Rocks, and Carbonates

Chemical dissolution, as well as the alteration of rock and soil minerals, are selective processes. Rock salt, gypsum and anhydrite, and other salt minerals are easily dissolved either direcly on the land surface or at some depth in the underground. Locally, they can contribute enormously to the dissolved river load and thus increase the denudation rate substantially (Fig. 9.2), even when the concentration of these salts in spring and river waters is mostly far below saturation.

During an active phase of *subsolution of rock salt* in a region with temperate climate, the land surface may be lowered by 1 to 5 mm/a (assuming an effective precipitation of 200 to 500 mm/a and a concentration of only 10 to 20 g/l $NaCl$ in the runoff, i.e., far below saturation). However, such a local phenomenon will last only for a short time period, because either the salt deposit is used up, or it is sealed by more or less impermeable residual clays released from the evaporitic sequence. Thereafter, the pore fluid of the neighboring rocks tends to become stagnant and saturated with rock salt, and thus subsolution more or less ceases.

Near-surface and subsurface dissolution of *sulfate rocks* may proceed in temperate humid regions at a rate of about 0.2 to 0.5 mm/a (200 to 500 mm/ka). These values do not necessarily mean that the land surface is lowered at this rate, because substantial dissolution occurs along fraction zones and in caves; nonetheless, in the long run, an entire sulfate layer can be removed in this manner.

Of particular interest in this context are *carbonate rocks*, because they often form thick and extensive deposits which may be exposed

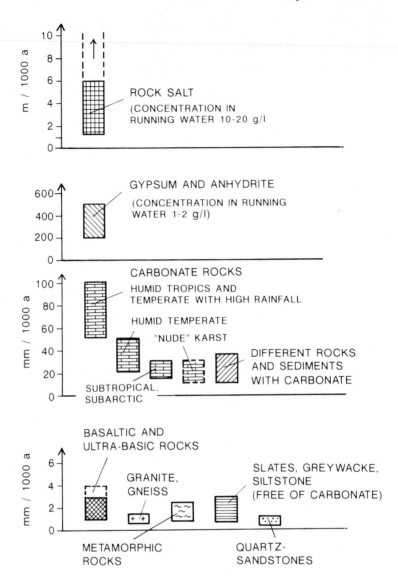

Fig. 9.2. Chemical denudation rates primarily controlled by rock type, mainly in temperate climates. Note the different orders of magnitude for evaporites (rock salt, sulfate rocks), carbonates, carbonate-bearing rocks, igneous, metamorphic, and siliciclastic sedimentary rocks. (After Priesnitz 1974; Hohberger and Einsele 1979; Ford and Williams 1989; and others). (1000 a = 1000 years = 1 ka)

on the land surface for comparatively long time periods. Since carbonate is chiefly dissolved as bicarbonate, the vegetation cover and production of CO_2 by decomposing organic matter in the soil zone are also important factors controlling the *carbonate denudation rate* (Fig. 9.2; Priesnitz 1974; Bögli 1978; Ford and Williams 1989). Therefore, regions with dense vegetation ("green" karst) and high effective precipitation, such as large parts of the tropical zone or mountainous areas with high rainfall in the more temperate regions,

enable denudation rates of between 50 and 150 mm/ka. In zones of temperate climate and medium rainfall (700 to 1000 mm/a), the carbonate dissolution reaches 20 to 50 mm/ka, and in subtropical and subarctic regions 15–30 mm/ka. Where soils and vegetation are absent (barren or "nude" karst), springs and stream waters have concentrations of bicarbonate only 25 to 50 % of those observed in green karst regions.

Although dissolution processes below the land surface (caves, etc., summarized as *endo-*

karst) can be very spectacular, most of the carbonate dissolution takes place near the surface *(exokarst)*, because the infiltrating waters become saturated with bicarbonate relatively fast. Therefore, subsolution is responsible for only a small percentage (up to a maximum of ca. 10 %) of the total carbonate dissolution (Bögli 1978). Despite the fact that lowering of the land surface in karst regions often proceeds very irregularly, one can assume and calculate from river output a certain mean denudation rate for such regions.

Silicate Rocks

Crystalline rocks free of carbonates and other easily soluble constituents yield lower quantities of dissolved matter than the rock types discussed so far, but they also show considerable variations in their weathering solutions. The characteristics of solutions from such silicate rocks or soils can be best evaluated by studying their groundwater chemistry (Fig. 9.3), because the establishment of equilibrium conditions in these rocks requires a fairly long residence time for the circulating water.

The mostly dark minerals of *ultrabasic, basic* (such as basalt and gabbro), and *intermediate volcanic rocks* (such as andesite), which are rich in iron, magnesium and calcium, break down more readily than most feldspars, muscovite, and quartz, which are abundant in granitic rocks. Among the plagioclase group, anorthite rich in calcium is less stable than albite (rich in sodium) or K-feldspar; biotite is altered more quickly than muscovite. From these and other observations we can draw the following conclusions:

- In the case of *basic and ultrabasic rocks,* all principal minerals are readily altered and therefore contribute substantially to the composition of groundwater and river water. The concentration of dissolved species tends to be relatively high (≥ 150 mg/l, Fig. 9.3), and the waters are fairly rich in Ca and Mg.
- Waters from *granitic rocks,* on the other hand, draw their ions essentially from the weathering of plagioclase (rich in Na), biotite, and K-feldspar. Since most of the potassium is fixed in clay minerals, such waters tend to be relatively rich in sodium, but poor in calcium and magnesium; their total dissolved species concentrations are low (50 to 100 mg/l, Fig. 9.3).

- SiO_2 released by the breakdown of minerals is partly retained in secondary minerals. Therefore, the silica content of natural waters from igneous rocks is limited and far below the values known for amorphous silica dissolved in water. In the first case, the aqueous silica content more or less reflects equilibrium conditions with newly formed aluminosilicates.
- Except for environmental conditions with very low pH-values, all the Al_2O_3 released by the weathering of primary minerals is retained to form secondary minerals.
- *Quartz* is dissolved in substantial quantities only under tropical to subtropical climatic conditions and is therefore of minor importance for the composition of natural waters outside such specific regions. As a result, it is enriched in the alteration products and thus in the sediments derived from these sources.
- *Iron,* although present in many rocks in considerable quantities, is rarely a substantial component in near-surface groundwater. It is generally fixed in the soil zone as oxide or hydroxide and may be transported as solid river load.

Meteoric waters passing *metamorphic rocks,* which do not contain substantial amounts of carbonate, have chemical characteristics similar to those of many claystones and shales, impure sandstones, and acidic and basic magmatic rocks. Figure 9.3 shows an example of mica schists, whose water does not deviate much from that of intermediate or basic igneous rocks.

Very low mineralization of near-surface groundwaters is usually observed in *quartzitic sandstones* and in *shales and slates* devoid of carbonate (Fig. 9.3). These rocks consist of the most stable minerals which are little dissolved under normal weathering conditions. Shales containing pyrite may deliver more concentrated waters rich in sulfate.

When sedimentary or metamorphic rocks contain carbonate (e.g., marly or dolomitic claystones, carbonate-cemented sandstones, and greenschists with lenses of marble), their groundwater chemistry approaches that of limestones or dolomites with high concentrations of earth alkali ions and bicarbonate.

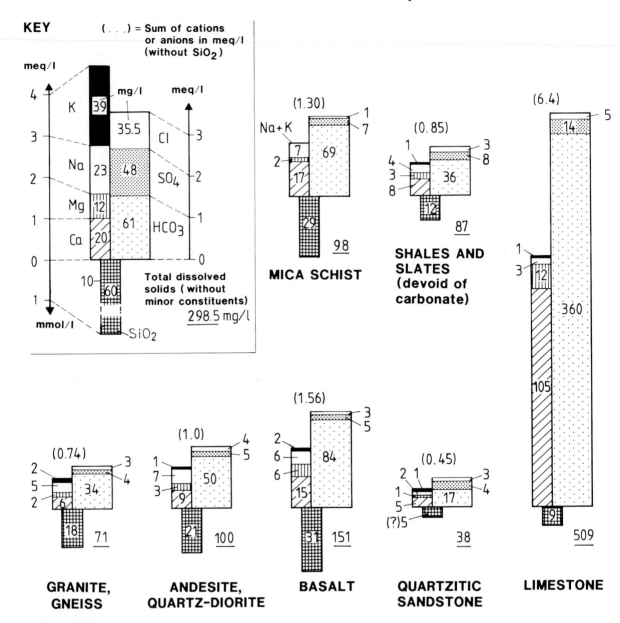

Fig. 9.3. Major constituents of near-surface natural groundwaters for some principal (nonevaporitic) rock types. All analyses come from humid temperate regions and represent mean values in mg/l, meq/l, or mmol/l (for SiO₂). Water chemistry of similar rock types may, however, vary considerably from location to location. (Data from various sources, e.g., Matthess 1973; Georgotas and Udluft 1978; Drever 1982; Hem 1983)

Denudation Rates in Various Rock Units and Morpho-Climatic Zones

Solute Transport in Rivers

In many drainage basins, the total dissolved river load is controlled mainly by the *ground-water chemistry* contributing significantly to river runoff.

For example in the river Main, a major tributary of the river Rhine in central Europe (in a temperate, humid climate), 86 % of the total dissolved constituents are provided by groundwater from different rock types (Fig. 9.4a). Of these lithologic units, carbonate and sulfate-bearing sedimentary rocks deliver approximately 90 % of the chemical load, whereas regions with crystalline rocks and quartzitic sandstones (comprising 22 % of the total drainage area) yield only 5 % of the dissolved load. The average transport rate of the entire drainage basin amounts

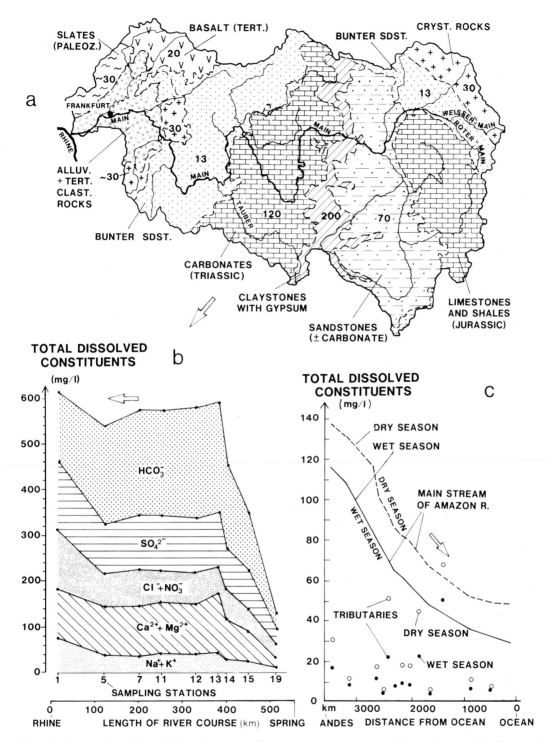

Fig. 9.4. a Drainage basin of river Main, Germany, with different rock types in humid temperate climate. *Numbers* refer to dissolved load (in t/(km²·a)) drained by groundwater. (Einsele and Hohberger 1978). **b** Changing concentrations of chemical species along the river Main passing areas of different rock types. Although the area is densely populated, only a small percentage of the dissolved river load (mainly NO_3^- and some Na^+, Cl^- and SO_4^{2-}) comes from anthropogenic sources. (Udluft 1978). **c** Changing water chemistry along the river Amazon caused by low-concentration tributaries from tropical regions. (Drever 1982)

to about 100 t/(km^2·a), which corresponds to a mean chemical denudation rate of 4 cm/ka or 40 m/Ma (see below). In reality, the denudation rate varies between approximately 5 m/Ma (Bunter sandstone) and 80 m/Ma (claystones with gypsum). The morphogenetic consequences of differing denudation rates are discussed in Chap. 9.5. Furthermore, the chemistry of the river water also varies significantly along its course (Fig. 9.4b) as a result of tributaries draining different rock types and industrial sites in the westernmost region.

Changing water chemistry is observed in many rivers. A more spectacular example is the Amazon River, where water from the Andes, relatively rich in dissolved constituents, is diluted by tributaries draining areas with deeply leached tropical soils (Fig. 9.4c).

Solute tranport in rivers all over the world has been studied for many years (Walling and Webb 1986). Figure 9.5 shows the water chemistry of major rivers as summarized by Meybeck (1979, 1983, 1987) and others; however, modern pollution problems (see, e.g., Meybeck 1989) are not discussed here. Most of the published data correspond to discharge-weighted mean values for one or several years. In these examples, which represent the most important drainage basins of the world, the concentrations of total dissolved solids (TDS) vary between 33 and 880 mg/l. If smaller catchments are included, TDS ranges from 5 to 8 mg/l (some tributaries of the Amazon river) to more than 20 000 mg/l (rivers draining sequences containing evaporites). Most rivers fall into the range of 30 to 300 mg/l (Walling and Webb 1986). In spite of this wide range, one can distinguish several groups with similar characteristics:

- Rivers draining either largely *tropical regions with thick soil cover* consisting of neo-minerals of low solubility (Amazon, Orinoco, Congo, Indonesian Rivers) or crystalline rocks (e.g., Scandinavia) exhibit low concentrations of TDS between 33 and 58 mg/l. The rivers Niger and Paraná show concentrations slightly above this range. In all these cases, the river water contains little bicarbonate and less TDS than the groundwater circulating in unweathered or only partly altered rocks, except quartzitic sandstones (Fig. 9.3). The concentration of SiO$_2$ in these river waters, however, approaches the norm (11 mg/l, average of world rivers).
- The Siberian rivers Lena, Ob, and Yenisei coming from *subarctic areas with sparse vegetation* also have low TDS values, although carbonate-bearing rocks are present in their drainage areas.
- Rivers originating in the *Alps and Himalayas* (including the Yangtze River), as well as the Mississippi River, range from about 170 to 300 mg/l of TDS. They all display medium values for Ca and bicarbonate due to the presence of carbonate rocks in their drainage areas.

- Two of the major drainage basins shown in Fig. 9.5 (Murray River in Australia and Rio Grande River in North America) are markedly affected by the dissolution of salt and/or gypsum and therefore reach unusually high TDS values.

According to Livingstone (1963) and Meybeck (1979, 1983, 1987) the *average river water of the world* contains about 115 mg/l TDS (load-weighted mean, i.e. total load transported by rivers) and has a composition as shown in Fig. 9.5. The most important constituent (by weight) is HCO$_3^-$ (about 58 mg/l) which, however, is derived partially from decaying plant material in the soil zone and recycled into the atmosphere. It is followed by Ca^{2+} (about 14 mg/l), SiO$_2$ (about 11 mg/l), and SO$_4^{-2}$ (10 mg/l). These four components generally make up more than 80 % of the total dissolved river load.

Specific Transport Rates

If the TDS concentrations are multiplied by the specific runoff (per unit area of the drainage basin, taking into account seasonal and short-term variations), one obtains the specific transport rate of dissolved constituents. Values frequently range between 5 and 80 t/(km^2·a), but can also be considerably lower or higher. Such data are plotted in Fig. 9.6 for major river basins of the world and grouped according to different climatic zones. In some cases, relief and rock type are also considered.

It is obvious that a high specific runoff, i.e., humid climate where precipitation substantially exceeds evapotranspiration, leads to high specific transport rates of TDS, even when the TDS concentrations are fairly low (e.g., Iceland or most of the tropical river basins). Under conditions of low specific runoff (arid to semi-arid climates), salt concentrations in river water must be extremely high in order to give rise to a high transport rate (e.g., catchments of the Aral Sea, the Caspian Sea, or the Shatt el Arab). Due to medium concentrations and high specific runoffs, many mountain ranges have high to very high TDS yields (e.g., the Alps, partially drained by the Danube River, or the Himalayas giving rise to the rivers Indus, Ganges, and Brahmaputra).

Mean Chemical Denudation Rates

The specific transport rates of TDS, expressed as the mass or weight of dissolved solids per unit area and time, can be converted into mean denudation rates (mm/ka or m/Ma) by dividing the weight by the basin area and

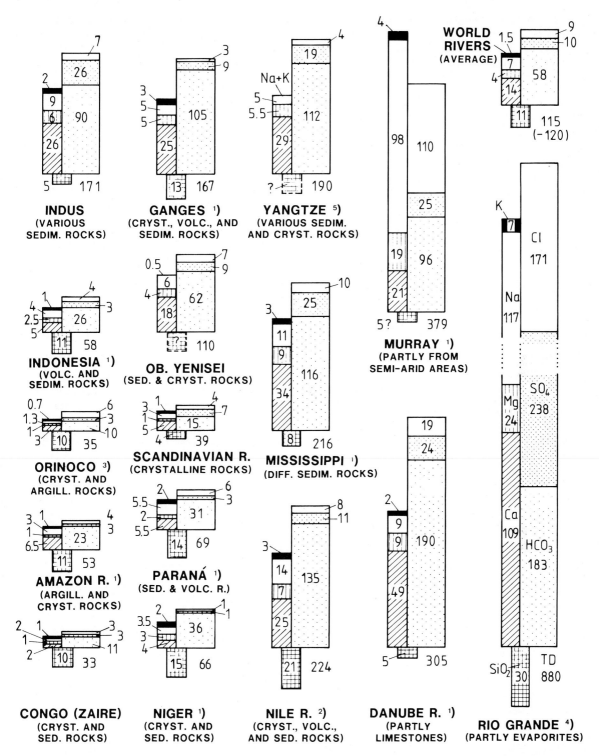

Fig. 9.5. Water chemistry of major rivers of the world, all values in mg/l, chemical species marked in Rio Grande plot and in Fig. 9.3. Total dissolved solids (TDS) increase with the proportion of carbonate and evaporitic rocks in the drainage basin (e.g., Danube and Rio Grande Rivers). Rivers draining tropical zones with deep soils are low in TDS and little affected by parent rock types (e.g., rivers Congo, Orinoco, etc.). Many large rivers drain areas of mixed rock types and zones of different climate and relief, and therefore show medium TDC values (e.g., Nile, Indian rivers, Yangtze, etc.). (Mainly after Livingstone 1963; Meybeck 1979, 1983, 1987; supplemented by Degens et al. 1982, 1983, 1985; and others)

convering from weight to volume. For the weight to volume conversion, an average rock density of 2.5 t/m^3 is assumed. The mean chemical denudation rate indicates the thickness of soil or rock which is removed from the land surface by solute transport after a certain time period. As shown in Fig. 9.6, chemical denudation rates are high in many drainage basins of the tropics, as well as in some river catchments of high mountain ranges with very high precipitation.

Globally, the mean (discharge-weighted) transport rate of dissolved solids is about 40 t/(km^2·a), which corresponds to a denudation rate of 16 mm/ka (Fig. 9.6).

Chemical Denudation in Specific Morpho-Climatic Zones

These interrelations are most obvious if TDS transport rates and chemical denudation rates are correlated with specific morpho-climatic zones (Fig. 9.7), rather than with the drainage areas of rivers, which are often composed of zones of different climate and relief. It then becomes evident that two maxima of solute transport and chemical denudation exist: the very humid temperate zone and the humid mountainous zone in the tropics. Minima of chemical denudation are found in arid zones and in tundra and taiga regions with low specific runoffs. Maximum and minimum values for chemical denudation thus differ by a factor of 20 to 30.

9.2.2 Mechanical Denudation from Solid River Load

General Aspects

Denudation rates on land can be evaluated in different ways. A common method is the measurement of *sediment discharge from rivers.*

Data gained by this method must be viewed with caution, however. Usually these measurements do not include rare events, for example catastrophic floods, which may occur once in 10 or 100 years and carry more suspended and bed load than a long period of normal river discharge. Therefore, average values based on measurements of a few years are often not representative of longer periods. In addition,

many of these measurements are strongly affected by the activities of humans (e.g., deforestation) and therefore often yield denudation rates much higher (by a factor of 2 to 10, but locally up to 100 to 1000!, see, e.g., Toy 1982) than those of prehistoric times. The sediment input from the Yellow River in China into the Yellow Sea has increased by a factor of about 10 since the land is agriculturally used (Milliman et al. 1987, cf. Chap. 11.3).

On the other hand, the solid load of rivers as measured today can also be too low in comparison with the prehistoric era. This is the case for most of those rivers in which man-made dams and reservoirs trap a large part of the sediment. Moreover, a considerable number of rivers draining previously glaciated areas pass natural, ice-formed lakes, where they deposit part of their load. Such sediment traps are short-lived and did not exist during long periods of the Earth's history. Hence, many measurements represent in fact delivery rates of sediment into the oceans and not denudation rates in the source areas (Leeder 1991). Furthermore, we should consider that normally only the *suspended river load* can be easily determined. It is true that the *bed load* of large rivers makes up only a few percent of the total solid load, but mountainous rivers may carry a much higher proportion of bed load (up to 50% or more), which can sometimes be measured in lakes or artificial reservoirs.

Finally, it should be mentioned that data collected at gauging stations upstream of the river mouth do not necessarily record the actual input of sediments into the depositional basins. Some of the load may be deposited on its way to the river mouth or, conversely, more sediment may be taken up down-stream from the gauging station. A striking example of the first case is the Yellow River in China which deposits as much as 76 % of its suspended load before it reaches the open sea.

In spite of these various sources of error, the large number of data collected all over the world allow us to infer some general rules and to estimate the orders of magnitude for mechanical denudation in different regions:

- Small drainage basins in *mountainous areas* yield more sediment than larger river systems, which also usually drain regions of lower relief. Steep stream-channel gradients and steep valley slopes, only partly protected by vegetation, favor erosion and sediment transport by running water. In mountain ranges, annual mean precipitation and rainstorm events are often greater than in lowlands. Most of the solid river load, especially the bed load (sand and gravel) is transported only during *peak flows*. Single, major floods with a recurrence interval of tens of years may accomplish more sediment transport than a span of many normal years.

Fig. 9.6. Chemical and mechanical denudation rates of major river basins of the world. *TDS* total dissolved solids; *TSS* total suspended solids. (Mainly after Meybeck 1979,

1987; Milliman and Meade 1983; supplemented by Emeis 1985; Subramanian 1985; Iriondo 1988; and others)

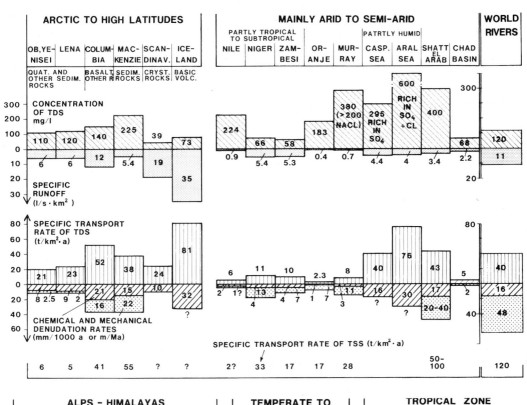

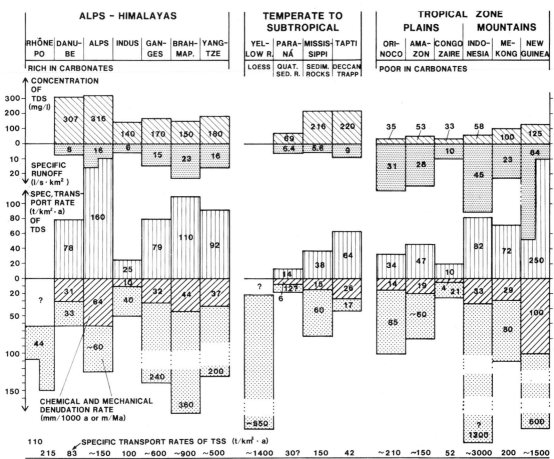

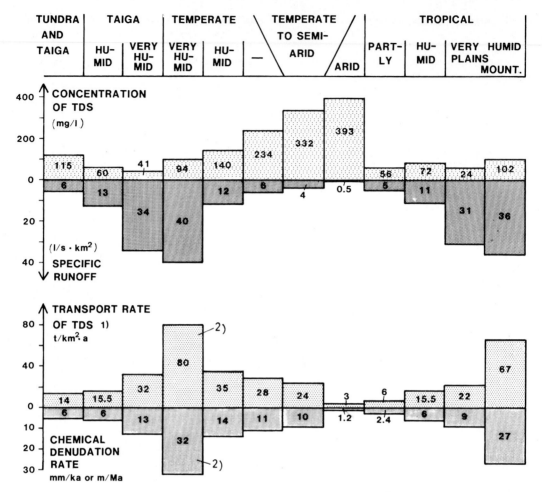

Fig. 9.7. Chemical denudation rates in different morpho-climatic zones derived from concentrations of total dissolved solids *(TDS)*, specific runoff, and specific transport rates of *TDS*. *1)* All measured rates are corrected for cyclic salts (e.g., NaCl) and atmospheric CO_2; rock density is assumed to be 2.5 t/m³. (Primarily after Meybeck 1979). *2)* Values for temperate, very humid zones are affected by the abundance of carbonate rocks and therefore appear too high as compared to the values of very humid tropic regions with predominantly crystalline rocks

- Dense, permanent *vegetation cover* prevents or strongly reduces soil erosion. Many studies in forested areas in the temperate humid zone have shown that even on rather steep slopes hardly any soil erosion can be observed, apart from sites where gravity mass movements occur. In these areas erosion is restricted mainly to the banks of stream channels. By contrast, if vegetation is sparse or completely absent, the soils and rocks are directly exposed to the agents of erosion. Consequently, not only linear erosion along the streams occurs, but also erosion on the surface of the slopes. Thus, even in regions of little rainfall, comparatively large amounts of soil and rock debris can be delivered to river systems or blown away by winds. Therefore, soil erosion and sediment yield from limited areas in arid to semi-arid zones are often greater than expected, especially if infrequent rain storms are taken into account.

Several authors have pointed out that sediment yield from some steppe and shrub desert regions with a mean annual rainfall of about 300 to 600 mm (precipitated seasonally or very irregularly) can reach greater values than those from temperate humid zones (mean annual precipitation up to 1000 mm and more). Generally, large river basins in arid to semi-arid regions appear to transport low amounts of suspended solids (see below and Fig. 9.6). As mentioned above, cultivation of land also exerts a strong influence on the denudation rate, which may be raised locally by a factor of 10 to 100, but regionally, for land of mixed use, normally by a factor of only 2 to 5. Because this text focuses on the sediment sources of ancient sedimentary basins, this problem is not discussed further here.

- If seasonal or annual rainfall surpasses a certain level characteristic of the temperate humid zone, the mechanical denudation rate can reach a second, yet higher peak than that mentioned above for semi-arid climates. The highest sediment yield of rivers is reported from *mountainous areas in the tropics*, where annual precipitation is on the order of 2000 to 3000 mm (e.g., Taiwan, New Guinea, Hawaii), but also regions with high seasonal monsoon precipitation fall into this category. In these cases, it is primarily the rapid weathering process (see Chap. 9.1) and the abundance of running water which enhance the denudation rates.

- Finally, denudation rates also are affected by the *rock types* exposed to weathering and erosion in the drainage area. Non-indurated and weakly indurated sediments are easily eroded, in contrast to granitic or metamorphic rocks. Quartzitic sandstones and cherts resist weathering far longer than carbonate-cemented or clayey silts and sands. Claystones and marlstones are rapidly disintegrated by mechanical processes if they are not protected by soil and vegetation cover. They may then form characteristic *badlands* incised by many rivulets and small streams which deliver relatively large amounts of fine-grained clastic sediments to larger river systems.

Outstanding examples of the influence of rock type on sediment yield are apparent in some large, present-day rivers of China. The Yellow River (Huanghe) delivers more sediment to its delta and the sea than the Yangtze River (Ying Wang et al. 1986), although its drainage area, and particularly its annual water discharge, are much smaller than those of the Yangtze River (Changjiang). This is due to the fact that the Yellow River enters a loess region in its lower reaches, where its course migrated southward and northward several times during its history. The high mean sediment yield of the Yellow River (1400 t/(km²·a) versus 250 t/(km²·a) of the Yangtze River) is, of course, also influenced by the more intensive cultivation of the land in the Yellow River drainage basin.

Mechanical Denudation Rates in Relation to Relief and Climate

Specific Transport Rates of Rivers

The following data (Figs. 9.6, 9.8, and 9.9) are based mainly on measurements of the suspended load in large rivers which usually drain regions of different relief, climate, and rock type and hence often cannot be associated with one particular morpho-climatic zone. Nevertheless, these measurements yield average values for the mechanical denudation of large regions which are of special interest here:

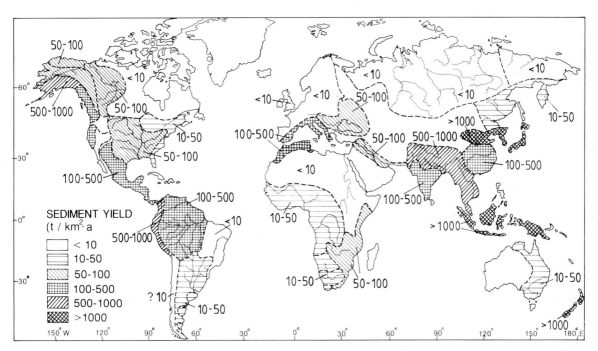

Fig. 9.8. Specific transport rates (in t/(km²·a)) of suspended solids from various drainage basins and larger regions of the world. (After Milliman and Meade 1983)

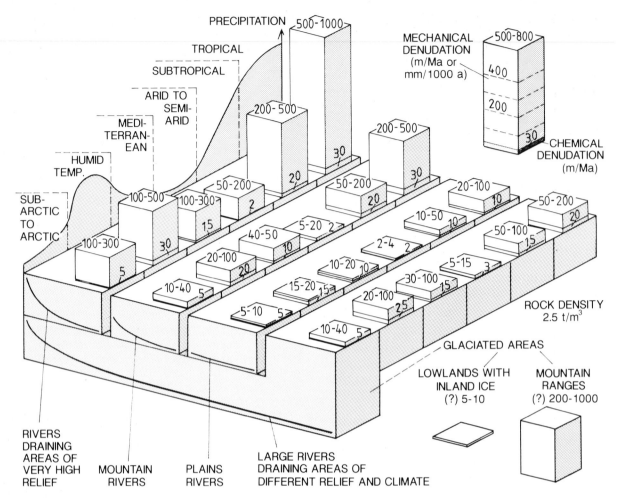

Fig. 9.9. Common range of mechanical and chemical denudation rates in relation to relief and climate (including vegetation) for moderate to large drainage areas. In small river basins, denudation may be largely controlled by rock types and therefore deviate considerably from the values shown in the graph. Specific transport rates (t/(km^2·a)) are converted into denudation rates (mm/ka or m/Ma) using an average rock density of 2.5 t/m^3. (Based on many sources, e.g., Selby 1974; Meybeck 1979, 1987; Milliman and Meade 1983; Chorley et al. 1984; Collins 1986; Hadley 1986; Tamrazyan 1989). See text for further explanation

- Rivers draining *arctic or high-latitude regions of mostly low relief,* such as the Siberian Ob, Yenisei, and Lena Rivers, have low specific transport rates for total suspended solids (TSS): approximately 5 t/(km^2·a). With TSS ≤10 t/(km^2·a), most of the rivers of the Canadian northern territories, as well as those of northwestern Europe, belong in this category (Fig. 9.8). If areas of high relief make up a large proportion of the drainage area, the sediment yield may increase to about 50 t/(km^2·a) (e.g., Mackenzie and Columbia Rivers).
- Similarly, in most *arid to semi-arid regions,* TSS is somewhat low. Apart from the Shatt el Arab, which drains nearby mountainous re-gions and reaches a sediment yield of 50 to 100 t/(km^2·a), all the other examples plotted in Fig. 9.6 range from only 2 to 33 t/(km^2·a). Of these, the highest values are from the Niger River and Murray River (Australia), which originate from humid tropical or temperate drainage basins, respectively. However, for limited, usually mountainous areas in semi-arid regions, very high TSS values (over 1000 t/(km^2·a)) are also found (e.g., Zachar 1982). In some of these cases, the high sediment yield may be due to human-induced soil erosion.
- In regions with *Mediterranean or subtropical climates,* the specific transport rate of sus-pended solids is usually moderate. Whereas the Paraná River and Tapti River (India) deliver

about 30 to 40 t/(km^2·a), the Mississippi River, which also drains areas of cooler climates, attains approximately 150 t/(km^2·a) (Fig. 9.6). Large areas in southern Europe, Middle America, India, and central China provide suspended solids within the range of 100 to 500 t/(km^2·a) (Fig. 9.8).

- *High mountain ranges* usually yield large amounts of suspended matter. The upper river Rhine draining central parts of the Swiss Alps provides about 1000 t/(km^2·a), which are deposited in glacial lakes along the northern rim of the Alps. The transport rates of the rivers Rhone and Danube measured farther downstream are much lower (110 and 150 t/(km^2·a), respectively) than that of the upper Rhine. The Alps in total are thought to attain 150 t/(km^2·a), a value which, by comparison with the Andes (500 to 1000 t/(km^2·a), Fig. 9.8), appears low. The large rivers draining the Himalayas have average sediment yields which differ considerably from one another. Whereas the Indus exhibits a comparatively low specific transport rate (100 t/(km^2·a)), the Ganges and Brahmaputra Rivers deliver 600 and 900 t/(km^2·a), respectively. The Yangtze River originating from the Tibetan Plateau also has a high sediment yield (500 t/(km^2 a)).

- *Lowland rivers of the tropical zone* have moderate TSS values of 100 to 500 t/(km^2·a) (Figs. 9.6, 9.8, and 9.9), although some may be less (e.g., the Congo River). However, mountainous regions in this climatic zone can provide a maximum TSS on the order of 1000 to 3000 t/(km^2·a) (e.g., New Guinea and Indonesia).

- On a *global scale*, the present-day, average river transport of suspended solids is 100 to 120 t/(km^2·a) (Milliman and Meade 1983; Tamrazyan 1989).

Mean Mechanical Denudation Rates

As previously mentioned (Chap. 9.2.2), TSS transport rates can be easily converted into average denudation rates (Figs. 9.6 and 9.9). These rates provide an indication of the mean thickness of soil or rock material removed from the total drainage area. The mean denudation rate, therefore, does not distinguish between (1) summit lowering, (2) slope denudation, and (3) rate of stream erosion in the river valleys or lowlands (Ahnert 1970; cf. Fig. 9.20b).

From the data discussed so far it is obvious that no simple law can be expressed for mechanical denudation, because the interrelationship between the various factors controlling this process is complex. Each region reacts in a somewhat different way, even if most of the variable factors are comparable. Therefore, an evaluation or prediction of mechanical denudation rates is more difficult than for chemical denudation rates, such as those shown in Figure 9.7 for different morpho-climatic zones.

Pinet and Souridu (1988) assume that the mechanical denudation rate (D_s) is largely correlated with the mean drainage basin elevation (H in m), but not with environmental factors. They give the following empirical formula:

$$D_s = 419 \times 10^{-6}\, H - 0.245 \quad (m/ka),$$

where 1 ka = 1000 years.

Using this formula, however, one obtains results which frequently differ from those depicted in Figures 9.6 and 9.9 (also see Leeder 1991). Some workers have also pointed out that the denudation rate is controlled by the slope gradient rather than by the absolute elevation of a region. An elevated plateau which is little dissected by deep valleys may display much less denudation than a coastal region with steep slopes, as in South Africa (Summerfield 1991).

In spite of the various problems mentioned above, some of the most important facts can be summarized in a simple, generalized diagram (Fig. 9.9). This diagram depicts the effects of relief and climate, and therefore, to some extent, the influence of soils and vegetation on denudation rate, although rock type is neglected. If we consider large drainage areas, which usually include many different rocks and soils, this omission may be permissible. If, however, only small drainage basins are evaluated, their rock types may strongly influence sediment yield, particularly in mountainous areas with limited soil cover and vegetation, as in semi-arid regions. In such areas, the denudation rate can deviate substantially from the rates shown in Fig. 9.9. Therefore, this simplified model should be used only to estimate average denudation rates for relatively large regions with mixed lithologic units.

Regions of Extremely Low Denudation

From Fig. 9.9, one might draw the conclusion that considerable denudation is taking place everywhere on the land surface, even in arid lowlands. This, however, is not quite true. As Fairbridge and Finkl (1980) have pointed out, some of the great Precambrian shields which were not affected by the Pleistocene glaciations have shown little evidence of denudation for very long time periods. These *tectonically stable areas* were leveled many hundred million years ago and since then have kept an elevation slightly above or below the base level of erosion. As a result, part of their almost planar erosion surface (peneplain) was repeatedly buried by a thin veneer of sediment and exhumed by subsequent erosion.

This is documented by the presence of relic sediments of different ages at the edges of these cratonic regions. In Africa and particularly in western Australia, such multiple "stacked" veneers are superimposed and subparallel not only to each other, but also to the original cratonic surface. They are often separated from the basement and from one another by paleosols and duricrusts (e.g., silcrete or lateritic crusts). In more central parts of the peneplain, which remained uncovered or were rarely buried by sediments, very deep, old weathering profiles may be preserved (also see, e.g., Firman 1988).

In such "cratonic regimes", the long-term net surface lowering can be extremely slow (in southwestern Australia 0.1 to 0.2 m/Ma, in the Chad Basin 0.5 to 2 m/Ma, see Chap. 9.4 and Fig. 9.11). However, in such cases there is no permanent denudation process, but alternating, shorter-term phases of accumulation and erosion.

Another important point is that substantial unloading of the crust by denudation commonly causes *isostatic rebound*. The amount of this uplift, however, is less than surface lowering by denudation. Nevertheless, this process enables denudation to continue for long time periods, even when thermotectonically generated uplift is excluded (cf. Chap. 9.5).

9.2.3 Chemical Versus Mechanical Denudation Rates

Global Denudation Rates

On a global scale, according to Milliman and Meade (1983) about 13.5×10^9 t/a (100 to 120 $t/(km^2 \cdot a)$) of suspended solids (TSS) are transported from the continents into the oceans. Likewise, according to Meybeck (1979) approximately $3.9 \cdot 10^9$ t/a (about 30 $t/(km^2 \cdot a)$) of dissolved materials (TDS) are transported. Thus the TSS/TDS ratio is approximately 3.5 (also see Fig. 9.6), and it is evident that, globally, the transport of particulate matter clearly exceeds solute transport.

Regional Variation and Trends

On a regional scale or for individual drainage basins, the situation may be quite different. In the river basins discussed earlier, the TSS/TDS ratios vary between 0.14 (Lena, Siberia) and 11 (Ganges, Brahmaputra), and, if smaller catchments are included, the deviation from the global mean is even greater. As Walling and Webb (1983) have pointed out, there is no clear relationship between the particulate and dissolved loads of rivers, which may be partly due to the significantly differing transport mechanisms for the two load components.

Most of the particulate load is tranported by peak flows during just a few days of the year, while the dissolved load moves downriver more or less continuously. TDS varies seasonally and the concentrations of the dissolved species normally decrease during storm events. Furthermore, the specific transport rate of suspended solids often declines with increasing basin area, while the transport rate of dissolved solids generally is not thus affected.

If the data plotted in Figs. 9.6 and 9.9 are considered together with information summarized by Walling and Webb (1986), the following trends can be identified:

- In *mountainous regions*, due to incomplete soil cover and high precipitation, usually both load components increase as relief increases, but the TSS transport rate increases much faster than the TDS rate. In the Alps the TSS/TDS ratio is about 1 to 2, in the Himalayas 4 to 8. In tropical mountain ranges, the TSS/TDS ratio reaches very large values of 10 to 30 (it should be noted that several of the areas studied are relatively small).
- In *tropical areas with low or mixed relief*, the TSS/TDS ratios vary considerably, between about 1 (Congo) and 3 to 5 (e.g., Amazon, Orinoco).
- In *large river basins in arid to semi-arid regions*, the TSS/TDS ratio appears to range between 2 and 4 (apart from the Nile River with 0.5). But as pointed out earlier, much higher (and lower) ratios may occur locally, which can exceed even those of high mountain ranges in other climatic zones.
- In *areas with temperate to subtropical climates and mixed relief* the TSS/TDS ratio varies between about 0.5 and 4 (Fig. 9.6). Most of the rivers in northwestern and central Europe (apart from those originating in the Alps) have ratios below 1, i.e., chemical denudation is more effective than mechanical denudation. With increasing annual runoff (and no change

in relief) the TDS transport rate is also usually enhanced, while the TSS load remains approximately constant, or even declines due to increasing soil protection by dense vegetation cover.
- The lowest reported TSS/TDS ratios are from the *arctic to high-latitude rivers* of Siberia which drain mainly lowlands. With growing relief, this ratio also increases and exceeds 1 (e.g., Mackenzie River, Fig. 9.6). Reliable data on the transport rates of solids from glaciated areas are sparse and greatly variable.

Very high mechanical denudation rates occur in areas with *wet-based, rapidly moving glaciers*, as in Alaska.

For the Glacier Bay area, an average denudation rate of 12 m/ka has been suggested (Powell and Molnia 1989). Pleistocene glacial denudation in central parts of Spitsbergen may have removed 3 km of Tertiary sediments (Svendsen et al. 1989). In the sea southwest of Alaska and on the Barents shelf (Vorren et al. 1988), Pleistocene diamicton accumulated at rates between 0.1 and 1 m/ka over large areas, and locally even faster. These data testify to the fact that denudation rates in the source areas of these accumulations, i.e., at the base of the eroding glaciers, must have reached about the same order of magnitude. In the Antarctic, however, glacial denudation operates at a much slower rate (Alley et al. 1988; Henrich 1990).

9.3 Mineralogical Composition of Suspended River Loads

General Aspects and Problems

Fine-grained sediments, comprising the *clay and silt grain size fractions*, are the most abundant group of all sedimentary rocks. To describe these rocks, various names are used, such as claystones, argillites, clayey siltstones, mudstones, marlstones, and shales, but slates and many schists also originate from this group. Their material is delivered mainly by the suspended load of rivers flooding alluvial plains, entering lakes, or reaching the sea. In addition, other fine-grained sediments, such as many biogenic deposits (hemipelagic and pelagic marls and marly limestones, siliceous oozes, etc.) also often contain considerable proportions of allochthonous, terrigenous material. Accordingly, the mineralogical (and chemical) composition of suspended river loads is of great interest to sedimentologists.

In Figure 9.10, the results are summarized of some recent studies of the solid loads of major world rivers.

It should be mentioned that these data display considerable scatter, when results gained by different authors on the same river system are compared. These discrepancies probably arise due to the investigation of samples of differing grain size distribution. Samples taken during peak floods contain a higher proportion of silt than those collected during normal or low runoff. If, for example, only the fraction containing particles ≤ 2 µm is analyzed, it is unlikely that much quartz or other non-altered rock minerals would be obtained, but, rather clay minerals would predominate (e.g., in Indian rivers draining the flood basalts of the Deccan Traps, Fig. 9.10). Furthermore, the different methods employed to analyze these fine-grained materials (which may be partially X-ray amorphous) add to the uncertainties regarding quantification of their mineralogical composition.

Results and Interpretation

As for the results, the following points are of interest:

- Small river basins have a more uniform mineralogical suspended solid composition, reflecting the influence of parent rocks and soils more than large drainage areas with mixed conditions.

A prominent example of limited and rather uniform river basins are the Deccan Plateau rivers in India, which are dominated by montmorillonitic clay due to the basaltic parent rocks and their alteration products in the subtropical climate. A second example of this type is the Caroni River in South America, which drains crystalline rocks in tropical conditions. Here illite, montmorillonite, and kaolinite are the most abundant clay minerals. All the other river basins listed in Fig. 9.10 represent large areas of mixed rock types, relief and climate, which is reflected in the mineralogical composition of their suspended load.

- The suspended solid flux consists essentially of a limited number of minerals which can be subdivided into two groups:

1. Primary detrital minerals inherited from the parent rocks.
2. Secondary minerals (neo-minerals) generated by chemical-mineralogical weathering in the soil zone in the drainage area (cf. Chap. 9.1).
3. An additional group of fine-grained minerals includes authigenic and diagenetic clay minerals, which form in the depositional environment or within the sediment (e.g., Eslinger and Pevear 1988). Generally, this last

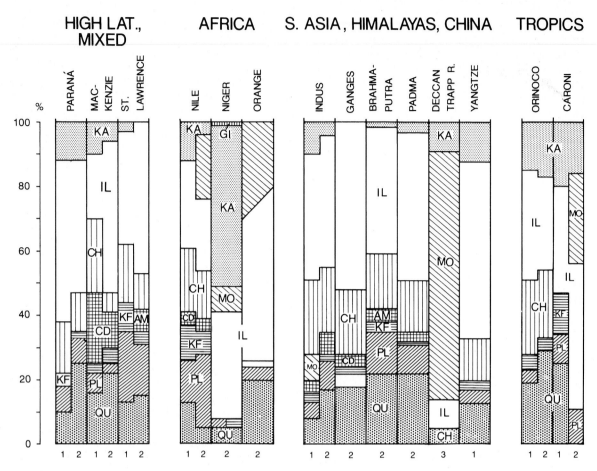

Fig. 9.10. Mineralogical composition of suspended matter of major world rivers. (Based on *1* Emeis 1985; *2* Konta 1985 and 1988; *3* Naidu et al. 1985). Differing results for the same river system may be caused by differences in grain size distribution of the investigated samples as well as by varying methods of analysis.

QU quartz; *PL* plagioclase; *KF* K-feldspar; *CD* calcite and dolomite; *AM* amphibole; *MO* montmorillonite, smectite; *IL* illite/muscovite; *CH* chlorite; *KA* kaolinite; *GI* gibbsite. Generally, the proportions of these minerals are not well established. See text for further explanation

group is of minor importance for the mineralogical composition of clayey sediments, and it is not discussed further here.

Limited Chemical Weathering

When physical weathering processes predominate and chemical weathering is limited, the minerals of group 1, i.e., both silt-size particles, such as *quartz, plagioclase, K-feldspar, calcite and dolomite,* and smaller particles, such as *chlorite and other parent rock-derived clay minerals,* contribute substantially to the river load (e.g., Konta 1985, 1988). The presence of carbonate minerals in the river load is a particularly sensitive indicator of the absence of deeply weathered soils in parts of the

drainage area. This situation is common in regions of high relief or under arid conditions (e.g., the rivers Nile, Mackenzie, the Himalayan Rivers, and the river Rhine entering Lake Constance).

The preservation and river transport of *feldspars,* especially the less stable plagioclase, are also strong indicators of limited chemical weathering conditions. Small proportions of plagioclase occur in nearly all the river loads shown in Fig. 9.10, but the Nile River (draining arid regions) and St. Lawrence River (from the Canadian shield) contain relatively large quantities. As already mentioned in Chapter 9.2, some of the clay minerals may also be derived directly from parent rocks, if claystones, shales, etc., are exposed to erosion, as for example in badlands.

Illite is a common weathering product from acidic crystalline rocks in humid temperate climates, but it is also abundant in many argillaceous bedrocks. Thus, sheet-silicates, including muscovite and sometimes mixed-layer minerals, may also originate directly from parent rocks. Illite predominates in the suspended load of the Indus and Ganges Rivers, Yangtze and Yellow Rivers, Paraná River, and in several major rivers of Europe (Fig. 9.10; Konta 1988).

Chlorite is a major constituent of slates and greenschists; therefore the presence of this mineral in considerable quantities, as observed in the suspended load of many large rivers (Fig. 9.10), results from the mechanical erosion of pre-existing minerals. However, chlorite is destroyed in warm, humid climates, because it is very sensitive to chemical weathering. Hence, the presence of detrital chlorite indicates a cold or possibly arid climate. On the other hand, some secondary chlorite may be formed under special conditions (acidic soils, humid temperate climates, basic parent rocks). In addition, it can form diagenetically at relatively low temperatures.

Limited Mechanical Denudation

If transport-limited conditions prevail in large parts of the drainage area, then *soil-formed neo-minerals* tend to predominate in the suspended river load. Under these conditions, thick soil profiles can develop, and fluvial erosion processes therefore remove mainly soils and little parent rock material. The neomineral group includes minerals of the *montmorillonite/smectite type* (mostly basic parent rocks, temperate to subtropical climates), the sheet-silicates of the *illite/muscovite type* (humid temperate climate), *kaolinite, gibbsite* (and the iron oxide *goethite;* semi-arid to tropical climates).

Montmorillonite is an important constituent in the Caroni and some other tributaries of the Amazon River, in the Orange River (SW Africa), and in the Deccan Plateau rivers draining basaltic rocks and their weathering products. It is also common in the Mississippi River near its mouth, where the ≤10 μm size fraction contains 40 % smectite or montmorillonite, 25 % illite, and 20 % kaolinite (Potter et al. 1975). Kaolinite and gibbsite are abundant in the soils of wet tropical climates, where the less stable chlorite disappears. Kaolinite is a principal component in the river loads of the Orinoco and Caroni, Niger, Congo, the Deccan rivers, and in some rivers of New Guinea; gibbsite was reported only from the Niger River.

There are more soil-formed clay minerals than those mentioned here, but a more detailed treatment of this topic is beyond the scope of this text (see e.g. Birkeland 1984; Eslinger and Pevear 1988; Chamley 1989; Weaver 1989). In any case, one should be cautious in deducing the paleoclimate of a specific region solely from the clay mineral composition of sediments derived from the area.

Chemical-mineralogical weathering is characterized by different *partitioning of elements* between the dissolved and the suspended phases. Generally, the solids tend to become enriched in aluminum, iron, and (in regions of humid, temperate climate) in silicon, while sodium, calcium, and (to a lesser extent) potassium and magnesium are carried away in solution (e.g., Stallard 1985). If the cation-poor minerals (kaolinite, gibbsite, goethite) are regarded as the end products of intensive weathering, and the chemical composition of the suspended load is identified, then a kind of *chemical-mineralogical maturity* of the soil-derived solids can be defined (Emeis 1985, Konta 1985 and 1988). Relatively high concentrations of Al^{3+} (and ferric iron), or relatively low SiO_2 contents and low concentrations of Na^+ and Ca^{++} (and Mg^{++}, K^+) indicate an advanced maturity (tropical rivers). Low maturity, expressed by a low Al/Si ratio, is typical of arid drainage basins. With growing maturity, the Al/Si ratio also increases.

Detrital Clay Minerals in Marine Sediments

Fine-grained marine sediments on *prodelta slopes* and in nearby regions usually exhibit clay mineral suites similar to those of the entering rivers.

The clay mineral composition of the sediments in the Gulf of Mexico is about the same as that of the Mississippi River at its mouth (Potter et al. 1975; cf. Chap. 11.3). The sediments off the Niger delta (cf. Chaps. 3.4.2 and 14.4) are rich in kaolinite delivered by the Niger River (Fig. 9.10). However, the kaolinite content in the sediment decreases seaward due to the specific transport and settling behavior of the various clay minerals (Chap. 3.4.2). The eastern Mediterranean accumulates sediments relatively rich in montmorillonite, which for the most part is supplied by the Nile River. A large portion of the suspended load in the Amazon River (predominantly montmorillonite and kaolinite) is transported to the northwest and deposited in the coastal region as mudflats (Weaver 1989).

The clay mineral distribution in *modern ocean sediments* reflects to some extent the global climatic zones. The largest amounts of kaolinite are found in low latitudes, where rivers from humid tropical regions enter the sea. The kaolinite content tends to decrease towards the poles and from coastal waters to the deep sea. Detrital chlorite is most abundant in the polar and subpolar regions, where chemical weathering is limited. Illite and smectite (montmorillonite) are the most important clay minerals in modern ocean sediments. Illite-rich sediments are found in temperate to cold climatic zones, whereas smectite is abundant in regions of extensive volcanic activity both below the sea and on land.

Authigenic clay minerals usually comprise only a minor proportion of the total composition of marine sediments. These and diagenetically formed clay minerals are not discussed in this text (see e.g. Weaver 1989).

9.4 Long-Term Denududation Rates from the Sediment Budget of Various Basins

Introduction

Because evaluating mechanical and chemical denudation rates from present-day river loads is problematic, geologists have repeatedly tried to derive long-term average denudation rates from the volume of sediment accumuluated in sedimentary basins during a certain time period. In these cases, the extent of the former drainage and depositional areas must be known, and the time period of the sediment slice or wedge (e.g., in a shelf sea, a submarine fan, or in front of a lake delta) should be sufficiently accurate. The general problems encountered in such a study are demonstrated in Fig. 9.11. This method is therefore more appropriate for the investigation of closed or semi-closed basins of young geologic age (Quaternary and Tertiary), where uncertainties about the drainage area can be minimized.

If we are dealing with *closed basins,* the dissolved river load should also be found, as chemical precipitate, in the skeletons and organic matter produced by organisms, or in the pore fluid of the deposited sediments. In *semi-closed basins,* part of the dissolved river load may be trapped by organisms and incorporated into the ultimate sink of the eroded material, but most is probably lost to regions beyond the depositional wedge under investigation.

Consequently, the "sediment budget method" should be used with caution for cases in which chemical denudation exceeded mechanical erosion. In addition, there are cases in which the exchange of water masses with larger basins exerted a strong influence on the character and volume of the investigated sediments (e.g., in the Gulf of California, see below).

Example of a Closed Basin (Chad Basin)

The Chad Basin in the central part of North Africa, described in detail by Gac (1980), is a completely closed basin with an internal drainage system (Fig. 9.12). River water, supplied mainly by a subtropical region to the southeast, flows into a large, shallow playa lake, Lake Chad, from which it evaporates. The total drainage area of the lake comprises an area of approximately 2×10^6 km^2, while the area of the lake itself varies in response to dry and wet years. Under average, present-day conditions it covers an area of 2.1×10^4 km^2. Six thousand years ago, when the climate in North Africa was more humid, the lake was much larger and had an outlet to the Niger River to the southwest. Today the climatic conditions range from hot and arid in the north (Sahara Desert) to subtropical in the southeast (Fig. 9.12 b). Evaporation from the lake is 2150 mm/a.

Mechanical denudation by rivers originating largely in mountainous regions to the southeast delivers an average of 10.5 t/(km^2·a) for the total drainge area; chemical denudation amounts to 3.6 t/(km^2·a). Thus, the present-day total *average denudation rate* is 5.6 mm/ka. This low value is a result of the low relief of large parts of the drainage area as well as of the thick, extensive soil cover on top of crystalline parent rocks (Fig. 9.12 c).

As climate changes, the intensity of mechanical erosion and chemical alteration of the parent rocks may vary significantly. When arid conditions in the region are more intense, and part of the soil cover is lost, denudation progresses more quickly than today, whereas more humid conditions only exert a minor influence on weathering.

If both mechanical and chemical river loads would be evenly spread over the present-day lake area, a solid sediment layer of about 550 mm/ka would result. Assuming a porosity of 60 %, the sedimentation rate of the uncompacted material in the lake should amount to about 1400 mm/ka. The actual thickness of wet Holocene sediments (10 ka) consisting of

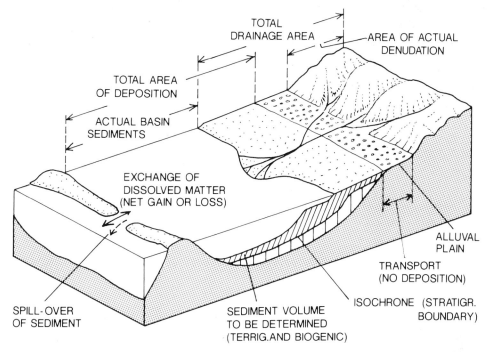

Fig. 9.11. Scheme demonstrating the general problems involved in the assessment of the denudation-basin fill budget. See text for further explanation

terrigenous silty clay, carbonate and biogenic silica ranges from 4 to 6 m and thus corresponds to an average uncompacted sedimentation rate of about 500 mm/ka or a sedimentation rate of compacted, solid material of 200 mm/ka. The latter value indicates an average denudation rate of only 2 mm/ka.

The discrepancy between the calculated and observed sediment thicknesses may be due to the fact that 6000 years ago the lake was larger, and therefore the sedimentation rate was lower than today. Furthermore, it should be mentioned that most of the dissolved and detrital material is derived from only about 30 % of the drainage area (in the southeast). Hence, the denudation rate there may be three times higher than that calculated for the area taken as a whole.

There are more recent examples of this type of closed basin with a salt water lake or playa lake in its center; well known are the Great Salt Lake in North America and Lake Eyre in Australia. A systematic study of the sediment volume accumulating in such lakes during a certain time span is not yet available, however.

Examples of Semi-Closed Basins (Black Sea, Gulf of California)

Adjacent seas can also be described as semi-closed basins, because they have an opening to the world oceans (Chap. 4). Some of these basins have been studied to determine the denudation rates in their drainage areas. Two examples include the Black Sea, with shallow, narrow opening to the Mediterranean Sea (also a semi-closed basin) and the Gulf of California, with a deep, wide connection to the Pacific. Whereas the Black Sea appears to be a relatively simple, straightforward case study, the Gulf of California incorporates a number of complications characteristic of the initial stage of a young, narrow ocean basin.

The Black Sea

The modern Black Sea is a gigantic catch basin for the river discharge of half of Europe and part of Asia (Fig. 9.13), where all incoming sediments are trapped. Degens et al. (1978) have calculated denudation rates in the drainage area from river loads and have compared these values with the sedimentation rates in the Black Sea (0.45×10^6 km^2). The total drai-

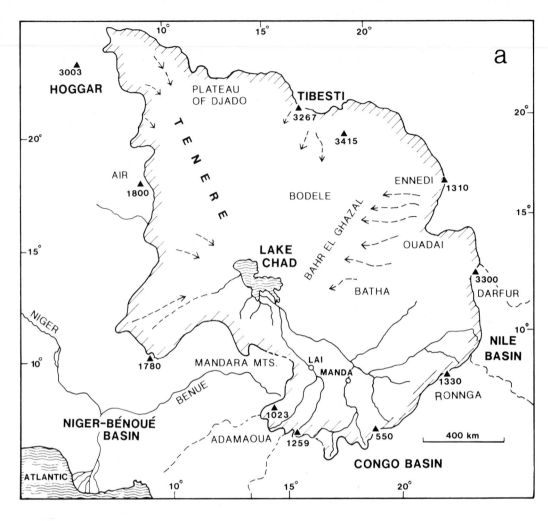

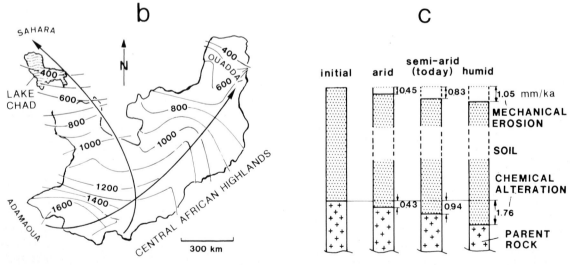

Fig. 9.12. a Closed internal drainage basin of Lake Chad, central North Africa. Note that the lake is fed primarily by rivers from the subtropical mountainous region in the southeast (**b**), while the Sahara desert in the north contributes hardly any water. **b** Annual precipitation (in mm/a) in the southeastern part of the Lake Chad drainage area. *Arrows* signify increasing aridity. **c** Progress of mechanical and chemical denudation (in mm/ka) under different climatic conditions. (After Gac 1980)

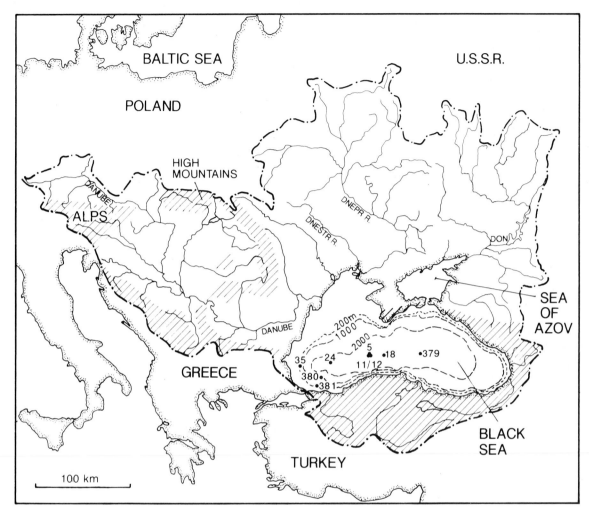

Fig. 9.13. Drainage area of the Black Sea and location of sediment cores and DSDP drilling sites (379 to 381) in the deep basin. (After Degens et al. 1978)

nage of the Black Sea comprises an area of 1.98×10^6 km²; this value does not include some limited regions where fluvial deposition takes place. The somewhat revised results of this study can be summarized as follows:

The total *suspended and dissolved river load* transported into the Black Sea amounts to 237 x 10⁶ t/a. The suspended load is slightly greater than the dissolved load. The Danube River delivers about 60 % of the total river load. The Don and Kuban Rivers discharge into the Sea of Asov, where most of their material is deposited. The Caucasian rivers from the east contribute about 20 % of the total detritus. As determined from their chemical and solid loads, the present-day denudation rates in the principal source areas vary from approximately 10 mm/ka (drainage area of the Dnepr Ri-

ver) to 80 mm/ka (Danube River). As above, an average rock density of 2.5 t/m³ is assumed for these calculations. The average denudation rate for the total drainage system is 50 mm/ka (or 120 t/(km²· a)), but of course neither the values of the individual river basins, nor the average value for the total area, represent the real denudation rates. In actual fact, about 30 % of the total drainage area is made up of mountain ranges, whereas the remainder is fairly flat. If we assume an average value of only 20 mm/ka for the lowlands, the denudation rate of the mountainous regions increases to 120 mm/ka, which seems realistic.

The estimation of the *sediment volume* in the Black Sea is based on data available for the deep part of the basin (Fig. 9.13), which comprises about two thirds of the total basinal area. In the past 5000 years, the sedimentation rate of uncompacted sediment (including coccolith ooze, sapropel, and intercalated turbidites), was approximately 1000 mm/ka. If we take into account an average porosity of 70 %, the sedimentation rate of compacted, solid material drops to 300 mm/ka. Normalized based on the drainage area, which is 4.4 times larger than the basin, this value yields an average denudation rate of about 70 mm/ka.

This result agrees fairly well with the rate found from the suspended and dissolved river loads. It is assumed that most of the dissolved constituents delivered by rivers is used up by organisms and deposited in the basin as biogenic carbonate and opaline silica. There is, however, also some exchange of dissolved material via the narrow passage of the Bosporus connecting the Black Sea with the Mediterranean, but outflow from the Black Sea is greater than inflow (Chap. 4.3). Due to this water exchange, the sedimentary environment of the present-day Black Sea is not far from normal marine conditions. Its sediment volume is probably not substantially affected by the inflow of sea water.

The *denudation rates* obtained in this study are characteristic of the steppe vegetation present in large parts of the drainage area. During times of more extensive forest growth, denudation was probably reduced by more than half. During such times, sapropels were preferentially deposited in the Black Sea basin. Similarly, in the course of Pleistocene glacial melting and loess mobilization in the hinterland, denudation and sedimentation rates increased considerably and may have reached values up to ten times greater than those found for the Holocene.

The Gulf of California

The Gulf of California is a young oceanic basin formed by strike-slip movements and therefore shows a rather complicated morphology with a number of small deep basins separated by higher areas (Fig. 9.14a, also see Fig. 4.5). Prior to the building of dams, the northern gulf was fed by the Colorado River draining a large hinterland. In the south the gulf has a wide, deep opening to the Pacific Ocean. To assess a hemipelagic sediment budget, the influence of the Colorado as a source of terrigeneous material was excluded by choosing some gulf islands as the northern boundary of the depositional study area (Fig. 9.14a), as these islands act as a sort of dam

which traps the Colorado sediments. In the south and southwest the boundary line of the study area separates a gulf region still influenced by hemipelagic sediments from the deeper ocean floor, where pelagic deposits (with a sedimentation rate of less than 20 mm/ka) prevail. The sediment volume of this part of the gulf has been determined for the last 4 Ma (Einsele and Niemitz 1982), i.e., for the so-called post-rift sediments which were deposited after the initiation of ocean spreading (Curray and Moore et al. 1982a).

Prior to this development a "protogulf" existed, in which *pre-rift sediments* of considerable thickness accumulated, particularly below the eastern shelf of the present gulf (Fig. 9.15a). The older *protogulf sediments* rest on subsiding continental crust, while the younger, post-rift sediments were deposited either on new oceanic crust in the deep main part of the gulf, or on top of protogulf sediments.

Because the pre-rift protogulf sediments are often somewhat faulted and tilted, the boundary between these and the post-rift sediments can be detected in seismic profiles. Thus, it is possible to determine the thickness and volume of sediments which accumulated during the last 4 Ma in the entire study area. Information on the nature of these sediments has also been gained from a number of deep holes drilled during Leg 64 of the Deep Sea Drilling Project. Based on these results, an isopach map for the post-rift sediments could be drawn, which shows extremely varying sediment thicknesses (Einsele and Niemitz 1982). In order to calculate the sediment volume of the depositional area, the different sediment thicknesses were subdivided into slabs of equal thickness and porosity. Wet, uncompacted sediment volume was then converted into volume of compacted, solid (dry) material using the porosity/depth relationship found in the boreholes (Fig. 9.15c and d).

In this way, the average sedimentation rate for dry solid material could be determined for the young gulf (25 m/Ma) as well as for its eastern shelf (58 m/Ma), where the young sediments lie on protogulf deposits. When the areal extents of both depositional areas are taken into account, the *average sedimentation rate* becomes 36 m/Ma (Fig. 9.15). Because 11 of the 36 m/Ma are made up of biogenic silica, carbonate, and organic matter, only 25 m/Ma originate from the influx of terrigenous material from the neighboring land masses. If the dissolved river load is excluded, and the sedimentation rate, drainage area (2.71×10^5 km$_2$), and the studied gulf area (1.92×10^5 km^2) are applied, an average denudation rate (TSS) of about 20 m/Ma for the last 4 Ma can be determined. This is, however, a minimum

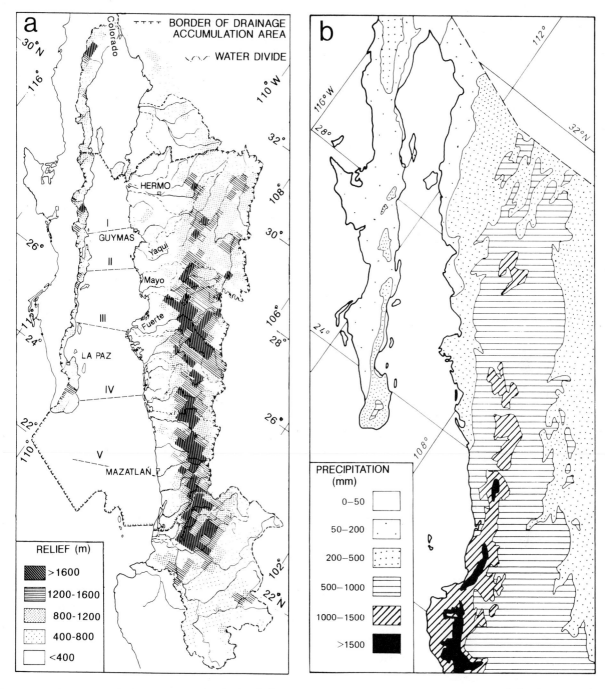

Fig. 9.14. a Relief of 20 x 20 km squares in the drainage area of the central and southern parts of the Gulf of California, as well as limits of the studied depositional area and special sections *(I - V)* across the gulf. **b** Mean annual precipitation in the drainage area of the gulf. (Einsele and Niemitz 1982)

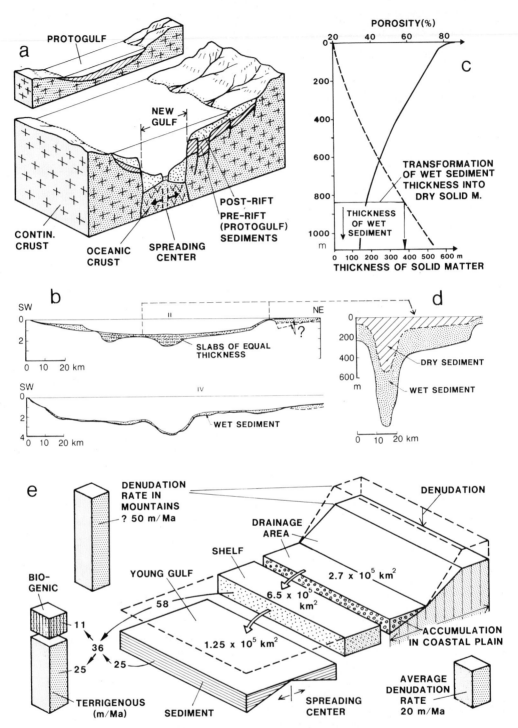

Fig. 9.15. a Simplified model for the development of an initial, Gulf of California-type ocean from (1) a shallow protogulf on continental crust to (2) a narrow, deep oceanic basin by rifting and accretion of new oceanic crust. Note the boundary (partly represented by an unconformity) between pre-rift and post-rift sediments. **b** Sections across the gulf (*II* and *IV* in Fig. 9.14a) with thicknesses of wet (uncompacted, porous) and dry (compacted, solid) sediment (shown only for part of the section); vertical exaggeration x 5. **c** Porosity/depth relationship representative of major parts of the Gulf of California and conversion from wet to solid sediment thickness: results are shown in **d**, vertical exaggeration x 50. **e** Average sedimentation rates in the studied gulf area and derived mean denudation rate in the corresponding drainage area. Formerly high sediment input from the Colorado river at northern end of gulf is excluded from this study. (Einsele and Niemitz 1982)

value, because some of the eroded material was deposited in alluvial plains before it reached the gulf.

Compared to drainage areas similar to that of the Gulf of California, the denudation rate, even if it were twice as high as the calculated value, appears to be rather low. The drainage area of the gulf lies in the transition zone between the hot dry climate to the northwest and the semiarid and increasingly wet climate of the lowlands and mountain ranges of mainland Mexico in the southeast (Fig. 9.14b). Here, some rivers drain regions of high relief, consisting largely of Cenozoic volcanics. It can be assumed that the size of the drainage area has not changed significantly during the last 4 Ma, since the principal physiographic features of this region, including the mountain range of the Sierra Madre Occidental, were formed earlier.

Average sedimentation rates calculated for *short time intervals* in the post-rift gulf have not been constant during the last 4 Ma, because its areal extent increased as a result of ocean spreading (extensional basin model, Chap. 11.3).

The Himalayas-Bengal Deep-Sea Fan Denudation-Accumulation System

Introduction and Previous Studies

In the present-day oceans, detritus of large rivers forms deep-sea sediment fans of various sizes and thicknesses (e.g., Barnes and Normark 1985; cf. Chap. 5.4.2). Some of these fans have been thoroughly studied using several methods, including seismic investigations and deep ocean drilling. If the volume and age of such a huge sediment wedge are known, as for the Mississippi Fan in the Gulf of Mexico (Chaps. 5.4.2 and 11.3), and the Indus Bengal Fans in the Indian Ocean, the mean denudation rate in the source area can be calculated. In addition, alluvial deposits on land and the outbuilding of river deltas have to be taken into account. This procedure is briefly demonstrated for the most prominent present-day example: the rivers Ganges and Brahmaputra, which drain a major part of the Himalayas and deposit most of their solid load in the Bengal Fan, a submarine sediment wedge approximately 3000 km long and, at its maximum, 1000 km wide (Fig. 9.16).

Curray and Moore (1971) made a first attempt to compare the volume of this sediment wedge with the sediment load of the two combined rivers (Table 9.1a and c). Because age control for both the base of the fan and an unconformity within the fan was poor, they calculated the age of the fan from the present-day load of the two rivers (Table 9.1c). Hence, their denudation rate for the Himalayas (620 or 720 mm/ka) is in fact based on the "river load method" and not on the sediment volume of the fan. Furthermore, it should be mentioned that their average denudation rate is determined for an area including lowlands, which is much greater than that occupied by the high mountain ranges.

New Mass Balance

In a new estimate, some of the results of Leg 116 of the Ocean Drilling Program were employed (The Leg 116 Shipboard Scientific Party 1987; Kassens and Wetzel 1989; Stow and Cochran 1989; Cochran 1990). It was found that the uplift of the Himalayas and the deposition of the deep-sea fan had already begun about 20 Ma ago, and the age of the unconformity within the fan sediments (caused by deformation of the underlying oceanic crust) is 7.5 to 8.0 Ma (in Table 9.1 an age of 7.7 Ma is used).

In addition to the huge sediment volume of the fan, large quantities of river detritus were deposited in the sub-Himalayan foredeep, an east-west trending molasse basin along the southern foot of the Himalayas, as well as in the Bengal Basin of Bangladesh (Alam 1989). The foredeep is essentially filled with coarse to fine fluvial sediments of mainly Middle Miocene to Holocene age. These sediments are derived directly from the Himalayan uplift and reach thicknesses of up to 10 km.

The Miocene to Quaternary sediments of the Bengal Basin underneath the modern deltaic flood plain of the Ganges River consist of proto-Bengal Fan deposits overlain by a seaward prograding delta complex. These two units form a wedge-like sedimentary body with thicknesses of a few kilometers or less in the northwest and up to more than 10 km (maximum of 15 km) near the present river mouth. The exact sediment volume deposited during the last 20 Ma in the sub-Himalayan foredeep, including the Bengal Basin is not known. An estimation based on available data on the sediment thicknesses and extents of these basins yields a volume of around 3×10^6 km^3.

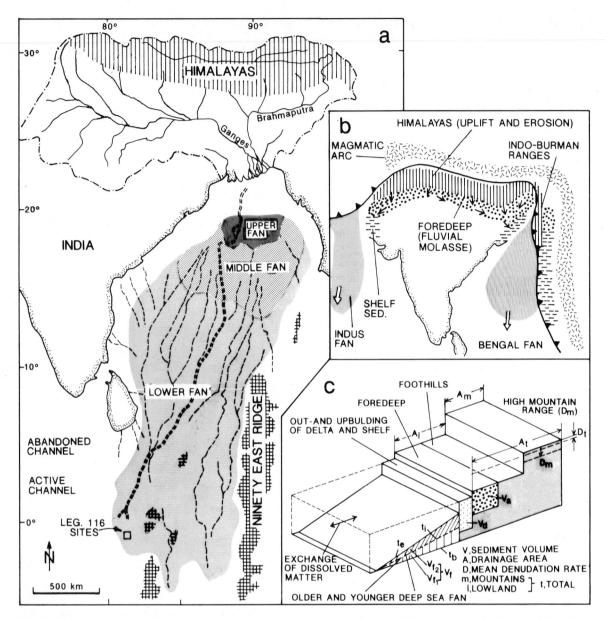

Fig. 9.16. a Drainage area of rivers Ganges and Brah-maputra in the Himalayas, their alluvial plain and delta, and the submarine Bengal Fan. (After Curray and Moore 1971; Alam 1989; Stow and Cochran 1989).

b Model for the different areas and volumes of denudation and deposition of the Ganges/Brahmaputra-Bengal Fan system (see Table 9.1 for results of calculations)

Average Denudation Rate in the Himalayas

Using the above mentioned value and the data listed in Table 9.1b, an average denudation rate of 950 mm/ka (950 m/Ma) can be determined for the high mountain ranges of the Himalayas. In this case, uniform erosion during the last 20 Ma is assumed.

The minor sediment contributions of mountainous regions in the southern part of the drainage area of the Ganges

River are neglected. Likewise, the sediment supply of rivers entering the Gulf of Bengal from the west are not taken into account. If the sediment volume of the Bengal Fan (as well as that of the plains) is subdivided into post-7.7 Ma (sediment volume V_{f2}) and pre-7.7 Ma (V_{f1}) portions, one obtains denudation rates for the Himalayas of 730 and 1070 m/Ma, respectively. This calculation takes into account the lower bulk density of the fan sediment (2.0 to 2.2 t/m³), as compared to the eroded solid rocks (2.6 t/m³) in the denudation area. Considering the many sources of error, all these values may deviate from the true rate by ± 20 to 30 %.

Table 9.1. Mechanical denudation rates of the Himalayas drained by the Ganges and Brahmaputra Rivers. Calculation based (a) on sediment volume of Bengal deep-sea fan, (b) on sediment volumes of Bengal Fan, molasse foredeep, alluvial plains, and subaerial delta, and (c) on present-day river loads. (For symbols see Fig. 9.16c). All values rounded

a) Curray and Moore (1971)

	$t_b - t_i$	$t_i - t_e$	$t_b - t_e$	
Time interval [a] (Ma B.P.)	8.8 - 1.8 = 7.0	1.8 - 0 = 1.8	8.8 - 0 = 8.8	
Volume of uncompacted fan sediment (x 10^6 km^3)	7.0	3.1	10.1	
		(Foredeep and subaerial delta neglected)		
Wet bulk density of fan sediment (t/m^3)	2.2	2.0	2.15	
Conversion factor for rock density in denudation area 2.6 t/m^3	0.846	0.762	0.827	
Eroded rock volume (x 10^6 km^3)	5.92	2.38	8.35	
Drainage area (x 10^6 km^2)				
Mountainous part	1.54 [b]	1.54	1.54	
Total area				2.05
Denudation rates (m/Ma)				
Mountainous part	550	860	620 [c]	
Total area				460

b) New estimate

	$t_b - t_i$	$t_i - t_e$	$t_b - t_e$
Time Interval (Ma B.P.)	20 - 7.7 = 12.3	7.7 - 0 = 7.7	20 - 0 = 20
Sediment volumes (x 10^6 km^3)			
Molasse foredeep, V_a, and delta plain, V_d	ca. 2	ca. 1	ca. 3
"Dry" fan sediments, V_{f1}, V_{f2}, and V_f (conversion wet/dry see a)	5.92	2.38	8.35
Denudation area (x 10^6 km^2)			
High mountains area	0.6	0.6	0.6
Total area	1.48 [d]	1.48	1.48
Average denudation rates			
High mountain area	1070	730	950
Total area (m/Ma)	435	300	380

Table 9.1 Continued

c) River load method (for rivers Ganges and Brahmaputra)

	After Holeman (1968) in in Curray and Moore (1971)	Milliman and Meade (1983)	
Total suspended solids, TSS (x 10^9 t/a)	2.2	1.67	
+ 25 % bedload	2.9	2.1	
Total denudation area	1.54	1.48	
High mountain area (x 10^6 km^2)			0.6
Average denudation rate [e] from TSS and bedload (m/Ma)	720	550	1350

[a] Calculated from present-day river load (no stratigraphic control)
[b] After Holeman (1968).
[c] Values should be equal, because both are based on present TSS (total suspended solids).
[d] After Milliman and Meade (1983).
[e] Density of eroded rocks 2.6 t/m^3.

The average denudation rate in the Himalayas is considerably lower if the Bengal Fan volume is only 4 x 10^6 km^3, as assumed by Barnes and Normark (1985), or if fan sedimentation initiated by crustal uplift in the Himalaya region began earlier. The denudation rate is somewhat higher if sediment loss to the Nicobar Fan (to the east of Ninety East Ridge) is considered (about 1 x 10^6 km^3, A. Wetzel, pers. commun.).

An average denudation rate on the order of 500 to 1000 m/Ma over a time period of 20 Ma signifies that a rock package 10 to 20 km thick has been removed. The present-day outcropping of 12 to 20 Ma old granitic intrusions in south Tibet, which were emplaced during to the main and late phases of tectonism in the Himalayas (Lefort et al. 1987), is compatible with such a large amount of erosion (also see Chap. 9.5.3).

Summary

Denudation rates found by the sediment budget method discussed here are summarized in Table 9.2, together with values taken from the literature. Most of these denudation rates are in relatively good agreement with those calculated from the suspended load of present-day rivers. Some of the long-term denudation rates, however, appear rather low. One of the reasons for this discrepancy was already mentioned in Chapter 9.2.3 (discontinuous denudation), but the ancient lower denudation rates may also be the result of differing climatic conditions or errors in the evaluation of both the sedimentary budget and/or the river bed load.

In addition, some authors have tried to establish sediment budgets for large individual ocean basins or budgets on a global scale for long time periods (cf. Chap. 11). Hay et al. (1989) have made an attempt to redistribute the sediments deposited in ocean basins to their former source areas.

9.5 Tectonic Uplift, Denudation, and Geomorphology

9.5.1 Long-Term Denudation Rates from Changes in Topography

A method similar to sediment budget assessment is the determination of the *rock volume* which was eroded from a *known topographic feature,* for example a volcanic cone, within a certain time period.

Table 9.2 Mean denudation rates derived from the budget of sedimentary basins, river deltas, and submarine fans

Study area		Time period	Average denudation rate (mm/ka)		Reference	Remarks
Denudation	Deposition	(Ma)	Total area	Mountain. part		
Chad drainage basin	Lake Chad N. Africa	0.01 Holocene	2 (Min. values)	6	Gac (1980)	Arid to semi-arid
Black Sea drainage	Black Sea	0.01 Holocene	70 (Steppe veget.)		Degens et al.(1978)	Modified
Mississippi drainage area	Miss. delta and fan	Pleistocene	46		Menard et al (1965)	
North. Appalachians South. App.	Volume of Cenozoic sediments	ca. 150	27 5		Mathews (1975)	Values probably too low
S. Californian coastal range	S. Californ. borderland	1 Pleistocene	120		Moore (1969)	
W. Mexico, coastal ranges	Gulf of California, central and southern part	4	20 (Min. values)	50	Einsele and Niemitz (1982)	
W. Alps	Rhone submarine fan	Pleistocene		400	Menard et al. (1965)	
Swiss Alps Upper Rhine	Rhine delta in Lake Constance	50 years		250	Müller (1966)	New calculation
Swiss Alps	Lake Zürich sediments	Holocene		1000	Schindler (1974)	
Himalayas, Ganges and Brahmaputra Rivers	Bengal deep sea fan, foredeep, coastal plain	8.8	460	720	Curray and Moore (1971)	
		20	380	950	This book	New calculation

In this way, Mills (1976) calculated a mean denudation rate of 1100 mm/ka (= 1100 m/Ma) for the slopes of the 4000 m high andesitic Mt. Rainier in the northwestern United States (for the last 0.32 Ma).

In the uplands of the Little Colorado Valley, Arizona, volcanic extrusions invaded the valley floor several times since the early Pliocene. A denudational lowering of the landscape in Permo-Triassic mudstones, siltstones, and sandstones on the order of 80 to 100 m/Ma has been deduced (Rice 1980) from the downcutting of the trunk river through these radiometrically dated lava flows.

In southern Germany, relics of Upper Jurassic limestones which formed the older landscape and have sunken into a Miocene volcanic pipe indicate that the land surface was lowered by about 600 m during the last 18 Ma (Geyer and Gwinner 1986). These data yield an average denudation rate of 33 m/Ma.

Extended post-Variscan peneplains at the base of the continental Triassic in Europe truncate granitic and metamorphic rocks and thus show that a thick rock cover has been removed by denudation subsequent to the Variscan orogeny.

Another method for estimating denudation is the enrichment of "insoluble" residues in the soils of karstic terrains or the formation of

lateritic crusts (duricrusts) in tropical "wet-dry" regions (e.g., Chorley et al. 1984).

Weathering of 30 m of granite is estimated to yield 1 m of laterite. The assessment of denudation rates from these observations is problematic, however, because lateral transport of soil and iron cannot be excluded. More reliable estimates can be gained from carbonate leaching in young loess, fluvial beds, or tills, the original carbonate contents of which are known.

In hilly areas, which were affected by Late Pleistocene solifluction processes forming a typical soil cover, the subsequent changes in topography by linear stream erosion can be evaluated.

Applying this method, Schmidt-Witte and Einsele (1986) found an average denudation rate of approximately 10 mm/ka for forested Keuper hills, consisting predominantly of claystones and sandstones, in the Holocene temperate humid climate of southern Germany.

9.5.2 Geomorphological Consequences of Differential Denudation

At the end of this chapter, a discussion of some of the consequences of chemical and mechanical denudation on landscape development appears appropriate. This is addressed in a semi-quantitative way, using simplified models of several characteristic morphogenetic regions containing subunits with various structures and lithologies. In order to simulate the *evolution of landforms* in such regions, it is assumed that (1) climatic conditions remain

constant, (2) denudation proceeds continuously at the same rate in each subunit, and (3) the investigated region is tectonically stable. These assumptions are, of course, unrealistic (cf. Chap. 9.5.3), but the purpose of this exercise is to get a feeling for the evolution of landforms in certain geological time periods.

Cuestas (Model)

Figure 9.17 presents a cross-section through a cuesta landscape similar to that of southern Germany. This type of landscape is found in many other regions of Europe and other continents. Tilted Mesozoic sedimentary rocks of different lithologies rest on a crystalline basement. It is assumed that in this case *only chemical denudation* takes place, which is by far the most important denudation process in humid temperate conditions with low to moderate relief. The denudation rates vary considerably, however, according to the rock type (Fig. 9.17). As a result, denudation proceeds very irregularly.

In this simplistic model, the outcropping carbonate rocks are more or less removed after 10 Ma, although they originally covered wide areas, unless they are protected by less soluble overlying beds or thick soils. The rocks least affected by chemical denudation are quartzitic sandstones. They maintain almost all their initial elevation and, in addition, their exposed surface area increases at the expense of less resistant rocks. This phenomenon can be observed in large areas of central Europe.

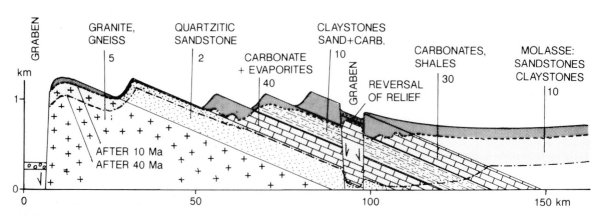

Fig. 9.17. Theoretical model for the development of a cuesta landscape with different rock types in persistently humid temperate climate. Only chemical denudation is considered. The geological setting resembles that in southern Germany. In reality, carbonate denudation in

particular may proceed differently if dissolution is hampered by primary clayey intercalations and/or soil formation protecting the underlying rocks. See text for further explanation. (After Hohberger and Einsele 1979)

After 20 Ma of continued erosion, the sandstones form the highest mountains or hills in a region largely eroded down to the base level of rivers. During this development, tectonic grabens may temporarily display inverted relief.

In reality, however, denudation in such regions is more complex and irregular. Mechanical denudation along the valley slopes of shifting river courses generates slope retreat, although the significance of this process in carbonate rocks is probably overestimated. The effect of progressive denudation on the landscape is also frequently modified by intercalated beds in the rock sequence which are particularly resistant to weathering. In addition, climatic changes, including periglacial processes which strongly promote mechanical denudation, play a role. A thorough discussion of this topic for particular regions is beyond the scope of this book.

Differential Denudation in the Amazon Drainage Area (Model)

Present-Day Situation

The second example deals with the drainage basins of the Amazon and Orinoco Rivers and is based on both Drever (1982) and especially on Stallard (1985). In terms of runoff, the Amazon is the largest river in the world, draining tropical forests as well as high, snow-covered mountain regions (Fig. 9.18a). Annual precipitation ranges mostly between 1000 and 2000 mm and in some parts is even higher. The relatively high load of solutes acquired in the Andes is diluted by *lowland rivers* with very low dissolved constituent concentrations (due to thick soils; Fig. 9.4c). Thus, 86 % of the dissolved load entering the Atlantic Ocean is supplied by 12 % of the total drainage area of the river, i.e., by the mountainous region of the Andes. Similarly, the *chemical denudation rates* of the different morphotectonic units of the Amazon and Orinoco drainage basins vary by a factor of nearly 100 and, if the evaporitic sediments are included, even up to a factor of 1000 (Figs. 9.18b and c). These chemical denudation rates are in fairly good agreement with the rates for the morphoclimatic zones shown in Fig. 9.7.

As for the *mechanical denudation rates*, the ratio of suspended/dissolved load is 3.1 for the Amazon and 4.2 for the Orinoco (Walling and Webb 1986, also see Fig. 9.6). The average ratio of mechanical/chemical denudation rates for the total drainage area of the Amazon is approximatly 3, i.e., 60/20 m/Ma. The mechanical denudation rates of the different morphotectonic units can only be estimated, however. According to Milliman and Meade (1983) and the data discussed in Chapter 9.2.2, the rates plotted in Fig. 9.18c (as mean denudation, MD, in m/Ma) are assumed.

Long-Term Landform Evolution

Starting with (unrealistically) idealized topography resembling the present-day in this region, and applying these chemical and mechanical denudation rates, one can predict the region's landform evolution after 1 and 5 Ma, respectively (Fig. 9.18c).

In addition to the assumptions mentioned above, it is assumed that the lithology of the units do not change with depth and thus bring about variations in denudation rate. Furthermore, isostatic rebound due to denudational unloading is neglected (see below).

Because both the chemical and mechanical denudation rates are high in parts of the drainage area, they produce marked changes in the landscape even within the relatively short time period of 1 Ma (Fig. 9.18c). Chemical denudation alone may considerably lower the elevation of the carbonate and evaporite sections in the Andes as well as in the foothills. After 5 Ma, the same units are strongly affected, whereas little change can be registered in the low-lying shield. The combined processes of chemical and mechanical denudation will erode in 5 Ma about three fourths of the Andean mountains and entirely remove the foothills. Both the low-lying and particularly the elevated shield are also significantly lowered, but the lowland sediments in the rivers remain adjusted to their erosional base, i.e., the Atlantic Ocean, and therefore more or less maintain their elevation by alternating erosion and deposition. In approximately 10 to 20 Ma (not shown in Fig. 9.18c) the crystalline rocks of the Andes and the elevated shield may have reached about the same (rather low) elevation above a very large lowland area, unless they undergo uplift coeval with denudation.

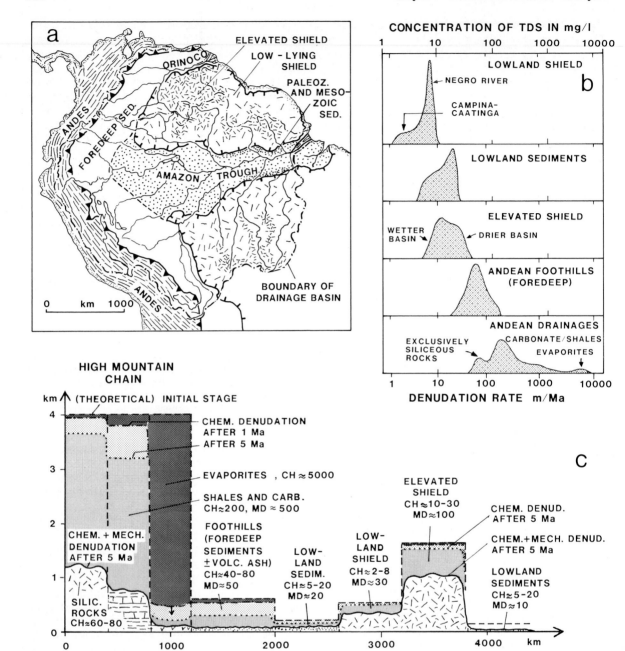

Fig. 9.18a-c. Model for predicting landscape development in region similar to that in the drainage basins of the rivers Amazon and Orinoco. **a** Morphotectonic map. **b** Total dissolved solids (TDS in mg/l) in river water and derived chemical denudation rates for the morphotectonic zones shown in **a**.

b Cross-section through drainage basin with different stages of prograding denudation, after 1 and 5 Ma, respectively. Note the extreme differences in denudation, leading to rapid destruction of great parts of the Andean cordillera. Isostatic rebound, however, is not taken into account. (After data of Stallard 1985, generalized)

9.5.3 Interrelationship Between Tectonic Uplift and Denudation

Uplift Rates

The examples in the previous section describe the progress of denudation in tectonically stable, hilly or mountainous areas which are assumed not to be affected by tectonic uplift or isostatic rebound as a result of denudation. This assumption is geologically unrealistic, because any kind of unloading, for example by substantial erosion of rock material or by melting ice, leads to isostatic rebound which partially compensates for denudation. This process will be taken into account in the following discussion. In addition, many regions are characterized by active tectonic uplift due to crustal processes or plate motions (cf. Chap. 12).

With *epeirogenic movements*, large areas are affected by moderate vertical uplift or subsidence on the order of 0.02 to 0.2 mm/a (20 to 200 m/Ma), which may persist for tens of millions of years. These movements originate from processes within the lithosphere and the mantle (cf. Chaps. 1 and 8). Higher rates of vertical movements result from isostatic response of the crust to rapid loading or unloading, for example by the growth or melting of icecaps. Post-glacial uplift in Scandinavia and other, formerly thickly glaciated areas is still in progress and reaches values as high as several mm/a and more, but is exponentially decreasing with time.

Orogenic uplift of mountain belts is, contrary to earlier hypotheses, a rather lengthy process lasting between 20 and 100 Ma (including isostatic rebound after initial isostatic equilibrium has been established). It evolves at plate boundaries as a result of plate convergence, subduction, and underthrusting of oceanic or continental crust. As long as these processes continue, uplift rates may be very high (up to several mm/a). Later, isostatic rebound of the thickened mountain root is the principal factor of slowly declining uplift.

From the cooling history and radiometric dating of metamorphic rocks, several authors (e.g., Grundmann and Morteani 1985; Selverstone 1985) derive an uplift rate of 0.2 to 1.0 mm/a (= 200 to 1000 m/Ma) for the last 30 to 50 Ma in the eastern Alps (Tauern Window). In the western Alps, uplift probably proceeded somewhat faster (e.g., Hurford 1991). According to Zhao and Morgan (1985), the Tibetan Plateau has experienced uplift for about 25 Ma (onset of collision) at an average rate of 0.2 to 0.3 mm/a, but for the High Himalayas an average rate of about 1.5 mm/a is assumed for the last 20 Ma (Burg et al. 1987).

During shorter periods and in limited areas, however, the uplift rates may be considerably higher in such tectonic settings.

For the Himalayas, dated, first-cycle detrital K-feldspar and muscovite samples from the distal Bengal Fan indicate that episodic rapid uplift and erosion must have taken place at rates greater than or equal to 5 mm/a (Copeland and Harrison 1990). Various radiometric measurements indicate that the uplift rates in the Nanga-Parbat-Haramosh massif have increased from ≤0.5 to more than several mm/a during the last 7 Ma (Hurford 1991).

Recent vertical movements in Japan, which are associated with an active subduction zone, amount to 1 to 2 mm/a. Parts of the North Island of New Zealand show young, subduction-related uplift of more than 1 mm/a, and locally more than 3 mm/a (Walcott 1987). Plio-Pleistocene, collision-related uplift of an accretionary wedge in Taiwan occurred at an average rate of 4 to 5 mm/a, as inferred from the unroofing history of this area. This is documented in adjacent sediments and derived from fission track ages, radiocarbon dates of coral reefs raised above sea level during the Holocene, and estimates on present-day denudation rates (Dorsey 1989). For the Caucasus, uplift rates of about 10 mm/a are reported (International Union of Geodesy and Geophysics 1975). Values of similar magnitude (4 to 12 mm/a) were measured in the Coastal Ranges of California. However, some of these high uplift rates may be caused by short-lived strike-slip movements and are therefore only of local importance.

In addition, melting of large, thick glaciers in the Holocene and subsequent crustal rebound may be responsible for the very high young uplift rates (≥ 3 mm/a) observed for several high mountain ranges with repeated geodetic surveys or with the aid of young, elevated fluvial terraces (Gansser 1984). These high values are probably not characteristic of longer periods.

Additional data of uplift rates in various tectonic settings are summarized in special articles (e.g., Leeder 1991).

Long-Term Epeirogenic Uplift and Denudation (Model)

The models in Fig. 9.19a and b demonstrate the interaction between comparatively strong, long-term epeirogenic uplift and regional denudation.

The rate of uplift (0.2 mm/a) is kept constant for a period of 40 Ma. Denudation is assumed to proceed uniformly over the entire area as is similarly observed in the formation of peneplains in tropical to subtropical climates. The influence of changing rock types, linear downcutting of deep valleys, as well as other minor geomorphological features such as benches, terraces, and pediments, are neglected.

Beginning at zero elevation (in relation to the base level of erosion), continuous uplift leads first to a rising land surface, because denudation lags behind uplift. With increasing eleva-

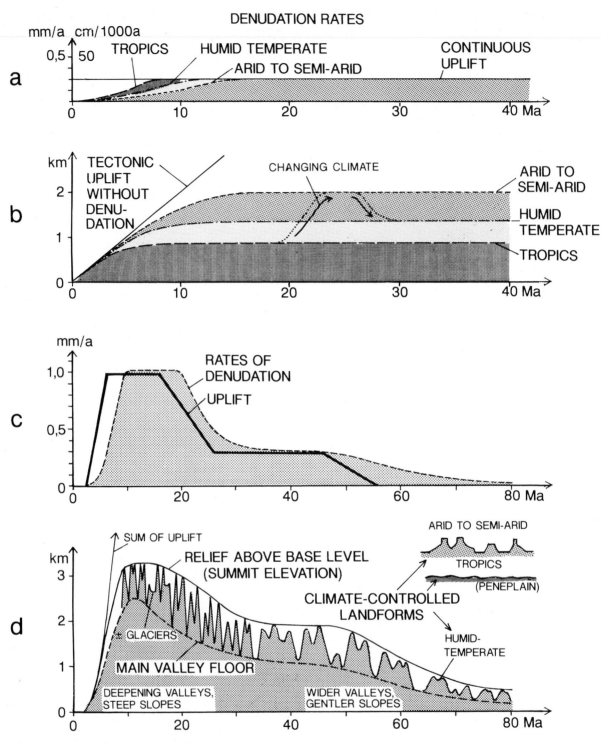

Fig. 9.19a-d. Interaction between tectonic uplift, denudation, and isostatic rebound, plotted semi-quantitatively for (**a** and **b**) a period of continuous, moderate uplift, or (**c** and **d**) one orogenic cycle. **a** and **c** Rates of uplift (including isostatic rebound) and long-term denudation; note that the denudation maximum lags behind **a** the strongest pulse of uplift, or **c** the onset of uplift. **b** and **d** Development of relief above base level with time. The resulting landforms also strongly reflect the influence of climate, which may change during the time span considered here

tion, which proceeds linearly versus time minus denudation loss, the denudation rate grows rapidly, more or less exponentially (cf. Figs. 9.9 and 9.20b) until it matches the uplift rate. Then, a kind of equilibrium between uplift and denudation is reached which tends to maintain a certain maximum land surface elevation under the given circumstances. This is also the case, when tectonic uplift is enhanced by isostatic rebound due to denudational unloading.

Depending on the climate, this *equilibrium elevation* is established after different time spans and at different heights above the base level. Tropical climates, which may cause high denudation rates even in lowlands, bring about equilibrium earlier and at lower altitude than temperate humid or arid to semi-arid conditions (Fig. 9.19b). With decreasing denudation rates (for the same elevation and relief), the

landscape is raised to increasingly high equilibrium levels (Fig. 9.20a), provided the rate of uplift remains constant. If the climate changes during continuous uplift, the *plateau* or *summit elevation* of the landscape will vary as shown in Fig. 9.19b.

Without denudation, an assumed uplift rate of 0.2 mm/a would cause an elevation of 8 km over a time period of 40 Ma. An increase in the uplift rate leads, of course, to higher altitudes (Fig. 9.20a). Because denudation grows rapidly with relief and therefore in very high mountain ranges always exceeds long-term uplift rates, there must be an *upper limit for the elevation* of the land surface above sea level (at present about 9 km) which cannot be surpassed on the Earth's surface. Again, the total uplift rate may include a proportion caused by isostatic rebound. In fact, uplift as a result of denudation may continue after the tectonic process has ceased. The maximum altitude of mountain belts or plateaus is also limited by isostacy, however, depending on rock densities and crustal thicknesses (cf. Chap. 8.1).

Uplift and Denudation of Orogenic Belts (Model)

Model Parameters

The model shown in Figure 9.19c and d represents a situation similar to that of large mountain belts. Because the long-term uplift and denudation history of orogenic belts, even of the best studied examples such as the Alps, the North American mountain ranges, or the Himalayas are not very well known and to some extent still controversially discussed (see above), the following simplifying assumptions are made:

- A phase of rapidly increasing, relatively strong uplift (1 mm/a), which persists for some time (10 Ma), is followed by a period (10 Ma) of decreasing uplift.
- Then a second phase of comparatively long-term (20 Ma), constant, moderate uplift (0.3 mm/a) is caused by isostatic adjustment of the thickened crust to previous or decelerating underplating, as well as by rebound resulting from erosional unloading.
- At the end of this phase, uplift drops to zero within another 10 Ma.

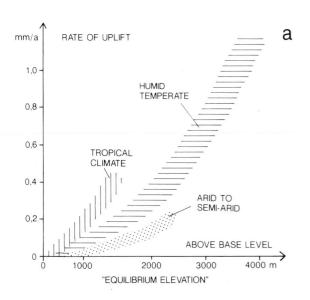

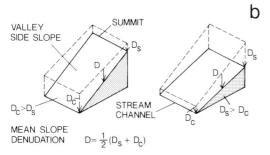

As in the example in Fig. 9.19a and b, the interaction between vertical tectonic movements and denudation,

Fig. 9.20. a Ultimate elevation above base level as a function of the rate of uplift and climate. **b** Development of valley-side slopes according to different rates of summit (D_s) and stream-channel denudation (D_c). (After Ahnert 1970; Chorley et al 1984; Gerrard 1988, greatly modified)

wearing down the increasing relief, is based primarily on W. Penck's fundamental hypothesis (cited and discussed, e.g., in Thornes and Brunsden 1977; Chorley et al. 1984). It suggests that there is a continuous interplay between endogenetic and exogenetic forces which shape the landscape. This idea is contrary to the once widely accepted hypothesis of W.M. Davis, who assumed short, pulse-like uplift and subsequent slow downwearing as the main factors in the evolution of mountain ranges.

Landform Evolution

At the start of uplift above base level, denudation once again lags behind uplift by 3 to 4 Ma (Fig. 9.19d). Similarly, when constant strong and moderate uplift ends, denudation still proceeds at a rate almost as high as before. This is a result of the high elevation the mountain belt has reached, even though the uplift rate has already decreased. During periods of constant uplift, an equilibrium elevation is established (at altitudes of ≥3000 m and about 2000 m, respectively) as discussed above (Fig. 9.19b).

These two levels are, of course, intersected by many deep valleys, which may generate a main valley floor considerably deeper than the *summit elevation* (Fig. 9.19 d). During certain times, for example when uplift surpasses denudation, channel lowering (D_c in Fig. 9.20 b) may predominate over summit lowering (D_s), or vice versa. In these cases, the mean denudation rate is approximately $D = 1/2 (D_c + D_s)$.

With the passage of time, the valleys usually become wider and their slopes gentler; in this way the areal extent and heights of the summits are reduced. Especially during the later stages of this development, when the summit elevation is already considerably lowered, the evolving landforms are significantly controlled by climatic factors. In the model shown here, uplift ceases completely after 80 Ma. (This is not quite realistic, because isostatic rebound due to erosional unloading is still possible). Denudation, however, continues at a low rate for a long time, showing only slow decline. Even 20 Ma later, a relief of about 500 m is still present.

In total, uplift during the preceding 80 Ma amounted to 23 km for this example. This appears to be realistic, since alpinotype mountain belts of this age now exhibit outcropping metamorphic rocks which were formed at burial depths of approximately 10 to 30 km.

Finally, it should again be stressed that the models described here are greatly simplified. Modern geomorphologists have become cautious regarding the usage of the term "equilibrium landforms" (see, e.g., Chorley et al. 1984). It seems that steady-state equilibrium, with only small fluctuations of certain landforms around a specific average condition, does not exist in nature. Rather, landforms resulting from uplift and denudation should be characterized as *dynamic metastable systems*. Such systems respond not only to external influences as discussed above, but may also be affected by internal processes such as the discontinuous storage or loss of large volumes of detrital sediments in or from its drainage area. During metastable equilibrium, channel patterns may change from straight and braided to sinuous, and slopes may retreat, become gentler, and evolve from convex to a more concave form.

10 Sedimentation Rates and Organic Matter in Various Depositional Environments

10.1 General Aspects

Sedimentation and Accumulation Rates

It has become common in the sedimentological literature to distinguish between *sedimentation rates* (sediment thickness per unit time) and *accumulation rates* (solid sediment mass per unit area and time), particularly in papers dealing with modern sediments and the results of ocean drilling (e.g., Van Andel 1983; Ehrmann and Thiede 1985). This discrimination is important when sediments of different porosities are compared, and it is indispensable for any kind of sediment budget.

Both sedimentation and accumulation rates are usually determined for sections taken between dated lower and upper boundaries. They then represent average values for these sec-tions, based on the assumption that during the corresponding time interval the sediments were deposited continuously at a constant rate.

Sedimentation Rates

The sedimentation rate, SR, is the thickness, z, of a certain vertical sediment section divided by the time span, t, necessary for its deposition:

$$SR = z/t \qquad (10.1)$$

It is calculated in mm/a (a = year), mm/ka (1 ka = 1000 years), cm/ka, or m/Ma (1 Ma = 10^6 years; 1 cm/ka = 10 m/Ma). In the following text, the latter two notations are used.

It is obvious that sedimentation rate is not an accurate scale to describe the deposition of solid particles quantitatively. Fresh sediments have a high porosity which is later reduced under the load of younger sediments by compaction (Chap. 13.2). Thus, sedimentation rates calculated without taking porosity into account become lower with decreasing porosity. However, if sediments have undergone deep burial and lithification, their differences in porosities become less pronounced. Nonetheless, the errors associated with dating of section boundaries and/or the assumption of steady state deposition within the measured section may be often substantial. For these reasons, the average sedimentation rate appears to be acceptable only as an approximation for many applications in sedimentology and geology.

Accumulation Rates

The accumulation rate, AR, is defined as the quantity (weight), W, of solid particles deposited during a certain time span, t, on unit area, A:

$$AR = W/(A \cdot t). \qquad (10.2)$$

It may be expressed in g/(m2·a),
kg/(m2·ka), g/(cm2·Ma), or t/(m2·Ma);
1 g/(m2·a) = 1 kg/(m2·ka)
= 100 g/(cm2·Ma) = 1 t/(m2·Ma).

One can distinguish between bulk accumulation rates, AR_b, comprising all constituents of a sediment, and specific accumulation rates for certain sediment components such as calcium carbonate, opaline silica, organic matter, etc. In order to calculate bulk accumulation rates, besides the requirements mentioned above, the porosity, n (as a portion of 1), as well as the wet and dry bulk densities, D_w and D_d (in g/cm^3 or t/m^3), of the sediment must be known. Note that D_d is related to the same volume as D_w; it must not be confused with average grain density, γ_s:

$$D_d = (1-n) \cdot \gamma_s . \qquad (10.3)$$

If the total pore space is filled with sea water (density $\gamma_{sw} = 1.025$ g/cm^3), wet and dry bulk densities are linked by the term

$$D_d = D_w - 1.025 \, n . \qquad (10.4)$$

For pore fluids differing from sea water, other correction factors have to be used. Porosity, n, derived from (5.3) is

$$n = (D_w - D_d)/1.025 . \qquad (10.5)$$

D_w and D_d are usually determined as follows:

$$D_w = W_w/V_t \quad \text{and} \quad D_d = W_d/V_t , \qquad (10.6)$$

where W_w and W_d are wet weight and dry weight, respectively, and V_t is total volume of the (water-saturated) sample being measured.

The porosity, n, is also found by

$$n = V_w/V_t = (V_w - W_d)/V_t , \qquad (10.7)$$

where V_w = the volume of water in the water-saturated sample.

Fresh fine-grained samples from drillholes or from the sea bottom usually are fully water-saturated. If part of the pore space is filled by gas, other formulae must be applied (see, e.g., special literature on soil mechanics or from the oil industry).

Having determined D_d, the bulk accumulation rate, AR_b, is

$$AR_b = SR \cdot D_d, \text{ or } AR_b = SR \cdot (1-n) \cdot \gamma_s \quad (10.8)$$

In order to calculate the accumulation rate of specific sediment components, AR_x, the con-

centration, C_x, of component x in the total sediment must be determined. If C_x is given in percent by weight, the accumulation rate, AR_x, of component x is

$$AR_x = AR_b \cdot C_x/100 . \qquad (10.9)$$

In order to convert sedimentation rates, SR, of sediments of different grain densities, γ_s, and porosities, n, into bulk accumulation rates, AR_b, and vice versa, the diagram in Fig. 10.1 may be used.

Short- and Long-Term Sedimentation Rates

As was pointed out by many authors (e.g., Wetzel and Aigner 1986; Anders et al. 1987; Tipper 1987; Dott 1988; Sadler and Strauss

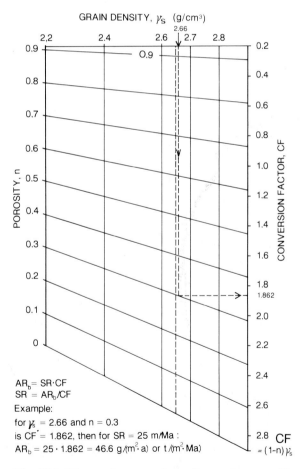

Fig. 10.1. Diagram for converting sedimentation rates, *SR*, into bulk accumulation rates, *AR*, and vice versa. τ_s mean grain density; *n* porosity; *CF* conversion factor. For $\gamma_s = 1$ g/cm^3 and n = 0 it is 1 m/Ma = 1 g/(m^2·a) or 1 t/(m^2·Ma)

1990; Ricken 1991), most sedimentary sequences have not accumulated continuously at a steady rate, but contain minor and major stratigraphic gaps (Fig. 10.2a). However, only the long intervals of non-deposition or erosion on the order of at least 0.1 to 1 Ma can be identified in terms of geologic time as missing in the stratigraphic record. Well known examples of *discontinuously accumulating sediments* are fluvial and tidal flat deposits displaying distinct erosional features, but their stratigraphic record is difficult to evaluate. Minor erosional events can be deduced, for example, from the truncation of primary bedding features below traction current beds, or from the truncation of burrows at the base of turbidites and tempestites (Wetzel and Aigner 1986). Discon-

tinuous sediment accumulation, such as that caused by a series of mud turbidites deposited at relatively long, irregular time intervals is, however, difficult to recognize. Similarly, foreshore sediments resulting from repeated wave and current reworking and redeposition contain many short episodes of erosion and sediment accumulation, which can also be barely defined.

Hence, short-term sedimentation rates determined for single beds or small proportions of the total sediment section are generally considerably higher than long-term rates (Fig. 10.2 a). However, this problem is not discussed further in the following chapters, because here long-term, large-dimensional basin fills are of primary interest.

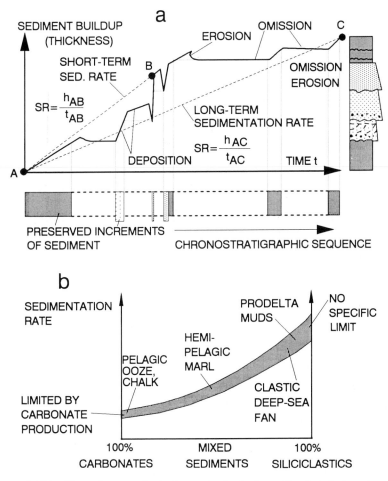

Fig. 10.2. a Sediment buildup/time diagram displaying discontinuous sedimentation with intervals of omission (nondeposition) and erosion. Note that long-term average sedimentation rates, SR_{AC}, may be lower than short-term rates, SR_{AB}, and rates of single beds, for example, those of tempestites. The stratigraphic gaps (hiatuses) are shown in the horizontally plotted chronostratigraphic sequence. b Mixed carbonate-siliciclastic sediments tend to produce higher sedimentation rates with increasing proportions of terrigenous material, as shown for some marine examples. (W. Ricken, pers. commun.)

Potential and Actual Sedimentation Rates

Another general problem is the distinction between potential and actual sedimentation rates. The term *potential sedimentation rate* as used here refers to the capacity of a sediment-delivering system to accumulate a maximum of sediment per unit time in a particular area. For example, the sediment supply of a river system can fill an open lake basin at a rate which is equal to the potential rate, as long as the lake can hold the total river sediment input. However, as soon as the lake is filled with sediment, the subsequent sedimentation rate in the lake basin is controlled by its subsidence, as well as by the base level and gradient of the river system. From then on, the *actual sedimentation rate* drops below the potential rate, because most of the river sediment is transported farther downstream out of the lake area. Similarly, a tectonic graben filled up to the gradient of a through-flowing stream can store sediment only in the space provided by subsidence. In a slowly subsiding, high-energy shelf basin, only part of the sediment supply can commonly settle permanently, while the remainder is swept into deeper water (cf. Chaps. 1.4 and 11). Additional examples of environments where the actual sedimentation rates are normally slower than the potential rates are delta plains, lagoons, tidal flats, and shallow carbonate platforms.

The sedimentation rate data discussed in the following sections are for scenarios in which either porous sediments or massive sediments with little pore space are considered. Most signify actual sedimentation rates observed in modern and ancient environments.

10.2 Average Sedimentation Rates

In Various Environments (Overview)

In a discussion of sedimentation rates in various depositional systems it appears useful to distinguish between two end members of sediments (Fig. 10.3):

- Allochthonous siliciclastic sediments, consistting mostly of terrigeneous material, but also including bioclastic components.
- Autochthonous biogenic sediments.

Siliciclastic Sediments and Evaporites

The sedimentation rates of siliciclastic sediments are controlled mainly by the size and characteristics of the source area (Chaps. 9 and 11), as well as by the distance of the depositional area from the site of sediment input. Lakes, marine prodelta slopes, and deep-sea fans fed with sediments from a major point source may accumulate large amounts of terrigenous material in short time periods and thus exhibit very high sedimentation rates. Similarly, proglacial lowlands, lakes, and marine basins reached by advancing glaciers and ice shields frequently display very high sedimentation rates.

On the Barents shelf glacigenic sediments accumulated at rates between 10 and 100 cm/ka; fjord troughs were filled at rates up to 6000 cm/ka (Vorren et al. 1989). During the Holocene, the rapidly moving, cool-temperate glaciers of southern and southeastern Alaska produced an average sedimentation rate of 500 cm/ka on the shelf, and in the last 6 Ma about 50 cm/ka on the shelf edge and in deeper water (Powell and Molnia 1989). In the Ross Sea of the Antarctic region, however, Pleistocene diamicton accumulated at a much lower rate (around 1 cm/ka, Alley et al. 1988).

In intracontinental tectonic grabens, delta plains, tidal flats, high-energy shallow seas, and in foreland basins the actual sedimentation rates are commonly high, but frequently lower than the potential rates (Fig. 10.3). In contrast, deep sea regions far away from the coast and any significant sediment source are characterized by extremely low sedimentation rates, as for example red, deep-sea clay. A major portion of these clays may consist of desert-derived eolian dust. The actual sedimentation rate is equal to the potential rate in these cases.

Extremely high sedimentation rates may be reached when concentrated brines precipitate *salts,* for example halite, with a rate up to about 5 000 cm/ka. For several reasons, however, such rates cannot be maintained for a long time (Chap. 6.4).

Biogenic Sediments

Biogenic sediments are formed from the skeletal remains of plants and animals. In addition, some carbonate may be precipitated from supersaturated surface waters, for example as a result of CO_2 uptake by subaqueous plants. Most of the biogenic sediment is produced by a few groups of organisms. In shal-

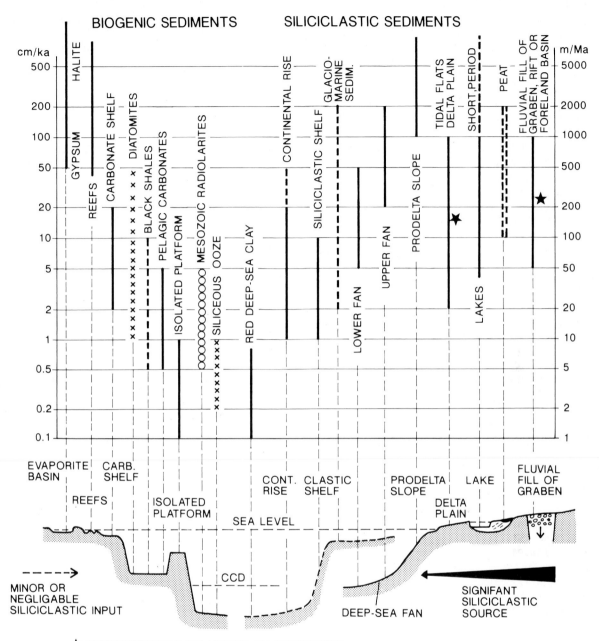

Fig. 10.3. Overview of sedimentation rates in various depositional environments which are either dominated by allochthonous, siliciclastic sediments, or autochthonous, biogenic materials.

(Data from many sources, including Seibold and Berger 1982; Scholle et al. 1983; Nelsen and Stanley 1984; Stow et al. 1985; Jenkyns 1986; Lützner 1989)

low marine waters, benthic organisms, such as various reef-builders, algae, sea grass, molluscs, brachiopods, and bryozoa, are most important. Since the Mesozoic, biogenic sediments in deeper basins consist mainly of the tiny remains of planktonic organisms including calcareous nannofossils, foraminifera, and pteropods, or siliceous tests of diatoms and radiolaria. In their freshly deposited state these deep-water sediments form water-rich oozes with porosities of around 80 %. Later, they compact and undergo chemical-mineralogical diagenesis, leading to chalk and pelagic limestone, or to porcellanite and chert.

The sedimentation rates of autochthonous biogenic sediments are limited by the production and dissolution of skeletal carbonate and opaline silica. (Phosphatic remains play a minor role in rock-forming processes and are therefore not discussed here.) These processes are controlled by a variety of factors (also see Chap. 5.3), the most important of which are favorable living conditions of the carbonate and silica secreting organisms, sufficient nutrient supply, and prevention of dissolution in the water column and on the sediment surface.

As regards dissolution in the oceans, calcareous and siliceous skeletal remains behave quite differently. While carbonate skeletons are preserved in shallow and medium deep waters and only dissolve in deep water regions below the lysocline and CCD, siliceous organism tend to dissolve in shallow waters which are undersaturated with respect to silica (cf. Chap. 5.3). This difference in carbonate and silica dissolution contributes to the general rule that shallow-water carbonates are commonly poor in opaline silica, and that pure siliceous sediments mostly occur below the carbonate compensation depth in relatively deep water.

The *highest biogenic sedimentation rates,* apart from reefs (see below), are observed below zones of coastal and equatorial upwelling, in estuaries, marginal basins with sufficient nutrient supply by rivers, and shelf seas. Such sediments may accumulate at rates up to 10 cm/ka and in special cases up to 50 to 100 cm/ka, for example Miocene diatomites in Sicily and uncompacted Recent diatomites in the Gulf of California (based on data from Martinson et al. 1987). Compacted Mesozoic radiolarites from the western Tethys (Jenkyns 1986) and the Miocene diatomites of the Monterey Formation, California Isaacs 1984), range between 0.2 and 2 cm/ka, in the eastern Pacific around 0.2 cm/ka (Gursky 1988).

Warm, low-latitude regions commonly promote organic productivity and thus the accumulation of skeletal remains, but high fertility of siliceous organisms also occurs in subpolar regions. Modern siliceous oozes of the North Pacific and Bering Sea reach values of about 1 cm/ka (when measured as massive beds without pore space). Cretaceous chalk formed in shelf seas reached sedimentation rates of about 5 to 15 cm/ka, while modern deep-sea carbonates commonly display lower rates of about 1 to 3 cm/ka.

Very low biogenic sedimentation rates are found below central parts of large ocean basins where fresh supply and recycling of nutrients are limited. The modern Sargasso Sea in the Atlantic is an example of this situation. When the CCD is elevated and silica concentration in near-surface sea water is low, the preservation of skeletal carbonate and silica may be markedly reduced, and sediments devoid of carbonate and/or siliceous remains may be deposited. The actual sedimentation rates of biogenic materials are mostly equal to the potential rates, except for carbonate deposition in very shallow water and tidal flats (see below).

Biogenic and biochemically precipitated *carbonate in fresh-water lakes* may form at rates of 10 to 100 cm/ka and more. The highest rates are observed in the littoral zones, due to the assimilating activity of macrophytes and the abundance of gastropods and molluscs (Kelts and Hsü 1978). Sedimentation rates similar to those of carbonates can be expected for lakes rich in diatoms (cf. Chap. 2.5).

Peat

The accumulation and preservation of peat depends on the terrestrial to lacustrine depositional environment and on the maintainance of a high groundwater level. Holocene peats in temperate and tropical regions have accumulated at rates of about 20 to 200 cm/ka (Cameron et al. 1989). Sedimentation rates reported in the literature for compacted Paleozoic to Cenozoic coals (Stephanian basins in France, upper Paleozoic to Mesozoic basins in China, Tertiary Rhine basin in Germany, African rift valleys) vary between 0.3 up to 200 cm/ka (Courel 1989). In all of these coal-bearing sequences, peat formation was sufficiently fast to keep pace with subsidence, which ranged, for example, in the Permian foreland basins of Australia from 5 to 40 cm/ka (Hunt 1989).

Mixed Sediments

Mixed sediments consisting of both siliciclastic and biogenic components commonly show higher sedimentation rates than do purely biogenic deposits. In calcareous sediments, the carbonate content tends to decrease with increasing sedimentation rate (Fig. 10.2b), reflecting dilution of the limited carbonate production by terrigenous material. Similarly, the percentage of opaline silica decreases with increasing sedimentation rate. Black shales frequently contain considerable amounts of carbonate and accumulate at a relatively slow rate (0.5 to 10 cm/ka). Up to a certain point, preservation of organic matter and thus the organic carbon accumulation rate in shales and marls is furthered by an increasing sedimentation rate (Chap. 10.3.3), but beyond this limit the effect of dilution prevails. The organic matter content also tends to decrease with growing carbonate content (Ricken 1991).

Sedimentation Rates in Carbonate Depositional Systems

Reefs and Shallow-Water Carbonates

Although the sedimentation and accumulation rates of biogenic carbonate are limited by organic productivity, they also display a wide range of values from about 0.1 to 1000 cm/ka (1 to 10 000 m/Ma). The highest values have been observed in young *coral reefs* which grew vertically upward with the rapid Holocene sea level rise. However, these extremely high rates should not be taken as average growth rates for longer time periods. *Reef mounds, calcareous tidal flats, ooid shoals, and coastal sabkhas* have the potential to grow upward at rates on the order of 50 to 100 cm/ka (Enos 1989; Hubbard et al. 1990; Strasser 1991). Thus, the average sedimentation rate of a very shallow carbonate platform may reach approximately 100 cm/ka (Fig. 10.4) provided there is an equal rate of subsidence maintaining steady state conditions. If subsidence is slower than this high potential sedimentation rate, the actual sedimentation rate drops below the potential rate.

Carbonate Shelves and Platforms

As pointed out by Sarnthein (1973) and confirmed by many other studies in tropical and subtropical regions, *benthic carbonate production* decreases with water depth. Coral reefs, algal structures, sea grass beds, and other organisms depending on photosynthesis can live only near the water surface where they receive sufficient sunlight. They may be accompanied by large foraminifera, molluscs, and gastropods growing on shallow banks. In such a faunal and floral community, biological and biochemical carbonate production reaches its maximum.

Below water depths of 30 to 40 m, benthic carbonate production drops by a factor of approximately 10 and is frequently on the order of 1 to 10 cm/ka. This signifies that outer, deeper portions of carbonate ramps exhibit benthic carbonate production which has approximately the same magnitude as planktonic production in this and deeper regions (for example chalk, Fig. 10.4). On the other hand, carbonate accumulation in shallower water may be inhibited by insufficient subsidence (see above) and lead to the *export of carbonate into deeper water*. In this way, carbonate accumulation in shallower and deeper water can be more or less equalized and make up in total an average of several cm/ka. Such rates are characteristic of the long-term buildup of carbonate platforms (cf. Chap. 12.2.2). An excess in carbonate production over the average subsidence rate is commonly used for the lateral progradation of the platform.

Deep-Water Carbonates

In deep water, benthic carbonate production becomes insignificant, but mass flows and turbidity currents may transport considerable volumes of shallow-water carbonates into the deep sea. There, they contribute to the buildup of deep-sea fans, slope aprons, and continental rises and may lead to sedimentation rates similar to those listed for their siliciclastic counterparts (Fig. 10.3). Sedimentation rates of plankton-produced carbonates are discussed above and plotted in Fig. 10.3.

Sedimentation Rates in Relation to Denudation, Subsidence, and Sea Level Change

The sedimentation rates in the specific environments discussed above represent relatively wide ranges and may be used as a general guideline. However, they do not include special cases which deviate from these common values. In addition, the fact that sedimentation rates may vary through time is not taken into account. During sea level highstands, terrigenous input into the sea usually slows, and carbonate production may also significantly vary as a result of sea level changes and other processes. Neglecting these complications, it is useful to compare common sedimentation rates with other rates relevant in geological processes.

It is interesting to note that *denudation rates* on land cover almost the same range (0.2 to more than 100 cm/ka, see Chap. 9.5) as the sedimentation rates discussed above (0.1 to

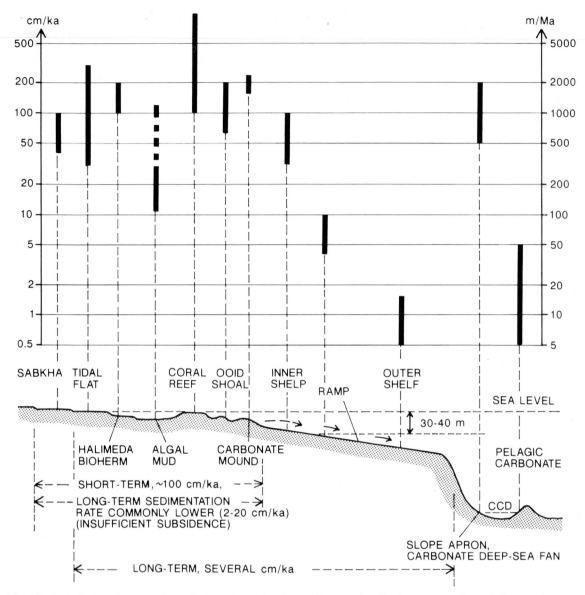

Fig. 10.4. Sedimentation rates in carbonate depositional environments. In shallow water, benthic carbonate production is dominant, while in deeper areas carbonate-secreting plankton prevails. Note that long-term rates are commonly much lower than short-term rates due to insufficient subsidence and sediment transport from shallow regions onto the outer shelf and into the deeper sea. (Compiled from several sources, e.g., Sarnthein 1973; Schlager 1981; Bosence 1989; Hubbard et al. 1990; Strasser 1991)

1000 cm/ka; also see Kukal 1990). However, a direct comparison of sedimentation and denudation rates makes sense only for such cases in which siliciclastic material predominates. In addition, the drainage area of the basin should be approximately equal to the area of the basin itself, or the ratio between drainage area and basin area should be known (cf. Chap. 11.2).

Furthermore, the *subsidence rates* in the various types of sedimentary basins have a range (about 0.2 to more than 100 cm/ka, Chap. 8) similar to that of the sedimentation rates. This signifies that, theoretically, any type of basin subsidence can be compensated for by sediment accumulation, i.e., even the most rapidly subsiding basin may be kept filled with sediments if an effective sediment source is available. Actually, all small basins, such as graben structures, foreland basins, and pull-apart basins, can be filled, for example with fluvial material, although they may display very high subsidence rates in their early evolutionary stage. However, when the basins become larger and continue to subside substantially, such as in growing ocean basins, sediment supply and sedimentation rates are normally no longer sufficient to keep the basins filled. These problems are discussed further in Chapter 11.4.

Finally, it is useful to know whether or not the sedimentation rates in specific environments could keep pace with pre-Quaternary and Quaternary *sea level rises*. In the early Holocene, the short-term sea level rise was so fast (6 to 10 m/ka) that carbonate platforms were drowned, and many individual reefs were not able to follow the rising water level (Schlager 1981). Pre-Quaternary, long-term sea level changes occurred at rates of about 1 cm/ka; short-term oscillations within the Milankovitch frequency band (Chaps. 7.1 and 7.2), which only had low amplitudes, may have caused maximum rates of sea level change of about 10 to 50 cm/ka. Even in the latter cases, the sedimentation rates of some environments, for example tidal and subtidal carbonates, were sufficiently high to keep pace with the sea level rise and to maintain approximately the same depositional system through long time periods (Strasser 1991).

10.3 Production and Preservation of Organic Matter

10.3.1 General Aspects

Production and Types of Organic Matter

Sediments with biogenic skeletal parts commonly also contain some organic matter. This chapter summarizes the most important facts about the production and preservation of organic matter. It is based to a large extent on publications by Durand (1980), Grant Gross (1980), Romankevich (1984), Tissot and Welte (1984), Brooks and Fleet (1987), Berger et al. (1989), and Stein (1991). The transformation of organic matter under increasing sediment load and temperature is described in Chapter 14.2.

All life on earth is based on photosynthesis, which began two to three billion years ago (Chap. 6.5). In this process carbon dioxide combines with water to form carbohydrates, at the same time releasing oxygen. The energy necessary for this synthesis is provided by the sun via light-absorbing pigments such as chlorophyll. This basic process can be accomplished by all *autotrophic organisms,* including primitive bacteria and blue-green algae, as well as by highly evolved plants. All of these organisms are the primary producers of the organic matter, whereas *heterotrophic organims* (animals) feed on this organic matter.

Carbohydrates contain stored energy which can be released when oxygen is taken up by animals and plants which thereby gain energy for their life processes (called "respiration"). Through respiration, the *gross productivity* of plants, which is defined as the mass of carbon fixed by photosynthesis per unit area and time, is reduced by 10 to 50 %. The remaining mass is the *primary net production,* which is discussed further below. Part of the energy gained by respiration is used to build up other organic compounds, such as proteins, lipids, lignin and tannin. Of these, the *carbohydrates* (including cellulose), which consist of carbon, oxygen, and hydrogen, are the most important organic matter constituent of plants, whereas animals (e.g., oceanic zooplankton or zoobenthos) and bacteria contain particularly high proportions of proteins. Some of the organic compounds incorporate small amounts of nitrogen, sulfur, and other elements. In the following discussion all of these organic compounds, including the products of biological activity such as excretions and secretions, are included in the term *organic matter.*

Primary Production in Various Environments

The primary production of organic matter (here described in a broader sense than defined in Chap. 10.3.2) is controlled by several factors. Light (for photosynthesis), temperature, and the availability of certain nutrients all play a large role. On the continents, climate and precipitation are of eminent importance. The most significant production per unit area is accomplished by tropical rain forests, while desert regions permit only very limited plant growth. In aquatic systems, of course, water is not the limiting factor. Here, primary production is a function of the amount of light received by the autotrophic community. Sun light also affects the surface temperature of the water body and thus promotes life processes. The growth rate of primary producers may be limited when nutrient supply (e.g., nitrate, phosphate, and so-called micro-nutrients such as vitamins, soluble organometallic

Table 10.1. Primary production in various Recent environments, dry organic matter[a] in g/m² per year. (Mainly after Huc 1980; for additional data see Romankevich 1984)

Continents	Short grasslands, deserts	50- 200
	Temperate forests	1300-3200
	Tropical rain forests	4000-9000
Fresh-water lakes		3000-6000
Salt marshes, mangroves		3300-6000
Marine	Open oceans	10- 400
	Continental shelves	200- 600
	Zones of upwelling	400-1200
	Estuaries	200-4000
	Algal beds	1300-2500
	Reefs	3500-9000

[a] Conversion factor for organic carbon/fresh organic matter: 0.31

compounds) is insufficient. Therefore, some lakes and low latitude oceanic regions only enable low growth rates. However, many swamps, lakes, estuaries, and nearshore zones with inflowing water rich in nutrients have high primary production rates (Table 10.1).

Standing Crop and Annual Productivity

So far we have discussed the primary net production of plants, which can be estimated by multiplying the average standing crop (the biomass present in a given area at a certain time) of a population by its rate of generation (doubling time). If the average standing crop doubles once every 2 months, the net production per year will be six times greater than the standing crop. Thus, the productivity of a population can be high even though its standing crop or biomass is small (e.g., in the oceans); similarly, the biomass or standing crop of an ecosystem can be considerably greater than the annual net productivity (e.g., in forests).

Preservation of Organic Matter (Overview)

Plants on the continents and in the oceans contribute approximately equal amounts to annual global primary production. However,

the situation *on land* is completely different from that in aquatic environments, if the consumption of primary production by other organisms and the preservation of organic matter in sediments are considered. Under aerobic conditions in soils, only the most resistant organic compounds, such as chitine or the cuticles of pollen and spores (sporopollenine), can be preserved over long time periods.

Therefore, with the exception of coal-forming environments (Chaps. 2.2.3 and 3.4.3), little organic matter is preserved in continental deposits.

In aquatic environments such as lakes, swamps, and anoxic marine basins, however, decomposition of organic matter is slower. Aerobic organisms may cease to destruct organic matter if the oxygen supply of the water body is used up and cannot be replaced rapidly enough. Thus, most of the organic matter preserved from earlier geologic times occurs in marine sediments; therefore the preservation of organic matter in the marine realm is dealt with here in some detail.

10.3.2 Organic Matter in the Oceans

Primary Production and Heterotrophic Organisms

Primary Production

In the modern oceans the main producers of *primary organic matter* are diatoms, coccolithophorids, and dinoflagellates. These phytoplankton can grow only in the *photic* (or "euphotic") zone where sunlight enables photosynthesis. As soon as the autotrophic organisms sink or are carried by currents into deeper water, their growth ceases or they die. The depth of the photic zone usually extends to between 50 and 100 m. At mid-latitudes, surface waters cool in winter, and the penetration of sunlight is limited. Consequently, primary production is low. As insolation increases with the onset of spring, the phytoplankton begin to grow and reproduce more rapidly, also developing in deeper water, and thus forming "plankton blooms". Such blooms, however, are restricted to areas with abundant nutrient supplies either from rivers or from deeper upwelling waters (Chap. 5.3.3). For this reason, most estuaries, continental shelves, and areas with upwelling water masses are regions of high productivity (cf. Fig. 10.13). About one half to two thirds of the present-day global oceanic primary production (30 gigatons, Gt, per year, where 1 Gt = 10^{15} g; Berger et al. 1989) is derived from these fertile areas of the ocean (approximately one third of the ocean surface). Such an unbalanced regional distribution of organic production is even more pronounced when new production and export production are considered (see below).

The Food Chain

The primary production by phytoplankton is used by *heterotrophic organisms* as food. In the present oceans we can observe the following food chain, or food web:

- Phytoplankton (primary producers).
- Zooplankton (herbivores, i.e., plant eaters).
- Zooplankton (carnivores, i.e., eaters of other zooplankton).
- Fish (plankton-eaters or fish-eaters).

Phytoplankton are eaten mainly by herbivorous zooplankton; they in turn are eaten by carnivorous zooplankton or fish. At each "trophic level" in the food chain, dissolved and suspended, dead organic matter is released from living organisms into the water. Part of this organic detritus is used up by other heterotrophic organisms living in the surface water, but some dead plant and animal matter also sinks into deeper water, thereby supporting life at all depths, including on the sea floor (Fig. 10.5).

Most of the energy the animals gain from their food is used for locomotion and other activities, and only a small part for growth and reproduction. In other words, the population at a higher trophic level must be much larger (often five to six times by weight) than the population at the lower trophic level which feeds on the higher one. Thus, the amount of organic matter produced by phytoplankton is reduced drastically from one trophic level to the next, and the biomass of the standing crop in the entire ecological system represents only a small fraction of the primary plant production. The organic matter, which is needed as energy by organisms in the food chain, is transformed into compounds with lower energy contents (through metabolism) and finally by bacterial activity into carbon dioxide and water.

At the same time, inorganic nutrients are released into the water. If these salts remain in the trophic zone or are recycled by currents into surface waters, they allow high primary production to be maintained.

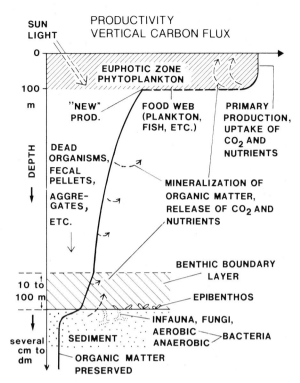

Fig. 10.5. Processes controlling vertical organic carbon flux from zone of primary productivity to burial in sediment. For further explanations see text.

New Production and Export Production

General Aspects

Primary productivity and life processes in the photic zone are not commonly a focus of geological and sedimentological studies. Geologists are more interested in the burial and preservation of organic matter below the sea floor. For this reason, marine biologists and sedimentologists have introduced the terms *new production* and *export production* (e.g., Berger et al. 1989). New production is the vertical organic carbon flux below the photic zone (at about 100 m below sea level, Figs. 10.5 and 10.6b).

Whereas primary production is difficult to quantify, new production, i.e., the downward flux of particulate organic matter, can be measured by free-floating particle traps. This quantity may also reflect the seasonal variation in primary production.

However, the vertical particle flux between 100 m and the sea floor is reduced by organisms in the deeper water. The remaining proportion of organic matter at a given depth, as on the sea floor, is called *export production*.

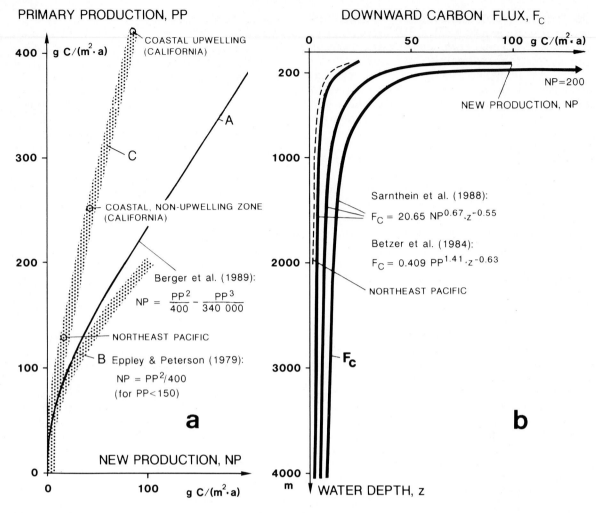

Fig. 10.6. a Relationship between primary net production, PP, and new production, NP, in the present-day oceans. Curve *A* after Berger et al. (1989); *B* after data from Eppley and Peterson (1979); *C* after Martin et al. (1987).

b Downward C flux, F_C (export production), as a function of new production, *NP,* and water depth, *z,* in the present-day oceans. Formula by Betzer et al. (1984) relates F_C, PP, and z

Thus, the vertical carbon flux finally reaching the sediment surface is often only a small fraction of the new production.

Adsorbed and Dissolved Organic Matter

Organic matter accumulates on the sea floor not only as pure organic particulate material, but also in combination with inorganic minerals, especially clay minerals. These adsorb dissolved organic compounds onto their surfaces during settling through the water column. Under specific conditions, particularly in front of deltas or in estuaries where fresh water mixes with sea water, organic com-

pounds and clay minerals can flocculate and form *organomineral aggregates* which slowly settle to the sea floor. In this and similar cases, the highest organic carbon contents correlate with the finest grain sizes. In addition, relatively large amounts of organic matter are dissolved in sea water as well as in fresh-water lakes and river runoff. The concentrations of dissolved organic compounds in sea water are on the order of 0.5 g C/m^3 and therefore are usually much higher than those of particulate organic matter. But only a small fraction of this pool can be adsorbed by clay and settle as coatings on the mineral particles.

Some of the dissolved organic compounds is also consumed by plankton organisms. Thus, the contribution of dissolved organic matter to the total vertical carbon flux is included in the data measured by the free-floating sediment traps.

Vertical Carbon Flux

In recent years considerable efforts have been made (see, e.g., Suess 1980; Betzer et al. 1984; Martin et al. 1987; Sarnthein et al. 1988; Berger et al. 1989) to quantify the vertical carbon flux in different parts of the modern oceans. Primary (total) production and new production at the 100 m water depth appear closely related (Fig. 10.6a). If primary production is low, most is consumed by heterotrophic organisms already in the photic zone, and little is left to sink as export production into deeper waters. However, as soon as primary production increases to 100 g/(m^2· a) and more, a large part of it is left over as new (and export) production. On the average, new production at 100 m is about 20 % of the primary production (Berger et al. 1989). It amounts to about 6 gigatons per year in the modern oceans.

Export production decreases regularly with depth, a pattern which can be found in all present-day ocean basins where the deep water oxygen content is at least 50 to 100 μmol/kg (1.6 to 3.2 g/m^3). Studies on settling dead foraminifers have shown that the number of tests devoid of organic matter increases rapidly with depth; below 1000 m most of the tests contain none.

From many trap studies it is now possible to make estimates for the downward vertical carbon flux below the photic zone (export production) as a function of either primary production, PP, or new production, NP, and water depth, z (Fig. 10.6b). Data from different regions reveal the same tendency. However, such calculations do not take into account terrestrial organic matter which may be transmitted into the sea in greatly varying quantities (see below).

Regional Variation

Primary (total) production and new production vary considerably in the different regions of the present-day oceans. In open ocean conditions, primary production of organic carbon is about 50 g/(m^2· a), whereas in coastal, non-upwelling regions about 150 g/(m^2· a) are typical (Table 10.2). Ocean regions with oligotrophic waters (e.g., areas of subtropical gyres

such as the Sargasso Sea) contribute less than 5 % of the global new production, although they constitute as much as 40 % of the ocean surface. Most new production occurs in narrow, high-fertility belts along the equator, the subpolar divergence zones, and in regions of coastal upwelling (Figs. 5.1 and 5.2, Chaps. 5.2.3 and 5.3.3).

The export production of organic carbon arriving on the sea floor is only about 0.5 g/(m^2· a) in the deep ocean (water depths of 5000 m), but some 15 g/(m^{2}a) in nonupwelling coastal waters (water depths of 250 m). Thus, a ratio of 1:3 in primary production is magnified to 1:30 on the sea floor. Consequently, the carbon flux to the deep-sea floor is less than that to the continental shelf floors, although the shelves comprise only about 10 % of the total surface area of the oceans. Below zones of coastal upwelling, the organic carbon fallout on the sea floor at depths of 250 m is about 250 g/(m^2· a).

Table 10.2. Primary production, new production, export production (organic carbon flux to the sea floor), and organic carbon accumulation in sediments from various, present-day ocean regions (all values in g/(m^2· a); from Stein 1991, based on Romankevich 1984; Betzer et al. 1984; Berger et al. 1989; for additional data also see Deuser 1986; Martin et al. 1987; Knauer et al. 1990)

	Open ocean	Coastal non-upwelling	Coastal up-welling
Primary production	50	150	250
New production	5	50	110
Export production (organic carbon flux, F$_c$, to the sea floor) at water depth of			
250 m	-	15	30
5000 m	0.5	-	-
Organic carbon accumulation in sediment, when sedimentation rate 1 cm/ka	0.01	-	-
10 cm/ka	-	3	-
20 cm/ka	-	-	13

These observations in the modern oceans agree with the geologic record, in which most of the organic matter preserved in ancient sediments accumulated in shallow seas. However, in certain regions and during specific periods organic carbon flux to the sea floor increased as a result of widespread anoxic deep-water environments (cf. Chaps. 5.2 and 7.5). In addition, considerable amounts of organic matter, including those in slope sediments, were transported by turbidity currents into deeper water (e.g., Degens et al. 1986; Stein et al. 1989; Stein 1991). Also important is the contribution of terrestrial organic matter to the organic carbon content of marine sediments.

Terrestrial Organic Matter in the Sea

In addition to the autochthonous production of organic matter in the oceans, input by rivers of both dissolved and particulate, *plant-derived material* from the continents is an important source. Furthermore, in the organic carbon-poor sediments of the central open-ocean regions, organic carbon supplied by winds may be a major proportion of the organic matter accumulated in these environments (Zafiriou et al. 1985; Prahl and Muehlhausen 1989; Stein 1991).

Detailed work on the composition of organic matter in many cores taken during the Deep Sea Drilling Project from Mesozoic and Cenozoic sediments has shown (Fig. 10.7) that a great portion of the kerogen (insoluble organic compounds, Chap. 14.1) must have come from neighboring continents (e.g., Stein et al. 1986, 1989).

The source of organic matter can be identified by different techniques (Section 14.1). As long as the sediments are immature, i.e., not deeply buried and substantially heated, marine organic matter (kerogen type II) derived from algae, phytoplankton, zooplankton and bacteria is characterized by a high *hydrocarbon index* (Fig. 10.8). These organisms contain a high proportion of hydrogen-rich lipid material, whereas terrestrial organic matter (kerogen type III), dominated by cellulose and lignin, renders low hydrocarbon index values. In the following paragraphs, only the quantitative importance of land-derived organic matter in marine sediments is considered.

According to Degens and Ittekkot (1985), the present-day continents deliver dissolved and particulate organic matter in large, but considerably varying quantities into the oceans (Fig. 10.9). Although these data only partially include the effects of peak floods (which transport large quantities of organic matter) and may often underestimate particle flux, they allow some important conclusions to be drawn:

- The carbon yield of the continents, in terms of dissolved and particulate matter delivered by rivers to the ocean, is almost of the same order of magnitute (per unit area of land surface) as the mean new productivity of the ocean (20 g/(m^2· a), see Table 3.2). The continents range between approximately 2.5 g/(m^2 a) (Africa and Europe) and 12 g/(m^2 a) (Asia and Oceania). The mean value of this type of "new productivity" for all the continents together is 8 g/(m^2· a), but it may actually be higher because peak floods with high organic carbon concentrations are not sufficiently taken into account, as mentioned above.
- The total carbon yield from the present-day continents to the oceans reaches at least 1.2 Gt/a; at least 20 % of this amount enters the oceans as dissolved organic compounds.
- Regions with high relief and forests, particularly tropical rain forests (for example Oceania) contribute larger amounts of organic matter per unit area to the oceans than do areas of low relief and dry climate (Africa).

10.3.3 Organic Matter Preservation and Black Shales

General Aspects

Present-day deep-sea sediments are generally poor in organic matter, because the vertical carbon flux originating from primary production is continuously reduced by remineralization in the water column as well as by the activity of benthic organisms. However, gravity mass movements can transport organic-rich sediments from the outer shelf or slope into deep water. Fine-grained mud turbidites, in particular, often contain relatively high amounts of organic carbon (Morris 1987; Stein 1991; also see Chap. 5.4.5). For that reason, the accumulation rate of organic matter can be fairly high in some deep water sediments, even when the vertical carbon flux from the overlying photic zone is negligible. In addition, some small, deep ocean basins with restricted water circulation (Chaps. 4 and 5) can store relatively large amounts of organic matter.

Most of the land-derived organic matter entering the ocean is mineralized, used up by heterotrophic organisms, or deposited in coastal and shallow-water zones. Here, the already fairly high autochthonous vertical carbon flux can be augmented, particularly in estuarine and prodelta environments.

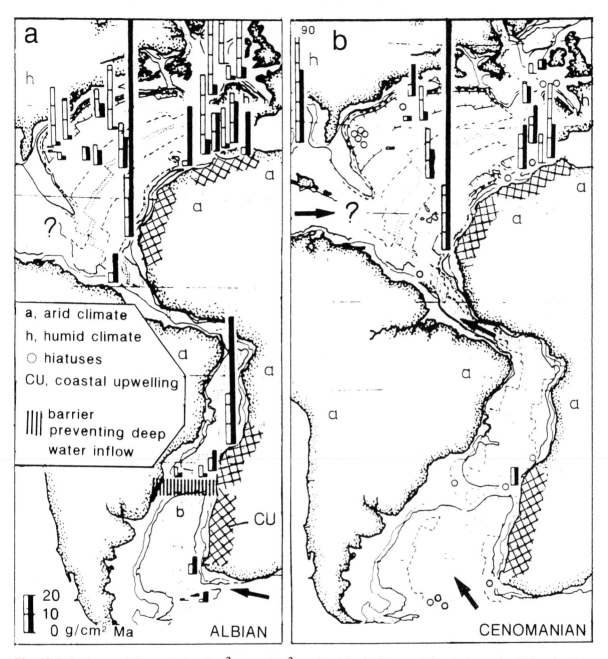

Fig. 10.7a,b. Accumulation rates in g/(cm²·Ma) or 10⁻² g/(m²·a) of marine *(solid bars)* and terrestrial organic matter *(open bars)* in **a** Albian and **b** Cenomanian sediments of the Atlantic Ocean. (After Stein et al. 1986 and 1989). Note that the land-derived allochthonous proportion of total organic matter often surpasses that of marine origin

Organic Carbon Loss on the Sea Floor

Processes

The export production which reaches the sea floor as particulate matter is commonly a small fraction of the primary and new production. It first enters the *benthic boundary layer* (Fig. 10.5), i.e., bottom water with a slightly higher concentration of suspended material and chemical characteristics which often deviate somewhat from those of the overlying water body. Increased activity by planktonic micro- and macro-organisms in this benthic boundary layer may reduce the vertical carbon flux more rapidly than it is in the higher water

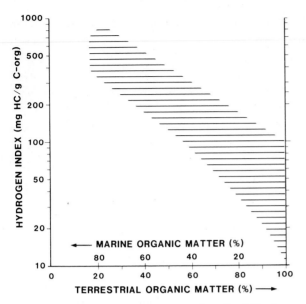

Fig. 10.8. Hydrocarbon index values (from Rock-Eval pyrolysis and maceral analysis) as a means of discriminating between marine and terrestrial organic matter. DSDP samples, vitrinite reflection R_o ≤0.5. (After Stein et al. 1986)

sediment (epibenthos, infauna, fungi and bacteria, Fig. 10.5). The sediment community with its micro-organisms growing rapidly in the uppermost millimeters or centimeters of the sediment, can accomplish further drastic reduction of the vertical carbon flux and thus release considerable quantities of CO_2 and nutrients.

Most of this activity is accomplished under *aerobic conditions*. Oxygen flux from the bottom water into the sediment is facilitated by burrowing organisms, but at some depth below the sediment/water interface (usually several centimeters) oxygen transfer becomes insufficient. Only *anaerobic organisms* can then persist, such as bacteria which reduce nitrate and sulfate, or bacteria which produce methane (cf. Chap. 13.6.3). They gain their energy from the mineralization of organic matter and thus further reduce the vertical carbon flux. In environments with oxygen-depleted bottom water and a high sedimentation rate, anaerobic bacteria decompose more organic matter than do aerobic organisms in the same situation. Ultimately, at the end of this long chain of consumers, only the most resistant organic particles and compounds are preserved in the sediment and have a chance of being fossilized. As a result of all these activities, most of the organic matter arriving at the sea floor is commonly decomposed above and below the sediment/water interface.

column. Although this process is linked with increased oxygen consumption, the bottom water usually still contains enough oxygen to support *bottom life* on top of and within the upper few centimeters to decimeters of the

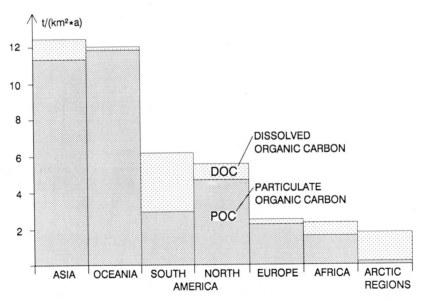

Fig. 10.9. Annual carbon yield of the present-day continents, transported by rivers into the ocean as particulate or dissolved organic matter. Oceania includes Australia. (After Degens and Ittekkot 1985)

Quantification

It is difficult to quantify these processes on the sea floor; nevertheless it is possible to measure the *accumulation rate of organic matter* (1) on the sediment surface and (2) at some depth below the zone of biological activity, i.e., to measure the difference between the input into and the output from the benthic life zone. In this case, however, the activity within the benthic boundary layer is neglected.

Therefore, attempts have been made to measure the biomass of the different communities living in this environment in order to estimate their reproduction and food requirements. One may also determine the oxygen consumption and CO_2 release of the living communities, e.g., the CO_2 flux from the sediment surface into the sea water. Oxygen consumption by benthic respiration, for example, is substantial in ocean regions with high organic carbon fallout to the sea floor. It contributes significantly to the strong oxygen minimum in many shallow-marine regions (Berger et al. 1989).

Applying these and other methods to the 1300 m-deep sea floor of the Santa Catalina basin off the coast of California, Smith et al. (1987) found that up to one half of the total mineralization of particulate and dissolved organic matter occurs in the benthic boundary layer, mainly by micro-plankton. At least the same quantity is decomposed within the top layers of the sediment.

In total, roughly 10 to 20 $g/(m^2 \cdot a)$ of carbon are mineralized. This is certainly more than the vertical carbon flux available in most of the ocean regions at this depth (Table 10.2), but it does correspond approximately with the flux of particulate matter typical in areas of upwelling, which is in fact the case for this basin. Furthermore, it should be noted that part of the organic matter mineralized on or near the sea floor is taken from the large pool of dissolved organic compounds in sea water. In the Catalina basin, the buried sediments below the benthic life zone still contain 6.5 % (by weight) of organic carbon.

A situation similar to that of the Santa Catalina basin has been studied in the zone of coastal upwelling off Peru (Suess et al. 1987). In the southern part of this region, average primary organic carbon production is on the order of 300 $g/(m^2 \cdot a)$, and the carbon flux arriving on the sea floor (water depth ≤500 m) is 88 $g/(m^2 \cdot a)$. Ultimately, 24 $g/(m^2 \cdot a)$ are buried, i.e., about 8 % of the primary production. This results in an organic carbon content of ≥5 % in the sediment which signifies that about 70 % of the organic carbon fallout are used up by benthic organisms on the sea floor and by processes of early diagenesis. At the same time, most of the biogenic carbonate reaching the sea floor (on the order of 150 $g/(m^2 \cdot a)$) is dissolved with the aid of CO_2 generated by the decomposition of organic matter. In order to dissolve a $CaCO_3$ flux of 142 $g/(m^2 \cdot a)$, an organic carbon flux of 34 $g/(m^2 \cdot a)$ is required.

In other regions, benthic degradation loss is frequently around 80 % of the carbon flux arriving on the sea floor (e.g., Stein 1991).

Organic Carbon Flux and Inorganic Accumulation Rates

The results of direct measurements, as reported from the Santa Catalina basin and off Peru, cannot be applied to other regions, but a number of studies compare organic carbon concentrations or accumulation rates (for definitions see Chap. 10.1) in young marine sediments with present-day production rates in the overlying water body (e.g., Müller and Suess 1979; Johnson Ibach 1982; Bralower and Thierstein 1984; Emerson 1985; Stein 1986, 1991). Although this information is still somewhat controversial, one can nonetheless draw some general conclusions on the interrelationship between the input of organic matter, the sedimentation or accumulation rate of carbon-free material, benthic life activity, and the preservation of organic matter (Fig. 10.10).

Flux Rates

In order to discuss this fairly complicated problem, we can distinguish between

- Flux rates of organic carbon, F_C, and inorganic material, F_I, from the water body into the benthic zone.
- Flux rates or accumulation rates of organic carbon, AR_C, and inorganic material, AR_I, in the deeper sediment below the zone of benthic life activity.

If erosional processes and dissolution on the sea floor are excluded, AR_I is equal to F_I, but AR_C becomes less than F_C due to *benthic degradation*.

When describing these relationships, one should not confuse accumulation rates, e.g., in $g/(m^2 \cdot a)$, linear sedimentation rates, e.g., in m/Ma, and concentrations in percent by dry weight.

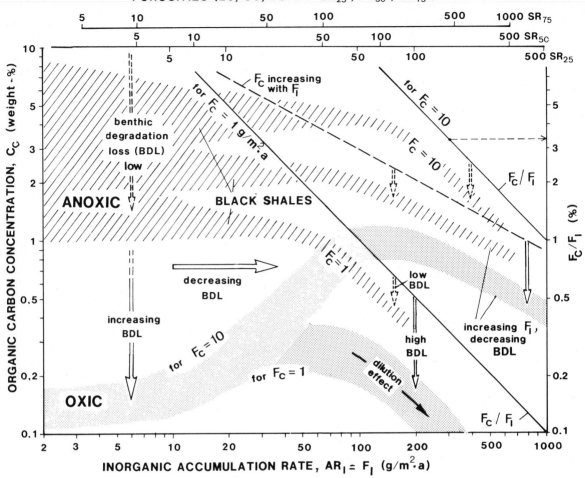

Fig. 10.10. Preservation of organic carbon, C_C (in % by weight), as a function of organic carbon flux, F_C, into the benthic degradation zone, flux of inorganic material, F_I, to the sea floor, and benthic degradation loss, *BDL*. *Solid straight lines* indicate changing F_C/F_I ratios, with F_C held constant. Dashed line indicates F_C/F_I ratio, with F_C increasing with F_I. *Both types of line* mark the theoretical upper limit of organic carbon concentration, C_C, for BDL = 0. *Stippled and hatched fields* show possible C concentrations in sediments of oxic and anoxic environments related to different carbon flux rates, F_C. Note that carbon contents increase with intermediate inorganic accumulation rates, F_I, but decrease with higher rates due to dilution

The relationship between the inorganic accumulation rate, AR_I, and the total sedimentation rate, SR (including biogenic skeletal material), is shown in Fig. 10.10 for three different porosities (25, 50, and 75 %). The concentration of organic carbon buried in deeper sediments is $C_C = AR_C/(AR_I + AR_C)$.

Flux Rates and Organic Carbon Concentration

A simple relationship exists between F_C and F_I (= AR_I) and thus between these parameters and C_C. If F_C remains constant, C_C decreases systematically with increasing F_I. This effect is usually referred to as *clastic dilution*. It is shown in Fig. 10.10 by solid lines for the (hypothetical) case that the sediment accumulation rate, AR_C, is equal to the fallout to the sea floor, F_C. In reality, AR_C is always smaller than F_C, even in anoxic environments. The concentration of organic carbon, C_C, in sediments is therefore generally lower than the theoretical upper limit shown for the various F_C/F_I ratios in Fig. 10.10. In areas of very high inorganic sedimentation rates, for example on prodelta slopes, C_C approaches a linear relationship in which it is negatively correlated with SR or AR_I, although the flux and accumulation rates of organic carbon, F_C and AR_C, into the depositional system are usually high.

Factors Controlling Benthic Degradation Loss

We define the difference between the input of organic carbon, F_C, into the benthic life zone and the accumulaation rate, AR_C, in the buried sediment as *benthic degradation loss*, BDL; ($BDL = F_C - AR_C$). This loss appears to be relatively low in environments where the residence time of organic matter in the zone of benthic life activity is short due to a high inorganic accumulation rate, AR_I (= F_I). Conversely, BDL is very high in areas of low AR_I (≤1 cm/ka = 10 m/Ma) and under oxic conditions, because the benthic community has a long time (hundreds to thousands of years) to decompose the organic material near the sediment surface. As a result, only the most resistant compounds settle into the deeper layers. Under these conditions, the organic carbon concentrations, C_C, in the buried sediments generally remain low (usually 0.1 to 0.3 % by weight), even if the carbon flux, F_C, into the benthic zone varies between low and medium values.

If, however, the inorganic accumulation rate, AR_I (= F_I), surpasses a certain lower limit (about 10 to 20 g/(m²·a) or roughly 10 to 20 m/Ma, assuming a porosity of 50 to 75 %, cf. Fig. 10.10), the residence time of organic matter in the benthic life zone is significantly reduced. Bioturbation begins to affect deeper layers, because some food is still available. This in turn introduces carbon into deeper portions of the sediment, where it more effectively depletes the still oxygenated pore water and thus may reduce degradation. With increasing inorganic accumulation rates, benthic degradation loss, BDL, is even more diminished. Better preservation of organic matter, AR_C, is thereby accomplished, and consequently the concentration of organic carbon, C_C, in the buried sediment may increase. Several authors (summarized by Stein 1986 and 1991) have found such a positive correlation between C_C and increasing sedimentation rates for oxic environments (Figs. 10.10 and 10.11).

It is not quite clear to what extent this somewhat surprising finding is caused by the processes mentioned above or by other phenomena. Higher sedimentation rates are also often associated with increasing carbon flux, F_C, into the benthic

zone, especially if part of the inorganic flux, F_I, consists of biogenic skeletal material. Such a case is indicated in Fig. 10.10 by the note "F_C increasing with F_I". Here it is obvious that C_C will grow with increasing F_I (= AR_I). Certainly the proportion of relatively resistant organic matter (e.g., land-derived plant material) in the total carbon flux also plays a significant role.

Clastic Dilution

Approaching the "dilution line" marked by solid lines in Fig. 10.10, C_C again decreases, as pointed out earlier. This bend in the band of organic carbon concentrations occurs when inorganic accumulation rates range from about 20 to 100 g/(m²·a), depending on the input rate, F_C.

Dilution of organic carbon also occurs in sediments deposited in *anoxic environments* (black shales). In contrast to oxic conditions, black shales can already have high organic carbon concentrations, C_C, in conjunction with low accumulation rates (Fig. 10.10), because the benthic degradation loss, BDL, is much lower under anaerobic conditions than under aerobic ones. Therefore, a decrease due to dilution rather than an increase in C_C with growing inorganic input, F_I, is expected. This is in agreement with actual data shown in Fig. 10.11.

Organic Carbon Preservation and Black Shales

Organic Carbon in Various Depositional Environments (Model Basins)

The simple models shown in Fig. 10.12 semi-quantitatively display the influence of the total carbon flux (F_C or AR_C) in both oxic and anoxic water conditions, as well as the resulting concentrations of organic carbon, C_C, in the buried sediment. Whereas they take into account the inorganic particle flux (F_I = AR_I), they neglect the different types of organic matter discussed in Chapter 14.1 as well as the changing proportions of bioclastic and terrigenous material in the inorganic flux. The models address the following environmental conditions:

- *Deep-sea oxic environments* (Fig. 10.12a) with either low or high primary and new production, very low or low to medium inorganic accumulation rate, AR_I, and low to medium benthic degradation loss, BDL. As a result, the organic carbon flux, AR_C, into the deeper

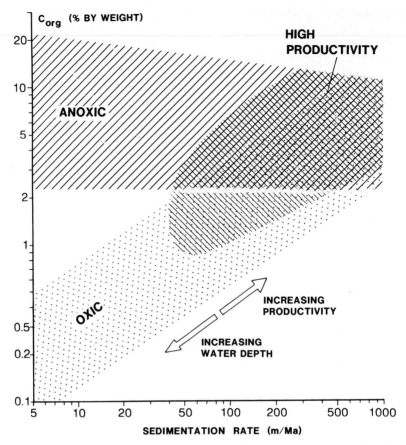

Fig. 10.11. Correlation between organic carbon concentrations and sedimentation rates of oxic and anoxic environments as well as areas of high organic productivity (e.g., regions of upwelling). Based on data from the present-day oceans (DSDP samples, including Jurassic and Cretaceous sediments) and adjacent seas, for example, Baltic Sea, Black Sea, Mediterranean. (After Stein et al. 1986)

sediment is very low to low; C_C is consequently very low or medium, respectively.

- *Deep-sea basins* (of various size) *with stratified water* and an *oxygen minimum* (Fig. 10.12 b), and either low or high primary and new productivity. Low AR_I enables high to very high values of C_C. Although BDL is very low, high AR_I only leads to low to medium C_C values.

- *Shallow seas* (e.g., shelves) *with oxic conditions,* low or high primary new production, and medium to very high inorganic accumulation rate, AR_I. Although organic carbon input, F_C, into the benthic zone may become very high (and thus the organic carbon accumulation rate, AR_C), the carbon concentration values, C_C, turn out to be low to medium due to high AR_I.

- *Shallow seas* (shelves or shallow basins) *with anoxic conditions* caused by restricted water circulation or coastal upwelling; low or high

primary and new production. The low to high values of C_C are mainly controlled by F_C and AR_I.

These models may be modified by additional factors, for example by the lateral input of organic carbon, including plant material, by bottom currents and gravity mass movements. The uppermost sediment layers may be reworked, and organic matter already buried may again be redeposited in the zone of intense benthic degradation.

The identification of the different types of black shale environments mentioned above and shown in Fig. 10.13 is often difficult in ancient rocks. Specific planktonic microfaunas of low diversity and with high amounts of biogenic opal may indicate upwelling conditions (e.g., Thiede and Suess 1983; Kuhnt et al. 1986, cf. Chap. 5.3.5). High concentrations of trace metals trapped from sea water by

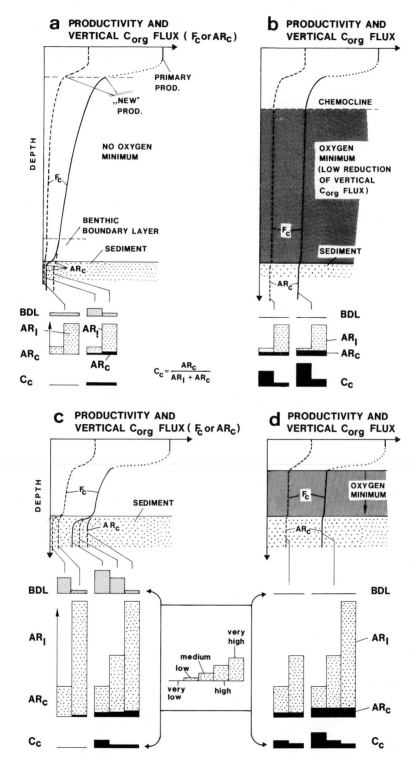

Fig. 10.12a-d. Semi-quantitative models showing the decrease in organic carbon flux, F_C, from surface waters through the deeper water column and due to benthic degradation loss, *BDL,* in the sediment (accumulation rate AR_C), in oxic and anoxic environments. BDL and concentration of organic carbon, C_C, in the buried sediment are controlled by the inorganic accumulation rate, AR_I

(cf. Fig. 10.10). **a,** Deep basin with deep-reaching water circulation. **b** Deep basin and stratified waters with oxygen minimum. **c** Shallow sea (e.g., shelf) with oxic conditions on the sea floor. **d** Shallow sea with oxygen minimum in shallow basin or under coastal upwelling. For further explanation see text

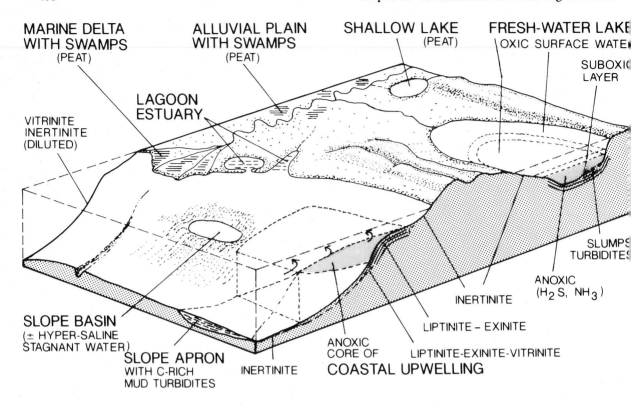

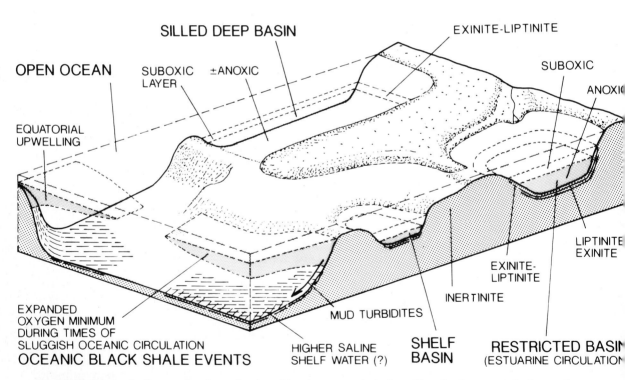

Fig. 10.13. Schematic diagram showing major depositional environments for the accumulation of organic-rich sediments, with their dominant types of kerogen. (After Brooks et al. 1987). Basins with stagnant bottom waters develop comparatively stable suboxic or completely anoxic zones, whereas zones of coastal and equatorial upwelling are characterized by spatial and temporal variations in their degrees of oxygen deficiency

hydrogen sulfide and/or organic matter are probably best explained by a stagnant anoxic basin and a low sedimentation rate (e.g., Brumsack 1986b).

Neglecting these complications, one can summarize the results of the models as follows:

1. Low concentrations of organic carbon, C_C, occur in sediments in many environments, including regions with high primary and new productivities.
2. Sediments with high to very high C_C values are formed only under anoxic or oxygen minimum conditions, in either deep or shallow water. Primary productivity can be fairly low, but inorganic accumulation rates must also be low or, at the most, intermediate (in shallow water). Under such conditions, organic carbon content does not increase with the sedimentation rate (Stein 1991).
3. High to very high accumulation rates of organic carbon, AR_C, are found in deep water if productivity is high and the water anoxic. In shallow water, AR_C can become high even under oxic conditions, if productivity is sufficiently developed.

Hydrocarbon Source Rocks and Black Shales

In the evaluation of petroleum source rocks (Chap. 14), the total organic carbon concentration, C_C, is a crucial factor. Carbonate source rocks should contain at least 0.3 % organic carbon and clastic source rocks at least 0.5 % (Tissot and Welte 1984). However, the nature of the buried organic matter also plays an important role. Oil-prone organic constituents are better preserved if the residence time of the organic particles in the zone of intense benthic activity is limited by high bulk accumulation rates. Therefore, if two marine source rocks have the same lithology and organic carbon content, the more rapidly deposited one usually has a greater potential of generating oil (Johnson Ibach 1982).

The most favorable terrestrial and aquatic environments for the formation of hydrocarbon source rocks are summarized in Fig. 10.13. They include:

- Swamps and the tidal flat zones of marine deltas (Chap. 3.4.2).
- Estuaries and lagoons with high nutrient supplies (Chap. 3.2.3).
- Basins with restricted circulation on the continental shelf and in adjacent seas (Chap. 4).

- Shelves and upper continental slopes under zones of coastal upwelling (Chap. 5.3.3).
- Widely extended deep-sea regions during periods of sluggish oceanic circulation.

In most of these environments, intermediate to relatively high sedimentation rates (up to several 100's of m/Ma) prevail.

In recent years, the term *black shale* has become rather popular. Black shales are fine-grained, dark-colored, compacted sediments which usually display distinct lamination and commonly contain ≥0.5 % of organic matter (Wetzel 1989). Their dark gray color is caused by finely dispersed iron sulfides and organic compounds. However, not all dark shales are true black shales with high organic matter contents, and not every black shale is a hydrocarbon source rock (cf. Chap. 14.1).

Organic Carbon Preservation Factor

Finally, we may ask the question: which proportion of the primary production, PP, in overlying surface waters is preserved in the sediments below the zone of benthic activity? The organic carbon accumulation rate, AR_C, in the sediment, expressed as a percentage of PP, is defined as the *organic carbon preservation factor*.

This factor can be determined for young sediments for which both values, PP and AR_C, can be measured with confidence. As can be seen in Fig. 10.14, most of the present-day pelagic and hemipelagic deep-sea sediments have a low to very low preservation factor (less than 1 %, often only about 0.01 to 0.1 %; also see Table 10.2). In contrast, near-shore sediments deposited more rapidly display much higher organic carbon preservation factors: up to 3 %, especially in cases of coastal upwelling. Under anoxic conditions, where burrowing organisms can no longer live on the sea floor and the sediments are laminated, the preservation factor can surpass 3 % and reach values as high as 10 % or more, for example on the Peru shelf (coastal upwelling), in some basins of the Gulf of California and in the Californian borderland, and in some bays and inlets along various coasts. Black Sea sediments under H_2S-bearing waters display an average preservation factor of 4.3 %, if land-derived organic carbon is neglected.

Taking into account such preservation factors and evaluating the paleoenvironment (oxic or anoxic conditions, bulk accumulation rate, etc.), some authors have recently tried to re-

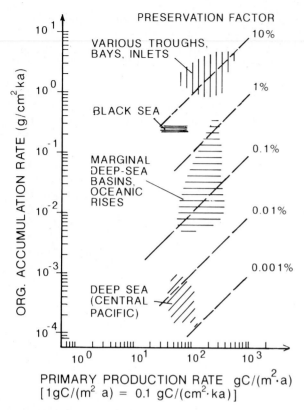

Fig. 10.14. Organic carbon preservation factors (accumulation rate, AR_C, expressed as a percentage of primary production rate, PP) of Holocene sediments. (After Bralower and Thierstein 1984)

calculate the primary productivity or, paleo-fertility, of ancient ocean basins from the organic carbon accumulation rate of their preserved sediments (summary in Berger et al. 1989).

Sarnthein et al. (1987), for example, demonstrate that the fertility (or in this case the new production) of low- and mid-latitude ocean regions during the last glacial period must have been about twice as high as in the Holocene. In their investigation of mid-Cretaceous, laminated black shales, Bralower and Thierstein (1984, 1987) come to the conclusion that the primary production rate may have been as much as one order of magnitude lower than that measured in the modern oceans.

Warm ocean basins with sluggish circulation and limited oxygen supply hamper the remineralization of organic matter and thus the recycling of nutrients. In addition, an increased loss of nitrate through denitrification in oxygen deficient waters may limit fertility and cause "starved ocean basins" (Berger et al. 1989).

In all estimates in which the organic carbon accumulation rate, AR_C, is determined for compacted and more or less lithified sediments, corrections for additional organic carbon losses from shallow to deep burial have to be introduced. In immature hydrocarbon source rocks (see Chap. 14) or nonsource rocks, however, these diagenetic carbon losses appear to be relatively small (Berner 1982; Bralower and Thierstein 1987).

11 The Interplay Between Sediment Supply, Subsidence, and Basin Fill

11.1 Introduction

Principal Factors Controlling Basin Fills

The general principles of mass balances between the drainage area of sedimentary basins and the corresponding basin fill were already discussed in Chapter 9.4. The examples described there were selected for the determination of long-term denudation rates. In this section, some basic laws of "basin-filling models" (e.g., Cross 1990; Angevine et al. 1990) are discussed. The filling of sedimentary basins is controlled by the interaction between specified, more or less independent factors, for example:

- Size and denudation characteristics of land areas delivering terrigenous sediments.
- Areal extent and geometry of corresponding basin.
- Tectonic subsidence of basin floor and compaction of sediments.
- Hydraulic regime of a river system or a water-filled basin.
- Eustatic sea level changes.
- General rules for autochthonous sediment production, including carbonate buildups.

Using these parameters and taking into account their interrelationships, other, dependent variables may be derived, such as:

- Grain size distribution and lateral and vertical sedimentary facies successions within the basin.
- Sedimentation rates at different locations in the basin.
- Occurrence and extent of carbonate buildups, etc.

Remarks to Basin-Filling Models

Basin-filling models or "quantitative dynamic stratigraphy" (Cross 1990) are still in an early state of development. In this state they can barely provide a means for exactly modeling many types of sedimentary basins, but they are a means of generalizing complex systems and of exploring the effects of varying parameters (Angevine et al. 1990). Basin-filling models can be subdivided into two groups:

1. *Geometric models* characterize basins which have a constant surface geometry, for example a fluvial basin which subsides but is filled with sediments all the time. In this case, sedimentation rate and subsidence are in balance and maintain an equilibrium. Similarly, a coastal plain and the adjacent foreshore zone may maintain a fixed slope which is restored all the time, regardless of sea level fluctuations and/or differential subsidence. Such models are relatively simple to construct.

2. *Dynamic models* take into account sediment transport and the rate of deposition. Both tend to change laterally, i.e., downstream a river, or from the coastline toward the center of a basin. As a result, the geometry of the sedimentary surface may undergo significant modification.

A precise treatment of this complex topic is beyond the intention of this book, but it may be useful to briefly discuss some general points related to basin modeling. Various working groups are trying to simulate the evolution of sedimentary basins with the aid of computers (e.g., Bitzer and Harbough 1987; Strobel et al. 1989; Cross 1990; Lawrence et al. 1990). This forward modeling is commonly carried out in two dimensions along transects and requires both a large data set and special training. The basic concept of this approach is founded on the fact that the factors mentioned above and additional relevant processes can be subdivided into two groups (Lawrence et al. 1990; cf. also Chap. 7.4):

- Processes controlling the creation and destruction of space in a basin.
- Processes controlling the introduction and removal of sediment.

Space is generated in a basin by tectonic subsidence of the basement, isostatic response of the crust to sediment and water loads, compaction of sediments, and rise in sea level. If these processes predominate over sediment accumulation, land-derived sediment can be deposited on the coastal plain and shelf. By contrast, an *excess of sediment supply* over space-generating processes causes the sediments to bypass the shelf and to settle in deeper water. This occurs in periods of slow subsidence or uplift and sea level fall. For a given rate of space creation, the volume of sediment introduced into the basin controls how far the sediments prograde seaward. If clastic sediment sources can be neglected, in situ biogenic sediment production is the principal factor interacting with space-creating processes.

For example, the buildup and drowning of reefs and carbonate platforms is frequently controlled by the rate of carbonate production, which in turn depends on a variety of other factors, such as carbonate-producing organisms at differing water depths, temperature, nutrient supply, and water turbidity (e.g., Schlager and Philip 1990).

Forward numerical simulation of these processes is only possible under the assumption that empirical growth rates at different locations within a basin can be used (e.g., Aigner et al. 1989). In addition, lateral prograding of carbonate buildups and their reactions to minor fluctuations of sea level have to be taken into account (Bosence and Waltham 1990; cf. Chap. 12.2.2).

Most of the factors mentioned above are subject to changes during the evolution of a basin. The characteristics of the source area on land, the size and geometry of the basin, and

processes within the basin may vary substantially with time. Such an evolution is referred to as *basin dynamics*.

The following sections deal (1) with simple models demonstrating some general principles and (2) a limited number of generalized case studies of the evolution of common basin types.

11.2 Simple Relationships Between Source Area on Land and Basin Fill

Predominantly Terrigenous Material

Model Parameters

First, some simple models demonstrating the *influence of the drainage area on land* and its clastic river load on the filling of sedimentary basins of various sizes are presented. The models of Fig. 11.1 are based on the following assumptions:

- Denudation area on land and depositional basin constitute a closed erosion-depositional system (Hay et al. 1989).
- The subsidence rate has a medium value (cf. Chap. 8) and is constant over the entire basin floor.
- The sediments are distributed evenly over the entire basin.
- The time period considered for denudation and sediment accumulation is 10 Ma.
- The initial elevation of the basin floor corresponds with the sea level which is held constant.

Over-Supplied Basins

If the drainage area, AD, on land is large in comparison to the area of the basin, AB, for example AB/AD = 0.1, the basin is frequently over-supplied with sediment and remains filled at all times (geometric basin-filling model, see above). This is also true for a low denudation rate, as long as the subsidence rate in our example does not exceed 100 m/Ma. In this case, the basin fill consists entirely of continental deposits, mostly fluvial sediments. Surplus river load not used for the basin fill is transported into other depositional areas. This is a common situation in many graben-like structures on the continents. Exceptions from

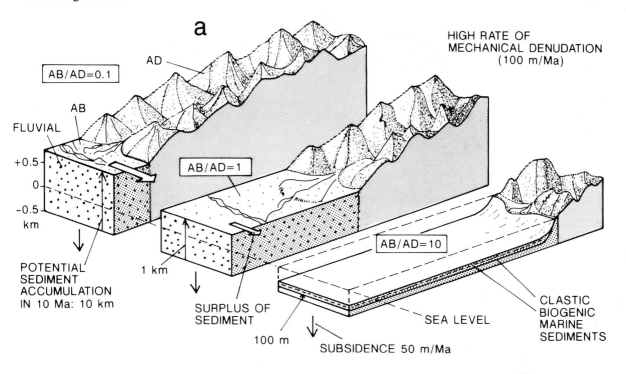

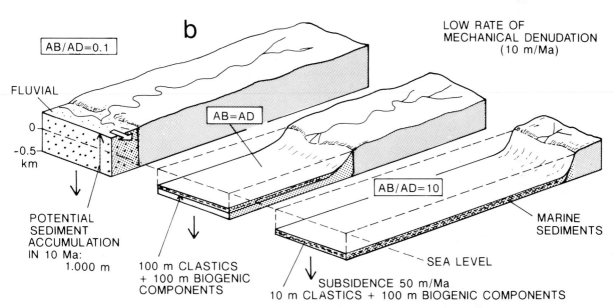

Fig. 11.1a,b. Simple scheme demonstrating relationship between mean mechanical denudation rates, ratio AB/AD of basin area, *AB*, and denudation area, *AD*, and subsidence rate. The initial basin floor corresponds to sea level. The rates of denudation and subsidence are kept constant. The time period for basin fill is assumed to be 10 Ma.
a High denudation rate and low AB/AD (= 0.1) cause rapid infilling of basin with fluvial deposits up to an elevation at which most sediment is carried away into other regions. AB/AD = 1 still provides more sediment than

necessary for (fluvial) basin fill. With AB/AD = 10, terrigenous clastics cannot fully compensate for subsidence: shallow sea, slowly deepening. **b** Low denudation rate still generates fluvial basin fill above sea level, as long as AB/AD is small (AB/AD = 0.1). With AB/AD = 1 and 10, the sub-siding basin is only partially filled with terrigenous clastics. For AB/AD = 10, autochthonous biogenic sediments (assumed sedimentation rate 10 m/Ma) predominate over very low clastic input

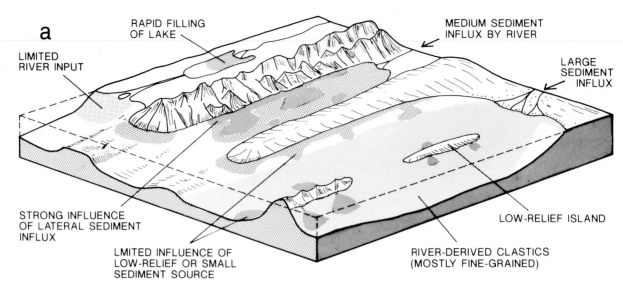

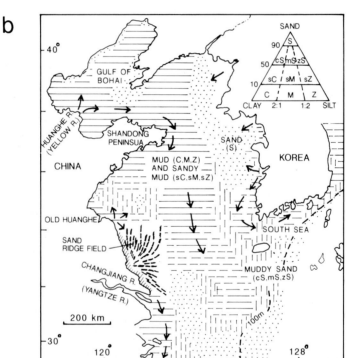

Fig. 11.2. a Qualitative model showing the influence of size and relief of denudation area on sediment fill of neighboring basins. **b** Holocene sediment distribution in the Yellow Sea controlled by suspended load of the rivers Huanghe and Changjiang; grain size classification after Folk 1954. (After Lee and Chough 1989, sand ridge field after Liu Zhenxia et al. 1989)

this rule are grabens or rift zones in a stage of initial rapid subsidence (on the order of 200 m/Ma, see Chap. 8.3) which, in addition, have a comparatively small drainage area with low denudation. In such cases, the graben may be partially filled by lake or sea water (for example in the East African rift zone), or the sediment surface may subside below sea level (Jordan graben with Dead Sea, Death Valley in California).

The modern Yellow Sea between mainland China and the Korean peninsula is an example of an epicontinental shelf sea oversupplied with terrigenous material from major rivers, especially the Huanghe and the Changjiang (Yangtze, cf. Chap. 9.2). Since post-glacial sea level rise had established approximately the present basin configuration about 7000 years ago, the Yellow Sea was largely controlled by siliciclastic mud deposition (mainly silty clays and clayey silts with minor proportions of sand, Fig. 11.2b; Lee and Chough 1989). In the northern portion (Gulf of Bohai) where the influence of the modern Huanghe river is greatest, the Holocene sedimentation rate attains 30 m/ka (solid dry material); in the central portion of the Yellow Sea, the sedimentation rate is around 3 m/ka, and in the south it is 0.5 to 1 m/ka. The highest accumulation rates have been found in the subaqueous delta areas with very gently inclined foresets and proximal bottomsets (Alexander et al. 1991). All these rates appear to be in good agreement with the suspended load which the rivers discharge into the sea. Under steady-state conditions, the Yellow Sea will be filled with sediments in a very short geological period. As a result of extensive agricultural use of the loess plateau, the Huanghe river load (about 1 km^3/a) was in the last 2300 years an order of magnitude higher than in the early and middle Holocene (Milliman et al. 1987).

Basins with Moderate to Low Sediment Supply

Given a ratio of AB/AD $\approx$ 1, the denudation rate on land predominantly controls basin fill and the nature of the sediment. High and medium denudation rates still deliver more sediment volume than required for basin filling due to moderate subsidence (50 m/Ma). Only a low sedimentation rate leads to insufficient sediment supply and, thus, to a deepening basin with marine sediment. Provided the biogenic sediment production reaches a medium value of 10 m/Ma, half of the marine sediment is autochthonous, the other half allochthonous (terrigenous).

With an increasing AB/AD (in Fig. 11.1: AB/AD = 10), even a very high rate of denudation is not able to fill the basin and compensate for subsidence. Under the assumptions of Fig. 11.1a, only one half of the resulting marine sediment is terrigenous which is typical of hemipelagic sediments in the modern oceans. AB/AD = 10 in combination with low denudation eventually leads to predominantly biogenic marine sediments which contain only a small fraction of land-derived material. This model approximates the conditions of pelagic sedimentation in large ocean basins some distance away from a continent (see below).

Influence of Relief and Several Sediment Sources

If the area of the basin is appoximately equal to the drainage area on land (AB/AD = 1), uplift and increasing relief of the drainage area lead to rising input of terrigenous material. The mechanical denudation rate is closely correlated with the mean relief (Chap, 9.2; Pinet and Souridu 1988). As a result, the depositional environment becomes shallower and finally passes into a fluvial system. Simultaneously, the proportion of terrigenous material increases at the expense of biogenic components (see example of Texas-Louisiana shelf, below, Fig. 11.6). On the contrary, a lowering of relief in the denudation area diminishes the terrigenous clastics/biogenics ratio and causes a deepening of the basin, provided the subsidence rate remains constant.

The influence of various terrigenous sediment sources on basin fills is shown qualitatively in Fig. 11.2. Rivers flowing through lakes before they reach the sea deposit their bed load and suspended load in the lake and therefore carry little material into the sea. A high-relief mountain range bordering the sea may exert a strong influence in terms of sedimentation rate and composition on the filling of a neighboring basin, even if a medium to major river from a large hinterland enters the same basin. Low-relief peninsulas and islands of limited extension commonly shed small amounts of clastic material into the sea in comparison to major rivers draining large areas of differing relief and climate. For this reason, the existence of islands in large, ancient ocean basins is often difficult to assess from the basin fill, unless the island differs significantly in its petrographic characteristics from the other, volumetrically predominating sediment sources.

Two modern examples (Gulf of Mexico and Atlantic Ocean) of the relationship between denudation area and basin filling are briefly discussed below (Chap. 11.3). In closed erosion-depositional systems, mass balances of sediments can also be used to reconstruct the size and topography of the drainage area of the basin (Hay et al. 1989). The sediment budgets of some closed and half-closed basins are described in Chapter 9.4; those of accretionary wedges are discussed in Chapter 12.5.

Chemical Sediments

General Aspects

The *dissolved river load* (Chap. 9.2) is carried into lakes and/or into the sea. Its nutrients (e.g., nitrate and phosphate) contribute, besides recycling of nutrients in the depositional area, to maintain and promote organic productivity in lakes, estuaries, and coastal waters (Chap. 10.3.2). Dissolved carbonate and silica are used for the production of skeletal material, but river supply is not sufficient to meet the demand of organisms in the present-day oceans. River-borne sodium chloride and sulfate more or less replenish the loss of the oceans to salt deposits; sulfate is partially also reduced to form metal sulfides; magnesium is required for the dolomitization of calcium carbonate.

Chemical Sediments in Closed Lakes

In the following simplified chemical budget of lakes, the above mentioned different sinks of the dissolved river load are not considered further. Instead it is assumed that the dissolved constituents are precipitated inorganically and form solid, pore-free layers in a lake basin. Such a model simulates to some extent the situation in salt lakes (Chap. 2.5.1) which are, however, not devoid of biota.

It is assumed that the depositional area of a closed lake remains constant and makes up one tenth of the denudation area feeding the lake with river water (Fig. 11.3). The climate in the river catchments is predominantly semiarid to humid, in the lake area it is arid. The rock types in the denudation area may be either granitic, basaltic, or carbonates. In relation to these prevailing rocks, the average denundation rates are assumed to be on the order of 5, 10, and 30 m/Ma, respectively. (In this case, the values for granite and basalt are higher than in most natural environments.) Taking into account the groundwater chemistry of these rock types, which mainly control the chemistry of river water (Chap. 9.2), one can calculate the thicknesses of the major chemical sediments precipitating in the lake during a period of 1 Ma. Provided all the dissolved material is reprecipitated in the lake, this model yields total sediment thicknesses ten times the denudation rates. This signifies that under the conditions of the model, lake sequences of 50, 100, and 300 m accumulate in a time period of 1 Ma.

The *thicknesses of the individual sediment types* (Fig. 11.3) are not only a function of the differing denudation rates, but also vary significantly in relation to the rock types. In terms of sediment volume, calcium carbonate is the most important lake sediment derived from a calcareous and basaltic drainage area, followed by magnesium carbonate which normally combines with calcium carbonate to form dolomitic limestones or dolomite. Granitic rocks, however, tend to deliver more silica than other constituents, but also in this case, calcium carbonate takes the second position. On the other hand, the highest silica supply comes from basaltic rocks, while carbonates deliver only small amounts, unless they are rich in opaline silica. Both carbonate and silica may be entirely or partially used up by organisms to form skeletal carbonate and opaline silica.

The thicknesses of the *gypsum and halite layers* produced in the model basin vary only slightly with rock types. In coastal areas, the rain water may be richer in NaCl and thus produce higher salt concentrations than found in the groundwater analyses used as standards for the model of Fig. 11.3. Nevertheless, the model basin demonstrates the orders of magnitude in terms of drainage area and time necessary to generate evaporitic layers of a certain thickness.

Using the data of Chapter 9.2 and varying the BA/DA ratio, other scenarios may be devised in order to evaluate the potential chemical sediment budget of closed lake-drainage systems. The fate of the dissolved river load entering the sea and its significance in geochemical cycles is discussed in special volumes, such as Holland (1964) and Gregor et al. (1988).

11.3 Different Modes of Basin Fill

Out- and Upbuilding of Sediment from a Point Source

Delta Sediments

Apart from the factors discussed in Chapter 11.2, the rate of sediment accumulation is also affected by the changing geometry and size of a basin, as well as sea (or lake) level changes. Given a point source with constant input of sediment, for example a major river building out a delta on an inclined basin floor, the

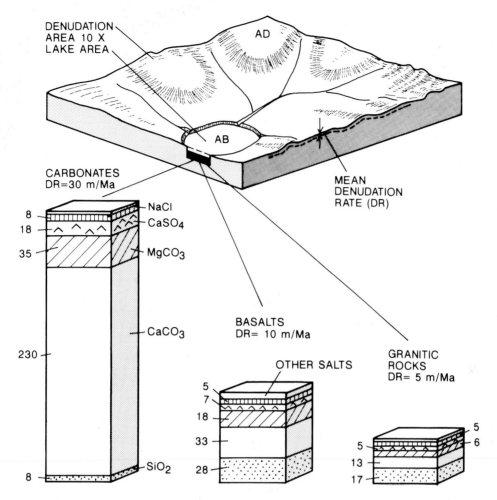

Fig. 11.3. Model of chemically precipitated lake sediments derived from chemical weathering in a closed lake-drainage system and deposited in 1 Ma. The ratio AB/AD of lake area, *AB,* and denudation area, *AD,* is 10 and kept constant; the climate in the drainage area is semiarid to temperate, around the lake it is arid. The rocks in the drainage area consist alternatively of carbonates, basalts, or granites and thus cause differing chemical denudation rates, *DR.* Note that part of the sediments may be biogenic and that $MgCO_3$ commonly is used to form dolomitic limestones and dolomites. See text for further explanation

progradation of the deltaic sediment wedge slows down through time (Fig. 11.4a and b; dynamic model, see Chap. 11.1). Falling sea level causes seaward progradation and rapid outbuilding of the delta, while rising sea level may result either in coastal retreat or seaward progradation, depending on the rates of both sea level rise and deposition (Fig. 11.4c and d).

The *Fraser River delta* of British Columbia, for example, continued to prograde during the Holocene relative rise in sea level. Progradtion amounted to 13 km in the last 9000 years (Fig. 11.4e and f, Williams and Roberts 1989). The sediment discharge of this river was sufficient to allow aggradation to keep pace with the rate of sea level rise and to maintain outbuilding of the delta. The delta growth was accomplished by both vertical accretion (about 2 to 5 mm/a) and lateral progradation (ranging from 1 to 6.5 m/a). During the most rapid Holocene sea level rise, progradation declined to less than 1 m/a. In other areas, the Holocene eustatic rise in sea level was not substantially reduced by rebound of the crust following deglaciation; thus, many former lowstand deltas were flooded and the rivers had to build new deltas adjusted to the present sea level (Fig. 11.4c).

Coastal Progradation

The frequently described drainage system of the **Mississippi river** delivering its sediment load into the Gulf of Mexico (Fig. 11.6a) is a classic example of the interplay between processes in the source area and deposition along

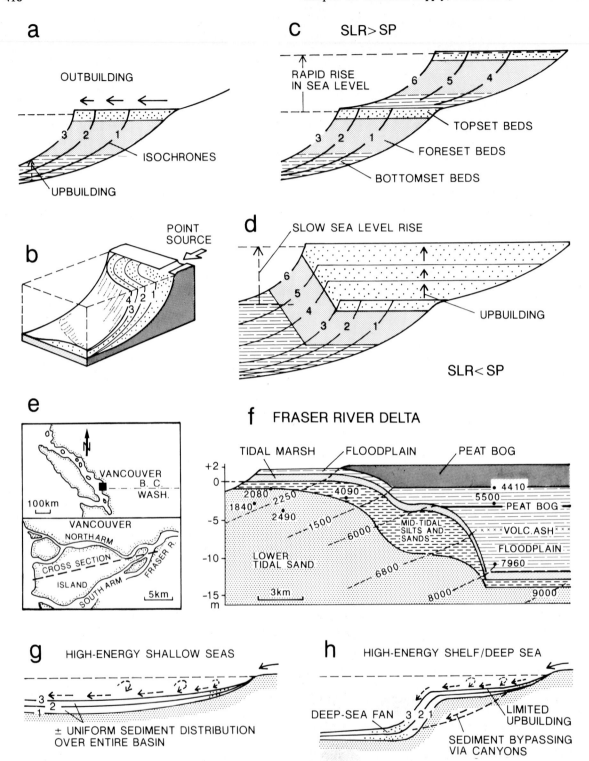

Fig. 11.4. a-d Simple theoretical models demonstrating outbuilding and upbuilding of delta during constant (**a** and **b**) or rising sea level (**c** and **d**). Note that outbuilding from a point source slows down due to both increasing water depth and lateral sediment dispersal (**b**). Outbuilding during rising sea level, *SLR*, can only be maintained by high sediment supply, *SP* (**d**). **e** and **f** Out- and upbuilding of Fraser River delta, British Columbia, during the last 9000 years (After Williams and Roberts 1989). **g** Tendency to uniform sediment distribution in high-energy shallow seas due to strong wave and current action. **h** Limited sediment build-up on high-energy shelf resulting in prograding slope and sediment bypassing via submarine valleys into the deep sea

a passive continental margin (summarized, e.g., by Matthews 1974, 1984a; mass balance see Hay et al. 1989). In contrast to the previous example, here the river-derived material is distributed over a wide shelf and slope region. With the onset of uplift of the Rocky Mountains and the Appalachians in the upper Cretaceous, terrigenous sediment input into the gulf increased significantly. As a result, Jurassic and lower Cretaceous carbonate deposition in a broad shelf region was replaced by the accumulation of a huge clastic wedge.

This wedge migrated about 200 km seaward since the upper Cretaceous. Uniform outbuilding was, however, modified by sea level fluctuations and regional differential subsidence in response to the increasing sediment load and diapirism of Jurassic salt. Cenozoic clastic sediments reach a thickness of up to 15 km. Mass balances indicate that extensive areas of the Rocky Mountains and High Plains have been uplifted as much as 1 to 3 km since the late Pliocene (Hay et al. 1989). On the western Florida shelf, which is an area outside of the influence of the Mississippi river, carbonate sedimentation continues to prevail up to the Present. The clay mineral composition of modern sediments in the gulf reflects the mineral suite of the Mississippi river (Chap. 9.3).

Using the present-day chemical and mechanical denudation rates of the Mississippi river also for the other drainage areas of the Gulf of Mexico, one can estimate the *average sedimentation rate of the gulf* if the incoming river load is distributed evenly over the entire basin. The gulf covers an area of about $1.5 \cdot 10^6$ km^2, the drainage area of the rivers approximately $4.5 \cdot 10^6$ km^2 (the drainage of the Mississippi river alone is $3.27 \cdot 10^6$ km^2); hence the AB/AD ratio is about 1/3. With a mechanical denudation rate of 60 m/Ma (Fig. 9.6), the average total sedimentation rate (solid, pore-free sediment) in the gulf can reach 180 m/Ma. The chemical denudation rate in the drainage area is 15 m/Ma, but only part of the dissolved river load is deposited in the basin. About 15 % of this load is Ca^{2+} (cf. Fig. 9.5, Mississippi and Rio Grande rivers), corresponding to a Ca denudation rate of around 2 m/Ma. After conversion into CaCO$_3$ (factor 2.5) and using AB/AD = 1/3, one obtains a (biogenic) calcium carbonate sedimentation rate of approximately 15 m/Ma.

These estimates do not yield the real sedimentation rates in the basin, which is filled irregularly by a large and several smaller point sources, but they provide the right order of magnitude. Most of the terrigenous material is deposited on submarine fans such as the Mississippi fan, where very high sedimentation rates occur (up to 1000 m/Ma, cf. Chap. 5.4.2). The calcium carbonate sedimentation rate calculated from the river load is about equal to rates found for pelagic carbonates in other ocean basins above the CCD (Chap. 10.2).

Apart from some shelf regions (western Florida, Yucatan), calcium carbonate deposition in the gulf, however, is too much diluted by siliciclastic material to generate carbonate-rich sediments. Finally, it should be mentioned that part of the dissolved river load is exchanged with sea water flowing into and out of the gulf.

Extensional and Closing Basin Filling

Extensional Basin Models

Similar to the declining rate of delta outbuilding into deepening water, the vertical sediment buildup of a basin growing in size will decrease with time, if the volume of sediment input is kept constant. Such a situation was investigated quantitatively by Schlische and Olsen (1990) for rift basins. Their extensional basin filling model is based on the following assumptions:

- The basin represents a continental full-graben or half-graben and is bounded by planar faults dipping at equal angles (Fig. 11.5a).
- Uniform extension causes uniform subsidence along the boundary faults and hence the depth of the basin increases linearly.
- The outlet of the basin is held at a constant level with respect to an external datum line.
- The volume of fluvial sediment added to the basin per unit time is constant; the sediment is distributed uniformly over the entire basin.

The principal results of such a model are briefly summarized here with a few modifications:

1. Early stage of basin evolution. Sediment supply is large enough to fill the small basin, and excess sediment and water leave the basin (Fig. 11.5a, Stage 1). The basin is *hydrologically and sedimentologically open*. During this phase, the sedimentation rate is constant and equal to the subsidence rate.
2. As the basin continues to grow, a point is reached at which the sediment input can no

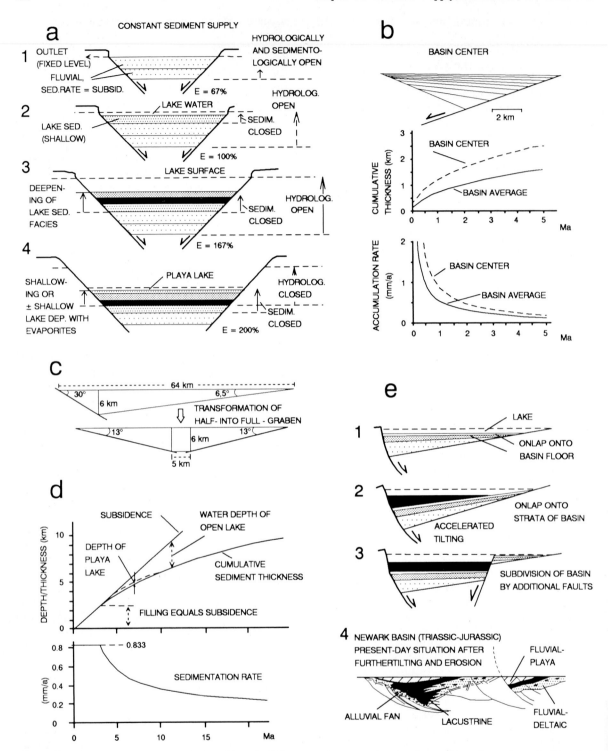

Fig. 11.5a-e. Simple model of extensional rift basin filling under the assumptions of constant volume input of sediment and uniform subsidence. **a** Full-graben filled with fluvial and lake deposits under hydrologically open (1 to 3) and closed conditions (4); E extension (%). **b** Two-dimensional model of progressive, complete fill of half-graben growing in size; cumulative sediment thickness in total basin (average) and basin center, and declining sedimen

tation rates through time. **c** Transformation of half-graben into full-graben to apply full-graben filling model (**d**) based on data from Triassic-Jurassic Newark basin (see Fig. 12.7). **e** Adjustment of deposition to asymmetric, uniform basin subsidence (1), accelerated tilting (2), additional faulting (3), and (4) situation after further tilting and erosion. (After Schlische and Olsen 1990; **e4** from Manspeizer 1988b)

longer completely fill the space made available through subsidence. Consequently, the basin is *sedimentologically closed* and becomes a lake (Stage 2 in Fig. 11.5a). The sedimentation rate of the lake deposits slows due to the increasing size of the basin. Younger strata progressively onlap the basement rocks of the hanging wall block.

3. Further development is controlled by climatic factors and the size of the hydrologic drainage area of the basin. If water supply into the basin remains high enough to keep the basin water-filled up to the fixed outlet, the *basin deepens* and the *sedimentation rate continuously decreases* (Fig. 11.5a, Stage 3).

4. However, if evaporation from the growing surface area of the lake exceeds water supply, the basin becomes *hydrologically closed*. In this case, lake depth decreases as a function of both the increasing size of the basin and increasing evaporation loss (Fig. 11.5a, Stage 4). As a result, the depositional environment is controlled by a playa lake which may record minor climatic variations.

Deposition in the rift basin can also begin with lake sediments, if sediment input is small. As long as rifting and subsidence continue at the initial rate, the sedimentation rate should decrease indefinitely (Fig. 11.5b and c). However, subsidence slows down and is finally only driven by sediment loading (cf. Chap. 8.3). Consequently, the basin is slowly filled up until the depositional surface reaches the level of the outlet. The subsequent sediments are again fluvial and accumulate at a rate equal to subsidence.

The results of two-dimensional modeling of a permanently filled half-graben, continuously growing in size (Fig. 11.5b), are similar to those found of a full-graben. The cross section of this model reflects the geometry of the early Mesozoic Atlantis basin in North America (cf. Fig. 12.7) and its time slices represent 0.55 Ma. If sediment input balances the space provided by subsidence, the depositional surface area of the basin ceases to grow. Accelerated tilting of the half-graben causes the basin fill to shift toward the border fault and young layers to onlap previously deposited strata as shown in Fig. 11.5e (Stage 2).

Modifications of the Models

These simple models cannot account for various complications realized in nature, such as nonuniform subsidence and sediment supply, lateral facies changes within the basin, long-term changes in climate and drainage areas, and compaction. In addition, several important

factors may significantly vary along-strike of the basin, i.e., one has to consider *three-dimensional problems*. Nevertheless, these models allow to predict some general trends in the sedimentary filling of such basins.

Modeling the situation of the early Mesozoic Newark basin (shown on map in Fig. 12.7), Schlische and Olsen (1990) found a fairly good agreement between their model (Fig. 11.5c and d) and the gross stratigraphic and depositional development of this basin (Fig. 11.5e, Stage 4). In order to apply the more simple full-graben model to the basin, the half-graben cross section of the basin was transformed into a full-graben. The subsidence rate was inferred from the sedimentation rate of time intervals during which both rates were equal. Over long periods, the cumulative sediment thickness and the sedimentation rate of the model approximates the true values. Major discrepancies between model and nature were caused by volcanic activity and climatic change. An idealized evolution of this basin is shown in Fig. 11.5e.

Consequences for Stratigraphic Sequences

In conclusion, this and similar extensional basin filling models can explain, to some extent, the tripartite stratigraphic sequences observed in numerous continental basins:

1. A basal, alluvial fan and fluvial unit indicates through-going drainage and open-basin conditions.
2. After an initial deepening trend, the subsequent lacustrine or shallow-marine unit may reflect gradually shoaling upward related to closed-basin conditions.
3. Overlying fluvial sediments once again testify to through-going drainage.

This general trend, however, may be modified by climatic and various other, mainly regional factors (see, e.g., Smoot 1991). Conditions similar to those of short-lived continental rift basins, i.e., graben and half-graben structures, are also found in many pull-apart basins (Chap. 12.8).

Rift-Drift Stage Transition

If the rift stage of a basin is followed by seafloor spreading, the extension of the basin commonly continues at an accelerated rate. For example, if a full-graben of 50 km width is formed during a time period of 20 Ma, the average rate of basin extension (but not crustal extension) is 0.25 cm/a. By contrast, ocean spreading frequently operates at a rate of one to several cm/a. Hence, the average sedimentation rate of an extensional basin passing from the rift to drift stage should drastically decrease. This is the case to some extent, but does not apply to the sediment accumulation on shelves. In large ocean basins most of the incoming sediment is trapped on the shelves and therefore cannot be distributed uniformly over the entire basin as assumed for the simple models discussed so far.

Example:
Sediments in the Growing Atlantic Ocean

Nevertheless, the sedimentary history of the Atlantic Ocean shows the expected trend of a decreasing clastic sedimentation rate with time. During its early stage of rifting the narrow South Atlantic displayed sedimentation rates for carbonate-free material of about 30 m/Ma (and more) and thus was dominated by terrigenous deposits (Van Andel et al. 1977). With continued sea-floor spreading, the sedimentation rate of clastics dropped in the Late Mesozoic and Early Cenozoic to values mostly under 10 m/Ma along the margins and to 2 to 5 m/Ma on the Mid-Atlantic ridge (Fig. 11.6c).

By contrast, the in situ biogenic carbonate production and preservation did not show such a distinct trend, but rather was controlled by changes in the oceanic circulation system (Fig. 11.6b). Similarly, the North Atlantic shows sedimentation rates of terrigenous material which were generally high during its early history and then declined with the widening of the basin up to the early Miocene (Ehrmann and Thiede 1985). However, this rate increased again from the middle Miocene up to the Quaternary as a result of climatic change.

An approximate mass balance of the present-day denudation on land and overall sedimentation in the Atlantic Ocean between 50°N and S yields the following results: The total drainage area in North and South America, Europe, and Africa is $30 \cdot 10^6$ km^2 and the basin area $75 \cdot 10^6$ km^2. Hence the AB/AD ratio is 75/30 = 2.5. Assuming the modern global mechanical denudation rate of 48 m/Ma (Chap. 9.2, Fig. 9.6) for the drainage area, the average sedimentation rate of clastic material in the Atlantic is about 48/2.5 = 19 m/Ma (solid, water-free material). As compared to Figure 11.6, this value appears to be realistic. However, this calculation does not take into account that much sediment is deposited on the relatively narrow shelves of this wide ocean basin.

As regards chemical denudation, the global rate is 16 m/Ma (Fig. 9.6) and approximately 12 % of the dissolved river load is Ca^{2+}, 10 % SiO$_2$. A calcium denudation of 1.9 m/Ma may lead to a calcium carbonate (factor 2.5) sedimentation rate in the Atlantic of 1.9 x 2.5/2.5 = 1.9 m/Ma or 1.9 mm/ka (AB/AD = 2.5).

This low value demonstrates that the calcium river supply is insufficient to maintain the present-day oceanic biogenic carbonate production and thus favors carbonate dissolution below the CCD (cf. Chap. 5.3.2). Similarly, the river-borne SiO$_2$ yields an average sedimentation rate of 1.5 m/Ma or 1.6 mm/ka, which is too low to provide diatoms, radiolaria, and sponges with enough silica to build their skeletons. Hence, most of their skeletons are dissolved in the water column (Chap. 5.3.5).

Closing Basins

In contrast to extensional basin filling, closing basins such as remnant basins, part of foreland basins, some forearc and backarc basins, as well as intramontane basins (see Chaps. 1.2, 12.5, and 12.6)) show the reverse development in their sedimentary fill. With decreasing area of sediment dispersal, the sedimentation rates grow and commonly cause shallowing-upward sequences until the basin is completely filled and subaqueous deposition is replaced by terrestrial environments. Some examples described in Chapters 12.5 and 12.6 reflect this situation.

Long-term, global sediment mass balances, as indicated in Chapter 9.2, are subject to further complications. Estimates of the mass-age distribution of Phanerozoic sedimentary rocks, preserved on the continents and beneath the sea, clearly show that only part of the older sediments is still present and can be identified (Mackenzie 1990). The older the rocks, the less of their original volume is preserved. The total mass of sediments deposited over geologic time has been recycled several times, because sediments are continuously subducted and incorporated into mountain building from where they are eroded again. However, this problem is not further discussed here.

The out- und upbuilding of an *accretionary wedge* along a subduction zone is also largely controlled by the sediment budget of the associated deep-sea trench. The specific problems of this topic are discussed in Chapter 12.5.2.

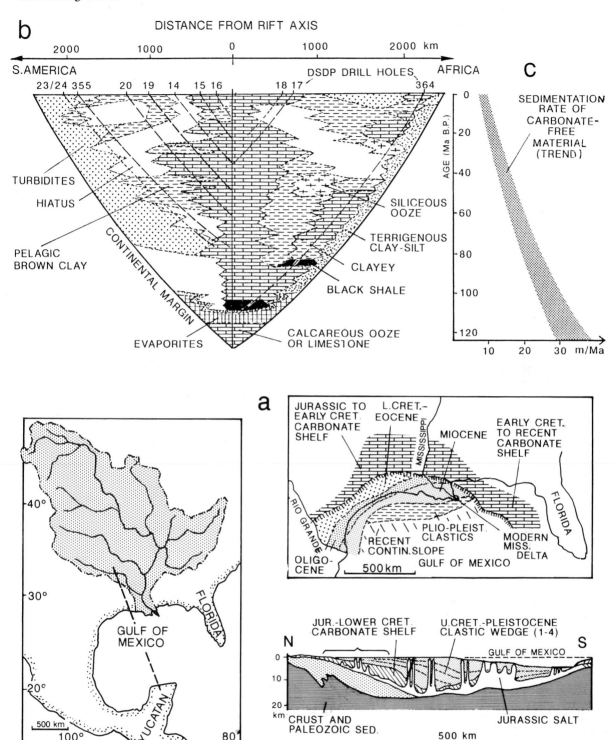

Fig. 11.6. a Present-day Mississippi drainage area and upper Cretaceous to Recent deposition of a huge, seaward migrating clastic wedge in the Gulf of Mexico. (After Matthews 1974). **b** Extension of South Atlantic through time and generalized chronostratigraphic cross section of its sediments between Brazil (Brazil basin) and Angola margin according to results from Deep Sea Drilling Project. Note that vertical drill holes appear as oblique lines in diagram. Calcareous oozes are almost restricted to mid-Atlantic ridge (rift axis), while sediments at greater distance from rift axis were mostly deposited below CCD (see Chap. 5.3.2). Asymmetry of sediment distribution is caused by circulation system. **c** Trend of carbonate-free sedimentation rate to diminish with increasing extension of ocean. (After Van Andel et al. 1977)

Variations in Subsidence and Sediment Supply

Variations in subsidence with time are another factor controlling the facies evolution of a sedimentary basin. In Fig. 11.7a through c, the sedimentation rate is approximately constant and of medium magnitude (20 m/Ma). In case (a), the basin remains shallow over a long time period, because sedimentation compensates for subsidence. In (b) and (c), time periods of rapid subsidence occur and therefore also of marked deepening of the basin with accompanying facies change. If the subsidence rate decreeases, the basin is filled up slowly, and the resulting vertical sequence will show a shallowing/coarsening trend.

If the rate of sedimentation substantially changes with time in a basin of constant subsidence, the depositional environment may vary drastically (Fig. 11.7d and e). In case (d), sedimentation starts with a very low rate and subsequently grows steadily. As a result, the basin is governed by marine conditions most of the time. If, in contrast, the development starts with a high sedimentation rate which gradually slows down (Fig. 11.7e), the basin remains filled up above sea level most of its lifetime and therefore collects continental deposits. The last example (Fig. 11.7f) shows a basin development in which both rates, subsidence and sedimentation, decrease with time and thus permit only a relatively short marine episode during the early basin history. A permanently high sedimentation rate would never enable the development of a water-filled, morphological basin, but maintain a continental depositional environment.

11.4 Vertical and Lateral Facies Associations (Overview)

Three Principal Types of Vertical Facies Evolution

General Trends

Summarizing the results of the interplay between denudation and depositional areas, sedimentation, and subsidence, three principal alternatives for the vertical facies evolution in sedimentary basins are established:

1. *Sediment supply ≥ subsidence (over-supplied basins):* shallowing basin, transition from marine or lake deposits into fluvial or another continental environment. Decrease in the proportion of biogenic sediment components. (Increase in benthic and reef-derived carbonate in the case of marine carbonates.)

2. *Sediment supply ≤ subsidence (sediment-starved basins):* deepening basin, transition from continental or shallow-marine into deep-marine environment. Increase in the proportion of biogenic sediment components, except for water depths below the lysocline (dissolution of carbonate and opaline silica).

3. *Sediment supply approximately compensates for subsidence* ("balanced" basins): long-persisting identical morphological basin configuration, maintainance of deep-marine, shallow-marine, or continental facies. No long-term drastic change in the proportions of terrigenous and biogenic sediment components.

Basins Over-Supplied with Sediment

Over-supplied basins are short-lived, but well represented in the geologic record. Their facies types are dominated by clastic sediments and comprise deep-sea to continental environments including

- Graben structures and rift basins with fluvial deposits (see examples of Chap. 12.1).
- Lakes fed by rivers from high-relief regions (Chaps. 2.2.2 and 2.5.1) and proglacial lake deposits (Chap. 2.1.1).
- Marine deltas and deep-sea fans (Chaps. 3.4 and 5.4.2).
- Remnant basins with flysch sedimentation (Chap. 12.6.1).
- Foreland basins with molasse deposition (Chap. 12.6.2).
- Forearc and backarc basins with high supply of volcaniclastics and other clastic material (Chaps. 12 5.3 and 12.5.4).
- Pull-apart basins in areas of high relief (Chap. 12.8.2).
- Some adjacent seas with high terrigenous sediment input, for example the northern Gulf of California (Chap. 4.3).

Sediment-Starved Basins

Sediment-starved basins tend to be deep and far away from high-relief terrigenous sediment sources. Their thin sediments make up only a small portion of the sedimentary record of a certain time period. Typical representatives include

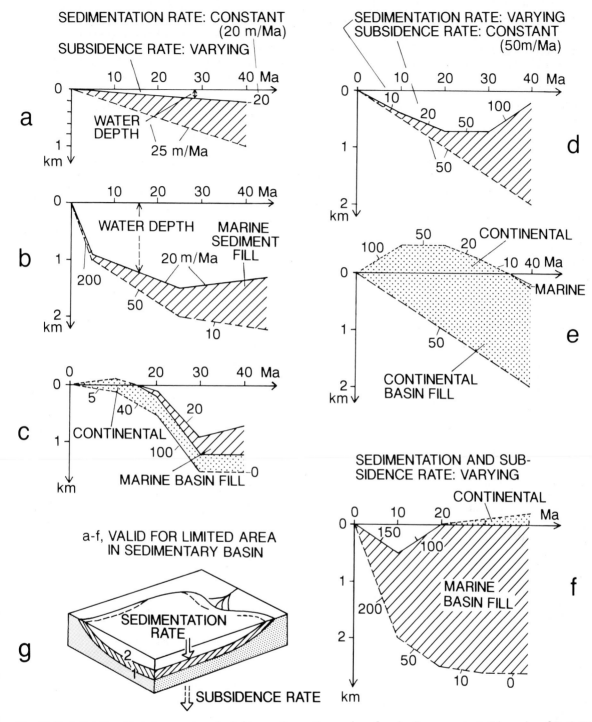

Fig. 11.7a-f. Relationship between rates of subsidence, *Su,* and sedimentation, *Sd,* versus time for a certain location within a sedimentary basin (all rates in m/Ma). **a** Both rates are constant, but Sd ≤ Su. **b** and **c** Sd = constant, Su varying. **d** and **e** Su = constant, Sd varying. **f** Both Sd and Su vary with time. Note that specific environmental conditions, for example a shallowing basin, may be created by different combinations of the factors Sd and Su

- Central parts of large oceanic basins and special smaller basins in the vicinity of mid-oceanic ridges (Chaps. 5.3.2 and 5.3.3).
- Isolated submarine platforms and highs (Chaps 5.3.7).
- Epicontinental seas surrounded by lowlands.
- Arid adjacent seas, for example the present-day Red Sea (Chap. 4.3).

Basins with Balanced Sediment Supply

These basins tend to maintain their depositional environment for considerable time periods. Characteristic examples are

- Shelf seas, particularly carbonate shelves and carbonate platforms, with and without various types of reefs (Chap. 3.3.2).
- Some areas subject to upwelling and high organic productivity, for example the modern shelves off Peru and South Africa (Chap. 5.3.4).
- Epicontinental seas with moderate influx of terrigenous sediment.
- Some forearc and backarc basins with limited sediment supply.

Modifying Factors

The three principal trends mentioned above, however, are modified in many ways by changes in oceanographic and climatic conditions, variations in organic productivity due to the faunal and floral evolution through time, as well as by sea level fluctuations and lateral tectonic movements, such as ocean spreading and subduction. The great variety of sedimentary facies through time is mainly a result of these modifying factors which obscure the simple basic principles.

The Abundance of Shallow-Marine Deposits

It is surprising that shallow-marine deposits constitute a great proportion of the total sediment volume in present-day environments as well as in the fossil record. One should expect that the principle condition (3), i.e., a balance between sediment supply and subsidence, is rarely realized for a time period of some geological significance. Hence, long-persisting shallow-water conditions should be an exception. However, the present-day, high-energy shelf environments demonstrate that the hydrodynamic conditions of the sea prevent ag-

gradation of sediment up to sea level. Instead, the incoming terrigenous sediment is widely distributed over the inner and outer shelf areas and partially transported into deeper water (Fig. 11.4g and h; Chap. 3.3). Outbuilding of the coastline is normally restricted to zones of very high sediment influx.

Thus, long persistence of shallow-marine conditions in the sedimentary record testifies to the fact that the former basin was probably oversupplied with sediment most of the time, but has lost part of the sediment influx to deeper, poorer supplied regions. Similarly, fluvial basins are commonly filled up to the level of their stream gradient. Any excess in river supply is carried farther downstream; thus a balance between sediment aggradation and basin subsidence is established for a significant time period. Such a process renders thick fluvial records. On the other hand, it should be mentioned that deep-sea sediments are under-represented in the fossil record, because they were largely subducted at convergent plate boundaries and transformed into metamorphic rocks.

The examples in Chapter 12 further illustrate the relationship between subsidence and sediment supply, but also take into account other modifying factors, including tectonic basin evolution. It is, however, not possible to deal with all basin types in detail. Once the general principles of the interplay between the various factors discussed above are demonstrated in a few representative cases, this approach can be applied to the great variety of other basin types in space and time.

Vertical and Lateral Facies Associations

Facies and Facies Change

As described in Part II, sediments deposited in special environments have distinct characteristics, i.e., texture, sedimentary structures, mineralogical and chemical composition, preserved fauna and flora, trace fossils, etc. All these characteristics are summarized under the term (sediment) *facies*, which is discussed in more detail in many textbooks (e.g. Dunbar and Rodgers 1957; Krumbein and Sloss 1963; Blatt et al. 1980). The facies of a sediment may indicate its particular depositional environment that distinguishes it from other facies in the same or another basin. Changing sedimentary environments cause facies changes both in vertical and lateral directions. One of the basic rules in sedimentary geology is that a special facies type can "migrate" obliquely through space and time, whereas the facies types change both in horizontal and vertical directions (*Walter's Law*, see, e.g., Teichert 1958). This means that a specific environment, producing a certain sediment facies, is established at different locations and times during the basin evolution.

Individual facies types can be combined to characteristic vertical facies successions or, more generally, to *vertical and lateral facies associations*. The morphological evolution of ancient sedimentary basins can be reconstructed mainly with the aid of such facies associations. The identification and interpretation of facies associations or, in other words, determining architectural elements (cf. Chap. 1.4.3) of a basin fill are one of the most important aims in sedimentary geology. For this reason, the major groups of *large-scale facies associations* are briefly mentioned here; smaller-scale trends in the facies evolution of specific depositional systems were described in Part II.

Vertical Facies Successions

Common vertical facies successions include:

1. *Deepening environments* with

High clastic input	*Lower clastic input*
Abyssal plain	Deep-sea pelagic sediments
Deep-sea fan	Deeper marine hemipel. sediments
Shallow marine clastics	Shallow marine (below wave base)
Delta and lake sediments	Coastal-marine (above wave base)
Braidplain	Delta, lagoon, tidal flats
Alluvial fan	Alluvial plain

2. *Shallowing environments* with

Siliciclastic sediments	*Carbonate sediments*
Fluvial plain	Alluvial plain (red beds)
Delta plain, lake, marsh	Coastal sabkha
Coastal, tidal-lagoonal	Carbonate lagoon, evaportites
River mouth bar sands	Reef and reef detritus

Prodelta, clastic shelf	Deeper carbonate platform
Deep-sea fan	Carbonate slope
Basin plain	Slope apron carbonates
	Deep-sea carbonate

Frequently, only part of these successions is realized in a particular basin (cf. Part II, e.g. Figs. 2.13 and 2.14, 2.28b, 3.15, 3.25a and b, 3.30 and 3.31, 4.6, 6.5e and 6.9).

In addition, there are frequently occurring *minor vertical facies associations* such as:

- Pelagic to hemipelagic shales, marls, limestones, and black shales (cf. Chap. 7.2).
- Pelagic carbonates, carbonaceous siliceous shales and chert, red shales, mostly associated with oceanic crust (cf. Chap. 5.3, Figs. 5.4 and 5.5).
- Skeletal lag deposits, consisting of cephalopods, crinoids, gastropods, etc., phosphorites, glauconitic minerals, crusts and nodules of iron and manganese oxyhydrates (condensed sections).
- Red and green shales and various evaporites (cf. Chap. 6.4, e.g., Fig. 2.32).

These and many other minor facies associations described in Part II are particularly useful for the identification of depositional environments.

Lateral Facies Associations

Similar successions also occur as lateral facies associations:

- Alluvial fan - braidplain - (playa) lake ± eolian sands (Figs. 2.8 and 2.32).
- Alluvial plain - tidal flats (lagoon) - shallow marine - deeper marine.
- Alluvial plain (red beds, eolian sands) - coastal sabkha - lagoonal carbonates/evaporites - reef - forereef - carbonate slope - slope apron. (Partially in Fig. 6.5e).
- Clastic shelf - slope and slope channels - deep-sea fan - basin plain (Fig. 5.18).
- Mid-oceanic ridge - deep-sea pelagic and hemipelagic environments (below and above CCD) - deep-sea trench - accretionary wedge with slope basins. (Chap. 12.5, Fig. 12.19).

Proximal-distal trends in relation to the sediment source also lead to distinct lateral facies associations, for example:

- Slope channel deposits - upper and lower deep-sea fan - overbank deposits - basin plain, displaying channel, proximal and distal turbidites (Chap. 5.4.2, Fig. 5.18).
- Lava flows - pyroclastic flows and ignimbrites - ash falls with decreasing grain size - fluvial-transported tephra - ash turbidites in lakes and in the deep sea (Chap. 2.4.2, Fig. 2.23).
- Relatively coarse fluvial sand - fine eolian sand - eolian dust (loess, Fig. 2.1).

Further examples of such facies associations in various tectonically defined basin types are described in Chapter 12.1 through 12.8.

Part IV
Basin Evolution

12 Basin Evolution and Sediments

12.1 Rift Basins

12.1.1 Rift Structures

Modern and ancient rift basins are amongst the most important geological structures in terms of their abundance and sediment accumulation. Geologically young, topographically expressed rift structures occur on all continents, as well as on thinned continental crust beneath the sea. Rift basins on both present-day and older, deformed passive continental margins signalize the initial stage (or rift stage) in the formation of a new ocean basin (drift stage). Other rift zones (i.e., failed rifts), including aulacogens, are those which have not evolved into an oceanic stage.

In Chapters 1.2 and 8.3 some simple models of rift basin evolution and subsidence were introduced. These models provide an initial approach for understanding the generation and tectonic evolution of such basins. In recent years, however, these models have been refined and modified considerably, and a wealth of new information on modern and ancient rift basins has become available (e.g., Frostick et al. 1986; Manspeizer 1988a; Price 1989; Allen and Allen 1990). These new results show that many rift basins are more complex and that their evolution is less predictable than was earlier assumed (Reading 1986b).

Pure Shear and Simple Shear

The extension of the crust by *pure shear* (Fig. 12.1a), as derived from "classic" graben structures (e.g., the Rhine graben in central Europe according to early interpretations, Fig. 12.2), is not as common as previously assumed. Pure shear implies instantaneous and uniform extension (cf. Chap. 8.3); if it reaches the drifting stage, the center of the rift graben coincides with the later spreading axis.

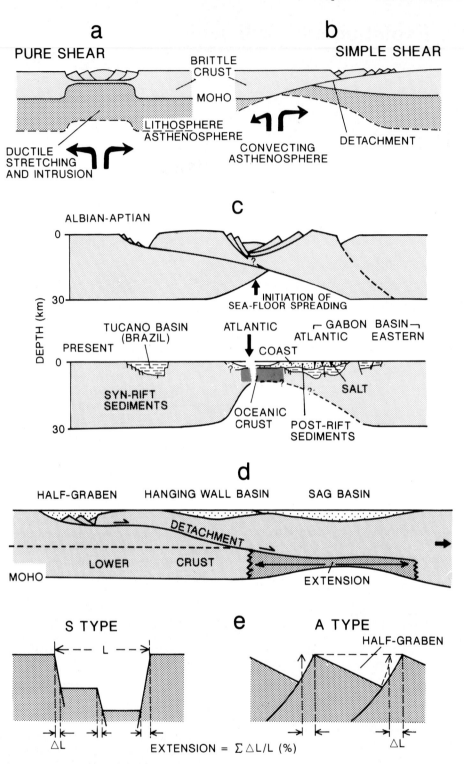

Fig. 12.1a-c. Pure shear (**a**) and simple shear (**b**) as alternative models for crustal extension and formation of rift basins (from Tankard and Welsink 1988). **c** Schematic section across lower Cretaceous basins of the south Atlantic continental margins and adjacent Tucano basin (Brazil) and eastern Gabon basin (Africa). All basins are assumed to result from one major extensional detachment zone. Note the lateral separation of the rift boundary fault from the basins produced by thermo-tectonic subsidence after continental breakup. (Kusznir et al. 1987). **d** Different styles of sedimentary basins developed from extension at various crustal levels. (Blundell et al. 1989). **e** S-type (limited extension) and A-type shear causing large extension (modified from Artyushkov and Baer 1989)

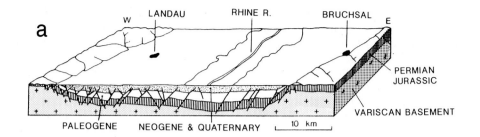

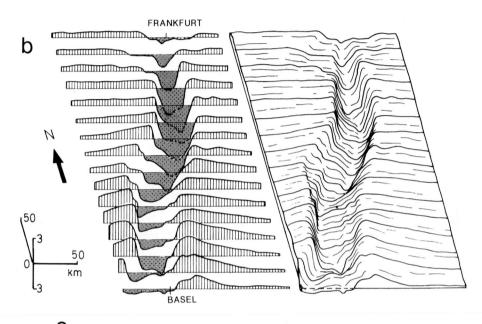

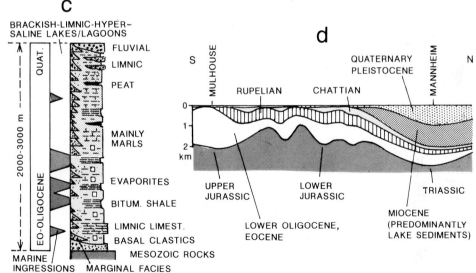

Fig. 12.2. a "Classic", approximately symmetric cross-section of Upper Rhine graben, north of Karlsruhe, central Europe, as an example for the pure shear model (Fig. 12.1a); no vertical exaggeration. **b** In reality, cross sections of this structure are more complicated and commonly asymmetric with changing polarity; various features, including graben morphology, variations in thickness of graben fill, and the geological structures of the graben shoulders (not shown here) indicate a left-lateral strike-slip component. **c** Simplified and idealized vertical facies sequence. **d** Schematic longitudinal section of Rhine graben fill. (After Pflug 1982; Geyer and Gwinner 1986; also see Illies and Greiner 1978; Fuchs et al. 1981)

However, many recent investigations of rift zones have revealed that crustal stretching is accomplished by slip along an *intra-crustal detachment surface (simple shear*, Fig. 12.1b). Such detachments were first observed in the Basin-and-Range Province of the western United States (Wernicke and Burchfiel 1982; Wernicke 1985) and confirmed by deep seismic reflection profiling in many other regions, for example on the Atlantic continental margins (e.g., Enachescu 1987; Boillot and Winterer 1988; Benson and Doyle 1988; Manspeizer 1988b; Tankard and Welsink 1988). In the detachment model, extension is nonuniform and decreases in the direction of shear. Simple shear allows the lateral separation of the zone of initial rifting and the zone of magma upwelling which may initiate ocean spreading (Fig. 12.1b and c). Low-angle detachments may explain the coeval initiation of different basin types (Fig. 12.1d).

To reconcile these two hypotheses, Barbier et al. (1986) proposed *combined shear*, i.e., a mechanism comprising thinning of the crust by both pure shear and simple shear (see also Ziegler 1988). Neugebauer (1987) emphasizes that lithospheric thinning is controlled not only by one principal parameter, but by several independent processes. Dome, rift, and basin structures represent stages of the thinning process and thus form a genetic sequence. Long-term, slow thinning may be superimposed by short-term episodes of increased tectonic and magmatic activity.

Other Structural Features

- Rift zones frequently evolve along pre-existing lineaments in the crust and may display a substantial strike-slip component (Fig. 12.3c and d, cf. Chap. 12.8). As long as extension exceeds strike-slip, the term rift is commonly used.
- Many rifts are not symmetrical, with the deepest basin floor located in the center, but may display the pronounced asymmetry of a half-graben (Figs. 12.1e and 8.10a).
- Rifts are not necessarily continuous, simple features as illustrated in small-scale maps; they may be broken into special segments by transverse structures (transfer or accommodation zones, Fig. 12.3a and e). Transfer zones in extensional regions display a wide range of geometries from discrete fault zones to zones of broad warping (Morley et al. 1990). The direction of asymmetry may alternate between neighboring segments, as observed, for example, in the region of Lake Tanganyika, East Africa (Burgess et al. 1988). The smaller the scale of observation, the more complex rifts become. Individual segments vary in their structural style and thus also in their basin fill. However, such rift structures may evolve into a full symmetric graben (Woldegabriel et al. 1990).
- Rifting along a rift zone does not necessarily occur simultaneously, but may proceed from one end to the other, as found in the East African rift zone (see below).
- One may find both thermally induced rift zones with initial updoming (Fig. 12.4b), as well as other rifts caused, for example, by continental collision and escape, which lack an early phase of uplift.

Rift Volcanism

Some of the rift zones and their sediments are strongly influenced by syn-rift volcanism (Fig. 12.4); others are little affected by these processes. When continental lithosphere is thinned by stretching, the underlying asthenosphere wells up to fill the space. As a result, the asthenosphere decompresses and generates partial melt (White 1989). The hotter the asthenosphere, the more melt is generated (and the more the crust is elevated). Relatively small variations in temperature cause major differences in the volume of magma generated. Large volumes of magma are produced if the lithosphere is underlain by abnormally hot mantle material (forming plumes or "hot spots"). The partial melt is buoyant and rises to be added to the overlying crust or to extrude at the surface. The phase prior to rifting may be characterized by widely extended flood basalts. Later, the rift valley may be filled by lava flows and tephra of variable composition, or volcanic structures are built up in the graben zone separating it into sub-basins.

Subsidence of Rift Zones

Finally, it should be mentioned that the principal mechanism of subsidence in rift zones is not always clear. The widely used stretching models (also see Chap. 8.3) are not generally accepted. According to Artyushkov and Baer (1989), there are two main modes of stretching (Fig. 12.1e): (1) S-type (synthetic) stretching, in which the fault blocks are not or only slightly tilted toward the down-thrown side of

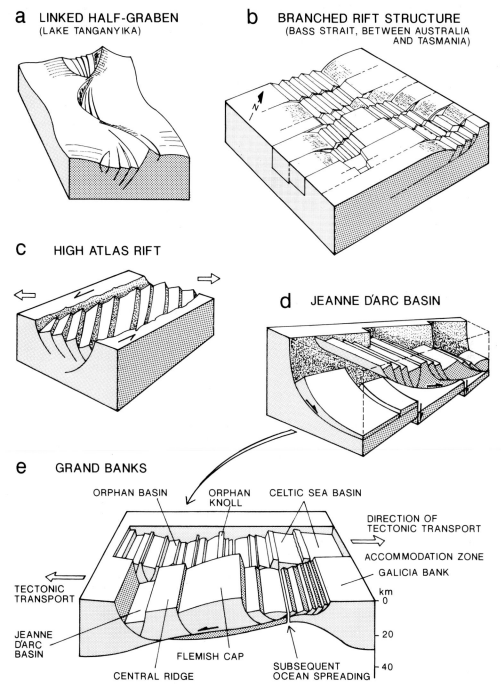

a LINKED HALF-GRABEN
(LAKE TANGANYIKA)

b BRANCHED RIFT STRUCTURE
(BASS STRAIT, BETWEEN AUSTRALIA
AND TASMANIA)

c HIGH ATLAS RIFT

d JEANNE D'ARC BASIN

e GRAND BANKS

ORPHAN BASIN ORPHAN KNOLL CELTIC SEA BASIN

DIRECTION OF TECTONIC TRANSPORT

ACCOMMODATION ZONE

GALICIA BANK

TECTONIC TRANSPORT

JEANNE D'ARC BASIN

FLEMISH CAP

CENTRAL RIDGE

SUBSEQUENT OCEAN SPREADING

km
0

20

40

Fig. 12.3. a Rift zone composed of half-grabens with alternating polarities. Adjacent half-grabens are separated by interbasinal ridges oblique to the rift axis. Model of Lake Tanganyika, Western Rift of the East African rift zone (see Fig. 12.4a; Rosendahl et al. 1986). **b** Cretaceous branched rift structure of the Bass Strait region between Australia and Tasmania. (Etheridge et al. 1987). **c** Combined syndepositional extension and strike-slip displacement. Pre-existing N 70°E faults are reactivated by strike-slip motion and produce secondary N 30°E normal faults within the graben structure. Model for Triassic rifting in the High Atlas, Morocco. (Beauchamp 1988; cf. Fig. 12.6). **d** Superposition of detachment above and perpendicular to previous extensional regime generates substantial dip-slip movement along listric faults and thus strike-slip parallel to boundary fault of extensional basin. Model for the development of secondary detachment in the Jeanne d'Arc basin, Grand Banks of Newfoundland (see **e**). **e** Simplified model of rifted continental margin between Newfoundland and Iberian Meseta, Spain (see Fig. 12.7a and c) prior to ocean spreading. (**d** and **e** Tankard and Welsink 1988)

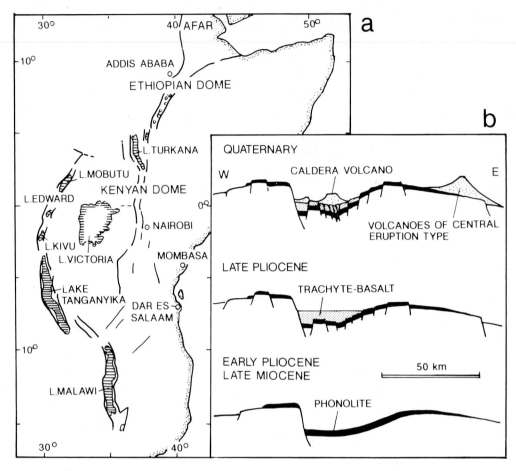

Fig. 12.4. a Simplified map of East African rift zone. **b** Different stages of rift evolution and volcanic activity in Eastern (Gregory) Rift. (After Williamson and Savage 1986; Uemura and Mizutani 1984, based on Baker et al. 1978)

a boundary fault; and A-type (antithetic) stretching, in which the blocks are markedly tilted and separated by listric faults. Substantial subsidence due to significant crustal thinning can be attained only by A-type processes. If S-type rift basins exhibit great subsidence but limited extension, the lowermost crust may have been altered or replaced by asthenospheric upwelling and injection of mantle-derived material. This process is also referred to as "basification" of the lower crust (e.g., Ziegler 1988). Then, the altered, moderately thinned crust may break into wide blocks under tensile stresses and subside isostatically along normal faults.

Consequences

From this brief discussion, it is obvious that one or even just a few rift basin prototypes

are not sufficient to define this basin type. Tectonic and volcanic processes in differing stress fields in the crust provide a great variety of rift basins. In addition, variations in the topography and drainage pattern of rift zones, climate, and source rocks contribute substantially to the great differences from one basin to another.

12.1.2 Examples of Young Rift Zones

The East African Rift Zone

The most prominent and frequently described present-day rift zone, the 4000 km long East African rift, is an example of the enormous variability of rift basins and their sediment fills (e.g., Uemura and Mizutani 1984; Frostick et al. 1986). It lies within a large, elongated dome structure (the Kenyan and Ethiopian dome) which has risen about 1800 m since the middle Tertiary (Fig. 12.4a). The narrow **Western Rift** is occupied by lakes arranged en échelon (cf. Fig. 12.3a) and displays a relief of approximately 3000 m between the rift shoulder and the floors of the deepest lakes (Lake Tanganyika and Lake Malawi). Here, volcanic activity is virtually absent. The 50 km wide *Eastern Rift* (or Gregory Rift) exhibits lower relief, and is subdivided by transverse ridges into several small, shallow basins, partially occupied by lakes. The **Eastern Rift** in particular was strongly affected by pre-rift and syn-rift volcanism of basaltic to more felsic nature (Fig. 12.4b). Where the volume of volcanic products exceeded the space created by rifting, the rift zone was totally filled by volcanics and volcaniclastic sediments.

In the main Ethiopian Rift, six major volcanic episodes were recognized from the late Oligocene to the Quaternary (Woldegabriel et al. 1990). Extruding carbonatitic magmas are at least partially responsible for the extremely high alkalinity (pH up to 10.3) and chemistry of Magadi-type sodium carbonate lakes in the Gregory Rift (Eugster 1986).

Furthermore, the climatic conditions of the East African rift zone range from a tropical, high-rainfall belt to desert areas. Open lake systems with fresh water alternate with closed lake basins where evaporites accumulate.

To the north, the East African rift system continues and forms the **Red Sea basin**. The evolution of this basin started 70 Ma ago and is more advanced than the East African rift zone. One can distinguish four phases of development (e.g., Guennoc et al. 1988): (1) A Cretaceous to Eocene pre-rift stage, (2) Oligocene rift formation, (3) Miocene subsidence (Red Sea depression), and (4) Pliocene to Recent sea-floor spreading. The sedimentary history of this basin is briefly discussed in Chapter 4.3.

Other Young Rift Zones

Other modern or Cenozoic rift zones include the Upper Rhine graben (Fig. 12.2), the Bresse-Rhone graben, and the Eger graben in the Alpine foreland of Europe; the Dead Sea rift, with its strong strike-slip component (see Chap. 12.8); the Baikal rift, with the very deep Lake Baikal in central Asia; several rift basins in China (Hsü 1989), and many other examples. A large number of greatly varying rift basins was formed during the Triassic and

early Jurassic as a result of the breakup of the Pangea supercontinent prior to the formation of the Atlantic Ocean and Indian Oceans (Manspeizer 1988b; Ziegler 1988).

A comparison of all these modern and ancient rift basins, with their great variations in tectonic style, volcanism, and sedimentary fill may lead to the conclusion that each rift basin should be regarded as an individual case. Nevertheless, some general trends in the evolution of these basins can be identified. These are briefly discussed in the following, with the principal aim of showing the interrelationship between tectonic basin evolution and sedimentation.

12.1.3 Sediments of Rift Basins

Introduction

The great variety of structural style and depositional environments mentioned above makes it difficult to define a specific facies evolution for rift basins. Although a succession from (1) fluvial deposits, (2) lake deposits with a tendency towards playa conditions, and finally a return to (3) fluvial deposits frequently occurs (as demonstrated in Chap. 11.3), many rift basins show other facies sequences (see, e.g., Smoot 1991). Pre-rift topography controls the positions of major rivers, which may or may not enter the basin and thus govern its hydrological and sedimentological budget. Rifting below sea level and at great distances from the continents may create, at least for some time, sediment-starved basins with marine sequences of limited thickness (cf. Fig. 12.14).

Facies Models for Half-Graben Basins

Some idealized facies models for half-graben basins exposed to various depositional environments are shown in Fig. 12.5 (Leeder and Gawthorpe 1987). All indicate a markedly asymmetric basin fill caused by differing sediment supply and depositional processes on the two sides of the basin, particularly if most of the material is supplied from the flanks of the half-graben. Sediments on the steep wall (or footwall) of the boundary fault tend to be coarse grained and form alluvial fans, fan deltas, and debris flows in deeper water. Sediments entering the basin from the gentler hanging wall usually come from more distant sources and are finer grained. If the basin is

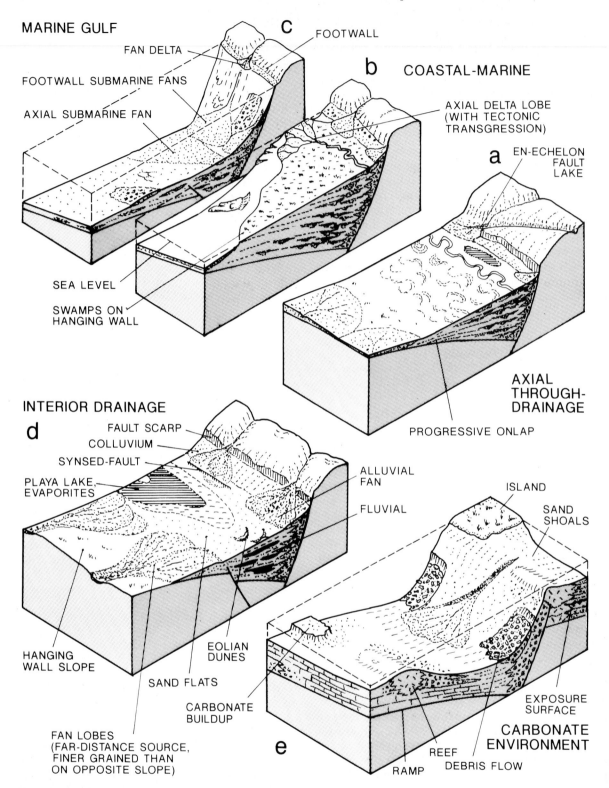

Fig. 12.5a-e. Simplified facies models of half-graben fill under continental (**a** and **d**) and coastal/marine conditions (**b**, **c**, and **e**). Note that the tectonic structure of the basins may be more complicated and that more synsedimentary faults occur. Lakes and the axial river tend to shift toward the boundary fault. **e** Half-graben morphology in shallow-marine carbonate environment may be modified by reef buildups and abundant reef detritus on both sides. Full discussion in text. (After Leeder and Gawthorpe 1987)

filled predominantly by a through-flowing axial river or by an axial delta lobe in conjunction with a prograding submarine fan, both the main trunk of the river and the submarine fan tend to shift towards the boundary fault. In a marine carbonate regime, the hanging wall represents a ramp setting where reefs and their associated facies may develop and finally create a morphology similar to that along the boundary fault.

In many cases, these "pure" facies associations will not persist throughout the entire evolution of the basin. For example, widening and deepening of the half-graben may lead to a deepening-upward facies succession a-b-c in Fig. 12.5 or, if tectonic movements have come to an end, to a shallowing-upward succession. Similarly, as mentioned in Chapter 11.3, axial through-drainage may be followed by internal drainage and thus result in succession a-d. Regional subsidence may allow the sea to enter and to turn a semi-arid to arid continental environment into a marine carbonate basin (succession d-e).

Volcanic activity is neglected in these facies models, but may modify them significantly. Well exposed middle Jurassic to early Cretaceous rift basins of eastern Greenland represent excellent examples of such halfgraben fillings consisting of alluvial fans, fan-deltas, shallow-marine sediments, and deeper marine fan to basin associations (Surlyk et al. 1981).

Examples

East African Rift Basins

The East African rift basins display a large variety of alluvial fan, fluvial, and lake deposits (Frostick et al. 1986). The latter reflect changing climatic conditions through time and from basin to basin. Diatomites and black shales occur in the more humid regions, while arid zones are characterized by playa lakes with evaporites, including sodium carbonate and lacustrine chert (Eugster 1986). The rift basins of the Eastern Rift in particular contain abundant volcaniclastics of varying composition. Repeated ashfalls favored the preservation of footprints, soft-bodied organisms, and all kinds of organic relics (Pickford 1986). Sedimentation rates in this rift zone range from about 10 cm/ka up to several 100 cm/ka (Tiercelin and Faure 1978).

Upper Rhine Graben

The rift basin of the Upper Rhine graben in central Europe is well explored by numerous drill holes (summaries by Roll 1979; Pflug 1982; Geyer and Gwinner 1986). Rifting began in the early Eocene and continued to the present. It was accompanied by left-lateral strike-slip motion, while the zone of maximum subsidence migrated from south to north (Fig. 12.2d, for subsidence curves see Fig. 8.12). The sedimentary fill in this basin is an example of the complexity of rift basin sediments. The simplified sections in Figures 12.2c and d omit most of the many stratigraphic details and boundaries known from this area. Nevertheless, they demonstrate the conspicuous vertical and lateral variations in facies and thicknesses of certain stratigraphic units.

The vertical section (Fig. 12.2c) reflects numerous changes in the depositional environment due to repeated marine ingressions and alternating limnic (lacustrine), brackish, and hypersaline conditions. Greenish to reddish marls, lacustrine limestones and dolomites, and various evaporites including potash salts characterize the more arid playa lake and lagoonal intervals. Bituminous shales are probably associated with more humid lake periods, but may also have accumulated during times of marine ingressions, which are normally recorded by gray marls and fossil-bearing sandy limestones and sands near the basin margin. The upper part of the section contains some layers of brown coal, indicating a general tendency towards a more humid climate. Volcanic activity during the Oligocene and Miocene was moderate and characterized by ashes of alkaline and bimodal mafic-felsic composition. With the onset of longitudinal drainage by the Rhine River, fluvial deposits accumulated in the entire graben. As a result of continuing differential subsidence, the youngest deposits show significant variations in thickness along strike and attained their maximum thickness in the northern part of the graben.

Remarks to Older Rift Zones

Many ancient rift zones which evolved into passive margins were later incorporated into orogenic belts by convergent plate movements. Good examples have been described from the Alps (Winterer and Bosellini 1981; Lein 1985; Eberli 1987; Lemoine and Trümpy 1987). Rifting in the Alps occurred in the early and middle Jurassic and pre-dated the opening of the Ligurian Tethys ocean by spreading and transform movements. The rift basins are bordered by normal faults with throws up to a few kilometers. However, these are often difficult to identify because of subsequent intensive deformation. The best evidence for former rift basins is therefore their basin fill, which is frequently characterized by mega-breccias, olistostromes, and turbidite fans of limited extent. In addition, such basin fills display marked lateral changes both in facies and in thicknesses.

The Inverted Rift Zone of the High Atlas

A quite different example of a rift zone evolution is the **High Atlas** in northwestern Africa, which represents an intracratonic mountain belt bordered by elevated plateaus, the Saharan craton to the south, and Oran Meseta microplate to the north (Fig. 12.6a; Beauchamp 1988; Jacobshagen 1988). The development of this structure may have resulted from northward directed crustal detachment (Fig. 12.6 b). It began with the breakup of Pangea during the Triassic (see below) after a period of erosion and peneplanation. Rifting was accompanied by strike-slip motion, and the rift zone extended from the Triassic proto-Atlantic to the Tethys ocean over a distance of more than 1000 km.

During the first phase of rifting (from the late Triassic to early Jurassic), the rift graben of the High Atlas accumulated continental red beds and evaporites, and was affected by syn-rift basaltic volcanism. Due to continuing rifting, subsidence, and eustatic high sea level stands, the sea entered the rift zone several times from the west (proto-Atlantic) and the east (Tethys) during the Jurassic, Cretaceous and Tertiary. In the central High Atlas, the widening and deepening rift zone was bordered by narrow carbonate shelves and subdivided by a central deep carbonate platform (Fig. 12.6c; Warme 1988). The northern and southern troughs were partially filled by turbidites and other mass flow deposits.

Beginning with the Toarcian high sea level rise, a middle Jurassic thick marly sequence was deposited in the central to eastern regions of the High Atlas, showing several shallowing-upward cycles. These cycles and several unconformities indicate continued differential subsidence and tilting of blocks within the rift zone. Periods in which sedimentation lagged behind subsidence and the basin deepened, alternated with phases in which the basin was filled up and the depositional environment changed from marine to continental. During the Paleogene, biogenic carbonates and reworked phosphorites accumulated in a marine, bay-like ramp setting in the western regions (Trappe 1991). Variations in subsidence and subsequent uplift along the strike of the rift brought about substantial changes in the depositional and erosional history of the rift zone. Transpression caused steep-angle overthrusting and gravitational gliding of rift sediments, locally including their

basement, onto molasse-like Tertiary sediments of the northern and southern Subatlas zones (Fig. 12.6d; Froitzheim et al. 1988).

In summary, the High Atlas rift zone represents an example of an inverted, uplifted, multi-phase, but relatively short-lived, failed rift complex (cf. Chap. 1.2). It shows a highly variable depositional history controlled by tectonics, eustatic sea level fluctuations, and climatic changes documented in continental and shallow-marine sediments.

12.1.4 Transition from Rift Basins to Continental Margin Basins

Introduction

In recent years a great number of buried rift basins have been detected by geophysical surveys and drilling below the present continental shelves, for example in the northern Atlantic (Fig. 12.7). These basins result from the breakup of Pangea and the stretching and thinning of a 500 to 1000 km wide zone of continental crust during the Mesozoic prior to continental breakup and sea-floor spreading. Some of these basins are exposed on land and commonly display simple structures and relatively short time periods of basin filling, as described in Chapter 11.3 (Fig. 11.4) and in the facies models for half-grabens (see above). The late Triassic-early Jurassic rift basins in the northwest Atlantic (in the regions of the former Appalachian orogeny) belong to this group (Fig. 12.7b). They are characterized by high relief and fluvial-lacustrine sediments (Manspeizer 1988b; Schlische and Olsen 1990).

Rift Basins of Continental Margins

Example: The North Atlantic

The rift basins on the continental shelves of the north Atlantic, for example, are buried under late Mesozoic and Cenozoic post-rift sediments and experienced longer and more complicated depositional histories than the rift basins mentioned above (Manspeizer 1988b; Tankard and Welsink 1988; Ziegler 1988; Welsink et al. 1989; Hiscott et al. 1990).

- *First phase of rifting*. The development of these basins also started in the late Triassic with continental siliciclastics, mostly red beds, but most of them lay in regions of low relief,

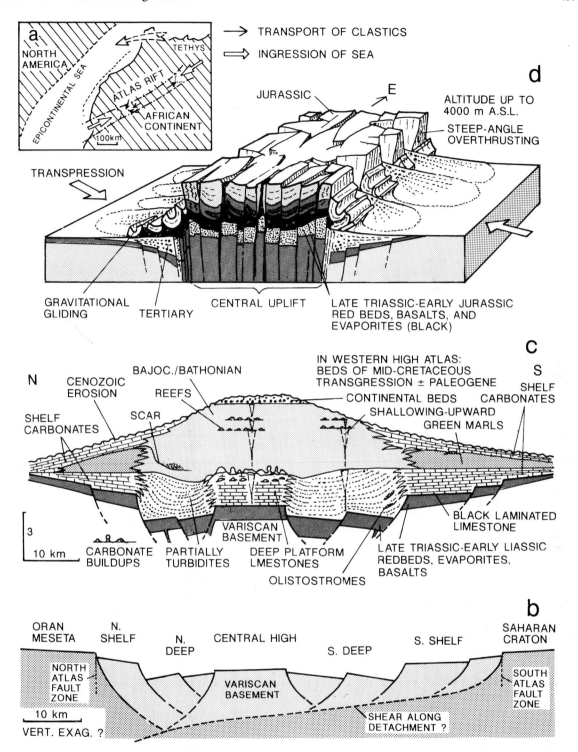

Fig. 12.6a-c. Evolution of High Atlas rift zone from extensional to compressional (transpressional) phase. **a** Paleogeographic reconstruction for the late Triassic, overview. (Beauchamp 1988). **b** Simplified cross section of High Atlas rift zone showing possible interpretation of structural features according to simple shear model with low-angle detachment. **c** Idealized stratigraphic cross section drawn on a Domerian-Toarcian datum (peak of transgression).

Note that younger sediments (Middle Jurassic, in the western High Atlas, to Turonian) indicate continued subsidence (**b** and **c** after Warme 1988). **d** Simplified block diagram of present structure of High Atlas caused by compression (transpression ?) and uplift of the central rift zone (inverted failed rift). (After Stets and Wurster 1982; Wurster and Stets 1982; Froitzheim et al. 1988)

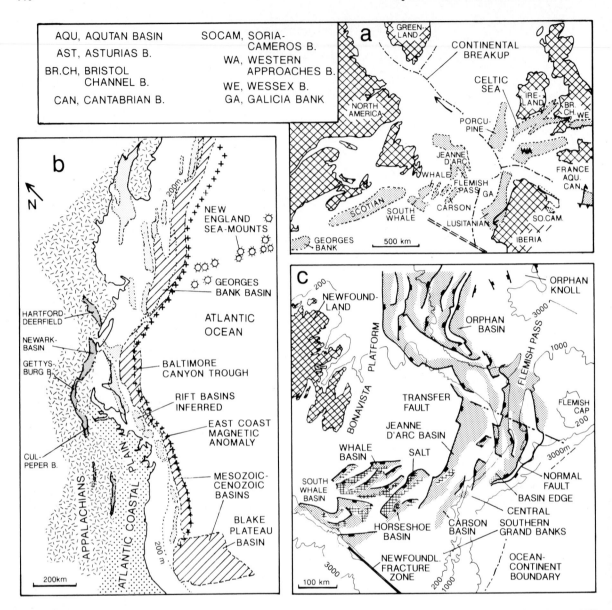

Fig. 12.7. a Main Jurassic and early Cretaceous basins on rifted continental crust of the North Atlantic ocean. (After Hiscott et al. 1990). **b** United States Atlantic continental margin between Nova Scotia and north Florida showing location of late Triassic to early Jurassic rift basins (black) exposed on land, rift basins below younger shelf sediments, and Mesozoic-Cenozoic basins along the shelf edge.

(Modified from Hutchinson and Klitgord 1988). **c** Rift basins of the Grand Banks shelf off Newfoundland, based on seismic and gravity data. Note subdivision of Grand Banks by transfer (strike-slip) faults and changing orientation of normal faults on southern Grand Banks. (After Tankard and Welsink 1988; Welsink et al. 1989)

were inundated by the sea, and accumulated evaporites of varying thickness. This first rifting phase was terminated before any oceanic crust was generated. It was followed in the middle Jurassic by a period of slow, thermal subsidence (Fig. 12.8a) which led to the formation of a wide epeiric sea and predominant deposition of shallow-marine shales and limestones.

- The *second phase of rifting* in the north Atlantic began in the late Jurassic (also see Winterer and Hinz 1984; Gradstein et al. 1990) and lasted approximately 50 Ma until the Aptian. After an initial episode of uplift accompanied by erosion and valley cutting, extension of a wide zone of crust was accomplished along a low-angle detachment (cf. Figs. 12.3e

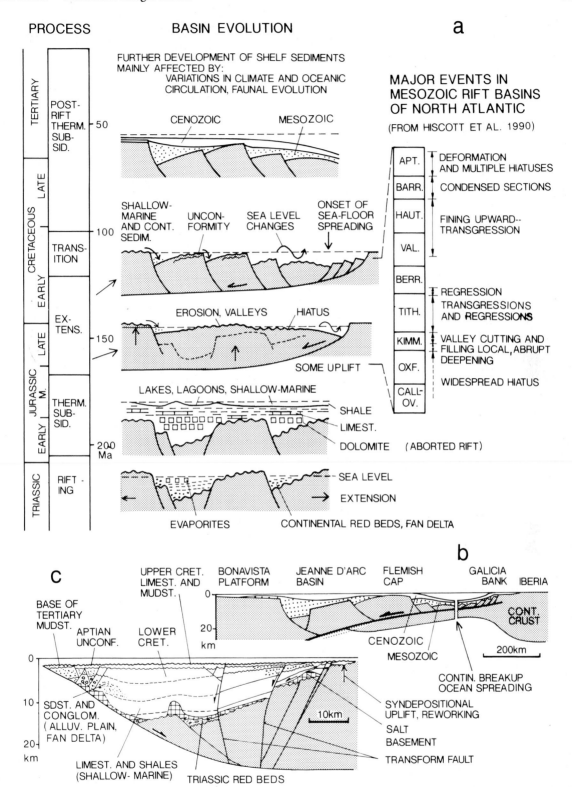

Fig. 12.8. a Main principles of Mesozoic rift basin evolution in the North Atlantic. (Based on Tankard and Welsink 1988; Hiscott et al. 1990). **b** Idealized cross-section between Newfoundland and Iberia showing boundary fault of aborted Triassic rift basin and Callovian-Aptian extension of continental crust along low-angle detachment. Size of listric faulted blocks and associated basins increases in direction of shear (distal). **c** Structure and sedimentary fill of Jeanne d'Arc basin based on seismic interpretation, also see Fig. 12.3d and e. (**b** and **c** after Tankard and Welsink)

and 12.8a and b). This mechanism reactivated pre-existing basins, for example the Jeanne d'Arc basin on the Grand Banks of Newfoundland (Fig. 12.8c), and generated a number of new basins. Fault-controlled synsedimentary subsidence created special depocenters separated by sills. The sedimentary fill in these basins consists predominantly of shallow-water limestones and shales with interbedded sandstones and organic-rich layers.

- At the *climax of rifting* in the late Jurassic, basins near the continent or adjacent to uplifted blocks were partially filled with fluvial and deltaic deposits.

- *Late-rift evolution* is characterized by decreasing local subsidence and a transition into regional thermal subsidence (for subsidence curves of some sites see Fig. 8.14). However, further adjustments of the broken, brittle upper crust to continuing extension and a changing stress field (see Fig. 12.3d and e) led to local variations in the sedimentary fill of the basins. A *basal Cretaceous transgression* created a widespread unconformity. The subsequent sediments include shallow-water limestones and are affected by additional unconformities.

- *Breakup unconformity*. The last and well developed unconformity during the Aptian is associated with the breakup of the stretched continental crust and the separation of Eurasia with Galicia Bank and the Iberian peninsula from the North American plate (Fig. 12.8b). This breakup unconformity indicates that up to this time the individual blocks of the continental crust had rotated, subsided differentially, or were uplifted and eroded.

Such a complex tectonic and stratigraphic history, with repeated regional rifting episodes and possibly more than one phase of breakup, is also known from other passive continental margins, for example from the Campos Basin of Brazil and the Gabon Basin of West Africa (Edwards and Santogrossi 1990). In northwestern Australia several depositional wedges were built out into the evolving marginal basin (Boote and Kirk 1989).

In spite of the great differences in their subsidence histories and sedimentary fills, the Mesozoic north Atlantic rift basins share a number of "sedimentological events" (Fig. 12.8 a; Hiscott et al. 1990). These features include more or less coeval hiatuses and periods of condensation. Unconformities are associated with the deposition of coarse-grained clastics. The rift phases in the Triassic and upper Jurassic are characterized by high rates of subsidence (50-100 m/Ma, up to 250 m/Ma in some basins). "Pulses" of extension and similarities in the evolution of these basins

result from their common origin by intracontinental rifting and subsequent separation of Europe and North America (see also Ziegler 1988).

General Trends in the North Atlantic

As crustal extension progressed, rifting tended to migrate toward the axial zone at the expense of peripheral graben and fault systems, which became inactive. The rifting stages of both aborted rifts (without crustal separation) and fully developed rifts (followed by sea floor spreading) lasted from about 50 Ma to more than 200 Ma. Continued subsidence and decreasing siliciclastic sediment supply commonly led to fining-upward sequences in the post-rift phase.

The effects of eustatic sea level variations and other long-term trends, such as climatic change and variations in the carbonate production, are superimposed on tectonic development. Shelf sediments from the post-breakup, mature ocean phase in the late Cretaceous and Tertiary are dominated by pelagic carbonates or silty clays. Their facies are controlled mainly by variations in climate, oceanographic circulation, sea level fluctuations, and faunal evolution (for a detailed description see, e.g., Gradstein et al. 1990).

12.2 Continental Margin and Slope Basins

12.2.1 General Aspects

Types of Continental Margins

A rift system evolves into a continental margin basin when two divergent plates are separated by a spreading center, creating new oceanic crust. Crustal extension and rifting *(syn-rift stage)* are then followed by drifting *(post-rift stage)*, but subsidence of the thinned, wedge-like continental crust continues due to cooling of the lithosphere and increasing sediment load. Syn-rift and post-rift sediments are frequently separated by a more or less distinct *breakup unconformity* caused by differential movements of the crustal blocks at the onset of drifting. Higher elevated blocks may undergo subaerial or submarine erosion. Three different classes of passive margins have been identified (Fig. 12.9, e.g., Roberts et al. 1984; Hinz et al. 1987):

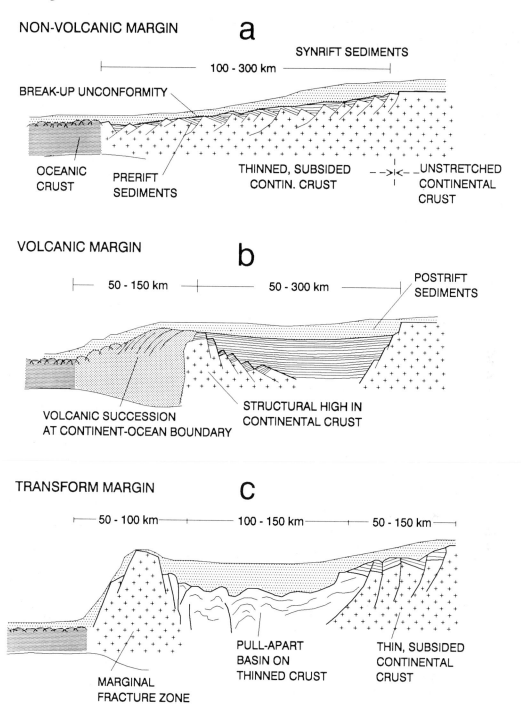

NON-VOLCANIC MARGIN **a**

SYNRIFT SEDIMENTS

BREAK-UP UNCONFORMITY

$\longmapsto$ 100 - 300 km $\longrightarrow$

OCEANIC CRUST   PRERIFT SEDIMENTS   THINNED, SUBSIDED CONTIN. CRUST   UNSTRETCHED CONTINENTAL CRUST

VOLCANIC MARGIN **b**

$\longmapsto$ 50 - 150 km $\longmapsto$ 50 - 300 km $\longrightarrow$ POSTRIFT SEDIMENTS

VOLCANIC SUCCESSION AT CONTINENT-OCEAN BOUNDARY   STRUCTURAL HIGH IN CONTINENTAL CRUST

TRANSFORM MARGIN **c**

$\longmapsto$ 50 - 100 km $\longmapsto$ 100 - 150 km $\longmapsto$ 50 - 150 km $\longrightarrow$

MARGINAL FRACTURE ZONE   PULL-APART BASIN ON THINNED CRUST   THIN, SUBSIDED CONTINENTAL CRUST

Fig. 12.9a-c. Three principal classes of passive continental margins. **a** Non-volcanic. **b** Volcanic. **c** Rift-transform margin.

(Report of Second Conference on Scientific Ocean Drilling, 1987, European Science Foundation, Strasbourg, France)

- *Nonvolcanic passive continental margins* (Fig. 12.9a), where lithospheric deformation is dominated by block faulting over a broad zone (100-300 km) and volcanic activity in the upper crust is of minor importance (examples: Red Sea, eastern United States, northwestern Africa, Galicia Bank off Iberia, Goban Spur to the southwest of Ireland, northwestern Australie, also see below). These margins may be either sediment-starved (≤1-2 km post-rift sediments) or heavily sedimented (up to ≥10 km post-rift sediments).

- *Volcanic passive margins* (Fig. 12.9b), where the boundary between continental and normal oceanic crust is characterized by a thick zone of seaward dipping volcanic units overlying continental crust. Margins of this type are narrower than nonvolcanic margins, and some of them appear to be associated with marginal plateaus (e.g., Rockall Bank, Vøring Plateau, see Fig. 12.10).

- *Rift-transform margins* (Fig. 12.9c), which originate from extensional deformation, including a significant strike-slip component. It is assumed, for example, that the early development of the South American and southwest African margins was affected by transtension. A young example is the Gulf of California (see below).

The following examples of basin evolution deal predominantly with the first and most common category of continental margins. The fact that most margins are segmented along their length on a scale of several hundred to more than a thousand kilometers (cf. Fig. 12.7c, Canadian shelf) is not further dealt with here. While the structure and subsidence history within individual segments appear to be relatively uniform, they may differ significantly from one segment to another.

Basin Morphology and Sediments (Overview)

Young continental margins with a limited sediment supply tend to have narrow shelves and to exhibit relatively thin sedimentary sequences. In contrast, the sediments on mature passive continental margins, such as those of the present-day Atlantic Ocean, can frequently store sediments more than 10 km thick. The accumulating sediments may show both upbuilding and outbuilding (cf. Figs. 12.11 and 12.15c), but they are frequently trapped by structural highs, reefs, or other dam-like structures at the outer margin (Fig. 12.11).

The width of continental margin basins is mainly a function of the continental crust, which was subjected to rifting prior to drifting. The circum-Atlantic basins are typically 200 to 300 km wide, but some are much narrower or considerably wider. In many cases, the present shelf edge or base of slope corresponds approximately with the boundary between subsided continental crust and oceanic crust. Narrower shelves may result from strong ocean currents preventing sediment accumulation (e.g., the Gulf Stream off Florida) or insufficient sediment supply and heavy storms (e.g., along the margins of islands exposed to the open sea). Then the subsided crust forms marginal plateaus, the surface of which lies several hundred to more than a thousand meters below sea level. The relatively young Norwegian-Greenland Sea, for example, exhibits continental margins of various widths and sunken plateaus, such as the Vøring Plateau (Fig. 12.10; Birkenmajer 1981; Ziegler 1988; Eldholm et al. 1989).

The *syn-rift sediments* of continental margin basins represent the deposits of buried rift graben systems, as already described in the previous sections. Along the margins of the Atlantic and other ocean basins resulting from the breakup of Pangea, these syn-rift sediments are of Triassic to early Jurassic age and consist predominantly of continental red beds, lake deposits, volcanics, evaporites, and some shallow-marine sediments of variable thickness and limited extent. Paleozoic and Tertiary passive continental margins may show other syn-rift sediment successions, but are commonly also characterized by fluvial, lacustrine, and shallow-marine sequences. The Miocene evaporites in the Red Sea were also deposited prior to the onset of sea floor spreading, and their changing thickness reflects a basin with nonuniform, locally rapid subsidence (see Chaps. 12.1.2 and 4.3).

Detailed descriptions of the structural features, stratigraphic sequences, and the hydrocarbon propects of the Atlantic continental margins have recently been published in several volumes (Emery and Uchupi 1984; Poag 1985; Sheridan and Grow 1988; Walsh 1988; Ziegler 1988; Tankard and Balkwill 1989; Edwards and Santogrossi 1990; Gradstein et al. 1990).

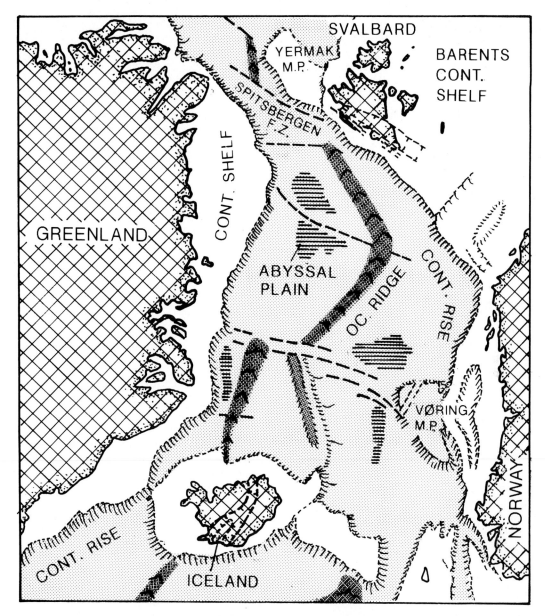

Fig. 12.10. Physiographic elements of the Norwegian-Greenland Sea, generalized. Note the variable width of the continental margin (shelf), continental platform of the Barents shelf, submarine plateaus, and rift structures along the eastern margin. (After Birkenmajer 1981; for more details see Eldholm et al. 1989)

12.2.2 Sediment Successions of Continental Margin Basins

Sediment Buildup on Atlantic-Type Margins

The post-rift sediments of continental margins overlie the rift basin fill and commonly cover wider areas. Rates of post-rift subsidence decay exponentially and are commonly lower than those of syn-rift subsidence of special rift basins (Chaps. 8.3 and 8.4).

On Atlantic-type, *"mature"* continental margins, where basins are filled with sediments up to near sea level (present water depth 100 to 200 m), the structure and nature of most of these sediments can be subdivided into four groups (Fig. 12.11; Kingston et al. 1983):

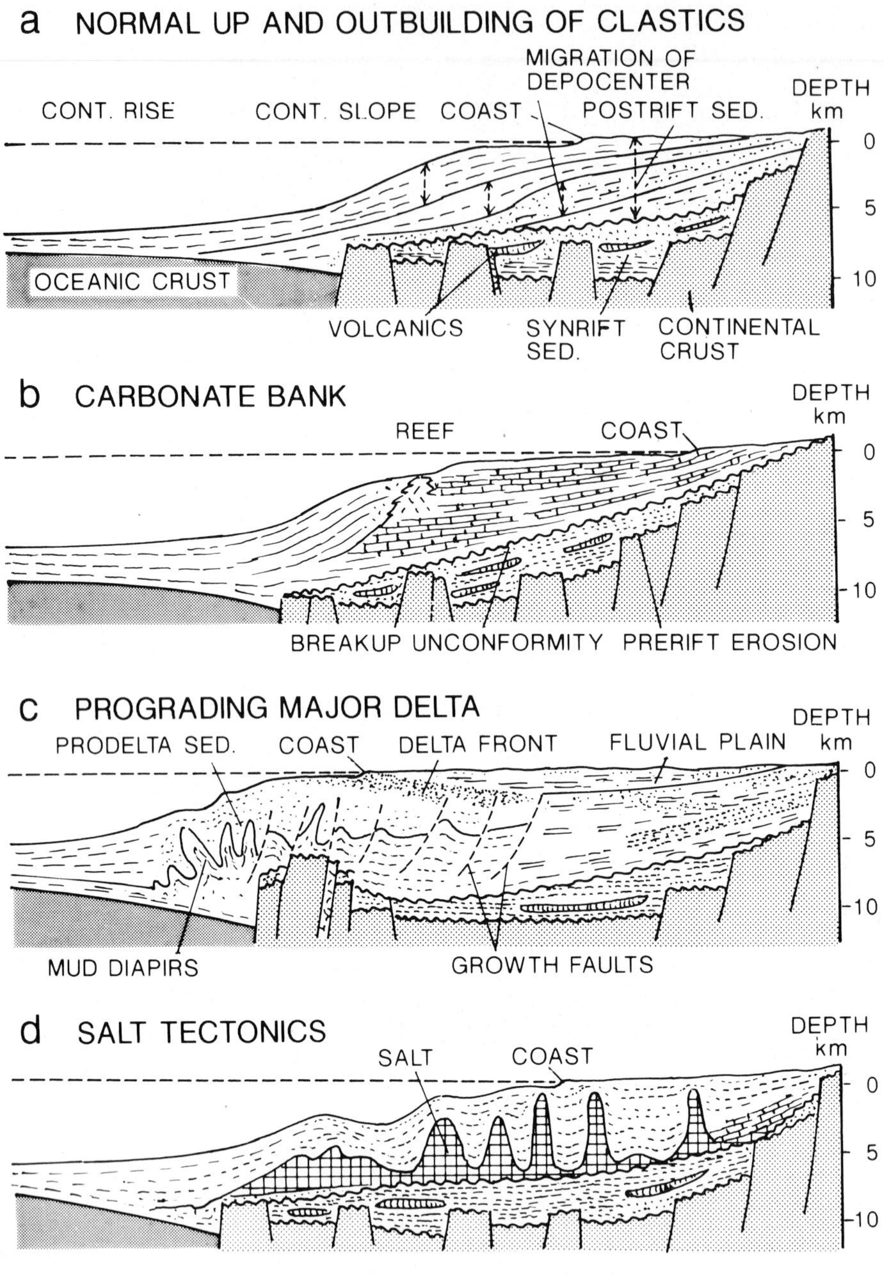

a NORMAL UP AND OUTBUILDING OF CLASTICS

b CARBONATE BANK

c PROGRADING MAJOR DELTA

d SALT TECTONICS

Fig. 12.11a-d. Generalized architecture and sediment successions of Atlantic-type passive continental margins (end-members).

Syn-rift sediments are predominantly continental, postrift sediments marine. (After Kingston et al. 1983). See text for explanation

1. *Upbuilding and Outbuilding of Siliciclastic Sediments*

The subsiding continental margin accumulates predominantly siliciclastic sediments at moderate rates. The isochrones show upbuilding and outbuilding of sediment. The rate of total subsidence of the basement (in places charac-terized by a breakup unconformity) increases seaward and thus produces a wedge-like sediment body. The deepest portion of the basement (basin floor) is commonly found below the continental slope or rise. The zone of maximum thickness (or *depocenter*) of a certain stratigraphic unit tends to migrate basinward with time (cf. Fig. 12.11a). The margins of

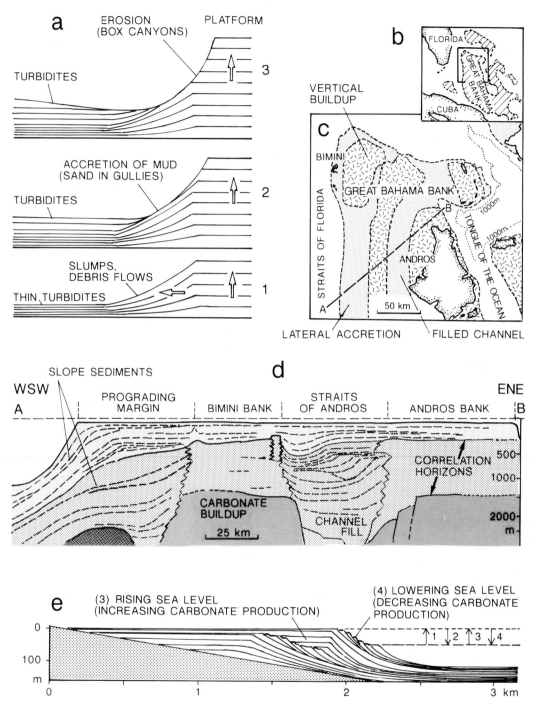

Fig. 12.12. a Model of lateral progradation and "self-erosion" of limestone escarpments in the Bahamas and Marshall Islands. Note oversteepening and various types of erosion with increasing vertical buildup (*1, 2, 3;* angles of upper one-third of slope up to about 30°). (Schlager and Camber 1986). **b** through **d** Location and internal structure of the northwest Bahama Bank displaying two buried banks (Bimini Bank and Andros Bank) with completely filled former trough (Straits of Andros) in between. Lateral progradation of Bimini western margin by 20 km as a result of sediment accretion on slope. Vertical upbuilding since the

mid-Cretaceous is about 2 km (lower correlation horizon); earlier platform consists of carbonates and evaporites. Vertical exaggeration about 12x. (Derived from seismic profiles, ocean drilling, and dredge hauls on the exposed flanks; Eberli and Ginsburg 1987; Schlager et al. 1988). **e** Simulated vertical and lateral growth of carbonate platform in response to sea level changes. During rising sea level, lateral progradation is rapid as a result of high carbonate production on shelf. Falling sea level leads to slow progradation due to reduced carbonate production on emerging platform. (Bosence and Waltham 1990)

southwest Africa, southern Morocco, and the northern Gulf of Mexico provide examples of this basin type. High-latitude continental margins may accumulate glacio-marine deposits of considerable thickness as known from the Barents shelf off northern Scandinavia.

2. *Carbonate Buildups*

Large portions of the modern and ancient continental margins are predominantly built up of carbonates (cf. Chap. 3.3.2; Stanton and Flügel 1989; Tucker et al. 1990). If the shelf edge is dammed by reefs (thereby forming a rimmed carbonate shelf), the slope can become very steep. Basinward of the reefs the slope may prograde due to the accumulation of reef talus (Fig. 12.11b) and debris flows, or shelf and terrigenous material may spill over the shelf edge. Modern examples of this type of shelf structure include the continental margin off the Senegal and Florida coasts. The upward growth of the Blake Plateau was largely terminated by the Gulf Stream, however (Chap. 5.3.7, Fig. 5.10). The marginal Mazagan plateau off Morocco shows an early, syn-rift phase of Jurassic carbonate buildup which was terminated due to insufficient subsidence in the early Cretaceous and then followed by post-rift hemipelagic and pelagic deposition (Chap. 5.3.7).

Isolated carbonate platforms on continental or intermediate crust as, for example, the present-day Bahamas Archipelago, also result entirely from carbonate accumulation.

The previously held view that carbonate buildup always occurred under shallow-water conditions and was accomplished predominantly by stationary upward growth has, however, to be modified (Schlager and Camper 1986; Eberli and Ginsburg 1987 and 1989; Leg 101 Scientific Party 1988). Seismic records and drilling have shown that this bank-and-basin archipelago was characterized by periods of lateral expansion and retreat as well as by phases of deeper water carbonate deposition (Fig. 12.12b through d). Upbuilding in the late Jurassic and early Cretaceous was followed by drowning and retreat during the mid-Cretaceous and by renewed lateral expansion in the Cenozoic. Lateral, discontinuous prograding of 500 to 1000 m thick carbonate platforms has also been described from the Triassic of the Dolomites in northern Italy (Bosellini 1984).

Generally, fine-grained, muddy carbonates produce gentle prograding slopes (2-4°), frequently characterized by erosional gullies, while coarser grained debris forms steep slopes (10-12°, up to more than 30° in the upper part), which commonly exhibit increased slumping, debris flows, and bypassing of sediment via canyons (Fig. 12.12a, also see Chap. 12.2.3 and Kenter 1990). In the latter case, sediment buildup is concentrated on debris aprons.

The result of a computer simulation for vertical and lateral carbonate platform growth in response to sea level changes is shown in Fig. 12.12e (Bosence and Waltham 1990). Such a depositional system is controlled largely by benthic carbonate production, which is high during rising sea level, but very limited during falling sea level. Hence, *lateral progradation of carbonate platforms* takes place predominantly during sea level rise and highstand (Eberli and Ginsburg 1989). Similarly, the high sediment input of platform sediments into deeper basins *is restricted to sea level highstands* (this is known as "highstand shedding", also see Chap. 7.4). Lowstands cause hiatuses and possibly karst horizons on emerged parts of the platform. Slope and periplatform carbonates have a high potential of being stabilized by early diagenesis.

3. *Prograding Deltaic Sediments*

High sediment influx from a major river delta causes continuous filling of the subsiding marginal basin and rapid progradation of the slope beyond the continental/oceanic crust boundary (Fig. 12.11c). As a result, the zone of maximum sediment thickness of a modern delta may lie landward of the present coast line. The low strength of muddy, underconsolidated sediments leads to the formation of growth faults, rollover structures, and mud diapirs (Figs. 3.30a and 12.16b). Prominent examples are the shelf off the Niger delta in West Africa (Doust and Omatsola 1990), and the northern margin of the Gulf of Mexico with the Mississippi delta (cf. Chap. 13.2), which is also modified by salt diapirism.

4. *Continental Margins Affected by Salt Structures*

The growing sedimentary wedge in the marginal basin is strongly affected by halokinesis of the underlying salt (Fig. 12.11d; cf. Chap. 6.4.4). Salt diapirs act as dams for shelf sediments which in addition are deformed by subsidence in special troughs and uplift over the rising salt structures. This type of continental margin basin is very common around the Atlantic. It is known, for example, from the shelf of Nova Scotia (Scotian basin) and the conjugate margin of West Africa (Essaouira-Agadir basin and Aaiun-Tarfaya basin; e.g., Jansa and Wiedmann 1982) or the Gabon basin (Kingston et al. 1983; Teisserene and Villemin 1990).

Many basins on passive continental margins also exhibit a combination of two or three of these end-member basin types. In the marginal basins of northwest Africa, for example, Triassic and Liassic salt may be overlain by Jurassic shelf carbonates and reefs, followed by early Cretaceous prodelta clastics. Generally, the sequence of post-rift sediments reflects the climatic and oceanographic history of the region, as well as the impact of the evolution of rock-forming calcareous microorganisms and eustatic or relative sea level changes.

Depositional History of Conjungate Marginal Basins

Particularly conjugate marginal basins which formed one common basin or lay close together prior to ocean spreading, but which are now separated by a large ocean (Fig. 12.13), may exhibit more or less *identical depositional histories*. For example, the Scotian basin along the North American margin and the Essaouira-Agadir basin off northwest Africa show the following common characteristics (from base to top of the basin fill):

- Late Triassic syn-rift graben infillings consisting of continental red beds and alkaline volcanics.
- Early Jurassic evaporites and dolomites, and some alkaline volcanics (late rifting stage, with an at least local marked decrease in subsidence).
- Mid-Jurassic latest rift sediments or early post-rift sediments consisting of clastics, deltaic sediments, red beds, and a first reef phase on the African side.
- Late Jurassic post-rift carbonate platforms with reefs (second reef phase followed locally by renewed rapid subsidence).
- Early Cretaceous platform drowning, "Wealden"-type fluvial clastics and marine prodelta deposits.
- Mid-Cretaceous siliciclastics and limestones and, on the African side, black shales and cherts resulting from coastal upwelling.
- Late Cretaceous to Paleogene sediments characterized by slow sedimentation rates and high proportions of biogenic material.

During mid-Cretaceous time, the central Atlantic had already reached the size of a large ocean basin and the depositional environments on the opposite margins of this ocean became increasingly dissimilar. This *asymmetric evolution* is documented by the mid-Cretaceous

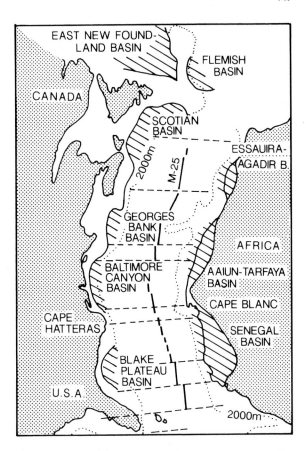

Fig. 12.13. Position of conjugate margins in the north Atlantic during the late Jurassic (magnetic line M-25) and major Mesozoic-Cenozoic marginal basins. (Jansa and Wiedmann 1982)

black shales in the Essaouira-Agadir basin, as well as by late Cretaceous shelf construction in the Scotian basin while the marginal basin on the African side was characterized by shallowing and the onset of emergence. In addition, the northwest Atlantic basins drifted into a region of increasingly cold climate and subarctic ocean currents, while the northwest African basins more or less remained in a zone of warm, relatively dry climate. Another example of a similar Mesozoic-Cenozoic tectonic history and depositional environment represent the eastern margin of the Grand Banks in the North Atlantic and the Galicia Bank margin of Spain (Grant et al. 1988; cf. Chap. 12.1.4 and Fig. 12.7). Another, well described example of conjugate marginal basins is the Campos Basin of Brazil and the Gabon Basin of West Africa (Edwards and Santogrossi 1990); one basin is the mirror image of the other.

As a result of differential subsidence and/or uplift of the landward side of the basin, many

Atlantic-type continental margin basins became partially emerged. These parts of the
basins are referred to as "*coastal basins*".

Sediment-Starved Marginal Basins

All the basin types discussed so far have in
common the tendency for being more or less
continuously filled with sediments up to the
depth of the continental shelves and therefore
generally display either shallow-water or slope
deposits along their outer margin. However,
subsidence can exceed vertical sediment aggradation and thus cause deepening, "sediment-starved" basins, or sediment supply is
greater than necessary to keep pace with subsidence. The latter case is common during late
basin evolution, when subsidence slows and
the basin becomes "over-supplied" with sediment. As a result, the excess sediment supply
will bypass the shelf and settle on the slope or
in deeper water (see below).

Examples of Young Continental Margins

After the onset of drifting, passive continental
margins tend to subside so rapidly that even
relatively high sedimentation rates are insufficient to keep pace with subsidence. Hence,
young continental margins frequently deepen.

For example, sea-floor spreading at the southern end of the
Gulf of California began 4 Ma B.P. and caused the passive, rifted continental margin adjacent to the tip of Baja
California to subside by about 2 km as shown in Fig.
12.14c and d (subsidence curve see Fig. 8.12). The present, post-rift sediment cover on top of this stretched and
sunken continental crust is only 100 to 200 m thick, although the sedimentation rate for diatomaceous ooze on
this slope was relatively high (30 to 80 m/Ma; Curray et
al. 1982b).

Deeply subsided, sediment-starved continental
margins also occur in other young ocean basins, for example in the Mediterranean (Burollet et al. 1978).

The young, recently drilled **passive continental margin
east of Sardinia** subsided rapidly with the onset of sea
floor spreading in the late Miocene or early Pliocene (Fig.
12.15a and b; Kasten, Mascle et al. 1988; Savelli 1988). It
borders the Tyrrhenian Sea, which is interpreted as a
backarc basin with young oceanic crust. The syn-rift sediments of the sunken continental margin (Site 654, Fig.
12.15a) partially show a typical transgressive sequence
beginning with subaerial conglomerates, overlain by oyster-
bearing sands and marine marls. During the Messinian
(latest Miocene) desiccation of the Mediterranean, the

landward Sardinian margin accumulated nannoplankton-
bearing clays interbedded with laminated gypsum, while
the seaward, more rapidly subsiding margin apparently
stood higher, as indicated by subaerial and lacustrine
facies. The post-rift sediments are about 200 m thick and
consist of nannofossil ooze with minor proportions of
calcareous mud, volcanic ash, and sapropel.

The evolution of the **Tyrrhenian Sea** as an extensional
basin from an early rifting stage prior to 8 Ma ago to the
present situation is shown in Fig. 12.15b. New oceanic
crust was generated during two spreading episodes (7 Ma
and 1.9 to 1.3 Ma). The center of magmatism migrated
from west to east and formed the Eolian Islands, Strom-
boli, and the volcanoes of central Italy (Savelli 1988).

Older, Sediment-Starved Outer Margins

If the outer part of a marginal basin does not
receive sufficient sediment to be filled up to
the normal shelf depth, a deep-water environment will be established. Then the post-rift
sediments tend to show a *deepening-upward
trend*, with hemipelagic and pelagic sediments
replacing shallow-water deposits. Figure 12.14
demonstrates such a situation in two differing
tectonic settings, i.e., on an old and on a
young continental margin.

In the first case, the **Goban Spur southwest
of Ireland**, the old continental margin of the
eastern Atlantic is too far away from continental sediment sources and therefore subsided to bathyal depths without sufficient
sedimentary cover (Fig. 12.14a and b; de Graciansky and Poag 1985).

Drilling of this sediment-starved margin has revealed a
half-graben within the rifted continental crust, which contains syn-rift, shallow-water marls and limestones of early
Cretaceous and possibly older ages. Post-rift development
began after the Albian after the formation of an unconformity. Bathyal to abyssal chalks indicate rapid subsidence
subsequent to the onset of sea floor spreading.

Further development is characterized by minor modifications due to a fluctuating CCD and periodic influx of
terrigenous detritus, especially in the Cenozoic. Additional
unconformities are associated with sea level lowstands.

In total, the thickness of the post-rift sediments is between 0.5 and 1.5 km and is thus in
striking contrast to the huge depositional
prism of the opposing North American margin, which displays post-rift sediment thicknesses in excess of 10 km (e.g., the Baltimore
Canyon Trough). The Galicia Bank off the
Iberian peninsula shows a similar situation,
with a very thin, incomplete post-rift sediment cover.

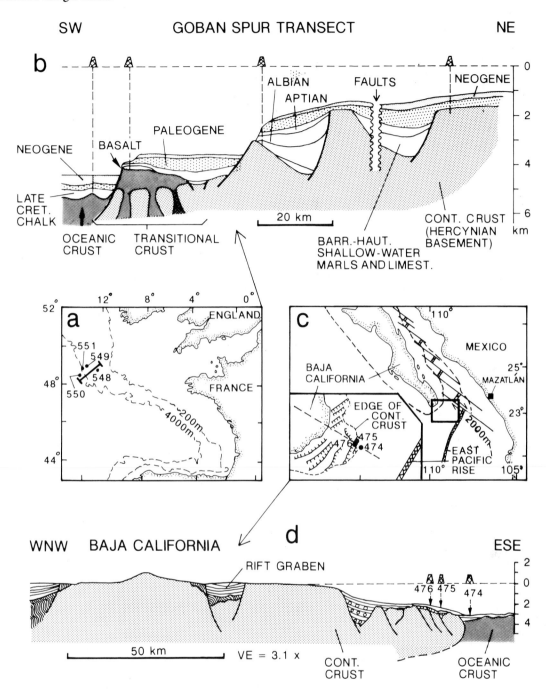

Fig. 12.14. a and **b** Sediment-starved passive continental margin of the eastern Atlantic. (After de Graciansky et al. 1985). **c** and **d** Young, rapidly subsided passive margin of the southern Gulf of California. (After Curray et al. 1982a). Note the different vertical and horizontal scales. See text for explanation

Submarine Marginal Plateaus

Submarine plateaus some 1000 to 3000 m below sea level occur in all large ocean basins. They are more or less separated from terrigenous sediment sources and their sediments show characteristics similar to those of sediment-starved outer margins mentioned above.

In the **southwest Pacific**, submarine plateaus occupy broad areas of the sea floor, including the Ontong-Java Plateau at the equator, several rises between Melanesia and

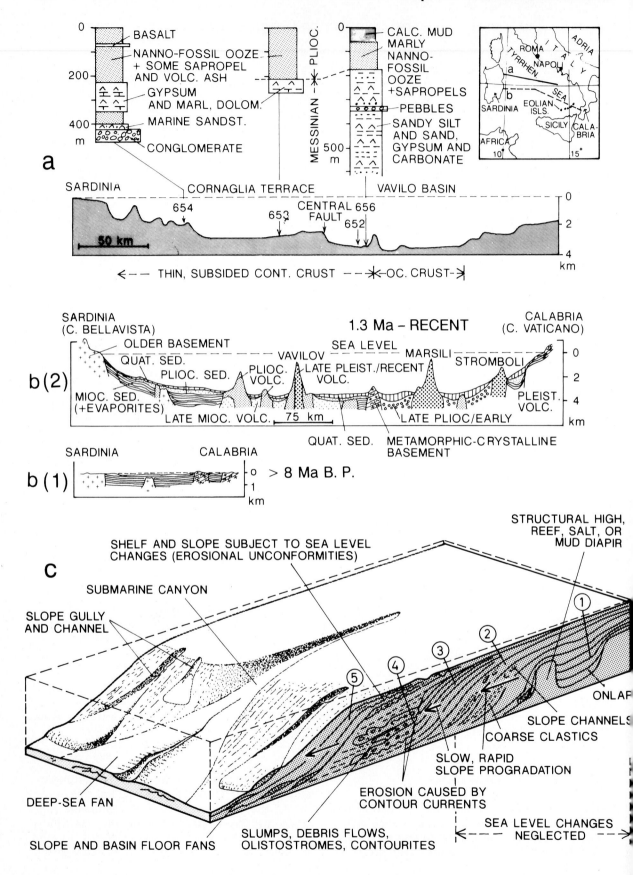

Australia, and some plateaus east of New Zealand (Kennett and von der Borch 1986). The tops of most of these plateaus lie well above the CCD. They are usually covered by relatively thick Neogene to Paleogene calcareous foraminiferal nannofossil oozes, which are little to not affected by carbonate dissolution. Deeper portions of such plateaus may have been affected by CCD fluctuations and therefore show distinct layering. Microfossils and carbonate contents of these sediments are useful not only for biostratigraphy, but also as indicators of the changing climatic and paleoceanographic conditions in the late Mesozoic and Cenozoic (e.g., Berger et al. 1991).

Parts of the **Exmouth Plateau off northwest Australia**, being an old passive continental margin of the Indian Ocean, lost their syn-rift Jurassic sediment cover due to uplift in conjunction with block faulting. After breakup of the Gondwana continent and the onset of sea floor spreading in the early Cretaceous, the entire Exmouth Plateau subsided below sea level. Outer parts of this marginal basin (e.g., the Wombat Plateau) became sediment-starved and therefore display only a thin post-rift sequence, including condensed horizons, pelagic limestones and black shales at the Cenomanian-Turonian boundary (Leg 122 Shipboard Scientific Party 1989; von Rad et al. 1989).

12.2.3 Sediment Successions on Continental Slopes

Introduction

A description of continental margin basins should also include their slopes leading from the shelf edge to the deep sea. Present-day continental slopes commonly have gentle gradients of a few degrees and therefore widths of 50 to 150 km. Hence, prograding slopes and slope aprons (Figs. 5.3 and 5.12) constitute a significant part of continental margin sedimentation. They accumulate sediments which have migrated over the shelf or are derived from the upper slope and are redeposited on the lower slope.

However, slopes may also undergo long-term nondeposition or erosion. For example, the postglacial rapid sea level rise may have caused oversteepening of some modern slopes and thus favored erosion. High temperature gradients between the poles and the equator during glacial periods, in conjunction with low sea level, have probably initiated particularly strong thermohaline oceanic circulation. As a result, contour currents were able to erode the foot of the slope and trigger large mass movements.

Construction and Destruction of Slopes

Some principles for the *upbuilding* and *outbuilding* (or so-called constructional phase) and destruction of mainly siliciclastic slope sediments are summarized in Fig. 12.15c (for carbonate slopes see above). Slopes *dammed by structural highs*, volcanic structures, reefs, salt or mud diapirs tend to remain sediment-starved until the shelf basin behind the dam is filled. Progradation of the slope may proceed either rapidly or slowly in relation to the influx of terrigenous or biogenic sediment. The upper slopes on modern continental margins are frequently characterized by sediment bypassing and local sediment ponding, progradation, and the filling of channels and gullies. Thus the outbuilding slope sediments frequently reflect the composition of the shelf deposits and their variation through time. The lower slope or slope apron collects slides, debris flows, and turbidites.

For example, seismic lines along the slopes of the western Atlantic off the United States have revealed that terrigenous material was transported across the early Cretaceous continental shelf to form several large shelf edge deltas (Poag et al. 1990). Each delta supplied sediment gravity flows to slope aprons and deep-sea fans on the lower slope and continental rise. In the northern Hatteras basin, one of these sediment distributory systems extended 500 km basinward.

Fig. 12.15. a Cross section of young, sediment-starved passive continental margin off Sardinia in the Thyrrhenian Sea with some drilling results. (After Kasten, Mascle, et al. 1988). **b** Sections similar to (a) displaying two stages of basin evolution and several magmatic episodes. (After Savelli 1988). **c** Theoretical model demonstrating principal types of outbuilding and upbuilding of slope sediments: *1* Mainly vertical aggradation with stratal onlap behind "dam". *2* Rapid slope progradation due to high sediment influx including infillings of slope channels and coarser grained redeposited material on lower slope. *3* Slow slope progradation. *4* Current-eroded and oversteepened slope with gravity mass flows and contourites. *5* Prograding slope affected by sea level changes (cf. Chap. 7.4). Note that several phenomena may be superimposed. (Based on von Rad and Wissmann 1982; Sheridan 1981; Pickering et al. 1989; Vail et al. 1990)

Erosion at the foot of the slope not only prevents further progradation, but also causes destruction and redeposition of older slope sediments. Figure 12.16 shows two examples of mid-Cretaceous to Cenozoic multi-phase slope destruction from the northwest African margin which are based on seismic profiles and information from boreholes (von Rad and Wissmann 1982).

Undercutting of the slope at various levels created long hiatuses and angular unconformities in the stratigraphic record. These processes initiated the cutting of major canyons into the slope sediments. In this case it is also assumed that intensified bottom currents occurred during times of cooler climate and sea level lowstand. Mass wasting was particularly active during the mid-Miocene and the late Pleistocene. Debris flows and huge olistostromes interfinger with contourite drifts. In addition, these cross sections display antithetic growth faults with giant rollover

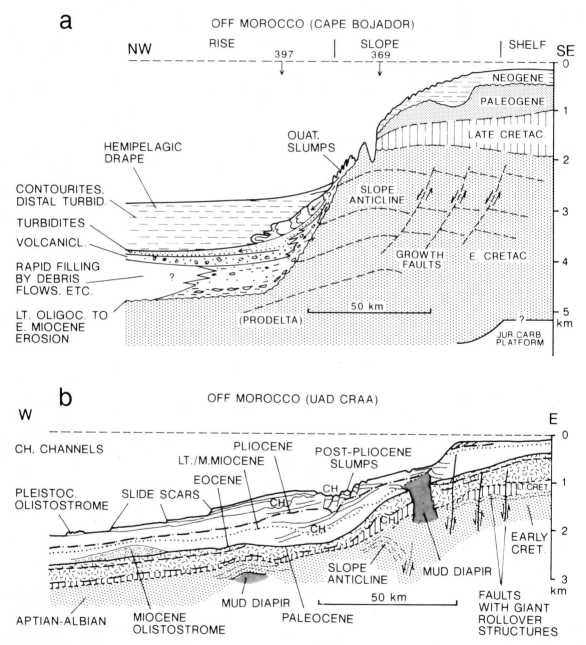

Fig. 12.16a,b. Schematic cross sections of continental slope off (**a**) Cape Bojador and (**b**) Uad Craa, Morocco. Note multi-phase erosion and mass wasting on huge scale. (After von Rad and Wissmann 1982)

structures, forming a "slope anticline", and mud diapirs. These structures are common in rapidly accumulating thick sedimentary sequences, as represented here by lower Cretaceous prodelta deposits.

Ancient Slope Deposits

The identification of slope sediments in the fossil record can be difficult. They may be characterized by predominantly siliciclastic, hemipelagic, or bioclastic carbonate sediments. Useful criteria are slope scars and gullies on the upper slope, canyons and mass flow deposits on the lower slope. At the foot of the slope contourites may accumulate, or the slope may be undercut by erosion. In any case, the large-scale facies association of the basin fill should be known. In addition, faunistic criteria may indicate the paleo-waterdepth.

If slope successions are incorporated into subduction and collision-related deposits, their identification is even more problematic. Slope sediments preserved due to rapid subsidence and burial under thick younger sequences tend to be affected by regional low-grade metamorphism.

Concluding Remarks

Continental margin sediments have been described from many places all over the world. They display a much greater variety in their structure and sedimentary fill than can be discussed here. Some margins show significant lateral changes (see, e.g., Ranke et al. 1982; Emery and Uchupi 1984). Continental margins affected by wrench tectonics and strike-slip motion may represent either a wide continental borderland (e.g., off southern California, cf. Fig. 5.10b) or a very narrow shelf, such as that of the young Gulf of California on its western side (Fig. 4.5). Finally, passive continental margins can migrate into zones of convergence or be transformed into active margins. Many ancient continental margin basins were incorporated into orogenic belts and were either subducted or obducted and strongly deformed. The sequences of continental margin basins may, for example, be overlain by undeformed and deformed sediments of a foreland basin (Chap. 12.6.3) or thrust sheets of the approaching mountain belt. Another possibility is that overthrusted and folded continental margin sequences are unconformably overlain by the deposits of an intramontane basin.

12.3 Intracratonic Basins Associated with Mega-Rifting

12.3.1 Permian to Mesozoic Basin Development in Europe (Overview)

Introduction

Central and western Europe, including its continental shelf and the North Sea, are characterized by a considerable number of sedimentary basins which developed since the Permo-Triassic and persisted through the Mesozoic or Cenozoic time. These basins represent well-known examples of intracratonic basins, the evolution of which is initiated by rifting and complicated by block faulting and wrench tectonics (Ziegler 1982, 1988). In all these cases, rifting did not lead to the formation of new oceanic crust (failed rifts). However, during and subsequent to rifting, these basins experienced long-lasting thermal subsidence similar to passive continental margins. Some of these basins may be regarded as continental sag basins (see below), although their development spans less time than that of typical sag basins.

Basin Evolution

During the Late Permian and Triassic, the Pangean supercontinent became increasingly unstable due to the reactivation of peripheral rift zones and the development of new interior rift systems. As a result of the northward drift of Pangea, Europe moved out of an equatorial position held during the Late Carboniferous, reached the northern trade-wind belt, and reached a zone of dry, hot climate. From the Triassic to Middle Jurassic, two principal, multi-directional rift systems transected the former Variscan fold belt and its foreland: from the north, the Norwegian-Greenland Sea rift prograded southward, and the Tethys rift system prograded westward. In the central and north Atlantic, these two mega-rift systems interfered with each other, but the neighboring areas, including western and central Europe, were also strongly affected by this plate reorganization. Pre-existing Permo-Carboniferous fracture systems were reactivated, and wide areas around the future continental margins underwent extensional stresses. These caused differential subsidence and thus basin fills of varying thicknesses between zones of higher elevation and moderate uplift.

The depositional environments and histories of these basins were affected by many local factors and therefore vary from basin to basin. Nevertheless, some general trends can be recognized which may also characterize other basins of similar origin. These trends predominantly reflect the interplay between subsidence and sediment accumulation, as well as the influence of climate and eustatic sea level changes. This is exemplified by the evolution of the European Mesozoic basin between the growing Tethys Ocean in the south and the North Sea in the north.

12.3.2 Mesozoic Sediments Between the North Sea and the Western Tethys

Triassic Sediments

Western Tethys, Alps

During the Triassic, the large shelf areas in the western Tethys Ocean were characterized by extensive carbonate platforms, some evaporite basins, and intervening troughs accumulating deep water carbonates and shales (Fig. 12.17a; e.g., Bechstädt et al. 1978; Ziegler 1988; Elmi 1990). The extensive, thick carbonate buildups of the eastern and southern Alps and other Alpine regions in southern Europe document subtropical climate and the efficiency of shallow-water carbonate production in keeping pace with the subsidence of the continental margins.

Germanic Facies Province

The fully marine facies in the Alpine-Mediterranean Triassic was replaced by predominantly clastic sediments in the Germanic facies province, i.e., on the western and northwestern Tethyan shelves and in the intracratonic basins of central and western Europe (Fig. 12.17a; e.g., Schröder 1982). During the Perm-

ian, red beds, evaporites, and volcanics were deposited in narrow troughs of the former Variscan fold belt. With the onset of the Triassic the relief was largely peneplained, but differential subsidence (20 to 50 m/Ma) of wide basins and uplift of large blocks continued. The term Triassic is derived from these regions, because its sequence can be subdivided into three distinctly different subunits: (1) The lower continental Buntsandstein clastics, (2) the marine Muschelkalk carbonates and evaporites, and (3) the upper coastal/playa lake/continental deposits of the Keuper. In Spain, for example, the Triassic sequence is quite similar to that in central Europe (García-Mondéjar et al. 1986).

1. *Buntsandstein.* The basins were filled by alluvial fans and extended red sandstones and claystones, which accumulated in braided and meandering fluvial systems (Tietze 1982; Mader 1985; Pérez-Arlucea and Sopena 1986). Locally, playa sediments with evaporites were deposited. Fluvial cycles and paleosols with silcrete and pedogenic carbonate reflect short-term climatic variations and changes in the drainage systems.

2. *Muschelkalk.* The shallow-marine Muschelkalk basin in central Europe is associated with a long-term eustatic sea level rise which caused a transgression of the Tethys Ocean through two graben systems, one to the east of the Bohemian massif and the other between the Bohemian massif and Massif Central in France (Fig. 12.17a). To the north the basin was connected with the Arctic Ocean by a seaway between Greenland and Fennosarmatia. This paleogeographic setting created a shallow-marine environment very sensitive to eustatic sea level variations. Warm and comparatively arid climate favored the production of biogenic and chemically precipitated carbonate, which in turn was slightly diluted by terrigenous siliciclastics. The carbonates show distinct lateral facies changes ranging from open marine ramp settings to marginal banks and protected, hypersaline back-bank regions (Aigner 1985). Mud flats and channels indicate moderate tidal activity (Schwarz 1975). Evaporites of limited thickness in the middle Muschelkalk reflect a sea level fall and arid climate. Cyclic, shallowing-upward tempestite/marginal bank sequences and the repeated occurrence of specific marker beds in the upper Muschelkalk are interpreted by small, short-term sea level oscillations superimposed on longer-term trends (Aigner 1985; Röhl 1990). The highstands in these oscillations are represented by thin, clay-rich beds which can be traced over long distances.

Fig. 12.17a-c. Triassic to Jurassic rift/sag basin evolution in central Europe, overview. **a** and **b** Paleotectonic and paleogeographic situation during the middle Triassic (**a** Muschelkalk) and late Jurassic (**b**). Abbreviations for land masses (massifs): *AM* Armorican; *BM* Bohemian; *EH* Ebro High; *IBM* Iberian; *IM* Irish; *LBM* London-Brabant; *RM* Rhenish. (Simplified from Ziegler 1988). **c** Greatly generalized Permian to lower Cretaceous sedimentary sequence.

(Based on Wurster 1968; Krömmelbein 1986; Anderton et al. 1979; Schröder 1982; Geyer and Gwinner 1986; and others) in relation to long-term sea level changes (after Haq et al. 1987, including time scale) and climatic trends (Schwarzbach 1974; Frakes 1979). Note that long-term trends are superimposed with shorter variations in climate and sea level, generating various cyclic phenomena. See text for further explanation

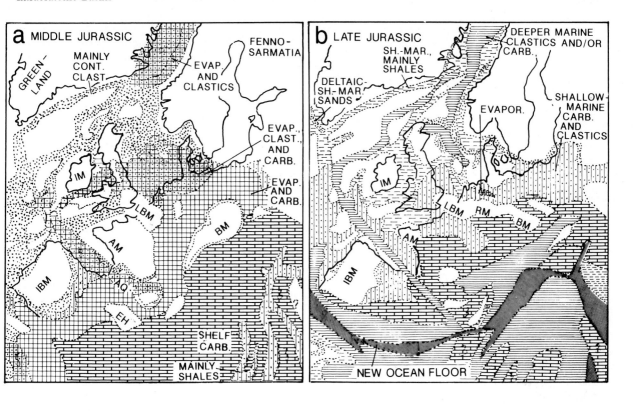

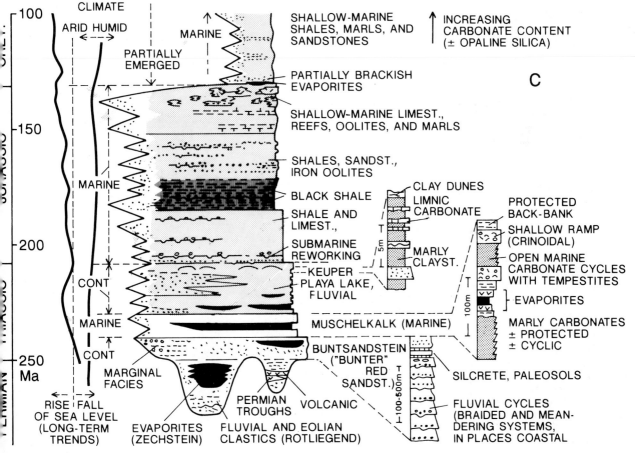

3. *Keuper.* The facies of the upper Triassic (Keuper) represent transitional or predominantly continental environmental conditions, although according to Haq et al. (1987) the eustatic sea level was still rising during the first half of this period (Fig. 12.17c). This indicates that, at least part of the time, the basin received a surplus of clastic sediments. Some short-term marine ingressions entered the central European basin and led to the deposition of tidal sediments, thin carbonate beds and evaporites. However, most of the time a huge playa lake occupied the region in which red and greenish clayey marls alternating with thin carbonate beds and some sabkha-type evaporites accumulated. Massive, red silty clays in the uppermost part of the sequence containing irregular layers of pedogenic carbonate nodules are interpreted as fossil clay dunes (Chap. 2.3.3). Throughout this development, the neighboring highlands shed sand and gravel into the basin from various sources, but mostly from Fennosarmatia southward.

The total thickness of the Triassic sequence in the Germanic facies frequently ranges from 800 to 2000 m, but is less adjacent to the uplifted massifs. By contrast, the thickness of the carbonates accumulated on the Tethyan shelves is much greater. These strata are also affected by sea level changes (e.g., Brandner 1985; also see Chaps. 7.2 and 7.4). Basin fills similar to those in the central European area resulted from Tethyan transgressions into other rift zones, for example into the Aquitaine basin and the Bay of Biscay rift up to the Paris basin and the Celtic Sea basin. Toward the west, the Gibraltar and Atlas rifts were affected by transgressions (Ziegler 1988). In most of these basins, the Triassic sequences are characterized by red beds and evaporites, Muschelkalk-type shallow-marine carbonates, and Keuper-type playa lake/continental deposits.

Jurassic Sediments

During the Rhaetian and early Jurassic, the pre-existing rift systems remained active, and differential subsidence in central and western Europe continued. This development led, in combination with long-term rising sea level, to a long-lasting seaway between the Tethys and Arctic Oceans (Fig. 12.17 b, Ziegler 1988). Ocean spreading in the Tethys prograded westward and connected the Tethys with the evolving central Atlantic.

In *Rhaetian time,* the sea again invaded the central European basin and established normal marine, epicontinental conditions until the end of the Jurassic and, in smaller areas, throughout the Cretaceous and part of the Tertiary. Due to the influence of both the mixing of colder Arctic water with warmer Tethys water and the increasingly humid and cooler climate, the sediments of the *lower* and *middle Jurassic* are predominantly shales with minor contributions of fine sand and carbonate (Fig. 12.17c). Sea level oscillations led to widespread submarine erosional unconformities characterized by reworked limestone and siderite concretions, redeposited iron ooids, and some layers containing glauconite and phosphorite.

The marked *Hettangian* and *Toarcian transgressions* favored the formation of relatively thin, but extensive black shale horizons. With the onset of increasing biogenic carbonate production in the *upper Jurassic,* possibly associated with a trend towards a warmer climate, marls, limestones, and reefal carbonates were deposited. Rhythmic and cyclic phenomena, including phases of reef emergence, again indicate variations in sea level.

At the same time, the Viking and Central graben systems in the **North Sea** became active and established comparatively deep water environments. The Kimmeridgian and Tithonian organic-rich shales are regarded as the principal hydrocarbon source rocks for the North Sea oil (cf. Chap. 14.3, Fig. 14.11). By contrast, the late Jurassic of the Lower Saxonian basin in Northern Germany is characterized by clastic sediments and evaporites indicating aridity.

Cretaceous Sediments

A land bridge running from the London-Brabant massif to the Bohemian massif, and wrench faulting in the North Sea and north of the land bridge, led to marked differences in the upper Jurassic and early Cretaceous basin evolution in central Europe.

While, for example, the southern (Franconian) basin went dry at the end of the Jurassic, subsidence continued in the Lower Saxonian basin, in the North Sea, and in the Paris basin. Evaporite accumulation came to an end due to a change to more humid climate.

The *lower Cretaceous Wealden facies* represents sandy to clayey nonmarine, brackish, deltaic, or coastal deposits. The overlying shallow-marine sediments consist mainly of shale, sand, and minor proportions of carbonate and opaline silica, but may vary significantly due to the influence of local sediment sources. In conjunction with the *mid-Cretaceous sea level rise,* and the evolution of carbonate secreting microorganisms, the sediments become increasingly rich in carbonate and show Milankovitch-type limestone-marl rhythms (Chap. 7.2).

The Jurassic sequence varies in thickness from about 400 to 600 m in southern Germany to approximately 1000 to 1800 m in northwestern Germany. The Cretaceous sediments in the Lower Saxonian trough may reach more than 1500 m in thickness. If regions with high sediment thicknesses are taken into account, the average rate of Mesozoic subsidence was on the order of 25 m/Ma.

Summary

This brief discussion of basin evolution in central Europe with a plate tectonic view demonstrates the following points.

- On the one hand, sedimentary sequences in neighboring basins evolving in a continent-wide common tectonic setting may develop striking similarities. In addition, the formation of easily correlatable sediment successions is favored by coeval variations in climate, eustatic sea level fluctuations, and changes in biogenic sediment production.
- On the other hand, the intracratonic Mesozoic basins in Europe and the north Atlantic show a great variation in both their tectonic and sedimentological histories.

Therefore it appears problematic to define certain "type basins". There are transitions between small, short-lived, simple rift basins and long-lasting, prograding, large rift systems which may in addition be affected by wrench faulting. Such processes may cause differences in regional subsidence over long time periods. Whether or not these basin types can be accurately discriminated from intracratonic sag basins as discussed below is not clear.

12.4 Continental or Intracratonic Sag Basins

General Aspects

As already mentioned in Chapter 1.2, continental or intracratonic sag basins are commonly *slowly subsiding, long-lived basins*. They lack major extensional faulting, because in these cases lateral tensile deviatoric stress is insufficient to overcome rock strength and cause brittle fracture (e.g., Allen and Allen 1990). Subsidence is brought about by thermal contraction without significant extension and fracturing. Intracratonic sag basins are known from all continents and may accumulate thick (up to 10 to 15 km), undeformed or little deformed sedimentary sequences over long periods of time (on the order of 200 to 1000 Ma).

Basins of this type with late Proterozoic to middle Paleozoic sediments have been described, for example, from northeastern Asia, around the Siberian platform, and the southern Urals (Artyushkov and Baer 1986). Other, mostly Precambrian to Paleozoic intracratonic sag basins are known from Africa, Australia, and North America (e.g., the Paleozoic Illinois, Michigan, and Williston basins, with 3,700 to 6,000 m of sediments; Klein and Hsui 1987; Fisher et al. 1988). The central parts of the Michigan basin subsided steadily at a an average rate of 24 m/Ma over a time span of 200 Ma (Fischer 1975). Several basins of North Africa, including the Tindouf basin in Morocco and Algeria, as well as the Murzuk and Kufra basins in Libya, existed since the Paleozoic and continued to subside until Neogene times.

"Modern" examples of intracratonic sag basins include the slowly subsiding, continental Chad basin in northern central Africa and the Eyre basin in Australia (Mitchell and Reading 1986); the North Sea and Hudson Bay in Canada are sag basins inundated by the sea.

The *crustal processes* forming these basins are not well understood. Some basins may be underlain by rift structures, but their post-rift subsidence appears to be much longer than that which can be attributed to a rift-produced thermal anomaly (Quinlan 1987). In the absence of significant stretching and thinning (or shortening) of the crust, a slow gabbro-eclogite transformation in the lowermost crust could be the principal mechanism causing subsidence (Artyushkov and Baer 1986; cf. Chap. 12.1.1). Other authors (e.g., Finlayson et al. 1989) emphasize that the earlier history of the crust (prior to the evolution of the sag basin) plays an important role in determining the subsidence and structure of this type of basin. Klein and Hsui (1987) assume that the North American cratonic basins resulted from the breakup of a late Precambrian supercontinent similar to the Triassic-Jurassic breakup of Pangea mentioned in Chapter 12.3.

Sediments

Due to slow subsidence, the sediments of intracratonic sag basins tend to be continental to shallow-marine. In addition, their stratigraphic record may display pronounced gaps. Special trends in their vertical facies successions are governed by climatic changes and sea level fluctuations, rather than by tectonic processes (see above). Some of the basins may reflect the evolution of land plants and rock-forming marine organisms from the Proterozoic era to Phanerozoic times. Basins isolated from sufficient terrigenous sediment sources may become sediment-starved and accumulate shallow-marine carbonates and evaporites.

The **Michigan basin** in North America, for example, collected sediments from the Cambrian to the Jurassic, and then again in the Quaternary, but it also experienced periods of erosion and sediment starvation (Fisher et al. 1988). After an initial stage of continental and shallow-marine deposition, the basin displays two upper Silurian, 600 m thick evaporite-carbonate cycles, which are replaced along the margins by reefs and skeletal shoals, while on the slopes and in deeper water pinnacle reefs occur (Schreiber 1988b).

The basin fills of the **Murzuk and Kufra basins** in north Africa are only 2 to 3 km thick and consist predominantly of fluvial deposits (Nubian sandstone), which were transported in northerly directions (Selley 1985a). In the late Cretaceous and early Tertiary, the basins were flooded by the sea.

12.5 Deep-See Trenches, Forearc and Backarc Basins

(Continued under the heading of Chap. 12.6)

12.5.1 General Features

The basic principles governing the structure and evolution of arc-trench systems have been outlined by several authors (e.g., Dickinson and Seely 1979), as well as in many modern text books on geodynamics, and therefore are not repeated here. Similarly, the highly variable geotectonic settings of these systems cannot be described in this book. The following discussion of sedimentary sequences in deep-sea trenches and forearc and backarc basins will concentrate on a few characteristic, well studied examples. The evolution of most arc-trench systems can be sufficiently defined by two-dimensional cross sections (Fig. 12.18) due to the well developed longitudinal continuity of these structures.

Arc-trench systems result either from the breakage of a previously intact oceanic plate or from the activation of a passive continental margin (cf. Fig. 1.2, A and B subduction). *Intraoceanic volcanogenic arc systems* have oceanic lithosphere in both the forearc and backarc regions. The island arc can be entirely volcanic (e.g., present-day Tonga-Kermadoc Islands and Mariana Islands) or comprise a continental sliver (e.g., Japan). "Normal" subduction of oceanic plate beneath an island arc is directed toward the continent following behind the island arc and backarc basin; reversed subduction (polarity reversal) affects backarc basins, whose oceanic crust is subducted oceanward, i.e., away from the continent. The minimum distance between arc and trench is about 100 km in modern oceanic arc systems; trenches are commonly smoothly arcuate, as for example the Sunda trench (see below).

By contrast, the edges of *rifted continental margins* tend to be jagged and to exhibit offsets and projections (Fig. 12.18a). Here, the subducting oceanic plate cannot trace all these irregularities but tends to adjust to the points of the crustal projections on the continental block. Hence, oceanic and transitional crust is caught between the projections. The presence of such thin, trapped crustal slivers is probably a prerequisite for the deep-water environment and/or large subsidence and sediment thickness within the forearc region. Similarly, backarc basins on oceanic or transitional crust have a higher potential for subsidence and sediment accumulation than basins in the region of continental crust. Additional subsidence may be caused by compressional downfaulting.

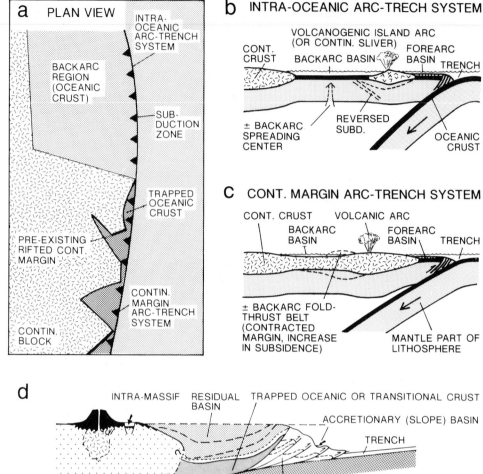

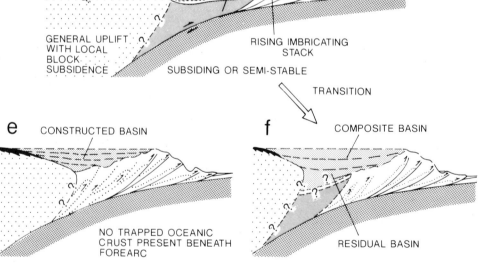

Fig. 12.18. a Plan view of intra-oceanic arc-trench system *(upper half)* with transition into continental margin arc-trench system *(lower half)*. Note that pre-existing rifted continental margin may be jagged and trap parts of the subducting oceanic crust (compare **d**). **b** Structure of intra-oceanic arc-trench systems with *(1)* the volcanic island arc consisting of a sliver of continental crust and/or *(2)* a spreading center developing in the backarc region, possibly giving rise to reversed subduction. **c** Continental arc-trench system with a fold-thrust belt in the backarc region, possibly leading to a "contracted" margin and causing additional subsidence. **d** through **f** Various types of forearc basins in relation to their basement, with or without trapped oceanic or transitional crust; see text for explanation. (After Dickinson and Seely 1979)

12.5.2 Deep-Sea Trenches

General Characteristics and Sediment Sources

Deep-sea trenches represent the deepest morphological features on Earth and reach water depths of 6 to 11 km. They evolve at the front of a subduction complex, where the sedimentation rate cannot balance the loss of sediment to the accretionary wedge as a result of subduction (see below). In regions with very high sedimentation rates, the site of a potential deep-sea trench remains permanently filled and morphologically resembles the lower slope or continental rise of a passive margin. Then, sediments delivered to the trench can spill over the outer trench margin and rest on the flat deep-sea floor (left-hand side of Fig. 12.19a).

Deep-marine trenches collect sediments from a number of different sources (Fig. 12.19a):

- *Lateral influx* of predominantly siliciclastic material from the side of the magmatic arc (inner slope and forearc region) via submarine canyons or gravity mass movements. These sediments mainly accumulate as trench fans along the inner side of the trench (Fig. 12.19b). Larger submarine fans remain in the confines of the trench, as long as the trench is not completely filled, and finally distribute their material in an axial direction.
- *Axial transport* of mainly siliciclastic material along the trench from distant areas with high sediment input. This mechanism can be active as long as the trench has a morphological expression. Sediment transport is accomplished by a channel system with levees (Fig. 12.19b, cf. Chap. 5.4.2), which may display depositional and erosional phases. At intermediate to long distances from the sediment source, the basin fill may become "sheeted" (Fig. 12.19d), as in distal fluvial plains, where fine-grained materials predominate. Contour currents can rework and redeposit part of these materials.
- Lateral migration of the pre-existing *sediment cover of the subducting oceanic plate* to the site of the trench. These sediments may consist of pelagic calcareous and/or siliceous oozes, red clay, and hemipelagic silts and clays. In regions adjacent to the deltas of major rivers, turbidite successions may rest on the incoming oceanic plate.
- *Pelagic* and *hemipelagic sediments* directly settling from the overlying water body to the trench floor. However, this sediment source is comparatively insignificant, because of the low sedimentation rate of pelagic and hemipelagic sediments in a basin whose sediments are permanently consumed by subduction (see below).

Trench Fills and Their Residence Times

Trench sections with high sediment input (*over-supplied systems*) become filled and have a wide cross section (Fig. 12.19a and b). If the deposits of a submarine fan spill over the outer trench margin, trench sediments are overlain by the sediments of a deep-sea fan association (Chap. 5.4.2). In contrast, *sediment-starved trenches* form narrow, ponded basins (Fig. 12.19a and f). Seamounts, oceanic ridges and faulted crust approaching the trench may create major irregularities in the trench fill and the ensuing subduction process.

The sediments in a deep-sea trench have little chance of being preserved in their original position, because they are continuously scraped off by the overriding accretionary wedge and incorporated into the subduction complex. The *residence time of sediments in a deep-sea trench* is a function of the convergence rate, CR, and the width, W, of the trench (Schweller and Kulm 1978). Given, for a example, an axial sediment source, the sediment wedge would be widest and thickest near the source and diminish distally (Fig. 12.20a). The maximum residence time of a sediment layer in the trench is W/CR.

Sediment deposited in a 20 km wide trench subject to a convergence rate of 5 cm/a, for example, would therefore have a maximum residence time of only 400 000 years.

Nevertheless, deep-sea trenches may collect relatively large volumes of material, in addition to the incoming sediment on the subducting oceanic plate, and therefore contribute greatly to the sediment stack of the accretionary wedge.

Relationship Between Trench Sedimentation and Accretion

Sedimentation and Convergence Rate

The relationship between new sediment accumulation in the trench and sediment loss by accretion is demonstrated by the models in Fig. 12.20a and b (based on Thornburg and Kulm 1987). If the sedimentation rate in the trench and the convergence rate of the sub-

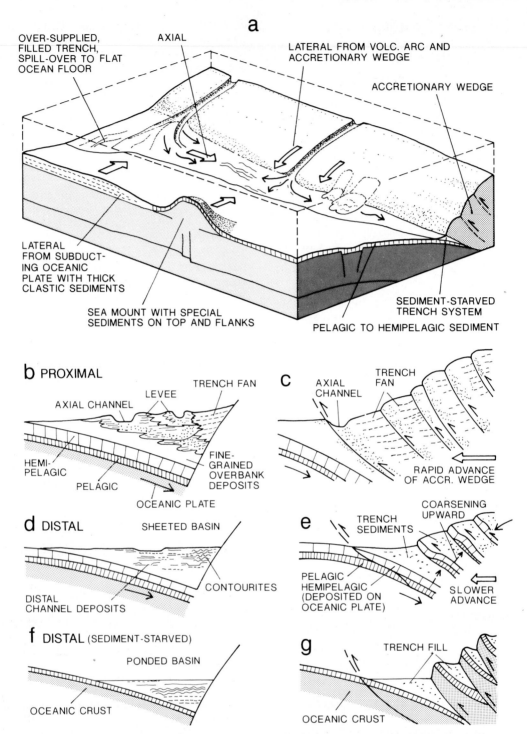

Fig. 12.19. a Scheme of the various processes which control the filling of deep-sea trenches. Over-supplied, filled trenches may be fed by axial channel systems (**b**), laterally from the side of the volcanic arc by submarine canyons or mass wasting on the slope, and by a thick clastic sediment cover from the subducting oceanic plate. A relatively distal, "sheeted" basin fill (**d**) may develop in the trench between submarine canyons. Sediment-starved, ponded trenches (**f**) accumulating mainly pelagic and hemipelagic mate rial from the overlying water body and from the subducting ocean plate, tend to be deeper and narrower than their over-supplied counterparts. Approaching seamounts (in **a**) may cause large irregularities in trench filling and the accretionary wedge. **b** and **c**, **d** and **e**, **f** and **g** Possible relationship between mode of trench fill and position of the décollement zone, controlling the lithofacies successions within the accretionary wedge. (Mainly after Thornburg and Kulm 1987; Pickering et al. 1989)

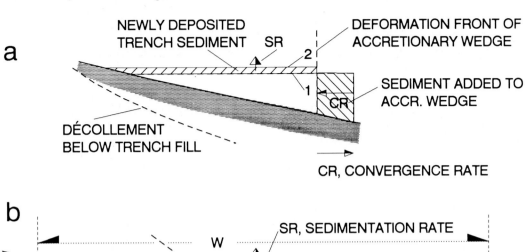

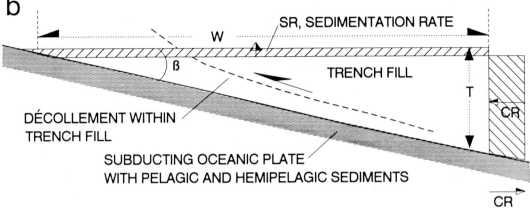

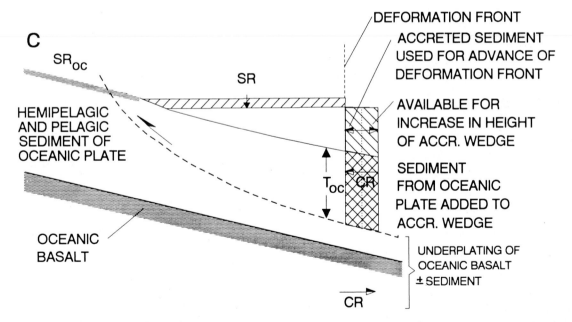

Fig. 12.20a-c. Sediment budget of accretionary wedges as a function of sedimentation rate, *SR,* in the trench, convergence rate, *CR,* and angle of dip, ß, of the subducting oceanic plate. *W* width of the trench. T_{oc} thickness of sediment scraped off from the subducting oceanic plate (in **c**). **a** and **b** Trench sediments only are considered. For steady-state conditions and ß and CR = constant, it is SR/CR = tanß. The volume of accreted sediment per unit length of trench is V_a = W·SR = W·CR·tanß. **c** Trench sediments and part of incoming sediment cover of oceanic plate (with thickness T_{oc}) are incorporated into accretionary wedge. Accreted sediment volume may be used for advancing deformation front and/or increasing height of the pre-existing accretionary wedge. See text for further explanation

duction process are kept constant, steady-state conditions at the front of the accretionary wedge can be achieved. This signifies that the sedimentation rate exactly balances the rate of sediment attrition caused by the seaward migration of the deformation front, i.e., the volume of newly deposited trench material is equal to the volume of sediment lost to the accretionary wedge (neglecting changes in porosity).

Under these conditions, the sedimentation rate, SR, is a function of the convergence rate, CR, and the angle of dip, ß, of the subducting oceanic plate:

$$SR = CR \cdot \tan ß , \qquad (12.1)$$

as can be seen from simple geometric considerations. In this case, SR is independent of the width, W, of the trench and the thickness, T, of the trench sediments.

Examples given by Thornburg and Kulm (1987) for the Chile Trench yield the following results for

Angle of dip of the subducting plate 3°,
Convergence rate 10 cm/a:

a) Sediment-starved trench off central Chile:
   Width W = 10 km, thickness T = 0.5 km.
b) Sediment-filled trench off south Chile:
   Width W = 30 km, thickness T = 1.5 km.

For both examples, SR is found to be about 0.5 cm/a (5000 m/Ma), which appears to be extremely high, particularly for the relatively sediment-starved central Chile trench. Hence, it is assumed that, in addition to trench material, incoming sediment from the subducting oceanic plate is being incorporated into the accretionary wedge (Fig. 12.20c). This assumption has been confirmed for several active subduction zones by seismic records and ocean drilling (see below).

Sediment Volume and Accretion

The volume of sediment derived from newly deposited trench deposits and incorporated into an accretionary wedge per unit time depends on the width of the trench fill. This means that trenches oversupplied with sediment provide more sediment for accretion than narrow trench fills, even when their sedimentation rate is equal. The volume of accreted trench sediment, V_t, per unit time and length of the trench is

$$V_t = W \cdot CR = W \cdot CR \cdot \tan ß \qquad (12.2)$$

and for accretion of both trench and oceanic sediment (Fig. 12.20c)

$$V_{t+oc} = W \cdot CR \cdot \tan ß + CR \cdot T_{oc} \qquad (12.3)$$

where T_{oc} = thickness of the sediment scraped off from the oceanic plate.

Part of the accreted sediment is used for the seaward advance of the accretionary wedge, the other part for the thickening of the accretionary wedge. Sediments below the décollement zone are atop basaltic oceanic crust.

In conclusion, the convergence rate and sediment supply from different sources are the dominant factors which control the configuration of a deep-sea trench and the buildup of an accretionary prism.

Consequences for Sediments in Accretionary Wedges

The considerations in the previous section and the assumption that a kind of equilibrium between sediment supply and sediment loss to accretion is established in arc-trench systems, lead to the qualitative models of sediment accretion shown in Fig. 12.19b through g:

1. **Oversupplied trenches**. Sediment fills tend to be sheared off along a décollement within the young trench fill. Failure by thrusting may be facilitated by increased pore pressures of the rapidly accumulated, underconsolidated sediments (von Huene 1986). As a result, the sequence within a slice of the accretionary wedge is built up entirely of trench sediments. These mainly represent the axial channel system and fan deposits from the foot of the landward trench slope (Fig. 12.19b and c). In this case, the sediments within a slice of the subduction complex are about the same age, but may contain some isolated, displaced older material transported by mass wasting from the landward slope into the trench.

2. **Trenches associated with low to medium sediment supply and a moderate convergence rate** (3 to 5 cm/a) usually develop medium-size trench wedges with discontinuous turbidite basins. The subduction décollement commonly develops below the young trench fill in pelagic or hemipelagic strata of the incoming oceanic plate (Fig. 12.19d and e). In cases where the pelagic sediment cover on the oceanic crust is scraped off, the idealized stratigraphic sequence within a little-disturbed slice of the accretionary wedge coarsens upward from older pelagic to hemipelagic to younger trench fill sediments (Fig. 12.19e; Lash 1985). This sequence may represent a comparatively long time period.

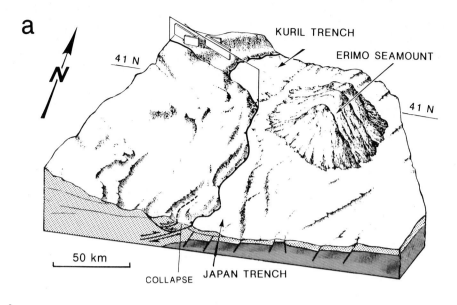

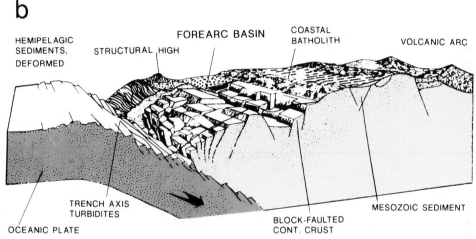

Fig. 12.21. a Northern Japan trench and its juncture with the Kuril trench at Erimo Seamount located on transform fault. Note collapse of lower slope due to tectonic erosion of overriding plate. (Cadet, Kobayashi, et al. 1987).

b Model of subduction-related tectonic erosion off Peru, causing continental crust to extend almost to the deep-sea trench. (Moberly et al. 1982)

3. **Sediment-starved trenches** which also receive little material from the incoming subducting plate are narrow, discontinuous, and display a partially empty axis. Modern examples of this type include trenches associated with volcanic island arcs in the western and southwestern Pacific. Starved trenches need a very long time span to build up an accretionary wedge of some size. Consequently, the deformation front progrades slowly, and the landward slope tends to steepen. In this case, thrust faults may propagate along weak zones in the upper oceanic crust and scrape off slices of basaltic crust, pelagic sediments and isolated bodies of ponded trench fill (Fig. 12.19f and g). Such settings allow the study of basal deep-water sediments on top of accreted oceanic crust in the subduction complex.

4. **Tectonic erosion.** When the subducting oceanic plate has a marked relief and the total sediment supply into the trench is low, sediment from the upper plate (accretionary wedge) can be scraped off and subducted as, for example, off southern Peru and northern Chile or in the northern portion of the Japan trench (Fig. 12.21). This process is referred to as "tectonic erosion" (Moberly et al. 1982; von Huene 1986; Cadet, Kobayashi, et al. 1987;

von Huene and Lallemand 1990). It creates stratigraphic gaps, reduces the volume of the accretionary wedge, and may finally lead to the collapse and retreat of this complex.

Off northwestern Peru, the edge of the continental crust may be as close as 20 to 40 km to the trench axis (Fig. 12.21b). Here, the subducting oceanic plate has scraped off the underside and edge of the adjacent continental crust.

Most ancient subduction complexes, whose internal structure has been exposed on land by deep erosion following uplift, are much more complex than may be inferred from the simple rules described above. Besides packets of bedded sequences, in which stratification is preserved in spite of faulting and folding, other, once deeply buried portions of the accretionary wedge display pervasive metamorphic fabrics which indicate low temperature and high pressure conditions. Zones of intensive shearing creating a scaly fabric are common.

Mélange, Wildflysch

Other portions of the subduction complex consist of a mixture of various rock types including large solid blocks in a more uniform, but sheared matrix. Such completely chaotic zones are referred to as *mélange* (see, e.g., Hsü 1974; Raymond 1984; Cowan 1985; Aalto 1989; Pickering et al. 1989). It is frequently difficult to decide whether a mélange results entirely from tectonic processes within the subduction complex, or whether it represents a former large mass flow (olistostrome) from the slope into the trench, which was later incorporated into the subduction complex and modified by tectonic deformation. Material from the accretionary wedge may be transported into the trench again and thus recycled within the arc-trench system. Where a genetic interpretation of a chaotic deposit is difficult, the descriptive term *wildflysch* may be used.

Specific Tectonic and Sedimentary Structures

The *internal structure* of modern accretionary wedges is poorly known. One of the best studied examples during deep-sea drilling is the accretionary wedge of the Barbados Ridge in the Caribbean Sea (Shipboard Scientific Party 1988b, Leg 110; Fig. 12.23d). The drilled cross section, about 30 km long, displays two generations of landward-dipping thrust faults which tend to steepen with increasing distance from the deformation front. The second generation disrupts the older thrust faults and thus promotes further deformation of the accreted sediments. As pointed out by Moore and Byrne (1987), initial fault surfaces are abandoned and deformation propagates into adjacent undeformed sediment. This is caused by sediment strengthening along the faults due to dewatering and release of excess fluid pressure. As a result, the fault and shear zones thicken and contribute to the intensive deformation of a tectonic mélange.

In addition to these major structures, deep-sea trenches, the lower slopes of accretionary prims, and some forearc basins may show a variety of medium to large-scale *sedimentary structures*, which partially result from liquefaction of water-rich, overpressured sediments. Such structures were observed in seismic records and by side-scan sonar mapping in modern environments. They include seepages on the sea floor (cf. Chap. 13.5), mud volcanoes and mud diapirs, wet sediment intrusions such as vertically and laterally injected sand dikes, and thick and widely extended "chaotic layers" generated in semi-lithified rocks. The latter can easily be misinterpreted as debris flows or tectonic mélange zones. A summary of these features is given by Pickering et al. (1989).

Example: The Sunda Arc-Trench System

A particularly instructive and large continental margin arc-trench system is the Sunda arc between Burma and the Sumba Island to the east of Java (Fig. 12.22; Moore GF et al. 1982). It represents a 5000 km long subduction zone of the Indian oceanic plate beneath the Eurasian plate, which has been intermittently active since the Permian. The youngest active phase began in the late Oligocene. During the Neogene, the northern part of this system was dominated by detritus from the Himalayas which was dispersed by mass flows and turbidity currents on the huge submarine Bengal fan, the Nicobar fan, and along the axis of the Sunda trench.

Whereas sediment supply to the Nicobar fan was cut off in the mid-Pleistocene as a result of the eastwardly migrating Ninetyeast Ridge, a major submarine channel still transports sediment today over the long distance from the northern end of the trench to the southeast into the region of the Sunda Strait (also see Velbel 1985). In the northwest, i.e., the Andaman section, transport of large volumes of material down the trench axis is indicated by a wide and relatively shallow trench floor (water depth about 3000 m,

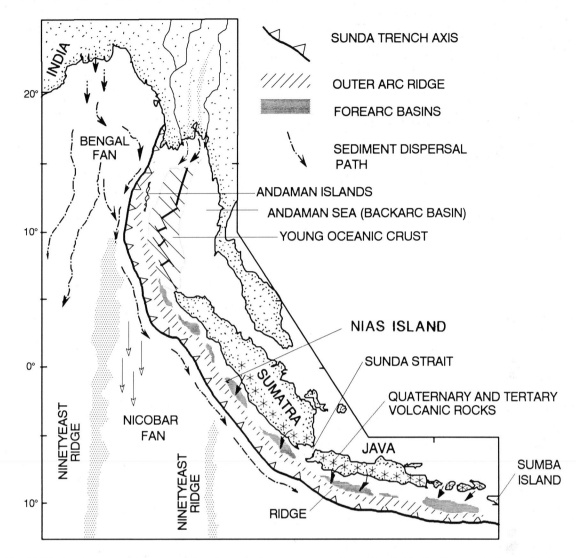

Fig. 12.22. Sunda arc-trench system. Map view with Bengal and Nicobar deep-sea fans accumulating sediments from the Himalayas; such material also migrates along the trench axis as far as Sunda Strait. At present, forearc basins and the outer arc ridge trap sediments from the island arc (Sumatra and Java)

Figs. 12.22 and 12.23a). West of Sumatra, the trench floor is still broad, but narrows and deepens to about 5000 m off Nias Island (Fig. 12.23b). South and southeast of Java, the trench is formed by narrow, isolated basins at water depths of 6000 to 6500 m.

Sediments on the subducting plate are thus primarily thick turbidite sequences in the northwestern and central portions of the system, but only thin (200 to 400 m) pelagic and hemipelagic sediments off Java. Consequently, accretion of thick Neogene fan and trench sediments created a wide, high outer arc ridge in the northwestern and central trench portions, while the ridge is narrower and less elevated in the southeastern part (Fig. 12.23).

Since this outer ridge formed, sediments from the volcanic arc, i.e., from Sumatra and Java, are trapped in forearc basins and therefore can no longer reach the trench. Prior to the mid-Tertiary, volcaniclastic material from the arc contributed significantly to the trench sediments and thus was incorporated into the accretionary wedge.

Between the arc and the outer ridge a series of *forearc basins* developed (Fig. 12.22). These basins are 150 to 500 km long and 50 to 100 km wide. In Burma, about 8 km of marine, deltaic, and fluvial sediments accumulated from the late Cretaceous to the Pliocene. Off Sumatra, water depths in the basins are 500 to

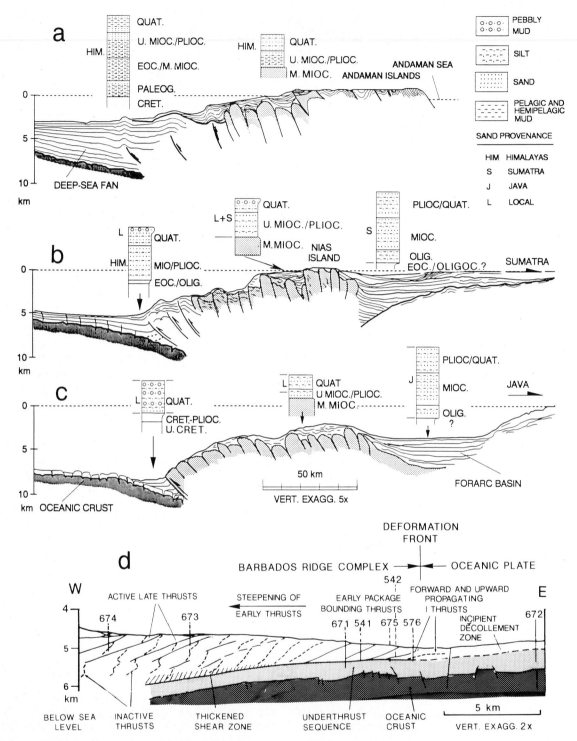

Fig. 12.23. a, b, and c Cross sections of arc-trench system from northwest (Andaman Islands) to southeast (Java). Note decreasing thickness of sediment on subducting oceanic plate, decreasing width and elevation of outer arc ridge, increasing water depth of forearc basin, and change in sand provenance from distal Himalayas to more local sources from the arc. See text for further explanation. (After Moore GF et al. 1982). **d** Schematic cross section at deformation front of northern Barbados Ridge, Caribbean Sea, as reconstructed from seismic records and ocean drilling. Note relatively shallow décollement zone on top of thicker, underthrust late Cretaceous to Oligocene strata, steepening of early thrusts in accretionary wedge, upslope formation of secondary thrusts disrupting early thrusts, and thickening basal shear zone. (Shipboard Scientific Party 1988, Leg 110; after Moore and Byrne 1987)

2000 m, while off Java the basins are 3500 to 4000 m deep; the sediment fill attains thicknesses of up to 6000 m. The basins are separated from each other by transverse structural highs, and therefore the fill in each basin differs in composition from adjacent basins. The sediments in the forearc basins are derived primarily from the arc terranes and are dispersed by submarine slope canyons and small deep-sea fans. Off north Sumatra, quartzose sediments accumulate, whereas off Java, volcaniclastic sediments predominate in the forearc basins. Shallow-water carbonates and their detritus also play a role.

Sediment Budget of Subduction Complexes and Depths of Décollement

The growth rate of accretionary wedges is commonly very high, because they take up huge masses of sediments from both the deep-sea trench and the subducting oceanic plate. As an initial approach, the rate of sediment uptake can be estimated from the average thickness of the accretionary prism, its average porosity, and the subduction rate.

In this manner, a sediment mass accumulation of around 350 km^3/Ma for a 1 km wide strip along strike was found for the Makran continental margin off southwestern Pakistan (White and Louden 1982). Here, the Arabian oceanic plate is subducting at a rate of 50 km/Ma beneath the Eurasian plate and creates an accretionary wedge about 7 km thick.

The Cascadia Subduction Zone

A more precise calculation was tried for the modern Cascadia subduction zone in the northeastern Pacific Ocean off British Columbia (Davis and Hyndman 1989). In this case, the mass of the accretionary wedge was balanced. This was done for two time periods: the Pleistocene accretionary prism (1.8 Ma) and the pre-Pleistocene prism which began forming in the Eocene. The geometry of the accretionary wedge is relatively well known from seismic records and the position of earthquake hypocenters (Fig. 12.24a through d). The décollement zone is situated at or near the top of the underlying oceanic crust, thus indicating that almost all of the incoming sediment is scraped off.

At the deformation front, the pre-Pleistocene section consists of fine-grained hemipelagic sediments about 1.5 km thick overlain by a 1.7 to 2 km thick Pleistocene turbidite

sequence, which was drilled and sampled during the Deep Sea Drilling Project. The high Pleistocene sedimentation rate of approximately 1000 m/Ma results from the high-relief, glaciated hinterland in the Rocky Mountains. The pre-Pleistocene/Pleistocene boundary can be traced in seismic records within the accretionary prism.

In this way, the Pleistocene portion of the wedge could be delineated. It contains about 170 km^3 of sediment per kilometer of its length. Taking into account an average sediment porosity of 15 to 20 %, the volume of the solid fraction is about 140 km^3. The total supply of incoming sediment from the trench and the oceanic plate is estimated from a subduction rate of 4.5 cm/a. This rate gives 80 km of convergence during the Pleistocene, but at the same time the deformation front advanced about 25 km seaward, thus yielding a relative, total convergence of 105 km. Assuming an average porosity of 35 % for the incoming sediment, a section about 2 km thick (1.3 km of solid material) is needed to build up the volume of solid material estimated in the accreted Pleistocene wedge. This calculated average thickness is within the range between the present sediment thickness at the deformation front (3.5 km) and the thickness at the beginning of the Pleistocene (about 1.5 km).

Hence, this balance supports the observation from seismic lines that, due to the deep position of the décollement zone, most of the incoming sediment was incorporated into the accretionary prism during the Pleistocene. The present-day trench structure has no topographic expression but is kept filled as a result of both high sedimentation rate in the trench itself and high sediment input to the accretionary prism by the obliquely subducting oceanic plate.

The sediment balance for the pre-Pleistocene proportion of the accretionary wedge yielded a 0.9 km thick sediment section (with 45 % porosity), which must have been permanently scraped off during the subduction process since the Eocene (Davis and Hyndman 1989).

Other Modern Examples

The thickness of undeformed sediment being underthrust with the subducting oceanic plate varies considerably.

Along the central **Aleutian trench** it is 1 km (McCarthy and Scholl 1985). Sediment accumulation in this trench is also rapid, and the convergence rate reaches 8 cm/a. At the northern **Japan trench**, the sediment thickness

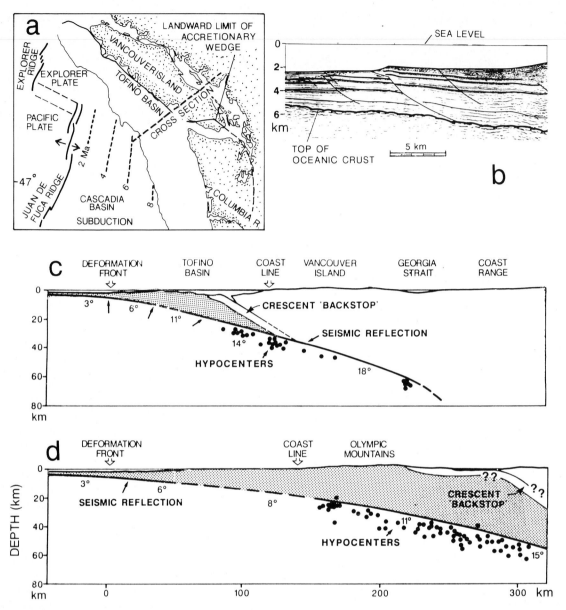

Fig. 12.24a-d. Depth of décollement and sediment budget of northern Cascadia subduction zone. **a** Map view indicating general plate tectonic setting, age of subducting oceanic plate, changing width of accretionary wedge, and position of cross section shown in **b** and **c**. **b** Multichannel seismic reflection line across the deformation front shown in **a**; travel time transformed to depth below sea level. **c** and **d** Generalized cross-sections through Cascadia continental margin (**c**) along line shown in **a** and **d** 150 km to the southeast through Olympic Mountains. (Davis and Hyndman 1989)

beneath the accretionary wedge is 1 to 1.5 km (Cadet, Kobayashi, et al. 1987). In both cases, the surface of the oceanic basement is rough due to faulting (Fig. 12.21a). Further south in the Nankai trench, 400 to 500 m thick oceanic sediments are underthrust (Karig 1986), but deeper below the accretionary prism the décollement zone moves down to the top of the oceanic crust (Taira et al. 1988, see below).

At the northern **Barbados ridge** in the region of the Lesser Antilles, the zone of décollement at the deformation front is rather shallow (about 300 m below the sea floor, Fig. 12.23d). Here, Neogene hemipleagic sediments are detached from thicker (about 600 m), little deformed Oligocene to Campanian underthrust strata.

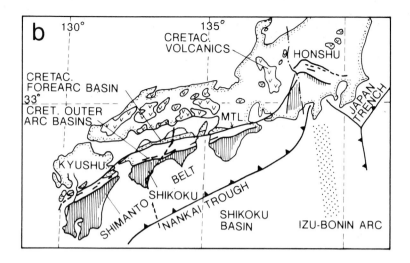

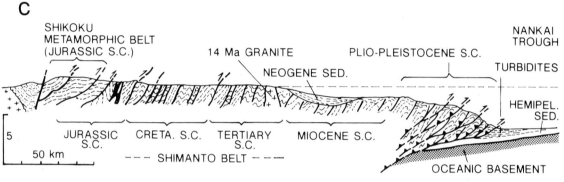

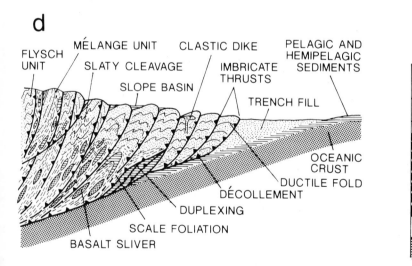

e STRATIGRAPHY OF MÉLANGE UNIT (IDEALIZED)

TURBIDITES AND SHALES (TRENCH FILL)

VARICOLORED HEMIPEL. SHALE AND ACIDIC TUFF

RED PELAGIC SHALE

BEDDED RADIOLARIAN CHERT

RED PELAGIC SHALE AND/OR NANNO-LIME-STONE

OCEANIC BASALT

Fig. 12.25a-e. Cretaceous to Miocene accretionary prism of the Shimanto belt, southeastern Japan. **a** and **b** Location maps. **c** Cross section from modern Nankai trough to Shikoku island, showing increasing age of subduction complexes *(S.C.)*.

d Idealized cross section through accretionary wedge, displaying principal features observed in Shimanto belt. **e** Idealized stratigraphic section of mélange zone. On top of oceanic crust, sulfide and/or manganiferous hematite deposits may occur. (After Taira et al. 1988)

Growth and Elevation of Accretionary Wedges

The seaward advance, thickness, and elevation of an accretionary wedge are controlled mainly by the rate of incoming sediment (in terms of volume) and the dip of the underthrusting oceanic crust (Davis et al. 1983). In the case of the Cascadia subduction zone, it is assumed that the deformation front advanced 60 km in pre-Pleistocene time and 25 km during the Pleistocene. The dip angle of the subducting oceanic plate normally increases landward or arcward (Fig. 12.24c and d). The Cascadia example demonstrates that the relatively steep dip ($\geq 11°$) of the subducting plate beneath the Vancouver island margin prevents the buildup of a highly elevated ridge. In contrast, the Olympic Mountains farther to the south (Fig. 12.24d) owe their existence to a more gentle dipping subduction zone ($\leq 11°$), as well as to particularly high sediment supply at the accretionary wedge front (Davis and Hyndman 1989).

Example of an Ancient Accretionary Wedge

The architecture of modern accretionary wedges is an important objective of current research. Ancient examples have been described from many regions (e.g., Leggett 1982), but many of them are not sufficiently exposed and therefore not well understood.

One of the best known examples of a Cretaceous to Miocene accretionary prism is the **Shimanto belt** in Shikoku, Japan (Taira et al. 1988). It lies landward of the Nankai trench (Figs. 12.25 and 12.21a) and resembles in its structure and sedimentary facies this modern subduction complex. The décollement of the young accretionary prism takes place above the pelagic and hemipelagic sediments of the downgoing oceanic plate within muddy sediments of the outer trench fill (Fig. 12.25c and d). Hence, the sediments of the young accretionary prism are dominated by muddy and sandy flysch from the trench fill. However, at a depth of about 5 km below the slope, the décollement zone moves down to the top of the oceanic crust and forms duplex structures into which sediments of the oceanic plate are incorporated.

Thus, the older thrust slices may contain two basic units:

- Mélange consisting of basalt slivers, pelagic and hemipelagic sediments, and fine-grained matrix from the trench fill.

- Flysch units from the trench fill (Fig. 12.25d).

Intact sections of pelagic and hemipelagic sediments show characteristic vertical facies successions deposited on pillowed basalt of the subducted oceanic plate (Fig. 12.25e). With increasing distance from the modern trench, increasing age and overburden pressure, the rocks of the thrust sheets develop scaly foliation, slaty cleavage, and are finally subject to metamorphism.

12.5.3 Forearc Basins

General Characteristics

Types of Forearc Basins

According to Dickinson and Seely (1979) four types of forearc basins can be distinguished (Fig. 12.18d through f):

- *Intramassif* or *intra-arc basins,* where the sediments lie unconformably on rocks of the island arc or continental margin arc. In the latter case, the basins collect arc or backarc-derived, commonly nonmarine sediments of limited extent and thickness. Intramassif sediments of island arcs tend to be predominantly marine. Several forearc basins along the Pacific coast of South America evolved on continental or transitional crust; for example, off Peru, crust with continental affinity extends to the shelf edge or lower slope of the margin (Fig. 12.21b, Moberly et al. 1982; Shipboard Scientific Party 1988c, Leg 112). After an earlier period of uplift, a series of elongated, narrow, coast-parallel forearc basins evolved on top of this crust, i.e., on the shelf and upper slope. These basins collected sediments which are in part strongly affected by coastal upwelling since the Miocene. Hence, some of these forearc sediments are characterized by black shale intervals, phosphorite horizons, widespread diagenetic dolomitization, and other special sedimentary features (cf. Chap. 5.3.4). The infillings of these basins are limited in thickness.

- *Accretionary basins,* i.e., basins within the accretionary wedge of the subduction zone (mostly slope basins), whose sediments lie unconformably on older, deformed strata. Their sediments are derived mainly from the uplifted subduction complex or the arc massif and reflect a comparatively deep marine environment. They may be deformed mainly by compressional forces exerted during the continuing growth of the underlying accretionary wedge.

- *Residual basins* accumulate sediments on oceanic or transitional crust trapped between the arc massif and the subduction zone (Chap. 12.5.1, Fig. 12.18d).

- *Constructed basins* collect sediments lying unconformably on both the arc massif (inner side of the basin) and deformed strata of the subduction complex (outer side of the basin).

The latter two basin types are most important and discussed further here.

Basin Evolution and Sediment Source

During its evolution, a forearc basin may go from a narrow, residual basin to a broader, composite basin that onlaps the arc massif on one flank and lies on the accretionary wedge on the other side. The initially deep residual forearc basins particularly have a high potential for great subsidence under increasing sediment load and can accumulate thick sedimentary sequences in comparatively short time periods.

The nearby volcanic arc acts as a sediment source of extreme efficiency. Continuing rapid subduction keeps the volcanic system active and thus provides both high relief and large amounts of easily erodable volcaniclastic material (cf. Chap. 2.4), which is transported into the forearc or backarc regions. The accretionary wedge frequently serves as a dam for ponding sediments in the forearc basin. In addition, this submarine or emerged high (shelf edge, outer ridge) may shed some sediment derived from the subduction complex, possibly including ophiolites, into the forearc basin.

The *subsidence histories* of forarc basins are more complicated and less predictable than those of rifted continental margins (Moxon and Graham 1987; Angevine et al. 1990). Several mechanisms play a significant role (cf. Chap. 8). (1) The sediment load of the depression between the topographic highs of the volcanic arc and the accretionary wedge. (2) Isostatic response to the emplacement of dense oceanic crust beneath the forearc region by subduction. (3) Rapid cooling of a relatively warm upper plate resulting from the subducted cooler plate. This third mechanism may apply to situations where hot, young oceanic crust is incorporated as upper plate into a forearc region. In addition, the subsidence

histories of forearc basins may be influenced signifiantly by variations in the subduction rates and, as in all basins, by the history of sediment loading.

Variations in Basin Evolution

The forearc basins in the modern oceans show considerable *variation in their configuration* (Fig. 12.26), which is partly due to their different stages of evolution. Sloped forearc regions (e.g., along New Britain, the Solomon Islands, and the south New Hebrides) may develop some small slope basins on top of the accretionary wedge, but do not provide larger basins for sediment accumulation. Similarly, terraced forearc regions, as known from the northeastern Pacific off Oregon and Washington, cannot accumulate sediments of substantial thickness on top of the subduction complex. Back-tilted forearcs (south Tonga, Mariana Islands, northern New Hebrides) are characterized by a partially submerged volcanic arc (Fig. 12.26a) and a kind of intra-arc basin directly adjacent to the arc.

Ridged and Shelved Forearc Basins

In terms of sediment accumulation, the ridged forarcs creating a ponded *basin* are most important. Their evolution depends on the volume of sediment which is collected by the deep-sea trench and accreted in the subduction complex per unit time (Chap. 12.5.2). Trench systems well supplied with sediments enable rapid progradation and upbuilding of the accretionary wedge and thus widening of the forearc basin or the outer ridge (structural high). Modern examples of ridged forearc regions include the western Aleutian trough, the Kermadoc trough, the Manila trough, and forearc basins bordered by an emerged outer ridge such as the Sumatra-Mentawai trough (Figs. 12.22 and 12.23 b) and the Lesser Antilles-Barbados ridge.

As a result of continuing subduction and accretion, such ponded basins tend to evolve from open, deep, narrow basins into broader basins with increasingly restricted environments and shallowing-upward sequences (Fig. 12.26b). Sediment-filled basins of this type are referred to as *shelved forearc basins*. Their basin fill may begin with abyssal sediments and finally end with terrestrial deposition, particularly on the side of the arc massif.

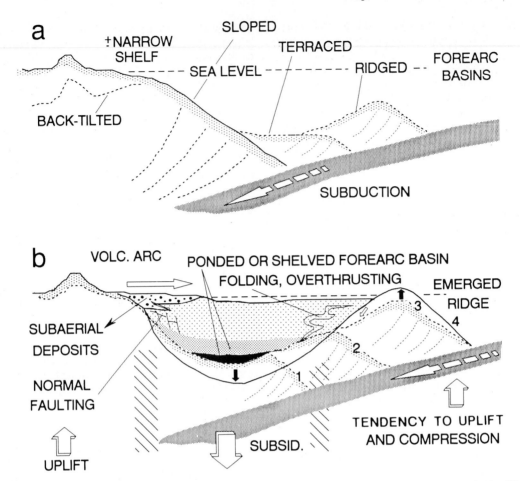

Fig. 12.26a,b. Variations in the configuration of forearc basins, partially related to their evolution. Note that in a back-tilted system (**a**) the volcanic arc may also be submerged below sea level; sedimentary fill (**b**) will subside more and become more deformed during basin evolution than shown in figure. In addition, the basin fill may be affected by extension or compression, with or without a strike-slip component. (Based on Karig and Sharman 1975; Dickinson and Seely 1979, simplified and modified)

Tectonic Overprint

A tendency for uplift of both the arc massif and the outer ridge and subsidence of the basin center favor the continuation of sedimentation for some time. Additional deformation of the strata can be caused by extension, compression, and wrench faulting of the basin fill. In the early stage of such deformation, normal or reverse faults are common on the side of the arc massif, while folding and overthrusting predominate on the seaward side of the basin fill (Fig. 12.26b). If the forearc basin fill is exposed to collision, as in ancient orogenic belts, it is further tectonically overprinted.

Sediment Successions of Forearc Basins

An idealized *residual forearc basin* may display the following succession (modified from Dickinson and Seely 1979, from top to bottom):

- *Fluvial* and *deltaic sands*, partially derived from the uplifted roots of the arc. On the inner edge of the basin, these sediments may interfinger with deposits from pyroclastic flows, relatively coarse grained ash fall, lava flows, and alluvial fans consisting of both volcanic and nonvolcanic materials (cf. Chap. 2.4).

- Sands and shales deposited in *shallow-marine environments* (shelf sea); high proportion of volcaniclastic material but a relatively low amount of biogenic components, resulting from a high influx of terrigenous material and rapid sediment buildup.
- *Flysch-like shale-sandstone sequence* with upwardly increasing proportion of coarser grained turbidites, largely composed of volcaniclastic material; shales may contain some carbonate (when deposited above the CCD) and partially have a relatively high organic carbon content.
- *Abyssal-plain sediments* containing montmorillonitic shales devoid of carbonate, fine-grained ash fall deposits, and fine-grained turbidites.

A large supply of terrigenous material commonly results in rapid infilling of the forearc basins, particularly of basins associated with continental margins.

The modern **forearc basin of Guatemala**, for example, has accumulated non-marine, predominantly volcaniclastic material during the past 20 000 to 30 000 years (Vessell and Davies 1981). These sediments reflect major volcanic eruptions, which have demonstrated a certain degree of cyclicity expressed by (a) a long quiescent phase with incision of meandering rivers and delta reworking, (b) an eruptive phase with airfall ash and pyroclastic flows, (c) a fan building phase dominated by debris flows and coarse grained fluvial deposits, and (d) a braided stream phase with the introduction of large sediment volumes into the fluvial system and delta progradation on the coast.

The **South American forearc basins** along the coast and offshore of Ecuador, Peru, and Chile contain Cretaceous and Tertiary sediments with thicknesses of several to ten kilometers (Moberly et al. 1982). Their facies range from deep marine sediments containing large proportions of turbidites, gravity mass flows and slides to paralic and continental deposits. They evolved on continental crust (intra-massif basins) and frequently show landward dipping strata in their lower sections, indicating syndepositional uplift of a seaward basement high. Syndepositional uplift of landward or seaward parts of the basins also caused the cutting of deep submarine channels. As a result of coastal upwelling, some of these basins accumulated substantial amounts of hydrocarbons.

The present-day **forearc basins of the Sunda arc** provide an example of basins unequally filled in relation to the sediment sources (Chap. 12.5.2, Figs. 12.22 and 12.23).

Similar to remnant basins and foreland basins (Chap. 12.6), forearc basins accumulate sediment successions of several to more than 10 km in thickness within a relatively short time span of 10 to 50 Ma. The contribution of volcanic material to the basin fill is normally high, but may also be low and insignificant, where a major river from the continent enters the basin, or where nonvolcanic uplands are nearby (e.g., in Peru). In *sediment-starved*, warm, equatorial regions, reefs and carbonate banks can grow along the shallow basin margins and build up thick carbonate sections (e.g., in the Mentawai trough).

The percentage and time equivalents of deep-water, shallow-water, and subaerial deposits in a forearc basin succession depend on the mode of basin filling (cf. Chap. 11.3). If a basin is filled up very rapidly during its early stage, as for example in continental margin settings, a large proportion of its sediments consists of shallow-water and nonmarine deposits which accumulate during the (sediment load-induced) later subsidence phase. In contrast, relatively slowly filling basins and coeval subsidence maintain deep-water conditions for longer time periods and thus lead to the buildup of a thick deep-water sequence.

The normal evolution of forearc basins may be modified significantly by jumps or shifts in the position of the subduction zone. Other modifications result from a change in the vector of plate convergence through time. Such processes complicate the interpretation of the structural evolution and facies successions of ancient forearc basins. The evolution of forearc basins as depositional systems ends when subduction of oceanic crust is followed by obduction and collision of plates, and the area is uplifted and exposed to erosion.

Example: The Great Valley Forearc Basin of California

The Great Valley forearc basin in California is one of the most thoroughly studied and best-understood ancient forearc basins in the world (Ingersoll 1982). All three major components of the arc-trench system, i.e., the (Franciscan) subduction complex, the (Great Valley) forearc basin, and the (Sierra Nevada) magmatic arc are exposed on land in their original relationship (Fig. 12.27a). This relationship is preserved, because further compression of the arc-trench system was terminated by strike-slip movement along the San Andreas Fault, converting the convergent margin into a transform margin.

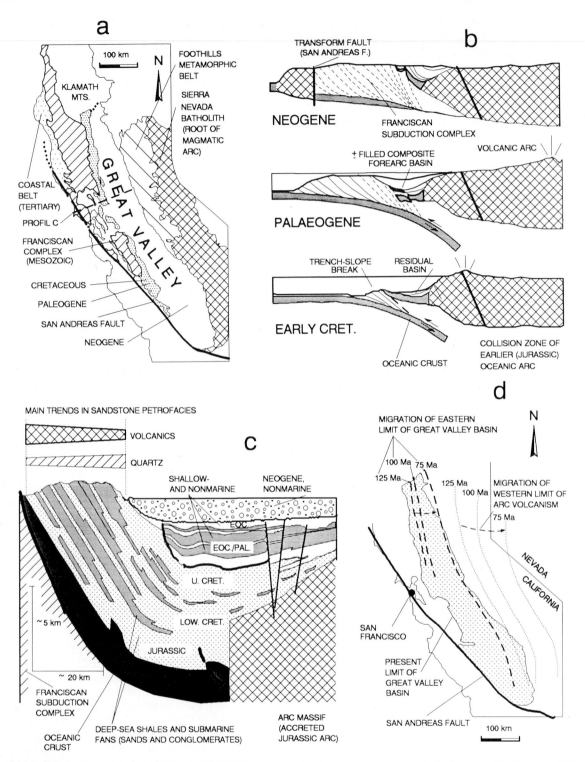

Fig. 12.27a-d. The Cretaceous-Tertiary Great Valley fore-arc basin of California. **a** Simplified map of major geotectonic units. **b** Evolution of convergent margin and forearc basin of central California in three stages; note transition from Cretaceous terraced, residual basin on oceanic crust to Paleogene shelved composite basin on both oceanic crust and subduction complex, and finally to "remnant" forearc basin, the western part of which was uplifted and partially eroded. **c** Simplified cross-section of Great Valley (location shown in **a** and **b**). **d** Map view of basin evolution, showing widening of basin as well as migration of volcanic arc with time. (After Dickinson and Seely 1979; Ingersoll 1982; Moxon and Graham 1987)

After the collision of an older intraoceanic arc with the North American plate in the Jurassic, a new trench and forearc basin formed west of the suture belt. The shape of the new continental margin was irregular and thus enabled the trapping of oceanic crust (Fig. 12.27) and the initiation of a residual basin in the Great Valley area, whereas to the north and south such basins are missing. Later, the basin evolved into a composite forearc basin resting on both oceanic and continental crust as well as partially on the newly formed accretionary wedge (Fig. 12.27b). In addition, the forearc region developed from a terraced or ponded, narrow basin with a deeply submerged ridge into a wider, shelved forearc basin behind a high, broad outer ridge.

Sedimentation in the newly created residual forearc basin began in the upper Jurassic and lower Cretaceous, with muddy slope deposits and conglomeratic channel fills. Sand and coarser grained components are derived from the underlying ophiolites, metamorphic and plutonic rocks of the former collision (suture) zone, and volcanic material from the magmatic arc. A great part of the suture-derived material came from the Klamath Mountains in the north (Fig. 12.27a). The petrofacies of the sandstones particularly reflect the development and erosion of the magmatic arc and the suture zone. For example, the proportion of volcanics decreases upsection, while the quartz content increases (Fig. 12.27c). Later, arkosic sands and sandstones point to the fact that the batholithic roots of the magmatic arc were exposed to erosion.

Concurrent with the westward growth and upbuilding of the Franciscan subduction complex, the forearc basin was enlarged and filled. Sands were dispersed by submarine slope channels and deposited in large fan systems. However, the Sierran magmatic arc migrated eastward (Fig. 12.27d) and lost its dominant role as a sediment source.

The total sediment thickness in the western part of the basin amounts to about 13 km; in the eastern part it is much less. The sedimentation rate during the upper Cretaceous was about 280 m/Ma in the basin center. By the end of the Cretaceous, the forearc basin had attained the morphology of a wide shelf, allowing arc-derived clastic material to migrate westward and to accumulate in the trench. Paleogene sediments were deposited predominantly under shallow-marine and nonmarine conditions; the Neogene is represented by primarily nonmarine strata, apart from some small, deep basins caused by Neogene wrench tectonics associated with the development of the San Andreas fault system. These movements finally terminated the evolution of the Great Valley basin and converted the subducting margin into a transform margin (Fig. 12.27b).

In the Great Valley basin, the western, oceanward parts of the basin first underwent strong subsidence and then uplift as a result of the growing subduction complex (Moxon and Graham 1987). In contrast, the eastern, landward parts of the basin display subsidence closely related to the thermal contraction and eastward migration of the Sierran magmatic arc.

12.5.4 Backarc Basins

General Characteristics

Backarc basins develop behind volcanic arcs either on continental, transitional, or oceanic crust. In *continental margin settings,* the early evolution of a backarc region is commonly characterized by subduction-related extension and rifting, which may be followed by accretion of new oceanic crust (Fig. 12.18c). In addition, the arcward side of the basin may be strongly affected by folding and overthrusting, similar to the tectonic evolution of foreland basins (see below). In *intra-oceanic arc-trench systems,* the backarc region is an oceanic basin of limited size which is subject to sedimentary processes similar to those in larger ocean basins with a volcanic sediment source.

Most of the modern backarc basins lie in the northwestern, western, and southwestern Pacific (e.g., Bering Sea, Okhotsk basin, Sea of Japan, South China basin, Celebes basin, Bismarck Sea, Solomon Sea, New Hebrides basin, and South Fiji basin, Fig. 12.28a) and developed behind volcanic island arcs on oceanic crust. However, several basins in this area cannot be classified absolutely as backarc basins and therefore are referred to as *small ocean basins* (e.g., Leitch 1984). Examples in the Mediterranean Sea are the Tyrrhenian (with new oceanic crust and active volcanism, Fig. 12.15; see, e.g., Savelli 1988) and Aegean Seas (continental crust, cf. Chaps. 12.2.2 and 12.7). The Black Sea and Caspian Sea are relics of a once larger ocean basin (see below).

The largest modern backarc basin associated with the Sunda trench in the northeastern Indian Ocean is the **Andaman Sea** (Fig. 12.22, Chap. 12.5.2). The center of this elongate basin is floored by young oceanic crust, but its northern and southern extensions into Burma and Sumatra, respectively, lie on continental crust. Thus, moving from north to south, the present-day depositional environment of the basin changes from continental (Burma) to deltaic and oceanic (Andaman Sea), and again resumes continental conditions in Sumatra (Stoneley 1981).

In some regions, intramontane troughs develop between the magmatic arc and backarc thrust belt. They may collect great thicknesses of volcaniclastic material as for example the Altiplano of Bolivia (Mitchell and Reading 1986).

Sediment Successions of Backarc Basins

From their variable tectonic setting, changes in sediment supply and oceanographic or other environmental conditions in the depositional area, it is evident that the sedimentary infillings of backarc basins also vary markedly. There are no generally applicable, simple facies models, even if backarc basins on continental and oceanic crust are considered separately. Therefore, ancient backarc basins are normally identified with the aid of their overall setting in a plate tectonic context. Some general rules, however, can be useful in recognizing such basins from their sedimentary infillings:

- Many backarc basins are typically *asymmetric*, with a steep, often thrust-faulted, outer margin against the volcanic arc and a gentler dipping, possibly block-faulted inner margin. The sediments include clastics derived from both margins, i.e., from the volcanic arc and drainage areas of the craton, but lateral sediment influx from distal sources may also play a role, as in the previously mentioned example of Burma and the Andaman Sea.
- On *continental crust*, the basin fill tends to evolve from an initial phase of coarse, clastic deposition into shallow-marine environments, and then to thick, molasse-type sediments derived largely from the volcanic arc and its basement. The backarc region on the eastern flanks of the Andes in South America (Subandean zone) is probably the most prominent example of this type of basin (e.g., Jordan et al. 1983; Mégard 1984; Reutter et al. 1988). These basins are 100 to 200 km wide, and their late Mesozoic to Neogene infillings frequently attain 5 to 10 km in thickness, locally up to 15 km. An example of such basin evolution is the Oriente basin east of the Andes in Ecuador and Peru (Fig. 12.28b, Stoneley 1981), where middle Cretaceous and older shelf sediments are overlain by thick upper

Cretaceous and Cenozoic molasse deposits. Cretaceous backarc basins in Chile contain thicknesses as great as 7 to 9 km of coarse clastics, including gravity mass flow deposits and turbidites (Mitchell and Reading 1986). Many of these basins experienced synsedimentary compressional deformation, mostly during the Tertiary, which created a thin-skinned fold-thrust belt.
- *Backarc basins* on oceanic crust or rift basins developing sea-floor spreading in the backarc region tend to become filled after shorter time periods than larger ocean basins due to their limited size and nearby clastic sediment sources. The different stages of a backarc basin resulting from the splitting of an oceanic arc are shown by the simplified model in Fig. 12.29. This model largely reflects observations in the backarc region of the Mariana arc characterized by several older remnant arcs and basins in the backarc region, and other examples from the western and southwestern Pacific (Carey and Sigurdsson 1984):

- *Stage 1.* Rifting of an island arc along a zone of crustal weakness and subsequent backarc spreading create a narrow trough between the active arc and inactive remnant arc. The basal fill in the trough is dominated by gravity mass flows, i.e., slides, slumps, debris flows, and turbidity currents, which transport basaltic pyroclastic and epiclastic material (eroded older pyroclastic deposits) into the basin (cf. Chaps. 2.4 and 5.4). Around the backarc spreading center, special hydrothermal deposits may accumulate. On the side of the remnant arc, epiclastic and some pelagic sediments are common. The basin attains only moderate depth, and sedimentation rates are generally high during this stage (at least 100 to 200 m/Ma).
- *Stage 2.* Continued backarc spreading causes basin widening and subsidence of previously formed oceanic crust as a result of cooling. Episodic volcanic activity in the arc region further provides large quantities of basaltic volcaniclastic material, which is deposited in aprons at the base of the arc or dispersed as ash falls over wider areas. Deposition of epiclastic material on the side of the remnant arc is more or less replaced by deposition of pelagic sediments, which also play some role on the active arc side. The basin floor is commonly still above the CCD (Chap. 5.3.2) and thus allows preservation of calcareous microfossils. Dissolution of siliceous fossils may be hampered by relatively high SiO_2 contents in

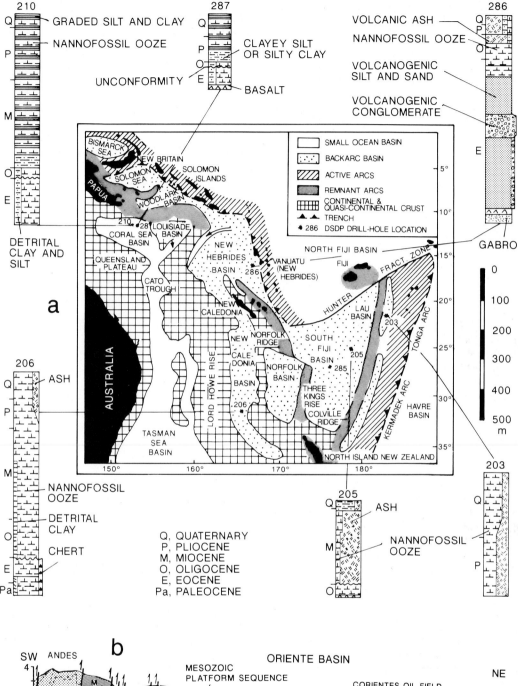

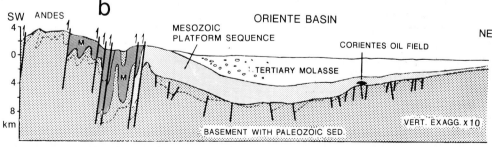

Fig. 12.28. a Modern backarc and small ocean basins of the southwestern Pacific and results of the Deep Sea Drilling Project shown in simplified sections for specific sites (numbers). (After Leitch 1984). **b** Simplified cross-section of Oriente backarc basin, Peru, filled with shallow-marine to molasse-type sediments. (Stoneley 1981)

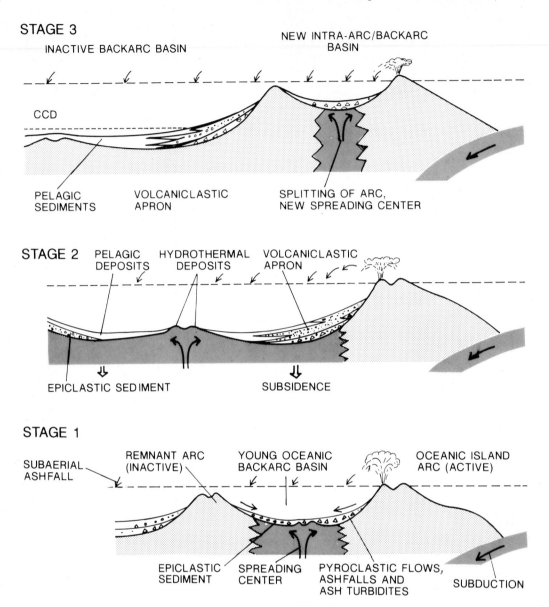

Fig. 12.29. Sedimentation model for backarc areas of oceanic island arcs in three stages of evolution. *1* Splitting of island arc by rifting, and initiation of backarc spreading into active arc and remnant arc; high influx of volcaniclastics. *2* Basin widening by continued backarc spreading; deposition of thick volcaniclasite apron at base of active arc flank and pelagic sediment farther away. *3* Backarc spreading ceases and renewed splitting of island arc may initiate a new basin. (After Carey and Sigurdsson 1984)

sea water due to the presence of volcanic ash and glass (cf. Chap. 5.3.5). In the more distal parts of the basin, the sedimentation rate drops to around 50 m/Ma, but may maintain high values arcward in the volcaniclastic apron. Increased arc volcanism leads to propagation of the areas previously dominated by pelagic sedimentation. If volcanism wanes, pelagic, ash-poor sedimentation will prevail. - *Stage 3.* Ocean spreading in the existing basin ceases, and splitting of the volcanic arc initiates the formation of a new backarc basin. The old basin still receives some epiclastic material and subaerially dispersed ash from the active arc, but is dominated by pelagic sedimentation. Continued subsidence may lower the basin floor below the CCD and thus cause carbonate dissolution, resulting in the deposition of brown clays. The sedimentation rate drops to values characteristic of normal pelagic deposits in the oceans (5-15 m/Ma for nannofossil ooze and around 5 m/Ma for brown abyssal clays). The subsequent sediment succession will be controlled mainly by factors unrelated to arc volcanism and backarc spreading.

The **sediment successions of small ocean basins** resulting from the rifting of continental crust or predating the formation of a volcanic island arc frequently begin with normal clastic sediments containing quartz, feldspar and mica. They are followed by pelagic and hemipelagic sediments and display a shallowing-upward trend, with increasing influence from volcaniclastic material, if a volcanic source is nearby. In addition, they show marked differences between their outer and inner margins.

Examples of Modern Backarc Basins

Modern backarc basins of the southwestern Pacific. Some examples of Tertiary sediment successions drilled in backarc basins during the Deep Sea Drilling Project more or less follow the rules outlined above (Fig. 12.28a, Leitch 1984; Klein 1985b). Young, narrow basins behind an active arc (e.g., west of the Tonga arc, Site 203) or older basins adjacent to an island arc (e.g., New Hebrides basin, Site 286, west of the Vanuatu islands, and the South Fiji basin, Site 205) display successions rich in various volcaniclastics and calcareous nannofossil ooze. The sedimentation rates of volcaniclastic sequences mostly range between 50 and 100 m/Ma. However, the drillholes (Sites 210 and 287) in the older Coral Sea basin, generated by rifting of continental crust, and in the New Caledonia basin (Site 206) are dominated by slowly deposited nannofossil oozes and silty clays, with minor proportions of volcaniclastic material.

The upper portion of Site 210 contains about 2000 muddy turbidite beds alternating with calcareous ooze, deposited during a time span of 10 Ma (sedimentation rate around 50 m/Ma). This increase in sediment supply can be explained by a rejuvenated sediment source in the remnant arc region. Early to mid-Oligocene unconformities observed in several basins are associated with accelerated bottom currents. According to seismic records, the narrow New Caledonia basin is filled with a 3000 m thick succession of submarine slumps derived largely off the Lord Howe Rise. The Tasman Sea basin adjacent to the Australian continent is characterized by a sequence rich in siliceous ooze and chert (not shown in Fig. 12.28a).

The same applies to the **Sea of Japan** which represents an approximately 20 Ma old, inactive backarc basin behind a volcanic arc associated with an active subduction zone. This basin is more than 3500 m deep and separated from the open sea by a shallow sill.

Its sediments and basement rocks were investigated during Legs 127 and 128 of the Ocean Drilling Project (Brumsack and Shipboard Scientific Party, Leg 127; Holler et al., Preliminary Reports 1990).

The basal sediments on top of young oceanic crust indicate hemipelagic conditions. At Site 799, the lower half of a 1000 m thick Miocene to Quaternary sequence (average sedimentation rate about 50 m/Ma) consists of siliceous claystones and porcellanites with minor proportions of calcareous ooze and rhyolite tuff. The organic carbon content of these sediments is mostly around 1 %. The upper half is dominated by diatom ooze with an upwardly increasing proportion of clay and silty clay, including volcanic ash.

In summary, the modern backarc basins of the western Pacific display a variety of sediment types including coarse-grained mass flow deposits, submarine fan turbidites and basinal turbidites, hemipelagic and pelagic silts and clays, biogenic carbonates and biosiliceous sediments, and various types of tephra deposits (Klein 1985b). Whereas volume and nature of the siliciclastic material are controlled by arc volcanism and the presence and relief of nearby land sources, the biogenic sediments depend on ocean circulation and surface productivity. Most of the sediments represent deep-water conditions, either above or below the CCD.

The deep basins of the **Black Sea and Caspian Sea** are assumed to be remnants of a much larger, 3000 km long, 900 km wide, Mesozoic to early Tertiary backarc basin (Steininger and Rögl 1984; Zonenshain and LePichon 1986).

This interpretation is based on seismic evidence indicating that these basins are floored by oceanic crust. In addition, the basins experienced increased rates of subsidence in conjunction with the approaching thrust belt from the south (the western and eastern Pontides, the Lesser Caucasus, and the Elburz Mountains). The area between the two basins was closed by collision in front of the northward moving Arabian promontory.

In the Black Sea, upper Cretaceous to Neogene sediments attain a maximum thickness of nearly 15 km. The oil-bearing middle Pliocene to Quaternary strata of the southern Caspian Sea are 3 to 6 km thick and thus were deposited at a rate on the order of 1000 m/Ma. They are underlain in the basin center by 8 to

10 km older sediments. The youngest development and present depositional environments of these basins are discussed briefly in Chapter 4.3.

12.5.5 Preservation and Recognition of Ancient Subduction-Related Basins

Ancient subduction-related basins are normally identified by evaluating their temporal and spatial relationship with magmatic arcs. Deep-sea trenches, forearc basins, and oceanic backarc basins are commonly closed by continued plate convergence. Their infills are incorporated into collision zones, where they are strongly deformed, disrupted and partially metamorphosed. Backarc basins behind continental margins, however, have a better chance of being preserved in a state which is only moderately modified by subsequent tectonic movements.

In many ancient examples, however, the differentiation of trench fills and sequences from forearc and backarc basins exposed in orogenic belts is difficult. Trenches are always filled in deep water, but their sediments include slumps and mass flows, which contain material from shallower or terrestrial environments. Both forearc and backarc basins display a shallowing-upward trend if the sediment supply is sufficient to fill the basins. The presence of oceanic basalts associated with deep-marine sediments in a subduction complex or mélange is taken as evidence for accreted sea floor, but may also indicate the former existence of a residual forearc basin or an oceanic backarc basin.

The determination of paleo-current directions and textural trends of subaerially and subaquatically deposited beds (e.g., fluvial transport, sediment gravity flows, turbidity currents) can be used to distinguish between forearc and backarc depositional systems. In addition, large proportions of volcaniclastic material and the occurrence of specific detrital minerals in sandstones (e.g., clinopyroxene and amphibole, Morris 1987) indicate nearby active arc volcanism.

The composition of volcaniclastic material derived either from continental margin arcs (dominated by intermediate and silicic, largely calc-alkaline magma types) or oceanic arcs (mainly basaltic and andesitic material) is frequently quoted as a means of discriminating these tectonic settings (e.g., Crook 1974; Dickinson 1982; Dickinson et al 1983; Maynard et al. 1982; Valloni and Mezzardi 1984; also cf. Chap. 2.4.3). Large proportions of feldspar and other minerals characteristic of granitic and granodioritic rocks may testify to the fact that the arc massif was already deeply eroded. Interpretations of ancient sedimentary successions based solely on composition should, however, be regarded with caution (e.g., Moore GF et al. 1982; Lash 1985; Velbel 1985; Zuffa 1987), since several modern examples of trench fills, forearc basins and backarc basins have shown that distant sediment sources and sediment recycling can play a decisive role.

12.6 Remnant and Foreland Basins

12.6.1 Remnant Basins with Flysch

General Aspects

The late phase of the development of an orogenic belt is commonly characterized by the transition from remnant basins of a closing ocean to a foreland basin. While remnant basins are largely filled with flysch-type sediments, foreland basins accumulate mainly molasse deposits. *Flysch* is typically a thick marine sequence of turbidite beds and gravity mass flows, contains an autochthonous deep-water fauna (apart from displaced shallow-water organisms), and is incorporated into the thrust and fold belt of an orogen. *Molasse* consists predominantly of shallow-marine to continental deposits including large alluvial fans and fan deltas. The evolution of both types of basin fills is closely related to the geodynamic development of an orogen. This is

shown in the simplified model of Fig. 12.30, which imitates to some extent the situation in the Swiss Alps (e.g., Matter et al. 1980; Stockmal and Beaumont 1987; Caron et al. 1989).

Remnant basins resulting from subduction are underlain by oceanic crust and therefore represent deep, narrowing basins (cf. Chap. 1.2). In Figure 12.30a, the remnant basin is confined on one side by a pre-existing passive continental margin with a wedge of older clastics and carbonate sediments, and on the other side by an approaching overthrust belt (accretionary wedge) shedding relatively large volumes of various clastics in the form of turbidites and mass flow deposits into the basin.

Some flysch basins may also develop in topographic depressions generated by tectonic deformation of continental crust and microcratons.

Basin Evolution and Sediments

Flysch sediments markedly vary in grain size distribution and composition in relation to the changing influence of intra-basinal and extra-basinal sediment sources (Fig. 12.31a). The wide spectrum of flysch deposits may include proximal conglomeratic beds, coarse-grained debris flows ("wildflysch", but also see Chap. 12.5.2) and olistostromes (Chap. 5.4.1), and shale-dominated sequences with distal fine-grained turbidites. In places also marly and calcareous flysch facies occur (Fig. 12.31a and b). Most of these deposits accumulate in deep water, but not all of them indicate water depths below calcite compensation, depending on the degree of basin fill and temporal and spatial fluctuations of the CCD (cf. Chap. 5.3.2).

Flysch sediments deposited along the front of the overthrust belt are successively incorporated into the overriding accretionary wedge (cf. Chap. 12.5.2). Older flysch is frequently overthrust onto younger flysch deposits. In a cross-section from the internal fold-

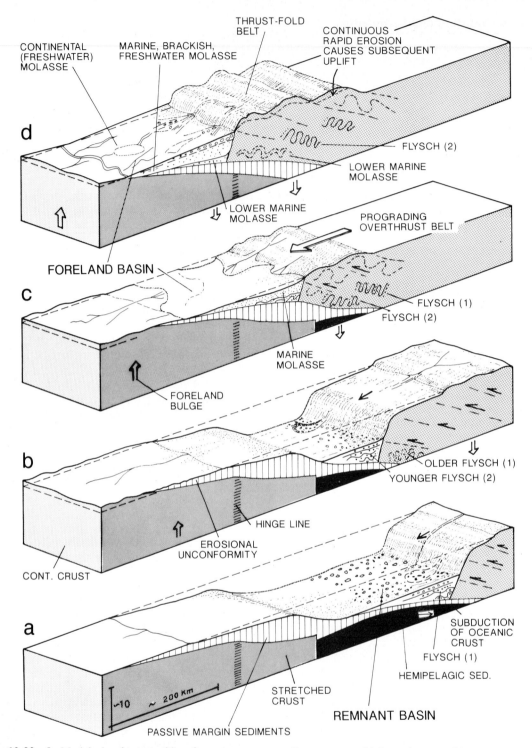

Fig. 12.30a-d. Model showing transition from remnant oceanic basin (**a** and **b**) to foreland basin (**c** and **d**) due to emplacement of overthrust belt as tectonic load onto pre-existing passive continental margin (collision). With *prograding overthrust belt,* zone of subsidence, basin axes, and flexural foreland bulge migrate outward toward the foreland. Successive flysch deposits (deep-sea fans and trench fills) may differ in composition due to changes in sediment source. *Molasse deposits* of foreland basin tend to evolve from marine (possibly rather deep, flyschoid) to continental sediments and to prograde over the shelf sequence of former passive margin. *Late phase, rapid erosion* of high elevated thrust-fold belt finally initiates uplift including the inner parts of the foreland basin. (Mainly after Matter et al. 1980; Stockmal and Beaumont 1987)

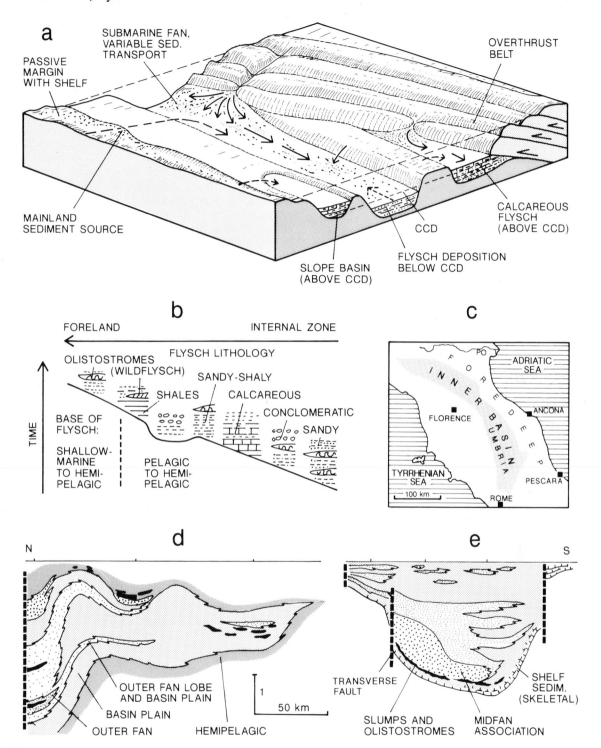

Fig. 12.31a-e. *Some general problems of flysch sediments in orogenic belts.* **a** *Different types of flysch basins with various sediment sources and transport directions, deposition above or below calcite compensation depth (CCD).* **b** *Simplified scheme of decreasing age toward the foreland and various lithologies of flysch basins in the Swiss Alps. (After Matter et al. 1980).* **c** *and* **d** *Location and megafacies distribution of longitudinal sections in flysch basins* of the northern Apennines, Italy. *(After Ricci-Lucchi 1975). Note coarsening-upward and progradational trend in Miocene Marnoso-arenacea flysch* **(d)** *over-supplied with sediment, while Upper Miocene-Lower Pliocene Laga formation* **(e)** *is characterized by fining-upward trend as a result of sediment supply insufficient to compensate for subsidence*

thrust belt to the foreland, the flysch deposits (in their original position) generally become younger toward the foreland (Fig. 12.31b). Younger flysch may differ in composition from older flysch due to variations in the sediment source provided by the growing accretionary wedge and its hinterland. Specific sediment sources such as ophiolites or overthrust metamorphics can be of particular diagnostic value (e.g., Winkler 1988). Over-supply of a narrowing basin (i.e., sedimentation ≥ subsidence) may lead, at least locally, to shallowing sequences (Fig. 12.31d). However, as long as oceanic crust continues to be subducted beneath the prograding overthrust belt at a relatively high rate (Chap. 12.5.2), the basin normally remains deep and may even exhibit a fining-upward trend.

The *megafacies distribution* in flysch and deep foreland basins (see below), as well as the directions of sediment transport may vary with time and from basin to basin (e.g., Ricci-Lucchi 1975 and 1986; Hesse and Butt 1976; Ackermann 1986). The most important facies associations include deep-sea midfan and outer fan sediments, outer fan sediments interfingering with those of the basin plain, basin plain deposits, slumps and olistostromes, and hemipelagic sediments (Fig. 12.31c through e, also see Chap. 5.4.2). The vertically and laterally changing megafacies result mainly from the varying efficiency and repeated shift of submarine canyon and fan systems funneling sediment into the elongate basins. In addition, the facies evolution is frequently controlled by synsedimentary tectonic basin deformation along the front of the overthrust belt (Homewood and Caron 1983; Ricci-Lucci 1990).

Some uplift of the foreland *(foreland bulge or forebulge)* may cause an erosional unconformity in the older sediment wedge of the passive margin sequence (Fig. 12.30c).

12.6.2 Foreland Basins with Molasse

General Aspects

The evolution of a peripheral foreland basin begins, when the former ocean is closed and the front of the overthrust belt has reached the stretched crust of the pre-existing continental margin (Chap. 1.2, Fig. 1.3a; Fig. 12.30c). Then the active subduction zone evolves into a *collisional over-thrust belt* as one continent overrides another.

The *subsidence history* of these basins reflects the rate of thrusting in the adjacent orogenic belt and sedimentation rate in the basin (Angevine et al. 1990). In addition, the nature and thickness of the lower plate plays a role. If the overriding tectonic load increases slowly with time and the continental crust beneath the thrust belt is thick, subsidence of the evolving foreland basin proceeds at a low rate and cannot exceed a specific limit. In contrast, rapidly overriding thick tectonic loads on thinned continental or transitional crust generate fast and deep subsidence in relatively narrow basins. The migration of thrust sheets into the basin results in an increasing rate of subsidence of the flexing crust in the basin in front of the thrust belt (convex-up subsidence curves, cf. Chap. 8.5 and Fig. 8.19). A discontinuous movement of the thrust sheets may lead to irregular subsidence curves. Sediments derived from the thrust belt are deposited in the basin and thus further cause subsidence to migrate outward toward the foreland. Finally, after thrusting has ceased, erosion of the thrust load and heating of the subducted crust generates *crustal rebound*. i.e., uplift.

A somewhat modified evolution of foreland basins was found in the southern Alps, the Po basin on the Apennine margin, and in the northern Apennines, Italy, as well as on the Pyrenean margin of the Ebro basin in Spain (Ori and Friend 1984; Massari et al. 1986; Ricci Lucchi 1986; Boccaletti et al. 1990). In these cases, *satellite foreland basins* also develop on top of deeper lying thrust sheets in addition to the main foreland trough. The basins may be split by prograding thrust sheets or carried "piggyback" toward the foreland (Fig. 12.33b). Furthermore, older portions of the molasse incorporated into the thrust belt are uplifted and eroded, i.e., their sediments are recycled.

Modern Examples

Some deep basins in modern oceans result from rapid emplacement of overthrust loads, such as the 1.5 km deep western Taiwan foreland basin (Covey 1986) and the 2 to 3 km deep Timor-Tanimbar Trough between Australia and the Banda Arc (Audley-Charles 1986).

The **Timor-Tanimbar Trough** is located to the east of the Sunda and Java Trench (Fig. 12.22a) where the Australian continental crust underplates the non-volcanic island chain

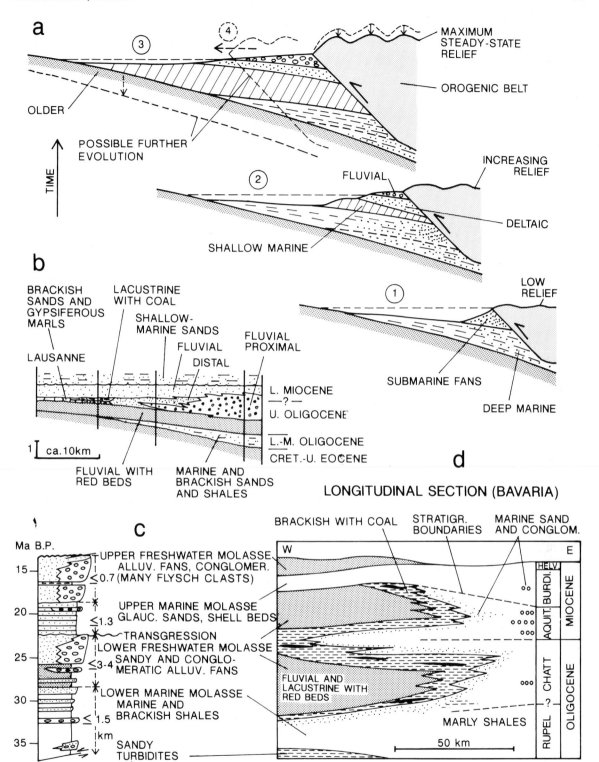

Fig. 12.32. a Evolution of foreland basin fill, stages *1* to *3 (4)*. (After Covey 1986). **b** Generalized, reconstructed cross-section of foreland (molasse) basin in western Switzerland. (Homewood et al. 1986). **c** Simplified stratigraphic section of Swiss molasse showing two major shallowing-upward cycles, each of them represented by marine and freshwater molasse. (After Matter et al. 1980). **d** Longitudinal section of foreland molasse in Bavaria, Germany, exhibiting transition from permanently marine conditions in the east to partial brackish and continental environments in the west. (After Bögli and Müller 1976)

of the Banda Arc. Subsequent to collision during the middle Pliocene, crustal shortening set in and continues to the Present. It operates at rates of 6 to 12 cm/a and causes both uplift of the island arc (1 to 3 mm/a) and subsidence of the foreland basin (1 mm/a along the present axis of the basin). In an upper Quaternary sediment section drilled in the Timor Trough, 400 m consist of fine-grained siliciclastics and biogenic carbonates. The total sediment thicknesses of the water-filled foreland basin ranges between 4 and 7 km.

Basin Evolution and Sediments

Early Stage of Evolution

Due to the emplacement of a tectonic load on thinned continental crust, the early stage of many ancient foreland basins tends to be fairly deep and thus may be still characterized by flysch-like marine sediments and submarine fans near the rising mountain range (Fig. 12.32a). The deformation front of the overthrust belt may advance at rates up to several cm/a onto the foreland and cause rapid sediment accumulation in the foredeep.

Excellent examples of this stage have been described from the southern Pyreneans (Mutti 1984; Mutti et al. 1985; Labaume et al. 1985), the northern Apennines and the southern Alps (Ricci Lucchi 1986; Boccaletti et al. 1990). In the northern Apennines, the Neogene deformation front advanced at a rate of 0.25 to 1 cm/a; further south the depocenters of Miocene foredeeps migrated outward at a rate of up to 7.5 cm/a, and sediments accumulated as fast as almost 1000 m/Ma. In the South Pyrenean foreland basin, a deformation rate of 3.5 cm/a is assumed. Rapid subsidence led to the accumulation of turbidite sequences up to 3.5 km in thickness. These were shed in an axial, western direction and also include carbonate turbidites derived from shelves to the north and south of the deep basin. In addition, the basin was segmented by synsedimentary structural highs transverse to the basin axis. The eastern portion of the foreland basin was largely filled with fluvial and deltaic sediments.

Foreland basins are markedly asymmetric. Thus, already in the early stage of evolution shallow-marine (marine molasse) or fluvio-deltaic sediments (freshwater molasse) predominate near the bordering platform.

Later Stages

During further development, the advance of nappes of the overthrust belt is increasingly hampered by the thickening, flexured and cratonward rising top of the underlying continental crust (Fig. 12.30d). Some cases have been described in which upper portions of the underplating crust and its shelf sediments were scratched off and incorporated into the overthrust belt (not shown in Fig. 12.30). Early molasse may be sheared off and folded in front of the overthrust belt, or overridden by the nappes. In this stage, the overthrust sheets are forced to move obliquely upward and reach elevations high above sea level, before they finally come to rest. This in turn promotes subsidence and additional flexure of the continental plate and thus expansion of the basin toward the foreland. The extension of the basin is, however, limited by the outward migrating foreland bulge.

Sediments filling the foreland basin are predominantly clastics derived from both sides of the basin, but particularly from the highly elevated parts of the folded overthrust belt. The general tendency is a *shallowing-upward sequence* which ends with continental sediments (Figs. 12.30a and 12.32a). During the early deep-water stage, the orogenic belt may still form a submarine ridge or a ridge of low elevation above sea level. Later, shallow-marine, deltaic, and fluvial sediments predominate. When uplift of the orogenic belt is balanced by erosion, the elevation of the mountain belt and its denudation rate have reached their maximum. This may lead to a kind of steady-state evolution of the foreland basin during its further migration toward the passive mainland (Covey 1986). Under these conditions, the basin's cross-sectional area remains more or less constant, and its sedimentary fill displays only minor changes.

Such a situation, however, was commonly superposed by additional processes. The depositional environment of many foreland basins frequently varied between shallow-marine, brackish, and continental as a result of significant spatial and temporal changes in sediment supply, sea level changes, and climate (Figs. 12.30b, c and 12.32; also see, e.g., Bachmann et al. 1987; Lemcke 1988; Barrio 1990). As mentioned above, a solely marine sequence may laterally pass into brackish, lacustrine or fluvial sediments (Fig. 12.32d). Coal seams of limited thickness and extension are common in such transitional regimes.

Example: The huge, about 5000 km long foreland basin east of the Rocky Mountains in North America, the so-called **Western Interior seaway**, was flooded from the Aptian to the late Cretaceous and formed a broad seaway, connecting the proto-Gulf of Mexico and the Circum-Boreal seaway for nearly 35 Ma (Weimer 1960; Jordan 1981; Kauffman 1984: Molenaar and Rice 1988). It experienced five major and numerous minor transgressive-regressive cycles (cf. Chap. 7.4), and accumulated marine

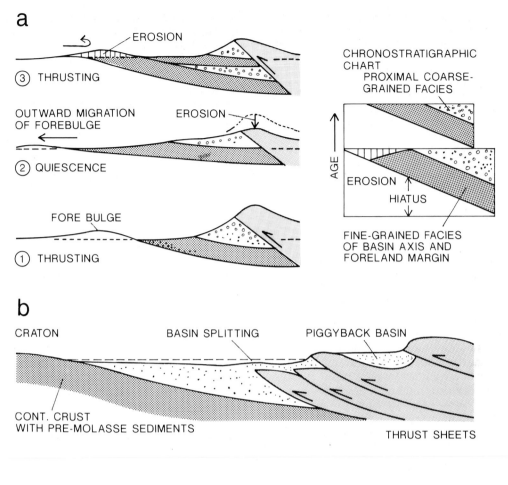

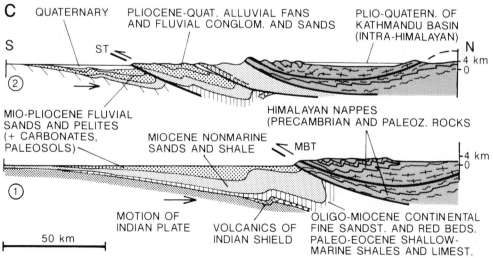

Fig. 12.33. a Model of foreland basin subject to episodic thrusting (stages *1* and *3*) causing migration of the fore-bulge toward the thrust belt. Periods of tectonic quiescence *(2)* lead to thrust erosion and outward migration of fore-bulge. Note stairstepped facies package and erosional unconformity. (After Flemings and Jordan 1990). **b** Scheme of foreland basin with deposition partially on top of fold-thrust belt and minor basin carried piggyback on thrust-sheets. (After Ori and Friend 1984; Ricci Lucchi 1986). **c** Evolution of Himalayan foreland basin in two stages (*1* and *2*) with sediment wedge of the Siwaliks. *MBT* Main boundary thrust; *ST* Siwaliks thrust. (After Mascle et al. 1986). See text for explanation

sediments up to about 6000 m in thickness. A special feature of this basin was its wide eastward extension onto the stable craton beyond the forebulge zone. This development was favored by the heavy sediment load deposited east of the thrust belt, as well as by the unusually high eustatic sea level during the middle Cretaceous. This basin displays a variety of sedimentary facies including deltaic sediments, coal seams, coastal barrier sandstones, offshore siltstones and shales, shelf sandstones and limestones, black shales, and even some turbidites.

Sediment Supply and Lateral Facies Change

Foreland basins receiving large volumes of clastic material during their evolution, as for example the widely extended Eocene to Quaternary molasse basin south of the High Himalayas, are filled predominantly with alluvial fans and fluvial deposits (Fig. 12.33c; Mascle et al. 1986; Delcaillau 1986). Such *over-supplied basins* develop a marine depositional environment only during their early stage.

Thick conglomerates deposited as *alluvial fans* frequently border the front of the mountain belt which undergoes rapid erosion. Fan deltas may prograde over shallow-marine sands. Some distance away from the mountain front, the distributory systems follow the longitudinal trend of the basin. A surplus of sediment can be transported longitudinally out of the basin into an adjacent ocean. The molasse sediments prograde onto the platform with time (Fig. 12.32 a); younger beds successively overlap older, pinching-out strata (e.g. Homewood et al. 1986). Finally, isostatic adjustment to the continuous erosion of the rising mountain range leads to exhumation of formerly deeply buried rocks and structures, as well as to some uplift of the neighboring parts of the foreland basin (cf. Chap. 8.5).

These general trends are modified in settings where thrusting occurs discontinually. *Episodic thrusting* with periods of tectonic quiescence in between may cause a stair-stepped facies package as modeled for a nonmarine foreland basin (Fig. 12.33a; Flemings and Jordan 1990). Thrusting and high elevation of the thrust belt (stage 1) induces thrustward migration of the forebulge and the deposition of proximal alluvial fans. The subsequent stage of quiescence (2) is characterized by continued erosion along the thrust belt, outward migration of the forebulge, and deposition of mainly fine-grained sediments in the center and distal portions of the basin. Renewed thrusting and thrustward migration of the forebulge (stage 3) causes

partial erosion of former distal sediments and the formation of a new proximal clastic wedge. The simplified chronostratigraphic chart of such an evolution shows periods of slowly prograding proximal and more distal facies, interrupted by intervals of erosion and retrogradation toward the thrust zone. Facies retrogradation marks the onset of a new thrusting event. Other variations in foreland basin evolution result from an irregular shape of a continental margin subject to collision (e.g. Lash 1988). In this case, the tectonic load of the thrust belt is emplaced at different times along strike of the basin and thus causes diachronous pulses of subsidence and sediment supply.

Sediment Composition and Sedimentation Rates

The mineralogical composition of molasse sandstones varies from quartz-rich, feldspar-poor types derived from intensely weathered continental blocks to varieties which are less quartz-rich but rich in rock fragments (Schwab 1986). The latter types constitute the bulk of sand in many foreland basins. Material derived from tectonically uplifted subduction complexes or magmatic arcs is found only in small quantities. Systematic variations in composition through time, related to unroofing of deeply buried rocks, are rarely observed.

The average sedimentation rates in foreland basins strongly differ between proximal (near the thrust belt) and distal areas. Average rates range between 50 and 1000 m/Ma; in the foreland basin south of the High Himalayas 200 to 500 m/Ma are common (Mascle et al. 1986).

12.7 Pannonian-Type Basins

General Characteristics

Relatively large, intramontane basins related to adjacent collision belts and subduction zones are referred to as Pannonian-type basins (cf. Chap. 1.2). Such basins are underlain by thinned continental crust, and they may originate from former backarc or retroarc basins which predominantly undergo extension behind fold-thrust belts migrating towards the foreland (Mitchell and Reading 1986). The prototype of these basins is the Pannonian basin of Hungary and its neighboring countries. Although recent studies (Royden and Horváth 1988) shed some doubt on the usefulness of the Pannonian basin as a model basin for the interpretation of other intramontane basins, some of the new results are briefly discussed here. The Pannonian basin is a young and relatively simple system which is not overprinted by subsequent tectonic processes. It is therefore well suitable for an example of basin analysis (Burchfiel and Royden 1988).

The **Pannonian basin** lies within the Alpine mountain belts of east-central Europe (Fig. 12.34a). From east to west it is about 800 km long and from north to south 400 km wide. It is currently interpreted by most authors as a back-arc extensional basin of middle Miocene age. However, it should more accurately be called the "Pannonian basin system" because it consists of several smaller individual basins separated by relatively shallow basement blocks. This structural pattern results from a combined backarc and escape tectonic setting during the Miocene (Fig. 12.34d). It is caused by the collision between the European plate and several smaller plates to the south including Africa. Deformation of the Pannonian region occurred during the final stage of thrusting and folding in the outer part of the Carpathians. Thrusting was directed outward from the Pannonian basin toward the European platform and the Adriatic region. Extension of the Pannonian plate fragment occurred along a conjugate system of faults and was accompanied by strike-slip (Fig. 12.34a). Size and shape of the resulting special troughs including pull-apart basins (see below) vary considerably, but their long axes more or less follow the directions of the dominant fault system.

In some of these special basins the Neogene-Quaternary sediments exceed 7 km in thickness. Part of the Neogene Pannonian basin sediments overlie the internal Mesozoic thrust sheets of the surrounding mountain belts, for example below the Vienna basin (Fig. 12.34b). Paleogene sediments occur only in limited areas.

The northernmost basins near the mountain belt (Vienna and Transcarpathian basins) subsided and were mainly filled in early to middle Miocene time, i.e., prior to the basins farther south (Danube, Zala, Drava, Makó and Békés basins) where rapid subsidence began in the middle Miocene.

This diachronous basin development was caused by two processes (Mattick et al. 1988; Royden 1988; Nikolaev et al. 1989):

1. Thermal subsidence of the basins located near the thrust belt occurred earlier and at a more rapid rate than in the central part of the Pannonian system. Hence, the basins adjacent to the mountain belt contain thick, fault-bounded sediments which were deposited dur-ing the extensional phase *(syn-extensional sediments)*. They are covered by thin *post-extensional strata* (see scheme of Fig. 12.34e and f). By contrast, basins located far from the thrust belt exhibit only thin fault-bounded syn-extensional sequences which in turn are overlain by thick, flat lying, unfaulted post-extensional sediments.

2. Due to the location of the principal sediment source to the north and east of the Pannonian system, the basins were successively filled from north or east to south and south-west.

Sediment Successions

The sedimentary fill strongly varies within individual basins and from one basin to another basin. Therefore only some general trends can be mentioned (Fig. 12.34g). Commonly, Neogen sedimentation started with continental or marine transgressive beds followed by shallow-marine shales and marls with sandy intercalations. During the early, marine deep-water stage of the basins, resulting from rapid subsidence and insufficient sedimentation to keep pace, predominantly marly, hemipelagic sediments with turbidites and submarine fans including gravity mass flows were deposited (Mattick et al. 1988). The depths of the troughs may have been on the order of 1000 m; stagnant water conditions occurred in several basins. As subsidence slowed down, the basins were slowly filled by prodelta and delta front sediments. The final stages were characterized by delta plains, brackish and freshwater lakes, and fluvial deposits of the evolving Danube river and its tributories.

Generally, the basins display a shallowing-upward tendency. However, the sedimentary facies may locally vary in type and thickness and display unconformities due to synsedimentary faulting, uplift and tilting of individual blocks, and small-scale horst and graben structures within the basins. In addition, the marginal facies of the basins differ from those of the more central parts.

Such complications are known, for example, from the Vienna basin (Fig. 12.34b and c), which is a significant oil and gas field and has therefore been intensively explored by numerous drill holes during the last 60 years (Wessely 1988, also see Chap. 12.8).

The subduction-related Neogene deformation of the Pannonian fragment was accompanied by Miocene calcalkaline volcanism and young-

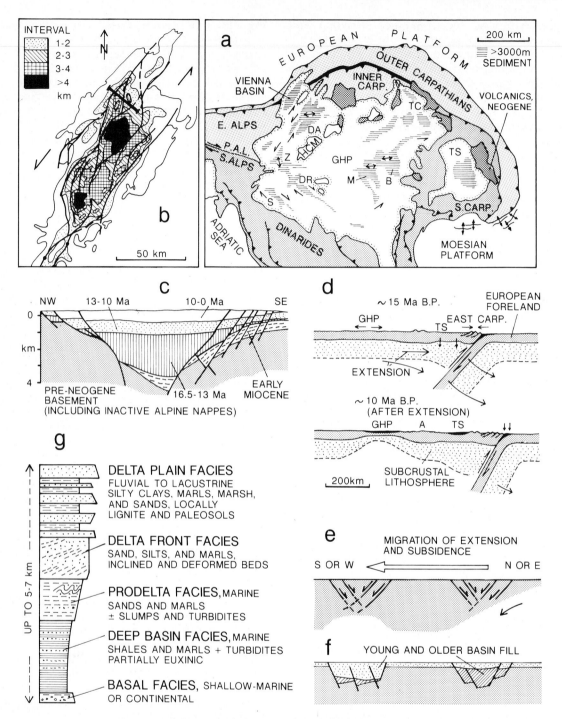

Fig. 12.34. a Pannonian basin system within the Alpine belts of central-eastern Europe. Hatched areas indicate smaller basins with Neogene to Quaternary sediment thicknesses ≥ 3000 m. *DA* Danube basin; *Z* Zala b.; *DR* Drava b.; *S* Sava b.; *TC* Transcarpathian b.; *TS* Transsylvanian b.; *M* Makó, and *B* Békés depressions; *GHP* Great Hungarian Plain; *HCM* Hungarian Central Mountains; *P.A.L.* Peri-Adriatic Line. **b** and **c** Simplified plan view and cross-section of Vienna pull-apart basin. (In realitiy the structure and sedimentary fill of the basin are much more complex;

Wessely 1988.) **d** Migration towards to the foreland and steepening of East Carpathian subduction zone causes extension of Pannonian plate fragment and formation of a number of small basins. **e** Simplified scheme demonstrating migration of specific zones of subsidence and basin fill away from thrust belt. **f** Idealized succession of sedimentary facies in special basins of the Pannonian system. (After Royden et al. 1982; Mattick et al. 1988; Royden 1988)

er alkali basalts. Their material contributed to the filling of the basins.

Other Examples

Pannonian-type intramontane back-arc basins which are deformed by continuing mountain building processes are also found in other regions. The **Aegean Sea** behind the Hellenic arc in the eastern Mediterranean may be regarded as a modern example of such a basin (Le Pichon 1982; Jacobshagen and Giese 1986). This area is underlain by continental crust, is subject to rapid extension (2-6 cm/a during the last 70 years, Jackson and McKenzie 1988), and started to subside in the Miocene.

12.8 Pull-Apart Basins

General Characteristics

Both continental and oceanic crust may be affected by strike-slip creating pull-apart basins (Chap. 1.2). These basins occur in different tectonic settings (Mann et al. 1983; Christie-Blick and Biddle 1985):

- Along strike-slip boundary zones between rigid, continental plates.
- In zones of oblique subduction along volcanic arcs.
- In zones of "continental escape" in response to collision.

In addition, many extensional continental graben and rift zones exhibit a significant strike-slip component.

Pull-apart basins are generated at the *releasing bend of transform faults* and display a certain evolution from a small initial stage (e.g., spindle shape) to a "lazy-S" and rhomboidal basin to a more mature, elongate trough stage (Fig. 12.35a). Furthermore, there are more complicated strike-slip systems with various basin morphologies (Ballance and Reading 1980; Biddle and Christie-Blick 1985), such as basins associated with the bifurcation of faults. Small basins are commonly underlain by thinned continental crust, while greater strike-slip motions lead to the formation of new oceanic crust in one or several special troughs within the larger basin, as for example in the Gulf of California (see below). Pull-apart basins are typically narrow (less

than 50 km wide). Extremely long pull-apart basins may develop a short spreading ridge comparable to mid-oceanic ridges.

The **Cayman trough** in the Caribbean is 1400 km long and probably the world`s longest pull-apart basin. It is characterized by such a spreading ridge, the 100 km long Cayman rise (Mann et al. 1983). This basin results from sinistral strike-slip motion between the North American and Caribbean plates.

Along a widely extended strike-slip zone, several pull-apart basins may be created simultaneously as, for example, known from the Gulf of Aqaba and Dead Sea Rift in the Middle East (Manspeizer 1985), the San Andreas Fault system in California, or the Anatolian Fault system in Turkey (Hempton and Dunne 1984; Sengör et al. 1985), and New Zealand.

The displacement along strike-slip faults may be accompanied either by extension (transtension) at a releasing bend or compression (transpression) at a restraining bend along the fault line in map view (Fig. 12.35b, cf. Chap. 1.2). *Transpression* leads to reverse fault separation, vertical uplift, thrusting, and folding (Fig. 12.35b and c) and is therefore ineffective in creating sedimentary basins of some size (Reading 1980). *Transtension*, however, and *simple, "sharp" pull-apart motion* (Fig. 12.37d, scheme for Gulf of California-type spreading troughs) form important basins with substantial sediment fills. The size of these basins is mainly controlled by fault separation and overlap of the master faults. Further prominent features of strike-slip faults are "en échelon" faults and folds which may occur within or adjacent to the principal displacement zone. In profile, strike-slip zones show characteristics as shown in Fig. 12.35c. The sediments may be segmented in wedge-shaped bodies with laterally variable facies and thickness. Both normal and reverse separation faults may be present in the same depth range and along the same fault at different levels.

Strike-slip movements commonly operate at very high rates (1 to 10 cm/a). With the aid of displaced alluvial fans, average slip rates of 2.2 to 3.4 cm/a were determined along the San Andreas fault in California for the last 30 000 years (Harden and Matti 1989). Maximum rates were as high as 6 to 8 cm/a.

Due to such high slip rates, *subsidence of pull-apart* basins is also relatively fast. The subsidence of narrow pull-apart basins is more rapid than that of rift basins of infinite length. In the former case, the heat provided

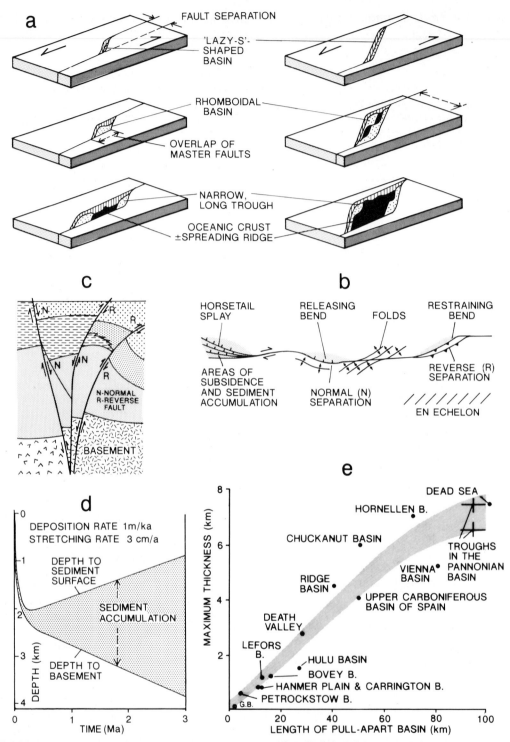

Fig. 12.35. a Generalized model of the evolution of simple pull-apart basins in relation to fault separation and overlap of master faults. Basins with large fault separation may develop two or several smaller, pull-apart troughs. Mature basins form narrow, elongate troughs partially underlain by newly accreted oceanic crust. (After Mann and Dunne 1984). **b** Plan view and **c** transverse profile of idealized strike-slip fault with associated structures.

(After Christie-Blick and Biddle 1985). **d** Calculated subsidence/time curve for crustal block of 10 km width and constant sedimentation rate. (Pitman and Andrews 1985, see text). **e** Relationship between maximum sediment thickness and length (overlap of master faults) of modern and ancient pull-apart basins. (After Hempton and Dunne 1984; data of Pannonian basin from Royden and Horváth 1988)

by the upwelling magma under the stretched crust is laterally lost and thus subsidence accelerated (Pitman and Andrews 1985; Sawyer et al. 1987; cf Chap. 8.3).

The Los Angeles basin was formed since the middle Miocene (11 Ma B.P.) and accumulated in its center a sequence of 9 to 10 km in thickness. About 4 km of this total subsidence result from purely tectonic subsidence. A pull-apart basin assumed below the northeastern Nile delta subsided at a rate of about 0.5 cm/a during the last 7000 years (Stanley 1988b).

A model calculation using the local Airy model for a crustal block of 10 km width subjected to a stretching rate of 3 cm/a and a continuous sediment loading of 1 m/ka (Fig. 12.35d) shows that subsidence during a period of 0.2 Ma can create a 2 km deep basin. In this initial phase the basin is sediment-starved in spite of the high sedimentation rate assumed. Later, sedimentation exceeds subsidence which is now only driven by sediment loading. When the basin is filled up, the sediments have reached a thickness of 4 to 5 km.

The empirical relationship between crustal stretching and maximum sediment thicknesses of pull-apart basins can also be seen from Fig. 12.35e. Here, the length of the basin, i.e., the amount of strike-slip displacement represented by the overlap of the master faults, is plotted versus maximum sediment thickness of the basin (Hempton and Dunne 1984). Short basins have a comparatively shallow basin floor, while long pull-aparts tend to be very deep and may display one or two depocenters on newly formed oceanic crust. Known basin depths range from a few hundred meters up to 7 to 10 km (Dead Sea, Los Angeles basin, and some troughs in the Pannonian basin, Royden and Horváth 1988). Subsidence is caused by both the local thinning of continental crust and the accumulating sediment load on top of continental or new oceanic crust.

Sediments of Pull-Apart Basins

In relation to the interaction of subsidence and sedimentation rate, the sedimentary facies of strike-slip basins may strongly vary. In their early stage of evolution, the deep basins tend to be filled by lake or sea water, because rivers or other sediment transporting systems are not yet adjusted to the newly formed morphologic depression. Later, modified or newly developed drainage systems of rivers or submarine canyons funnel larger volumes of sediment into the pull-aparts and keep pace with

or overcome subsidence. Mean sedimentation rates of known basin fills are very high ranging from 0.5 to 4 m/ka.

Another characteristic feature of many pull-aparts is their *asymmetric basin fill* displayed in both longitudinal and lateral sections. The basin margin bounded by a strike-slip fault is usually characterized by a steep fault scarp of high relief, while the opposite side is more irregular and morphologically less pronounced. Hence, lateral cross-sections mostly show deep downfaulting on the side of the dominant fault and less subsidence on the opposite side along the subordinate fault system (Fig. 12.36a and b).

Several of the modern examples on land (e.g., the Dead Sea, the Death Valley in California, the Erzincan Basin in Turkey) show a relief between 2000 and more than 4000 m on the strike-slip bounded side. Pull-apart basins below the sea display similar morphological features (e.g., the troughs of the Californian borderland, see below).

Intercratonic Pull-Apart Basins

These basins are frequently filled with fluvial and lake deposits, but these facies have a limited lateral extent (Fig. 12.36b). If an asymmetric basin is fed predominantly from its sides, the deeply downfaulted, high-relief side of the basin commonly displays marginal sediments consisting of locally derived coarse-grained alluvial fans and debris flows. By contrast, the opposite, subordinate margin is characterized by a wider zone of finer materials of alluvial plains built up by gentler streams. The remaining, more central part of the basin tends to be filled with deltaic sediments and various lake deposits depending on the climate of the region. Pull-aparts in humid climates may accumulate some black shales and peat. *External drainage* in conjunction with an open lake system (Chap. 2.5) or a through-flowing river may be replaced by an *internal drainage system* dropping its total dissolved and suspended load in the basin (cf. Chap. 12.1.3, Fig. 12.5).

Strike-slip motion is frequently accompanied by *volcanism* of variable nature including silicic and intermediate magmas. Hence, volcaniclastic sediments can contribute to the infill of such basins.

The central part of the 400 km long, but narrow Dead Sea-Jordan rift system provides an example of rapidly accumulated, up to 4000 m thick Pliocene salts, mostly halite. They indicate that the rift was connected with the open sea during that time. The evaporites interfinger with fan deltas,

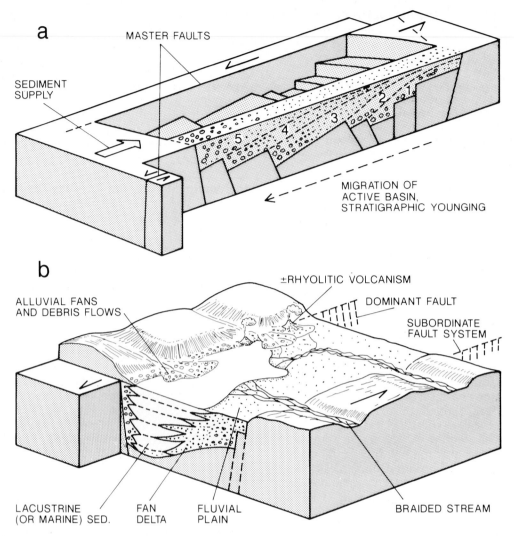

Fig. 12.36. a Generalized longitudinal section of pull-apart basin formed by sinistral strike-slip; major part of basin fill is removed to show basin floor. Depocenter migrates away from the direction of strike-slip motion and sediments young in direction opposite to sediment influx. **b** Cross section of pull-apart basin demonstrating asymmetry of basin floor and basin fill. (After Crowell 1974; Steel and Gloppen 1980)

fluvial sediments, lacustrine limestones and marls in a series of strike-slip-induced depressions (Manspeizer 1985; Kashai 1988). In the northern part of the transform zone, a 4000 m deep Pliocene section of alternating salts and gabbroid to basaltic flows was drilled. The deepest basin fills may reach a thickness of 8 to 10 km. The surface of the Dead Sea lies approximately 400 m below sea level because neither salts nor fluvial sediments and volcanic rocks could fill up the basin to a higher level.

If a pull-apart basin is fed from one end by a river sufficiently competent to keep it filled all the time, a depositional architecture similar to that shown in Fig. 12.36a may result. In this case, the depocenter migrates away from the direction of the strike-slip motion, and the sediments young upstream as shown by the dip of the strata (Crowell 1974; Steel and Gloppen 1980; Nilsen and McLaughlin 1985). Hence, the thickness of the dipping strata may turn out to be greater than the true basin depth. However, if a major river enters the basin from the opposite end, the depocenter tends to migrate downstream.

Marine Pull-Apart Basins

Most of these basins evolve along continental margins or on subsided continental blocks. Here, deposition may begin with comparatively deep-marine facies, including mass flows and turbidites. As they enlarge their drainage area and fill, they pass into shallow-marine and finally continental facies.

Neogene basins of this type are known from **southern California** (for example the Ridge basin, Crowell 1974; Link and Osborne 1982) and the **Californian borderland** (see Fig. 5.10 in Chap. 5.3.7 and Gorsline and Douglas 1987; Teng and Gorsline 1989). Further examples were described from the **Pannonian basin system** in Hungary and its neighboring countries (Chap. 12.7, Royden and Horváth 1988) and many other regions (Biddle and Christie-Blick 1985).

The intensively drilled **Vienna basin** is an example of the complex structure of a pull-apart basin caused by extension and left-handed shear with a strike-slip displacement of a few tens of kilometers (Fig. 12.34b and c; Royden et al. 1982). This basin is superimposed on early Miocene and older Alpine thrust sheets and has a maximum depth of 6 km (cf. Chap. 12.7).

Larouziere et al. (1987) and Montenat et al. (1987) describe Neogene basins of southeastern Spain which experienced, besides strike-slip motion and mafic to calcalkaline volcanism, a main phase of extension followed by compression.

Marine pull-apart basins far away from efficient sediment sources form special troughs which are considerably deeper than the neighboring sea floor. These basins tend to remain unfilled or only partially filled for considerable time spans. Their sedimentary facies may include slowly deposited pelagic sediments, submarine fans, and gravity mass movements. Such basins exist in the Gulf of Aden, the Red Sea, and the Gulf of California (see below and Chap. 4). The previously mentioned large Cayman trough in the Caribbean Sea has a mean depth of 5 km below sea level and contains a sediment cover of significant thickness only in its eastern part.

The Guaymas Basin Model (Gulf of California)

One of the submarine pull-apart basins of the Gulf of California, the Guaymas Basin, was investigated during the Deep Sea Drilling Project (Curray, Moore, et al. 1982a) and the results used to develop a model for this type of basin (Fig. 12.37; Einsele 1986).

After an initial shallow-marine "proto-gulf" stage, the present Gulf of California was formed along the strike-slip boundary between the North American and the Pacific plates. The dextral, somewhat transtensional plate motions started about 4 Ma B.P. and operated at an average rate of several cm/a. As a result, a series of pull-apart basins was generated (Fig. 12.37a and d) which are 2000 to 3000 m deep and separated by higher elevated sills (cf. Fig. 4.5).

During the first stage of basin evolution *(rift stage)*, the continental crust reacted by thinning, block faulting and listric faulting, which was accompanied by the injection of magmatic dikes. The subsiding crust was covered by shallow-marine to hemipelagic rift-stage sediments. This stage was followed by the accretion of new oceanic crust in the center of the pull-apart basins *(stage of early drifting)*. In contrast to the mid-oceanic spreading ridges of mature oceans, the spreading centers of these pull-apart basins are mostly situated in the morphological basin deep, thus forming *spreading troughs*. In spite of the fact that these troughs collect large amounts of hemipelagic sediments, turbidites, and mass flows from nearby slopes and marine deltas, they remain open due to the continuing growth of the basins.

Only in the northern part of the gulf and in the Salton Trough in southern California is the sedimentation rate so high that the underlying pull-apart basins and their spreading troughs are filled up (Fig. 12.37e; Fuis and Kohler 1984) and no longer have a topographic expression.

The *interrelationship between spreading rate (strike-slip) and sedimentation rate* in such basins is demonstrated by the idealized models of Fig. 12.37b and c.

Assuming a 4 km wide spreading trough and a (realistic) sedimentation rate as high as 3000 m/Ma in this trough, then a 4 km wide and 3 km high sediment body can accumulate during 1 Ma. However, due to simultaneous accretion of oceanic crust at a rate of 6 cm/a = 60 km/Ma, this sediment volume is spread out to a width of 60 km. Thus, its thickness is reduced to 200 m. In other words: A sediment volume element E of 200 m thickness (Fig. 12.37c) migrates laterally with the half-spreading rate and reaches a position 30 km away from the spreading center after 1 Ma. On its path to the new position, it will be covered by coeval flank sediments which are assumed to accumulate at a rate of 1000 m/Ma (Fig. 12.37b).

Further evaluations of this problem utilizing various other assumptions indicate that Gulf of California-type spreading troughs can be maintained under sedimentation rates of up to several km/Ma, if the rate of strike-slip and crustal accretion is also high. In this case, the basaltic magma rarely extrudes on the sea floor, but rather generates a sill-sediment complex of several hundreds of meters in thickness between sheeted dikes (oceanic layer 2) and the sediment cover (layer 1, also see Chap. 13.5).

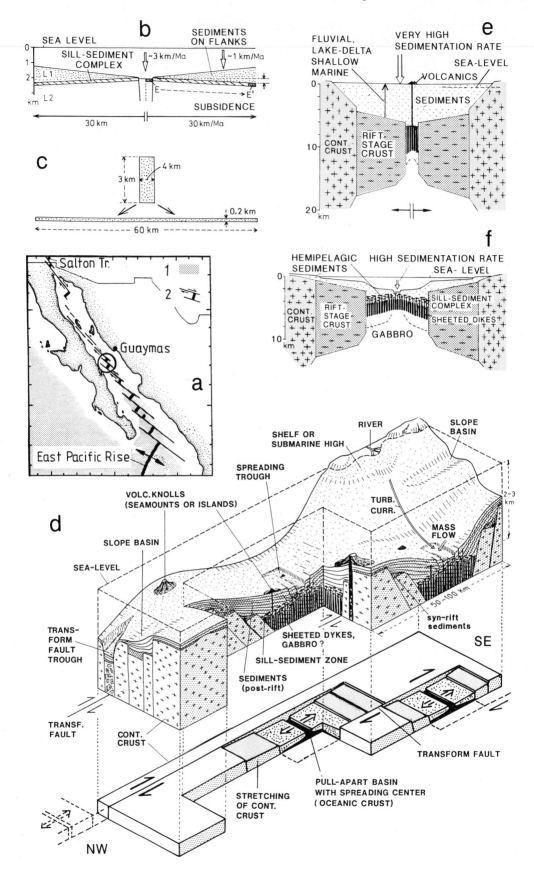

b

SEA LEVEL

SILL-SEDIMENT
COMPLEX

SEDIMENTS
ON FLANKS

~3 km/Ma   ~1 km/Ma

L1

L2

SUBSIDENCE

30 km   30 km/Ma

E   E'

c

4 km

3 km

0.2 km

60 km

e

FLUVIAL,
LAKE-DELTA
SHALLOW
MARINE

VERY HIGH
SEDIMENTATION RATE

SEA-LEVEL

VOLCANICS

SEDIMENTS

CONT.
CRUST

RIFT-
STAGE
CRUST

f

HEMIPELAGIC
SEDIMENTS

HIGH SEDIMENTATION RATE

SEA-LEVEL

CONT.
CRUST

RIFT-
STAGE
CRUST

GABBRO

SILL-SEDIMENT
COMPLEX

SHEETED DIKES

a

Salton Tr.

1

2

Guaymas

East Pacific Rise

d

RIVER   SLOPE
BASIN

SHELF OR
SUBMARINE HIGH

SPREADING
TROUGH

TURB.
CURR.

MASS
FLOW

VOLC.KNOLLS
(SEAMOUNTS OR ISLANDS)

SLOPE BASIN

SEA-LEVEL

50–100 Km

syn-rift
sediments

SE

TRANS-
FORM
FAULT
TROUGH

SHEETED DYKES,
GABBRO?

SILL-SEDIMENT ZONE

SEDIMENTS
(post-rift)

TRANSF.
FAULT

CONT.
CRUST

NW

STRETCHING
OF CONT.
CRUST

PULL-APART BASIN
WITH SPREADING CENTER
(OCEANIC CRUST)

TRANSFORM FAULT

If the *sedimentation rate is extremely high* and surpasses the spreading rate, the pull-apart basins are filled up and become continental (Fig. 12.37e).

The Pliocene to Quaternary infilling of the **Imperial Valley depression** to the north of the Gulf of California is 5 to 6 km thick and begins with an approximately 1000 m thick shallow-marine to brackish sequence overlain by very thick, mainly lacustrine and fluvial deposits (Quinn and Cronin 1984). Their basement, consisting of rift-stage crust and newly formed oceanic crust, is depressed under the high sediment load. Magma of basaltic to acidic composition rarely extrudes at the surface of the alluvial plain.

By contrast, low sedimentation rates in conjunction with normal spreading rates (1 to 10 cm/a) lead to very thin sediment covers on top of newly accreted crust.

The tectonic setting, depositional environment of the **Bündnerschiefer in the Alps**, and the intrusion of basaltic magma in the Mesozoic Northern Penninian (Valais) trough may have been similar to the conditions and processes in the modern Gulf of California (Kelts 1981).

Concluding Remarks

The number of known modern and ancient strike-slip zones and pull-apart basins is steadily increasing with the progress of geological investigations on land and below the sea. Similar to the Pannonian basin system mentioned above, other intramontane basins may have been affected by extension and transform faulting. A new evaluation of the sedimentary basins of China (Hsü 1989; Zhu 1989), for example, reveals that several of these basins or parts of them can be interpreted as pull-apart basins. Strike-slip motions also occur in conjunction with oblique rifting and on divergent continental margins. They are probably common on convergent margins with oblique subduction as for example observed in New Zealand (Lewis 1980), but the resulting basins and their sedimentary fills

are complex. Ancient examples of this category are usually strongly overprinted by subsequent compressional tectonic events and metamorphism, or the former basin sediments were eroded upon uplift.

The sedimentary fills of many pull-apart basins, particularly those accumulated in continental environments, may be distinguished from rift basin sediments by higher sedimentation rates, greater thicknesses in limited areas, and greater structural complexity (Hempton and Dunne 1984). Displaced fan/ source relationships, the migration of the depocenter parallel to the master faults, and special tectonic features provide evidence for strike-slip motion.

Most of the pull-apart basins have a short life span on the order of a few million years, because their growth is terminated by superimposed tectonic movements unrelated to the basin-forming processes. Their length is therefore mostly limited to several tens to hundreds of kilometers.

If their depositional environment was favorable for the accumulation of organic matter and subsequent tectonic overprint was moderate, pull-apart basins are interesting targets for hydrocarbon exploration in spite of their limited size. Rapid subsidence, sediment accumulation, and burial of source rocks favor the preservation and maturation of organic matter in these basins more effectively than in many other basin types (cf. Chap. 10.3.3).

12.9 Basin-Type Transitions (Polyhistory Basins)

General Aspects

The tectonic style of a sedimentary basin may change over time, as already indicated in the previous chapters. Rift basins, for example,

Fig. 12.37a-e. Interaction of sedimentation and strike-slip-induced spreading in Gulf of California-type pull-apart basins. **a** Location of Guaymas Basin (*1* water depth ≥2000 m; *2* transform faults and spreading centers). **b** Idealized cross-section of submarine pull-apart basin with spreading trough and flanks showing its evolution during 1 Ma (*thick arrows* indicate sedimentation rates; spreading rate is 6 cm/a; vertical scale exaggerated 4x). Note that after 1 Ma the basin will have grown in length by 60 km; sediment volume element *E* of the spreading trough will have moved to *E'* and have subsided under the load of

1 km thick flank sediments. **c** As **b**, during 1 Ma a body of trough sediments, 3 km thick and 4 km wide, is distributed by basin spreading to form a 60 km wide blanket of 200 m thickness (sediment compaction neglected). **d** Simplified plan view and block diagram of gulf-type pull-apart basins in stage of early drifting. Formation of sill-sediment complexes in spreading troughs. **e** and **f** Pull-apart basins in stage of early drifting subject to high (**e**) and very high (**f**) sedimentation rates. See text for further explanation. (After Einsele 1986)

frequently evolve into continental margin basins (Chaps. 8.1 and 8.2), or remnant basins
(with flysch deposits) exhibit transitions into
peripheral foreland basins (with molasse deposits; Chap. 12.6). Basins displaying such transitions are referred to as "polyhistory basins"
(e.g., Kingston et al. 1983; Klein 1987). The
early and even young sediments of these basins are commonly deformed by syndepositional or subsequent basin transformation (cf.
Chap. 1.3).

The evolution of polyhistory basins follows
the general principles of plate tectonics. Ocean
basins open along a rift zone, grow by accretion of new oceanic crust to form a deep
ocean basin, and close again, when the subduction rate of oceanic crust exceeds the
spreading rate. Subduction may finally result
in a continent-continent collision (cf. Figs. 1.1
and 1.2). This life cycle of an ocean basin is
called *Wilson cycle* (Wilson 1965, 1968; Dewey
and Bird 1970) and may last for a few hundred million years. A specific region may undergo several Wilson cycles during its evolution. The most important stages of this concept are summarized in Table 12.1

This simplistic scheme of the Wilson cycle
does not take into account several modifications, for example: (1) the "mature ocean" can
also be relatively narrow before it starts to
close again, as assumed for the Alpine orogeny
(Frisch and Loeschke 1986). (2) Micro-continents and terranes (cf. Chap. 1.2) may produce various regional complications, partially
associated with plate rotations, as known from
the Mediterranean. (3) There is an increasing
amount of evidence that strike-slip movements and wrench faulting play a significant
part in basin evolution related to the stages of
the Wilson cycle.

Basin-Transition Models

Passive Continental Margins

Figure 12.38 shows some straightforward
models of basin-transitions on a passive continental margin, i.e., from the embryonic to
the mature stages of the Wilson cycle.

The early rift stage with predominantly
continental deposits (Chap. 12.1.3) may be
followed by a shallow-marine protogulf stage,
in which tectonic basin subsidence is still controlled by thinned continental or transitional
crust (Fig. 12.38a). With the onset of ocean
spreading (b), the basin develops an elongate
central trough (as in the Red Sea) or several

individual small troughs (Gulf of Aden, Gulf
of California), where hemipelagic sediments
and sediments of various gravity mass movements accumulate. Ascending basaltic magma
may form dikes and sills in these rapidly accumulating first deep-water sediments (Chaps.
4.3 and 12.8.3).

In the subsequent stages (Fig. 12.38c
through e), the sedimentary history of the
passive margin basin is characterized by the
aggradation and outbuilding of sediments on
the shelf, the continental slope and rise (Chap.
12.2). In the early stage of this evolution the
shelf tends to be narrow and its sediments
largely consist of clastic material including
sands. Later, frequently shales and carbonate-
buildups take over depending on the climatic
situation and presence of large rivers. The
depositional area may include coastal regions
which in earlier stages acted as sediment source or were sediment-free. In a final stage, a
wide, very thick embankment may prograde
seaward beyond the oceanic/continental crust
boundary and deeply bury older sediments
under its load.

Basin Transitions on Convergent Margins

Basins evolving from the declining to the relic
scar (suture zone) stage of the Wilson cycle,
are more diversified and complicated. Figure
12.39 displays, for example, the closing of an
ocean basin between two continental plates
with a subduction-related szenario at the
right-hand side. The terminal stage of the
ocean (Fig. 12.39a) exhibits basin-type transitions from

1. Passive margin basin with a shelf-continen-
   tal slope-rise association to remnant basin
   with flysch-type sediments.
2. Shelf basin along active margin to forarc
   basin, and possibly
3. Deformed sediments of accretionary wedge
   to undeformed slope basin sediments.

Terminal and Relic Basin Development

After the closure of the remnant basin and the
formation of fold-thrust belts both on top of
the downgoing passive margin plate and possibly in the backarc region, several types of
narrow, shallow-marine or continental basins
are left. One can identify the following basin-
type transitions (Fig. 12.39b):

Table 12.1. Stages of the Wilson cycle and basin evolution. (After Wilson 1968, modified)

Stage	Basin type (modern examples)	Motions, volcanism, formation of terrig. sediment sources	Characteristic depositional environments and sediments (depending on climate)
Embryonic	Continental rift (East African Rift)	Frequently updoming and block uplift, extension and strike-slip; alkaline, mostly basic, but also more silicic volcanism	Continental and lacustrine, partially also marine, clastics, volcaniclastics, (red beds and evaporites)
Young	Narrow ocean (Red Sea, Gulf of Aden, Gulf of California)	Incipient ocean spreading, rapid subsidence of basin, continued uplift of nearby continental regions, tholeiitic basalts	Marine, narrow shelves, one central trough or several small spreading troughs, variable sediments depending on connection to world ocean, entering rivers, etc.
Mature	Wide ocean (Atlantic Ocean)	Spreading, mid-oceanic ridges, continued subsidence of basin flanks, tholeiitic ocean floor, alkali basalt islands, distant terrigenous sediment sources	Wide shelves with thick clastic and carbonate sediments, deep-sea fans and basin plains, large areas with pelagic clay and biogenic oozes
Declining	Narrowing of ocean basin (Pacific Ocean)	Convergence and marginal uplift, island arcs or or Anden-type active margin, basaltic and andesitic volcanics	Deep-sea trenches, forarc and backarc basins, flysch-type and pelagic sediments, volcaniclastics
Terminal	Ocean basin strongly reduced in size (Mediterranean Sea)	Convergence and uplift, young orogenic belts, andesitic volcanics and rising granodiorite-gneis complexes	Remnant basins (flysch) and foreland basins (marine and continental) accumulate clastic wedges, volcaniclastics, (red beds and evaporites)
Relic scar	Suture zone (Indus-Tsangpo suture, Himalayas)	Collision and continued uplift of orogenic belt, exposed granitic and metamorphic rocks	Final stage of foreland basins with continental clastic wedge, intramontane molasse (red beds)

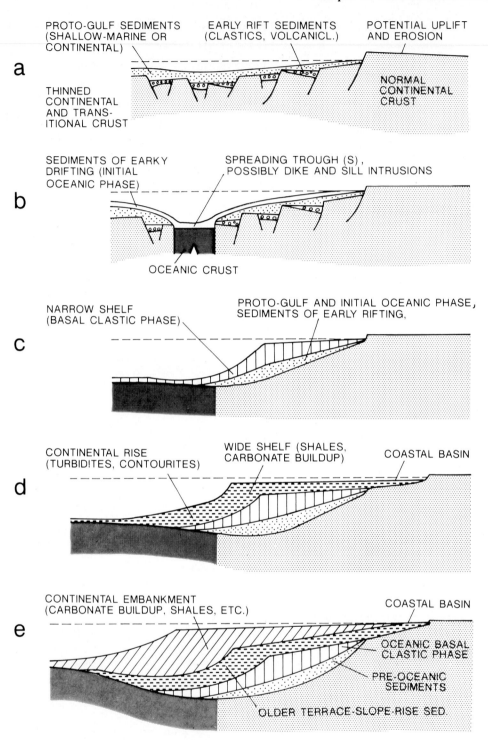

Fig. 12.38a-e. Basin evolution from **a** early rift stage and proto-gulf to **b** initial oceanic phase and **c, d, e** several stages on passive continental margin. Note increase in shelf width (including coastal basin) and sediment thickness, as well as onset of slope and continental rise deposition. Not to scale (crustal thicknesses reduced), tectonic modifications omitted. (Largely after Dickinson and Yarborough 1976)

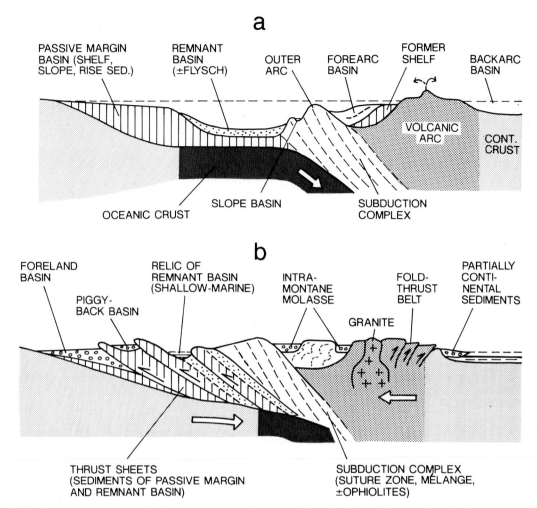

Fig. 12.39a,b. Basin evolution during two stages of a closing ocean. **a** Remnant basin between passive margin with wide shelf and subduction-related basins. **b** Closed oceanic basin and onset of collision of two continental (or transitional) plates. Note various basin-type transitions; folding, strike-slip, and other structural modifications are omitted; not to scale. See text for further explanation. (Based on various sources)

4. Transition from deep remnant basin into shallow-marine *relic basin*.
5. Passive margin basin (inner shelf) to peripheral foreland basin, or deformed shelf basin (overthrust) to piggy-back basin (Chap. 12.6.2).
6. Forearc basin to intramontane or intra-arc basin (continental molasse).
7. Marine backarc basin to predominantly continental retroarc basin with molasse.

These few examples of basin-type transitions do not take into account syn-depositional and post-depositional deformations, as well as wrench tectonics, which may strongly modify basin types and their sedimentary fill (e.g., Kingston et al. 1983).

Polyhistory basins may be subdivided further into the following groups (Klein 1987):

- *Superposed basins* which develop on a suture zone.
- *Successor basins* consisting of changing basin types along a zone of long-persisting tectonic "weakness".
- *Resurgent basins* in situations, where the original basin style is repeated and maintained along older tectonic boundaries and trends.

The unraveling of the "polyhistories" of such basins is one of the tasks of regional and structural geology, but this can only be done by using and testing basic concepts.

Sediment Preservation Potential of Various Basin Types

It is important to note that mountain belts resulting from continental collision preserve only part of their sedimentary history. The sediments of early rifting and drifting as well as post-rift sediments on continental crust have a relatively good chance of being preserved. In addition, part of the record of the terminal and final stages, i.e., sediments of remnant basins, foreland basins, and intramontane basins, may escape erosion and trans-

formation. However, the sediments of the intervening oceanic period are largely vanished or obscured by subduction, accretion, and metamorphism.

Below the present-day oceans, the oldest oceanic crust with oceanic sediments is Jurassic in age. It is found only in marginal regions, whereas most deep-sea sediments and their underlying crust are younger. This signifies that older *deep-sea sediments* only occur in large accretionary complexes and mountain belts, where they are commonly highly deformed and altered by metamorphism.

Part V
Diagenesis and Fluid Flow

13 Mechanical and Chemical Diagenesis

13.1 General Aspects of Mechanical and Chemical Diagenesis

Introduction

Freshly deposited sediments are commonly unconsolidated, have a relatively low bulk density and high permeability and, if accumulated under water, a high water content. However, with increasing burial depth under younger sediments, and occasionally shortly after deposition, the sediments become denser and more solid or lithified. All the processes involved in such a change of sediment state are summarized under the term *diagenesis*. It comprises both mechanical and chemical-mineralogical processes.

However, if sediments, for example clays, become very deeply buried and/or are affected by relatively high temperatures, they start to re-crystallize and form new minerals. In addition, their porosities drop to values around 3 % and less. This occurs to a large extent in the so-called *low-grade greenschist facies* ($\geq$200 °C). This phenomenon is referred to as *metamorphism* (see below). A well-defined boundary between diagenesis and metamorphism, however, cannot be set.

The phenomena at the transition from diagenesis to metamorphism also depend on the tectonic setting of a basin, i.e., whether baric or thermal effects predominantly control this evolution (Robinson 1987).

Some authors, particularly those dealing with the transformation of organic matter into hydrocarbons and coal (Chaps. 14.2 and 14.4), distinguish a zone of *catagenesis* or an *anchimetamorph zone (anchizone)* between diagenesis and metamorphism. At this depth range, most sediments have become lithified, shales and claystones are transformed into slates, smectite and mixed-layer clay minerals are replaced by well-crystallized illite and chlorite, and vitrinite reflectance is high, but temperatures are only 100 to 150 °C (e.g., Krumm et al. 1988).

Mechanical diagenesis results from vertical and accompanying lateral stresses caused by the overburden load of younger sediments, and possibly by additional stresses due to compressional tectonic movements. It expels pore water and leads to a rearrangement of the sediment particles (fabric).

Chemical-mineralogical or, simply, *chemical diagenesis* includes dissolution and recrystallization of primary sediment particles, as well as repre-cipitation of dissolved matter as cement in the pore spaces. Dissolution and cementation may take place at different levels within the sediment and thus involve mass transfer from one layer to another. Chemical diagenesis is driven by the tendency of solid matter to reach thermodynamic stability under the changing conditions in the subsurface (increasing temperature and pressure, possibly changing pore water chemistry). As a result, a limited number of *stable minerals* are formed at the expense of various unstable phases. In old and deeply buried shales, for example, the illite-chlorite-quartz mineral assemblage is common, regardless of the initial clay mineral association. Calcite and dolomite replace less stable skeletal carbonates, such as aragonite and high Mg clacite. Furthermore, all sorts of minerals having a large surface area tend to dissolve and reprecipitate into more compact particles and thus to achieve a state of lower energy.

All these processes promote the *reduction of porosity and permeability*, enhance the bulk density, and affect the strength of the rock. The loss in porosity and thus in sediment thickness is referred to as *compaction*. Compaction of a water-saturated sediment is always associated with the expulsion of pore water, which is frequently called *compaction flow*.

The effects of both mechanical and chemical diagenesis in various sediment types are very complex and can be studied and described in several ways, as for example by Engelhardt (1973), Bathurst (1975), Magara (1978), Larsen and Chilingar (1979), Parker and Sellwood (1983), McDonald and Surdam (1984), Gautier (1986), Mumpton (1986), and Füchtbauer (1988). In this text, only a brief summary is given, in which the aspects of compaction and pore water flow are emphasized, while the various chemical and mineralogical processes are little addressed; the latter aspects have been frequently and extensively described in many special articles and textbooks. An introduction to carbonate and silica diagenesis is given in Chapters 3.3.2 and 5.3.5.

Principles of Mechanical Diagenesis

Subaqueous mechanical diagenesis starts immediately below the sediment-water interface and progressively alters all the mass physical properties of the sediments versus depth. It controls or influences the porosity, water content, bulk density, permeability and pore water flow, shear strength and rheologic behavior, sensitivity to liquefaction and, finally, mass transfer via molecular diffusion and advection.

In *normally consolidated* sediments, the fabric strength or compaction state is in equilibrium with the overburden pressure at all depths below the sediment-water interface (Fig. 13.1a). The overburden pressure can be expressed as total lithostatic pressure (see below) or as effective vertical stress (reduced by buoyancy) per unit area of a horizontal plane within the sediment. In reality, such an effective vertical stress acts as a grain-to-grain stress at the contact of individual particles (see below). The state of compaction can be defined in terms of porosity, water content, or bulk density. The pore water pressure is equal to the hydrostatic pressure at the corresponding depth; hence, if sediment accumulation has ended, there is no upward or laterally directed pore water flow. However, if a new increment of sediment is added on top of such a sedimentary column (Δ h in Fig. 13.1a), this equilibrium is disturbed. In order to establish a new equilibrium between the increased vertical stress and the compaction state, some pore water must be expelled (see below). In a fairly permeable sediment, this pore water expulsion takes place more or less simultaneously with the growth of the sedimentary column, i.e., the sediment maintains its state of normal consolidation. Normal consolidation can also be maintained in fine-grained, low-permeable sediments if the sedimentation rate is low and the pore water has sufficient time to escape.

If, by contrast, the sedimentation rate is high and the growing sedimentary column has a low permeability and becomes less and less permeable with depth, the pore water to be released by prograding compaction cannot escape readily to the sediment-water interface. Hence, compaction is delayed (Fig. 13.1b), and the sediment is in a state of *underconsolidation*, in which fabric strength is less than that which would result from the total overburden pressure. In this state, excess pore water pressure builds up, which may approach and counteract the effective vertical pressure induced by the (buoyancy-reduced) weight of solid matter in the sediment. Such a state signifies that the weight of the overlying sedimentary column is no longer carried alone by the solid grain structure (grain-to-grain stress), but to a great part or nearly entirely by the practically incompressible pore water. This condition causes a drastic reduction in shear strength (friction) and is therefore one of the most important factors generating gravity mass movements (Chap. 5.4.1).

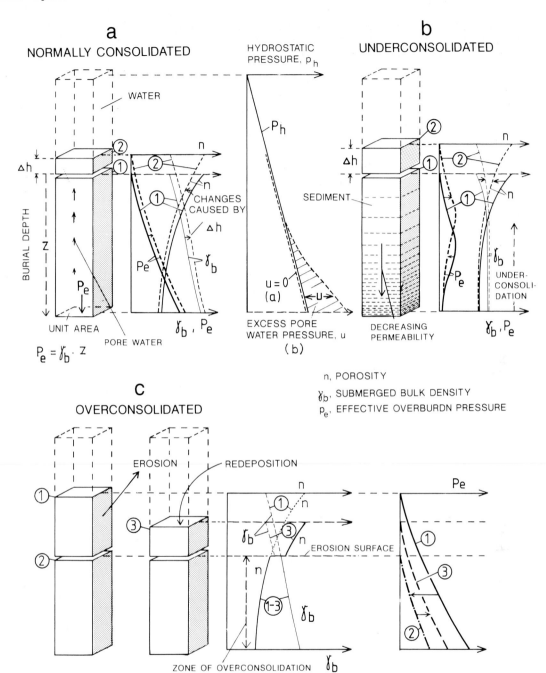

Fig. 13.1a-c. Normal consolidation (**a**), underconsolidation (**b**), and overconsolidation (**c**) of subaqueous fine-grained sedimentary columns. *Encircled numbers* indicate subsequent stages of evolution. **a** Newly added sediment increment, Δh, disturbs equilibrium between vertical effective stress, p_e, and compaction state of sediment at all depths, z. However, rapid escape of pore water adjusts new equilibrium, exemplified by shifting curves of porosity (or water content), and bulk density, γ_b. **b** Downward decreasing permeability prevents the escape of pore water released by additional sediment load, Δh, and thus hampers further compaction. Consequently, porosity and bulk density barely decrease versus depth, and most of the sediment load is carried by the pore water. The resulting excess pore water pressure, u, counteracts the vertical pressure exerted by the sediment load. **c** After erosion (Stage *2*) the remaining sedimentary column is overconsolidated. Further consolidation can only take place if redeposition (Stage *3*) surpasses the former top of the sequence. Note that during Stages *1* to *3*, porosity (or water content) and bulk density remain approximately constant in the zone of overconsolidation, but the vertical stress varies. See text for further explanation

Cohesive, commonly clay-bearing sediments, may exhibit a state of mechanical *overconsolidation*, i.e., a state in which their fabric strength is higher than under equilibrium with overburden pressure. This frequently occurs when the upper part of the sedimentary column is removed by erosion (Fig. 13.1c). Due to a kind of "memory effect", the underlying sediment maintains its compaction state achieved under the former higher overburden pressure, i.e., too high a bulk density and too low a porosity with respect to the lowered sediment-water interface. The erosion-induced reduction in vertical stress has practically no influence on the compaction state. Furthermore, renewed deposition on top of

the erosion surface (Stage 3 in Fig. 13.1c) does not affect the compaction state of the underlying material as long as the newly added sediment does not surpass the former top of the sequence.

Recent marine sediments frequently show apparent overconsolidation, discernible, for example, by shear strength values higher than normal for this depth, at or slightly below the sea floor. This phenomenon is not related to submarine erosion, but to the onset of chemical diagenesis (the beginning of cementation).

As sediment builds up over time, a fine-grained sequence may undergo changes in its consolidation state, as demonstrated in the model in Fig. 13.2. Emergence of sediments

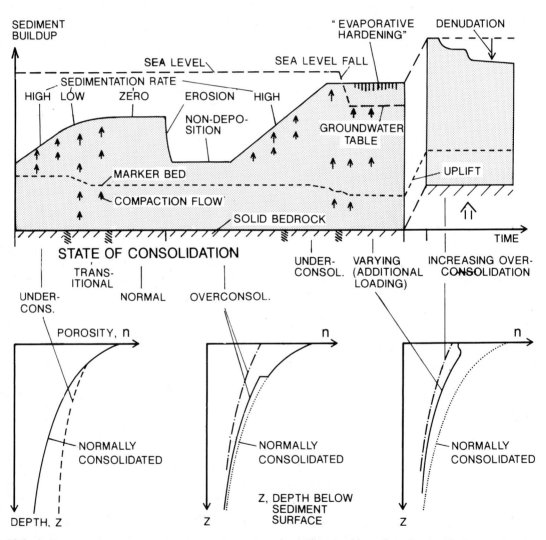

Fig. 13.2. Sediment buildup versus time, displaying different consolidation states for fine-grained material in relation to sedimentation rate, erosion, emergence caused by falling sea level, and tectonic uplift with denudation on land. Note position of marker bed indicating the progress of compaction associated with compaction flow. All the porosity-depth curves represent sediment of equal, homogeneous composition

above sea level and the groundwater table leads to an increase in effective overburden pressure and therefore initiates renewed compaction or underconsolidation. Water loss by evaporation near the land surface, associated with high capillary pressures, can produce "evaporative hardening" of fine-grained materials. Present-day erosion on land generally leads to overconsolidated shales and mudrocks, while earlier erosional events followed by subsequent deposition mostly become masked. Underconsolidation and excess pore water pressures in the subsurface are well known phenomena in the exploration of hydrocarbons and deep groundwater (see below).

Mechanical and Chemically Modified Compaction (Overview)

The aim of this brief overview is to show that there is hardly a case in nature where the porosity-depth relationship of a thick, usually nonuniform sedimentary column is controlled only by gravitational, mechanical compaction. The pronounced compaction of fine-grained sediments, for example, is caused by both mechanical and chemical-mineralogical processes, while coarser-grained materials (sand and gravel) are but little affected by purely mechanical diagenesis.

Purely mechanical (or gravitational) compaction of a fine-grained, mainly siliciclastic sediment may lead to a porosity-depth relationship as indicated in Fig. 13.3a. As stated by many authors (e.g., Chilingarian 1983, Brückmann 1989), the absolute values of porosity at certain burial depths are reversely related to grain size (Fig. 13.6, siliciclastic sediments). Clays rich in smectite, montmorillonite, and mixed-layer clay minerals generally exhibit higher porosities than illitic or kaolinitic clays. Clays rich in exchangeable Na^+, for example montmorillonitic material, usually bind more pore water and hence have higher porosities than Ca^{++}-rich clays. The porosity-depth curve of a *normally consolidated*, fine-grained sequence can, as mentioned above, be modified by *underconsolidation* or *overconsolidation* (Fig. 13.3c and d).

In Figure 13.3a, the regularly decreasing porosity versus depth is only interrupted by intercalated sands (thick beds or relatively thin sandy turbidites) which cause a drastic drop in the porosity curve in the upper part of the sequence. However, at greater burial depth, as long as the sands are not affected by pressure solution and cementation (see below),

they may maintain a higher porosity than their neighboring shales and mudrocks.

If the same sequence is also affected by chemical diagenesis, the porosity-depth relationship may be modified in such a way as that shown in Fig. 13.3b. At a burial depth of several hundreds of meters, the less stable portions of the fine-grained material, for example various bioclasts, start to dissolve and reprecipitate as pore cement. At greater depths, unstable clay minerals "mature" into stable ones and lose their crystal water (dehydration). As a result, the porosity curve shifts to lower values. Sandy interbeds in particular undergo shallow-burial cementation and, in addition, deep-burial pressure solution at the contacts of individual grains. Both processes strongly reduce the pore space and permeability of such beds.

Pelagic carbonates, consisting mainly of calcareous nannofossils, tend to display a step-like decrease in porosity and strength (and other properties) versus depth (Fig. 13.3e). Partial dissolution and recrystallization of calcium carbonate as cement already starts in the soft nanno ooze (Fig. 13.3f; Schlanger and Douglas 1974; Matter et al. 1975), but becomes more important in the deeper regions where firm chalk and hard limestone are encountered. While the nanno chalk is still highly porous, the pelagic limestone has lost most of the porosity of the initial nanno ooze at a depth, where a siliciclastic mud would maintain a considerably higher porosity (Fig. 13.3a). The depths of the transition zones between ooze, chalk, and limestone also depend on the temperature gradient within the sedimentary column (Wetzel 1989). A high gradient furthers chemical diagenesis and thus a shift of the boundaries into shallower burial depths. Normally, calcareous ooze is converted into chalk at burial depths of at least 150 to 300 m (Garrison 1981). The chalk-limestone transition commonly occurs at burial depths between 500 and 1000 m (e.g., Neugebauer 1974) and thus takes a very long time.

Siliceous sediments rich in opaline silica exhibit a behavior similar to that of nanno oozes, but commonly have a higher initial porosity. Opal A is generally transformed into opal CT, in conjunction with a marked decrease in porosity, at depths below 100 to 300 m (Kastner 1981), but high geothermal gradients lead to lithification of these sediments at considerably shallower burial depths (see also Chap. 5.3.5 and Fig. 5.6). Beginning shallow-burial lithification of layers rich in diatoms, which originate from short upwelling periods

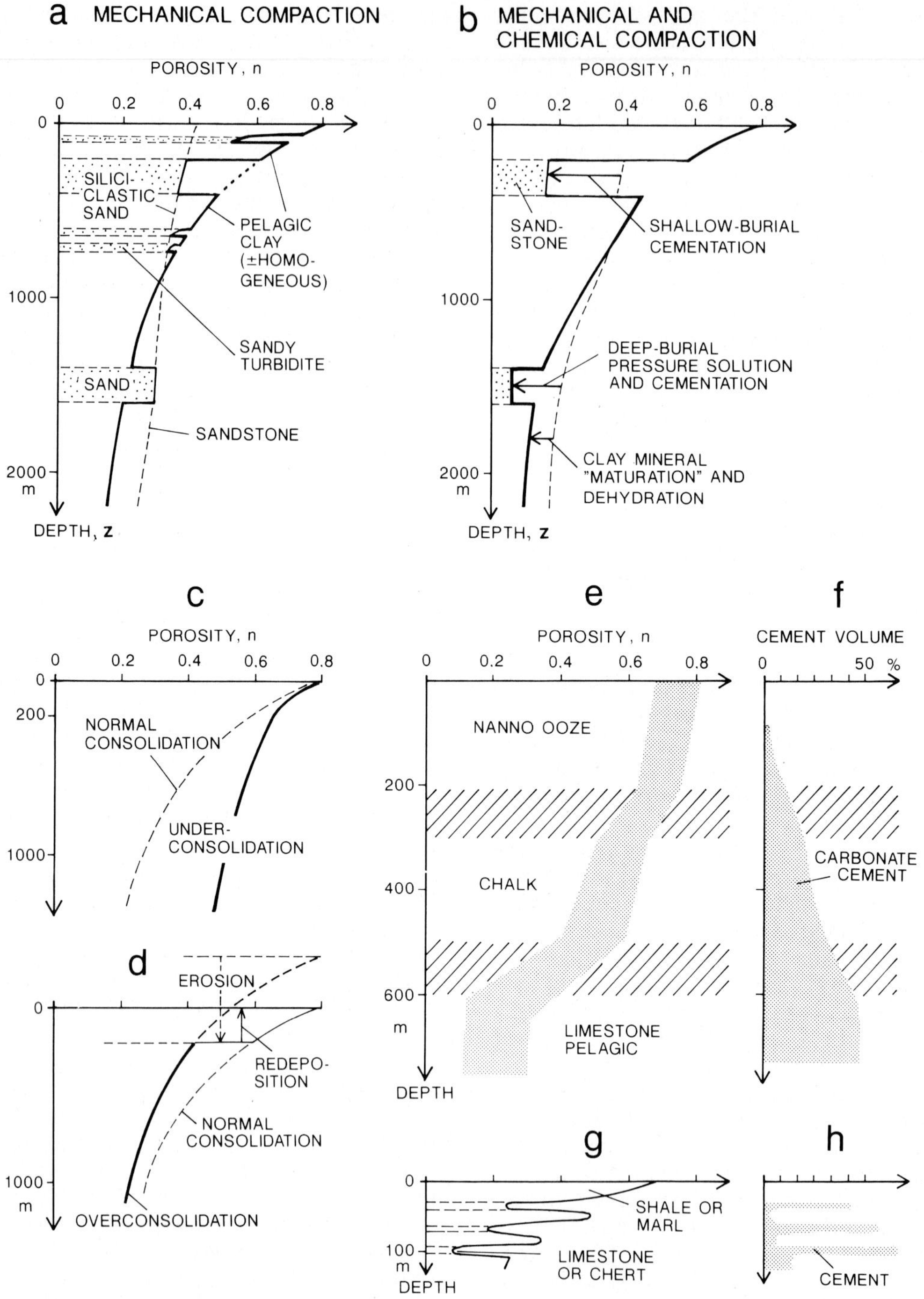

Fig. 13.3a-h. General porosity-depth relationships of various marine and lacustrine sediments; depth (z) below sediment-water interface. **a** Mechanical (gravitational) compaction of silty clay and sands. **b** Mechanical compaction (equal to **a**) is enhanced by chemical-mineralogical reduction of porosity. **c** Underconsolidation of low-permeable sediments resulting from delayed escape of pore water, as compared to normally consolidated material. **d** Overconsolidation due to erosion of thicker overlying sediments prior to the accumulation of the present (thinner) sediment column. **e** and **f** Transition from pelagic nanno ooze to chalk and limestone accompanied by mechanical and chemical compaction (dissolution and reprecipitation of carbonate as cement; after Matter et al. 1975; Wetzel 1989; Hobert and Wetzel 1989); cement volume in **f** and **g** is related to total sediment volume. **g** and **h** Shallow-burial onset of carbonate (or silica) diagenesis in shelf and platform sediments (containing unstable mineral phases), generating extreme fluctuations in porosity versus depth

during the late Pleistocene, have also been found in the Norwegian Sea (Kassens 1990).

In shallow-marine sediments, chemical diagenesis frequently starts at shallow burial depths and proceeds very irregularly (Fig. 13.3g and h). Layers containing higher proportions of unstable, relatively reactive components, such as *aragonitic bioclasts, high Mg calcite, and opaline silica,* become cemented and lithified earlier than their clay-rich interbeds which maintain a rather high porosity. Instead of continuous hard beds, layers of isolated carbonate concretions or chert nodules can be formed. The role of microbial decomposition of organic matter on the formation of carbonate nodules is briefly described in Chapter 13.3.

There are numerous other factors influencing sediment diagenesis: Sediments rich in finely dispersed organic matter, for example lake gyttjas or Black Sea muds, have particularly high initial porosities. Clayey sediments deposited in freshwater lakes tend to develop higher initial porosities than marine clays, but they can be surpassed by marine diatomaceous and radiolarian oozes (see below). When exposed to freshwater circulation, carbonates especially tend to become rapidly lithified. In slowly reactive and low permeable sediments, such as shales devoid of carbonate, age is an important factor in diagenesis, because the escape of both the excess pore water and water released by dehydration of clay minerals as a result of their transformation into stable phases takes a long time.

In most cases, the effects of mechanical compaction are more or less overprinted by chemical diagenesis, particularly in sediments with large proportions of relatively highly soluble, unstable components. Variations in temperature gradients and differences in age further modify the porosity-depth relationship in sediments of similar composition. For all these reasons, the porosity-depth curves summarized in many articles and special volumes mentioned above display a wide scatter, even within sediments of approximately identical composition. It is therefore frequently not possible to successfully describe the porosity-depth relationship and compaction of natural sediments in a general way by simple rules or equations (see below). This can only be done for compaction or consolidation experiments in the laboratory, and to some extent for young, shallow-buried sediments below the present sea or lake floors.

Some examples of such *porosity-depth curves* which reflect primarily mechanical compaction are shown in Fig. 13.4. The highest initial porosities are in sediments rich in diatoms and radiolarians, as long as the delicate opal skeletons resist mechanical and chemical destruction. They can maintain porosities higher than 0.7 (70 %) up to several hundreds of meters below the sea floor. In addition, other biogenic oozes may display abnormally high initial porosities as a result of their intratest porosity. This may increase the total sediment porosity by approximately 35 to 45 % (Bachman 1984), as compared to a sediment consisting simply of solid grains of the same size. Pelagic carbonates consisting mainly of nannofossils and foraminifers, however, exhibit about the same initial porosities as siliciclastic marine muds with minor proportions of microfossils and organic matter.

13.2 Compaction, Compaction Flow, and Other Flow Mechanisms

Analytical Equations to Describe Compaction

In spite of the complexity of sediment compaction and the constraints mentioned above, several authors have made an attempt to describe the porosity-depth relationship or consolidation of sediments by simple analytical equations. They were encouraged by the observation that the reduction of porosity versus depth takes place in a more or less exponential fashion, i.e., the porosities of succeeding depth intervals decrease systematically. Figure 13.5a through d illustrates four different ways to approach this problem: (1) Soil engineers (Terzaghi 1925, Fig. 13.5a) have learned from laboratory consolidation tests that the compaction of clayey materials renders straight line relationships, if the void ratio, e, is plotted versus the logarithm of the effective overburden pressure, p_e (see above and Fig. 13.1). The void ratio, e, is the ratio between the volumes taken by the voids and the solid material of a sediment sample (cf. Fig. 13.6a, Eq. 3). Because p_e is not a linear function of depth, z, and therefore not very convenient to use in the study of sedimentary columns with downward-increasing bulk density, Athy (1930) proposed plotting porosity versus log z according to the equation

$$n_z = n_o \cdot e^{-\beta z} \quad , \qquad (13.1)$$

where n_o = initial porosity
            n_z = porosity at the depth z .

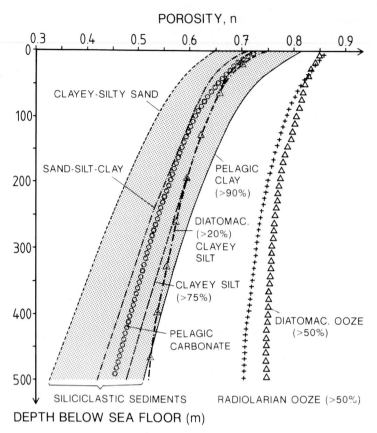

POROSITY, n

Fig. 13.4. Generalized porosity/depth type-curves for young marine sediments little affected by chemical diagenesis. Curves represent most important groups of hemipelagic to pelagic marine sediments according to DSDP data.

(Einsele 1989). Each curve is based on 20 to approximately 70 DSDP-sites. The oozes contain more than 50 % diatoms or radiolarians

Athy's exponential equation (cf. Fig. 13.5b) is characterized by the initial porosity, n_o, and an exponent, ß, the value of which depends on the sediment type and the dimension of z (m or km). Young marine sediments frequently have ß values between 5 and 15 x 10^{-4}, if z is measured in meters (Brückmann 1989). However, while both the Terzaghi and the Athy equations describe the true situation fairly well in medium depth ranges (Fig. 13.5b, Wetzel 1986), they render differing results for shallow burial depths, and the "Terzaghi curve" cannot be used for deeply buried sediments. The reason for this is that the straight lines approach zero porosity with depth, which is on the whole unrealistic.

To overcome this difficulty, Beaudoin et al. (1985) and Maillart (1989) set certain limits for the initial porosity, n_o, and the residual porosity, n_r, at great depth. (several kilometers). Between these two boundaries, the porosity changes exponentially (Fig. 13.5c).

The specific parameters, α and ß, are empirical and may include effects of chemical diagenesis.

Instead of using an exponential equation, Baldwin and Butler (1985) emphasize that a power-law equation (Fig. 13.5d) may suit the porosity-depth relationship better than an exponential equation. Their equation is based on the average porosity-depth relationship of a number of fine-grained sediments. In this case, however, the sediments start with an (unrealistic) porosity of 1 (100 %) at the sediment surface. The power-law equation plots as a straight line on log-log paper (Fig. 13.5d). In addition, these authors replace porosity by "solidity" (S = 1 - n), which is the complement of porosity.

Calculation of burial depths. If the compaction of a sedimentary column follows one of these equations, one can calculate from porosity or solidity the burial depth, z_x, for a spe-

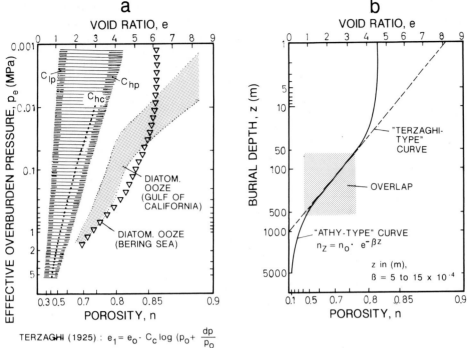

a

VOID RATIO, e

C_{lp}, C_{hp}, C_{hc}

DIATOM. OOZE (GULF OF CALIFORNIA)

DIATOM. OOZE (BERING SEA)

EFFECTIVE OVERBURDEN PRESSURE, p_e (MPa)

POROSITY, n

b

VOID RATIO, e

"TERZAGHI-TYPE" CURVE

OVERLAP

"ATHY-TYPE" CURVE
$n_z = n_o \cdot e^{-\beta z}$

z in (m),
$\beta = 5$ to 15×10^{-4}

BURIAL DEPTH, z (m)

POROSITY, n

TERZAGHI (1925) : $e_1 = e_0 \cdot C_c \log (p_0 + \dfrac{dp}{p_0}$

C_c = COEFFICIENT OF COMPRESSIBILITY (DEPENDS ON MATERIAL AND USE OF log OR ln)
p_0 = INITIAL, EFFECTIVE VERTICAL STRESS

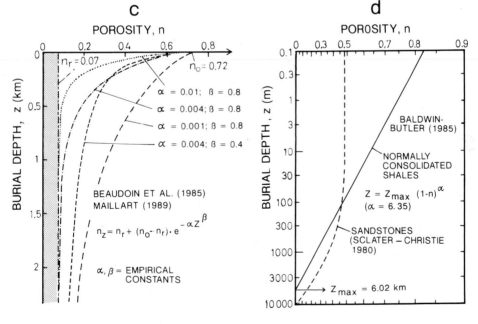

c

POROSITY, n

$n_r = 0.07$

$n_o = 0.72$

$\alpha = 0.01$; $\beta = 0.8$

$\alpha = 0.004$; $\beta = 0.8$

$\alpha = 0.001$; $\beta = 0.8$

$\alpha = 0.004$; $\beta = 0.4$

BEAUDOIN ET AL. (1985)
MAILLART (1989)

$$n_z = n_r + (n_o - n_r) \cdot e^{-\alpha z^\beta}$$

α, β = EMPIRICAL CONSTANTS

BURIAL DEPTH, z (km)

d

POROSITY, n

BALDWIN-BUTLER (1985)

NORMALLY CONSOLIDATED SHALES

$$Z = Z_{max} (1-n)^\alpha$$
$(\alpha = 6.35)$

SANDSTONES (SCLATER – CHRISTIE 1980)

$Z_{max} = 6.02$ km

BURIAL DEPTH, z (m)

Fig. 13.5a-d. Approximations to describe the porosity-depth relationship of fine-grained sediments. **a** Void ratio versus logarithm of effective vertical stress after Terzaghi; C_{lp}, low plasticity, normally consolidated clay; C_{hp}, high plasticity clay; C_{hc}, highly colloidal clay (after Skempton 1970) and young diatomaceous oozes (Einsele 1982a, simplified). **b** Comparison of Athy-type and Terzaghi-type curves using logarithmic scales for depth and porosity, but linear scale for void ratio.

Note the field of good agreement between both curves. (Wetzel 1986). **c** Porosity-depth curves between fixed upper and lower limits, n_o and n_r, respectively. Shape of curve is controlled by empirical constants, α and β (absolute values depend on dimension of z; after Maillart 1989). **d** Porosity-depth curve in log-log diagram according to a power-law equation. (Baldwin and Butler 1985, solidity replaced by porosity). See text for further explanation

cific site (e.g., Magara 1976). Using the Athy equation, one obtains

$$z_x = - \ln \frac{n_z}{n_0} \cdot \frac{1}{\beta} , \qquad (13.2)$$

where z_x is found in m, when β values are, for example, between 5 and 15 x 10^{-4} as mentioned above. The depth values calculated may differ considerably from the true values, when n_0 and β are not known exactly.

According to Baldwin and Butler (1985), the burial depth, z_x (in km), for normally consolidated shales is:

$$z_x = 6.02 \, (1 - n)^{6.35} \text{ or } 6.02 \, S^{1.35} \qquad (13.3)$$

and for underconsolidated shales:

$$z_x = 15 \, (1 - n)^8 \text{ or } 15 \, S^8 . \qquad (13.4)$$

Whether this method can also be applied to sandstones as quoted by Baldwin and Butler (1985) is, however, not sufficiently ascertained. Lithic arenites and wackes containing a significant volume of ductile grains display a reduction in porosity from approximately 40 % at the surface to 15 % at a depth of 500 to 800 m (Smosna 1989). This is caused by simple grain rearrangment, whereas plastic grain deformation operates at greater depths and proceeds more slowly. Evaluating a large data base, Scherer (1987) concludes that, in addition to the factors mentioned above, sandstone porosity also significantly decreases with the age of the rocks considered. This finding indicates that long-term processes, such as pressure solution and various types of slow, chemical processes are involved (see below).

There are other and more sophisticated approaches to the mathematical modeling of compaction (summarized, e.g., by Brückmann 1989; Bayer 1989b; Smosna 1989). All of them try to understand and interpret this process in order to predict situations which cannot be directly measured.

Compaction Ratio and Differential Compaction

Compaction Ratio

In the following we consider horizontal layers or slices of a compacting sedimentary column at various depths. All the layers contain the same solid mass per unit area or the same height, h_s, of pore-free, solid material. The reduction of porosity versus increasing burial depth causes individual layers or beds of an initial thickness, h_1, and mean porosity, n_1, to compact to a secondary thickness, h_2, with a mean porosity, n_2, at the average depth, z_2, below the top of the sequence (Fig. 13.6a and b).

The ratio between h_1 and h_2 is the *compaction ratio*, CR (or τ), which directly indicates the reduction in thickness between a higher and lower interval in a sedimentary column. Since h_1 is usually not known, the compaction ratio is determined from the porosities, n_1 and n_2 (Fig. 13.6a, Eq. 1).

In this case, n_2 is measured from rock samples, and n_1 is assumed from published porosity-depth curves for the corresponding sediment type. One has, however, to take into account that the value of n_1 and thus the compaction ratio significantly depend on the height, h_1, chosen for the topmost (initial) layer as compared to that of the deeper buried layer (Fig. 13.6b). A thick top layer renders a lower n_1 value than a thin one, and therefore yields a lower compaction ratio for the same sequence and depth range considered.

Under favorable conditions, compaction ratios can also be determined or estimated directly (Fig. 13.6c through e). For example, horizontal, originally cylindrical, sediment-filled burrows, generated several to tens of centimeters below the sediment-water interface, are deformed under higher burial load to bodies with elliptical cross sections. Shallow-burial concretions preserve more or less the original thickness of parallel sediment laminae, while the same laminae become compacted outside. Sediment-filled body fossils, the shells of which are dissolved on or slightly below the sea floor, also register subsequent sediment compaction.

Differential Compaction

Except for the above mentioned special observations and changes in mass physical properties, the effects of compaction are difficult to detect in homogeneous, widely extended sediments. Compaction becomes evident, however, in lithologic sequences with varying compaction ratios, as well as in basins with inclined or irregular basin floors (Fig. 13.7). For example, mechanically resistant reef structures undergo little or no compaction, particularly if their pore space is cemented early. By contrast, the adjacent interreef sediments usually

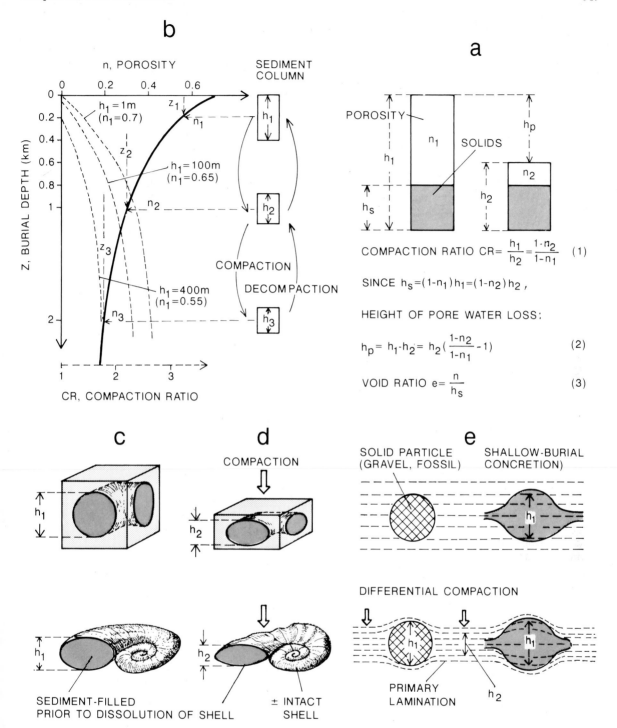

b Relationship between porosity-depth curve and compaction ratio, CR, at different depths, z, below top of sedimentary sequence.

a Demonstration of compaction ratio, CR, void ratio, e, and thickness or height of expelled pore water by comparing two sedimentary columns.

COMPACTION RATIO $CR = \dfrac{h_1}{h_2} = \dfrac{1-n_2}{1-n_1}$ (1)

SINCE $h_s = (1-n_1)h_1 = (1-n_2)h_2$,

HEIGHT OF PORE WATER LOSS:

$h_p = h_1 - h_2 = h_2 \left(\dfrac{1-n_2}{1-n_1} - 1 \right)$ (2)

VOID RATIO $e = \dfrac{n}{h_s}$ (3)

Fig. 13.6. a Demonstration of compaction ratio, *CR*, void ratio, *e*, and thickness or height of expelled pore water by comparing two sedimentary columns. These have the same height, h_s, of compact, solid matter, but differing porosities, n_1 and n_2. **b** Relationship between porosity-depth curve and compaction ratio, *CR*, at different depths, *z*, below top of sedimentary sequence. Note that CR depends on the height, h_1, of the uppermost sediment column and on its average porosity, n_1, which decreases with increas-ing thickness of h_1. **c, d,** and **e** Direct measurement of compaction ratio using: **c** horizontal, initially cylindrical sediment-filled burrows, **d** compacted part of sediment-filled fossil, the shell of which was dissolved or broken early within the sediment, and **e** initial, parallel lamination deformed by differential compaction adjacent to solid gravel, compact fossil (e.g., belemnite, bone of vertebrae), and early diagenetic concretion

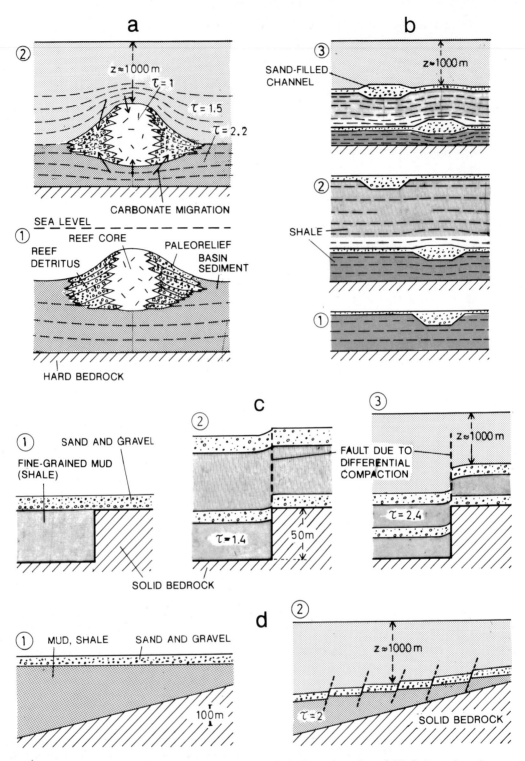

Fig. 13.7a-d. Frequent examples of medium to large-scale sedimentary structures caused by differential compaction at a burial depth of z ≈ 1000 m. **a** Primary relief of bioherm with talus (Stage *1*) is strongly enhanced *(2)*, assuming compaction ratios, τ, of 1.0 (reef core) up to 2.2 (basin sediments). Mechanical compaction may be modified by carbonate redistribution. (After Shaver 1977).

b Deformation of sand-filled channels and accompanying shales. **c** Differential compaction in three stages of development on top of a step-like structure on the basin floor generating a nontectonic fault. **d** Deformation of primary horizontal bed on top of inclined basin floor. (**b, c** and **d** after Maillart 1989). τ = CR = compaction ratio (see Fig. 13.6a

display a high compaction ratio. Such differences in compaction, i.e., differential compaction of sediments not far apart, may enhance the primary relief of a bioherm (Fig. 13.7a), or accentuate and modify the shape of sand-filled channels within a shale sequence (Fig. 13.7b). Differential compaction may generate faults on top of a step-like relief in the underlying bedrock (Fig. 13.7c) and lead to a synthetic fault system on top of an inclined basin floor (Fig. 13.7d).

Differential compaction is also a common feature in marine deltaic environments. Where fluvial sands of distributary channels, river mouth bars and the delta front (see Chap. 3.4) accumulate on top of thick prodelta muds, they cause greater compaction than in adjacent areas less loaded. The resulting slight depressions in the topography are immediately filled up with more sand, which adds to the already existing higher overburden pressure at this site. In this way, so-called growth faults are generated (Fig. 13.8a and b), which may cause a rotation of the more loaded sediment bodies. High sand input into the basin and a seaward prograding delta front or shelf edge during sea level lowstand may lead to a series of basinward prograding minor to medium-sized faults (Fig. 13.8a), while a balance between subsidence and deposition tends to form a single or only a few large growth faults (Fig. 13.8b).

Some of these initially steeply inclined faults may be affected by the lateral sliding of sediment masses, such that they then continue basinward as gently dipping reflectors (Bruce 1973). As a result, a rather complicated extensional fault pattern can develop, as observed in seismic records of the Texas coast (Fig. 13.8c). Cross-sections of the continental margin in the northwestern Gulf of Mexico, which is little affected by salt diapirs, show Tertiary sands and sandy shales subdivided by growth faults into a series of coastal-parallel troughs. These rest on the irregular top of older, thick shales which form "ridge-like" structures between the sand-filled troughs. The growth faults die out near the deepest parts of the troughs. The underlying shales appear to be more compacted than the shales forming the ridges, which are underconsolidated and exhibit excess pore fluid pressures from subsurface depths of approximately 1500 m downwards.

The true origin of all these structures is revealed by *decompacting* the sediments with the aid of the (downward increasing) compaction ratio. In this way, the state and thickness of a sediment body prior to compaction can be reconstructed. As demonstrated in Figs. 13.6b and 13.7, decompaction can be carried out in several steps to show the evolution of a specific structure. Using this method, the compaction-induced structures can also be dated.

Compaction Flow

Another important aim of the study of compaction is the evaluation of compaction-induced pore water flow, frequently referred to as *compaction flow* or *advection* (e.g., Engelhardt 1973). The reduction of the water-filled pore space of sediments versus depth is accompanied by the expulsion of considerable volumes of water. This water moves either vertically upward through the entire sedimentary column, or uses inclined, highly permeable layers to escape laterally. In this way, pore waters can transport dissolved inorganic matter and hydrocarbons.

Model Calculations

In the following consideration, the amount of pore water is expressed in terms of height of a pure water column. The height or volume of expelled pore water can be calculated from the reduction in porosity and the thickness of the sedimentary column considered (Fig. 13.6a, Eq. 2). An example of such a calculation for vertically ascending water, using the Baldwin-Butler (1985) porosity-depth curve for normally consolidated argillaceous sediments, is shown in Fig. 13.9.

First, the volume of expelled pore water is determined for specific depth intervals (1000 m in thickness, in the uppermost 250 to 500 m of the section), which was released by the deposition of the last 1000 m increment of sediment. For this purpose, the difference in mean porosity between succeeding intervals is evaluated. In a second step, the volume increments of the intervals are summed from bottom to top of the sequence (right-hand side of diagram). As expected, the water volumes expelled from the single intervals increase upward, and thus, the volume of the total pore water flow through given horizontal cross-sections or marker beds in the sedimentary column grows substantially toward the top. A new increment of 1000 m of sediment would release the same amounts of pore water from the underlying intervals and generate an equal cumulative compaction flow. Values derived by Bjørlykke (1983) from a somewhat different porosity-depth curve are similar to those reported here.

However, these values represent neither the total pore water loss from individual sediment intervals nor the total water loss from the entire sequence during its burial history. If, for example, the mean porosity of the lowermost sediment interval in Fig. 13.9 is compared with the porosity of the uppermost 1000 m thick section (mean porosity $n = 0.35$), one gets (using Eq. 2 in Fig. 13.6a, as before) a volume of released pore water of 515 m^3/m^2. If the initial porosity at the sediment-water interface ($n_o = 0.72$) is taken into account, one obtains 2520 m^3/m^2. The latter value corresponds to a water column higher (or thicker) than the present thickness of the deeply buried sediment interval, which is reasonable, because 72 % of the surface sediment consisted of water.

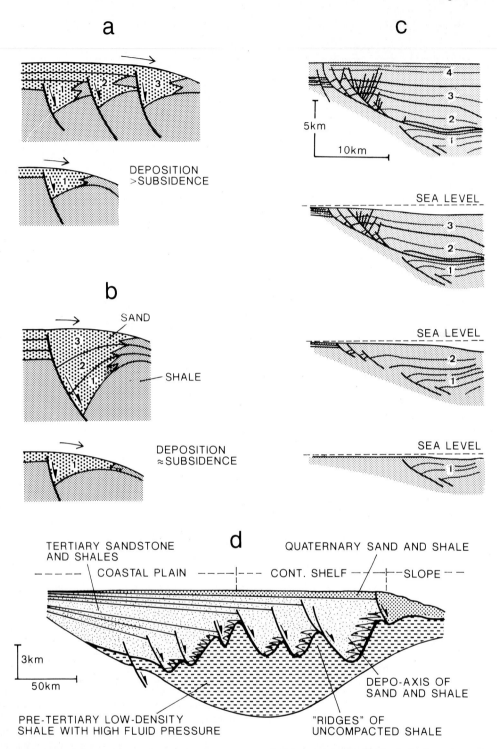

Fig. 13.8a-d. Large-scale sediment deformation by differential compaction and development of syndepositional fault systems. (After Bruce 1973). **a** and **b** Growth faults in marine deltas or along edge of continental shelf as controlled by subsidence and rate of deposition in three stages of development (*1, 2, 3*). **c** Reconstruction of the evolution (Stages *1* through *4*) of individual fault system due to a combination of differential compaction and laterally direct- ed slide over gently inclined slip face (Stage *4* constructed after seismic record). **d** Generalized cross section of passive continental margin on the Gulf coast of Texas. Note seaward migrating depocenters of sand-shale sedimentation on top of a thick shale sequence, forming wide "ridges" of underconsolidated sediment in between the sand-filled "troughs"

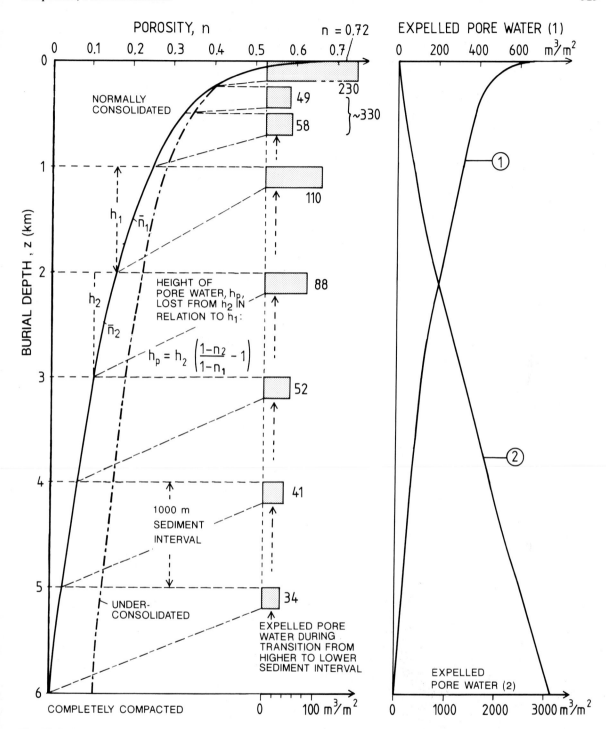

Fig. 13.9. Upward-directed compaction flow calculated for a standard porosity-depth curve for normally consolidated, argillaceous sediments (after Baldwin and Butler 1985); curve for underconsolidated material for comparison. Heights, or rather volumes, of expelled pore water are based on equation in Fig. 13.6a (Hamilton 1976). *Columns* Water released from individual lower to higher sediment intervals. *Right-hand curves: 1* Cumulative compaction flow, released by an additional 1 km increment of sediment, from all six succeeding 1 km-thick sediment sections. (2) Total volume of pore water released from all six, 1 km-thick sediment sections during their burial history, summed from top to bottom. 1 m³/m² = 0.1 l/cm². (See also Bjørlykke 1983 and Galloway 1984)

The overlying sediment intervals have increasingly higher porosities and therefore have lost smaller total amounts of water. In addition, they had to take up the ascending pore water from the compacting underlying sections, i.e., part of their original pore water was replaced by older pore water.

This is demonstrated in the three-layer model in Fig. 13.10b, using the same porosity-depth curve as in Fig. 13.9. As a result of compaction caused by the new sediment increment (Layer 1, h_1), the original pore water in the middle layer (Layer 2) moves upward beyond its base and top and penetrates far into the uppermost layer (Layer 1).

This process is particularly interesting when this pore water differs in chemistry from that of the other layers. Even more conspicuous is the case in which the lowermost layer (Layer 3) deviates in its pore water chemistry from the higher layers. If the oldest layer was a fresh water sediment or was deposited under hypersaline conditions, and was then overlain by two normal marine sections, its pore water can later influence a large part of the younger sediments (Fig. 13.10c).

In order to determine the total water loss of the entire sequence during its burial history, the cumulative water loss of all intervals can be summed from top to bottom (Fig. 13.9, Curve 2). It is found for this example that a water volume of approximately 3150 m^3/m^2 was released. This means, in other words, that a water column of about one half the present thickness of the sequence was expelled.

One can deduce from this example, as well as by simple reasoning, that the height of the total compaction flow is always less than the thickness of the sequence. As long as sedimentation continues and compaction flow occurs uniformly and solely vertically, the ascending pore water from deeper sections can never reach the sediment-water interface (Einsele 1976). In this case, the total amount of water confined in the pores over all depths also increases and therefore must be complemented by newly added sea water.

Ascending pore water can only cross the sediment-water interface if sediment accumulation has ended and an underconsolidated sequence is still in the process of compaction. In addition, irregular dewatering of compacting sediments may cause local pore water discharge at the sediment-water interface.

Compaction Flow as Function of the Sedimentation Rate

During deposition, the range and velocity of ascending pore water flow and its potential solute transport are mainly a function of the sedimentation rate. This relationship is exemplified in Fig. 13.10 (based on the same porosity-depth curve as Fig. 13.9).

Here, a two-layer sequence with the layers h_1 and h_2 is overlain, in the course of time, by a new increment of sediment with thickness h_1. Under conditions of normal consolidation, the former sediment layers, h_1 and h_2, are compacted to h_2 and h_3, respectively, and their former porosities, n_1 and n_2, are reduced to n_2 and n_3, until a new state of equilibrium is established. In contrast to the model in Fig. 13.9, the thickness, h_s, of compact solid matter is kept constant in all three succeeding sections, i.e., they become thinner with increasing burial depth. The loss of pore water in terms of the height, h_p, from the freshly compacted sections h_2 and h_3 can easily be found graphically. Because of the preservation of mass, the sum of h_3 and h_{p2}, for example, is equal to the thickness, h_2, of the former, less compacted section.

The height, h_p, of expelled pore water can also be calculated with the aid of the compaction ratio, CR (Fig. 13.6a, Eq. 1). If $h_1/h_2 = CR_{1,2}$ and $h_2/h_3 = CR_{2,3}$ etc., we find

$$h_{p2} = h_2 - h_3 = \frac{h_1}{CR_{1,2}} - \frac{h_2}{CR_{2,3}} \qquad (13.5)$$

and

$$h_{p1} = h_1 - (h_2 - h_{p2}) = h_2 \cdot CR_{1,2} - \frac{h_2}{CR_{2,3}}$$

$$= h_1 - h_3 . \qquad (13.6)$$

Sequences subdivided into more sections can be treated in the same way.

The h_p values of expelled pore water signify the heights of columns or the thicknesses of layers of pure water. In reality, the ascending pore water passes through pores which occupy only a fraction of the sediment volume. Hence, the range of ascending pore water flow, h_f, crossing the boundaries of the sediment sections in question is greater than h_p. It is generally

$$h_f = \frac{h_p}{n} . \qquad (13.7)$$

For our example we find

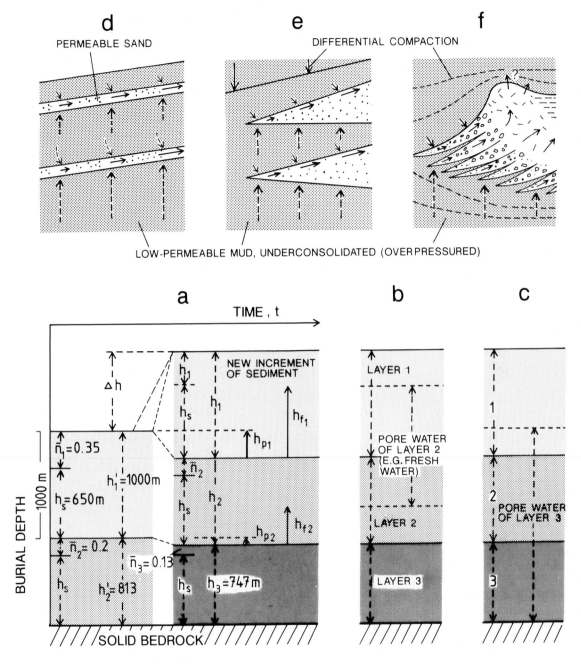

Fig. 13.10. a Two-layer sequence (h_1' and h_2') based on the same porosity-depth curve as in Fig. 13.9, but later loaded by new increment of sediment (h_1). This causes compaction and pore water flow (h_p or h_f) from underlying Layers 2 and 3. The thickness, h_s, of compact solid matter is the same in all layers, but mean porosities, n, vary with depth. **b** Upward migration of pore water from Layer 2 and Layer 3, respectively, caused by additional compaction under the load of Layer 1 (h_1), with dispersion and mixing neglected. **c** Total upward migration of pore water from lowermost layer (h_3) resulting from sediment load of both overlying layers (h_2 and h_1). Note the especially wide, secondary depth range of pore water from the lowermost layer. **d through f** Refraction of vertically ascending pore water flow through highly permeable, inclined individual sand layers, sand wedges, and reef detritus. See text for further explanation

$$h_{f1} = \frac{h_{p1}}{n_1} \quad \text{and} \quad h_{f2} = \frac{h_{p2}}{n_2} .$$

Velocity of Ascending Pore Water

The velocity, v, of the pore water ascending in the porous sediment is then approximately (t = time interval)

$$v = \frac{h_f}{t} \quad \text{or} \quad v = \frac{h_f \cdot SR}{h} , \qquad (13.8)$$

because the sedimentation rate is
SR = h/t.

Equation (13.8) shows that the velocity of the upstreaming pore water is controlled by the sedimentation rate and is inversely related to the thickness, h, of the uppermost sediment layer considered. For the example in Fig. 13.10 we find (using n_1 = 0.3)

$$v_1 = \frac{h_{f1} \cdot SR_w}{h_1} = 0.84 \, SR_w ,$$

where SR_w = sedimentation rate of wet sediment in the uppermost section with h_1 = 1000 m.

For SR_w = 10, 100, 1000 m/Ma, the velocity of ascending pore water in our example is v_1 = 8.4, 84, and 840 m/Ma, or 2.7×10^{-13}, 2.7×10^{-12}, and 2.7×10^{-11} m/s, respectively. These extremely low values will be discussed further in Chapter 13.3. In the underlying sediment intervals, the velocity of compaction flow becomes successively less with depth.

It should be mentioned here that one can distinguish between different sedimentation rates (measured, for example, in m/Ma; also cf. Chap. 10.1): The sedimentation rate of wet sediment calculated from h_1, the sedimentation rate for dry, compact sediment derived from h_s, and the elevation, Δ h, of the sediment-water interface above the basin floor (Fig. 13.10).

Pore Water Pressure

As discussed earlier, compaction flow is delayed or prevented if the permeability of the overlying beds drops to such low values that the released pore water cannot escape. It then becomes *overpressured* with respect to the normal hydrostatic pressure, p_h, because it has

to carry part of or almost the entire load of solid matter exerting *lithostatic pressure* (Fig. 13.11a). Hence, the state of underconsolidation of subsurface layers can be determined from the pressure of their pore fluids. The ratio between the pressure of the pore fluid, p_f (= p_h + u; u = excess pore pressure), and the lithostatic pressure, p_l, is called *geostatic ratio*. Values for this ratio vary between 0.47 (no excess pore pressure) and in excess of 0.9, indicating very high overpressure.

In the latter case, the fluid pressure can exceed the minimum horizontal stress and thus cause *fracturing* (Fig. 13.11b; Rouchet 1978). This preferentially occurs in the depth range of 1000 to 3000 m, where horizontal stresses are considerably lower than vertical stress. The existence of overpressure signifies that the section is more or less sealed by overlying low permeability layers. Consequently, there is a high likelihood that hydrocarbons, possibly released from the sequence, are not lost by migration to the land surface or sea bottom. Similarly, one can assume that the primary pore water (connate water or chemically altered formation water) was not recently mixed with meteoric water (see below).

Some sedimentary basins also exhibit *subnormal pressures* in drillholes of some depth. These pressures are less than the hydrostatic pressure. Such low pressures can only occur in porous rocks that are separated from circulating groundwater by a low permeable barrier. Under this condition, subnormal pressure can be generated by extracting fluids or gas from the reservoir. Other causes for the reduction in pressure may be an increase in pore volume by fracturing and decompression, a decrease in the subsurface temperature, or osmotic effects.

Refraction of Pore Water Flow

So far, only vertically ascending pore water in homogeneous sediments has been considered. The direction of compaction flow can, however, be significantly modified by the intercalation of highly permeable layers, sediment wedges, and irregular sediment bodies, which in addition may give rise to differential compaction and inclined bedding (see above). If vertically upstreaming pore fluid in low-permeable muds reaches, for example, an inclined porous sand bed or sand wedge, its direction of flow is refracted and thus forced to follow more or less the dip of the sandy bed (Fig. 13.10d and e).

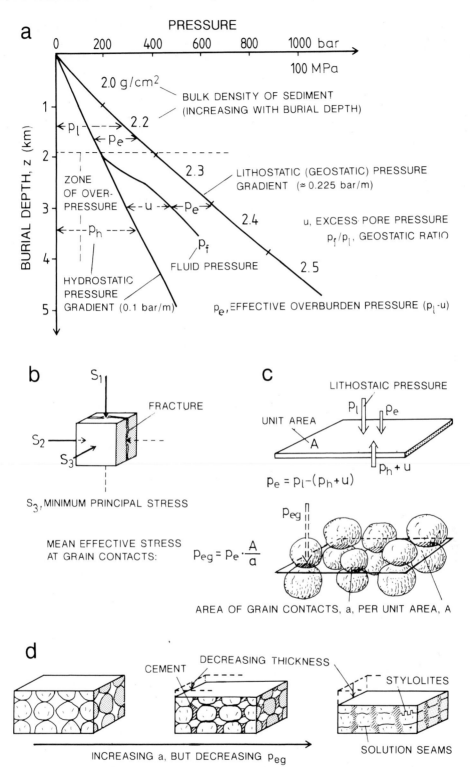

Fig. 13.11. a Relations between hydrostatic pressure, lithostatic (or geostatic) pressure, excess pore water pressure, and burial depth (from Gretener 1979; Gretener and Feng 1985). Note the buildup of excess pore water pressure in the zone of "overpressure" and the associated reduction in effective stress. **b** Possible fracturing normal to direction of minimum principal stress in zone of overpressure. (Rouchet 1978). **c** Comparison of effective stress on unit area of horizontal plane and mean effective stress at grain contacts. **d** Increasing compaction due to pressure solution. See text for explanation

In the South Caspian basin subsurface pressure data indicate that sands are acting as drains not only for the underlying, but also for the overlying compacting shales (Bredehoeft et al. 1988). The highly permeable sands therefore display flow velocities considerably higher than those calculated for vertically ascending pore waters. The lateral movement of compaction water may allow subdivision of a sequence into several horizontal units which are hydraulically more or less isolated from one another. Similar to sand layers and sand wedges, reef detritus acts as conduit for compaction flow (Fig. 13.10f).

In conclusion, the compactional flow regime is characterized by upward and outward expulsion of pore waters from growing basin fill. Lithostatic loading or compressive tectonic stress provides the driving mechanism for flow. The zone of compaction flow may also include burial depths where the fluid pressure exceeds hydrostatic pressure (Fig. 13.11). Compaction flow is substantial in thick sequences and may contribute significantly to diagenetic processes. Nevertheless its volume is finite and limited by the water content of the sediments involved. Growth faults frequently act as barriers for the updip migration of compaction water (including hydrocarbons).

Other Hydrogeologic Flow Regimes in Sedimentary Basins

In growing sedimentary sequences, compaction flow is the most obvious result of pore space reduction with increasing burial depth. Sedimentary basins are, however, also affected by other flow mechanisms. We have to consider at least four different types of hydrogeologic flow regimes (Bjørlykke 1984; Galloway 1984; Hanor 1987) (Fig. 13.12):

- Meteoric flow (gravity-driven flow).
- Compaction flow (pressure-driven flow, discussed above).
- Thermobaric flow (temperature/pressure driven flow).
- Convection (density driven flow).

All these flow regimes may occur simultaneously in an active, repeatedly prograding shelf-slope depositional system as is common on passive continental margins; this is shown in the model of Fig. 13.12.

Meteoric water circulation primarily occurs in shallow portions of the basin fill near the coast. Groundwater infiltrated on land from precipitation or surface water tends to move down the topographic gradient in the direction of decreasing hydraulic potential and discharges at or below sea level. A drop in sea level or the emergence of parts of the sequence permit deeper and more widely extended *fresh water flushing* of sediments which originally contained marine pore water.

In addition to local circulation cells, larger regional circulation systems may develop (Fig. 13.12d). Along the Atlantic continental margin of North America, fresh, artesian groundwater was found far offshore by ocean drilling (Kohout et al. 1988). If special water conduits such as fault systems are present, meteoric water circulation can also affect deeply buried strata. However, the volume of water passing through deeper parts of the basin fill during a certain time interval is usually much less than that of shallow meteoric circulation systems. Compared to other flow mechanisms, shallow meteoric flow involves large volumes of water which not only replace the primary pore water, but also leach soluble and unstable minerals from the sediments.

Thermobaric flow occurs in the deepest parts of basin fill and is related to compaction flow. Under increasing temperature and pressure, significant volumes of water may be released by dehydration of clay minerals and other hydrous mineral phases, besides the expulsion of residual pore water. Similar to pure compaction flow, the fluids tend to move slowly upward until they reach permeable layers allowing lateral flow. However, the extremely low permeability of the intensely compacted sediments in this depth zone commonly restricts water movement over long time periods. As a result, this zone is characterized by abnormally high pore water pressures. Thermobaric waters frequently contain methane, carbon dioxide, and hydrogen sulfide derived from the thermal alteration of organic matter within the sediment.

Convective or density-driven flow can operate in different ways:

- Near the surface when water of higher density (salinity) forms a kind of reflux system through sediments. This is known from some modern salt lagoons separated from the open sea (cf. Chap. 6.4). Circulation of sea water with slightly elevated salinity may also occur in carbonate platforms. Beneath the North Island of the Great Bahama Bank, for example, sea water with salinities between 38 and 42 ‰ flows eastward, mixes with colder groundwater of normal salinity, and enters

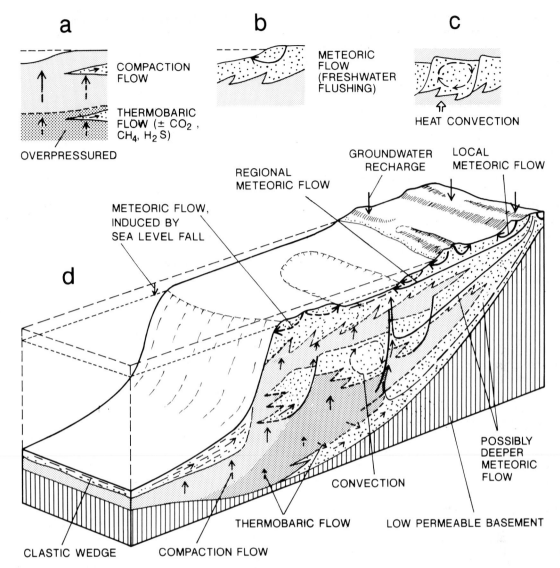

a

COMPACTION FLOW

THERMOBARIC FLOW (± CO_2, CH_4, H_2S)

OVERPRESSURED

b

METEORIC FLOW (FRESHWATER FLUSHING)

c

HEAT CONVECTION

REGIONAL METEORIC FLOW

GROUNDWATER RECHARGE

LOCAL METEORIC FLOW

METEORIC FLOW, INDUCED BY SEA LEVEL FALL

d

POSSIBLY DEEPER METEORIC FLOW

CONVECTION

THERMOBARIC FLOW

LOW PERMEABLE BASEMENT

CLASTIC WEDGE

COMPACTION FLOW

Fig. 13.12. a,b,c Principal types of flow regimes possible in sedimentary basins. **d** Simplified model of repeatedly prograding, partially deltaic shelf-slope facies on passive continental margin, illustrating various contemporaneous hydrogeologic flow regimes. A sea level fall may extend the influence of meteoric water circulation, MF. (Based on Bjørlykke 1983; Galloway 1984; Bethke 1989, greatly modified)

again the sea at water depths of 200 to 300 m (Whitaker and Smart 1990). Such flow systems are thought to be important for pervasive dolomitization of carbonate buildups (see below).

- Pore water flow in a circle or loop within a permeable sediment body may also occur without uptake of water from outside or loss of water to an adjacent flow system. Density differences may arise as a result of differences in temperature and/or salinity (Chap. 5.2). However, convection of fluid can only occur in relatively large, highly permeable rock bodies if the difference in density is sufficiently high and, in addition, the viscosity of the fluid is low enough to overcome friction within the porous medium.

Theoretical calculations and some field observations concerning temperature gradients and pore water chemistry have presented evidence that such flow systems are possible in several sedimentary basins (Wood and Hewett 1984; Hanor 1987). For example, density inversions in pore fluids on the Gulf coast of Texas are sufficient to drive large-scale convective flow at rates of meters per year.

Convective mass transport can redistribute relatively large quantities of mass in comparatively short times (Schwartz and Longstaffe 1988). A quantitative evaluation of these various large-scale flow systems is difficult, because the three-dimensional structure of basin fill and its heterogeneities are seldom sufficiently known. The permeabilities of the different sedimentary facies in particular vary markedly in space and time (Bethke 1989, Bjørlykke et al. 1989).

There are several other, so-called *nonhydraulic* mechanisms, such as osmosis and ultrafiltration, which may cause or influence fluid flow and pore water chemistry in argillaceous sediments (Neuzil 1986; Hanor 1987).Although several experimental studies have been carried out in this field, the significance of these phenomena in large-scale natural systems is still poorly known.

Significance of Various Flow Regimes

The importance of various hydrogeologic regimes depends on the type of sedimentary basin and its age and evolution through time (Bjørlykke 1983; Galloway 1984). *Compactional and thermobaric* regimes are associated with the *active depositional phase* of many marine basins, particularly basins on continental margins, marine rift zones, or delta-fed basins which subside significantly and are filled rapidly. If this active phase comes to an end and is followed by emergence of the basin fill due to sea level fall or the onset of uplift, compactional and thermobaric flow slow down and finally cease.

Depositionally inactive basins or *hydrogeologically maturing* basins are then increasingly subjected to gravity-driven, meteoric water circulation and therefore flushed by fresh water. Recharge commonly occurs along the uplifted margins of the basin, and meteoric flow is directed toward the less elevated basin center. Pressure gradients versus depth commonly record hydrostatic conditions. Many basins all over the world have evolved to the stage when gravity flow predominates, for example the Aquitaine basin and Paris basin in France, the Pannonian basin in Hungary, and several basins in North America.

Pore Water Flow in Various Depositional Systems

The young, slowly deposited sediments in *deep-sea basins* far away from the continents, as well as most sediments of slowly subsiding epicontinental seas, exhibit only minor compaction flow and are barely thick enough to develop overpressured thermobaric zones.

The emerged, more or less horizontally layered sediments of an epicontinental sea, for example, provide little possibility for deeply circulating meteoric waters (Fig. 13.13a), as long as significant hydraulic gradients are excluded due to low relief. Hence, the evolving circulation systems are limited in extent both laterally and vertically, except for emerged carbonates which may provide more effective conduits for ground water circulation. Sandstones in such sequences are usually thin and contain small proportions of unstable minerals which could be later leached and allow the formation of secondary porosity (see below).

Thick, *fluvial, and deltaic sediments* along the margins of a continental rift basin are usually well and deeply flushed by meteoric waters during and after the time of their deposition (Fig. 13.13b). Their frequently immature sediments derived from nearby uplifted rift shoulders can thus be strongly affected by leaching and alteration, as in the example of kaolinitization of feldspar. Some compaction flow may originate in the basin center and migrate some distance outward.

Deep-water turbidites, which have been buried by pelagic and hemipelagic marine sediments (Fig. 13.13c), are normally not affected by early freshwater flow. If the turbidites are relatively fine-grained and poorly sorted, they lose most of their primary porosity due to compaction, or their pores are filled by carbonate cement provided by compaction flow from carbonate-bearing host sediments (Fig. 13.13d). Only thicker turbidites with thin intercalations of host sediment devoid of carbonate or easily soluble opaline silica, have a chance of preserving some of their primary porosity, or they may gain some secondary porosity by the leaching effect of compaction water (Fig. 13.13e).

The flow regimes of other basin types and their evolution over time can be inferred from these general rules. Many basins also experience changes in the development of certain flow regimes during their depositional history.

For example, Lower and Middle Jurassic deltaic sands, representing the rifting stage in the central North Sea, were

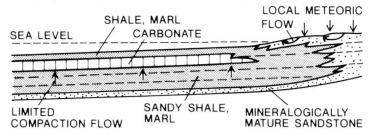

a EPICONTINENTAL SEA

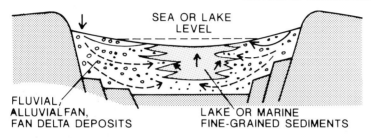

b CONTINENTAL RIFT BASIN

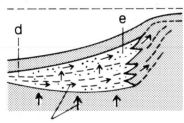

c BURIED DEEP-SEA FAN

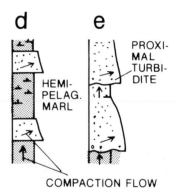

Fig. 13.13a-e. Dominant flow regimes in various types of sedimentary basins. **a** Epicontinental sea in stage of emergence. **b** Continental rift basin with central lake or marine environment. **c** Buried deep-sea fan and **d, e** sections of special turbidite sequences. See text for further explanation. (After Bjørlykke 1983)

subjected to meteoric water circulation during and immediately after their deposition (Fig. 13.14a). As a result, feldspar and mica in the sands were converted to kaolinite. Then the sandy deposits (Brent Sandstone of the Viking graben) were covered by transgressive shales and carbonates, which cut them off from the meteoric water supply and initiated some compaction flow (Fig. 13.14b). Grain-to-grain pressure was, however, not sufficiently high to produce pressure solution (see below). In a third stage during the Upper Jurassic, part of the sequence emerged due to tectonic uplift. Thus, the clay cover on top of the sandstones was removed, allowing the renewal of meteoric water circulation within certain portions of the sandstone (Fig. 13.14c).

Later development (Fig. 13.14d) is characterized by significant subsidence and deposition of more than 2000 m of younger beds, which again sealed the Jurassic sandstones and caused overpressure, limited compaction flow, and possibly thermobaric flow and hydrocarbon accumulation within the deeply buried part of the basin fill. Further compaction and cementation of the sandstone by pressure solution (see below) was inhibited by the buildup of very high excess pore pressure equivalent to approximately 3000 m of water column. Today, about 70 to 80 % of the overburden load is carried by pore water and only 20 to 30 % by the sandstone framework (Bjørlykke 1983).

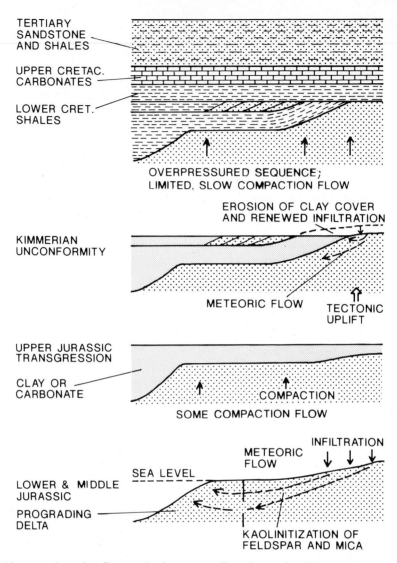

Fig. 13.14. Burial history and varying flow mechanisms controlling diagenesis of Jurassic Brent sandstone in the North Sea, burial depth 2.5 to 3 km. (After Bjørlykke 1983)

13.3 Principles
of Chemical Diagenesis

Introduction

Chemical diagenesis is an extremely complex, but very interesting and important topic in sedimentology and economic geology. There is an immense, never ending literature on the different aspects of chemical diagenesis in various rock types from many regions all over the world. Chemical diagenesis affects siliciclastic and biogenic materials (skeletal remains and organic matter) of greatly varying solu-

bility and thermodynamic stability, and it includes mineral reactions with pore fluids ranging from meteoric waters to highly concentrated brines. Primary minerals may be replaced to a great extent by secondary minerals; pore cements often display a characteristic evolution of the diagenetic system with several cement generations and secondary leaching. Primary minerals may be depleted in trace elements, which are enriched in secondary mineral phases or may migrate in solution to other sites. Some inorganic processes are closely related to the decomposition and diagenesis of organic matter. Stable isotopes, such as ^{13}C and ^{18}O, may undergo fractionation relative to ^{12}C and ^{16}O, respectively, during diagenesis and thus reveal the origin and nature of the secondary minerals in which they are found. Fluid inclusions in cement minerals reflect the composition, temperature, and pressure of the former diagenetic environment.

Chemical diagenesis is commonly accompanied by compaction. The term "chemical compaction" may refer to pressure solution, but is also used in a broader sense to describe the total compaction effect of chemical diagenesis. We may, however, distinguish between processes retarding the onset and efficiency of chemical compaction (e.g., early cementation, pore filling with oil) and situations accelerating chemical compaction (presence of metastable mineral phases, aggressive pore fluids, effective flow systems; Moore 1989). Carbonates normally undergo earlier and more intensive diagenesis than siliciclastic sediments.

Sediments rich in volcaniclastic material (Chap. 2.4) tend to be altered rapidly during diagenesis, because they contain high porportions of very unstable minerals which are easily converted into clay minerals. These form rims and coatings around the grains and more or less fill the pore space, thus reducing the primary porosity and permeability of sands.

Due to all these and further complications, it is not possible to present a simple system including and summarizing all important problems of chemical diagenesis. In the following, only a few of these various aspects can be discussed.

Pressure Solution

Mechanical compaction includes ductile deformation and fracturing of minerals which contribute to denser grain packing. However, more important in diagenesis is pressure solution at the contacts of individual grains. This can be regarded as a process transitional between mechanical and chemical diagenesis. The principle mechanism has long been known, but its efficiency as a provider of pore cement for various types of sandstones, orthoquartzites, and carbonates is somewhat controversial (Bjørlykke 1983 and 1984; Bathurst 1987 and 1991; Pettijohn et al. 1987; Füchtbauer 1988; Mitra 1988; Bayer 1989b; McBride 1989; Tada and Siever 1989).

According to Riecke's principle, the solubility of minerals is enhanced under compressional stress, i.e., at the contact between two grains or bedding planes. If the composition of adjacent particles is different, one usually maintains its shape at the expense of the other, which is partially dissolved. Pressure solution has frequently been described from carbonate rocks and quartzitic sandstones, but it also occurs in other rock types.

The *effective vertical stress* at the grain contacts is a function of the effective overburden pressure and the area, a, which the grain contacts occupy per unit area, A, of a horizontal plane within the rock (Fig. 13.11c). The smaller the grain contacts, the higher is the effective stress at these contacts.

For example, at a burial depth of 2000 m, the effective vertical stress exerted on a plane with area A is

$$p_e = p_l - p_h = 25 \text{ Mpa} (= 250 \text{ bar})$$

(using the hydrostatic pressure and lithostatic pressure gradients listed in Fig. 13.11a, and assuming that there is no excess pore pressure). Then, the mean effective stress at the grain contacts, p_{eg}, is

$$p_{eg} = p_e \cdot \frac{A}{a} . \qquad (13.9)$$

If a = 0.1 A, the effective stress at the grain contacts amounts to 250 MPa (= 2500 bar).

Such a high stress is, however, strongly reduced if excess pore water pressure builds up. In this way, significant pressure solution is prevented. This is the reason why many reservoir rocks can maintain a great part of their primary porosity at an advanced stage of their burial history and thus allow the migration and accumulation of hydrocarbons.

Fine-grained quartzose sandstones appear to develop more intergranular pressure solution than coarser counterparts (Houseknecht 1988). Pressure solution of quartz may be enhanced by the presence of clay as grain coatings. Chemical reactions involving clays probably yield high pH in the pore water and thus favor the dissolution of quartz. Therefore, clean quartz sands have a higher potential to preserve their porosity with depth than "dirty" sands. At burial depths of 2.5 to 3.5 km microstylolites between quartz grains are common (see below).

Ongoing pressure solution will lead to an increase in the area of grain contacts (Fig. 13.11d) and thus to a reduction in stress at these contacts. As a result, the solution process slows down. Finally, the grain contacts often merge and display *dissolution seams, sutures, or stylolites*. The dissolved matter is usually precipitated immediately next to the grain contacts on free grain surfaces and in the open pore space, where minimum stress conditions prevail. The pressure-solved material forms *mineral overgrowths and pore cement* (Trurnit and Amstutz 1979; Mitra 1988). If all pores are filled, pressure solution tends to end (Park and Schot 1968). It may continue during early stages of rock deformation, when fractures can be filled by pressure-solved material (Mitra 1988).

The question whether or not the cement of orthoquartzites or carbonate-cemented strata has been derived entirely from the same layer by pressure solution is controversial. Several authors assume that both carbonate and silica can also be delivered from neighboring mudstones and shales (e.g., Füchtbauer 1974a and b; Ricken 1986; McBride 1989). However, it was also inferred from petrographic evidence that, in the case of quartz sandstones, pressure solution during deep burial diagenesis may provide more silica than needed for complete cementation. In other words, deeply buried sandstones can export silica to less deeply buried strata.

While pressure solution and cementation of quartz appear to take place predominantly at deep burial depths of at least 1.5 to 2 km (Füchtbauer 1988) or in the "silica mobility window" at 80 to 120 °C (McBride 1989), pressure solution of carbonates largely occurs at depths of a few hundred meters (Bathurst 1975, Ricken 1986).

**Various Diagenetic Realms
and Solute Transport Mechanisms**

Pore fluids are the main mediator of all reactions in chemical diagenesis. Solid-solid reactions are negligible under the low temperature conditions characterizing diagenetic processes. Therefore, the chemistry of pore fluids is one of the most important criteria for distinguishing between various diagenetic realms (Fig. 13.15). Furthermore, the mechanisms of solute transport within the sediments play a great role in the type and intensity of chemical diagenesis. Both factors, the chemistry of pore fluids and the types of solute transport, are closely related and operate in different ways at shallow and deep burial depths.

There are *open systems*, which exchange dissolved matter with sea water or other parts

of the sedimentary basin (*allochemical diagenesis*), and more or less *closed systems*, which preserve their bulk chemical composition (*isochemical diagenesis*), although they exhibit internal solution and recrystallization processes.

*Solute Transport
by Meteoric Water Circulation*

Dissolution, transport, and reprecipitation of solid matter by meteoric water circulation is a very important process in carbonate sediments (Chap. 3.3.2), but also affects siliciclastic material as will be demonstrated in the following.

On the continents, for example in fluvial systems or along lake shores, chemical diagenesis is controlled by meteoric water circulation with recharge in elevated areas and discharge in lowlands. The unsaturated, *vadose zone* above the groundwater table is commonly well aerated, but oxidation also affects the upper part of the saturated, *phreatic zone*. As a result, primary sediment particles containing ferrous iron and manganese are oxidized and partially replaced by oxyhydrates of iron and manganese (e.g., limonite) or hematite. Unstable minerals in sands, for example biotite and feldspar, are slowly dissolved or transformed into clay minerals. In humid climates, kaolinite predominates as a secondary mineral, while arid conditions may lead to the neoformation of potassium feldspar (Füchtbauer 1974 a). The carbonate fraction in sands tends be be dissolved, but part of the dissolved carbonate may be reprecipitated either as overgrowths on the surfaces of various grains (meniscus cement) in the vadose zone, or as pore-filling cement (equant low Mg calcite) in the deep phreatic zone where water circulation is very slow. Both these effects and those of brackish and hypersaline waters on carbonate diagenesis in the coastal zone are briefly described in Chapter 3.3.2 (Fig. 3.25).

Similar to sands on land, emerging coastal sands are subjected to relatively strong *meteoric leaching*. Thus, their unstable components also tend to be altered and dissolved. Cementation only occurs locally, for example along beaches, when unstable carbonate is present, then partially dissolved and reprecipitated as aragonite or high Mg calcite cement (*beach rock*). Under semiarid conditions dissolved carbonate or silica can be precipitated near the land surface as nodules or crusts (*calcrete, silcrete*). Wood and pre-existing carbonate, including calcareous fossils, are fre-

a

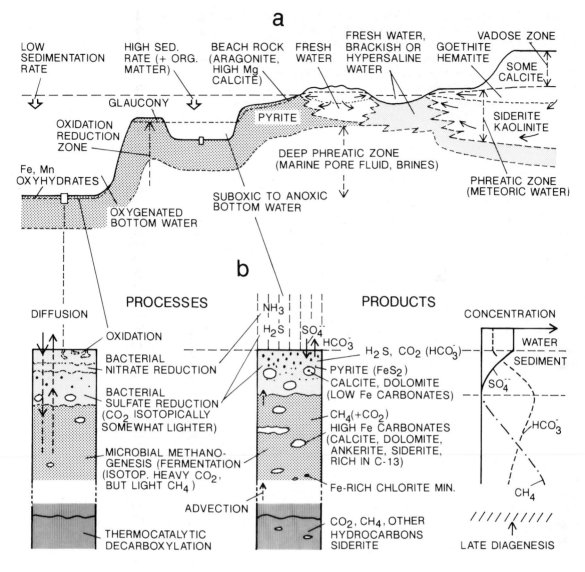

LOW SEDIMENTATION RATE

HIGH SED. RATE (+ ORG. MATTER)

BEACH ROCK (ARAGONITE, HIGH Mg CALCITE)

FRESH WATER

FRESH WATER, BRACKISH OR HYPERSALINE WATER

GOETHITE HEMATITE

VADOSE ZONE

SOME CALCITE

GLAUCONY

PYRITE

OXIDATION REDUCTION ZONE

SIDERITE KAOLINITE

Fe, Mn OXYHYDRATES

DEEP PHREATIC ZONE (MARINE PORE FLUID, BRINES)

PHREATIC ZONE (METEORIC WATER)

OXYGENATED BOTTOM WATER

SUBOXIC TO ANOXIC BOTTOM WATER

b

PROCESSES

PRODUCTS

CONCENTRATION

DIFFUSION

NH_3

H_2S SO_4

HCO_3

WATER

SEDIMENT

OXIDATION

BACTERIAL NITRATE REDUCTION

H_2S, CO_2 (HCO_3)

PYRITE (FeS_2)

CALCITE, DOLOMITE (LOW Fe CARBONATES)

SO_4

BACTERIAL SULFATE REDUCTION (CO_2 ISOTOPICALLY SOMEWHAT LIGHTER)

CH_4 (+ CO_2)

HIGH Fe CARBONATES (CALCITE, DOLOMITE, ANKERITE, SIDERITE, RICH IN C-13)

HCO_3

MICROBIAL METHANO-GENESIS (FERMENTATION (ISOTOP. HEAVY CO_2, BUT LIGHT CH_4)

Fe-RICH CHLORITE MIN.

CH_4

ADVECTION

THERMOCATALYTIC DECARBOXYLATION

CO_2, CH_4, OTHER HYDROCARBONS SIDERITE

///////////

LATE DIAGENESIS

IRON-POOR SED. (CARBONATES) : LITTLE OR NO AUTHIGENIC CARBONATES IN ZONES OF SULFATE RED. AND METHANOGENESIS

C   FRESH WATER ENVIRONMENTS (POOR IN SO_4)

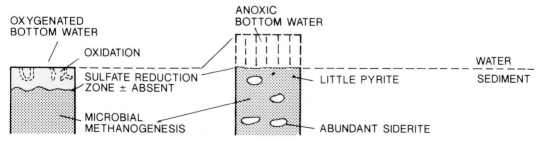

OXYGENATED BOTTOM WATER

ANOXIC BOTTOM WATER

OXIDATION

WATER

SEDIMENT

SULFATE REDUCTION ZONE ± ABSENT

LITTLE PYRITE

MICROBIAL METHANOGENESIS

ABUNDANT SIDERITE

Fig. 13.15. a Overview of various realms of diagenesis. **b** and **c** Processes and products of early diagenesis in **a** marine sediments with oxygenated bottom water and slow burial (*left column*), and with anoxic bottom water and rapid burial (*right column*). **c** Same as **b** but for fresh water environments. (Based on several sources, e.g., Bjørlykke 1983; Curtis 1987). See text for further explanation

quently silicified (see, e.g., Hesse 1989b). The effects of meteoric water circulation may also control large proportions of shallowly buried shelf sands, if there is sufficient recharge on land and a hydraulic connection between the emerged coastal and submerged shelf sands (cf. Chap. 3.3.1 and 13.2). All these meteorically controlled systems are "open" and may transfer considerable quantities of solutes from the recharge to the discharge zone.

Solute Transport by Compaction Flow

In contrast, a sedimentary sequence below a lake or the sea may be affected only by compaction flow (upward directed advection) and diffusion (see below). Because these mechanisms of solute transport operate relatively slowly, such systems are commonly thought of as more or less "closed" (Fig. 13.17a and b). As pointed out in Chapter 13.2, the velocity of vertical compaction flow, v, and the filter velocity, $v_f = v \cdot n$, are commonly less than the sedimentation rate, SR_w.

Using the example in Chapter 13.2 (velocity of compaction flow $v_1 = 0.84\ SR_w$, $SR_w = 40$ m/Ma, porosity $n_1 = 0.3$), the filter velocity of upstreaming pore water is $v_{fl} = 10$ m/Ma. If, for example, the pore fluid contains 20 mg/l or 20 g/m^3 dissolved silica, then the transport rate of silica through a horizontal plane within the sediment is 200 g/m$^2 \cdot$Ma. This amount corresponds to a 0.074 mm thick layer of solid quartz or chert. If a water column 1000 m high, with the same SiO_2 concentration as before, passes through a unit area of the horizontal plane, it may leave behind a quartz layer of 7.4 mm at most.

This solute transport is not enough to provide cement for a sand layer of a few to tens of meters in thickness (Bjørlykke 1984). Similarly, limited compaction flow can barely dissolve enough preexisting cement in order to create secondary porosity (see below). Even the relatively well soluble calcium carbonate (100 to 1000 g/m^3) cannot be transported by compaction flow in sufficient quantities to account for the large volumes of carbonate cement observed in many sequences. However, where thin sandy beds act as lateral drains of upstreaming compaction flow, the volume of pore fluid and the concentration of solutes may match the requirements for significant cementation.

Solute transport by diffusion

Molecular or ionic diffusion in aqueous solutions is one of the most important processes of mass transport in diagenesis. It tends to overcome differences in concentration or chemical potential of a specific ion or pair of ions, for example Na^+ and Cl^- at the surface of buried rock salt, between two locations (distance, d) within the solution. These locations are usually the source and sink (or reaction zone) of the specific ions in question. A thorough treatment of diffusion in sediments is a rather difficult topic (see, e.g., Berner 1980; Ranganathan and Hanor 1987).

Under simplified steady state conditions, the flux, J_i, of a dissolved chemical species, i, through a unit area, A, is directly proportional to the concentration gradient, $(C_2-C_1)/d$, between the two locations (distance, d) and the time, t, the diffusion process operates (Fig. 13.16a; based on Garrels and Mackenzie 1971; Pingitore 1982; Hesse 1986):

$$J_i = \frac{q}{t} = D \cdot (C_2 - C_1) \cdot \frac{A}{d} \quad [g/(m^2 \cdot a)]. \tag{13.10}$$

The quantity, q, of the transported material is expressed here in g, the ion concentration in the aqueous solution, C, in g/m^3, A in m^2, and d in m. The factor of proportionality, D, is the diffusion coefficient for the specific ion in water. It is here expressed in m^2/a, but values given in the literature are usually in cm^2/s or cm^2/day (10^{-6} cm^2/s = 0.0864 cm^2/day = 0.00315 m^2/a). The diffusion coefficient signifies the mass of material transported through a unit area per unit time, if the distance corresponds to one unit length and the concentration gradient is also 1 (in terms of the mass units used, Fig. 13.16a).

If, for example, the diffusion coefficient of Ca^{2+} in water is $D = 7.9 \cdot 10^{-6}$ cm^2/s ≈ 0.025 m^2/a (at 25° C, Lerman 1980), the diffusion flux under steady state conditions (as shown in Fig. 13.16a) is $J = 0.025$ g/(m$^2 \cdot$ a) or 25 kg/(m$^2 \cdot$ Ma).

Solute transport by diffusion increases with growing concentration gradient, decreasing distance between the ion source and the reaction zone, and increasing temperature. An increase in temperature on the order of 30 °C doubles the effectiveness of diffusion. However, this does not necessarily mean that diffusion increases with burial depth. The significance of diffusion is reduced by low con-

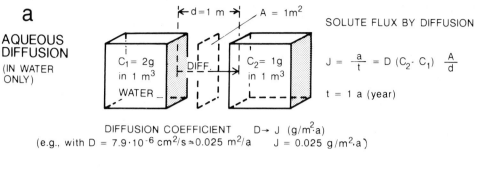

a

AQUEOUS DIFFUSION

(IN WATER ONLY)

SOLUTE FLUX BY DIFFUSION

$C_1 = 2g$ in $1 m^3$ WATER

$C_2 = 1g$ in $1 m^3$

$\leftarrow d = 1 m \rightarrow$ $A = 1m^2$

$J = \dfrac{a}{t} = D (C_2 - C_1) \dfrac{A}{d}$

$t = 1$ a (year)

DIFFUSION COEFFICIENT $D \rightarrow J$ $(g/m^2 \cdot a)$

(e.g., with $D = 7.9 \cdot 10^{-6}$ cm^2/s $\approx$ 0.025 m^2/a $J = 0.025$ g/m$^2 \cdot$a)

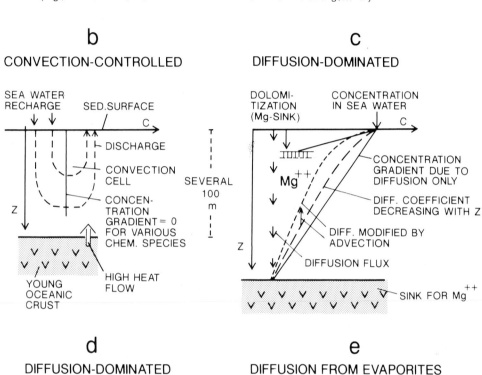

b

CONVECTION-CONTROLLED

SEA WATER RECHARGE   SED.SURFACE

DISCHARGE

CONVECTION CELL

CONCEN-TRATION GRADIENT = 0 FOR VARIOUS CHEM. SPECIES

SEVERAL 100 m

YOUNG OCEANIC CRUST

HIGH HEAT FLOW

c

DIFFUSION-DOMINATED

DOLOMI-TIZATION (Mg-SINK)

CONCENTRATION IN SEA WATER

Mg^{++}

CONCENTRATION GRADIENT DUE TO DIFFUSION ONLY

DIFF. COEFFICIENT DECREASING WITH Z

DIFF. MODIFIED BY ADVECTION

DIFFUSION FLUX

SINK FOR Mg^{++}

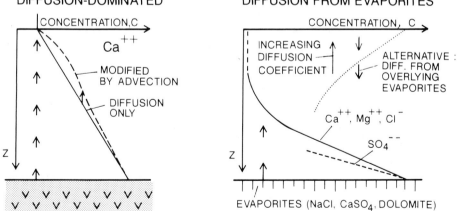

d

DIFFUSION-DOMINATED

CONCENTRATION, C

Ca^{++}

MODIFIED BY ADVECTION

DIFFUSION ONLY

e

DIFFUSION FROM EVAPORITES

CONCENTRATION, C

INCREASING DIFFUSION COEFFICIENT

ALTERNATIVE: DIFF. FROM OVERLYING EVAPORITES

Ca^{++}, Mg^{++}, Cl^-

SO_4^{--}

EVAPORITES (NaCl, CaSO$_4$, DOLOMITE)

Fig. 13.16. a Meaning of the diffusion coefficient in terms of diffusion transport under simplified conditions.
b through **d** Various types of vertical concentration gradients in marine sediments. **b** In young sediments of limited thickness on top of the oceanic crust of mid-oceanic ridges. Differences in heat flow cause convective flow, which prevents chemical gradients. **c** Diffusion-dominated Mg^{2+} gradient, somewhat modified by depth-dependent diffusion coefficient or advection, in sediments on top of older oceanic crust. **d** Same as **c** but for Ca^{2+}. **e** Diffusion-controlled, strong vertical gradients of chemical species associated with underlying (or overlying) evaporites. (Largely based on Hesse 1986)

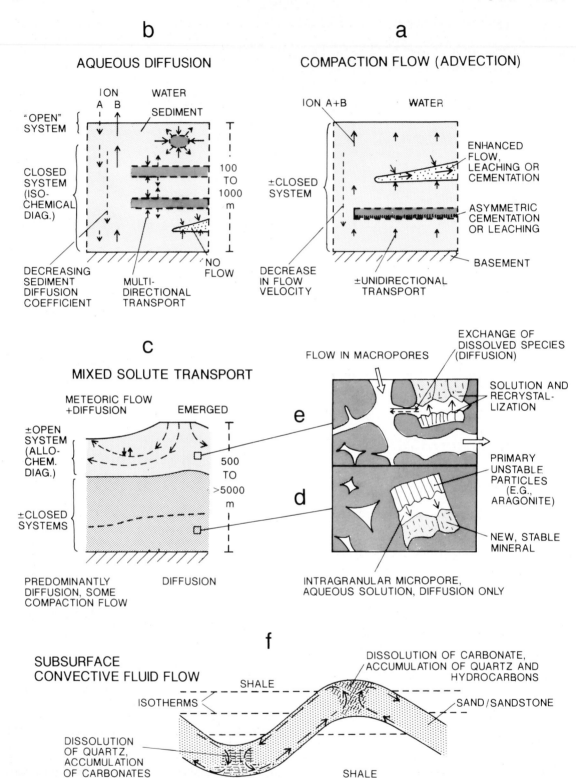

b

AQUEOUS DIFFUSION

a

COMPACTION FLOW (ADVECTION)

c

MIXED SOLUTE TRANSPORT

f

SUBSURFACE CONVECTIVE FLUID FLOW

Fig. 13.17. a-c Various mechanisms of solute transport in sediments reflecting more or less "closed" or "open" chemical systems. **d** Solution-recrystallization via thin film of pore fluid in micropore of closed (isochemical) system. **e** Same as **d** but in open (allochemical) system.

f Convective pore fluid flow in folded sand sheet intercalated with shales. (After Wood and Hewett 1984). Quartz is dissolved in the hotter synclinal trough and precipitated in the cooler anticline, while carbonate behaves vice versa

centration gradients and long transport distances. In addition, diffusion is generally hampered in sediments as compared to water, because sediments consist for the most part of solid matter and have rather irregular, changing pore space and pore geometry. Therefore, both the porosity, n, and the tortuosity (length/length) of the pores, i.e., the actual path a dissolved species has to travel between two points within a sediment, have to be taken into account. Thus, the diffusion coefficients for sediments, D_s, are commonly smaller than those for water, D, only. The diffusion transport in sediments becomes

$$J_s = \frac{q}{t} = n \cdot D_s \cdot (C_2 - C_1) \cdot \frac{A}{d} . \quad (13.11)$$

For n = 0.2 and a tortuosity of 1.33, or D_s = 0.75 D, the diffusion transport for the example in Fig. 13.16a is reduced to 15 % of the amount calculated for diffusion in water only.

With increasing burial depth and continued loss in porosity, the diffusion factor, $n \cdot D_s$, may drop to values of 10^{-7} to 10^{-8} cm^2/s (Manheim 1970), and thus the efficiency of diffusion declines further (Fig. 13.16c). Nevertheless, solute transport by molecular diffusion is frequently more effective than compaction flow.

Using the example in Fig. 13.16a and taking into account a reduction in the diffusion factor, $n \cdot D_s$ = 0.1 D, as discussed above, the diffusion flux in sediments is J_s = 0.1 x 25 = 2.5 kg/(m$^2 \cdot$ Ma) (for distance d = 1 m). For d = 100 m, J_s drops to 0.025 kg/(m$^2 \cdot$ Ma), for d = 0.1 m it increases to 25 kg/(m$^2 \cdot$ Ma). The transport rate of compaction flow, calculated in the previous section for a concentration (or difference in concentrations between the inflowing and outflowing pore water) of 10 g/m^3 was only 0.2 kg/(m$^2 \cdot$ Ma).

This means that molecular diffusion is the main transport mechanism over short distances in both closed and open sedimentary systems.

The significance of time in diffusion processes may also be defined in terms of a *mean diffusion path length*, z_m (Gieskes 1975):

$$z_m = (D_s \cdot t)^{0.5} \quad . \quad (13.12)$$

This is the distance over which diffusion will dissipate concentration anomalies for a given period of time.

The mean diffusion path length, z_m, is 100 to 200 m for a time period of 1 Ma and values of D_s between approximately 4 and 10 x 10^{-6} cm^2/s (= 0.0126 to 0.0315

m^2/a), which include the bulk sediment diffusion coefficients of Na$^+$, Ca^{2+}, Cl$^-$, and SO$_4^{2-}$ in young soft muds with n = 0.71 at a temperature of 20 to 25 °C (Berner 1980).

This signifies, for example, that solute transport by diffusion from sea water can penetrate about 100 m downward into the sediment, as long as the sedimentation rate does not exceed 100 m/Ma. Long diffusion pathways on the order of several hundred to more than 1000 m were found in drill holes of the Deep Sea Drilling Project (Fig. 13.16c and d). Here, diffusion is reflected by vertical concentration gradients between sea water and oceanic basement (e.g., for Ca^{++} and Mg^{++}) in the pore fluids of the sediments over great depth ranges. Only where convective flow causes relatively fast, upward and downward directed solute transport is the effect of diffusion largely eliminated (Fig. 13.16b).

Evaporites greatly influence their overlying or underlying sediments, because they create very steep concentration gradients for several chemical species (e.g., Cl$^-$, SO$_4^{2-}$, Ca^{2+}, Mg^{2+}, Na$^+$; Fig. 13.16e).

For example, in the Mediterranean, the pelagic to hemipelagic sediment cover on top of Late Miocene (Messinian) evaporites (Chap. 6.4) forms a section of about 200 m with abnormally high, but upwardly decreasing pore water salinity. The pore fluids are characterized by the easily soluble constituents of the underlying halite, anhydrite, and dolomite deposits. In this case, diffusion transport has covered a vertical distance of 200 to 300 m in a time interval of approximately 5 Ma. Similar observations were reported from the North and South American Atlantic margin, offshore West Africa, and the Gulf coast of the United States (Holmes 1986; Bjørlykke 1988). As known from several other examples, sandy sediments in the neighborhood of evaporites may contain salt minerals as cement (e.g., anhydrite).

The diffusion coefficients of different chemical species vary slightly. Chloride and carbonate ions have relatively high diffusion coefficients and can therefore migrate somewhat faster than most of the common cations.

In a sediment undergoing chemical diagenesis, the concentration gradients of the various chemical species often vary considerably. At a single site, a strong gradient may exist for one or a pair of species, but weak or no gradients at all for other species. Hence, only the species affected by a significant concentration gradient will exhibit diffusion transport. Such conditions are encountered, for example, along the *boundary between an oxidation and reduction zone*, where SO$_4^{2-}$ is used up by anaerobic bacteria and ferrous iron precipitated as iron monosulfide or pyrite (Fig. 13.15, reaction-controlled diffusion).

In contrast to advective mass transport through pore water flow, mass transfer of specific ions by diffusion can occur toward or away from the reaction zone simultaneously ("two-way traffic"). In relation to the concentration gradients, one species may enter a certain reaction zone, while another species leaves (Pingitore 1982).

For example, during the transformation of skeletal aragonite (e.g. corals) to calcite, the aragonite may release strontium to its environment, while the new calcite takes up magnesium from outside (compare Fig. 13.17d).

Redistribution of Solutes by Convective Fluid Flow

As already mentioned in Chapter 13.2, convection of pore fluids in a relatively large, closed circulation system appears to be a rather common phenomenon in the subsurface (Wood and Hewett 1984), even under conditions of normal geothermal gradients. Such long-term operating *circulation systems* have been postulated by several other authors in order to explain the volume of dissolved and redistributed cement in thick sandstone bodies, which otherwise are difficult to explain (e.g., Bjørlykke 1984).

For example, the slowly moving pore fluid (rate of convection about 1 m/year) may circulate in a thick sand layer in a folded sequence in a so-called toroidal cell (Fig. 13.17f). It is heated during its downward flow to the synclinal trough and cools as it approaches the anticlinal crest. During its circulation, the pore fluid has the tendency to maintain chemical equilibrium with the adjacent rock matrix. Because the solubility of minerals is, among other factors, a function of temperature, the cooling pore fluid precipitates minerals which have "prograde solubilities" (for example quartz) and dissolves minerals whose solubility decreases with increasing temperature ("retrograde solubility", for example carbonates). As a result, quartz will move from hot source zones to cooler sinks, but carbonates from cool sites to hot ones. Dissolution and reprecipitation are particularly intense in anticlinal crests and synclinal troughs where the pore fluids are most rapidly cooled or heated. In this way, a continuous tranfer of solutes is accomplished in the permeable rock layer.

It is interesting to note that from 60 to 150 °C the solubilities of common hydrocarbons in water are similar to that of quartz. Hence, both components can be transferred and accumulated in comparable quantities under these conditions.

Combination of Various Processes of Solute Transport

Finally, there are many sedimentary basins where several of the above mentioned solute transport systems operate simultaneously (Fig. 13.17c). While the meteoric, open system is realized in emerged, coastal and shallow-buried parts of the basin fill, water-covered and deeply buried portions of the basin fill are characterized by compaction flow, diffusion, and convective flow in a more or less closed system.

As a result of differing solute transport, the *mineral reactions* occurring in individual pores and micropores in the sediments also vary fundamentally. In a closed system, an unstable mineral may dissolve and be replaced by one or several stable minerals, which in total have the same chemical composition as the parent mineral. For example, high Mg calcite may be transformed into low Mg calcite and some dolomite. The new mineral phase(s) are formed directly adjacent to the dissolving mineral; solute transport takes place by diffusion via a very thin film of pore water (Fig. 13.17d).

In an open system, the micropores between a dissolving and newly crystallizing mineral are more or less in contact with the macropores (Fig. 13.17d). These serve as conduits for pore fluid flow and thus for the exchange of chemical species between the reaction zone of the micropore and zones far away from this location. In this case, the new mineral phase(s) and their trace element contents frequently differ in their bulk composition from the primary mineral(s) (allochemical diagenesis).

Early Diagenesis

Most geologists and sedimentologists dealing with diagenesis distinguish between processes of "early" *(shallow-burial)* and "late" *(deep-burial)* diagenesis, but some experts have described more stages of diagenesis for particular rock sequences or regions. A sharp boundary between early and late diagenesis is difficult to define, because it depends on the nature of the process which is considered most important.

If, for example, the transition from biogenic decomposition of organic matter to thermocatalytic reactions (see Chap. 14.2) and the onset of hydrocarbon migration are taken as particularly significant in diagenesis, the lower boundary of early diagenesis coincides with a temperature of 60 to 75 °C (e.g., Edman and Surdam 1984; Hesse 1986). Assuming a normal temperature gradient of 25 °C per 1000 m and a surface temperature of 10 °C, the transition to catalysis corresponds with a burial depth of at least 2000 m. However, if the peak of carbonate diagenesis is thought to characterize early diagenesis, a much shallower burial depth appears to be appropriate.

Therefore, many authors (e.g., Berner 1980) leave this question rather open. Thus, early diagenesis may refer to changes occurring during burial to a few hundred meters, where elevated temperatures are not encountered.

The processes and products of early diagenesis vary strongly according to the depositional environment and the type of sediment.

Marine Muds under Oxic and Anoxic Conditions

Fine-grained marine sediments deposited slowly under *oxic bottom waters* display the following diagenetic subzones (Fig. 13.15b). The upper sediment layer below the interface is kept in an oxygenated state by downward diffusion of dissolved oxygen from the overlying water column. Hence this layer can be colonized by *benthic organisms and aerobic bacteria*, which consume and oxidize organic matter reaching the sea floor. Burrowing organisms in turn promote mass transfer from the sea water into the sediment and vice versa. If all the organic matter is used up and iron and manganese are oxidized to form oxyhydrates, *brown or reddish brown sediment* may result which is no longer affected by organic processes (see Chap. 5.3.3). If some reactive organic matter is left over below the oxidation zone, a very thin zone of bacterial nitrate reduction and a somewhat thicker zone of anaerobic sulfate reduction develop. In this way, CO_2 and H_2S (and small amounts of NH_3) are produced, which may migrate to the sea floor by upward diffusion or react with other ions to form new minerals, for example pyrite.

Sediments *deposited more rapidly under anoxic conditions* are commonly rich in organic matter and have a stronger tendency to form early diagenetic authigenic minerals (Irwin et al. 1977; Goldhaber and Kaplan 1980; Gautier and Claypool 1984; Curtis 1987). *Nitrate reduction* occurs in the water column and *sulfate reduction* operates both above and below the sediment-water interface (Fig. 13.15b). Since normal benthic organisms are absent, comparatively large amounts of organic matter are available for bacterial sulfate reduction. A shortage in the sulfate content in the pore water can to some extent be replenished by downward diffusion of sulfate from the sea water reservoir. Hence, considerable quantities of CO_2 and H_2S can be produced. Iron released from minerals and reduced to Fe^{2+} is precipitated as (metastable) monosulfide and stable pyrite, particularly in the upper part of the sulfate reduction zone. Commonly, early diagenetic pyrite occurs as fine-grained framboidal grains and aggregates, but it may also form internal moulds of fossils (e.g., Hudson 1982). Calcareous fossil shells, particularly those consisting of aragonite, are frequently dissolved or replaced by pyrite.

Somewhat deeper, but still within the sulfate reduction zone, the decomposition of organic matter causes the *formation of authigenic calcite or dolomite* (Lippmann 1973; Kelts and McKenzie 1982), often in the form of concretions. Diffusive transport of methane from below may stimulate anaerobic sulfate reduction and provide the alkalinity for the precipitation of carbonate ($CH_4 + SO_4^{2-} = HCO_3^- + HS^- + H_2S$; Raiswell 1987). A slow sedimentation rate or a break in sedimentation favors the formation of concretions. In particular cases, salinity variations in the overlying water body may be recorded in the mineral phases of concretions and their isotopic composition (e.g., Carpenter et al. 1988).

Normally, the early formed carbonate concretions are nonferroan, since the pore fluids are depleted in iron by the preceding iron sulfide precipitation. Diagenetically early iron carbonate *(siderite)* only forms under specific conditions when the rate of iron reduction exceeds the rate of sulfate reduction. Then, the dissolved sulfide is insufficient to precipitate all the available dissolved ferrous iron, as observed in late Holocene marsh and sandflat sediments on the Norfolk coast, England (Pye et al. 1990).

The lowered ^{13}C content of carbonate concretions reflects some bacterial fractionation of carbon isotopes. The lower boundary of this zone may lie several meters up to 10 m below the sea floor. Early diagenetic phosphogenesis as briefly discussed in Chapter 5.3.6, is related to the processes taking place in the sulfate reduction zone.

If sulfate is exhausted and can no longer be sufficiently supplied by downward diffusion, the precipitation of pyrite and isotopically light carbonate ceases. Further decomposition of organic matter is now accomplished by *microbial fermentation* reactions, which produce isotopically light methane and heavy carbon dioxide. Carbon dioxide may be reduced to provide oxygen for anaerobic respiration. These processes are frequently summarized under the term *methanogenesis* (Fig. 13.15b). Characteristic authigenic minerals in this zone are *carbonates rich in ferrous iron and* ^{13}C (ferroan calcite, dolomite, ankerite, and siderite).

Siderites formed in such marine environments are commonly impure (in contrast to freshwater siderites) and are rich in Mg and Ca (Mozley 1989). Even the occurrence of rhodochrosite ($MnCO_3$) has been reported (e.g., von Rad and Botz 1987). Carbonate precipitation often occurs in the form of concretions, which tend to develop a flatter shape than in the sulfate reduction zone. Some of the carbonate in the concretions may be derived from dissolution and reprecipitation of detrital, biogenic carbonate.

In addition, iron-rich chlorite minerals can form in the zone of methanogenesis. However, iron-poor marine muds, for example pelagic carbonates, can only form small quantities of pyrite in the sulfate reduction zone and little or no authigenic carbonates in both the sulfate reduction and the methanogenesis zones.

In *fresh water environments* where pore fluids are poor in sulfate, the sulfate reduction zone and the precipitation of pyrite are subordinate or absent (Fig. 13.15c). As a result, the zone of methanogenesis expands upward. Under anoxic conditions, it may extend into the water body. The principal authigenic mineral is siderite, which may occur in large quantities, while pyrite is rare.

The zone of microbial methanogenesis extends deep below the sea floor, but a specific depth range is difficult to define. In organic matter-rich sediments on the continental margins of the present oceans, methane has been observed in the form of *gas hydrates* (*clathrates;* Hesse et al. 1985; Hesse 1986) in several drill holes.

Gas hydrates are ice-like substances, which form with high methane concentrations, elevated hydrostatic pressures (at water depths of at least 500 m), and relatively low temperatures (less than +10 to +30 °C). The vertical range of gas hydrates may be up to 1000 m within a sedimentary column. When the gas hydrates thaw, for example due to increasing temperature with depth, they release large quantities of methane and fresh water, thus influencing sediment stability and pore water chemistry.

Biogenic Sediments

Fine-grained sediments rich in carbonate (e.g., hemipelagic oozes and marls), but also grainstones, show additional early diagenetic features, which result from the relatively good solubility of the primary carbonate minerals. Aragonitic and high Mg calcitic skeletal particles in particular are easily dissolved and reprecipitated as carbonate cement in neighboring layers (Figs. 13.17b and 13.18). With increasing burial depth, pressure solution may become an important mechanism providing carbonate cement.

It occurs, however, only in those layers, which have not been lithified by previous cementation (e.g., Hird and Tucker 1988; James and Bone 1989). The dissolved carbonate is transported by diffusion over short distances (over few to tens of centimeters).

In this way, indistinct primary bedding is enhanced by diagenetic overprint (Ricken 1986; Ricken and Eder 1991; Bathurst 1987, 1991). Even minor variations in primary composition and pore space are sufficient to cause significant diagenetic modifications. Typical examples demonstrating this internal carbonate redistribution process are limestone-marl successions, layered carbonate concretions, and nodular limestones (Fig. 13.18). Hardgrounds may result from both the reaction of surface sediment with the overlying sea water and internal carbonate redistribution. Marked differences in compaction between limestone concretions, limestones, and adjacent marls testify to the fact that carbonate is commonly redistributed at relatively shallow burial depths (several to tens of meters). In contrast, pelagic oozes consisting predominantly of calcite skeletons usually dissolve and recrystallize at deeper burial depths (Chap. 13.1, Fig. 13.3e and f). The diagenesis of shallow-water carbonate buildups and reefs is briefly described in Chapter 3.3.2.

A frequently discussed problem is the *dolomitization* of thick, deeply buried calcium carbonate deposits. The Mg content of normal marine pore waters and compaction flow is not sufficient for transforming large volumes of calcium carbonate into secondary dolomite (*pervasive dolomitization*). Mg-rich bittern salts are rare in the subsurface. Thus, pervasive dolomitization normally appears to require a system which allows downward Mg transport from the sea water reservoir or other Mg sources by diffusion, convective flow, or meteoric water circulation (Land 1985; Moore

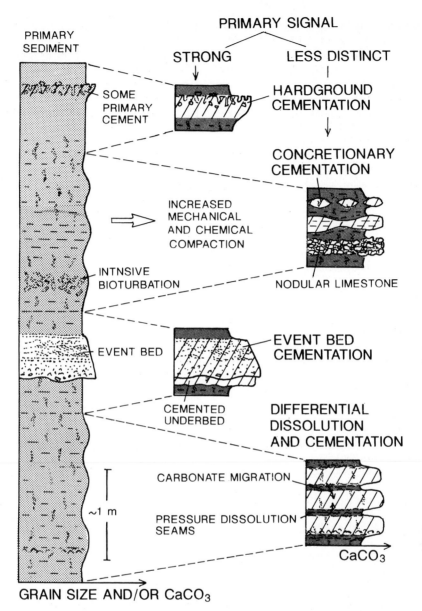

Fig. 13.18. Various types of diagenetic overprint and accentuation of indistinct primary bedding features in fine-grained calcareous sediments. *Left-hand side* Uncompacted primary sediment. *Right-hand side* Selected sections after mechanical and chemical compaction, as well as cementation of particular beds. Several of these processes may operate simultaneously in the same sediment. (After Ricken and Eder 1991)

1989). Smaller quantities of Mg can be released in conjunction with the transformation of unstable to stable clay minerals (see below), but at least part of the available Mg may be consumed in the formation of chlorite. More details on this difficult and somewhat controversial topic, including dedolomitization (the process of calcite replacing dolomite), are given, for example, by Chilingar et al. (1979), Lee and Friedman (1987), Füchtbauer (1988), Moore (1989), and Tucker and Wright (1990).

The diagenesis of *siliceous*, predominantly deep-sea sediments is described in Chapter 5.3.5 (see also summary by Iijima 1988). The phenomena of silica diagenesis in shelf carbonates are in many ways similar to those observed in pelagic carbonates. Due to more pronounced undersaturation of shallow waters with respect to silica, however, opal skeletons, such as sponge spicules, are frequently dissolved and replaced by carbonate as long as they are in diffusive contact with sea water (e.g., Brachert et al. 1987). Opaline silica preserved within the sediment usually forms early diagenetic chert nodules which are little affected by compaction. Under special conditions, calcareous fossil remains become silicified. This may result, for example, from silica-rich alkaline pore fluids below hypersaline lagoonal waters. Silicification of carbonate may also occur at the contact with fresh pore water, when part of the dissolved silica in saline pore fluids is precipitated due to lowered pH.

In highly saline, alkaline, sodium carbonate-bicarbonate lakes, a special type of early diagenetic chert may be precipitated (Sheppard and Gude 1986). This was first observed in Upper Pleistocene deposits of Lake Magadi in Kenya, but since then has been found in many lake deposits of Jurassic to Pleistocene age. This Magadi-type chert occurs as thin, discontinuous beds or as plates and nodules of irregular shape. Typically, the chert is dense and translucent and has a white rind. It probably formed from a primary precipitate of hydrous sodium silicate. The conversion of the precursor mineral to chert is accompanied by a loss of sodium and water and a reduction in volume. Magadi-type chert is an excellent indicator of past environmental conditions.

Both fine-grained, deep-sea carbonate and siliceous sediments commonly lithify as a result of recrystallization and cementation at burial depths of a hundred to several hundred meters. Shallow-water carbonates containing high proportions of unstable mineral phases, however, tend to lithify earlier. Hypersaline conditions frequently promote the formation of hardgrounds at or slightly below the sea floor. In contrast, siliciclastic shales and marls maintain their plastic, ductile state and undergo compaction up to great burial depths (several kilometers).

Late Diagensis

At deep burial depths the sediments are subject to *increased temperature and effective stress*, but solute transport by the various mechanisms described in the previous section generally slows down as a result of *decreasing permeability and sediment diffusion coefficients*. Thus, on the one hand, the solubilities of many mineral species (e.g., quartz) increase considerably due to elevated temperature and pressure solution. Other mineral species that are stable at surface temperature become unstable and tend to transform into new minerals (Fig. 13.19). On the other hand, the macro-scale exchange of dissolved matter

is hampered, which frequently leads to more highly concentrated pore fluids. These may significantly deviate in their composition from pore waters near the sediment-water interface. However, when freshwater or brackish water deposits are overlain by marine sediments, the deeply buried pore fluids may have lower salinities than marine pore waters.

With increasing temperature and burial depth, the zone of microbial methanogenesis is followed by the zone of *thermocatalytic decarboxylation* (cf. Chap. 14.2). This process is generally ascribed to late diagenesis. Its products are methane, carbon dioxide, and various hydrocarbons. In addition, siderite may form in minor quantities. If the decomposition of kerogen releases some sulfur, small quantities of late diagenetic pyrite may also result.

The solubility of quartz markedly increases with temperature, while carbonates show little effect or even decreasing solubilities (compare Fig. 13.17f). At deep burial depth, hydrous mineral phases tend to transform into denser, less hydrous phases (Fig. 13.19). Thus, for example smectite (i.e., montmorillonite) forms illite at temperatures between 60 and 100 °C and releases water. Similarly, kaolinite is converted to illite and quartz (120 to 150 °C).

In conjunction with continued porosity reduction, possibly enhanced by lateral tectonic stress, shales transform into slates. Oxyhydrates of iron and manganese may form less hydrous compounds or, after reduction, may be incorporated into carbonates.

Volcanic glass (tuffs) take up water when shallowly buried and form zeolites (Iijima 1988). These and other unstable minerals may later transform into chlorite, low-temperature feldspar (preferentially potash feldspar or albite), and quartz. Most of these various reactions release water and dissolved species and thus modify the ionic strength and composition of the pore fluids.

Cementation and Secondary Porosity of Sandstones (and Calcarenites)

Sandstone diagenesis (as well as diagenesis of relatively coarse-grained carbonate buildups, Chap. 3.3.2) is of particular interest in hydrocarbon exploration and hydrogeology, because these rocks provide reservoirs and conduits for water, oil, and gas. One of the principal problems is the question whether the sandstones maintained or regained significant porosity, when hydrocarbons started to mi-

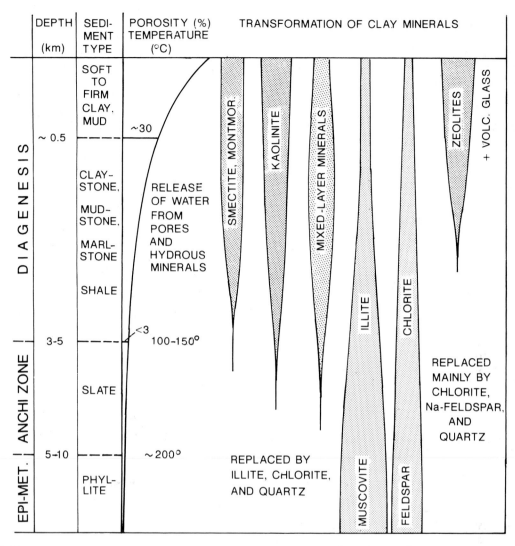

Fig. 13.19. Diagenesis of pelitic sediments with most important mineral transformations. (After Frey et al. 1980; Heling 1988)

grate. It is also important to know whether meteoric water could have circulated in the deeply buried sandstone. Sands and sandstones are permeable and therefore act as favored conduits for all kinds of flow (Chap. 13.2), as long as they are not completely cemented. During a long burial history they may be exposed to the influence of pore fluids of varying composition. For these reasons and many other factors, sandstone diagenesis is a very complex topic.

The initial porosity of sands (frequently between 25 and 45 %) is just moderately reduced by mechanical compaction, i.e., rearrangement and deformation of grains (Fig. 13.20a; Chap. 13.2). More important are the

following processes (e.g., Bjørlykke 1983, 1988; Füchtbauer 1988):

- Precipitation of cement in the primary pore space and overgrowth of grains.
- Transformation of primary minerals (e.g., feldspar) into expanded hydrous mineral phases (kaolinite and other clay minerals).
- Pressure solution.

Early Diagenesis of Sands

Significant early diagenesis of sands occurs where they are circulated by meteoric waters with low ionic strength. Then their unstable

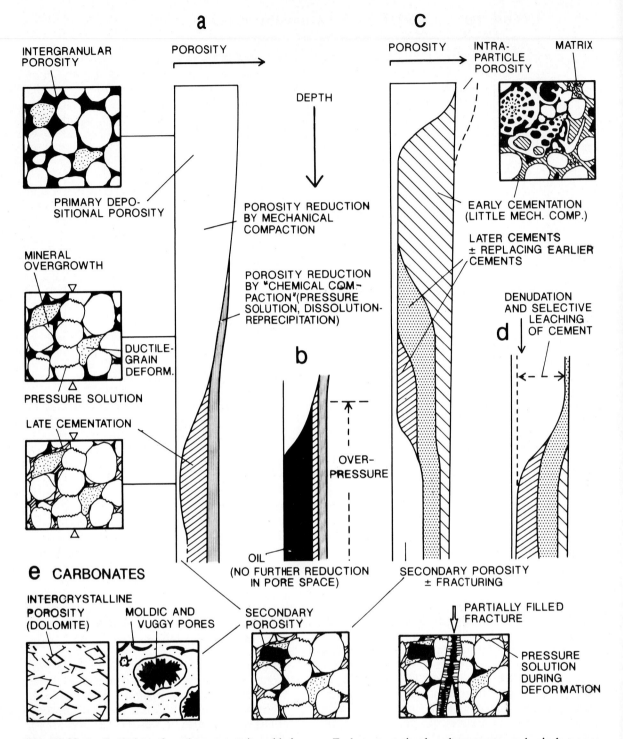

Fig. 13.20a-e. Evolution of sandstone porosity with increasing burial depth and time. **a** Reduction of porosity by mechanical and (iso)chemical compaction predates late (allochemical) cementation. **b** Same as **a** but prograding pore space reduction is hampered or terminated by overpressure and oil fill.

c Early cementation largely prevents mechanical compaction and pressure solution; secondary porosity by leaching of cement and unstable grains. **d** Selective leaching of cement and unstable grains after uplift and denudation. **e** Some features of secondary porosity in carbonates. (Partially based on Selley 1976; Mitra 1988)

and easily soluble mineral phases are removed or altered, reflecting to some extent the climatic conditions of the region. Some of the solutes may be precipitated in zones of very slow water movement or stagnant pore water (Chap. 13.3). In warm climate, shallow-marine waters are frequently supersaturated with respect to calcium carbonate. Hence, coastal and foreshore sands which are in contact with these waters tend to precipitate carbonate cement (Fig. 13.20c).

Sands in deeper marine environments, on the other hand, are affected neither by meteoric waters nor by carbonate supersaturated sea water. Since most of their common mineral phases (quartz, feldspar, clay minerals) are more or less in equilibrium with sea water, there is little cause for reactions between the sand particles and the marine pore fluid during the early stages of diagenesis (Fig. 13.20a). Above or near the redox boundary (Fig. 13.15a and b), some glauconitic minerals and iron-rich chlorites and chamosites may form (cf. Chap. 6.1). In the sulfate reduction and methanogenesis zones, pyrite and various carbonates may locally fill the pore space.

Where relatively thin sand layers and sand bodies act as drains for upstreaming compaction flow (Chap. 13.2) which is supersaturated with respect to carbonate and silica, small amounts of quartz cement and possibly larger volumes of carbonate may precipitate. However, the solubility of carbonate may increase with the cooling of the pore fluids and thus prevent precipitation. Sands in direct or indirect contact with evaporative brines frequently exhibit sulfates, halite, and zeolites as pore cement.

The Cement Source of Deeper Buried Sandstones

In spite of these various reactions releasing water and dissolved constituents in the deep subsurface, many authors have pointed out that the mass of dissolved matter and the volume of pore fluids are insufficient to account for the large volumes of cement observed in many sandstones.

Even in the case of relatively easily soluble carbonates, about 100 000 volumes of pore water are needed to fill one volume of pore space (Bathurst 1975). Therefore, the release of carbon dioxide from the decomposition of organic matter and kerogen is often quoted as a mechanism for dissolving carbonate, which can be precipitated elsewhere.

Particularly the source of silica for quartz cement, which is the most important cement type besides carbonate, is frequently discussed in the literature. Several sources of silica have been proposed (e.g., Bjørlykke 1988; Füchtbauer 1988): (1) Pressure solution, (2) transformation of feldspar and mica into kaolinite, (3) smectite-illite transformation, (4) meteoric water rich in silica due to intensive chemical weathering, etc. The last process may account for the cementation of shallowly buried sandstones, but is normally excluded for deeply buried strata. Therefore, in the deep subsurface, pressure solution within the sands and mass transfer by diffusion from directly underlying and overlying shales probably provide the major part of silica necessary for large volumes of quartz cement. Finally, in some regions, convective flow in large circulation cells (Fig. 13.17f) may transfer silica in considerable quantities from deeper and warmer to more elevated and cooler portions of a sand body.

Cement Mineral Sequences

The previous discussion on the various realms of diagenesis and systems of mass transport in different rocks and rock sequences demonstrates that a common rule for the succession of cement minerals (diagenetic mineral paragenesis) cannot exist. In addition, the numerous mineral reactions involved, including the replacement of primary grains and early cement by secondary mineral phases (Fig. 13.20c), render it difficult in many cases to establish the true, complete cement sequence. Great, small and larger scale spatial variations in porosity and permeability, as found for example in siliciclastic delta plain sediments or in carbonate buildups (Dreyer et al. 1990; Schroeder 1988) enhance this problem. They may lead to differing coeval cements and cement successions which are caused by primary differences in grain size and fabric, skeletal structures, bitoturbation and bioerosion, oceanographic parameters, etc.

For certain depositional environments or individual sedimentary basins, however, fairly regular cement sequences have been found (e.g., Schmidt and McDonald 1979; Edman and Surdam 1984; Füchtbauer 1988). According to observations of thin sections (Fig. 13.20), which may be supplemented by evidence from geochemical, unstable and stable isotope studies (Chap. 13.5), the following cement sequences occur frequently in *sandstones:*

1. Aluminium silicates
2. Quartz
3. Carbonates
4. Sulfates and chlorides.
   Or:
1. Quartz
2. Calcite or dolomite
3. Anhydrite or calcite.

Other cement minerals of some importance include goethite, hematite, and zeolites. Later cements can replace earlier ones and thus obliterate the diagenetic history. In many sandstones calcite is found to replace quartz. Late diagenetic cements may include small quantities of dolomite and pyrite. In volcaniclastic sands, glass may be followed by zeolites which later transform into feldspar or mica. Greywackes and sands rich in feldspar and lithic components develop a matrix consisting of altered primary particles and cement which was precipitated later (clay minerals, chlorite, and other silicates).

Shallow-marine carbonates or mixed siliciclastic-carbonate sediments with different grain size distributions tend to show the following (simplified) sequence of calcite cements:

1. High Mg calcite (in addition to aragonite).
2. Ankeritic calcite, Fe, Mn ≥ Mg.
3. Ferroan calcite, high Fe.
4. Low Mg calcite.

Such cement sequences frequently reflect the changing pH and Eh conditions evolving within the sediments. There are, however, many exceptions from this trend. In each of these stages, calcite may be replaced by dolomite.

Secondary Porosity

Many sandstones exhibit secondary porosity in contrast to shales, which commonly maintain their state of compaction and cementation they have achieved during diagenesis. This is frequently observed in thin sections as voids after the leaching of clastic grains (e.g., feldspar, bioclastic carbonate) or carbonate cement (Fig. 13.20). In addition, dissolution and recrystallization of mineral phases in place can change the distribution and geometry of pores, but it hardly creates a significant increase in overall porosity. Generally, such an increase in porosity and permeability can be caused by

- Dissolution of primary siliciclastic or bioclastic grains
- Dissolution of cement.
- Transformation of unstable, poorly crystallized, water-rich minerals (e.g., smectite, kaolinite) into well crystallized, less hydrous minerals (illite), accompanied by loss of water.
- Dolomitization of calcium carbonate.
- Fracturing by overpressure (Chap. 13.2) or tectonic movements.

One of the most important processes for providing CO_2 for the dissolution of carbonate appears to be the maturation of kerogen (Schmidt and McDonald 1979; Bjørlykke 1988). It is, however, only the humic type of kerogen with a high O/C ratio, which enables substantial leaching of carbonate. Other mechanisms for the dissolution of carbonate and possibly other types of cement are the circulation of acid meteoric water and convective flow (Fig. 13.17f).

For the migration and storage of hydrocarbons it is crucial that sufficient primary (or secondary) porosity is preserved at critical burial depths (approximately 3 to 5 km). Thus, the time and depth range of pore space reduction (or secondary increase in pore space) control the quality of sands as reservoir rocks. Once the pores have become filled with oil (Fig. 13.20c), the residual water is no longer effective as means of mass transfer. Thus, the reservoir can more or less maintain its porosity, whereas neighboring rocks with water-filled pores are further affected by processes changing their porosity.

13.4 Thermal History of Sedimentary Basins and Onset of Metamorphism

Assuming a normal geothermal gradient, a specific rock unit must be buried very deeply in order to attain the temperature needed for the onset of low-grade metamorphism (around 200 °C). With a vertical temperature gradient of 25°/km, for example, the sediment unit should be covered by a younger sequence 8 km in thickness. From this simple calculation, one can deduce that regional metamorphism normally occurs only in deep subsiding basins accumulating great thicknesses of sediments. Otherwise, metamorphisms is caused by specific factors, such as abnormally high temperature gradients, lateral compressional stress, and overthrusting in an accretionary wedge.

Thermal History of Basin Fills

The temperature distribution in a sedimentary basin is controlled by *conductive heat flow* from the deeper crust and, in particular cases, by *advective heat transfer* due to meteoric water circulation. Figure 13.21 demonstrates examples of the reconstruction of the thermal history and temperature distribution of basins assuming that meteoric flow was absent or insignificant. Rift basins evolving into passive continental margin basins experience an early

phase of enhanced conductive heat flow (Chap. 8.3), i.e., high temperature gradients, which is followed by cooling and decreasing temperature gradients. Such a situation has been modeled for several cross-sections of the North American Atlantic margin (Sawyer 1988).

In the model of the **Georges Bank basin** off Newfoundland (Fig. 13.21a), it is assumed that the stretching factor (lithospheric attenuation) was $\beta = 2.5$ (see Chap. 8.3). During the initial rifting stage, 200 Ma before present, the temperature gradient may have been about 50 °C/km, but later it dropped to approximately 30° C. Because of the rapid subsidence and sediment accumulation during the Jurassic, the base of these sediments reached a temperature of 125° C within a time period of about 40 Ma (subsidence curve crossing temperature lines in Fig.

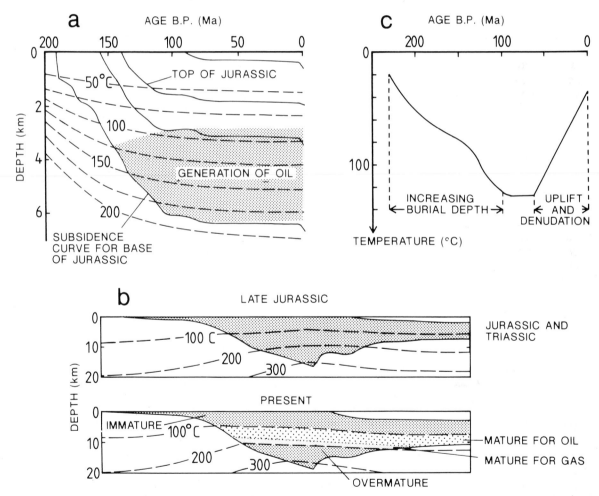

Fig. 13.21a-c. Temperature-depth history of sediments not affected by meteoric water circulation. **a** On passive Atlantic continental margin of North America, Georges Bank basin, COST well G-2. **b** Cross section of Baltimore Canyon trough, Atlantic continental margin of North America, with results of thermal model for Late Jurassic and present time. (a and b Sawyer 1988). **c** Thermal history of Permian sediments of the Colorado Plateau, reflecting increasing burial and subsequent uplift and denudation. (Meyer et al. 1989)

13.21a). About 110 Ma B.P., the maximum temperature was established at a burial depth of approximately 6 km. Then the temperature at the base of the Jurassic (as well as in younger strata) remained more or less constant.

During rapid subsidence, a temperature of approximately 125 °C is thought to be necessary for the generation of oil. Later, the minimum temperature for oil generation can drop (Fig. 13.21a), because oil generation is controlled by both the temperature and the time elapsed since deep burial (Waples 1980; Tissot and Welte 1984).

For the deeply filled **Baltimore Canyon trough** farther south, thermal modeling predicts a subsurface temperature distribution as shown in Fig. 13.21b.

During the Late Jurassic, thermal gradients in the center of the trough (containing Triassic sediments at burial depths of up to 15 km) and farther offshore reflect the rifting and early drifting stage and therefore were greater than at present. For a specific sediment layer, however, the temperature still increased somewhat during the last 140 Ma, because subsidence and sediment accumulation in the passive margin basin continued. The sediments below the 200 °C isoline are considered overmature with respect to oil generation. They should display the characteristics of low-grade metamorphism.

Thermal Effects of Subsidence Followed by Uplift

If a phase of subsidence is followed by uplift and significant denudation, the thermal history of a specific sediment body exhibits both a phase of conductive heating and a phase of cooling. This is a common feature in fold belts (Chap. 9.10), but here the reconstruction of the thermal and stress history of certain rock units is complex due to tectonic implications. A prominent example of a simple case is the **Colorado Plateau** in North America (Fig. 13.21c, based on Meyer et al. 1989).

After their deposition, Permian sediments were buried under 2700 to 4000 m of younger sediments and heated to 100-140 °C, assuming a geothermal gradient of 30° C/km. After a short period of nondeposition, the Colorado Plateau sediments were uplifted, and about 2500 m of the overlying sediments were removed. To verify the estimation of maximum temperature and its date, three independent additional methods were applied (see below): (1) investigation of fission tracks in apatite, (2) study of fluid inclusions, and (3) determination of the so-called vitrinite reflectance of the organic matter.

All three methods showed that the maximum temperature reached during the burial history

of the Permian sediments was between 120 and 140 °C. Thus, significant changes in the evolution of the thermal gradients or strong effects of meteoric water circulation can be ruled out. During the last 65 Ma, the Permian sediments underwent cooling down to about 30 °C (Fig. 13.21c).

Thermal Effects of Meteoric Water Circulation

As discussed in Chapters 13.2 and 13.3, many sediments which have partially or entirely emerged, are affected by meteoric water circulation. In these cases, the subsurface temperature distribution caused by conductive heat flow may be greatly disturbed or completely modified. The direction of meteoric flow is controlled mainly by the topography of the land surface. In an elevated former foreland basin sloping away from a thrust belt, groundwater circulation tends to develop a pattern as generalized by Jones and Majorowicz (1987) for the western Canadian Prairies Basin (Fig. 13.22a).

Below the area of regional recharge (A), the geothermal gradient and thus the vertical heat flow densitiy in the upper part of the basin fill (Mesozoic-Cenozoic sediments) are less than in the underlying Paleozoic strata. In the area of regional discharge (B), temperature increases rapidly with depth in the basin fill.

Local areas not affected by this basin-wide redistribution of heat by groundwater flow may be of interest for the exploration of oil and gas. Such areas display the original temperature field and thus indicate the presence of impermeable rocks favorable for the retention of hydrocarbons. A subsurface temperature distribution similar to that in Fig. 13.22a was calculated for the Paleozoic Illinois Basin (Fig. 13.22b, Bethke 1989). It is also caused by regional, long-distance groundwater circulation.

The marked *influence of the permeability* of the rock or sediment mass on the distortion of conductive heat flow was demonstrated by Smith et al. (1989) for a situation with high relief (Fig. 13.22d and e). If hydraulic conductivity is relatively high, the thermal gradients in the basin fill are significantly affected by groundwater circulation.

In the model, the difference in elevation between groundwater recharge and discharge zones was assumed to be 500 m over a distance of a few km. If the sediment mass is homoge-neous and has a very low permeability ($k = 10^{-18}$

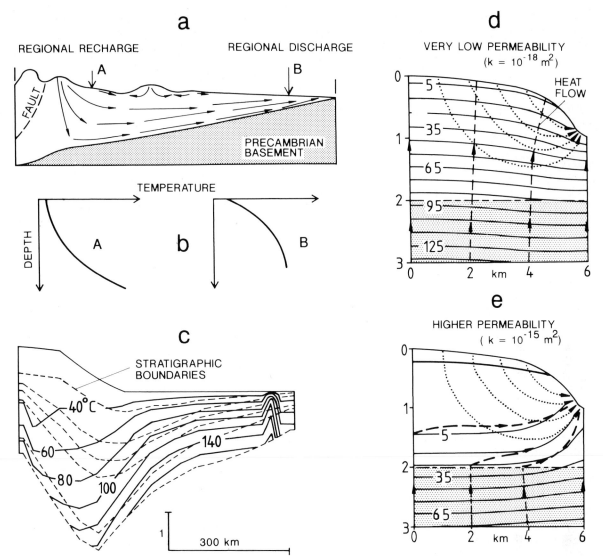

Fig. 13.22a-e. Subsurface temperature distribution disturbed by meteoric water circulation. **a** and **b** Foreland basin of the Canadian Rocky Mountains, displaying marked differences in thermal gradients of the recharge and discharge areas. (Jones and Majorowicz 1987). **c** Temperature distribution calculated for Mesozoic groundwater flow system in the Illinois basin. (Bethke 1989). **d** and **e** Disturbance of conductive thermal regime in areas of high relief (e.g., entirely or partially elevated sediments on narrow shelf) in relation to sediment permeability. (Smith et al. 1987)

m^2, corresponding with a hydraulic conductivity of about 10^{-6} m/d or 0.4 mm/a), conductive heat flow lines and the geothermal gradient are only slightly disturbed (Fig. 13.22d). However, if the permeability increases by three orders of magnitude (hydraulic conductivity around 0.4 m/a), conductive heat flow lines and thermal gradients are strongly affected (Fig. 13.22e). Other examples are reported, for example, by Chapman and Rybach (1985).

All these thermal effects have to be taken into account in the interpretation of inorganic and organic processes in diagenesis.

13.5 Special Methods and Processes in Diagenesis

Age, Temperature, and Degree of Diagenetic Reactions

As already indicated, there are several methods to determine the *maximum temperature* a sediment body has experienced. If particulate organic matter is present, its state of diagenesis, defined by the vitrinite reflectance, is a function of temperature and time (e.g., Teichmüller and Teichmüller 1979). Fluid inclusions frequently occur in secondary crystals precipitated in rock fractures or in large pores. At normal temperatures they consist of a fluid phase and a gas phase. By heating them under the microscope, one can observe that the fluid and gas are transformed into a homogeneous fluid. From the homogenization temperature, corrected for the effects of pressure and considering other sources of error, one can reconstruct the temperature and pressure at the time the inclusion was formed (Roedder 1984).

Stable isotopes, for example the $\delta^{18}O$ values of carbonate cements, can be used to determine the temperature at the time the cement mineral was formed. However, one has to be cautious, because these values may be affected by the mixing of pore fluids derived from sea water and meteoric water. In addition, one must take into account that the isotopic composition of ocean water changed from the Paleozoic to the present time (Garrels 1986; Veizer et al. 1986; Carpenter and Lohmann 1989). For example, the $\delta^{18}O$ values increased, but carbon and sulfur isotopes fluctuated during Phanerozoic time (Schidlowski et al. 1984). To investigate such a question, one must search in ancient sediments for primary minerals or early cements which formed in equilibrium with the former ocean water and were preserved unaltered.

The *timing of thermal events* can be determined by dating cement or recrystallized minerals containing unstable isotopes. Another method is the study of fission tracks in apatite, other minerals, and volcanic glass (e.g., Naeser 1979; Odin 1982; Faure 1986). Fission tracks are caused by nuclear particles emitted from uranium-238 impurities. They are obliterated if the temperature exceeds a certain limit (for example 100 to 120 °C for apatite) and start to form again and grow with time if the temperature drops below this boundary. Hence, fission tracks indicate that the rock sample has been exposed to elevated temperatures a certain time ago.

The transition from shales to slates, i.e., from deep-burial diagenesis to low-grade or "sub-greenschist facies" metamorphism is characterized by increasing *crystallinity of illite/muscovite* (e.g., Dunoyer de Segonzac et al. 1968; Frey 1987; cf. Chap. 13.4).

Diagenetically Controlled Ore Deposits

From the processes discussed above in this chapter, it is obvious that the generation, migration, and accumulation of oil and gas (Chap. 14), as well as the precipitation of ore deposits from pore fluids, are controlled by the various mechanisms operating during diagenesis. Of particular interest with respect to hydrocarbons and metal deposits is deep burial diagenesis. In this regime the elevated temperatures needed for the generation of oil and gas are established, and many of the deep pore fluids are highly saline and contain metals in concentrations sufficient to be precipitated when the pore fluids move to areas of lower temperature.

Many *sediment-hosted ore deposits* result from diagenetic processes and can be explained without the aid of magmatic exhalations or metal-bearing hydrothermal fluids derived from the zone of metamorphism (see e.g. Amstutz et al. 1982; Maynard 1982; Guilbert and Park 1986; Friedrich and Herzig 1988). Hot brines, particularly those rich in chlorine, have the capacity to leach metals (Zn, Pb, Ba, Cu, U, Ag, and other elements such as Ba and F) from deeply buried strata. Such strata may consist of continental, fluvial or shallow-marine clastics derived from igneous and volcanic rocks. The metal content of such siliciclastic sediments may become relatively enriched by weathering processes, either prior to deposition or in situ.

A prominent example for such a possibility is the Permian red beds (Rotliegendes) below the Zechstein evaporites in Europe (Chap. 6.4.1), which provided the metals in the overlying, copper-bearing black shales of the "Kupferschiefer". Other metal-rich black shales gained their metal content from sulfides precipitated in normal sea water as well as from organic matter.

In Mesozoic extensional basins, "immature" syn-rift sediments frequently alternate with evaporites and are overlain by post-rift, thick marine shales. Such basins appear to provide a favorable tectonic setting for the formation of strata-bound, diagenetically-formed ore deposits of the *Mississippi Valley type* (Maynard 1982; Large 1988). Due to the low-permeable shales acting as caprocks in more central parts of the basin, sediment compaction forces the hot, Cl-rich, metal-bearing brines to migrate laterally upward. As they pass black shales, the brines may take up additional metals, before they escape along more permeable layers toward the basin margins. Alternatively, the brines may migrate upward along syngenetic (i.e., post-rift) faults.

Cooling, changing Eh and pH conditions, and mixing with shallow pore fluids from other sources cause the precipitation of metals, mostly as low soluble sulfides (for example Pb, Zn, and Fe); however, several elements, such as Ba, Mn, and U, behave differently. Variations in the chemical conditions over time lead to special successions of ore minerals (ore mineral paragenesis), as in the case of the common cement minerals.

Diagenetically generated ore deposits are found in siliciclastic and carbonate host rocks. The ores may fill macropores and voids, as well as faults and fracture systems. *Carbonates* with initially high porosities, such as oolitic wackestones, packstones, and bioherms, and rocks with karst cavities, are particularly promising host rocks. One can distinguish between several phases and types of ore genesis, for example between early diagenetic cementation of pores (disseminated ores), replacement of carbonate, sulfidization of carbonates, secondary concentration of primary ores (formation of discontinuous flat beds and nests), and filling of fractures.

The precipitation of *metal sulfides* in areas outside of the sulfate reduction zone is difficult to understand, possibly because two different types of pore fluids are needed. One pore fluid has to provide the metals in solution (in part as chloride complexes at high temperatures or as organic complexes), while the other (probably from shallow sources) has to deliver sulfur in the form of HS^- from the sulfate reduction zone. However, an alternative model has also been proposed (see, e.g., Roberts and Sheahan 1988), in which sulfur migrates with the metals to the site of deposition. If ore deposition takes place within the shallowly buried sulfate reduction zone, the upstreaming metal-rich brine may mix with more or less stagnant water in a reduced state, with sulfur being supplied by diffusion from the sea bottom.

Redox conditions can easily change under such conditions. Thus, precipitation of metal sulfides may be followed by the formation of barite and other compounds characteristic of oxidizing conditions. The deposition of metal ores in carbonates is frequently associated with dolomitization and/or brecciation, and occasionally with silicification. This indicates that various diagenetic processes may operate in the same rock body, either more or less simultaneously, or successively.

The special conditions of *ore precipitation at the contact with evaporites* are related to the phenomena discussed above. For example, in the Gulf Coast basin of the United States, the intrusion of salt into the overlying argillaceous sediments had several significant effects (Holmes 1986). As in the Mediterranean, a 125 m thick Tertiary/Quaternary sediment section on top of drilled salt diapirs displays upwardly decreasing pore water salinity. In addition, the increased heat flow through salt dehydrated the clay minerals at the salt-sediment interface at relatively shallow burial depth. The chemical reactions between the pore water containing sulfate and the sediments rich in organic matter created a diagenetic sulfide-rich front. Within this zone, metals such as zinc and lead were concentrated. Anhydrite residue from evaporite dissolution, transformation of anhydrite to gypsum, and newly precipitated carbonate form the cap rock of the salt domes.

Pore Water as a Source of Special Hydrothermal Systems

Most of the hydrothermal systems discussed in the literature, including those at active spreading centers on mid-oceanic ridges (e.g., Fyfe and Lonsdale 1981), are fed either by sea water cr meteoric water circulating in fracture zones or highly permeable layers through various rocks. If there is no difference in hydraulic head, such systems are driven by a magmatic heat source or by local differences in the geothermal gradients. Hot water is forced to move upwards to the sea floor or land surface to vents or springs, but it has to be simultaneously replaced by downward moving sea water or meteoric water.

However, there also are hydrothermal systems which are partially or entirely fed by the expulsion of pore water from young, soft sediments. In the last ten years two types of such systems were detected in the present oceans. Because their origin is closely related to the topics of this chapter, they are briefly described here. Whether or not they significantly contribute to the formation of mineral deposits is not yet sufficiently known.

Expulsion of Pore Water from Young Spreading Troughs

In young, narrow ocean basins and pull-apart basins (see Chaps. 1, 12.1, and 12.8), the accretion of new oceanic crust does not take place at sediment-starved oceanic ridges, but in spreading troughs collecting large quantities of sediment. This can be observed, for example, in the central graben of the Red Sea and in special basins in the Gulf of California (Guaymas basin). Since these spreading troughs represent the deepest depressions in the total depositional area and are situated close to terrigeneous sediment sources, they are commonly characterized by high sedimentation rates. In such cases, the upwelling basaltic magma intrudes into soft sediments and tends to generate horizontal sills as soon as the magma pressure exceeds the lithostatic pressure and tensile strength of the sediments (Fig. 13.23a).

During Leg 64 of the Deep Sea Drilling Project in the Gulf of California, such basaltic sills were encountered in several drillholes in the Guaymas basin at burial depths between approximately 50 and 350 m (Curray, Moore, et al. 1982; Einsele 1982b and 1986). Two of the shallowly buried sills were still hot. Downhole measurements showed that the porosity of the sediments was reduced by 20 to 40 % within several tens of meters above and especially below the sill-sediment contacts (Fig. 13.23b).

Sediment porosity is reduced by "thermal tamping" (Walker and Francis 1987), which is caused by heat transfer from the hot magma into the neighboring sediment, as well as by the emplacement of heavy magmatic loads on top of highly porous sediments.

In addition, siliceous and calcareous microfossils may dissolve, and sediments below thick sills and not far above the sheeted dike zone of young oceanic crust (Fig. 13.23d) may be further altered by high-temperature, low-pressure metamorphism ranging from anhydrite-dolomite to chlorite-pyrite-albite-epidote-sphene associations. Such an interaction between the injection of basaltic dikes and sills and soft sediments leads to the expulsion of great volumes of water (in terms of height on the order of 30 m). This water moves upward to layers of relatively high permeability which act as lateral conduits to fractures (Fig. 13.23c). Along such growth faults in the spreading center, the hot water escapes to the sea floor where it reacts with sea water and precipitates part of its dissolved constituents.

This type of hydrothermal system is controlled by special magmatic pulses and is thus subject to considerable variation in space and time. Dike and sill intrusions may also cause hydrothermal convection systems, due to the marked thermal anomalies they create, and thus produce metal deposits (Tarkian and Garbe 1988). The interaction between rapidly accumulating sediments, ongoing ocean spreading, and discontinuous large basalt injections in young oceanic basins may create special sill-sediment complexes (Fig. 13.23d; Einsele 1986: Gibb and Kanaris-Sotiriou 1988).

Expulsion of Pore Water Along Subduction Zones

A second possibility for expelling pore water from soft sediments was found along convergent continental margins. When the sediments on a subducting oceanic plate reach the seaward edge of an accretionary wedge (Fig. 13.24), they are deformed by two processes: (1) increasing overload due to underthrusting, and (2) additional lateral stress originating from compressional forces. As a result, the sediment porosity will be reduced and pore water expelled (e.g., Bray and Karig 1985; Pickering et al. 1989; Suess and Whiticar 1989).

At the Barbados Ridge accretionary complex, the porosity in the uppermost 300 m of pelagic to hemipelagic sediments with initial values between 50 and 75 % was reduced by 2 to 9 % at the deformation front (Mascle, Moore, et al. 1988; Brückmann 1989). Assuming an average porosity reduction from 63 to 57 % for this 300 m sediment section, the expelled pore water is equivalent to a water layer of about 50 m in thickness (compare Fig. 13.6). Using this value and assuming a subduction rate of 4 cm/a, one has a volume of 2000 m^3 of pore water expelled from a 1 km length of the subduction zone per year. Submarine springs venting pore fluids at the toe of the Cascadia accretionary complex off the Oregon coast have been observed to discharge approximately 70 m^3/a.

In total, the porosity reduction of deeply underthrusted sediments within an accretionary wedge, and thus their water loss are much higher than estimated here for the onset of deformation at the toe of the wedge.

In the case of the Cascadia subduction zone (Fig. 13.24) and along convergent plate boundaries in the northwestern Pacific (Boulegue et al. 1987), the pore fluid vents were observed and sampled with deep-sea submersibles and additional devices.

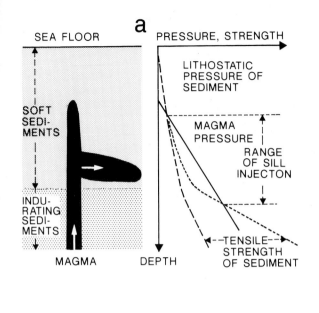

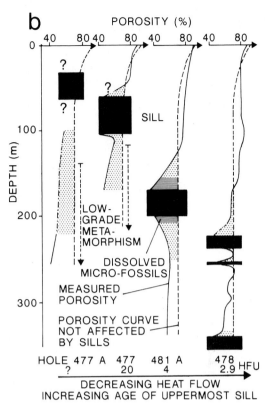

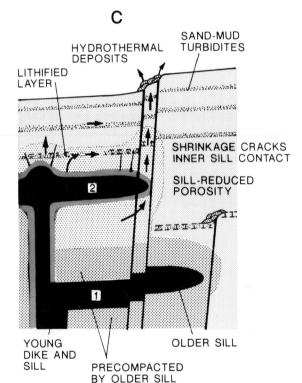

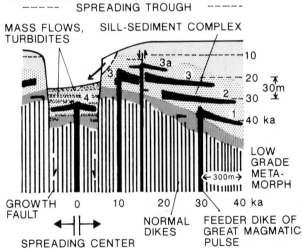

Fig. 13.23a-d. Sill-induced expulsion of pore water and hydrothermal deposits in young, Gulf of California-type spreading center. a Emplacement of basaltic sill in soft sediment. The magma pressure has to overcome the lithostatic pressure and tensile strength of the sediment. b Porosity-depth curves of sill-affected sediments in the Guaymas Basin (DSDP, Leg 64). *Hatched areas* Sill-induced reduction of porosity as compared to drillhole in similar sediments not affected by basaltic sills. Older, more deeply buried sills are associated with lower values of heat flow

(*HFU*), because sill-induced thermal anomaly dissipates with time. c Model of successive sill intrusions and expulsion of pore water from neighboring sediments. Hot water migrates along sand layers and growth faults and may deposit minerals on the sea floor. d Buildup of sill-sediment complex on top of sheeted dike zone by special magmatic pulses (*1* through *4*) in time intervals of about 10 ka; fixed spreading center, half-spreading rate 3 cm/a. (Einsele 1982b, 1986).

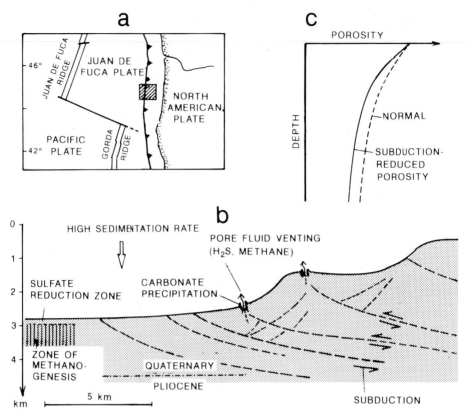

Fig. 13.24a-c. Simplified model of Cascadia accretionary complex off the Pacific coast of North America, showing vents with discharge of hot pore water containing hydrogen sulfide and/or methane, and other dissolved constituents. **a** Location map. **b** Cross-section with vents. **c** Dewatering of sediment caused by reduction in porosity due to increased vertical and lateral stress. (After Suess and Whiticar 1989; Han and Suess 1989)

The submarine vents may discharge not only water with elevated temperatures, but also pore fluids reflecting the chemical properties of the sulfate reduction zone and the zone of methanogenesis mentioned in Chapter 13.3. If the subducting sediments are deposited relatively rapidly and are rich in organic matter, the expelled pore water may contain *hydrogen sulfide and/or methane* in considerable quantities (Fig. 13.24; Suess and Whiticar 1989; Han and Suess 1989). Part of the methane is oxidized near the surface and gives rise to the precipitation of calcium carbonate as cement or in the form of chimneys at the vents. Near these vents, faunal communities, including giant clams and large tube worms adapted to methane, hydrogen sulfide, and warm water, were observed.

Such phenomena may also be found in ancient subduction zones. *Overpressuring of sediments* may also cause liquefaction and fluidization. As a result, sediment is forced upward to form mud diapirs, or it is injected vertically and laterally as dikes and sills into neighboring deposits (Pickering et al. 1989). In this way, *chaotic structures* of large dimension can form which resemble phenomena known from thick debris flows or tectonic mélange zones. Fluidized sediments reaching the sea floor may also create mud volcanoes. The dewatering of accretionary prisms is an important process in the deformation, stress release, diagenesis and subsequent metamorphism of sediments at convergent plate boundaries.

14 Hydrocarbons and Coal

14.1 Source Rocks, Kerogen Types, and Hydrocarbon Potential

Source Rocks

Hydrocarbon source rocks may be defined as fine-grained sediments which in their natural setting have generated, are generating, or will generate and release enough hydrocarbons to form a considerable accumulation of oil or gas (Brooks et al. 1987). Whether such an accumulation is commercial largely depends on economic considerations. A potential source rock has the capacity to generate hydrocarbons in substantial quantities, but has not yet reached sufficient organic maturation.

Oil shales contain thermally degradable organic material, usually about 20 % of the total organic matter, TOC, the remainder consisting of insoluble compounds (kerogen, see below). Oil shales and coal must have a very high organic matter content to be of economic interest. In contrast, hydrocarbon source rocks may release only very small proportions of oil or

gas per unit rock volume to form an important accumulation in reservoir rocks. This is possible if the source rocks represent a large and sufficiently subsided rock mass. The following brief discussion focuses on hydrocarbon source rocks, but also includes to some extent other types of organic-rich sediments. The depositional environment of these sediments is discussed in Chapters 5.3 and 10.3.

As already mentioned in Chapter 10.3.3, the preservation of organic matter is, among other factors, a function of the oxygen content of bottom waters, the sedimentation rate, and the intensity of benthic life. The influence of these factors is summarized in Fig. 14.1. With water oxygenation and benthic activity decreasing, the zone of methane fermenting bacteria expands upward at the expense of the sulfate reduction zone. As a result, more and also less stable organic matter can be preserved in the sediment.

Kerogen Types and Hydrocarbon Potential

During burial and under increasing temperature, the organic matter undergoes a series of geochemical reactions leading from "biopolymers" to "geopolymers" (Fig. 14.4), often collectively called *kerogen* (Tissot and Welte 1984; Brooks et al. 1987). Besides kerogen, the organic matter of buried sediments still contains a small fraction of organic compounds which are similar to the compounds originally produced by organisms. Such "geochemical fossils or markers" provide evidence for the source of organic matter and, to some extent, also for the depositional environment (cf. Fig. 10.13; Tissot and Welte 1984). They are also found in crude oil and thus testify to the origin of petroleum from the remnants of animals and plants.

Carbonate sediments containing at least 0.3 to 0.6 % organic carbon and shales with 0.5 to 1 % organic carbon may already be regarded as hydrocarbon source rocks. These lower li-

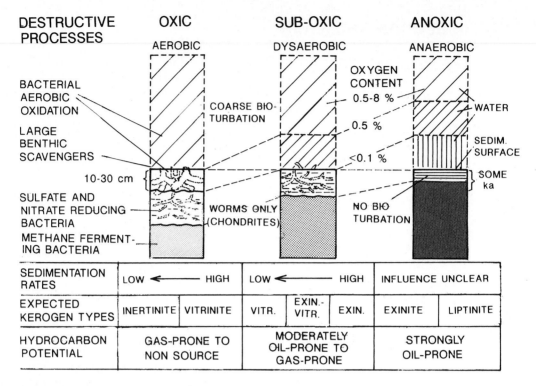

Fig. 14.1. Dominant kerogen types and hydrocarbon potential of aquatic petroleum source rocks in relation to benthic environment and sedimentation rate. (After Demaison and Moore 1980; Brooks et al. 1987)

mits of potential source rocks depend, however, on the kerogen type present in these rocks. The composition of kerogen in the different source rocks is significantly controlled by the processes shown in Fig. 14.1. Low sedimentation rates under well oxygenated conditions only allow the preservation of the so-called *inertinite* (usually in very small amounts), whereas anoxic depositional environments enable the preservation of comparatively large quantities of the H-rich *liptinite* (see below) and thus strongly enhance the hydrocarbon potential.

Principally, the organic matter of hydrocarbon source rocks is subdivided into two groups (Fig. 14.2):

- *Bitumen:* organic matter soluble in organic solvents which represents only a small proportion of total organic carbon, TOC.
- *Kerogen:* organic matter which is insoluble in organic solvents, nonoxidizing mineral acids, and aqueous alkaline solvents. Keregon always represents the bulk of TOC.

Whereas the extractable bitumen is already in a migratable state, the kerogen is fixed in the sediment, but has the potential to generate

migratable crude oil and gas. A thorough study of the amount and composition of kerogen in source rocks is therefore an important modern tool in evaluating the hydrocarbon potential of lacustrine and marine sediments. For this purpose optical methods (organic petrography), physicochemical methods, and organic geochemical analyses including certain geochemical (biological) markers are used (see, e.g., Durand 1980; Tissot and Welte 1984; Brooks et al. 1987; Tissot et al. 1987). With the aid of such methods, kerogen can be classified into four main groups (Fig. 14.2):

1. *Liptinite-type kerogen.* Derived mainly from the lipid components of algal material after partial bacterial degradation, altered by decomposition, condensation, and polymerization. Liptinite-rich deposits are typically dark, finely laminated or structureless, and rich in TOC. They commonly form in lakes and lagoons, but liptinite is also an important constituent in the organic matter of marine environments (Fig. 10.13). Liptinite is relatively rich in hydrogen and therefore exhibits a high H/C ratio; it has a low oxygen content and low O/C ratio (Fig. 14.3). Liptinite-dominated kerogen is classified as *type I kerogen* (Tissot

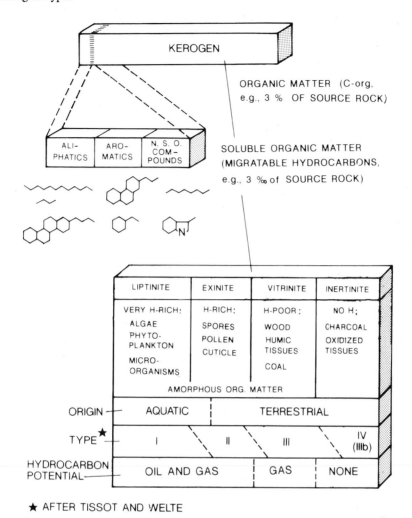

Fig. 14.2. Soluble and insoluble (kerogen) organic matter of hydrocarbon source rocks, kerogen type after Tissot and Welte (1984), and hydrocarbon potential. (After Brooks et al. 1987).

and Welte 1984). Liptinite-rich source rocks have a very good potential for the generation of oil.

2. *Exinite-type kerogen.* Derived mainly from relatively resistant membraneous plant debris such as spores, pollen, leaf cuticles etc. Resins and waxes also belong to this group. Such plant particles are not only produced on land and swept into lakes and swamps later forming certain varieties of coals, but they also grow in lakes and in the ocean (e.g., dinoflagellates and other phytoplankton). Exinite has a high H content and H/C ratio (but lower than liptinite) and a medium O content and O/C ratio (Fig. 14.3). Many marine sediments and hydrocarbon source rocks contain a mixture of liptinite, exinite, and vitrinite (see below) which is classified as *type II kerogen.* Exinite-

rich source rocks have a good potential for the generation of oil, condensate, and wet gas (see below).

3. *Vitrinite-type kerogen.* Derived mainly from woody material of higher plants, more or less degraded. Vitrinite has a relatively low H content and H/C, but a high initial O/C ratio (Fig. 14.3). It is the main constituent of most coals. However, it also occurs in marine and lake sediments in varying quantities. Vitrinite-dominated organic matter corresponds to *type III kerogen,* which has a high potential for the generation of gas, but only a limited potential for oil and condensate.

4. *Inertinite-type kerogen* ("dead carbon", *type IIIb* or *type IV kerogen*). Black opaque debris of highly altered, frequently resedimented older organic matter, mostly derived from

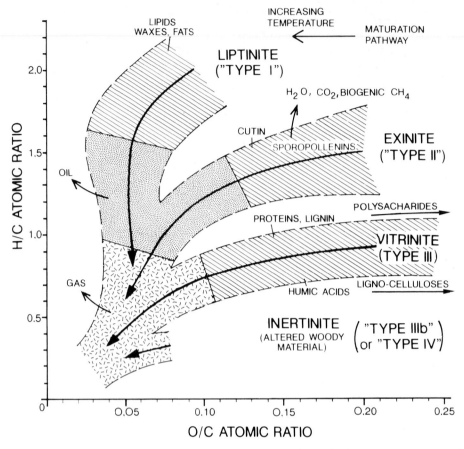

Fig. 14.3. Kerogen-types in van Krevelen-type diagram showing H/C versus O/C and pathways of organic maturation. (After Brooks et al. 1987). With increasing burial depth and temperature, kerogen composition moves to lower H/C and O/C values (carbon enrichment, loss of H_2O and CH_4). All kerogen types tend to approach more or less constant C, H, and O values

plants. Due to previous oxidation and/or high levels of carbonization, the H content and the H/C ratio of inertinite are very low. Rocks containing only inertinite practically have no potential for oil or gas.

The van Krevelen-type diagram (Fig. 14.3), based on the ratios of the three most important elements of kerogen, C, H, and O, clearly displays the different kerogen types and is therfore widely used. However, this is applicable only for immature organic matter. With increasing maturity, i.e., under growing temperature due to subsidence and burial beneath younger sediments, the elemental composition of the initial kerogen types gradually changes, and the curves of the different kerogen types tend to merge. All types of kerogen become relatively richer in C, but lose H and O, because they release H_2O, CH_4, and other hydrocarbons.

14.2 Generation of Hydrocarbons

The Evolution of Organic Matter

The evolution of organic matter from biopolymers to geopolymers with increasing burial depth (and temperature) is shown in the overview of Fig. 14.4. As a result of different biochemical, chemical, and physicochemical processes, the primary organic compounds are transformed into insoluble kerogen or, in the case of coal evolution, into sub-bituminous brown coal. This first stage of evolution is referred to as *diagenesis;* it ends as soon as extractable humic acids are more or less used up.

In the zone of *catagenesis*, part of the kerogen is converted into hydrocarbons. It is the main zone of oil and wet gas generation ("oil kitchen"). The coal evolution proceeds to the

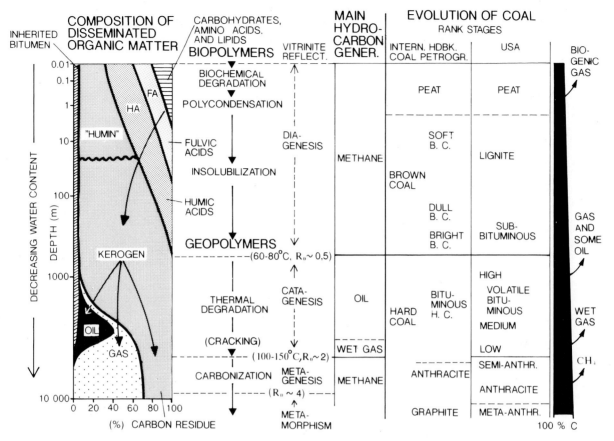

Fig. 14.4. Overview of the evolution of organic matter from young, organic-rich sediment or peat to the deep-burial metamorphic zone. The source rocks generate hydrocarbons (oil and gas) or are transformed to coal (simul-taneously releasing gas and some oil). Note the logarithmic depth scale; R_o signifies vitrinite reflectancy (optical measure for maturity). (After Tissot and Welte 1984; coal ranks after Stach et al. 1982)

development of bituminous hard coal, which also releases gas and some oil. In the next zone, namely that of *metagenesis,* both hydrocarbon source rocks and hard coal mainly release gas. In oil source rocks the carbon-rich residue remains disseminated in the shale as a minor constituent. In coal deposits (Chap. 14.4), the carbon-enriched residue forms anthracite and, after the onset of metamorphism, graphite.

The relationship between increasing kerogen maturity with temperature (and depth) and the release of migratable organic compounds, i.e., the generation of hydrocarbons, is demonstrated in more detail in Fig. 14.5.

- *At shallow burial depths* the immature organic matter can only release (*biogenic*) *methane* gas produced by methane fermenting bacteria (cf. Fig. 14.1 and Chap. 13.6.2) and small quantities of heavier hydrocarbons.

- *Early to mid-mature stage.* Large quantities of *oil* are generated within the temperature range between 60 to 80 °C and 120 to 150 °C. The relatively *heavy oil* is predominantly composed of molecules with 15 and more C atoms (C_{15}+ hydrocarbons) and contains *condensates of lighter molecules* (C_{8-15} hydrocarbons), such as paraffins and aromatic compounds.
- *Late mature zone for oil generation.* At temperatures higher than about 130 °C, part of the large organic molecules are cracked to form *light hydrocarbons* of the C_{2-7} fraction (wet gas) and methane (dry gas).

Transformation Factor

The percentage of oil and gas that the kerogen can generate if subjected to adequate temperature during a sufficient time period, is referred to as *genetic potential* or *transformation*

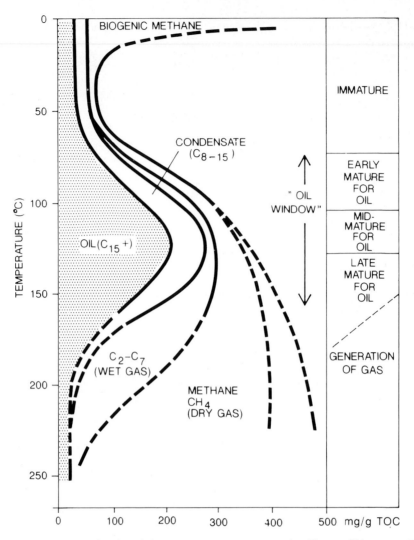

Fig. 14.5. Hydrocarbon generation in relation to temperature (burial depth) leading to different stages of source rock "maturity". Horizontal axis shows amount of generated hydrocarbons in mg/g TOC (total organic matter) in source rock with type II kerogen. Temperature scale adjusted to North Sea conditions; the trends for temperatures greater than about 160 °C are not well established. (After Brooks et al. 1987)

ratio. The optimal conditions, in terms of thermal and burial history, to reach a maximum transformation ratio vary between different source rocks (Tissot et al. 1987).

Type I kerogen has a genetic potential up to 80 to 90 % (e.g., the lacustrine bituminous shales of the Eocene Green River Formation of North America). The common *type II kerogens* of many marine shales have a genetic potential of about 60 %, but carbonaceous shales rich in *type III kerogen* (vitrinite, coal) only reach about 25 % (Tissot and Welte 1984).

In the example of Fig. 14.5, the horizontal axis indicates the amount of oil and gas which can be generated by 1 gram of total organic matter, TOC. The fraction of the primary production which, after losses in the water column and in the benthic degradation zone of the sediment, can finally be converted into hydrocarbons, is shown schematically in Fig. 14.6. These values vary greatly from site to site, but one can again see the enormous difference between the preservation of organic matter and the hydrocarbon potential of oxic and anoxic sediments.

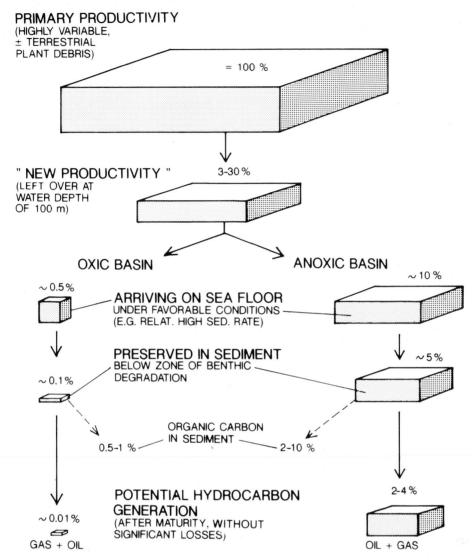

Fig. 14.6. Percentage of primary production of organic matter (= 100 %) in oxic and anoxic environments which can be converted ultimately into oil and gas after losses in the water column and on the sea floor. Note that these values represent orders of magnitude and vary greatly from location to location

Expulsion and Migration of Oil

Fig. 14.7 demonstrates two stages of the burial history of a hydrocarbon source rock in a subsiding basin. At stage (a) only the lowermost part of the source beds have reached early maturity and can begin to generate and expel heavy oil. After deeper burial (stage b), most of the source bed has become early, mid-, or late mature and can yield heavy and light oil including condensate. The released hydrocarbons migrate first through narrow pores and capillaries of the fine-grained source bed (*primary migration*). Later they reach the water-saturated wider pores and more permeable carrier beds and final reservoir rocks (*secondary migration*). Because the densities of oil and gas are lower than that of (usually highly mineralized) formation water, the hydrocarbons tend to move upward into reservoirs located in structural highs.

The oil generated in source rocks migrates either in *molecular solution* in pore waters or as a *hydrocarbon phase* (Magara 1978; Roberts and Cordell 1980; Tissot and Welte 1984; Durand 1987). Because of the low solubility of hydrocarbons in water, the first process can only be effective if a relatively large volume

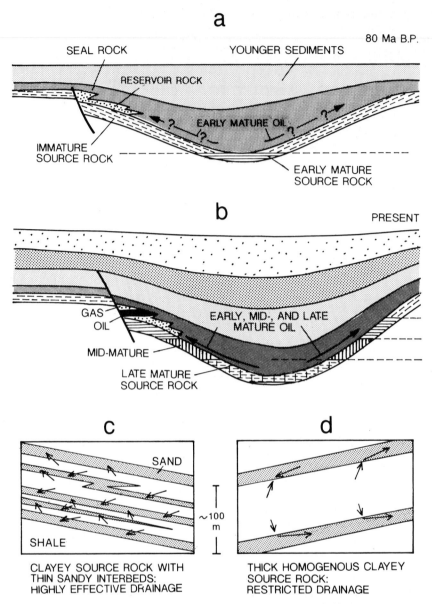

Fig. 14.7a-d. Progressive maturation of hydrocarbon source rock and expulsion of oil with increasing subsidence and burial beneath younger sediments. **a** Only deepest part of structural basin reaches stage of early mature source rock with generation of heavy oil. **b** Large parts of source rock have reached different stages of maturity (including generation and expulsion of light oil and condensate). **c** and **d** Differences in hydrocarbon drainage and migration due to the presence or lack of interbedded permeable layers. (After Brooks et al. 1987)

of water moves from the source bed to the reservoir. The volume of pore water expelled from sediments by compaction is, however, limited, particularly at the depths where the source rocks become thermally mature (cf. Chap. 13.4). Therefore many experts assume that the movement of hydrocarbons in liquid phase is the major factor in primary oil migration. To enable this process, a substantial part of the pores must be filled with oil (oil saturation on the order of 20 %, i.e., in excess of the residual saturation).

Oil movement occurs mainly as *two-phase flow* and is driven by buoyancy ("buoyancy drive", e.g., Davis 1987), which is caused by the fact that oil has a lower density than water. According to this concept, water drive is not important except in those cases when the tops of the petroleum transporting strata have very low dips (generally less than 5°).

The expulsion of oil is promoted by *high fluid pressures* which evolve mainly from continuing compaction. Assuming that the source rock and part of the overlying deposits are more or less impermeable, the excess pore water cannot escape and thus develops high geostatic pressures. Such a pressure-driven, discrete hydrocarbon phase movement can occur in different rock types and has to be maintained for long time periods in order to accomplish significant hydrocarbon accumulations.

Hydrocarbon expulsion from a source rock already becomes effective in an early stage if *thin carrier beds* are interbedded (Fig. 14.7c). They allow the migration of heavy, early mature oil, as well as a high oil expulsion efficiency (50 % and more), even if the total organic carbon content is relatively low. As a consequence, after further burial, the yield of lighter hydrocarbons, condensate, and gas, however, is reduced (Brooks et al. 1987).

In contrast, *thick homogeneous shale* source rocks (Fig. 14.7d) require a higher organic carbon content and an advanced level of maturity to expel oil in a certain period of time. Hence they tend to retain a large proportion of heavy oil products until they reach higher maturity in the course of further subsidence. Then they expel light oils, condensate, and gas, but their expulsion efficiency is lower (5 to 25 %) than that of the interbedded source rocks.

Secondary oil migration occurs as multi-phase flow. Oil globules or gas bubbles in porewater tend to move upward solely due to buoyancy or driven by hydrodynamic conditions (Tissot and Welte 1984). Oil and also gas are trapped if they can no longer be pressed through fine rock pores. The pore space of traps is, however, never fully occupied by petroleum, but always contains some residual water which cannot be replaced by hydrocarbons. For secondary migration in permeable rocks, some authors stress the importance of large-dimension, regional subsurface flow systems (e.g., Toth 1980), where water is the principal medium for carrying oil and gas (mostly in solution).

14.3 Examples of Hydrocarbon Habitats

The general concept of hydrocarbon generation in a subsiding basin (Fig. 14.7) is demonstrated here in a few examples of well known oil and gas fields:

Lacustrine Source Beds

The large, intracratonic, rift-style **Songliao Basin** in northeastern China (Fig. 14.8) is filled with fluviatile and lacustrine sediments of Cretaceous and Tertiary age (Demaison 1985; Ma Li 1985).

During the early Cretaceous highly oil-prone black shales (kerogen type I), 500 to 700 m thick, were deposited in a lake. The main river entering the lake formed a delta with elongated distributary channel sands and lenticular subaqueous mouth bar sands (see Chap. 3.4.2) extending into and building up with the shale in the subsiding basin. In the basin center, these deposits were buried under 1500 to 2000 m younger sediments, tectonically deformed, and split up into fault blocks.

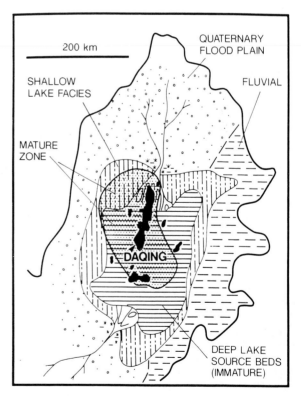

Fig. 14.8. Oil fields associated with bituminous lake sediments and delta sands in the Songliao Basin, northeastern P.R. of China. (After Demaison 1985)

The oil released by the thermally mature source rock migrated into the anticlines and other structural traps and accumulated in the deltaic sandstones. The giant Daqing oil field is located immediately updip from the center of the oil-generating basin. It represents the largest known oil accumulation in a non-marine basin fill. The maximum distance of oil migration is less than 40 km.

Marine Deltas

Large, young deltas are favorable settings for the formation of hydrocarbon source rocks, coal seams, and the accumulation of oil and gas. Whereas peat-derived coal frequently originates from swamps located on the deltaic plain (Fig. 10.13 and Chap. 3.4.3), hydrocarbon source rocks commonly accumulate on the delta slope in marine environments. Although such sites receive large amounts of terrigenous material, delta slope sediments may become sufficiently rich (0,5 to 1 %) in land-derived and marine organic matter to act as hydrocarbon source rocks. The disadvantage of their relatively low C content is compensated for by their large rock volume accumulated in a short time period. The delta front and river mouth sands provide reservoirs which are sealed by the overlying, fine-grained deposits of the delta plain. As a result of delta front prograding and rapid sediment accumulation, the delta complex subsides considerably. Hence, the source rocks reach thermal maturity after a relatively short time span and can generate hydrocarbons.

The well explored Tertiary **Niger delta** is one of the best known examples for this type of petroleum habitats (Evamy et al. 1985; cf. Chap. 3.4.2).

The prograding delta complex has reached a thickness of 9000 to 12 000 m on top of down-faulted oceanic crust (Fig. 14.9). The huge sediment wedge can be subdivided from bottom to top into three main units: marine shales (principal source rocks with an organic C content of about 1 %, mainly type III kerogen), a paralic sequence (shales and sands of the delta plain with distributary channels), and fluvial sands and gravels of the alluvial plain. Deep-reaching basement faults and shallower gravity-induced growth faults created separate structural units for the generation and accumulation of hydrocarbons. The marine shales of these units are generally over-pressured. They release oil and gas at depths in excess of 3000 m, but the temperature gradient varies considerably (from about 2.7 to 5.5 °C/100 m) due to the cooling effect of locally deeply circulating groundwater.

The hydrocarbons accumulate in sands of the overlying paralic sequence sealed by shales. The giant oil fields (largely light waxy paraffinic oils, partially transformed bacterially to heavier crude oil) and gas fields are concentrated in the area of maximum thickness of the deltaic sediment wedge where sufficient sand traps are available in the subsurface. Within a given structure the gas-bearing rocks become more important downdip.

Shallow Epicontinental Seas and Rift Basins

The well explored **Paris Basin** (Fig. 14.10) represents a classic example of a slowly subsiding, wide basin on continental crust, which was filled by alternating continental and shallow-marine sediments of Triassic to Tertiary age (Schönenberg and Neugebauer 1987). The center of the basin appears to have been affected by rifting during Permian/Triassic time. The younger basin fill reaches a maximum thickness of 3000 m. The moderate oil production of the basin is provided by Middle Jurassic limestones; the source rocks are Liassic (Toarcian) shales cropping out as bituminous marlstones along the rim of the basin (Fig. 14.10). They are also known from other parts of Western Europe and represent a kind of anoxic event with kerogen type II. Although source beds and reservoirs are separated by a thick shale sequence, the producing fields lie within or very close to the central part of the basin where the Toarcian source rocks are most deeply buried (2000 to 2500 m) and mature for the generation of oil (Huc 1980; Tissot and Welte 1984; Demaison 1985).

Outside this central depression no oil or gas could be found. According to geochemical modeling, the fraction of kerogen which has been converted to petroleum decreases from the center of the basin outward (from about 150 to less than 50 g petroleum per kg organic carbon). The evolution of the Toarcian kerogen with increasing burial depth is shown in Fig. 14.10b in comparison to values from deeper burial sites in northwestern Germany.

The **failed rift zone of the North Sea**, which has not evolved into a wide, deep ocean basin, has become one of the major oil and gas provinces of the world (Fig. 14.11). The principal source rocks for oil are marine shales (largely with kerogen type II) of Late Jurassic age.

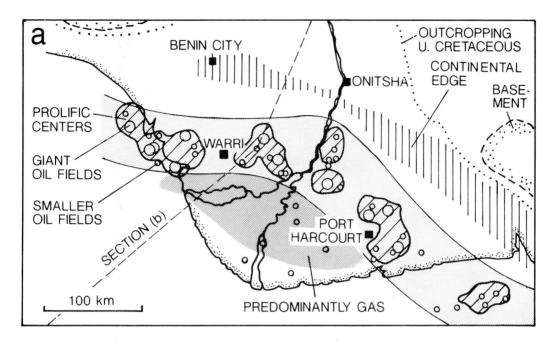

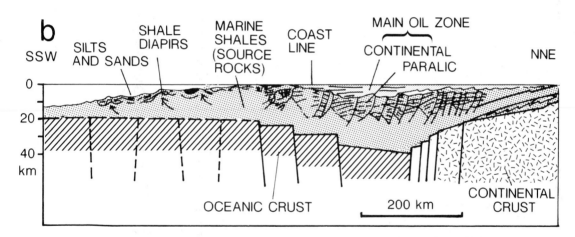

Fig. 14.9. a Major oil fields of the Tertiary Niger Delta on top of downwarped oceanic basement, seaward of the continental edge. **b** Cross section demonstrating the prograding of the delta complex, the down-faulting of continental and oceanic crust, and the subsided and buried growth faults generated along the former delta front. (After Evamy et al. 1978; Tissot and Welte 1984)

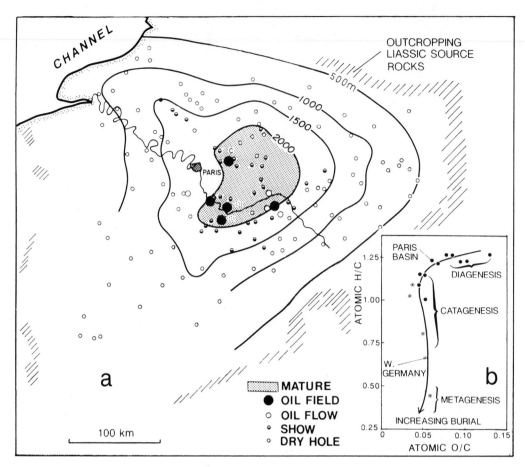

Fig. 14.10. a Petroleum zone of the Paris Basin restricted to the basin center where Liassic (Toarcian) shales have reached maturity for the generation of oil. **b** Type II kerogen development of Toarcian shale with increasing burial depth, based on samples from the Paris Basin and deeper buried samples from northern Germany. (After Tissot and Welte 1984; Demaison 1985)

In the center of the rift zone, these source beds are found at depths in excess of 3000 m. At subsurface temperatures of at least 90 °C, the source rocks are fully mature and actively generating and expelling oil (Demaison 1985). In areas of lower burial depth, where temperatures are lower than 90 °C, no significant oil generation has taken place.

The oil moves to the nearest available reservoirs (e.g., Middle Jurassic sands or Upper Cretaceous chalk) in tilted fault blocks or associated with salt swells (Fig. 14.11b). Virtually all the oil and gas fields lie within or very near the oil generating depression containing mature Kimmeridgian source rocks. The fields with the largest reserves tend to be close to the center of the generative depression (also see Nielsen et al. 1986; Ziegler 1982 and 1988).

Late Jurassic bituminous shales of large areal extent were also deposited in the Western Sibirian Basin, an intracratonic downwarp. Where they are thermally mature, they act as source rocks of the giant fields in this very important petroleum province. All the oil fields are located close to the center of this large hydrocarbon generating basin.

Carbonate Shelves with Evaporites

This hydrocarbon habitat is demonstrated here by an example which was later incorporated into a foredeep. It represents the richest hydrocarbon province of the world, the Middle East oil province around the **Arabian/Persian Gulf** (Fig. 14.12a).

During the Mesozoic this area was characterized by a broad stable platform on the southwestern side of the Tethys ocean (Murris 1985; Stoneley 1987; Watts and Blome 1990).

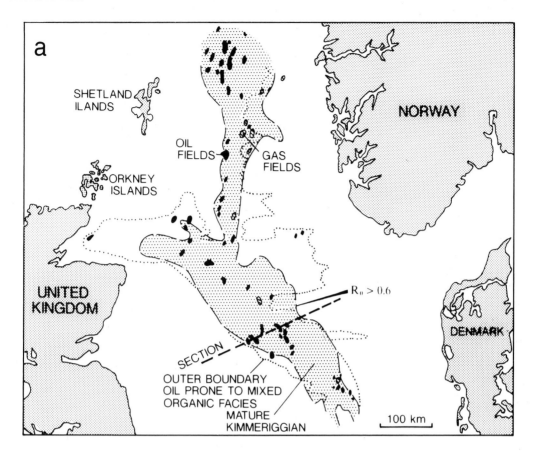

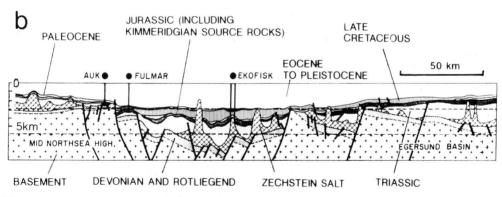

Fig. 14.11. a Oil and gas fields in the rift zone of the North Sea with mature Kimmeridgian source rocks. (After Demaison 1985, modified).

b Cross section of the central North Sea showing fault blocks and oil fields in structural highs or on flanks of salt dome. (After Ziegler 1982)

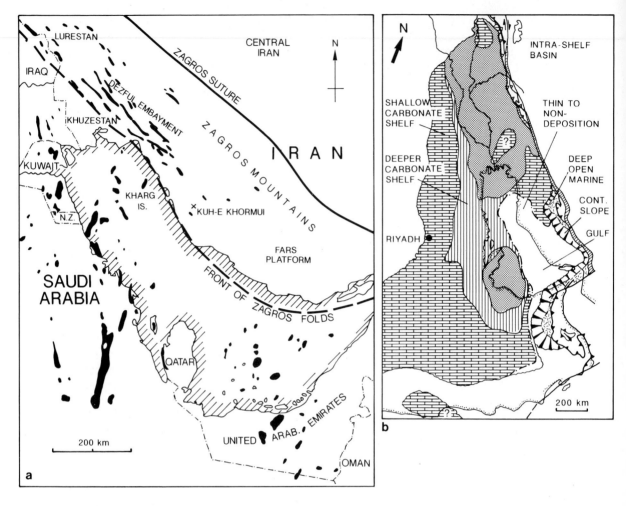

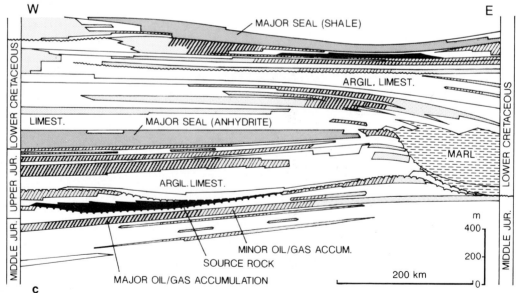

Fig. 14.12. a Principal oil fields of the Middle East, associated with source beds deposited in intra-shelf basins of a very large Jurassic-Creataceous carbonate platform (after Stoneley 1987). **b** Depositional environment during the Upper Jurassic (Late Oxfordian to Early Kimmeridgian). **c** Simplified, undeformed cross section of central Gulf displaying the relationship between major hydrocarbon source beds, reservoirs, and seal rocks. (After Murris 1985)

On the landward side of the platform, ramp-type, mixed carbonate-clastic units were deposited during regressive phases. The central and northeastern, seaward parts of the shelf were differentiated into broad regional highs, horsts, tilted faultblocks, and salt swells. These areas were dominated either by carbonate sediments accumulated under euxinic conditions in intra-shelf basins, or shallow-water carbonates and evaporites on the higher parts of the platform (Fig. 14.12b).

During the Upper Cretaceous the outer shelf region was affected by the onset of orogenic movements creating a foredeep along the Tethys margin. The deposition of platform sediments continued up to the late Alpine orogeny in the Miocene, when the northeastern part of this area was overthrust by the Zagros chains and anticlinal traps formed in the foredeep now occupied by the Gulf (Fig. 14.12a).

The hydrocarbon source rocks accumulated from middle Jurassic to late Cretaceous times in the starved shelf basins, especially during transgressive periods. The richest basinal source rocks (thinly laminated lime-mudstones, kerogen type II) are of Oxfordian to Kimmeridgian age. They usually contain between 3 and 5 % organic carbon, and their oil yield can be as high as 2 % of the total rock volume. The neighboring marginal reef mounds, ooidal grainstones and other shallow-water sands were used as reservoirs.

The porosity of these rocks was improved in places by subaerial leaching during times of low sea level stand. The reservoirs are sealed by shales and supratidal evaporites (Fig. 14.12c). The temperature gradient from Jurassic to present time varied only slightly around 3.5 °C/100 m. At 2000 m depth about 90 °C and at 2700 m approximately 110 °C are reached in drillholes. Upper Cretaceous and Tertiary source rocks remain immature.

Several giant oil fields on the south side of the Gulf are associated with relatively shallow burial depths of the older source rocks. The oil has a relatively high sulfur content indicating the marine euxinic depositional environment and possibly the influence of the intercalated sulfate rocks. Cross-strata migration of oil from deeper source beds to higher reservoirs appears to have been facilitated by (oil-filled) microfractures. Gas fields coincide with those areas where the source rocks are buried below 5000 m and have reached the zone of metagenesis.

The exceptionally prolific habitat for oil and gas of the Gulf region is not a result of extraordinary rich and/or thick source rocks, but rather originates from the vast horizontal dimensions of the depositional platform area (2000 to 3000 km wide and 5000 km long). In addition, abundant highly porous and permeable reservoirs as well as efficient seals lie in close proximity to the source rocks. Tectonic structures are simple and of very large scale, allowing comparatively long-distance migration and a high degree of reservoir fill. Finally, the area was not affected by a major phase of post-depositional erosion and intensive, deep meteoric water circulation.

Summary

All these examples have in common that the hydrocarbon source rocks reached a burial depth on the order of 1500 to 3000 m with a temperature of at least 60 to 90 °C. In this depth range part of the kerogen is or was transformed to petroleum and/or gas under the influence of both subsurface temperature and time. Areas underlain by source rocks undergoing such processes are called *petroleum generative depressions* or "hydrocarbon kitchens" (Demaison 1985). A *generative basin* may contain one or more petroleum generative depressions.

The largest petroleum accumulations tend to occur in reservoir rocks located close to the center of the generative basins or on structural highs neighboring deeper generative depressions. Large volumes of hydrocarbon generating sediments and long time spans for the drainage are a prerequisite for giant oil and gas accumulations. Evaporites and fine-grained, water-wet clayey shales and mudrocks, which must not have open fractures, act as efficient seals on top of the porous oil and gas traps. Such seal rocks prevent oil (lower density than water!) and gas from migrating upward to the surface over long time periods, if the generative basin is not affected by subsequent tectonic activity. Large oil and gas fields still exist in Paleozoic basins, for example in North America, North Africa, and in the Soviet Union.

Oil usually migrates over distances not more than a few tens of km in the updip drainage area of individual tectonic structures. Most of the producible hydrocarbons and the largest oil fields occur within the ranges of generative depressions. These conclusions apply to different types of basins (e.g., intracratonic basins, rift basins, passive margin basins). Long-distance oil migration is documented from some foreland basins. Migration of gas, however, is frequently more difficult to predict.

Subsurface mapping of hydrocarbon generative basins, including kerogen analyses and an evaluation of the thermal history of the basin, is one of the most important tasks in petroleum exploration. In addition, the three-dimensional facies distribution and architecture of the depositional system, reservoir and seal rocks in the neighborhood of the source rocks, and the tectonic structure of the basin fill have to be investigated.

14.4 Evolution of Coal

Introduction

In contrast to oil and gas, coal remains in its primary position within a sedimentary body, but with continued burial it undergoes processes similar to the diagenesis and catagenesis of hydrocarbon source rocks (Teichmüller and Teichmüller 1979; Tissot and Welte 1984).

Important prerequisites for the formation of substantial volumes of coal are the evolution of plant life on the continents, a relatively high plant productivity under favorable climatic conditions with sufficient nutrient supply, and a depositional environment where plant remains can be preserved. Most coals originate from the detritus of terrestrial plants forming peat either in shallow freshwater lakes and swamps (Fig. 10.13, limnic coals and coals of fluvial plains, Chap. 2.2.3), or in brackish transitional environments such as those observed in coastal and marine deltaic areas (paralic coals, Chap. 3.4.3). Summaries on coal-forming environments are given, for example, by Teichmüller (1973) Galloway and Hobday (1983), and Lyons and Alpern (1989).

Regional studies are published in special journals. An overview on the Permo-Carboniferous coals in China, for example, is given by Liu Guanghua (1990). These coals accumulated either in *lakes* (coals with *low sulfur content*), in backswamp areas of marine deltas with switching lobes (coal seams of limited extent with intermediate sulfur content), or in supratidal marshes of coastal regions. The latter type is characterized by wide lateral extent, *high sulfur content*, and *coal cycles* resulting from sea level changes.

Preservation of Peat

In both limnic and paralic environments, the plant debris has to be protected from rapid microbial decomposition under oxic conditions. This is accomplished by a water level standing just at or above the surface of the deposited plant material. To maintain such conditions for longer time periods, the depositional area has to subside; only that part of the plant accumulation can be preserved which is compensated for by subsidence and is therefore kept under the protection of a water cover or water-saturated sediment.

Therefore, long-term *sedimentation rates of peat and coal* commonly correspond with overall sedimentation and subsidence rates of the basin considered (cf. Chap. 10.2). Thus, peats of different thicknesses, composition, and purity are formed. Some of the peat may still be subject to aerobic decomposition by bacteria and fungi, while deeper-seated parts undergo changes by anaerobic bacteria. In this biochemical phase of early diagenesis, some *biogenic methane* is produced.

Coalification and Composition of Coal

After burial, geochemical processes take over to convert the peat into brown coal and, later, into bituminous hard coal and anthracite (Fig. 14.4). These processes are summarized under the term *coalification*. Special investigations, such as the determinations of the moisture and volatile matter contents, carbon and hydrogen contents, the reflectance of vitrinite (see below), and the calorific value, allow to distinguish between several stages of coalification (e.g., Alpern 1987). These stages are referred to as *rank levels* which indicate the maturity of coal.

Peat and coal contain various organic components (*macerals*) which, similar to the components of the hydrocarbon source rocks, are affected by the coalification process in different ways. The three main groups of macerals are (Tissot and Welte 1984; Robert 1985):

1. *Huminite and vitrinite.* Remains of woody and humic substances which are the major components of brown coals and bituminous hard coals.
2. *Liptinite or exinite.* Remains of lipid-rich plant relics (resins, waxes, spores, cuticles, algae) which are summarized under the terms liptinite or exinite.

3. *Inertinite.* Oxidized or reworked older carbon-rich particles.

Over all, the organic components of coal resemble kerogen type III described earlier for hydrocarbon source rocks. Their potential to generate hydrocarbons is similar to that of this kerogen type.

Prograding coalification leads to the following physical and chemical changes:

- Decrease in water content and porosity, increase in density.
- Condensation of organic molecules, polymerization, aromatization, and loss of functional groups.

The net result of these processes is a *relative enrichment of carbon* with increasing burial depth and temperature (or coal rank). These changes are accompanied by the expulsion of H_2O and the generation and release of carbon dioxide and hydrocarbons (Fig. 14.4):

- *At shallow burial depth:* carbon dioxide and biogenic methane.
- *At medium burial depth* (range of medium volatile bituminous coal): large amounts of methane and minor volumes of heavier hydrocarbons including oil.
- *At deep burial depth:* methane.

Part of the generated hydrocarbons can be stored in coal under high in-situ pressure and temperature. However, especially the generation of methane greatly exceeds the storage capacity of coal. Therefore, the bulk of methane migrates into structural traps with reservoir rocks or escapes into the atmosphere.

Example of a Coal Basin

The **Carboniferous in northwestern Germany** extending from the Rhenish Massif with the Ruhr district to the North Sea (Fig. 14.13). is one of the best explored coal basins. In conjunction with the exploration of gas fields, the top of the coal-bearing strata was drilled at numerous sites, and vitrinite reflectance (% Rm) was determined (Teichmüller et al. 1984). The cross section in Fig. 14.13 shows several characteristic features:

- The predominantly pre-orogenic, relatively *high-rank coalification* of the folded Upper Carboniferous is found at the present land surface in the south. This area was uplifted, and part of the Carboniferous was eroded.
- The *less intense coalification* in the deeply buried northern part of the section is explained by young subsidence of this area. The temperature gradient is relatively low, and burial time was not long enough for the coal to reach a higher rank. Most gas fields occur where the vitrinite reflectance varies between 1 and 2 % Rm. Reservoir rocks are sands of the upper Carboniferous, the Permian (Rotliegend), and Lower Triassic (Bunter sandstone). They are sealed by evaporites of Permian and Triassic age.

- The striking *maximum in the coal rank* of the Ibbenbüren mine district (4-5 % Rm) results from an Upper Cretaceous magmatic intrusion, causing a very high temperature gradient. Local strong uplift and subsequent erosion brought the coal-bearing strata to or close to the present land surface.

The results of a model calculation for the degassing of Ruhr coal is shown in Fig. 14.14 (after Tissot and Welte 1984, based on Jüntgen and Klein 1975).

It is assumed that the initial shallow-burial, highly volatile bituminous coal subsided to a burial depth of 3000 m during a time period of 20 Ma (Fig. 14.14a). Simultaneously, the temperature rose from 20 to 140 °C, which corresponds to a heating rate of 6 °C/Ma or about 1.6×10^{-8} °C/day. Using the results of experimental heating of coal and extrapolating these data to the natural heating rate, the history of degassing of the initial coal as well as the amounts of released gas were found (Fig. 14.14b).

During the first 16 Ma of heating, only H_2O, CO_2, and CO were released according to this calculation. Then methane, hydrocarbons, and also some nitrogen were produced. This process continued until about 40 Ma after the onset of heating. Over all, the coal has lost more than 30 % of its weight and evolved from a highly volatile into a less volatile bituminous coal and later into C-rich anthracite. Finally, the area was uplifted and exposed to erosion as mentioned above. As a result, burial depth and temperature of the coal seams decreased. Because high maturation of the coal was already established prior to this process, however, this development had no further significant effect on the coal and its degassing.

These general principles can also be applied to other coal deposits of various burial and uplift histories.

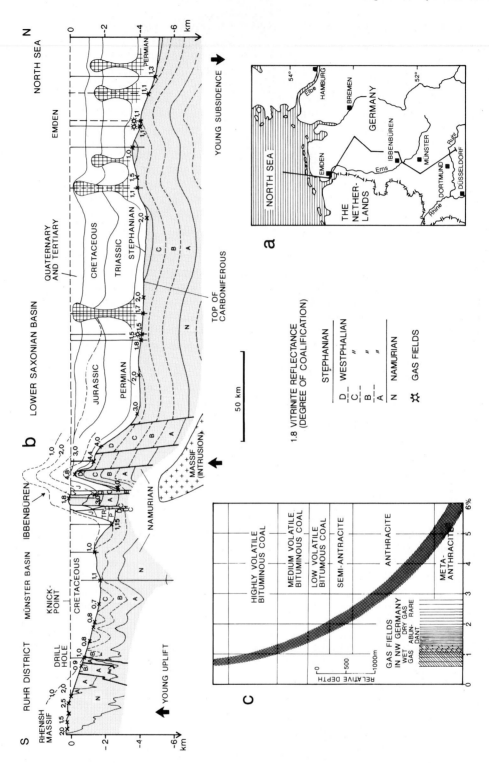

Fig. 14.13. a Location of section from the Ruhr coal district (Rhenish Massif), NW Germany, via the Ibbenbüren coal mine to the North Sea. **b** Cross section in- dicated in **a** displaying degree of coalification at the surface of the Carboniferous. Note influence of post-orogenic uplift in the south, young subsidence in the north, and magmatic intru- sion in the Ibbenbüren area. For further explanation see text. **c** Relationship between burial depth, mean reflectance, coal rank, and generation of gas. (After Teichmüller et al. 1984; in **b** approximate position of isolines for mean refraction, % Rm, was added)

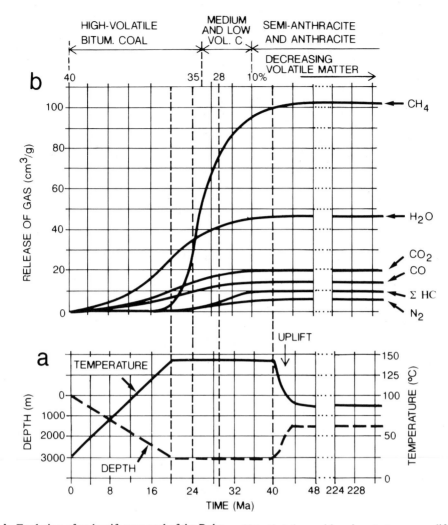

Fig. 14.14a,b. Evolution of carboniferous coal of the Ruhr area, western Germany, and release of different types of gas (b) as a result of increasing burial depth and temperature (a). Note that the coal-bearing strata were uplifted about 40 Ma after their subsidence. Geothermal gradient is 4 °C/100 m. (After Tissot and Welte 1984). For further explanation see text

References

Aalto KR (1989) Franciscan complex olistostrome at Crescent City, northern California. Sedimentology 36: 471-495

Ackermann A (1986) Le flysch de la nappe du Niesen. Eclogae Geol Helv 79: 641-684

Adey WH, Burke RB (1977) Holocene bioherms of Lesser Antilles: geologic control of development. In: Frost SH, Weiss MP, Saunders JB (eds) Reefs and related carbonates: ecology and sedimentology. Am Assoc Petrol Geol, Studies in Geology, 4: 67-81

Agterberg FP, Gradstein FM (1988) Recent developments in quantitative stratigraphy. Earth Sci Rev 25: 1-73

Ahnert F (1970) Functional relationship between denudation, relief, and uplift in large mid-latitude drainage basins. Am J Sci 268: 243-263

Ahr WM (1973) The carbonate ramp: an alternative to the shelf model. Trans Gulf Coast Assoc Geol Soc, 23rd Annu Conv, p 221-225

Aigner T (1982) Event-stratification in nummulite accumulations and in shell beds from the Eocene of Egypt. In: Einsele G, Seilacher A (eds) Cyclic and event stratification, Springer, Berlin Heidelberg New York, p 248-262

Aigner T (1985) Storm depositional systems. Lecture Notes Earth Sciences, 3, Springer, Berlin Heidelberg New York, 174 p

Aigner T, Dott RH (eds) (1990) Patterns and processes in epeiric basins. Sediment Geol 69 (Spec Issue): 165-334

Aigner T, Reineck HE (1982) Proximity trends in modern storm sands from the Helgoland Bight (North Sea) and their implications for basin analysis. Senckenb Marit 14: 183-215

Aigner T, Doyle M, Lawrence D, Epting M, Van Vliet A (1989) Quantitative modeling of carbonate platforms: some examples. Soc Econ Paleontol Mineral Spec Publ 44: 27-37

Akou AE (1984) Subaqueous debris flow deposits in Baffin Bay. Geomar Lett 4/2: 83-90

Alam M (1989) Geology and depositional history of Cenozoic sediments of the Bengal Basin of Bangladesh. Paleogeogr Paleoclimatol Paleoecol 69: 125-139

Alexander CR, De Master DJ, Nittrouer CA (1991) Sediment accumulation in a modern epicontinental-shelf setting: the Yellow Sea. Marine Geol 98: 51-72

Allen JRL (1965) Late Quaternary Niger delta, and adjacent areas: sedimentary environments and lithofacies. Am Assoc Petrol Geol Bull 49: 547-600

Allen JRL (1982) Sedimentary structures, their character and physical basis, v. II. Developm Sedimentol 30 B, Elsevier, Amsterdam, 663 p

Allen JRL (1983) Studies in fluviatile sedimentation: bars, bar-complexes and sandstone sheets (low sinuosity braided streams) in the Brownstones (L. Devonian), Welsh Borders. Sediment Geol 33: 237-293

Allen PA, Allen JR (1990) Basin analysis - principles and application. Blackwell, Oxford, 451 p

Allen PA, Collinson JD (1986) Lakes. In: Reading HG (ed) Sedimentary environments and facies. Blackwell, Oxford, p 63-94

Allen PA, Homewood P, Williams G (eds)(1986) Foreland basins: an introduction. Spec Publ Int Assoc Sedimentol 8: 3-12

Alley BB, Blankenship DD, Rooney ST, Bentley CR (1988) Sedimentation beneath ice shelves - the view from ice stream B. Marine Geol 85: 101-120

Alpern B (1987) Application de la pétrographie des organoclastes à l´histoire géologique et thermique des bassins sédimentaires carbonés. Mém Soc Géol Fr NS 151: 55-75

Amstutz GC, El Goresy A, Frenzel G, Kluth C, Moh G, Waschkuhn A, Zimmermann RA (1982) Ore genesis, the state of the art. Soc Geology Applied to Mineral Deposits, Spec Publ 2, Springer, Berlin Heidelberg New York, 804 p

Anders MH, Krueger SW, Sadler PM (1987) A new look at sedimentation rates and the completeness of the stratigraphic record. J Geol 95: 1-14

Anderson FE, Black L, Watling SE, Mook W, Mayer LM (1981) A temporal and spatial study of mudflat erosion and deposition. J Sediment Petrol 51: 729-736

Anderson JB, Molnia BF (1989) Glacial marine sedimentation. Short course in geology 9, Am Geophys Union, Washington D.C., 127 p

Anderson RY (1982) A long geoclimatic record from the Permian. J Geophys Res 87, C9: 7285-7294

Anderton R, Bridges PH, Leeder MR, Sellwood BW (1979) A dynamic stratigraphy of the British Isles. Allen & Unwin, London, 301 p

Andreyev SN, Kulikov AN (1987) Sedimentation rates in areas of manganese-nodule formation in the Pacific. Trans (Doklady) USSR Acad Sciences, Earth Science Sections, 297, 6: 75-77

Angevine CL, Heller PL, Paola C (1990) Quantitative basin modeling. Continuing Education Course Note Series, 32, Am Assoc Petrol Geol, Tulsa, 133 p, 108 figs

April R, Newton R, Truettner Coles L (1986) Chemical weathering in two Adirondeck watersheds: past and present-day rates. Geol Soc Am Bull 97: 1232-1238

Arthur MA, Dean WE (1991) An holistic geochemical approach to cyclomania: examples from Cretaceous pelagic limestone sequences. In: Einsele G, Ricken W, Seilacher A (eds) Cycles and events in stratigraphy. Springer, Berlin Heidelberg New York, p 126-166

Arthur MA, Dean WE, Pratt LM (1988) Geochemical and climatic effects of increased marine organic carbon burial at the Cenomanian/Turonian boundary. Nature 335: 714-717

Arthur MA, Dean WE, Schlanger SO (1985) Variations in the global carbon cycle during the Cretaceous related to climate, volcanism, and changes of atmospheric CO_2. In: Sundquist ET, Broecker WS (eds) The carbon cycle and atmospheric CO_2: natural variations

578

Archean to Present. Am Geophys Union, Geophys Monogr 32, p 504-530, Washington DC

Arthur MA, Schlanger SO, Jenkyns HC (1987) The Cenomanian-Turonian oceanic anoxic event, II. Palaeoceanographic controls on organic matter production. In: Brooks J, Fleet AJ (eds) Marine petroleum source rocks. Geol Soc Spec Publ 26: 401-420

Artyushkov EV, Baer MA (1986) Mechanism of formation of deep basins on continental crust in the Verkhoyanak fold belt: miogeosynclines and cratonic basins. Tectonophysics 122: 217-245

Artyushkov EV, Baer MA (1989) The mechanism of formation of the North Sea basin. In: Price RA (ed) Origin and evolution of sedimentary basins and their energy and mineral resources. Geophys Monograph, 48 (IUGG v 3): 109-123, Am Geophys Union, Washington DC

Artyushkov EV, Sobolev SV (1983) Mechanism of passive margins and inland sea formation. In: Watkins JS, Drake CL (eds) Studies in continental margin geology. Am Assoc Petrol Geol Mem 34: 689-701

Ashley GM, Symposium chairperson (1990) Classification of large-scale subaqueous bedforms: a new look at an old problem. J Sediment Petrol 60: 160-172

Athy LF (1930) Density, porosity, and compaction of sedimentary rocks. Am Assoc Petrol Geol Bull 14: 1-24

Audley-Charles MG (1986) Timor-Tanimbar Trough: the foreland basin of the evolving Banda orogen. In: Allen PA, Homewood P (eds) Foreland basins. Int Assoc Sedimentol Spec Publ 8: 91-102, Blackwell, Oxford

Auffret G, Malinverno A, Pautot G, Ryan WBF (1988) Sonar images of the path of recent failure events on the continental margin off Nice, France. In: Clifton HE (ed) Sedimentologic consequences of convulsive geologic events. Geol Soc Am Spec Pap 229: 59-75

Bachman R (1984) Intratest porosity in foraminifera. J Sediment Petrol 54: 257-262

Bachmann GH, Müller M, Weggen K (1987) Evolution of the Molasse Basin (Germany, Switzerland). Tectonophysics 137: 77-92

Baker BH, Crossley R, Goles GG (1978) Tectonic and magmatic evolution of the southern part of the Kenya rift valley. In: Neumann ER, Ramberg IB (eds) Petrology and geochemistry of continental rifts, Reidel, Dordrecht, p 29-50

Baldwin B, Butler CO (1985) Compaction curves. Am Assoc Petrol Geol Bull 69: 622-626

Ballance PL, Reading HG (1980) Sedimentation in oblique-slip mobile zones. Int Assoc Sedimentol Spec Publ 4, Blackwell, Oxford, 265 p

Bally AW (1987) Atlas of seismic stratigraphy. AAPG Studies in Geology, 27, vol 1, 125 p

Bally AW, Snelson S (1980) Realms of subsidence. Canad Petrol Geol Bull, Jan 1980: 9-75

Barbier F, Duvergé J, Le Pichon X (1986) Structure profonde de la marge Nord-Gascogne. Implications sur le méchanisme de rifting et dee la formation de la marge continentale. Bull Cent Rech Explor-Prod Elf-Aquitaine 10: 105-121

Barnes NE, Normark WR (1985) Diagnostic parameters for comparing modern submarine fans and ancient turbidite systems. In: Bouma AH, Normark WR, Barnes NE (eds) Submarine fans and related turbidite systems. Springer, New York Heidelberg, p 13-14 with maps and table

Barrett TJ (1982) Stratigraphy and sedimentology of Jurassic bedded chert overlying ophiolites in the North Apennines, Italy. Sedimentology 29: 353-373

Barrio CA (1990) Late Cretaceous-Early Tertiary sedimentation in a semi-arid foreland basin (Neuquén Basin, western Argentina). Sediment Geol 66: 255-275

Barron EJ, Peterson WH (1991) The Cenozoic ocean circulation based on General Ocean Circulation Model results. Palaeogeogr Palaeoclimat Palaeoecol 83: 1-28

Barron EJ, Washington WM (1985) Warm Cretaceous climates: high atmospheric CO_2 as a plausible mechanism. In: Sundquist ET, Broecker WS (eds) The carbon cycle and atmospheric CO_2. Geophys Monogr 32: 546-553, Am Geophys Union

Basu A (1985) Influence of climate and relief on compositions of sands released at source areas. In: Zuffa GG (ed) Provenance of arenites. NATO ASI Series, Reidel, Dordrecht, p 1-18

Bathurst RGC (1975) Carbonate sediments and their diagenesis, 2nd edn. Developm Sedimentol 12, Elsevier, Amsterdam, 658 p

Bathurst RGC (1987) Diagenetically enhanced bedding in argillaceous platform linestones: stratified cementation and selective compaction. Sedimentology 34: 749-778

Bathurst RGC (1991) Pressure dissolution and limestone bedding: the influence of stratified cementation. In: Einsele G, Ricken W, Seilacher A (eds) Cycles and events in stratification. Springer, Berlin Heidelberg New York, p 1-19

Baturin GN (1982) Phosphorites on the sea floor. Developm Sedimentol 33, Elsevier, Amsterdam, 330 p

Baturin GN (1983) Some unique sedimentological and geochemical features of deposits in coastal upwelling regions. In: Thiede J, Suess E (eds) Coastal upwelling - its sediment record. Part B, Sedimentary record of ancient coastal upwelling. NATO Conference Ser 4: Marine Sci, 10, Plenum Press, New York London, p 11-27

Baum GR, Vail PR (1987) Sequence stratigraphy, allostratigraphy, isotope stratigraphy and biostratigraphy: putting it all together in the Atlantic and Gulf Paleogene. Eighth Annual Res Conf, Gulf Coast Section, Soc Econ Paleont Mineral Foundation, Earth Enterprises, Austin, p 15-23

Bayer U (1989a) Stratigraphic and environmental patterns of ironstone deposits. In: Young TP, Taylor WE (eds) Phanerozoic ironstones. Geol Soc Spec Publ 46: 105-117

Bayer U (1989b) Sediment compaction in large scale systems. Geol Rundsch 78: 155-169

Bayer U, Wetzel A (1989) Compactional behavior of fine-grained sediments - examples from Deep Sea Drilling Project cores. Geol Rundsch 78: 807-819

Bayer U, Altheimer E, Deutschle W (1985) Sedimentary cycles and bed formation. In: Bayer U, Seilacher A (eds) Sedimentary and evolutionary cycles. Lecture Notes Earth Sci 1: 347-381

Beach A, Bird T, Gibbs A (1987) Extensional tectonics and crustal structure: deep seismic reflection data from the northern North Sea Viking graben. In: Coward MP, Dewey JF, Hancock PL (eds) Continental extensional tectonics. Geol Soc Spec Publ 28: 467-476

Beauchamp J (1988) Triassic sedimentation and rifting in the High Atlas (Morocco). In: Manspeizer W (ed) Triassic-Jurassic rifting, parts A and B. Developm Geotectonics 22, Elsevier, Amsterdam, p 477-497

Beaudoin B, Cojan I, Fries G, Pinoteau B (1985) Lois de décompaction et approche de évolution du taux de sédimentation dans les forages pétroliers du Sud-Est de la France. Programme Géologie Profonde de la France. Doc BRGM 95/11, thème 11: 133-148

Bechstädt (1975) Zyklische Sedimentation im erzführenden Wettersteinkalk von Bleiberg-Kreuth (Kärnten, Österreich). N Jahrb Geol Paläontol Abh 149: 73-95

Bechstädt T, Brandner R, Mostler H, Schmidt K (1978)

Aborted rifting in the Triassic of the Eastern and Southern Alps. N Jahrb Geol Paläontol Abh 156: 157-178

Beerbower JR (1964) Cyclothems and cyclic depositional mechanisms in alluvial plain sedimentation. In: Merriam DF (ed) Symposium on cyclic sedimentation, Kansas Geol Surv Bull 169,1: 31-42

Bender ML (1984) On the relationship between ocean chemistry and atmospheric CO_2 during the Cenozoic. In: Hansen JE, Takahashi T (eds) Climate processes and climate sensitivity. Geophys Monogr 29: 352-359, Am Geophys Union

Ben Ismail MH, M'Rabet A (1990) Evaporite, carbonate, and siliciclastic transitions in the Jurassic sequences of southeastern Tunisia. Sediment Geol 66: 65-82

Benson RN, Doyle RG (1988) Early Mesozoic rift basins and the development of the United States middle Atlantic continental margin. In: Manspeizer W (ed) Triassic-Jurassic rifting. Developm Geotectonics 22, Elsevier, Amsterdam, p 99-127

Bentor YK (1980) Phosphorites - the unsolved problems. In: Bentor YK (ed) Marine phosphorites. Soc Econ Paleontol Mineral Spec Publ 29: 3-18.

Berger A, Imbrie J, Hays J, Kukla G, Saltzman B (eds) (1984) Milankovitch and climate. NATO series C 126, I, II, Reidel, Dordrecht, 895 p

Berger WH (1970) Biogenous deep-sea sediments: fractionation by deep-sea circulation. Geol Soc Am Bull 81: 1385-1402

Berger WH (1974) Deep-sea sedimentation. In: Burk CA, Drake CL (eds) The geology of continental margins. Springer, New York Heidelberg, p 213-241

Berger WH (1976) Biogenous deep-sea sediments: production, preservation and interpretation. In: Riley JP, Chester R (eds) Chemical oceanography, vol 5, 2nd edn. Academic Press, London, p 265-388.

Berger WH (1981) Paleoceanography: the deep-sea record. In: Emiliani C (ed) The sea, v 7, The oceanic lithosphere. Wiley-Interscience, New York, p 1437-1519

Berger WH (1985) CO_2 increase and climate prediction: clues from deep-sea carbonates. Epsodes, 8, 3: 163-168

Berger WH, Kroenke LW, Mayer LA, and Shipboard Scientific Party (1991) Ontong Java Plateau, Leg 130: synopsis of major drilling results. In: Kroenke LW, Berger WH, Janecek TR, et al. Proc ODP Init Repts 130: 497-537, College Station, TX (Ocean Drilling Program)

Berger WH, von Rad U (1972) Cretaceous and Cenozoic sediments form the Atlantic ocean. In: Hayes DE, Pimm AC, et al, Init Repts DSDP, 14: 787-954, US Govt Print Office, Washington

Berger WH, Smetacek VS, Wefer G (1989) Ocean productivity and paleoproductivity - an overview. In: Berger WH, Smetacek VS, Wefer G (eds) Productivity of the ocean: present and past. Dahlem Konferenzen, Wiley, Chichester, pp 1-34

Berner HD, Gebhard G, Wiedmann J (1982) Kondensationserscheinungen in der marokkanischen und alpinen Mittelkreide (Apt, Alb). N Jahrb Geol Paläontol Abh 165: 102-124, Stuttgart

Berner RA (1980) Early diagenesis. Princeton Univ Press, 237 p

Berner RA (1982) Burial of organic carbon and pyrite sulfur in the modern ocean: its geochemical and environmental significance. Am J Sci 282: 451-473

Bernoulli D, Jenkyns HC (1974) Alpine, Mediterranean, and central Atlantic Mesozoic facies in relation to the early evolution of the Tethys. In: Dott RH, Shaver RH (eds) Modern and ancient geosynclinal sedimentation. Soc Econ Paleontol Mineral Spec Publ 19: 129-160

Bernoulli D, Lemoine M (1980) Birth and early evolution of the Tethys: the overall situation. In: Auboin J, Debelmas J, Latreille M (eds) Geology of the Alpine chains born of the Tethys. Mem BRGM 115: 168-198

Bethke CM (1989) Modeling subsurface flow in sedimentary basins. Geol Rundsch 78: 129-154

Betzer PR, Showers WJ, Laws EA, Winn CD, DiTullio GR, Kroopnik PM (1984) Primary productivity and particle fluxes on a transect of the equator at 153° W in the Pacific Ocean. Deep-Sea Res 31: 1-11

Beyth M (1980) Recent evolution and present stage of Dead Sea brines. In: Nissenbaum A (ed) Hypersaline brines and evaporitic environments. Developm Sedimentol 28, Elsevier, Amsterdam, p 155-165

Biddle KT, Christie-Blick N (eds)(1985) Strike-slip deformation, basin formation, and sedimentation. Soc Econ Paleontol Mineral Spec Publ 37, 386 p

Bigarella JJ (1972) Eolian environments: their characteristics, recognition and importance. In: Rigby JK, Hamblin WK (eds) Recognition of ancient sedimentary environments. Soc Econ Paleontol Mineral Spec Publ 16: 12-62

Birkeland PW (1984) Soils and geomorphology. Oxford Univ Press, New York

Birkenmajer K (1981) The geology of Svalbard, the western part of the Barents Sea, and the continental margins of Scaninavia. In: Nairn AEM, Churkin MJr, Stehli FG (eds) The ocean basins and margins, vol 5, the Arctic ocean. Plenum Press, New York London, pp 265-329

Bitzer K, Harbough JW (1987) DEPOSIM: a Macintosh computer model for two-dimensional simulation of transport, deposition, erosion, and compaction of clastic sediments. Computers & Geosci 13: 611-637

Bjørlykke K (1983) Diagenetic reactions in sandstones. In: Parker A, Sellwood BW (eds) Sediment diagenesis. NATO ASI Series C 115: 169-213, Reidel, Dordrecht

Bjørlykke K (1984) Formation of secondary porosity: how important is it? In: McDonald DA, Surdam RC (eds) Clastic diagenesis. Am Assoc Petrol Geol Mem 37: 277-286

Bjørlykke K (1988) Sandstone diagenesis in relation to preservation, destruction and creation of porosity. In: Chilingarian GV, Wolf KH (eds) Diagenesis, I. Developm Sedimentol 41, Elsevier, Amsterdam, p 555-588

Bjørlykke K, Ramm M, Saigal SC (1989) Sandstone diagenesis and porosity modification during basin evolution. Geol Rundsch 78: 243-268

Blatt H, Middleton GV, Murray RC (1980) Origin of sedimentary rocks, 2nd edn. Prentice-Hall, Englewood Cliffs, 782 p

Blueford JR (1989) Radiolarian evidence: Late Cretaceous through Eocene ocean circulation patterns. In: Hein JR, Obradovic J (eds) Siliceous deposits of the Tethys and Pacific regions. Springer, New York Heidelberg, p 19-29

Blundell DJ, Reston RJ, Stein AM (1989) Deep crustal structural controls on sedimentary basins geometry. In: Price RA (ed) Origin and evolution of sedimentary basins and their energy and mineral resources. Geophys Monograph, 48 (IUGG vol 3), Am Geophys Union, Washington DC, p 57-64

Boccaletti M, Calamita F, Deiana G, Gelati R, Massari F, Moratti G, Ricci-Luchi F (1990) Migrating foredeepthrust belt system in the northern Apennines and southern Alps. Palaeogeogr Palaeoclimatol Palaeoecol 77: 3-14

Bögel H, Schmidt K (1976) Kleine Geologie der Ostalpen. Ott Verlag, Thun, 231 p

Bögli A (1978) Karsthydrographie und physische Speläologie. Springer, Berlin Heidelberg New York, 292 p

Boer PL de, (1991) Pelagic black shale-carbonate rhythms: orbital forcing and oceanographic response. In: Einsele G, Ricken W, Seilacher A (eds) Cycles and events in stratigraphy. Springer, Berlin Heidelberg New York, p 63-78

Boer PL de, Oost AP, Visser MJ (1989) The diurnal inequality of the tide as a parameter for recognizing tidal influence. J Sediment Petrol 59: 912-921

Boer PL de, van Gelder A, Nio SD (eds) (1988) Tide-influenced sedimentary environments and facies. Reidel, Dordrecht, 530 p

Boersma JR, Terwindt JHJ (1981) Neap-spring tide se quences of intertidal shoal deposits in a mesotidal estuary. Sedimentology 28: 151-170

Boggs SJr (1987) Principles of sedimentology and stratigraphy. Merril, Columbus, Ohio, 784 p

Bohrmann G (1988) Zur Sedimentationsgeschichte von biogenem Opal im nördlichen Nordatlantik und dem Europäischen Nordmeer. Berichte Sonderforschungsbereich 313 Sedimentation im Europäischen Nordmeer. Univ Kiel 9, 221 p

Bohrmann G, Henrich R, Thiede J (1990) Miocene to Quaternary paleooceanography in the northern North Atlantic: variability in carbonate and biogenic opal accumulation. In: Thiede J, Bleil U (eds) Geological history of the polar oceans: Arctic versus Antarctic. NATO ASI Series C, Kluwer, Dordrecht, p 647-675

Boillot G, Winterer EL (1988) Drilling on the Galicia margin: retrospect and prospect. In: Boillot G, Winterer EL, et al., Proc ODP Sci Results 103: 809-829; College Station, TX (Ocean Drilling Program)

Bond GC, Kominz MA, Grotzinger JP (1988) Cambro-Ordovician eustacy: evidence from geophysical modeling of subsidence in Cordilleran and Appalachian passive margin. In: Klienspahn K, Paola C (eds) New perspectives in basin analysis. Springer, New York Heidelberg, p 129-161

Boote DRD, Kirk RB (1989) Depositional wedge cycles on evolving plate margin, western and northwestern Australia. Am Assoc Petrol Geol 73: 216-243

Bosellini A (1984) Prograding geometries of carbonate platforms: examples from the Triassic of the Dolomites, northern Italy. Sedimentology 31: 1-24

Bosellini A, Mutti E, Ricci Lucchi F (1989) Rocce e successioni sedimentarie. Unione Tipografico-Editrice Torinese (UTET), Torino, 395 p

Bosence D (1989) Biogenic carbonate production in Florida Bay. Bull Marine Sci 44: 419-433

Bosence D, Waltham D (1990) Computer modeling the internal architecture of carbonate platforms. Geology 18: 26-30

Bott MHP (1982) The interior of the earth: its structure, constitution and evolution, 2nd edn. Arnold, London, 403 p

Boulegue J, Iiyama JT, Charlou JL, Jedwab J (1987) Nankai trough, Japan trench and Kuril trench: Geochemistry of fluids sampled by submersible "Nautile". Earth Planet Sci Lett 83: 363-375

Boulton GS, Deynoux M (1981) Sedimentation in glacial environments and the identification of tills and tillits in ancient sedimentary sequences. Precambrian Res 15: 397-422

Bouma AH (1962) Sedimentology of some flysch deposits. Elsevier, Amsterdam, 168 p

Bouma AH (1990) Clastic depositional styles and reservoir potential of Mediterranean basins. Am Assoc Petrol Geol Bull 74: 532-546

Bouma AH, Coleman JM, Meyer AW, et al (1986) Init Repts DSDP, 96, Washington, US Govt Print Office

Bouma AH, Normark WR, Barnes NE (eds) (1985) Submarine fans and related turbidite systems. Springer, New York Heidelberg, 351 p

Bourrouilh R (1987) Evolutionary mass flow-megaturbidites in an interplate basin: example of the North Pyrenean basin. Geo-Marine Lett 7: 69-81

Bowler JM (1973) Clay dunes: their occurrence, formation and environmental significance. Earth Sci Rev 9: 315-338

Bowler JM (1986) Spatial variability and hydrologic evolution of Australian lake basins: analogue for Pleistocene hydrologic change and evaporite formation. Palaeogeogr Palaeoclimatol Palaeoecol 54: 21-41

Bowler JM, Teller JT (1986) Quaternary evaporites and hydrological changes, Lake Tyrell, north-west Victoria. Aust J Earth Sci 33: 43-63

Bowler JM, Huang Qi, Chen Kezao, Head MJ, Yuan Baoyin (1986) Radiocarbon dating of playa-lake hydrologic changes: examples form northwestern China and central Australia. Palaeogeogr Palaeoclimatol Palaeoecol 54: 241-260

Brachert TC (1987) Diagenesis of siliceous sponge limes tones from the Pleistocene of the Tyrrhenian Sea (Mediterranean Sea). Facies 17: 41-50, Erlangen

Braitsch O (1971) Salt deposits, their origin and com position. Springer, New York Heidelberg, 297 p

Bralower TJ, Thierstein HR (1984) Low productivity and slow deep-water circulation in Mid-Cretaceous oceans. Geology 12: 614-618

Bralower TJ, Thierstein HR (1987) Organic carbon and metal accumulation rates in Holocene and mid-Cretaceous sediments: palaeoceanographic significance. In: Brooks J, Fleet AJ (eds) Marine petroleum source rocks. Geol Soc Spec Publ 26: 345-369

Brandner R (1985) Meeresspiegelschwankungen und Tektonik in der Trias der NW-Tethys. Jahrb Geol Bundesanst 126: 435-475, Wien

Brandt K (1985) Sea-level changes in the upper Sinemurian and Pliensbachian of Southern Germany. In: Bayer U, Seilacher A (eds) Sedimentary and evolutionary cycles. Lecture Notes in Earth Sciences, 1, Springer Berlin Heidelberg New York, p 113-126

Bray CJ, Karig DE (1985) Porosity of sediments in accretionary prisms and some implications for dewatering processes. J Geophys Res 90: 768-778

Bredehoeft JD, Djevanshir RD, Belitz KR (1988) Lateral fluid flow in a compacting sand-shale sequence: South Caspian basin. Am Assoc Petrol Geol Bull 72: 416-424

Breitkopf JH (1988) Iron formations related to mafic vol canism and ensialic rifting in the southern margin zone of the Damara Orogen, Namibia. Precambrian Res 38: 111-130

Brett CE, Seilacher A (1991) Fossil lagerstaetten: a taphonomic consequence of event sedimentation. In: Einsele G, Ricken W, Seilacher A (eds) Cycles and events in stratigraphy. Springer, Berlin Heidelberg New York, p 283-297

Bristow CS (1987) Brahmaputra River: Channel migration and deposition. In: Ethridge FG, Flores RM, Harvey MD (eds) Recent developments in fluvial sedimentology, Soc Econ Paleontol Mineral Spec Publ 39: 63-74

Brodzikowski K, van Loon AJ (1991) Glacigenic sediments. Developm Sedimentol 49, Elsevier, Amsterdam, 674 p

Brongersma-Sanders MB (1957) Mass mortality in the sea. In: Hedgpeth JW (ed) Treatise on Marine Ecology and Paleoecology. Geol Soc Am Mem 67 (1): 941-1010

Brooks J, Fleet AJ (eds)(1987) Marine petroleum source rocks. Geol Soc Spec Publ 26, Blackwell, Oxford,

444 p Brooks J, Cornford C, Archer R (1987) The role of hydrocarbon source rocks in petroleum exploration. In: Brooks J, Fleet AJ (eds) Marine petroleum source rocks. Geol Soc Spec Publ 26: 17-46

Bruce CH (1973) Pressured shale and related sediment deformation: contemporaneous faults. Am Assoc Petrol Geol Bull 57: 878-896

Brückmann W (1989) Typische Kompaktionsabläufe mariner Sedimente und ihre Modifikation in einem rezenten Akkretionskeil (Barbados Ridge). Tübinger Geowiss Arbeiten, A 5, Geol Inst Univ Tübingen, 135 p

Brumsack HJ (1986a) The inorganic geochemistry of Cretaceous black shales (DSDP Leg 41) in comparison to modern upwelling sediments from the Gulf of California. In: Summerhayes CP, Shackleton NJ (eds) North Atlantic palaeoceanography. Geol Soc Spec Publ 21: 447-462

Brumsack HJ (1986b) The geochemical facies of black shales from the Cenomanian/Turonian Boundary Event (CTBE). Mitt Geol-Paläontol Inst Univ Hamburg, SCOPE UNEP Sonderb, H 60: 247-265

Brumsack HJ and Shipboard Scientific Party ODP Leg 127 (1990) Erste Ergebnisse von ODP-Leg 127 (Japan-See I). In: Beiersdorf H (ed) Deutsche Forschungsgemeinschaft, Protokoll Kolloquium im SPP "Ocean Drilling Program/Deep Sea Drilling Project" in Bremen, Bonn, p 31-33

Brunner CA, Ledbetter MT (1987) Sedimentological and micropaleontological detection of turbidite muds in hemipelagic sequences: an example from the late Pleistocene levee of Monterey Fan, central California continental margin. Marine Micropaleontol 12: 223-239

Brunsden D (1979) Weathering. In: Embleton C, Thornes J (eds) Process in geomorphology. Arnold, London, p 73-129

Bryant WR, Bennett RH (1988) Origin, physical and mineralogical nature of red clays: the Pacific Ocean basin as a model. Geo-Marine Lett 8: 189-249

Budyko MI, Ronov AB, Yanshin AL (1987) History of the Earth's atmosphere. Springer, Berlin Heidelberg New York, 139 p

Buggisch W (1991) The global Frasnian-Famennian "Kellwasser Event". Geol Rundsch 80: 49-72

Buick R, Dunlop JSR (1990) Evaporitic sediments of Early Archaean age from the Warrawoona Group, North Pole, Western Australia. Sedimentology 37: 247-277

Bull WB (1977) The alluvial fan environment. Progress in physical geography, vol 1, Arnold, London, p 222-270

Burchfiel BC, Royden LH (1988) A general approach to basin analysis (afterword). In: Royden LH, Horváth F (eds) The Pannonian basin: a study in basin evolution. Am Assoc Petrol Geol Mem 45: 373-375

Burckle LH, Abrams N (1987) Regional late Pliocene-early Pleistocene hiatuses of the southern ocean - diatom evidence. Marine Geol 77: 207-218

Burg JP, Leyreloup A, Girardeau J, Chen GM (1987) Structure and metamorphism of a tectonically thickened continental crust: the Yalu Tsangpo suture zone (Tibet). In: Oxburgh ER, Yardley BWD, England PC (eds) Tectonic settings of regional metamorphisms. Philos Trans R Soc London A 321: 67-86

Burgess CF, Rosendahl BR, Sander S, Burgess CA, Lambiase J, Derksen S, Meader N (1988) The structural and stratigraphic evolution of Lake Tanganyika: a case study of continental rifting. In: Manspeizer W (ed) Triassic-Jurassic rifting, parts A and B. Developm Geotectonics 22, Elsevier, Amsterdam, p 859-881

Burnett WC, Riggs SR (eds) (1990) Phosphate deposits of the world, vol 3, Genesis of Neogene to Recent phosphorites. Cambridge Univ Press

Burollet PF, Mugniot JM, Sweeney P (1978) The geology of the Pelagian Block; The margins and basins off southern Tunisia and Tripolitania. In: Nairn AEM, Kanes WH, Stehli FG (eds) The ocean basins and margins, v 4 B: the Western Mediterranean. Plenum Press, New York, p 331-359

Burton R, Kendall CGSC, Lerche J (1987) Out of our depth: on the impossibility of fathoming eustacy from the stratigraphic record. Earth Sci Rev 24: 237-277

Busson G (1990) Le Messinien de la Méditerranée - vingt ans après. Géologie de la France 1990, 3-4: 3-58

Butler GP, Harris PM, Kendall CGSC (1982) Recent evaporites from the Abu Dhabi coastal flats. In: Handford CR, Loucks RG, Davies GR (eds) Deposition and diagenetic spectra of evaporites. Soc Econ Paleontol Mineral Core Workshop 3, Calgary, p 33-64

Buxton MWN, Pedley HM (1989) A standardized model for Tethyan Tertiary carbonate ramps. J Geol Soc London 146: 746-748

Cadet J-P, Kobayashi K, et al. (1987) The Japan Trench and its juncture with the Kuril Trench: cruise results of the Kaiko project, Leg 3. Earth Planet Sci Lett 83: 267-284

Calvert SE (1966) Origin of diatom-rich varved sediments from the gulf of California. J Geol 76: 546-565

Calvert SE (1974) Depostion and diagenesis of silica in marine sediments. In: Hsü KJ, Jenkyns HC (eds) Pelagic sediments on land and under the sea. Int Assoc Sedimentol Spec Publ 1: 273-299

Calvert SE, Fontugne MR (1987) Stable carbon isotopic evidence for the marine origin of the organic matter in the Holocene Black Sea sapropel. Chem Geol (Isotope Geosci Sect) 66: 315-322

Calvert SE, Piper DZ (1984) Geochemistry of ferromanganese nodules from DOMES Site A, Northern Equatorial Pacific: multiple diagenetic metal sources in the deep sea. Geochim Cosmochim Acta 48: 1913-1928

Cameron CC, Esterle JS, Palmer CA (1989) The geology, lbotany and chemistry of selected peat-forming environments from temperate and tropical latitudes. In: Lyons PC, Alpern B (eds) Peat and coal: origin, facies, and depositional models. Elsevier, Amsterdam, p 105-156

Campell CV (1967) Laminae, lamina set, bed and bedset. Sedimentology 8: 7-26

Carey SN, Sigurdsson H (1984) A model of volcanogenic sedimentation in marginal basins. In: Kokelaar BP, Howells RE (eds) Marginal basin geology: volcanic and associated sedimentary and tectonic processes in modern and ancient marginal basins. Geol Soc Spec Publ 16: 37-58

Caron C, Homewood P, Wildi W (1989) The original Swiss flysch: a reappraisal of the type deposits in the Swiss Prealps. Earth-Sci Rev 26: 1-45

Carpenter SJ, Lohmann KC (1989) $\delta^{18}O$ and $\delta^{13}C$ variations in Late Devonian marine cements from the Golden Spike and Nevis reefs, Alberta, Canada. J Sediment Petrol 59: 792-814

Carpenter SJ, Erickson JM, Lohmann KC, Owen MR (1988) Diagenesis of fossiliferous concretions from the upper Cretaceous Fox Hills Formation, North Dakota. J Sediment Petrol 58: 706-723

Carson MA, Kirkby MJ (1972) Hillslope, form and process. Cambridge Univ Press, 475 pp

Carter RM (1988) The nature and evolution of deep-sea channel systems. Basin Res 1: 41-54

Carter RWG (1988) Coastal environments. Academic Press, London, 617 p

Cas RAF, Landis CA (1987) A debris-flow deposit with multiple plug-flow channels and associated side accretion deposits. Sedimentology 34: 901-910

Cas RAF, Wright JV (1987) Volcanic successions: modern and ancient. Unwin Hyman, London, 528 p

Chambre de syndicale de recherche et de la production du pétrole et du gaz naturel (1980) Evaporite deposits, illustration and interpretation of some environmental sequences. Editions Technip, Paris, 266 p

Chamley H (1989) Clay sedimentology. Springer, Berlin Heidelberg New York, 623 p

Chapin CE, Elston WE (eds) (1979) Ash-fall tuffs. Geol Soc Am Spec Pap 180, Boulder, 211 p

Chapman DS, Rybach L (1985) Heat flow anomalies and their interpretation. J Geodynamics 4: 3-37

Chenet P, Montadert L, Gairaud H, Roberts D (1983) Extension ratio measurements on the Galicia, Portugal, and northern Biscay continental margins: implications for evolutionary models of passive continental margins. In: Watkins JS, Drake CL (eds) Studies in continental margin geology. Am Assoc Petrol Geol Mem 34: 703-715

Chilingar GV, Zenger DJ, Bissel HJ, Wolf KH (1979) Dolomites and dolomitization. In: Larsen G, Chilingar GV (eds) Diagenesis in sediments and sedimentary rocks. Developm Sedimentol 25A, Elsevier, Amsterdam, p 423-536

Chilingarian GV (1983) Compactional diagenesis. In: Parker A, Sellwood BW (eds) Sediment diagenesis. NATO ASI Series C 115, Reidel, Dordrecht, p 57-168

Chorley RJ, Schumm SA, Sugden DE (1984) Geomorphology. Methuen, London New York, 605 p

Chough SK (1984) Fine-grained turbidites and associated mass flow deposits in the Ulleung (Tsushima) back-arc basin, East Sea (Sea of Japan). In: Stow DAV, Piper DJW (eds) Fine-grained sediments: deep-water processes and facies. Geol Soc Spec Publ 15: 185-196

Christie-Blick N, Biddle KT (1985) Deformation and basin formation along strike-slip faults. In: Biddle KT, Christie-Blick N (eds) Strike-slip deformation, basin formation, and sedimentation. Soc Econ Paleontol Mineral Spec Publ 37: 1-34

Christie-Blick N, Grotzinger JP, von der Borch CC (1988) Sequence stratigraphy in Proterozoic successions. Geology 16: 100-104

Cita MB, Ricci Lucchi F (eds)(1984) Seismicity and sedimentation. Marine Geol 55, No 1/2, 161 p

Clemmensen LB, Abrahamsen K (1983) Aeolian stratification and facies association in desert sediments, Arran basin (Permian) Scotland. Sedimentology 30: 311-339

Clemmey H, Badham N (1982) Oxygen in the Precambrian atmosphere: an evaluation of the geologic evidence. Geology 10: 141-146

Cliff RA, Droop GTR, Rex DC (1985) Alpine metamorphism in in south-east Tauern Window, Austria:2. Rates of heating, cooling and uplift. J Metamorphic Geol 3: 403-415

Clifton HE (ed)(1988) Sedimentologic consequences of convulsive geologic events. Geol Soc Am Spec Pap 229, 157 p

Clifton HE, Hunter RE, Phillips RL (1971) Depositional structures and processes in the non-barred, high-energy nearshore. J Sediment Petrol 41: 651-670

Cloetingh S (1986) Intraplate stresses: a new tectonic mechanism for fluctuations of relative sea level. Geology 14: 617-620

Cloud P (1973) Paleoecological significance of the banded iron formation. Econ Geol 68: 1135-1143

Cochran JR (1981) The Gulf of Aden: structure and evolution of a young ocean basin and continental margin. J Geophys Res 86: 263-287

Cochran JR (1983a) A model for the development of the Red Sea. Am Assoc Petrol Geol Bull 67: 41-69

Cochran JR (1983b) Effects of finite rifting times on the development of sedimentary basins. Earth Planet Sci Lett 66: 289-302

Cochran JR (1990) Himalayan uplift, sea level, and the record of Bengal fan sedimentation at the ODP Leg 116 sites. In: Cochran JR, Stow DAV, et al., Proc ODP, Sci Results, 116: 397-414. College Station, TX (Ocean Drilling Program)

Cohen AS (1989) Facies relationships and sedimentation in large rift lakes and implications for hydrocarbon exploration: examples from Lakes Turkana and Tanganyika. In: Talbot MR, Kelts K (eds) The Phanerozoic record of lacustrine basins and their environmental signals. Palaeogeogr Palaeoclimatol Palaeoecol 70, 65-80

Collins MB (1986) Process and controls involved in the transfer of fluviatile sediments in the deep ocean. J Geol Soc London 143: 915-920

Collinson JD (1986a) Alluvial sediments. In: Reading HG (ed) Sedimentary environment and facies. Blackwell, Oxford, p 20-62

Collinson JD (1986b) Deserts. In: Reading HG (ed) Sedimentary environments and facies. Blackwell, Oxford, p 95-112

Collinson JD, Lewin J (eds) (1983) Modern and ancient fluvial systems, Int Assoc Sedimentol Spec Publ 6

Collinson JD, Thompson DB (1982) Sedimentary structures. Allen & Unwin, London, 194 p

Colman SM, Dethier DP (eds)(1986) Rates of chemical weathering of rocks and minerals. Academic Press, Orlando, 603 p

Conway Morris S (1990) Late Precambrian and Cambrian soft-bodied faunas. Ann Ev Earth Planet Sci 18: 101-122

Cook HE, Enos P (eds) (1977) Deep-water carbonate environments. Soc Econ Paleontol Mineral Spec Publ 25, 336 p

Cooke RV, Warren A (1973) Geomorphology in deserts. Batsford, London, 374 p

Copeland P, Harrison TM (1990) Episodic rapid uplift in the Himalaya revealed by $^{40}Ar/^{39}Ar$ analysis of detrital K-feldspar and muscovite, Bengal fan. Geology 18: 354-357

Costa JE (1984) Physical geomorphology of debris flows. In: Costa JE, Fleisher PJ (eds) Developments and applications of geomorphology. Springer, Berlin Heidelberg New York, p 268-317

Cotillon P (1985) Les variations à différentes echelles du taux d'accumulation sédimentaire dans le séries pélagiques alternantes du Crétacé inférieur, conséquences de phénomènes globaux. Bull Soc Géol Fr 8: 59-68

Cotillon P (1991) Varves, beds, and bundles in pelagic sequences and their correlation (Mesozoic of SE France and Atlantic). In: Einsele G, Ricken W, Seilacher A (eds) Cycles and events in stratigraphy. Springer, Heidelberg New York, p 820-839

Courel L (1989) Organics versus clastic conditions necessary for peat (coal) development. In: Lyons PC, Alpern B (eds) Peat and coal: origin, facies, and depositional models. Elsevier, Amsterdam, p 19-207

Covey M (1986) The evolution of foreland basins to steady state: evidence from the western Taiwan foreland basin. In: Allen PA, Homewood P (eds) Foreland basins. Int Assoc Sedimentol Spec Publ 8: 77-90, Blackwell, Oxford

Cowan DS (1985) Structural styles in Mesozoic and Cenozoic melanges in the western Cordillera of North America. Geol Soc Am Bull 96: 451-462

Crevello PD, Schlager W (1980) Carbonate debris sheets and turbidites, Exuma Sound, Bahamas. J Sediment Petrol 50: 1121-1147

Cronan DS, Damiani VV, Kinsman DJJ, Thiede J (1974) Sediments from the Gulf of Aden and western Indian Ocean. In: Fisher RL, Bunce ET, et al, Init Repts DSDP, 24, US Govt Print Office, Washington, p 1947-1109

Cronan DS, Galácz A, Mindszenty A, Moorby SA, Polgari M (1991) Tethyan ferromanganese oxide deposits from Jurassic rocks in Hungary. J Geol Soc London 148: 655-668

Crook KAW (1974) Lithogenesis and geotectonics: the significance of compositional variation in flysch arenites (graywackes). Soc Econ Paleontol Mineral Spec Publ 19: 304-310

Cross TA (1986) Tectonic controls of foreland basin subsidence and Laramide style deformation, western United States. In: Allen PA, Homewood P, Williams G (eds) Foreland basins: an introduction. Spec Publ Int Assoc Sedimentolog 8: 15-39

Cross TA (ed) (1990) Quantitative dynamic stratigraphy. Prentice Hall, Englewood Cliffs

Crowell JC (1974) Origin of late Cenozoid basins in southern California. In: Dott RH, Shaver RH (eds) Modern and ancient geosynclinal sedimentation. Soc Econ Paleontol Mineral Spec Publ 19: 292-303

Curray JR, Moore DG (1971) Growth of the Bengal deep-sea fan and denudation in the Himalayas. Geol Soc Am Bull 82: 563-572

Curray JR, Moore DG (1974) Sedimentary and tectonic processes in the Bengal deep-sea fan and geosyncline. In: Burk CA, Drake CI (eds) The gelogy of continental margins. Springer, New York Heidelberg, p 617-627

Curray JR, Moore DG, et al. (1982a) Init Repts DSDP 64, Parts 1 and 2, US Govt Print Office, Washington, 1313 p

Curray JR, Moore DG, Kelts K, Einsele G (1982b) Tectonics and geological history of the passive continental margin at the tip of Baja California. In: Curray JR, Moore DG, et al., Init Repts DSDP 64, part 2: 1089-1116, US Govt Print Office, Washington

Currey DR (1990) Quaternary palaeolakes in the evolution of semidesert basins, with special emphasis on Lake Bonneville and the Great Basin, U.S.A. Palaeogeogr Palaeoclimatol Palaeoecol 76: 189-214

Curtis C (1987) Mineralogical consequences of organic matter degradation in sediments: Inorganic/organic diagenesis. In: Leggett JK, Zuffa GG (eds) Marine clastic sedimentology, Graham and Trotman, p 108-123

Dahanayake K, Krumbein WE (1985) Microbial structures in oolitic iron formations. Miner Deposits 21: 85-94

Damuth JE, Flood RD, Kowsmann RO, Belderson RH, Gorini MA (1988) Anatomy and growth pattern of Amazon deep-sea fan as revealed by long-range side-scan sonar (GLORIA) and high-resolution seismic studies. Am Assoc Petrol Geol Bull 72: 885-911

Davis DM, Suppe J, Dahlen FA (1983) Mechanics of fold-and-thrust belts and accretionary wedges. J Geophys Res 88 B2: 1153-1172

Davis EE, Hyndman RD (1989) Accretion and recent deformation of sediments along the northern Cascadia subduction zone. Geol Soc Am Bull 101: 1465-1480

Davis RA Jr (1978) Coastal sedimentary environments. Springer, New York Heidelberg, 420 p

Davis RA Jr (1983) Depositional systems: a genetic approach to sedimentary geology. Prentice-Hall, Englewood Cliffs, 669 p

Davis RA, Fox WT, Hayes MP, Boothroyd JC (1972) Comparison of ridge and runnel systems in tidal and nontidal environments. J Sediment Petrol 42: 413-421

Davis RW (1987) Analysis of hydrodynamic factors in petroleum migration and entrapment. Am Assoc Petrol Geol Bull 71: 643-649

Dean WE (1981) Carbonate minerals and organic matter in sediments of modern north temperate hard-water lakes. Soc Econ Paleontol Mineral Spec Publ 31: 213-231

Dean WE, Arthur MA (1986) Origin and diagenesis of Cretaceous deep-sea, organic carbon-rich lithofacies in the Atlantic Ocean. In: Mumpton FA (ed) Studies in diagenesis. US Geol Survey Bull 1578: 97-128

Dean WE, Fouch TD (1983) Lacustrine environment. In: Scholle PA, Bebout DG, Moore CH (eds) Carbonate depostional environments. Am Assoc Petrol Geol Mem 33: 98-130

Decker K (1991) Rhythmic bedding in siliceous sediments - an overview. In: Einsele G, Ricken W, Seilacher A (eds) Cycles and events in stratigraphy. Springer, Heidelberg New York, p 464-479

Decima A, McKenzie JA, Schreiber BC (1988) The origin of "evaporative limestones: an example from the Messinian of Sicily (Italy). J Sediment Petrol 58: 256-272

Degens ET (1989) Perspectives on biogeochemistry. Springer, Berlin Heidelberg New York, 423 p

Degens ET, Ittekot V (1985) Particulate organic carbon: an overview. In: Degens ET, Kempe S, Herrera R (eds) Transport of carbon and minerals in major world rivers. Part 3. SCOPE/UNEP Sonderband, Mitt Geol-Paleont Inst Univ Hamburg, H 58: 7-27

Degens ET, Ross DA (eds) (1969) Hot brines and recent heavy metal deposits in the Red Sea. Springer, Berlin Heidelberg New York, 600 p

Degens ET, Ross DA (1974) The Black Sea - geology, chemistry, and biology. Am Assoc Petrol Geol Mem 20, 633 p

Degens ET, Stoffers P (1980) Environmental events recorded in Quaternary sediments of the Black Sea. J Geol Soc London 137: 131-138

Degens ET, Kempe S, et al. (eds) (1982, 1983, 1985) Transport of carbon and minerals in major world rivers, parts 1, 2, and 3. SCOPE/UNEP Sonderband, Mitt Geol-Paleontol Inst Univ Hamburg, H 52, 764 p, H 55, 535 p, H 58, 645 p

Degens ET, von Herzen R, Wong HK (1971) Lake Tanganyika: water chemistry, sediments, geological structure. Naturwissenschaften 58: 229-241

Degens ET, Emeis KC, Mycke B, Wiesner MG (1986) Turbidites, the principle mechanism yielding black shales in the early deep Atlantic ocean. In: Summerhayes CP, Shackleton NJ (eds) North Atlantic palaeogeography. Geol Soc Spec Publ 21: 361-376, Blackwell, Oxford

Degens ET, Stoffers P, Golubic S, Dickman MD (1978) Varve Chronology: estimated rates of sedimentation in the Black Sea deep basin. In: Ross DA, Neprochnov YP et al, Init Repts DSDP 42, 2: 499-508

de Graciansky PC, Poag CW, et al. (1985) The Goban Spur transect: geologic evolution of a sediment-starved passive continental margin. Geol Soc Am Bull 96: 58-76

Delcaillau B (1986) Dynamique et évolution morphostructurale du piémont frontal de l'Himalaya: les Siwaliks du Népal oriental. Revue Géol Dynam Géogr Phys 27: 319-337

Demaison G (1985) The generative basin concept. In: Demaison G, Murris RJ (eds) Petroleum geochemistry and basin evaluation. Am Assoc Petrol Geol Bull Mem 35: 1-14

Demaison GJ, Moore GT (1980) Anoxic environments and oil source bed genesis. Am Ass Petrol Geol 64: 1179-1209

584

Deuser WG (1986) Seasonal and interannual variations in deep-water particle fluxes in the Sargasso Sea and their relation to surface hydrography. Deep-Sea Res 33: 225-246

De Visser JP, Ebbing JHJ, Gudjonsson L, Hilgen FJ, Jorissen FJ, Verhallen PJJM, Zevenboom D (1989) The origin of rhythmic bedding in the Pliocene Trubi formation of Sicily, southern Italy. Palaeogeogr Palaeoclimatol Palaeoecol 69: 45-66

Dewey JF, Bird JM (1970) Mountain belts and the new global tectonics. J Geophys Res 75: 2625-2647

Dickinson WR (1982) Composition of sandstones in circum-Pacific subduction complexes and forearc basins. Am Assoc Petrol Geol Bull 66: 121-137

Dickinson WR, Seely DR (1979) Structure and stratigraphy of forearc regions. Am Assoc Petrol Geol Bull 63: 2-31

Dickinson WR, Yarborough H (1976) Plate tectonics and hydrocarbon accumulation. Am Assoc Petrol Geol Educ Course Note Series 1, Tulsa Oklahoma

Dickinson WR, Beard IS, Brakenridge GR, Erjavec JL, Ferguson RC, Inman KF, Knepp RA, Lindberg FA, Ryberg PT (1983) provenance of North American Phanerozoic sandstones in relation to tectonic setting. Geol Soc Amer Bull 93: 95-107

Dickinson WR, Armin RA, Beckvar N, Goodlin TR, Janecke SU, Mark RA, Norris RD, Radel G, Wortman AA (1987) Geohistory analysis of rates of sediment accumulation and subsidence for selected California basins. In: Ingersoll RV, Ernst WG (eds) Cenozoic basin development of coastal California. Rubey vol 6: 1-23, Prentice-Hall, Englewood Cliffs

Diester-Haass L (1973) No current reversal at 10,000 B.P. in the Strait of Gibraltar. Marine Geol 15: M1-M9

Diester-Haass L (1991) Rhythmic carbonate content variations in Neogene sediments above the oceanic lysocline. In: Einsele G, Ricken W, Seilacher A (eds) Cycles and events in stratigraphy. Springer, Heidelberg New York, p 94-109

Dietrich G (1975) Allgemeine Meereskunde. Eine Einführung in die Ozeanographie, 3. Aufl. Bornträger, Berlin Stuttgart, 593 p

Dingler JR, Anima RJ (1989) Subaqueous grain flows at the head of Carmel submarine canyon, California. J Sediment Petrol 59: 280-286

Donovan DT, Jones EJW (1979) Causes of world-wide changes in sea-level. J Geol Soc London 136: 187-192

Dorsey RJ (1989) Provenance evolution and unroofing history of a modern arc-continent collision: evidence from petrography of Plio-Pleistocene sandstones, eastern Taiwan. J Sediment Petrol 58: 208-218

Dott RH Jr (1988) An episodic view of shallow marine clastic sedimentation. In: de Boer PL, van Gelder A, Nio SD (eds) Tidal-influenced sedimentary environments and facies. Reidel, Dordrecht, p 3-12.

Dott RH Jr, Bourgeois J (1982) Hummocky stratification: significance of its variable bedding sequences. J Sediment Petrol 53: 5-23

Doust H, Omatsola E (1990) Niger Delta. In: Edwards JD, Santogrossi PA (eds) Divergent/passive margin basins. Am Assoc Petrol Geol Mem 48: 201-238

Dreesen R (1989) Oolitic ironstones as event-statigraphical marker beds within the Upper Devonian of the Ardenno-Rhenish Massif. In: Young TP, Taylor EG (eds) Phanerozoic ironstones. Geol Soc Spec Publ 46: 65-78

Drever JI (1982) The geochemistry of natural waters. Prentice-Hall, Englewood Cliffs, 388 p

Drever JI, Li Y-H, Maynard JB (1988) Geochemical cycles: the continental crust and the oceans. In: Gregor CB, Garrels RM, Mackenzie FT, Maynard JB (eds) Chemical cycles in the evolution of the Earth. Wiley, New York, p 17-53

Dreyer T, Schleie A, Walderhaug O (1990) Minipermeameter-based study of permeability trends in channel sand bodies. Am Assoc Petrol Geol Bull 74: 359-374

Dronkert H (1985) Evaporite models and sedimentology of Messinian and Recent evaporites. GUA Papers of Geology, Series 1, No 24, Univ Amsterdam, 283 p

Duff PM, Hallam A, Walton EK (eds)(1967) Cyclic sedimentation. Developm Sedimentol 10, Elsevier, Amsterdam, 280 p

Duke WL (1990) Geostrophic circulation or shallow marine turbidity currents? The dilemma of paleoflow patterns in storm-influenced prograding shoreline systems. J Sediment Petrol 60: 870-883

Dunbar CD, Rodgers J (1957) Principles of stratigraphy. Wiley, New York, 356 p

Dunoyer de Segonzac G (1970) The transformation of clay minerals during diagenesis and low-grade matamorphism: a review. Sedimentology 15: 281-346

Durand BM (ed) (1980) Kerogen, insoluble matter from sedimentary rocks. Editions Technip, Paris, 519 p

Durand BM (1987) Du kérogène au pétrole et au charbon: les voies et les méchanismes des transformations des matières organiques sédimentaires au cours de l'enfouissement. Mém Soc Géol Fr NS 151: 77-95

Eberli GP (1987) Carbonate turbidite sequences deposited in rift-basins of the Jurassic Tethys Ocean (eastern Alps, Switzerland). Sedimentology 34: 363-388

Eberli GP (1991) Calcareous turbidites and their relation ship to sea-level fluctuations and tectonism. In: Einsele G, Ricken W, Seilacher D (1989) Cycles and events in stratigraphy. Springer, Heidelberg New York, p 340-359

Eberli GP, Ginsburg RN (1987) Segmentation and coalescence of Cenozoic carbonate platforms, northwestern Great Bahama Bank. Geology 15: 75-79

Eberli GP, Ginsburg RN (1989) Cenozoic prograding of northwestern Great Bahama Bank, a record of lateral platform growth and sea level fluctuations. Soc Econ Paleontol Mineral Spec Publ 44: 339-351

Ebinger CJ (1989) Tectonic development of the western branch of the East African rift system. Geol Soc Am Bull 101: 885-903

Eder FW, Engel W, Franke W, Sadler PM (1983) Devonian and Carboniferous limestone-turbidites of the Rheinisches Schiefergebirge and their tectonic significance. In: Martin H, Eder FW (eds) Intracontinental fold belts. Springer, Berlin Heidelberg New York, p 93-124

Edman JD, Surdam RC (1984) Diagenetic history of the Phosphoria, Tensleep and Madison Formations, Tip Top Field, Wyoming. In: McDonald DA, Surdam RC (eds) Clastic diagenesis. Am Assoc Petrol Geol Mem 37: 317-345

Edwards JD, Santogrossi PA (eds) (1990) Divergent/passive margin basins. Am Assoc Petrol Geol Mem 48, 252 p

Edwards M (1986) Glacial environments. In: Reading HG (ed) Sedimentary environments and facies. Blackwell, Oxford, p 445-470

Ehrmann WU, Thiede J (1985) History of Mesozoic and Cenozoic sediment fluxes to the North Atlantic. Contributions to Sedimentology 15, Schweizerbart, Stuttgart, 109 p

Einsele G (1976) Range, velocity, and material flux of compaction flow in growing sedimentary sequences. Sedimentology 24: 639-655

Einsele G (1982a) Mass physical properties of Pliocene and Quaternary sediments in the Gulf of California, Deep Sea Drilling Project Leg 64. In: Curray JR, Moore DG et al., Init Repts DSDP, 64: 529-542, US Govt Print Office, Washington

Einsele G (1982b) Mechanism of sill intrusion into soft sediment and expulsion of pore water. In: Curray JR, Moore DG et al., Init Repts DSDP, 64, 2: 1169-1176, US Govt Print Office, Washington

Einsele G (1982c) Remarks about the nature, occurrence, and recognition of cyclic sequences. In: Einsele G, Seilacher A (eds) Cyclic and event stratification. Springer, Berlin Heidelberg New York, p 3-7

Einsele G (1985) Response of sediments to sea-level changes in differing subsiding storm-dominated marginal and epeiric basins. In: Bayer U, Seilacher A (eds) Sedimentary and evolutionary cycles. Lect Notes Earth Sci 1: 68-97

Einsele G (1986) Interaction between sediments and basalt injections in young Gulf of California-type spreading centers. Geol Rundsch 75: 197-208

Einsele G (1989) In-situ water contents, liquid limits, and submarine mass flows due to a high liquefaction potential of slope sediment (results from DSDP and subaerial counterparts). Geol Rundsch 78: 821-840

Einsele G, Hohberger K (1978) Zusammensetzung und Bilanzierung der Lösungsfracht im Maingebiet. In: Das Mainprojekt, Schriftenreihe Bayer Landesamt Wasserwirtschaft H 7: 278-289, München

Einsele G, Kelts K (1982) Pliocene and Quaternary mud turbidites in the Gulf of California: sedimentology, mass physical properties and significance. In: Curray JR, Moore DG et al., Init Repts DSDP 64, 2, US Govt Print Office, Washington, p 511-528

Einsele G, Niemitz JW (1982) Budget of postrifting sediments in the Gulf of California and calculation of the denudation rate in neighboring land areas. In: Curray JR, Moore DG, et al., Init Repts DSDP 64, Part 2, US Govt Print Office, Washington, p 571-592

Einsele G, Ricken W (1991) Limestone-marl alternations - an overview. In: Einsele G, Ricken W, Seilacher A (eds) Cycles and events in stratigraphy. Springer, Berlin Heidelberg New York, p 23-47

Einsele G, Seilacher A (1982) (eds) Cyclic and event stratification. Springer, Berlin Heidelberg New York, 536 p

Einsele G, Seilacher A (1991) Distinction of tempestites and turbidites. In: Einsele G, Ricken W, Seilacher A (eds) Cycles and events in stratification. Springer, Heidelberg New York, p 377-382

Einsele G, Werner F (1972) Sedimentary processes at the entrance Gulf of Aden/Red Sea. Meteor-Forschungs-Ergebnisse C, 10: 39-62, Berlin Stuttgart

Einsele G, Ricken W, Seilacher D (1991) Cycles and events in stratigraphy: introduction. In: Einsele G, Ricken W, Seilacher A (eds) Cycles and events in stratigraphy. Springer, Heidelberg New York, p 1-19

Einsele G, Ricken W, Seilacher A (eds) (1991) Cycles and events in stratigraphy. Springer, Berlin Heidelberg New York, 955 p

Einsele G, Elouard P, Herm D, Kögler FC, Schwarz HU (1977) Source and biofacies of Late Quaternary sediments in relation to sea level on the shelf of Mauritania, West Africa. Meteor-Forschungs-Ergebnisse C 26: 1-43, Berlin Stuttgart

Ekdale AA, Bromley RG, Pemberton SG (1984) Ichnology: trace fossils in sedimentology and stratigraphy. Short Course 16, 316 p, Soc Econ Paleontol Mineral, Tulsa

Eldholm O, Thiede J, Taylor E (1987) Summary and preliminary conclusions, ODP Leg 104. In: Eldholm O, Thiede J, Taylor E, and Shipboard Scientific Party. Proc Init Repts ODP 104 A: 751-771, College Station, TX (Ocean Drilling Program)

Eldholm O, Thiede J, Taylor E (1989) The Norwegian continental margin: tectonic, volcanic, and paleoenvironmnetal framework. In: Eldholm O, Thiede J,

Taylor E, et al., Proc ODP, Sci Results, 104: 5-24, College Station, TX (Ocean Drilling Program)

Elliott T (1986a) Siliciclastic shorelines. In: Reading HG (ed) Sedimentary environments and facies. Blackwell, Oxford, p 155-188

Elliott T (1986b) Deltas. In: Reading HG (ed) Sedimentary environment and facies, 2nd ed. Blackwell, Oxford, pp 113-154

Elmi S (1990) Stages in the evolution of late Triassic and Jurassic carbonate platforms: the western margin of the Subalpine Basin (Ardèche, France). In: Tucker ME, Wilson JL, Crevello PD, Sarg JR, Read F (eds) Carbonate platforms: facies, sequences and evolution. Int Assoc Sedimentol Spec Publ 9, Blackwell, Oxford, p 109-144

Elmore RD, Pilkey OH, Cleary WJ, Curran HA (1979) Black shell turbidite, Hatteras abyssal plain, western Atlantic ocean. Geol Soc Am Bull 90: 1165-1176

Elorza JJ, Bustillo MA (1989) Early and late diagenetic chert in carbonate turbidites of the Senonian flysch, northeast Bilbao, Spain. In: Hein JR, Obradovic J (eds) Siliceous deposits of the Tethys and Pacific regions. Springer, New York Heidelberg, p 93-105

Emeis K (1985) Particulate suspended matter in major world rivers - II: results on the rivers Indus, Waikato, Nile, St. Lawrence, Yangtse, Parana, Orinoco, Caroni, and Magdalena. Mitt Geol-Paläontol Inst Univ Hamburg, H 58: 593-617

Emelyanov EM (1972) Principal types of recent bottom sediments in the Mediterranean Sea: their mineralogy and geochemistry. In: Stanlea DJ (ed) The Mediterranean Sea: a natural sedimentation laboratory. Dowden, Hutchinson & Ross, Stroudsburg, p 355-386

Emerson S (1985) Organic carbon preservation in marine sediments. In: Sundquist ET, Broecker WS (eds) The carbon cycle and atmospheric CO_2: Natural variations Archean to Present. AGU Geophys Monogr 32: 78-87

Emery KO (1960) The sea off southern California. Wiley, New York London, 366 p

Emery KO, Uchupi E (1984) The geology of the Atlantic Ocean. Springer, New York Heidelberg, 1050 p and map set

Emiliani C (ed) (1981) The oceanic lithosphere. The sea, v 7, Wiley-Interscience, New York, 1738 p

Enachescu ME (1987) Tectonic and structural framework of the northeast Newfoundland continental margin. In: Beaumont C, Tankard AJ (eds) Sedimentary basins and basin-forming mechanisms. Can Soc Petrol Geol Mem 12: 117-146

Engelhardt W von (1973) Sedimentpetrologie, Teil III: Die Bildung von Sedimenten und Sedimentgesteinen. Schweizerbart, Stuttgart, 378 p

Enos P (1989) Islands in the bay - a key habitat of Florida Bay. Bull Marine Sci 44: 365-386

Eppley RW, Peterson BJ (1979) Particulate organic matter flux and planktonic new production in the ocean. Nature 282: 677-680

Eschner TB, Kocurek G (1986) Marine destruction of eolian sand seas: origin of mass flows. J Sediment Petrol 56: 401-411

Eslinger E, Pevear D (1988) Clay minerals. Soc Econ Paleontol Mineral Short Course Notes 22: 4-1 to 4-22

Etheridge MA, Branson JC, Stuart-Smith PG (1987) The Bass, Gippsland and Otway basins, Southeast Australia: a branched rift system formed by continental extension. In: Beaumont C, Tankard AJ (eds) Sedimentary basins and basin-forming mechanisms. Can Soc Petrol Geol Mem 12: 147-162

Ethridge FG (1985) Modern alluvial fans and fan deltas. In: Flores RM, Ethridge FG, Miall AD, Galloway WE, Fouch TD (eds) Recognition of fluvial systems and their resource potential. Soc Econ Paleontol

Mineralog Short Course 19: 101-126

Ethridge FG, Flores RM, Harvey MD (1987) Recent developments in fluvial sedimentology. Soc Econ Paleontol Mineralog Spec Publ 39: 389 p

Eugster HP (1985) Oil shales, evaporites and ore deposits. Geochim Cosmochim Acta 49: 619-635

Eugster HP (1986) Lake Magadi, Kenya: a model for rift valley hydrochemistry and sedimentation? In: Frostick LE, Renaut RW, Reid I, Tiercelin JJ (eds) Sedimentation in African rifts. Geol Soc Spec Publ 25: 177-189

Eugster HP, Hardie LA (1978) Saline lakes. In: Lerman A (ed) Lakes: chemistry, geology, physics. Springer, Berlin Heidelberg, p 237-293

Eugster HP, Kelts K (1983) Lacustrine chemical sediments. In: Goudie AS, Pye K (eds) Chemical sediments and geomorphology. Academic Press, London, p 321-368

Evamy BD, Haremboure J, Kamerling P, Knaap WA, Molloy FA, Rowlands PH (1985) Hydrocarbon habitat of Tertiary Niger Delta. Am Assoc Petrol Geol Bull 62: 1-39

Eyles CH, Eyles N, Miall AD (1985) Models of glacio-marine sedimentation and their application on ancient glacial sequences. Palaeogeogr Palaeoclimatol Palaeoecol 51: 15-84

Eyles N (ed) (1984) Glacial Geology - an introduction for engineers and earth scientists. Pergamon Press, Oxford, 409 p

Eyles N, Miall AD (1984) Glacial facies. In: Walker RG (ed) Facies models, 2nd edn. Geosci Canada Reprint Ser 1, Geol Assoc Canada, p 15-37

Fairbridge RW, Finkl CW (1980) Cratonic erosional unconformities and peneplains. J Geol 88: 69-86

Farell JW, Prell WL (1987) Climate forcing of calcium carbonate sedimentation: a 4.0 my record from the central equatorial Pacific Ocean. EOS Trans 68: 333, Am Geophys Union

Farrow CEG, Mossman DJ (1988) Geology of Precambrian paleosols at the base of the Huronian Supergroup, Elliot Lake, Ontario, Canada. Precambrian Res 42: 107-134

Faupl P, Sauer R (1978) Zur Genese roter Pelite in Turbiditen der Flyschgosau in den Ostalpen (Oberkreide-Alttertiär). N Jahrb Geol Paläontol 1978, 2: 65-86

Faure G (1986) Principles of isotope geology, 2nd edn. Wiley, New York, 589 p

Feeley MH, Moore TC Jr, Loutit TS, Bryant WR (1990) Sequence stratigraphy of Mississippi fan related to oxygen isotope sea level index. Am Assoc Petrol Geol Bull 74: 407-424

Feldhausen PH, Stanley DJ, Knight RJ, Maldonado A (1981) Homogenization of gravity-emplaced muds and unifites: models from the Hellenic Trench. In: Wezel FC (ed) Sedimentary basins of Mediterranean margins. CNR, Italian Project of Oceanography, Tecnoprint, Bologna, p 203-226

Field ME, Gandner JW, Jennings AE, Edwards BE (1982) Earthquake-induced sediment failures on a 0.25° slope, Klamath river delta, California. Geology 10: 542-546

Finlayson DM, Wright C, Leven HJ, Collins CDN, Wake-Dyster KD, Johnstone DW (1989) Basement features under four intra-continental basins in central and eastern Australia. In: Price RA (ed) Origin and evolution of sedimentary basins and their energy and mineral resources. Geophys Monogr 48 (IUGG vol 3): 43-55, Am Geophys Union, Washington DC

Firman J (ed) (1988) Landscapes of the southern hemisphere. Earth Sci Rev 25, 5,6: 343-519

Fischer AG (1964) The Lofer cyclothems in the Alpine Triassic. Kansas Geol Surv Bull 169: 107-149

Fischer AG (1975) Origin and growth of basins. In: Fischer AG, Judson S (eds) Petroleum and global tectonics. Princeton Univ Press, Princeton, New Jersey, p 47-79

Fischer AG (1982) Long-term climatic oscillations recorded in stratigraphy. In: Climate in earth history. Studies in Geophysics, Nat Academy Press, Washington DC, p 97-104

Fischer AG (1991) Orbital cyclicity in Mesozoic strata. In: Einsele G, Ricken W, Seilacher A (eds) Cycles and events in stratigraphy. Springer, Heidelberg New York, p 48-62

Fischer AG, Sarnthein M (1988) Airborne silt and dune-derived sands in the Permian of the Delaware basin. J Sediment Petrol 58: 637-643

Fischer AG, de Boer PL, Premoli-Silva I (1989) Cyclostratigraphy. In: Ginsburg RN, Beaudoin B (eds) Cretaceous resources, rhythms, and events. NATO ASI Ser C 304: 139-172, Kluwer, Dordrecht

Fisher JH, Barratt MW, Droste JB, Shaver RH (1988) Michigan basin. In: Sloss LL (ed) The geology of North America, v D-2: Sedimentary cover - North American craton: U.S. Geol Soc America, Boulder, p 361-382

Fisher RV, Schmincke HU (1984, 1990) Pyroclastic rocks, 1st and 2nd edn. Springer, Berlin Heidelberg New York, 472 p

Flemings PB, Jordan TE (1990) Stratigraphic modeling of foreland basins: Interpreting thrust deformation and lithosphere rheology. Geology 18: 430-434

Flemming RW (1980) Sand transport and bedform patterns on the continental shelf between Durban and Port Elizabeth. Sediment Geol 26: 179-206

Flores RM (1981) Coal deposition in fluvial paleoenvironments of the Paleocene Tongue River Member of the Fort Union Formation, Powder River area, Powder River basin, Wyoming and Montana. Soc Econ Paleontol Mineralog Spec Publ 31: 169-190

Flores RM, Ethridge FG, Miall AD, Galloway WE, Fouch TD (1985) Recognition of fluvial systems and their resource potential. Soc Econ Paleontol Mineralog Short Course 19, 290 p

Flügel E (1982) Microfacies analysis of limestones. Springer, Berlin Heidelberg New York, 633 p

Flügel E (1988) Problems with reef models: the late Triassic Steinplatte "reef" (northern Alps, Salzburg/Tyrol, Austria). Facies 20: 1-138, Erlangen

Flügel E (1989) Typen und wirtschaftliche Bedeutung von Riffkalken. Archiv f Lagerstättenforsch, Geol Bundesanst Wien 10: 25-32

Föllmi KB, Garrison RN, Grimm K (1991) Stratification in phosphatic sediments. In: Einsele G, Ricken W, Seilacher D (eds) Cycles and events in stratigraphy, Springer, Heidelberg New York, p 492-507

Folk RL (1971) Longitudinal dunes of the northwestern edge of the Simpson Desert, Northern Territory, Australia. 1. Geomorphology and grain-size relationships. Sedimentology 16: 5-54

Ford D, Williams P (1989) Karst geomorphology and hydrology. Unwin Hyman, London, 601 p

Foster NH, Beaumont EA (eds) (1987) Geologic basins I: Classification, modelling, and predictive stratigraphy. Treatise of Petroleum Geology, Reprint Series 1, Am Assoc Petrol Geol, Tulsa, 458 p

Frakes LA (1979) Climate throughout geologic times. Elsevier, Amsterdam, 310 p

Franke W, Paul J (1980) Pelagic redbeds in the Devonian of Germany - deposition and diagenesis. Sediment Geol 25: 231-256

Franke W, Walliser OH (1983) "Pelagic" carbonates in the Variscan Belt - their sedimentary and tectonic environments. In: Martin H, Eder W (eds) Intracontinental fold belts. Springer, Berlin Heidelberg New York, p 77-92

Fraser GS (1989) Clastic depositional sequences. Prentice Hall, Englewood Cliffs, 459 p

Frazier DE (1967) Recent deltaic deposits of the Mississippi river: their development and chronology. Gulf Coast Assoc Geol Soc Trans 17: 287-315

Frey M (1987) Low-temperature metamorphism. Blackie, Glasgow

Frey M, Teichmüller M, Teichmüller R, Mullis J, Kunzi B, Breitschmid A, Gruner U, Schwizer B (1980) Very low-grade metamorphism in the external parts of the Central Alps: Illite crystallinity, coal rank, and fluid inclusion data. Eclogae Geol Helv 73: 173-203

Friedman GM, Sanders JE (1978) Principles of sedimentology. Wiley, New York, 792 p

Friedrich GH, Herzig PM (1988) Base metal sulfide deposits in sedimentary and volcanic environments. Soc Geology Applied to Mineral Deposits, Spec Publ 5, Springer, Berlin Heidelberg New York, 290 p

Frisch W, Loeschke J (1986) Plattentektonik. Erträge der Forschung, Band 236, Wissensch Verlagsbuchgesellschaft, Darmstadt, 190 p

Froelich PN, Arthur MA, Burnett WC, Deakin M, Hensley V, Jahnke R, Kaul L, Kim KH, Roe K, Soutar A, Vathakanon C (1988) Early diagenesis of organic matter in Peru continental margin sediments: phosphorite precipitation. Marine Geol 80: 309-343

Froitzheim N, Stets J, Wurster P (1988) Aspects of western High Atlas tectonics. In: Jacobshagen VH (ed) The Atlas system of Morocco. Lecture Notes Earth Sci 15, Springer, Berlin Heidelberg New York, 219-244

Frostick LE (ed) (1987) Desert environments: ancient and modern. Geol Soc Spec Publ 35, 401 p

Frostick LE, Renaut RW, Reid I, Tiercelin JJ (eds)(1986) Sedimentation in African rifts. Geol Soc Spec Publ 25, Blackwell, Oxford, 382 p

Fryberger SG, Ahlbrandt TS, Andrews S (1979) Origin, sedimentary features, and significance of low-angle eolian "sand sheet" deposits, Great Sand Dunes National Monument and vicinity, Colorado. J Sediment Petrol 49: 733-746

Fuchs K, Bonjer KP, Prodehl C (1981) The continental rift system of the Rhinegraben - Structure, physical properties and dynamical processes. Tectonophysics 73: 79-90

Füchtbauer H (1974a) Zur Diagenese fluviatiler Sandsteine. Geol Rundsch 63: 904-925

Füchtbauer H (1974b) Some problems of diagenesis in sandstones. Bull Cent Rech Pau SNPA 8: 391-403

Füchtbauer H (1988) (ed) Sedimente und Sedimentgesteine, 4th edn. Schweizerbart, Stuttgart, 1141 p

Füchtbauer H, Peryt T (eds) (1980) The Zechstein basin with emphasis on carbonate sequences. Contributions to Sedimentology 9, Schweizerbart, Stuttgart, 328 p

Füchtbauer H, Richter DK (1983) Relations between submarine fissures, internal breccias and mass flows during Triassic and earlier rifting periods. Geol Rundsch 72: 53-66

Füchtbauer H, Richter DK (1988) Karbonatgesteine. In: Füchtbauer H (ed) Sedimente und Sedimentgesteine. Schweizerbart, Stuttgart, p 233-434

Fuis GS, Kohler WM (1984) Crustal structure and tectonics of the Imperial Valley. In: Rigsby CA (ed) The Imperial Basin - tectonics, sedimentation, and thermal aspects. Soc Econ Paleontol Mineral Pacific Section, p 1-13

Fulthorpe CS (1990) Fourth order sequences and the response of a basin to eustatic sea-level variation. 13th Int Sedimentol Congress, Aug 1990, Nottingham, Int Assoc Sedimentol, Abstracts of Papers, p 177

Funk HP (1985) Mesozoische Subsidenzgeschichte im Helvetischen Schelf der Ostschweiz. Eclogae Geol Helv 78: 249-272

Fürsich FT (1979) Genesis, environments, and ecology of Jurassic hardgrounds. Neues Jb Geol Paläontol Abh 158: 1-63

Fyfe WS, Lonsdale P (1981) Ocean floor hydrothermal activiy. In: Emiliani C (ed) The oceanic lithosphere. The sea, vol 7, Wiley-Interscience, New York, p 589-638

Gac JY (1980) Géochemie du bassin du Lac Tchad, bilan de alteration, de l'érosion et de la sédimentation. Travaux et Documents de l'ORSTOM 123, 251 p

Galloway WE (1981) Depositional architecture of Cenozoic Gulf coastal plain fluvial systems. Soc Econ Paleontol Mineral Spec Publ 31: 127-155

Galloway WE (1984) Hydrogeologic regimes of sandstone diagenesis. In: McDonald DA, Surdam RC (eds) Clastic diagenesis. Am Assoc Petrol Geol Mem 37: 3-13

Galloway WE (1985) Meandering streams - modern and ancient. In: Flores RM, Ethridge FG, Miall AD, Galloway WE, Fouch TD (eds) Recognition of fluvial systems and their resource potential. SEPM Short Course 19: 145-166

Galloway WE (1989) Genetic stratigraphic sequences in basin analysis I: architecture and genesis of flooding-surface bounded depositional units. Am Assoc Petrol Geol Bull 73: 125-142

Galloway WE, Hobday DK (1983) Terrigenous clastic depositional systems - applications to petroleum, coal, and uranium exploration. Springer, New York Heidelberg, 423 p

Gansser A (1984) The morphogenetic phase of mountain building. In: Hsü KJ (ed) Mountain building processes, 2nd ed. Academic Press, London, p 221-228

García-Mondéjar J, Pujalte V, Robles S (1986) Caracteristicas sedimentologicas, secuenciales y tectoestratigraficas del Triasico de Cantabria y norte de Palencia. Cuardernos Geología Ibérica 10: 151-172, Madrid

Garrels RM (1986) Sediment cycling and diagenesis. In: Mumpton FA (ed) Studies in diagenesis. US Geol Surv Bull 1578: 1-11, US Govt Print Office, Washington

Garrels RM, Mackenzie FT (1971) Evolution of sedimentary rocks. Norton, New York, 397 p

Garrison RE (1981) Diagenesis of oceanic carbonate sediments: a review of the DSDP perspective. In Warme JE, Douglas RG, Winterer EL (eds) The Deep Sea Drilling Project: a decade of progress. Soc Econ Paleontol Mineral Spec Publ 32: 181-207

Garrison RE, Kastner M, Kolodny Y (1987) Phosphorites and phosphatic rocks in the Monterey Formation and related Miocene units, coastal California. In: Ingersoll RV, Ernst WG (eds) Cenozoic basin development in coastal California. Rubey vol 6: 348-381, Prentice-Hall, New Jersey

Gautier DL (ed) (1986) Roles of organic matter in sediment diagenesis. Soc Econ Paleontol Mineral Spec Publ 38, 203 p

Gautier DL, Claypool GE (1984) Interpretation of methanic diagenesis in ancient sediments by analogy with processes in modern diagenetic environments. In: McDonald DA, Surdam RC (eds) Clastic diagenesis. Am Assoc Petrol Geol Mem 37: 111-123

Geister J (1983) Holozäne westindische Korallenriffe: Geomorphologie, Ökologie und Fazies. Facies 9: 173-284, Erlangen

Georgotas N, Udluft P (1978) Inhaltsstoffe des oberflächennahen Grundwassers und des Oberflächenwassers. Anorganische Inhaltsstoffe. In: Das Mainprojekt,

Schriftenreihe Bayer Landesamt f Wasserwirtschaft H 7: 258-265, München

Gerdes G, Krumbein WE, Reineck H-E (1985) Verbreitung und aktuogeologische Bedeutung mariner mikrobieller Matten im Gezeitenbereich der Nordsee. Facies 12: 75-96, Inst Paläontol Univ Erlangen-Nürnberg

Gerdes G, Krumbein WE, Reineck H-E (1991) Biolamination: ecological versus depositional dynamics. In: Einsele G, Ricken W, Seilacher A (eds) Cycles and events in stratigraphy. Springer, Heidelberg New York, p 592-607

Gerrard AJ (1988) Rocks and landforms. Unwin Hyman, London, 319 p

Geyer OF, Gwinner MP (1986) Geologie von Baden-Württemberg. Schweizerbart, Stuttgart, 472 p

Gibb FGF, Kanaris-Sotiriou R (1988) The geochemistry and origin of the Faeroe-Shetland sill complex. In: Morton AC, Parson LM (eds) Early Tertiary volcanism and the opening of the NE Atlantic. Geol Soc Spec Publ 39: 241-251

Gibbs RJ (1985) Settling velocity, diameter and density of flocs of illite, kaolinite, and montmorillonite. J Sediment Petrol 55: 65-68

Gieskes JM (1975) Chemistry of interstitial waters of marine sediments. Annu Rev Earth Planet Sci 3: 433-453

Ginsburg RN (ed)(1975) Tidal deposits: a casebook of Recent examples. Springer, Berlin Heidelberg New York, 428 p

Ginsburg RN, Hardie LA (1975) Tidal and storm deposits, northern Andros Island, Bahamas. In: Ginsburg RN (ed) Tidal deposits, p 201-208. Springer, Berlin Heidelberg New York

Girdler RW, Whitmarsh RB (1974) Miocene evaporites in Red Sea cores, Their relevance to the problem of the width and age of oceanic crust beneath the Red Sea. In: Whitmarsh RB, Weser OE, Ross DA, et al., Init Repts DSDP 23, US Govt Print Office, Washington, p 913-921

Girty GH, Mossman BJ, Pincus SC (1988) Petrology of Holocene sand, Peninsular Ranges, California and Baja Norte, Mexico: implications for provenance-discrimination models. J Sediment Petrol 58: 881-887

Glenn CR (1990) Pore water, petrology and stablef carbon isotopic data bearing on the origin of modern Peru margin phosphorites and associated authigenic phases. In: Burnett WC, Riggs SR (eds) Phosphate deposits of the world, v 3, Genesis of Neogene to Recent phosphorites. Cambridge Univ Press, p 46-61

Glenn CR, Arthur MA (1985) Sedimentary and geochemical indicators of productivity and oxygen contents in modern and ancient basins: the Holocene Black Sea as the "type" anoxic basin. Chemical Geol 48: 325-354

Glenn CR, Arthur MA (1988) Petrology and major element geochemistry of Peru margin phosphorites and associated diagenetic minerals: authigenesis in modern organic-rich sediments. In: Burnett WC, Froelich PN (eds) The origin of marine phosphorite. The results of the R.V. Robert D. Conrad Cruise 23-06 to the Peru shelf. Marine Geol 80: 231-267

Glenn CR, Arthur MA (1990) Anatomy and origin of a Cretaceous phosphorite-greensand giant, Egypt. Sedimentology 37: 123-154

Glenn CR, Kelts K (1991) Sedimentary rhythms in lake deposits. In: Einsele G, Ricken W, Seilacher A (eds) Cycles and events in stratigraphy. Springer, Heidelberg New York, p 188-221

Glennie KW (1970) Desert sedimentary environments. Developm Sedimentol 14, Elsevier, Amsterdam, 222 p

Gloppen TG, Steel RJ (1981) The deposits, internal structure and geometry in six alluvial fan-fan delta bodies (Devonian Norway) - a study in the significance of braiding sequence in conglomerates. In: Ethridge FG, Flores RM (eds) Recent and ancient nonmarine depositional systems: models for exploration. Soc Econ Paleontol Mineral Spec Publ 31: 49-69

Goldhaber MB, Kaplan R (1980) Mechanisms of sulfur incorporation and isotope fractionation during early diagenesis in sediments of the Gulf of California. Marine Chemistry 9: 95-143

Goldhammer RK, Dunn DA, Hardie LA (1987) High-frequency glacio-eustatic sea-level oscillations with Milankovitch characteristics recirded in Middle Triassic platform carbonates in northern Italy. Am J Sci 287: 853-992

Goodwin PW, Anderson EJ (1985) Punctuated aggradational cycles: a general hypothesis of episodic stratigraphic accumulation. J Geol 93: 515-533

Goodwin RH, Prior DB (1989) Geometry and depositional sequences of the Mississippi canyon, Gulf of Mexico. J Sediment Petrol 59: 318-329

Gordon A (1966) Caribbean Sea - oceanography. In: Fairbridge RW (ed) Encyclopedia of earth sciences series, vol 1, The encyclopedia of oceanography. Reinhold, New York, p 175-181

Gore PJW (1989) Toward a model for open- and closed-basin deposition in ancient lacustrine sequences: The Newark supergroup (Triassic-Jurassic), eastern North America. In: Talbot MR, Kelts K (eds) The Phanerozoic record of lacustrine basins and their environmental signals. Palaeogeogr Palaeoclimatol Palaeoecol 70: 29-51

Görler K, Reutter KJ (1968) Entstehung und Merkmale der Olisthostrome. Geol Rundsch 57: 484-514

Gorsline DS, Douglas RG (1987) Analysis of sedimentary systems in active-margin basins: California continental borderland. In: Ingersoll RV, Ernst WG (eds) Cenozoic basin development of coastal California. Rubey vol 6, Prentice-Hall, Englewood Cliffs, p 64-80

Gorsline DS (1984) Review of fine-grained sediment origins, characteristics, transport and deposition. In: Stow DAV, Piper DJW (eds) Fine-grained sediments: deep-water processes and facies. Geol Soc Spec Publ 15: 17-22, Blackwell, Oxford

Gradstein FM, Agterberg FP, Brower JC, Schwarzacher WS (1985) Quantitative stratigraphy. Reidel, Dordrecht, 598 p

Gradstein FM, Jansa LF, Srivastava SP, Williamson MA, Bonham Carter G, Stam B (1990) Aspects of North Atlantic paleo-oceanography. Chapter 8. In: Keen MJ, Williams GL (eds) Geology of the continental margin of eastern Canada. Geolog Survey Canada, Geology of Canada, no 2, p 351-389

Grandstaff DE, Endelman MJ, Foster RW, Zbinden E, Kimberley MM (1986) Chemistry and mineralogy of Precambrian paleosols at the base of the Dominion and Pongola Groups (Transvaal, South Africa). Precambrian Res 32: 97-131

Grant AC, Jansa LF, McAlpine KD, Edwards A (1988) Mesozoic-Cenozoic geology of the eastern margin of the Grand Banks and its relation to Galicia Bank. In: Boillot G, Winterer EL et al., Proc ODP, Sci Results, 103: 787-808, College Station, TX (Ocean Drilling Program)

Grant Gross M (1980) Oceanography, a view of the earth, 2nd edn. Prentice-Hall, Englewood Cliffs, 475 p

Grasshoff K (1975) The hydrochemistry of land-locked basins and fjords. In: Riley JP, Skirrow G (eds) Chemical Oceanography, vol 2, 2nd edn. Academic Press, London, p 456-597

Greenlee SM, Schroeder FW, Vail PR (1988) Seismic stratigraphy and geohistory analysis of Tertiary strata from the continental shelf off New Jersey; calculation of eustatic fluctuations from stratigraphic data. In: Sheridan RE, Grow JA (eds) The Geology of North America, vol 1/2, The Atlantic continental margin. US Geol Soc Am, p 437-444

Gregor CB, Garrels RM, Mackenzie FT, Maynard JB (eds) (1988) Chemical cycles in the evolution of the Earth. Wiley, New York

Gretener PE (1979) Pore pressure: fundamentals, general ramifications and implications for structural geology. Am Assoc Petrol Geol, Cont Ed Course Note Ser 4, 2nd edn, 131 p

Gretener PE, Feng ZM (1985) Three decades of geopressures - insights and enigmas. Bull Verein Schweiz Petrol-Geol Ing 51/120: 1-34

Grobe H (1987) Facies classification of glacio-marine sediments in the Antarctic. Facies 17: 99-109

Grötsch J, Wu G, Berger WH (1991) Carbonate cycles in the Pacific: reconstruction of saturation fluctuations. In: Einsele G, Ricken W, Seilacher A (eds) Cycles and events in stratification. Springer, Heidelberg New York, p 110-125

Grotzinger JP (1986) Cyclicity and paleoenvironmental dynamics, Rocknest platform, northwest Canada. Geol Soc Am Bull 97: 1208-1231

Grundmann G, Morteani G (1985) The young uplift and thermal history of the central Eastern Alps (Austria/Italy), evidence from apatite fission track ages. Jahrb Geol Bundesanst (Wien) 128: 197-216

Guennoc P, Pouit G, Nawar Z (1988) The Red Sea: history and associated mineralization. In: Manspeizer W (ed) Triassic-Jurassic rifting, parts A and B. Developm Geotectonics 22: 957-982, Elsevier, Amsterdam

Guidish TM, Kendall CG, Lerche SC, Toth DJ, Yarzar RF (1985) Basin evaluation using burial history calculations: an overview. Am Assoc Petrol Geol Bull 69: 92-105

Guilbert JM, Park CF Jr (1986) The geology of ore deposits. Freeman, New York, 985 p

Gursky H-J (1988) Gefüge, Zusammensetzung und Genese der Radiolarite im ophiolitischen Nicoya-Komplex (Coasta Rica). Münstersche Forschungen Geol Paläontol 68, 189 p, Geol-Paläontol Inst Münster

Gygi RA (1981) Oolitic iron formations: marine or not marine. Eclogae Geol Helv 74: 233-254

Haas J (1991) A basic model for Lofer cycles. In: Einsele G, Ricken W, Seilacher A (eds) Cycles and events in stratification. Springer, Heidelberg New York, p 722-732

Hadley RF (ed) (1986) Drainage basin sediment delivery. Proc Symp Albuquerque, IAHS Int Comm on Continental Erosion, IAHS Publ 159

Hallam A (1984) Pre-Quaternary sea-level changes. Ann Rev Earth Planet Sci 12: 205-243

Hallam A (1988) A reevaluation of Jurassic eustacy in the light of new data and the revised Exxon curve. In: Wilgus CK, Hastings BS, Posamentier H, Van Wagoner J, Ross CA, Kendall CGSC (eds) Sea-level changes: an integrated approach. Soc Econ Paleontol Mineral Spec Publ 42: 261-273

Hallam A, Bradshaw MJ (1979) Bituminous shales and oolithic ironstones as indicators of transgressions and regressions. J Geol Soc London 136: 157-164

Halbach P, Friedrich G, Stackelberg U von (1988) The manganese nodule belt of the Pacific ocean: geological environment, nodule formation, and mining aspects. Enke, Stuttgart, 254 p

Hallock P, Schlager W (1986) Nutrient excess and the demise of coral reefs and carbonate platforms. Palaios 1: 389-398

Hamilton EL (1976) Variations of density and porosity with depth in deep-sea sediments. J Sediment Petrol 46: 280-300

Hampton MA (1972) The role of subaqueous debris flow in generating turbidity currents. Jour Sediment Petrol 42: 775-793

Hampton MA (1979) Buoyancy in debris flows. J Sediment Petrol 49: 753-758

Han MW, Suess E (1989) Subduction-induced pore fluid venting and the formation of authigenic carbonates along the Cascadia continental margin: implications of the global Ca-cycle. Palaeogeogr Palaeoclimatol Palaeoecol 71: 97-118

Hancock JM (1989) Sea-level changes in the British region during the Late Cretaceous. Proc Geol Assoc 100: 565-594

Hanor JS (1987) Origin and migration of subsurface sedimentary brines. Lecture Notes for Short Course 21, Soc Econ Paleontol Mineral, 247 p

Haq BU, Hardenbol J, Vail PR (1987) Chronology of fluctuating sea levels since the Triassic. Science 235: 1156-1167

Harden JW, Matti JC (1989) Holocene and late Pleistocene slip rates on the Sand Andreas fault in Yucaipa, California, using displaced alluvial fan deposits and soil chronology. Geol Soc Am Bull 101: 1107-1117

Hardie LA (1987) Dolomitization: a critical view of some current views. J Sediment Petrol 57: 166-183

Harms JC, Fahnestock RK (1965) Stratification, bed forms, and flow phenomena (with an example from the Rio Grande). Soc Econ Paleontol Mineral Spec Publ 12: 84-115

Harms JC, Southard JB, Walker RG (1982) Shallow marine environments - a comparison of some ancient and modern examples. In: Structures and sequences in clastic rocks. Soc Econ Paleontol Mineral Short Course 9: 8.1-8.51

Hart BS, Plint AG (1989) Gravelly shoreface deposits: a comparison of modern and ancient facies sequences. Sedimentology 36: 551-557

Hart BS, Vantfoort RM, Plint AG (1990) Is there evidence for geostrophic currents preserved in the sedimentary record of inner to middle shelf deposits? - Discussion. J Sediment Petrol 60: 633-635

Hattin DE (1986a) Interregional model for deposition of Upper Cretaceous pelagic rhythmites, U.S. Western Interior. Paleoceanography 1: 483-494

Hattin DE (1986b) Carbonate substrates of the Late Cretaceous Sea, Central Great Plains and Southern Rocky Mountains. Palaios 1: 347-367

Hattin DE (1989) Global perspective on Cretaceous shelf-sea chalk deposits. 28 th Int Geol Congr, Abstr 2: 38-39, Washington DC

Hay WB (1987) The past and future of scientific ocean drilling. In: Kozlowsky EA (ed) 27th Int Geol Congr, Moscow 1984, General Proc 27-40

Hay WW, Shaw CA, Wold CN (1989) Mass-balanced paleogeographic reconstructions. Geol Rundsch 78: 207-242

Hayes MO (1980) General morphology and sediment patterns in tidal inlets. Sediment Geol 26: 139-156

Heckel PH (1986) Sea-level curve for Pennsylvanian eustatic marine transgressive-regressive depositional cycles aalong mid-continent outcrop belt, North America. Geology 14: 330-334

Heezen BC, Hollister CD (1971) The face of the deep. Oxford Univ Press, New York, 659 p

Hegarty KA, Weissel JK, Mutter JC (1988) Subsidence history of Australia's southern margin. Am Assoc Petrol Geol Bull 72: 615-635

Hein JR, Obradovic J (1989) Siliceous deposits of the Tethys and Pacific regions. In: Hein JR, Obradovic J (eds) Siliceous deposits of the Tethys and Pacific regions. Springer, New York Heidelberg, p 1-17

Hein JR, Parrish JT (1987) Distribution of siliceous deposits in space and time. In: Hein JR (ed) Siliceous sedimentary rock-hosted ores and petroleum. Van Nostrand Reinhold, New York, p 10-57

Heling D (1988) Ton- und Siltsteine. In: Füchtbauer H (ed) Sedimente und Sedimentgesteine. Schweizerbart, Stuttgart, p 185-231

Heller F, Liu Tungsheng (1984) Magnetism of Chinese loess deposits. Geophys J Royal Astron Soc 77: 125-141

Heller PL, Angevine C (1985) Sea-level cycles during the growth of Atlantic-type oceans. Earth Planet Sci Lett 75: 417-426

Hem JD (1983) Study and interpretation of the chemical characteristics of natural water, 3rd ed. US Geol Surv Water-Supply Pap 2254: 263 pp

Hemleben C, Swinburne NHM (1991) Cyclical deposition of plattenkalk facies. In: Einsele G, Ricken W, Seilacher A (eds) Cycles and events in stratigraphy. Springer, Heidelberg New York, p 572-591

Hempton MR, Dunne LA (1984) Sedimentation in pull-apart basins: active examples and eastern Turkey. J Geology 92: 513-530

Henderson-Sellers B, Henderson-Sellers A (1989) Modelling the ocean climate for the early Archean. Palaeogeogr Palaeoclimatol Palaeoecol (Global and Planetary Change Section 1/3) 75: 195-221

Henrich R (1990) Cycles, rhythms, and events in Quaternary Arctic and Antarctic glaciomarine deposits. In: Bleil U, Thiede J (eds) Geological history of the polar oceans: Arctiv versus Antarctic. Kluwer, Dordrecht, p 213-244

Herbert TD, D'Hondt SL (1990) Precessional climatic cyclicity in Late Cretaceous-Early Tertiary marine sediments: a high resolution chronometer of Cretaceous-Tertiary boundary events. Earth Planet Sci Lett 99: 263-275

Herbert TD, Stallard RF, Fischer AG (1986) Anoxic events, productivity rhythms, and the orbital signature in a mid-Cretaceous deep-sea sequence from central Italy. Paleoceanography 1: 495-506

Hesse R (1974) Long-distance continuity of turbidites:Possible evidence for an Early-Cretaceous trench-abyssal plain in the East Alps. Bull Geol Soc Am 85: 859-870

Hesse R (1975) Turbiditic and non-turbiditic mudstone of Cretaceous flysch sections of the East Alps and other basins. Sedimentology 22: 387-416

Hesse R (1986) Early diagenetic pore water/sediment interaction: modern offshore basins. Geosci Canada 13, 3: 165-196

Hesse R (1989a) "Drainage systems" associated with mid-ocean channels and submarine yazoos: alternative to submarine fan depositional systems. Geology 17: 1148-1151

Hesse R (1989b) Silica diagenesis: origin of inorganic and replacements cherts. Earth Sci Rev 26: 253-284

Hesse R, Butt A (1976) Paleobathymetry of Cretaceous turbidite basins of the East Alps relative to the calcite compensation level. J Geol 34: 505-533

Hesse R, Chough SK (1980) The Northwest Atlantic mid-ocean channel of the Labrador Sea: II. Deposition of parallel laminated levee-muds from the viscous sublayer of low density turbidity currents. Sedimentology 27: 697-711

Hesse R, Lebel J, Gieskes JM (1985) Interstitial water chemistry of gas hydrate bearing sections on the Middle-America Trench Slope Deep Sea Drilling Project, Leg 84. In: Von Huene R, Aubouin J, et al., Init Repts DSDP 84: 727-737, US Govt Print Office, Washington

Heward AP (1978) Alluvial fan sequence and megasequence models: with examples from Westphalian D-Stephanian B coalfields, northern Spain. In: Miall AD (ed) Fluvial Sedimentology, Can Soc Petrol Geol Mem 5: 669-702

Hieke W (1984) A thick Holocene homogenite from the Ionian abyssal plain (eastern Mediterranean). Marine Geol 55: 63-78

Higgs R (1990) Is there evidence for geostrophic currents preserved in the sedimentary record of inner to middle shelf deposits? - discussion. J Sediment Petrol 60: 630-632

Hill PR (1984) Facies and sequence analysis of Nova Scotian slope muds: turbidite vs "hemipelagic" deposition. In: Stow DAV, Piper DJW (eds) Fine-grained sediments: deep-water processes and facies. Geol Soc Spec Publ 15: 311-318

Hinz K, Kögler FC, Richter I, Seibold E (1971) Reflexionsseismische Untersuchungen mit einer pneumatischen Schallquelle und einem Sedimentecholot in der westlichen Ostsee. Meyniana (Kiel) 21: 17-24

Hinz K, Mutter JC, Zehnder CM, and the NGT Study Group (1987) Symmetric conjugation of continent-ocean boundary structures along the Norwegian and East Greenland margins. Marine and Petrol Geol 4: 166-187

Hinze C, Meischner D (1968) Gibt es rezentes Rotsedimente in der Adria? Marine Geol 6: 53-71

Hird K, Tucker ME (1988) Contrasting diagenesis of two Carboniferous oolites from South Wales: a tale of climatic influence. Sedimentology 35: 587-602

Hiscott RN, Wilson RCL, Gradstein FM, Pujalke V, Garcia-Mondéjar J, Boudreau RR, Wishart HA (1990) Comparative stratigraphy and subsidence history of Mesozoic rift basins of North Atlantic. Am Assoc Petrol Geol Bull 74: 60-76

Hobert LA, Wetzel A (1989) On the relationship between silica and carbonate diagenesis in deep-sea sediments. Geol Rundsch 78, 3: 765-778

Hoffmann A (1989) Mass extinctions: the view of a sceptic. J Geol Soc London 146: 21-35

Hoffmann PF (1988) United plates of America, the birth of a craton: early Proterozoic assembly and growth of Laurentia. Annu Rev Earth Planet Sci 16: 573-603

Hoffmann PF (1989) Speculations on Laurentia's first gigayear (2.0 to 1.0 Ga). Geology 17: 135-138

Hohberger KH, Einsele G (1979) Die Bedeutung des Lösungsabtrags verschiedener Gesteine für die Landschaftsentwicklung in Mitteleuropa. Z Geomorph NF 23: 361-382

Holland HD (1984) The chemical evolution of the atmosphere and oceans. Princeton Univ Press, Princeton, 582 p

Holler P, Krumsiek K, Stein R (1990) Erste Ergebnisse von ODP-Leg 128 (Japan See). In: Beiersdorf H (ed) Deutsche Forschungsgemeinschaft, Protokoll Kolloquium im SPP "Ocean Drilling Program/Deep Sea Drilling Project" in Bremen, Bonn, p 47-49

Holmes CW (1986) Salt-induced diagenesis of argillaceous sediments. In: Mumpton FA (ed) Studies in diagenesis. US Geol Surv Bull 1578: 347-362

Holser WT (1984) Gradual and abrupt shifts in ocean chemistry during Phanerozoic time. In: Holland HD, Trendall AF (eds.) Patterns of change in Earth evolution, Springer, Berlin Heidelberg New York, p 123-143.

Homewood P, Allen PA (1981) Wave-, tide-, and current-controlled sandbodies of Miocene molasse, western Switzerland. Am Assoc Petrol Geol Bull 65: 2534-2545

Homewood P, Caron C (1983) Flysch of the western Alps. In: Hsü KJ (ed) Mountain building processes. Academic Press, London, p 155-186

Homewood P, Allen PA, Williams GD (1986) Dynamics of the molasse basin of western Switzerland. In: Allen PA, Homewood P, Williams G (eds) Foreland basins: an introduction. Spec Publ Int Assoc Sedimentol 8: 199-217, Blackwell, Oxford

Horie S (1978) Lacustrine sedimentation. In. Fairbridge RW, Bourgeois J (eds) The Encyclopedia of sedimentology. Dowden, Hutchinson & Ross, Stroudsburg, p 421-427

Houseknecht DW (1988) Intergranular pressure solution in four quartzose sandstones. J Sediment Petrol 58: 228-246

Houten FB van (1973) Origin of red beds: a review - 1961-1971. Annu Rev Earth Planet Sci 1: 39-61

Howard JD, Reineck H-E (1981) Depositional facies of high energy beach to offshore sequence: comparison with low energy sequence. Am Assoc Petrol Geol Bull 65: 807-830

Howell DG, Crouch JK, Greene HG, McCulloch DS, Vedder JG (1980) Basin development along the late Mesozoic and Cainozoic California margins: a plate tectonic margin of subduction, oblique subduction and transform tectonics. In: Ballance PF, Reading HG (eds) Sedimentation in oblique-slip mobile zones. Spec Publ Int Assoc Sedimentol 4, Blackwell, Oxford, p 43-62

Hsü KJ (1972) Origin of saline giants: a critical review after the discovery of the Mediterranean evaporite. Earth Sci Rev 8: 371-396

Hsü KJ (1974) Mélanges and their distinction from olistostromes. Soc Econ Paleontol Mineral Spec Publ 19: 321-333

Hsü KJ (1978) When the Black Sea was drained. Scient American 238/3: 52-63

Hsü KJ (1989) Origin of sedimentary basins of China. In: Zhu X (ed)(1989) Chinese sedimentary basins. Elsevier, Amsterdam, p 207-227

Hsü KJ, Jenkyns HC (eds) (1974) Pelagic sediments on land and under the sea. Int Assoc Sedimentol Spec Publ 1, Blackwell, Oxford

Hsü KJ, Kelts K (eds) (1984) Quaternary geology of Lake Zürich: An interdisciplinary investigation by deep lake drilling. Contributions to Sedimentology 13, Schweizerbart, Stuttgart

Hsü KJ et al. (1977) History of the Mediterranean salinity crisis. Nature 267: 399-403

Huang TC, Stanley DJ (1974) Current reversal at 10,000 years B.P. at the Strait of Gibraltar - a discussion. Marine Geol 17: M1-M7

Hubbard DK, Miller AI, Scaturo D (1990) Production and cycling of calcium carbonate in a shelf-edge reef system (St. Croix, U.S. Virgin Islands): applications to the nature of reef systems in the fossil record. J Sediment Petrol 60: 335-360

Hubbard RJ (1988) Age and significance of sequence boundaries in Jurassic and Early Cretaceous rifted continental margins. Am Assoc Petrol Geol Bull 72: 49-72

Huc AY (1980) Origin and formation of organic matter in recent sediments and its relation to kerogen. In: Durand B (ed) Kerogen. Editions Technip, Paris, p 445-474

Hudson JD (1982) Pyrite in ammonite-bearing shales from the Jurassic of England and Germany. Sedimentology 29: 639-667

Hughes CJ (1983) Igneous petrology. Elsevier, Amsterdam

Hughes Clarke JE, Shor AN, Piper DJW, Mayer LA (1990) Large-scale current-induced erosion and deposition in the path of the 1929 Grand Banks turbidity current. Sedimentology 37: 613-629

Hunt JW (1989) Sedimentation rates and coal formation in the Permian basins of eastern Australia. In: Lyons PC, Alpern B (eds) Peat and coal: origin, facies, and depositional models. Elsevier, Amsterdam, p 259-274

Hunter RE (1977) Basic types of stratification in small eolian dunes. Sedimentology 24: 361-388

Hunter RE (1981) Stratification styles in eolian sandstones: some Pennsylvanian to Jurassic examples from the western Interior, U.S.A. In: Ethridge FG, Flores RM (eds) Modern and ancient nonmarine depositional environments. Soc Econ Paleontol Mineral Spec Publ 31: 315-329

Hurford AJ (1991) Uplift and cooling pathways derived from fission track analysis and mica dating: a review. Geol Rundsch 80: 349-368

Hutchinson DR, Klitgord KD (1988) Evolution of rift basins on the continental margin off southern New England. In: Manspeizer W (ed) Triassic-Jurassic rifting, parts A and B. Developm Geotectonics 22, Elsevier, Amsterdam, p 81-98

Hynes A (1990) Two-stage rifting of Pangea by two different mechanisms. Geology 18: 323-326

Ibbeken H (1983) Jointed source rock and fluvial gravel controlled by Rosin's law: a grain size study in Calabria, South Italy. J Sediment Petrol 53: 1213-1231

Iijima A (1988) Diagenetic transformations of minerals as exemplified by zeolites and silica minerals - a Japanese view. In: Chilingarian GV, Wolf KH (eds) Diagenesis, II. Developm Sedimentol 43, Elsevier, Amsterdam, p 147-211

Iijima A, Matsumoto R, Tada R (1985) Mechanism of sedimentation of rhythmically bedded chert. Sediment Geol 41: 221-233

Illies H, Greiner G (1978) Rhinegraben and the Alpine system. Bull Geol Soc Am 89: 770-782

Ingersoll RV (1982) Initiation and evolution of the Great Valley forearc basin of northern and central California, U.S.A. In: Leggett JK (ed) Trench-forearc geology: modern and ancient active plate margins. Geol Soc Spec Publ 10: 458-467

Ingersoll RV, Ernst WG (eds) (1987) Cenozoic basin development in coastal California. Rubey, vol 6, Prentice-Hall, New Jersey, 496 p

Iriondo MH (1988) A comparison between the Amazon and Paraná river systems. In: Degens ET, Kempe S, Naidu S (eds) Transport of carbon and minerals in major world rivers, lakes and estuaries, Part 5, Mitt Geol-Paläontol Inst Univ Hamburg, SCOPE/UNEP Sonderb 66: 77-92

Irwin H, Curtis C, Coleman M (1977) Isotopic evidence for source of diagenetic carbonates formed during burial of organic-rich sediments. Nature 269: 209-213

Isaacs CM (1984) Hemipelagic deposits in a Miocene basin, California: toward a model of lithologic variation. In: Stow DAV, Piper DJW (eds) Fine-grained sediments: deep-water processes and facies. Geol Soc Spec Publ 15: 481-496

Isaacs CM, Pisciotto KA, Garrison RE (1983) Facies and diagenesis of the Miocene Monterey Formation, California: a summary. In: Iijima A, Hein JR, Siever R (eds) Siliceous deposits in the Pacific region. Elsevier, Amsterdam, p 247-282

Jackson J, McKenzie D (1988) Rates of active deformation in the Aegean Sea and surrounding regions. Basin Res 1: 121-128

Jackson MPA, Talbot CJ (1986) External shape, strain rates, and dynamics of salt structures. Geol Soc Am Bull 97: 305-323

Jacobshagen VH (ed) (1988) The Atlas system of Morocco. Lecture Notes Earth Sci 15, Springer, Berlin Heidelberg New York, 499 p

Jacobshagen V, Giese P (1986) Bau und geodynamische Entwicklung der Helleniden - ein Gesamtbild. In: Jacobshagen V (ed) Geologie von Griechenland. Beiträge zur Regionalen Geologie der Erde 9, Bornträger, Berlin, 257-279

James NP (1983) Reef environment. In: Scholle PA, Bebout DG, Moore CH (eds) Carbonate depositional environments. Am Assoc Petrol Geol Mem 33:345-440

James NP (1990) Cool water carbonate sediments: viable analogues for Paleozoic limestones? Abstracts of Papers. 13th Int Sedimentol Congress, Nottingham, England, Int Assoc Sedimentol, p 245-246

James NP, Bone Y (1989) Petrogenesis of Cenozoic, temperate water calcarenites, South Australia: a model for meteoric/shallow burial diagenesis of shallow water calcite cements. J Sediment Petrol 59: 191-203

Jansa LF, Steiger TH, Bradshaw M (1984) Mesozoic carbonate deposition on the outer continental margin off Morocco. In: Hinz K, Winterer EL et al., Init Repts DSDP 79: 857-891, US Govt Print Office, Washington

Jansa LF, Wiedmann J (1982) Mesozoic-Cenozoic development of the eastern North American and northwest African continental margin: a comparison. In: Rad U von, Hinz K, Sarnthein M, Seibold E (eds) Geology of the Northwest African continental margin. Springer, Berlin Heidelberg New York, p 215-269

Jarvis GT (1984) An extensional model of graben subsidence - the first stage of basin evolution. Sediment Geol 40: 13-31

Jarvis I (1980) The initiation of phosphatic chalk sedimentation - the Senonian (Cretaceous) of th Anglo-Paris Basin. In: Bentor YK (ed) Marine phosphorites. Soc Econ Paleontol Mineral Spec Publ 29: 167-192

Jauzein A (1984) Sur la valeur de quelques hypothèses relatives à la genèse de gramdes séries salines. Rev Géol Dynam Géogr Phys 25: 149-156

Jenkyns HC (1986) Pelagic environments. In: Reading HG (ed) Sedimentary environments and facies. Blackwell, Oxford, p 343-398

Jenkyns HC, Winterer EL (1982) Palaeoceanography of Mesozoic ribbon radiolarites. Earth Planet Sci Lett 60: 351-375

Jenyon MK (1986) Salt tectonics. Elsevier Applied Science Publ, London New York, 191 p

Johnson AM (1984) Debris flows. In: Brunsden D, Prior DB (eds) Slope instability. Wiley, Chichester, p 257-361

Johnson HD, Baldwin CT (1986) Shallow siliciclastic seas. In: Reading HG (ed) Sedimentary facies and environment, 2nd edn. Blackwell, Oxford, p 229-252

Johnson SY (1989) Significance of loessite in the Maroon Formation (Middle Pennsylvanian to Lower Permian), Eagle Basin, northwest Colorado. J Sediment Petrol 59: 782-791

Johnson Ibach LE (1982) Relationship between sedimentation rate and total organic carbon content in ancient marine sediments. Am Assoc Petrol Geol Bull 66: 170-188

Johnsson MJ (1990) Tectonic versus chemical-weathering controls on the composition of fluvial sands in tropical environments. Sedimentology 37: 713-726

Jones FW, Majorowicz JA (1987) Some aspects of the hydrodynamicx of the western Canadian sedimentary basin. In: Goff JC, Williams BPJ (eds) Fluid flow in sedimentary basins and aquifers. Geol Soc Spec Publ 34: 79-85

Jonsson P, Carman R, Wulff F (1990) Laminated sediments in the Baltic - a tool for evaluating nutrient mass balances. AMBIO 19/3: 152-158

Jordan T (1981) Thrust loads and foreland basin evolution, Cretaceous, Western United States. Am Assoc Petrol Geol Bull 65: 2506-2520

Jordan T, Isacks B, Allmendinger RW, Brewer JA, Ramos VA, Ando CJ (1983) Andean tectonics related to geometry of subducted Nazca plate. Geol Soc Am Bull 94: 341-361

Kälin O, Bernoulli D (1984) Schizosphaerella Deflandre and Dangeard in Jurassic deeper-water carbonate sediments, Mazagan continental margin (Hole 547B) and Mesozoic Tethys. In: Hinz K, Winterer EL, et al., Init Repts DSDP 79: 411-448, US Govt Print Office, Washington

Karig DE (1986) The framework of deformation in the Nankai Trough. In: Kagami H, Karig DE, Coulbourn WT, et al., Init Repts DSDP 87: 927-940, US Govt Print Office, Washington

Karig DE, Sharman III GF (1975) Subduction and accretion in trenches. Geol Soc Am Bull 86: 377-389

Karner GD, Steckler MS, Thorne JA (1983) Long-term thermo-mechanical properties of the continental lithosphere. Nature 304: 250-253

Kashai EL (1988) A review of the relations between the tectonics, sedimentation and petroleum occurrences of the Dead Sea-Jordan rift system. In: Manspeizer W (ed) Triassic-Jurassic rifting, parts A and B. Developm Geotectonics 22: 883-909, Elsevier, Amsterdam

Kassens H (1990) Verfestigte Sedimentlagen und seismische Reflektoren: Frühdiagenese und Paläo-Ozeanographie in der Norwegischen See. Doct thesis, Math-naturwiss Fak, Univ Kiel, 100 p

Kassens H, Wetzel A (1989) Das Alter des Himalaya. Die Geowissenschaften 7, 1: 15-20, VCH Verlag, Weinheim

Kassler P (1973) The structural and geomorphic evolution of the Persian Gulf. In: Purser BH (ed) The Persian Gulf. Springer, Berlin Heidelberg New York, p 11-32

Kasten K, Mascle J, et al. (1988) ODP Leg 107 in the Tyrrhenian Sea: insights into passive margin and back-arc basin evolution. Geol Soc Am Bull 100: 1140-1156

Kasting JF (1989) Long-term stability of the Earth's climate. Palaeogeogr Palaeoclimatol Palaeoecol (Global and Planetary Change Section 1) 75: 123-136

Kastner M (1981) Authigenic silicates in deep-sea sediments: formation and diagenesis. In: Emiliani C (ed) The sea, v 7, The oceanic lithosphere. Wiley-Interscience, New York, p 915-980

Keefer, D.K., & Johnson, A.M. (1983): Earth flows: morphologbilization, and movement. US Geol Surv Prof Pap 1264: 1-56

Kauffman EG (1983) A geological and paleoceanographic overview of Cretaceous history in the Western Interior seaway of North America. In: Kauffman EG (ed) depositional environments and paleoclimates of the Greenhorn tectono-eustatic cycle, Rock Canyon anticline, Pueblo, Colorado. Geol Soc Am, Penrose Conf Guidebook, p 3-32

Kauffman EG (1984) Paleobiogeography and evolutionary response dynamic in the Cretaceous Western Interior seaway of North America. In: Westermann GEG (ed) Jurassic-Cretaceous biochronology and paleogeography of North America. Geol Assoc Canada Spec Pap 27: 273-306

Kauffman EG (1988) Concepts and methods of high-resolution event stratigraphy. Annu Rev Earth Planat Sci 16: 605-654

Kauffman EG, Walliser OH (eds)(1990) Extinction events in earth history. Lecture Notes in Earth Sciences, 30, Springer, Berlin Heidelberg New York, 432 p

Kauffman EG, Elder WP, Sageman BB (1991) High-resolution correlation: a new tool in chronostratigraphy. In: Einsele G, Ricken W, Seilacher A (eds) Cycles and events in stratigraphy. Springer, Heidelberg New York, p 795-819

Keen CE, Beaumont C, Boutilier R (1983) A summary of thermo-mechanical model results for the evolution of continental margins based on three rifting processes. In: Watkins JS, Drake CL eds) Studies in continental margin geology. Am Assoc Petrol Geol Mem 34: 725-728

Kehle RO (1988) The origin of salt structures. In: Schreiber BC (ed) Evaporites and hydrocarbons. Columbia Univ Press, New York, p 345-404

Kelts K (1981) A comparison of some aspects of sedimentation and translational tectonics from the Gulf of California and the Mesozoic Tethys, Northern Penninic margin. Eclogae Geol Helv 74: 317-338

Kelts K, Arthur MA (1981) Turbidites after ten years of deep-sea drilling - wringing out the mop? In: Warme JE, Douglas RG, Winterer EL (eds) The Deep Sea Drilling Project: a decade of progress. Soc Econ Paleontol Mineral Spec Publ 32: 91-127

Kelts K, Hsü KJ (1978) Freshwater carbonate sedimentation. In: Lerman A (ed) Lakes - chemistry, geology, physics. Springer, New York Heidelberg, p 295-323

Kelts KR, McKenzie JA (1982) Diagenetic dolomite formation in Quaternary anoxic diatomaceous muds of Deep Sea Drilling Project Leg 64, Gulf of California. In: Curray JR, Moore DG et al., Init Repts DSDP 64: 553-569, US Govt Print Office, Washington

Kempe S, Degens ET (1985) An early soda ocean? Chemical Geol 53: 95-108

Kempe S, Kazmierczak J (1991) Chemistry and stromatolites of the sea-linked Satonda Crater lake, Indonesia: a recent model for the Precambrian sea? Chemical Geol (in press)

Kempe S, Kazmierczak J, Degens ET (1989) The soda ocean concept and its bearing on biotic evolution. In: Crick RE (ed) Origin, evolution and modern aspects of biomineralisation in plants and animals. Proc Internat Symp Biomineralisation, Arlington, Texas, May 1986. Plenum Press, New York, p 29-43

Kemper E (1987) Das Klima der Kreide-Zeit. Geol Jahrb (Hannover) A 96: 5-185

Kendall AC (1983) Subaqueous evaporites. In: Walker RG (ed) Facies models, Geosci Canada Reprint Ser 1, p 159-174

Kendall AL (1988) Aspects of evaporite basin stratigraphy. In: Schreiber BC (ed) Evaporites and hydrocarbons. Columbia Univ Press, New York, p 11-65

Kendall CGSC, Schlager W (1981) Carbonates and relative changes in sea level. Marine Geol 44: 181-212

Kennett JP (1982) Marine geology. Prentice-Hall, Englewood Cliffs, 813 p

Kennett JP, von der Borch CC et al. (1986) Init Repts DSDP 90, parts 1 and 2, US Govt Print Office, Washington

Kenter JAM (1990) Carbonate platform flanks: slope angle and sediment fabric. Sedimentology 37: 777-794

Kern JP (1980) Origin of trace fossils in Polish Carpathian flysch. Lethaia 13: 347-362

Kidwell SM (1991) Taphonomic feedback (live/dead interactions) in the genesis of bioclastic beds: keys to reconstructing sedimentary dynamics. In: Einsele G, Ricken W, Seilacher A (eds) Cycles and events in stratigraphy. Springer, Berlin Heidelberg New York, p 682-695

Kimberley MM (1979) Origin of oolitic iron formations. J Sediment Petrol 49: 111-132

Kineke GC, Sternberg RW (1989) The effect of particle settling velocity on computed suspended sediment concentration profiles. Marine Geol 90: 159-174

King CAM (1972) Beaches and coasts, 2nd edn. Edward Arnold, London, 570 p

Kingston DR, Dishroon CP, Williams PA (1983) Global basin classification system. Am Assoc Petrol Geol Bull 67: 2175-2193

Kinsman DJJ (1975a) Rift valley basins and sedimentary history of trailing continental margins. In: Fischer AG, Judson SS (eds) Petroleum and global tectonics. Princeton, Univ Press, 322 p

Kinsman DJJ (1975b) Salt floors to geosynclines. Nature 5: 375-378

Kipphut GW (1988) Sediments and organic carbon cycling in an Arctic lake. Mitt Geol-Paläontol Inst, Univ Hamburg, SCOPE/UNEP Sonderb H 66: 129-135

Kirkland DW, Evans R (eds)(1973) Marine evaporites: origin, diagenesis, and geochemistry. Benchmark Papers in Geology. Dowden, Hutchinson & Ross, Stroudsburg, 426 p

Klein G deV (1970) Depositional and dispersal dynamics of intertidal sand bars. J Sediment Petrol 40: 1095-1127

Klein G deV (1971) A sedimentary model of determining paleotidal range. Geol Soc Am Bull 82: 1095-1127

Klein G deV (1985a) The frequency and periodicity of preserved turbidites in submarine fans as a quantitative record of tectonic uplift in collision zones. Tectonophysics 119: 181-193

Klein G deV (1985b) The control of depositional depth, tectonic uplift, and volcanism on sedimentation processes in the back-arc basins of the western Pacific ocean. J Geol 93: 1-25

Klein G deV (1985c) Sandstone depositional models for exploration of fossil fuels, 3rd edn. IHRDC Publ, Boston, 209 p

Klein G deV (1987) Current aspects of basin analysis. Sediment Geol 50: 95-118

Klein G deV (1989) Origin of coal-bearing cyclothems of North America. Geology 17: 152-155

Klein G deV (1990) Pennsylvanian time scales and cycle periods. Geology 19: 455-457

Klein G deV, Hsui AT (1987) Origin of cratonic basins. Geology 15: 1094-1098

Klein G deV, Ryer TA (1978) Tidal circulation patterns in Precambrian, Paleozoic and Cretaceous epeiric and mioclinal shelf seas. Geol Soc Am Bull 89: 1050-1058

Knauer GA, Redalje DG, Harrison WG, Karl DM (1990) New production at the VERTEX time-series site. Deep-Sea Res 37: 1121-1134

Knight RJ, McLern JR (eds)(1986) Shelf sands and sandstones. Canad Soc Petrol Geol Mem 11, 347 p

Koenigswald W von, Michaelis W (1984) Fossillagerstätte Messel - Literaturübersicht der Forschungsergebnisse aus den Jahren 1980-1983. Geol Jahrb Hessen 112: 5-26

Kögler FC, Larsen B (1979) The West Bornholm basin in the Baltic Sea: geological structure and Quaternary sediments. Boreas 8: 1-22, Oslo

Koerschner III WF, Read JF (1989) Field and modelling studies of Cambrian carbonate cycles, Virginia Appalachians. J Sediment Petrol 59: 654-687

Kohout FA, Meisler H, Meyer FW, Johnston RH, Leve GW, Wait RL (1988) Hydrogeology of the Atlantic continental margin. In: Sheridan RE, Grow JA (eds) The geology of North America, vol 1-2, The Atlantic continental margin, US. Geol Soc Am, p 463-483

Kolla V, Coumes F (1987) Morphology, internal structure, seismic stratigraphy, and sedimentation of Indus fan. Am Assoc Petrol Geol Bull 71: 650-677

Kolodny Y (1981) Phosphorites. In: Emiliani C (ed) The oceanic lithosphere. The sea vol 7, Wiley-Interscience, New York, p 981-1023

Komar PD (1969) The channelized flow of turbidity currents with application to Monterey deep sea fan channel. J Geophys Res 74: 4544-4558

Komar PD (1970) The competence of turbidity current flow. Geol Soc Am Bull 81: 1555-1562

Komar PD (1985) The hydraulic interpretation of turbidites from their grain sizes and sedimentary structures. Sedimentology 32: 395-408

Kominz MA, Bond GC (1986) Geophysical modelling of the thermal history of foreland basins. Nature 320: 252-256

Konta J (1985) Mineralogy and chemical maturity of suspended matter in major rivers sampled under the SCOPE/UNEP Project. Mitt Geol-Paläontol Inst Univ Hamburg H 58: 569-592

Konta J (1988) Minerals in rivers. In: Degens ET, Kempe S, Naidu S (eds) Transport of carbon and minerals in major world rivers, lakes and estuaries, Part 5, Mitt Geol-Paläontol Inst Univ Hamburg, SCOPE/UNEP Sonderb 66: 341-365

Kotwicki V, Isdale P (1991) Hydrology of Lake Eyre, Australia. Palaeogeogr Palaeoclimatol Palaeoecol 84: 87-98

Krömmelbein K (1986) Brinkmanns Abriß der Geologie, Band 2, Historische Geologie, 12/13th edn. Enke, Stuttgart, 400 p

Krumbein WC, Sloss LL (1963) Stratigraphy and sedimentation, 2nd edn. Freeman, San Francisco, 660 p

Krumm H, Petschick R, Wolf M (1988) From diagenesis to anchimetamorphism, upper Austroalpine sedimentary cover in Bavaria and Tirol. Geodynam Acta 2, 1: 33-47

Kuenen PH (1950) Marine geology. Wiley, New York London, 551 p

Kuenen PH, Migliorini CI (1950) Turbidity currents as a cause of graded bedding. J Geol 58: 91-127

Kuhnt W, Thurow J, Wiedmann J, Herbin JP (1986) Oceanic anoxic conditions around the Cenomanian/Turonian boundary and the response of the biota. Mitt Geol-Paläontol Inst Univ Hamburg, SCOPE UNEP Sonderb H 60: 205-246

Kukal Z (1971) Geology of recent sediments. Academia Publ House, Prague, Academic Press, London, 490 p

Kukal Z (1990) The rate of geological processes. Earth Sci Rev 28: 1-284

Kusznir NJ, Karner GD, Egan S (1987) Geometric, thermal and isostatic consequences of detachments in continental lithosphere extension and basin formation. In: Beaumont C, Tankard AJ (eds) Sedimentary basins and basin-forming mechanisms. Can Soc Petrol Geol Mem 12: 185-203

Labaume P, Mutti E, Seguret M (1985) Megaturbidites: a depositional model from the Eocene in the SW-Pyrenean foreland basin. Geo-Marine Lett 7: 91-101

Lancaster N (1990) Palaeoclimatic evidence from sand seas. Palaeogeogr Palaeoclimatol Palaeoecol 76: 279-290

Lancelot Y (1973) Chert and silica diagenesis in sediements from the Central Pacific. In: Winterer EL, Ewing JI et al, Init Repts DSDP 17: 377-405, US Govt Print Office, Washington

Land LS (1967) Diagenesis of skeletal carbonates. J Sediment Petrol 37: 914-930

Land LS (1985) The origin of massive dolomite. J Geol Educ 33: 112-125

Langbein R (1987) The Zechstein sulphates: The state of the art. In: Peryt TM (ed) The Zechstein facies in Europe. Lecture Notes Earth Sci 10, Springer, Berlin Heidelberg New York, p 143-188

Langford RP (1989) Fluvial-eaolian interactions: part I, modern systems. Sedimentology 36: 1023-1035

Langford RP, Chan MA (1989) Fluvial-aeolian interaction: part II, ancient systems. Sedimentology 36: 1037-1051

Large D (1988) The evaluation of sedimentary basins for massive sulfide mineralization. In: Friedrich GH, Herzig PM (1988) Base metal sulfide deposits in sedimentary and volcanic environments. Soc Geol Applied to Mineral Deposits, Spec Publ 5, Springer, Berlin Heidelberg New York, p 3-11

Larsen G, Chilingar GV (1979) Diagenesis in sediments and sedimentary rocks. Developm Sedimentol 25A, Elsevier, Amsterdam, 519 p

Larouzière FD de, Montenat C, Ott D'Estevou P, Griveaud P (1987) Évolution simultanée de bassins Néogènes en compression et en extension dans un couloir de décrochement; Hinojar et Mazarrón (Sudest de l'Espagne). Bull Cent Rech Explor Prod Elf-Aquitaine 11: 23-38

Larsen H (1980) Ecology of hypersaline environments. In: Nissenbaum A (ed) Hypersaline brines and evaporitic environments. Developm Sedimentol 28, Elsevier, Amsterdam, p 23-40

Lash GG (1985) Recognition of trench fill in orogenic flysch sequences. Geology 13: 867-870

Lash GG (1988) Along-strike variations in foreland basin evolution: possible evidence for continental collision along an irregular margin. Basin Res 1: 78-83

Last WM (1990) Lacustrine dolomite - an overview of modern, Holocene, and Pleistocene occurrences. Earth Sci Rev 27: 221-263

Lawrence DT, Doyle M, Aigner T (1990) Stratigraphic simulation of sedimentary basins: concepts and calibration. Am Assoc Petrol Geol Bull 74: 273-295

Leckie DA (1988) Wave-formed, coarse-grained ripples and their relationship to hummocky cross-stratification. J Sediment Petrol 58: 607-622

Leckie DA, Krystinik LF (1989) Is there evidence for geostrophic currents preserved in the sedimentary record of inner to middle shelf deposits? J Sediment Petrol 59: 862-870

Lee HJ, Chough SK (1989) Sediment distribution, dispersal, and budget in the Yellow Sea. Marine Geol 87: 195-205

Lee YL, Friedman GM (1987) Deep-burial dolomitization in the Ordovician Ellenburger Group carbonates, west Texas and southeastern New Mexico. J Sediment Petrol 57: 544-557

Leeder MR (1982) Sedimentology - process and product. Allen & Unwin, London, 344 p

Leeder MR (1991) Denudation, vertical crustal movements and sedimentary basin infill. Geol Rundsch 80: 441-458

Leeder MR, Gawthorpe RL (1987) Sedimentary models for extensional tilt-block/half-graben basins. In: Coward MP, Dewey JF, Hancock PL (eds) Continental extensional tectonics. Geol Soc Spec Publ 28: 139-152

Lees BG, Cook PG (1991) A conceptual model of lake barrier and compound lunette formation. Palaeogeogr Palaeoclimatol Palaeoecol 84: 271-284

Lefort P, Cuney M, Deniel C, France-Lanord C, Sheppard SMF, Upreti BN, Vidal P (1987) Crustal generation of the Himalayan leucogranites. Tectonophysics 134: 39-57

Leg 101 Scientific Party (1988) Leg 101 - an overview. In: Austin JA, Schlager W et al., Proc ODP, Sci Results 101: 455-472, College Station TX (Ocean Drilling Project)

Leg 122 Shipboard Scientific Party (1989) Breakup of Gondwanaland. Nature 337: 209-210

Leggett JK (ed) (1982) Trench-forearc geology: modern and ancient active plate margins. Geol Soc Spec Publ 10, 576 p

Lein R (1985) Das Mesozoikum der Nördlichen Kalkalpen als Beispiel eines gerichteten Sedimentationsverlaufes infolge fortschreitender Krustenausdünnung. Archiv f Lagerstättenforsch Geol Bundesanst (Wien) 6: 117-128

Leinen M, Sarnthein M (eds)(1989) Paleoclimatology and paleometeorology: modern and past patterns of global atmospheric transport. NATO ASI Series C, 282, Reidel, Dordrecht, 909 p

Leitch EC (1984) Marginal basins of the SW Pacific and the preservation and recognition of their ancient analogues: a review. In: Kokelaar BP, Howells RE (eds) Marginal basin geology: volcanic and associated sedimentary and tectonic processes in modern and ancient marginal basins. Geol Soc Spec Publ 16: 97-108

Lemcke K (1974) Vertikalbewegungen des vormesozoischen Sockels im nördlichen Alpenvorland vom Perm bis zur Gegenwart. Eclogae Geol Helv 67: 121-133

Lemcke K (1988) Geologie von Bayern I. Das bayerische Alpenvorland vor der Eiszeit. Schweizerbart, Stuttgart, 175 p

Lemoalle J, Dupont B (1976) Iron-bearing oolites and the present conditions of iron sedimentation in Lake Chad. In: Amstutz GC, Bernard AJ (eds) Ores in sediments. Int Union Geol Sci A3, Springer, Berlin Heidelberg New York, p 167-178

Lemoine M, Trümpy R (1987) Pre-oceanic rifting in the Alps. Tectonophysics 133: 305-320

Le Pichon X (1982) Land-locked oceanic basins and continental collision: the eastern Mediterranean as a case example. In: Hsü KJ (ed) Mountain building processes. Academic Press, London, p 201-211

Le Pichot X, Cochran JR (eds) (1988) The Gulf of Suez and the Red Sea rifting. Tectonophysics 153, 1-4 (Spec Iss), 320 p

Le Pichon X, Sibuet JC (1981) Passive margins, a model of formation. J Geophys Res 86: 3708-3720

Lerman A (1980) Geochemical processes. Wiley, New York, 481 p

Lever A, McCave IN (1983) Eolian components in Cretaceous and Tertiary North Atlantic sediments. J Sediment Petrol 53: 811-832

Lewis KB (1980) Quaternary sedimentation in the Hikurangi oblique-subduction and transform margin, New Zealand. In: Ballance PL, Reading HG (eds) Sedimentation in oblique-slip mobile zones. Int Assoc Sedimentol Spec Publ 4, Blackwell, Oxford, p 171-189

Limarino CO, Spalletti LA (1986) Eolian Permian deposits in west and northwest Argentina. Sediment Geol 49: 109-127

Link GJ, Osborne RH (1982) Sedimentary facies of Ridge Basin. In: Crowell JC, Link GJ (eds) Geologic history of Ridge Basin, southern California. Soc Econ Paleontol Mineral, Pacific Sect p 63-78

Lipman PW, Mullineaux DR (eds) (1981) The 1980 eruptions of Mount St. Helens Washington. US Geol Surv Prof Pap 1250, 844 p

Lippmann F (1973) Sedimentary carbonate minerals. Springer, Berlin Heidelberg New York, 226 p

Lisitzin AP (1972) Sedimentation in the world ocean. Soc Econ Paleontol Mineral Spec Publ 17

Liu Guanghua (1990) Permo-Carboniferous paleogeography and coal accumulation and their tectonic control in the North and South China continental plates. Int J Coal Geol 16; 73-117

Liu Tungshen (1988) Loess in China, 2nd edn. Springer, Berlin Heidelberg New York, China Ocean Press

Beijing, Springer Series in Physical Environments, 224 p

Liu Zhenxia, Huang Yichang, Zhang Qinian (1989) Tidal current rideges in the southwestern Yellow Sea. J Sediment Petrol 59: 432-437

Livingstone DA (1963) Chemical composition of rivers and lakes. US Geol Surv Prof Pap 440, 64 pp

Locke S, Thunell RC (1988) Paleoceanographic record of the last glacial/interglacial cycle in the Red Sea and Gulf of Aden. Palaeogeogr Palaeoclimatol Palaeoecol 64: 163-187

Logan BW (1987) The MacLeod evaporite basin western Australia. Am Assoc Petrol Geol Mem 44, 140 p

Long B, Sala M, Durand J, Michaud L (1989) Géométrie d'un lobe deltaique en contexte régressif. Bull Cent Rech Explor-Prod Elf-Aquitaine 13: 189-213

Lotze E (1957) Steinsalz und Kalisalze, 2nd edn, Band 1. Bornträger, Berlin, 466 p

Loutit TS, Hardenbol J, Vail PR, Baum GR (1988) Condensed sections: the key to age determination and correlation of continental margin sequences. In: Sea-level changes - an integrated approach. Soc Econ Paleontol Mineral Spec Publ 42: 183-213

Lowe DR (1976) Grain flow and grain flow deposits. J Sediment Petrol 46: 188-199

Lowe DR (1979) Sediment gravity flows: their classification and some problems of application to natural flows and deposits. Soc Econom Paleontol Mineral Spec Publ 27: 75-82

Lowe DR (1982) Sediment gravity flows: II. Depositional models with special reference to the deposits of high-density turbidity currents. J Sediment Petrol 52: 279-297

Lucas J, Prévot L (1984) Synthèse de l'apatite par voie bactériénne à partir de matière organique phosphatée et de divers carbonates de calcium dans des eaux douce et marine naturelles. Chemical Geol 42: 101-118

Lucia FJ (1972) Recognition of evaporite-carbonate shoreline sedimentation. In: Rigby JK, Hamblin WK (eds) Recognition of ancient sedimentary environments. Soc Econ Paleontol Mineral Spec Publ 16: 160-191

Lützner H (1989) Sedimentation rates of Variscan molasse basins in central Europe. Z Geol Wissensch (Berlin) 17-9: 859-868

Lyon-Caen H, Molnar P (1985) Gravity anomalies, flexure of the Indian plate, and the structure, support and evolution of the Himalaya and the Ganga basin. Tectonics 4: 513-538

Lyons PC, Alpern B (eds) (1989) Peat and coal: origin, facies, and depositional models. Elsevier, Amsterdam, 882 p

Ma Li (1985) Subtle oil pools in Xingshugang Delta, Songliao Basin. Am Assoc Petrol Geol Bull 69: 1123-1132

Macdonald DIM (1986) Proximal to distal variation in a linear turbidite trough: implications for the fan model. Sedimentology 33: 243-259

Mackenzie FT (1990) Sea level change, sediment mass and flux and chemostratigraphy. In: Ginsburg RN, Beaudoin B (eds) Cretaceous resourceds, events and rhythms. NATO ASI Series C 304, Kluwer, Dordrecht, p 289-304

MacLeod N, Keller G (1991) Hiatus distribution and mass extinctions at the Cretaceous/Tertiary boundary. Geology 19: 497-501

Mader D (ed) (1985) Aspects of fluvial sedimentation in the lower Triassic Buntsandstein. Lecture Notes Earth Sci 4, Springer, Berlin Heidelberg New York, 626 p

Magara K (1976) Thickness of removed sedimentary rocks, paleopressure, and paleotemperature, southwestern part of western Canada Basin. Am Assoc Petrol Geol Bull 60: 554-565

Magara K (1978) Compaction and fluid migration - practical petroleum geology. Elsevier, Amsterdam, 319 p

Maillart J (1989) Differentiation entre tectonique syn sédimentaire et compaction differentielle. Thèse l'Ecole Nationale Superieure des Mines, Paris, 193 p

Maizels J (1989) Sedimentology, paleoflow dynamics and flood history of Jökulhlaup deposits: paleohydrology of Holocene sediment sequences in southern Iceland sandur deposits. J Sediment Petrol 59: 204-223

Malouta DN, Gorsline DS, Thornten SE (1981) Processes and rates of recent (Holocene) basin filling in an active transform margin: Santa Monica Basin, California continental borderland. J Sediment Petrol 51: 1077-1096

Manheim FT (1970) The diffusion of ions in unconsolidated sediments. Earth Planet Sci Lett 9: 307-309

Manheim FT, Pratt RM, McFarlin FP (1980) Composition and origin of phosphorite deposits of the Blake Plateau. In: Bentor YK (ed) Marine phosphorites - geochemistry, occurrence, genesis. Soc Econ Paleontol Mineral Spec Publ 29: 117-137

Mann P, Hempton MR, Bradley DC, Burke K (1983) Development of pull-apart basins. J Geol 91: 529-554

Manspeizer W (1985) The Dead Sea Rift: impact of climate and tectonism on Pleistocene and Holocene sedimentation. In: Biddle KT, Christie-Blick N (eds) Strike-slip deformation, basin formation, and sedimentation. Soc Econom Paleontol Mineral Spec Publ 37: 143-158

Manspeizer W (ed) (1988a) Triassic-Jurassic rifting, parts A and B. Developm Geotectonics 22, Elsevier, Amsterdam, 998 p

Manspeizer W (1988b) Triassic Jurassic rifting and opening of the Atlantic: an overview. In: Manspeizer W (ed) Triassic-Jurassic rifting, parts A and B. Developm Geotectonics 22, Elsevier, Amsterdam, p 41-79

Martin JH, Knauer GA, Karl DM, Broenkow WW (1987) VERTEX: carbon cycling in the northeast Pacific. Deep-Sea Res 34, 2A: 267-285

Martinsen OJ (1989) Styles of soft-sediment deformation on a Namurian (Carboniferous) delta slope, Western Ireland Namurian Basin, Ireland. In: Whateley MKG, Pickering KT (eds) Deltas; sites and traps for fossil fuels. Geol Soc Spec Publ 41: 167-177

Martinson DG, Pisias NG, Hays JD, Imbrie J, Moore TC, Shackleton NJ (1987) Age dating and orbital theory of ice ages: development of a high-resolution 0-3,000,000-year chronostratigraphy. Quat Res 27: 1-29

Marzo M (1986) Secuencias fluvio-eolicas en el Buntsandstein de Macizo de Garraf (Provincia de Barcelona). Cuad Geol Ibérica (Madrid) 10: 207-233

Marzo M, Nijman W, Puigdefabregas C (1988) Architecture of the Castissent fluvial sheet sandstones, Eocene, South Pyrenees, Spain. Sedimentology 35: 719-738

Mascle A, Moore JC, et al. (1988) Proc. ODP, Init Repts 110 A. College Station, Texas (Ocean Drilling Project), 603 p

Mascle G, Hérail G, Van Haver T, Delcaillau B (1986) Structure et évolution des bassins d'épisuture et de périsuture liés à la ch^aine Himalayenne. Bull Cent Rech Explor-Prod Elf-Aquitaine 10: 181-203

Massari F, Grandesso P, Stefani C, Jobstraibitzer PG (1986) A small polyhistory foreland basin evolving in a context of oblique convergence: the Venetian basin (Chattian to Recent, Southern Alps, Italy). In: Allen PA, Homewood P, Williams G (eds) Foreland basins: an introduction. Spec Publ Int Assoc Sedimentol 8: 141-168

Mathews WH (1975) Cenozoic erosion and erosion surfaces of eastern North America. Am J Sci 275: 818-824

Matter A, Douglas RG, Perch-Nielsen K (1975) Fossil preservation, geochemistry, and diagenesis of pelagic carbonates from the Shatsky Rise, Northwest Pacific. In: Larson RL, Moberly R, et al., Init Repts DSDP 32: 891-921, US Govt Print Office, Washington

Matter A, Homewood P, Caron C, Rigassi D, Stuijvenberg J van, Weidmann M, Winkler W (1980) Flysch and molasse of western and central Switzerland. In: Trümpy R (ed) Geology of Switzerland, a guide-book, part B: 261-293. Schweiz Geol Kommission, Wepf, Basel

Matthess G (1973) Die Beschaffenheit des Grundwassers. Lehrbuch der Hydrologie, Bd 2. Bornträger, Berlin Stuttgart, 324 p

Matthews RK (1974) Dynamic stratigraphy. Prentice Hall, Englewood Cliffs, 370 p

Matthews RK (1984a) Dynamic stratigraphy, an introduction to sedimentation and stratigraphy, 2nd edn. Prentice-Hall, Englewood Cliffs, 489 p

Matthews RK (1984b) Oxygen isotope record of ice-volume history: 100 million years of glacio-eustatic sea-level fluctuation. In: Schlee JS (ed) Interregional unconformities and hydrocarbon accumulation. Am Assoc Petrol Geol Mem 36: 97-107

Mattick RE, Phillips RL, Rumpler J (1988) Seismic stratigraphy and depositional framework of sedimentary rocks in the Pannonian basin in southeastern Hungary. In: Royden LH, Horváth F (eds) The Pannonian basin: a study in basin evolution. Am Assoc Petrol Geol Mem 45: 117-145

Mayer L (1987) Subsidence analysis of the Los Angeles basin. In: Ingersoll RV, Ernst WG (eds) Cenozoic basin development of coastal California. Rubey vol 6, Prentice-Hall, Englewood Cliffs, p 299-318

Maynard JB (1983) Geochemistry of sedimentary ore deposits. Springer, New York Heidelberg, 305 p

Maynard JB, Valloni R, Yu H-S (1982) Composition of modern deep-sea sands from arc-related basins. In: Leggett JK (ed) Trench-forearc geology: modern and ancient active plate margins. Geol Soc Spec Publ 10: 551-561

McArthur JM, Thomson J, Jarvis I, Fallick AE, Birch GF (1988) Eocene to Pleistocene phosphogenesis of western South Africa. Marine Geol 85: 41-63

McBride EF (1989) Quartz cement in sandstones: a review. Earth Sci Rev 26: 69-112

McCarthy J, Scholl DW (1985) Mechanism of subduction accretion along the central Aleutian Trench. Geol Soc Am Bull 96: 691-701

McCave IN (1984) Erosion, transport and deposition of fine-grained marine sediments. In: Stow DAV, Piper DJW (eds) Fine-grained sediments: deep-water processes and facies. Geol Soc Spec Publ 15: 35-64

McDonald DA, Surdam RC (eds)(1984) Clastic diagenesis. Am Assoc Petrol Geol Mem 37, 434 p

McGhee GR, Bayer U (1985) The local signature of sea-level changes. In: Bayer U, Seilacher A (eds) Sedimentary and evolutionary cycles. Lecture Notes Earth Sci (Springer) 1: 98-112

Mchargue TR, Webb JE (1986) Internal geometry, seismic facies, and petroleum potential of canyons and inner fan channels of the Indus submarine fan. Am Assoc Petrol Geol Bull 70: 161-180

McIlreath JA, James NP (1984) Carbonate slopes. In: Walker RG (ed) Facies models, 2nd edn. Geosci Canada Reprint Ser 1: 245-257

McKee ED (ed) (1979) A study of global sand seas. US Geol Sur Prof Pap 1052, 429 p

McKee ED, Ward WC (1983) Eolian environments. In: Scholle PA, Bebout DG, Moore CH (eds) Carbonate depositional environments. Am Assoc Petrol Geol Mem 33: 131-170

McKenzie DP (1978) Some remarks on the development of sedimentary basins. Earth Planet Sci Lett 40: 25-32

McQueen HWS, Beaumont C (1989) Mechanical models of tilted block basins. In: Price RA (ed) Origin and evolution of sedimentary basins and their energy and mineral resources. Geophs Monograph 48, IUGG vol 3, Int Union Geodesy Geophysics, p 65-71

Mégard F (1984) The Andean orogenic period and its major structures in central and southern Peru. J Geol Soc London 141: 893-900

Meischner KD (1964) Allodapische Kalke, Turbidite in riff-nahen Sedimentations-becken. In: Bouma AH, Brouwer A (eds) Turbidites. Elsevier, Amsterdam, p 156-191

Menard HW, Smith SM, Pratt RM (1965) The Rhone deep sea fan. In: Whitard WF, Bradshaw R (eds) Submarine geology and geophysics. Butterworths, London, p 271-284

Meulenkamp JE, Wortel MJR, Van Wamel WA, Spakman W, Hoogerduyn Strating E (1988) On the Hellenic subduction zone and the geodynamic evolution of Crete since the Middle Miocene. Tectonophysics 146: 203-215

Meybeck M (1979) Concentration des eaux fluviales en élements majeurs et apports en solution aux océans. Rev Géol Dynam Géogr Phys 21: 215-246

Meybeck M (1983) Atmospheric inputs and river transport of dissolved substances. In: Dissolved loads of rivers and surface water quantity/quality relationships. Int Assoc Hydrol Sci Publ 141: 173-192

Meybeck M (1987) Global chemical weathering of surficial rocks estimated from river dissolved loads. Am J Sci 287: 401-426

Meybeck M (1989) The Quality of rivers: from pristine stage to global pollution. Palaeogeogr Palaeoclimatol Palaeoecol 75: 283-309

Meyer AJ, Landais P, Brosse E, Pagel M, Carisey JC, Krewdl D (1989) Thermal history of the Permian formations from the Breccia Pipes area (Grand Canyon region, Arizona). Geol Rundsch 78: 427-438

Miall AD (1978) Fluvial sedimentology. Can Soc Petrol Geol Mem 5, 859 p

Miall AD (1980) Cyclicity and the facies model concept in fluvial deposits. Bull Can Petrol Geol 28: 59-80

Miall AD (1981) Alluvial sedimentary basins: tectonic setting and basin architecture. In: Miall AD (ed) Sedimentation and tectonics in alluvial basins. Geol Soc Canada Spec Pap 23: 1-33

Miall AD (1984) Principles of sedimentary basin analysis. Springer, New York Heidelberg, 490 p

Miall AD (1985) Architectural-element analysis: a new method of facies analysis applied to fluvial deposits. Earth Sci Rev 22: 261-308

Middleton GV, Hampton MA (1976) Subaqueous sediment transport and deposition by gravity flows. In: Stanley DJ, Swift DJP (eds) Marine sediment transport and environmental management. Wiley, New York, p 197-218

Miller PM, Barakat H (1988) Geology of the Safaga Concession, northern Red Sea, Egypt. Tectonophysics 153: 125-136

Milliman JD, Meade RJ (1983) World-wide delivery of river sediment to the oceans. J Geol 91: 1-21

Milliman JD, Qin YS, Ren ME, Saito Y (1987) Man's influence on the erosion and transport of sediment by Asian rivers: the Yellow River (Huanghe) example. J Geol 95: 751-762

Millot G (1964) Géologie des argiles. Masson, Paris, 500 p

Mills HH (1976) Estimated erosion rates on Mount Rainier, Washington. Geology 4: 401-406

Mitchell AHG, Reading HG (1986) Sedimentation and tectonics. In: Reading HG (ed) Sedimentary environments and facies, 2nd edn. Blackwell, Oxford, p 471-519

Mitra S (1988) Effects of deformation mechanisms on reservoir potential in central Appalachian overthrust belt. Am Assoc Petrol Geol Bull 72: 536-554

Moberly R, Shepherd GL, Coulbourn WT (1982) Forearc and other basins, continental margin of northern and southern Peru and adjacent Ecuador and Chile. In: Leggett JK (ed) Trench-forearc geology: modern and ancient active plate margins. Geol Soc Spec Publ 10: 171-189

Molenaar CM, Rice DD (1988) Cretaceous rocks of the Western Interior Basin. In: Sloss LL (ed) The geology of North America, vol D-2: Sedimentary cover - North American craton: US Geol Soc Am, Boulder, p 77-82

Molnia BF (ed) (1983) Glacial-marine sedimentation. Plenum Press, New York, 844 p

Montadert L, de Charpal O, Roberts D, Guennoc P, Sibuet JC (1979) Northeast Atlantic continental margins: rifting and subsidence processes. In: Talwani M, Hay W, Ryan WBF (eds) Deep drilling results in the Atlantic ocean: continental margins and paleoenvironment. Am Geophys Union, Maurice Ewing Ser 3, Washington DC, p 154-184

Montenat C, Ott d'Estevou P, Masse P (1987) Tectonic-sedimentary characters of the Betic Neogene basins evolving in a crustal transcurrent shear zone (SE Spain). Bull Cent Rech Explor-Prod Elf-Aquitaine 11: 1-22

Montenat C, et al (1988) Tectonic and sedimentary evolution of the Gulf of Suez and the northwestern Red Sea. In: Le Pichot X, Cochran JR (eds) The Gulf of Suez and the Red Sea rifting. Tectonophysics 153, 1-4 (Spec Iss): 161-177

Moore CH (1989) Carbonate diagenesis and porosity. Developm Sedimentol 46, Elsevier, Amsterdam, 338 p

Moore DG (1969) Reflection profiling studies of the California continental borderland: structure and Quaternary turbidite basins. Geol Soc Am Spec Pap 107, 142 p, 18 plates

Moore DG, Buffington EC (1968) Transform faulting and the growth of the gulf of California since late Miocene. Science 161: 1238-1241

Moore DG, Curray JR, Einsele G (1982) Salado-Vinorama submarine slide and turbidity current off southeast tip of Baja California. In: Curray JR, Moore DG et al, Init Repts DSDP 64, US Govt Print Office, Washington, p 1071-1082

Moore GF, Curray JR, Emmel FJ (1982) Sedimentation in the Sunda Trench and forearc region. In: Leggett JK (ed) Trench-forearc geology: modern and ancient active plate margins. Geol Soc Spec Publ 10: 245-258

Moore JC, Byrne T (1987) Thickening of fault zones: a mechanism of mélange formation in accreting sediments. Geology 15: 1040-1043

Moore TC, Van Andel TH, Sancetta C, Pisias N (1978) Cenozoic hiatuses in pelagic sediments. Micropaleontology 24: 113-138

Morales C (ed) (1979) Saharan dust-mobilization, transport, deposition. Wiley, New York, 316 p

Morgan JP, Shaver RH (eds)(1970) Deltaic sedimentation modern and ancient. Spec Publ Soc Econ Paleontol Mineral 15, 312 pp

Morgenstern NR (1967) Submarine slumping and the initiation of turbidity currents. In: Richards AF (ed) Marine geotechnique. Urbana, Univ Illinois Press, p 189-220

Morley CK, Nelson RA, Patton TL, Munn SG (1990) Transfer zones in the East African rift system and their relevance to hydrocarbon exploration in rifts. Am Assoc Petrol Geol Bull 74: 1234-1253

Morris PA (1988) Volcanic arc reconstruction using discriminant function analysis of detrital clinopyroxene and amphibole from the New England fold belt, eastern Australia. J Geol 96: 299-311

Morris RJ (1987) Turbidite flows as a source of organic matter in deep water marine deposits: evidence from Quaternary sediments on the Madeira abyssal plain. Mém Soc Géol Fr NS 151: 43-53

Morton RA (1988) Nearshore responses to great storms. Geol Soc Am Spec Pap 229: 7-22

Morton RA, Nummedal D (eds)(1989) Shelf sedimentation, shelf sequences and related hydrocarbon accumulation. Gulf Coast Section, Soc Econ Paleontol Mineral 220 p

Moxon IW, Graham SA (1987) History and controls of subsidence in the late Cretaceous-Tertiary Great Valley forearc basin, California. Geology 15: 626-629

Mozley PS (1989) Relation between depositional environment and the elemental composition of early diagenetic siderite. Geology 17: 704-706

Müller G (1966) The new Rhine delta in Lake Constance. In: Shirley ML (ed) Deltas in their geologic framework. Houston Geol Soc 107: 107-124

Müller G (1988) Salzgesteine (Evaporite). In: Füchtbauer H (ed) Sedimente und Sedimentgesteine. Schweizerbart, Stuttgart, p 435-500

Müller G, Irion G, Förstner U (1972) Formation and diagenesis of inorganic Ca-Mg carbonates in the lacustrine environment. Naturwissenschaften 59: 158-164

Müller PJ, Suess E (1979) Productivity, sedimentation rate and sedimentary organic carbon content in the oceans. Deep-Sea Res 26: 1347-1362

Mullins HT, Cook HE (1986) Carbonate apron models: alternatives to submarine fan model for paleoenvironmental analysis and hydrocarbon exploration. J Sediment Geol 48: 37-79

Mullins HT, Gardulski AF, Hinchey EJ, Hine AC (1988) The modern carbonate ramp slope of central West Florida. J Sediment Petrol 58: 273-290

Mumpton FA (ed) (1986) Studies in diagenesis. US Geol Surv Bull 1578, US Govt Print Office, Washington, 368 p

Murat A, Got H (1987) Middle and late Quaternary depositional sequences and cycles in the eastern Mediterranean. Sedimentology 34: 885-899

Murris RJ (1985) Middle East: Stratigraphic evolution and oil habitat. In: Demaison G, Murris RJ (eds) Petroleum geochemistry and basin evaluation. Am Assoc Petrol Geol Mem 35: 353-372

Mutti E (1977) Distinctive thin-bedded turbidite facies and related depositional environments in the Eocene Hecho Group (South-central Pyrenees, Spain). Sedimentology 24: 107-131

Mutti E (1984) The Hecho Eocene submarine fan system, south-central Pyrenees, Spain. Geo-Marine Lett 3: 199-202

Mutti E, Normark WR (1987) Comparing examples of modern and ancient turbidite systems: problems and concepts. In: Legget JK, Zuffa GG (eds) Marine clastic sedimentology. Graham and Trotman, p 1-38

Mutti E, Ricci Lucchi F (1978) Turbidites of the northern Apennines: introduction to facies analysis. In: Nilsen T (ed) Am Geol Inst Reprint Ser 3: 127-166 (translation of 1972 article in Italian)

Mutti E, Ricci Lucchi F, Seguret M, Zanzucchi G (1984) Seismoturbidites: a new group of resedimented deposits. Marine Geol 55: 103-116

Mutti E, Remacha E, Sgavetti M, Rosell J, Valloni R, Zamorano M (1985) Stratigraphy and facies characteristics of the Eocene Hecho Group turbidite systems, south-central Pyrenees. Field-trip guide book, 6th European Meeting Int Assoc Sediment, Lleida, Spain, Excursion 12, p 521-600

Naeser CW (1979) Thermal history of sedimentary basins: fission track dating of subsurface rocks. In: Schlolle PA, Schluger PR (eds) Aspects of diagenesis. Soc Econ Paleontol Mineral Spec Publ 26: 109-112

Naidu AS, Mowatt TC, Somayajulu BLK, Sreeramachandra Rao K (1985) Characteristics of clay minerals in the bed loads of major rivers in India. Mitt Geol-Paläontol Inst Univ Hamburg, SCOPE/UNEP Sonderb 58: 559-568

Nairn AEM, Stehli FG (eds) (1975) The ocean basins and margins, vol 3. The Gulf of Mexico and the Caribbean. Plenum Press, New York London, 706 p

Neev D, Emery KO (1967) The Dead Sea. Depositional processes and environments of evaporites. Geol Surv Israel Bull 41, 147 p

Nelsen TA, Stanley DJ (1984) Variable deposition rates on the slope and rise off the Mid-Atlantic States. Geo-Marine Lett 3: 37-42

Nelson CH, Maldonado A (1988) Factors controlling depositional patterns of Ebro turbidite system, Mediterranean Sea. Am Assoc Petrol Geol Bull 72: 698-716

Nelson CH, Normark WR, Bouma AH, Carlson PR (1978) Thin-bedded turbidites in modern submarine canyons and fans. In: Stanley DJ, Kelling G (eds) Sedimentation in submarine canyons, fans and trenches. Dowden, Hutchinson & Ross, Stroudsburg, p 177-189

Nemec W, Steel RJ (eds)(1988) Fan deltas - sedimentology and tectonic setting. Blackie, Glasgow, 444 p

Neugebauer HJ (1987) Models of lithospheric thinning. Annu Rev Earth Planet Sci 15: 421-443

Neugebauer J (1974) Some aspects of cementation in chalk. In: Hsü KJ, Jenkyns HC (eds) Pelagic sediments: on land and under the sea. Int Assoc Sedimentol, Spec Publ 1: 149-176

Neuzil CE (1986) Groundwater flow in low-permeability environments. Water Resour Res 22: 1183-1195

Newell ND, Rigby JK, Fischer AG, Whiteman AJ, Hickox JE, Bradley JS (1953) The Permian reef complex of the Guadelupe Mountains region, Texas and New Mexico. Freeman, San Francisco, 236 p

Nicols MM (1989) Sediment accumulation rates and relative sea-level rise in lagoons. In: Ward LG, Ashley GM (eds) Physical processes and sedimentology of siliciclastic-dominated lagoonal systems. Marine Geol (Spec Iss) 88: 201-219

Nielsen OB, Sorensen S, Thiede J, Skarbo (1986) Cenozoic differential subsidence of North Sea. Am Assoc Petrol Geol Bull 70: 276-298

Nikolaev VG, Vass D, Pogacsas D (1989) Neogene-Quaternary Pannonian basin: a structure of labigenic type. In: Price RA (ed) Origin and evolution of sedimentary basins and their energy and mineral resources. Geophys Monograph 48 (IUGG vol 3), Am Geophys Union, Washington DC, p 187-196

Nilsen TH (1982) Alluvial fan deposits. In: Scholle PA, Spearing DR (eds) Sandstone depositional environments. Am Assoc Petrol Geol Mem 31: 49-66

Nilsen TH, McLaughlin RJ (1985) Comparison of tectonic framework and depositional patterns of the Hornelen strike-slip basin of Norway and the Ridge and little Sulphur Creek strike-slip basins of California. In: Biddle KT, Christie-Blick N (eds) Strike-slip deformation, basin formation, and sedimentation. Soc Econ Paleontol Mineral Spec Publ 37: 79-103

Nissenbaum A (ed) (1980) Hypersaline brines and evaporitic environments. Developm Sedimentol 28, Elsevier, Amsterdam, 270 p

Nitecki MH (ed)(1981) Biotic crises in ecological and evolutionary time. Academic Press, New York, 301 p

North CP, Todd SP, Turner JP (1989) Alluvial fans and their tectonoc controls. J Geol Soc London 146: 507-508

Notholt AJG, Jarvis I (eds)(1990) Phosphorite research and development. Geol Soc Spec Publ 52, Blackwell Oxford

Nottvedt A, Kreisa RD (1987) Model for the combined flow origin of hummocky cross-stratification. Geology 15: 357-361

Nummedal D (1991) Shallow marine storm sedimentation: the oceanographic perspective. In: Einsele G, Ricken W, Seilacher A (eds) Cycles and events in stratification. Springer, Heidelberg New York, p 227-248

Nummedal D, Pilkey OH, Howard JD (eds) (1987) Sea level fluctuation and coastal evolution. Soc Econ Paleontol Mineral Spec Publ 41, 267 p

Nummedal D, Swift DJP (1987) Transgressive stratigraphy at sequence-bounding unconformities: some principles derived from Holocene and Cretaceous examples. In: Nummedal D, Pilkey OH, Howard JD (eds) Sea level fluctuation and coastal evolution. Soc Econom Paleontol Mineral Spec Publ 41: 241-260

Nur A, Ben-Avraham Z (1983) Displaced terranes and mountain building. In: Hsü KJ (ed) Mountain building processes. Academic Press, London, p 73-84

Oberhänsli H, Allen PA (1987) Stable isotope signature of Tertiary lake carbonates, eastern Ebro Basin, Spain. Palaeogeogr Palaeoclimatol Palaeoecol 60: 59-75

Oberhänsli R, Stoffers P (eds)(1988) Hydrothermal activity and metaliferous sediments on the ocean floor. Marine Geol (Spec Iss) 84, 3/4: 145-284

O'Brien GW, Harris JR, Milnes HR, Veeh HH (1981) Bacterial origin of East Australian continental margin phosphorites. Nature 294: 442-444

Odin GS (ed) (1982) Numerical dating in stratigraphy. Wiley, Chichester, 1094 p

Odin GS (ed) (1988) Green marine clays. Developm Sedimentol 45, Elsevier, Amsterdam, 445 p

Odin GS, Matter A (1981) De glauconiarum origine. Sedimentology 28: 611-641

Odin GS, Morton AC (1988) Authigenic green particles from marine environments. In: Chilingarian GV, Wolf KH (eds) Diagenesis, II. Developm Sedimentol 43: Elsevier, Amsterdam, p 213-264

Odin GS, Debenay JP, Masse JP (1988a) The verdine facies deposits identified in 1988. In: Odin GS (ed) Green marine clays. Developm Sedimentol 45, Elsevier, Amsterdam, p 131-158

Odin GS, Knox RW O'B, Gygi RA, Guerrak S (1988b) Green marine clays from the oolitic ironstone facies: Habit, mineralogy, environment. In: Odin GS (ed) Green marine clays. Developm Sedimentol 45, Elsevier, Amsterdam, p 29-52

Ouwehand PJ (1986) Werden Phosphorite wirklich frühdiagenetisch gebildet? Beispiele aus der helvetischen "Mittelkreide" der Ostschweiz: In: Bechstädt T, Knitter H (eds) Erstes Treffen deutschsprachiger Sedimentologen, Freiburg i.B., p 86-89

Ogg JG, Haggerty J, Sarti M, von Rad U (1987) Lower Cretaceous pelagic sediments of Deep Sea Drilling Project Site 603, western North Atlantic: a synthesis. In: van Hinte JE, Wise SW, et al., Init Repts DSDP 93: 1305-1331, US Govt Print Office, Washington

Ollier C (1969) Weathering. Elsevier, New York

Olsen PE (1986) A 40-million year lake record of Early Mesozoic orbital climatic forcing. Science 234: 842-848

Opdyke BN, Wilkinson BH (1990) Paleolatitude distribution of Phanerozoic marine ooids and cements. Palaeogeogr Palaeoclimatol Palaeoecol 78: 135-148

Ori GG, Friend PF (1984) Sedimentary basins formed and carried piggyback on active thrust sheets. Geology 12: 475-478

Ori GG, Roveri M (1987) Geometries of Gilbert-type deltas and large channels in the Meteora Conglomerate, Meso-Hellenic basin (Oligo-Miocene), central Greece. Sedimentology 34: 845-859

Orson RA, Simpson RL, Good RE (1990) Rates of sediment accumulation in a tidal freshwater marsh. J Sediment Petrol 60: 859-869

Oschmann W (1991) Anaerobic - poikiloaerobic - aerobic: a new facies zonation for modern and ancient neritic redox facies. In: Einsele G, Ricken W, Seilacher A (eds) Cycles and events in stratigraphy. Springer, Heidelberg New York, p 565-571

Park WC, Schot EH (1968) Stylolites: their nature and origin. J Sediment Petrol 38: 175-191

Parker A, Sellwood BW (eds) (1983) Sediment diagenesis. NATO ASI Series, C, 115, Reidel, Dordrecht, 427 p

Parrish JT (1987) Palaeo-upwelling and the distribution of organic-rich rocks. In: Brooks J, Fleet AJ (eds) Marine petroleum source rocks. Geol Soc Spec Publ 26: 199-205

Parrish JT, Curtis RL (1982) Atmospheric circulation, upwelling, and organic-rich rocks in the Mesozoic and Cenozoic Eras. Palaeogeogr Palaeoclimatol Palaeoecol 40:31-66

Parrish JT, Ziegler AM, Humphreville RG (1983) Upwelling in the Paleozoic era. In: Thiede J, Suess E (eds) Coastal upwelling - its sediment record. Part B, Sedimentary record of ancient coastal upwelling. NATO Conference Ser 4: Marine Sci 10, Plenum Press, New York London, p 553-578

Paul J (1982) Zur Rand- und Schwellenfazies des Kupferschiefers. Z Deutsch Geol Ges (Hannover) 133: 571-605

Paul J (1987) Der Zechstein am Harzrand: Querprofil über eine permische Schwelle, Stop 1-23. (Zechstein at the Harz Mountains: a geotraverse across a Permian platform.) In: Kulick J, Paul J (eds) Int Symp Zechstein 1987, Exkursionsführer (Wiesbaden) II: 193-276

Payton CE (ed) (1977) Seismic stratigraphy - application to hydrocarbon exploration. Am Assoc Petrol Geol Mem 26

Pedersen TF, Calvert SE (1990) Anoxia vs. productivity: what controls the formation of organic-carbon-rich sediments and sedimentary rocks? Am Assoc Petrol Geol Bull 74: 454-466

Penland SP, Boyd R, Suter JR (1988) Transgressive depositional systems of the Mississippi delta plain: a model for barrier shoreline and shelf sand development. J Sediment Petrol 58: 932-949

Pérez-Arlucea M, Sopena A (1986) Estudio sedimentologico del Saxoniense y del Buntsandstein entre Molina de Aragon y Albarracin (Cordillera Iberica). Cuad Geol Ibérica (Madrid) 10: 117-150

Perrodon A (1988) Bassins sédimentaires, provinces pétrolières et tectonique globale. Bull Cent Rech Explor-Prod Elf-Aquitaine 12, 2: 493-512

Peryt TM (ed)(1987a) The Zechstein facies in Europe. Lecture Notes Earth Sci 10, Springer, Berlin Heidelberg New York, 272 p

Peryt TM (ed) (1987b) Evaporite basins. Lecture Notes Earth Sci 13, Springer, Berlin Heidelberg New York, 188 p

Peryt TM, Hoppe A, Bechstädt T, Köster J, Pierre C, Richter DK (1990) Late Proterozoic aragonitic cement crusts, Bambui Group, Minas Gerais, Brazil. Sedimentology 37: 279-286

Pethick J (1984) An introduction to coastal geomorphology. Edward Arnold, London

Pettijohn FJ, Potter PE, Siever R (1987) Sand and sandstone, 2nd edn. Springer, New York Heidelberg, 553 p

Pflug R (1982) Bau und Entwicklung des Oberrheingrabens. Erträge der Forschung, Bd 184, Wissensch Buchgesellschaft, Darmstadt, 145 p

Picard MD, High LRJr (1981) Physical stratigraphy of ancient lacustrine deposits. Soc Econ Paleontol Mineral Spec Publ 31: 233-259

Pickard GL, Emery WJ (1982) Descriptive physical oceanography, 4th edn. Pergamon Press, Oxford, 249 p

Pickering KT, Hiscott RN, Hein FJ (1989) Deep marine environments: clastic sedimentation and tectonics. Unwin Hyman, London, 416 p

Pickering KT, Stow DAV, Watson MP, Hiscott RN (1986) Deep-water facies, processes and models: a review and classification scheme for modern and ancient sediments. Earth Sci Rev 23: 75-174

Pickford M (1986) Sedimentation and fossil preservation in the Nyanza rift system, Kenya. In: Frostick LE, Renaut RW, Reid I, Tiercelin JJ (eds) Sedimentation in African rifts. Geol Soc Spec Publ 25: 345-362

Pierre C (1988) Application of stable isotope geochemistry to the study of evaporites. In: Schreiber BC (ed) Evaporites and hydrocarbons. Columbia Univ Press, New York, p 300-344

Pilkey OH, Locker SD, Cleary WJ (1980) Comparison of sand layer geometry on flat floors of 10 modern depositional basins. Am Assoc Petrol Geol Bull 64: 841-856

Pinet P, Souridu M (1988) Continental erosion and large-scale relief. Tectonics 7: 563-582

Pingitore NE Jr (1982) The role of diffusion during carbonate diagenesis. J Sediment Petrol 52: 27-39

Piper DJW, Normark WR (1982) Effects of the 1929 Grand Banks earthquake on the continental slope off eastern Canada. Geol Surv Canada, Current Res B, Pap 82 - 1B

Piper DJW, Normark WR (1983) Turbidite depositional patterns and flow characteristics, Navy Submarine Fan, California Borderland. Sedimentology 30: 681-694

Piper DJW, Shor AN (1988) The 1929 "Grand Banks" earthquake, slump, and turbidity current. In: Clifton HE (ed) Sedimentologic consequences of convulsive geologic events. Geol Soc Am Spec Pap 229: 77-92

Piper DJW, Stow DAV (1991) Mud turbidites. In: Einsele G, Ricken W, Seilacher A (eds) Cycles and events in stratification. Springer, Heidelberg New York, p 360-376

Pisciotto KA (1981) Distribution, thermal histories, isotopic compositions and reflection characteristics of siliceous rock recovered by the Deep Sea Drilling Project. Soc Econ Paleontol Mineral Spec Publ 32: 129-147

Pitman WC (1978) Relationship between eustacy and stratigraphic sequences of passive margins. Geol Soc Am Bull 89: 1389-1403

Pitman WC III, Andrews JA (1985) Subsidence and thermal history of small pull-apart basins. In: Biddle KT, Christie-Blick N (eds) Strike-slip deformation, basin formation, and sedimentation. Soc Econ Paleontol Mineral Spec Publ 37: 43-49

Pitman WC III, Golovchenko X (1983) The effect of sea level change on the shelf edge and slope of passive margins. In: Stanley DJ, Moore GT (eds) The shelf-break: Critical interface on continental margins. Soc Econ Paleontol Mineralog Spec Publ 33: 41-58

Pitman WC, Golovchenko X (1988) Sea-level changes and their effect on the stratigraphy of Atlantic-type margins. In: Sheridan RE, Grow JA (eds) The Atlantic continental margin, U.S., The Geology of North America, vol 1/2, Geol Soc Am, Boulder, p 429-436

Plint AG (1988) Sharp-based shoreface sequences and "offshore bars" in the Cardium Formation of Alberta: their relationship to relative changes in sea level. In: Wilgus CK, Hastings BS, Posamentier H, Van Wagoner J, Ross CA, Kendall CGSC (eds) Sea-level changes: an integrated approach. Soc Econ Paleontol Mineral Spec Publ 42: 357-370

Poag CW (ed) (1985) Geologic evolution of the United States Atlantic margin. Van Nostrand Reinhold, New York, 369 p

Poag CW, Swift BA, Schlee JS, Ball MM, Sheets LL (1990) Early Cretaceous shelf-edge deltas of the Baltimore Canyon Trough: principal sources for sediment gravity deposits of the northern Hatteras basin. Geology 18: 149-152

Porada H (1989) Pan-African rifting and orogenesis in southern to equatorial Africa and eastern Brazil. Precambrian Res 44: 103-136

Posamentier HW, Vail PR (1988) Eustatic controls on clastic deposition II - Sequence and system tract models. In: Sea-level changes - an integrated approach. Soc Econ Paleontol Mineralog Spec Publ 42: 125-154

Postma G (1986) Classification of sediment gravity-flow deposits based on flow conditions during sedimentation. Geology 14: 291-294

Potter PE, Heling D, Shimp NF, VanWie W (1975) Clay mineralogy of modern alluvial muds of the Mississippi River Basin. Bull Cent Rech Pau-SNPA 9: 353-389

Powell RD, Molnia BF (1989) Glacimarine sedimentary processes, facies and morphology of the south-southeast Alaska shelf and fjords. Marine Geol 85: 359-390

Prahl FG, Muelhausen LA (1989) Lipid biomarkers as geochemical tools for paleoceanographic study. In: Berger WH et al (eds) Productivity of the ocean: past and present. Life Sci Res Rept 44, Wiley, New York, 271-290

Pratt LM (1984) Influence of paleoenvironmental factors on preservation of organic matter in Middle Cretaceous Greenhorn Formation, Pueblo, Colorado. Am Assoc Petrol Geol Bull 68: 1146-1159.

Pretorius DA (1981) Gold and uranium in Quartz-pebble conglomerates. In; Skinner BJ (ed) Economic geology: seventy-fifth anniversary vol, Econ Geology, El Paso, p 107-138

Price RA (ed) (1989) Origin and evolution of sedimentary basins and their energy and mineral resources. Geophs Monograph 48, IUGG vol 3, Int Union Geodesy and Geophsics

Price SP, Scott B (1991) Pliocene Burdur basin, SW Turkey: tectonics, seismicity and sedimentation. J Geol Soc London 148: 345-354

Priesnitz K (1974) Lösungsraten und ihre geomorphologische Relevanz. In: Poser H (ed) Geomorphologische Prozesse und Prozeßkombinationen in der Gegenwart unter verschiedenen Klimabedingungen. Abh Akad Wissensch Göttingen, Math-Physik Klasse, Dritte Folge 29, Vandenhoeck & Ruprecht, Göttingen, p 68-85

Prior DB, Coleman JM (1984) Submarine slope instability. In: Brunsden D, Prior DB (eds) Slope instability. Wiley, New York, 455 p

Puckett TM (1991) Absolute paleobathymetry of Upper Cretaceous chalks based in ostracodes - evidence from the Demopolis Chalk (Campanian and Maastrichtian) of the northern Gulf Coastal Plain. Geology 19: 449-452

Pulham AJ (1989) Controls in internal structure and architecture of sandstone bodies within Upper Carboniferous fluvial-dominated deltas, County Clare, Western Ireland. In: Whateley MKG, Pickering KT (eds) Deltas: sites and traps for fossil fuels. Geol Soc Spec Publ 41: 167-177

Purdy EG (1963) Recent carbonate facies of the Great Bahama Bank II. Sedimentary facies. J Geol 71: 472-497

Purser BH (ed) (1973) The Persian Gulf: Holocene carbonate sedimentation and diagenesis in a shallow epi continental sea. Springer, Berlin Heidelberg New York, 471 p

Purser BH (1980) Sédimentation et diagenèse des carbonates néritiques recents, tome 1. Publ Inst Francais du Petrole, Édition TECHNIP, Paris, 366 p

Purser BH (1985) Coastal evaporite systems. In: Friedman GM, Krumbein WE (eds) Hypersaline ecosystems. Springer, Berlin Heidelberg New York, p 72-102

Purser BH, Schroeder JH (1986) The diagenesis of reefs: a brief review of our present understanding. In: Schroeder JH, Purser BH (eds) Reef diagenesis. Springer, Berlin Heidelberg New York, p 424-446

Purser BH, Soliman M, M'Rabet A (1987) Carbonate, evaporite, siliciclastic transitions in Quaternary rift sediments of the northwestern Red Sea. Sediment Geol 53: 247-267

Pye K (1987) Aeolian dust and dust deposits. Academic Press, London, 334 p

Pye K, Dickson JAD, Schiavon N, Coleman ML, Cox M (1990) Formation of siderite-Mg-calcite-iron sulphide concretions in intertidal marsh and sandflat sediments, north Norfolk, England. Sedimentology 37: 325-343

Quinlan G (1987) Models of subsidence mechanism in intracratonic basins and their applicability to North American examples. In: Beaumont C, Tankard AJ (eds) Sedimentary basins and basin-forming mechanisms. Can Soc Petrol Geol Mem 12: 463-481

Quinlan GM, Beaumont C (1984) Appalachian thrusting, lithosphere flexure, and the Paleozoic stratigraphy of the eastern interior of North America. Can J Earth Sci 21: 973-996

Quinn HA, Cronin TM (1984) Micropaleontology and depositional environment of the Imperial and Palm Spring formations, Imperial Valley, California. In: Rigsby CA (ed) The Imperial Basin - tectonics, sedimentation, and thermal aspects. Soc Econ Paleontol Mineral Pacific Section, p 71-85

Rad U von (1972) Zur Sedimentologie und Fazies des Allgäuer Flysches. Geologica Bavarica (München) 66: 92-147

Rad U von, Botz R (1987) Authigenic Fe-Mn carbonates in Cretaceous and Tertiary sediments of the continental rise off eastern North America, Deep Sea Drilling Project Site 603. In: Hinte JE van, Wise SW Jr, et al., Init Repts DSDP, 92: 1061-1077, US Govt Print Office, Washington

Rad U von, Kudrass R (eds) (1984) Geology of the Catham Rise phosphorite deposits east of New Zealand. Geol Jahrb Reihe D, H 65, Hannover, 252 p

Rad U von, Wissmann G (1982) Cretaceous-Cenozoic history of the West Saharan continental margin (NW Africa): Development, destruction and gravitational sedimentation. In: Rad U von, Hinz K, Sarnthein M, Seibold E (eds) Geology of the Northwest African continental margin. Springer, Berlin Heidelberg New York, p 106-131

Rad U von, Thurow J, Haq BU, Gradstein F, Ludden J, and ODP Leg 122/123 shipboard scientific parties (1989) Triassic to Cenozoic evolution of the NW Australian continental margin and the birth of the Indian Ocean (preliminary results of ODP Legs 122 and 123). Geol Rundsch 78: 1189-1210

Raiswell R (1987) Non-steady state microbiological diagenesis and the origin of concretions and nodular limestones. In: Marshall JD (ed) Diagenesis of sedimentary sequences. Geol Soc Spec Publ 36: 41-54

Raiswell R, Berner RA (1987) Organic carbon losses during burial and thermal maturation of normal marine shales. Geology 15: 853-856

Ramberg H (1981) Gravity, deformation and the Earth's crust, 2nd edn. Academic Press, London, 452 p

Ranganathan V, Hanor JS (1987) A numerical model for the formation of saline waters due to diffusion of dissolved NaCl in subsiding sedimentary basins with evaporites. J Hydrol 92: 97-120

Ranke U, von Rad U, Wissmann G (1982) Stratigraphy, facies and tectonic development of the on- and offshore Aaiun-Tarfaya basin - a review. In: Rad U von, Hinz K, Sarnthein M, Seibold E (eds) Geology of the Northwest African continental margin. Springer, Berlin Heidelberg New York, p 86-105

Rao CP (1981) Criteria for recognition of cold-water carbonate sedimentation: Bermiedale limestone (lower Permian), Tasmania, Australia. J Sediment Petrol 51: 491-506

Raup DM, Jablonski D (eds) (1986) Patterns and processes in the history of life. Dahlem Workshop Reports, Life Sciences Res Repts, 36, Springer, Berlin Heidelberg New York, 447 p

Raymond LA (1984) Classification of melanges. In: Ray mond LA (ed) Melanges: their nature, origin and significance. Geol Soc Am Spec Pap 198: 7-20

Read JF (1982) Carbonate platforms of passive (extensional) continental margins: types, characteristics, and evolution. Tectonophysics 81: 195-212

Read JF (1985) Carbonate platform facies models. Am Assoc Petrol Geol 69: 1-21

Reading HG (1980) Characteristics and recognition of strike-slip fault systems. In: Ballance PL, Reading HG (eds) Sedimentation in oblique-slip mobile zones. Int Assoc Sedimentol Spec Publ 4, Blackwell, Oxford, p 7-26

Reading HG (ed) (1986a) Sedimentary environments and facies. Blackwell, Oxford, 615 p

Reading HG (1986b) African rift tectonics and sedimentation: an introduction. In: Frostick LE, Renaut RW, Reid I, Tiercelin JJ (eds) Sedimentation in African rifts. Geol Soc Spec Publ 25: 3-7, Blackwell, Oxford

Reineck H-E (1980) Sedimentationsbeträge und Jahres schichtung in einem marinen Einbruchsgebiet/Nordsee. Senckenberg Marit 12: 281-309

Reineck H-E (1984) Aktuogeologie klastischer Sedimente. W. Kramer, Frankfurt a.M., 348 p

Reineck H-E, Singh IB (1980) Depositional sedimentary environments, 2nd edn. Springer, Berlin Heidelberg New York, 549 p

Reinson GE (1984) Barrier island and associated strandplain systems. In: Walker RG (ed) Facies models, 2nd edn. Geosci Canada Reprint Ser 1, Geol Soc Canada, p 119-140

Reiss Z, Hottinger L (1984) The gulf of Aqaba: ecological micropaleontology. Ecolog Studies, 50, Springer, Berlin Heidelberg New York, 354 p

Reiss Z, Luz B, Almagi-Labina A, Halicz E, Winter A (1980) Late Quaternary paleoceanography of the Red Sea. Quat Res 14: 294-308

Remane J (1960) Les formations bréchiques dans le Tithonique du sud-est de la France. Trav Lab Géol Fac Sci Grenoble 36: 6-114

Retallack GJ (1990) Soils of the past. Unwin Hyman, Boston, 520 p

Reutter KJ, Giese P, Götze H-J, Scheuber E, Schwab K, Schwarz G, Wigger P (1988) Structural and crustal development of the Central Andes between 21° and 25° S. In: Bahlburg H, Breitkreuz C, Giese P (eds) The southern Central Andes. Lecture Notes Earth Sci 17, Springer, Berlin Heidelberg New York, p 231-261

Ricci-Lucchi F (1975) Depositional cycles in two turbidite formations of northern Apennines (Italy). J Sediment Petrol 45: 3-43

Ricci Lucchi F (1986) The Oligocene to Recent foreland basins of the northern Apennines. In: Allen PA, Homewood P (eds) Foreland basins. Int Assoc Sedimentol Spec Publ 8: 105-139, Blackwell, Oxford

Ricci-Lucchi F (1990) Turbidites in foreland and on-thrust basins of the northern Apennines. Palaeogeogr Palaeoclimatol Palaeoecol 77: 51-66

Ricci Lucchi F, Valmori E (1980) Basin-wide turbidites in a Miocene "over-supplied" deep-sea plain: a geometrical analysis. Sedimentology 27: 241-270

Rice RJ (1980) Rates of erosion in the Little Colorado valley, Arizona. In: Cullingford RA, Davidson DA, Lewin J (eds) Timescales in geomorphology. Wiley-Interscience, Chichester, p 317-331

Richter DK (1985) Die Dolomite der Evaporit- und der Dolcrete-Playasequenz im mittleren Keuper bei Coburg (NE-Bayern). N Jahrb Geol Paläontol Abh 170: 87-128

Richter-Bernburg G (1960) Zeitmessung geologischer Vorgänge nach Warven-Korrelationen im Zechstein. Geol Rundsch 49: 132-149

Richter-Bernburg G (1968) Salzlagerstätten. In Bentz A, Martini HJ (eds) Lehrbuch der Angewandten Geologie, 2, Teil 1, Bornträger, Berlin, p 918-1061

Richter-Bernburg G (1985) Zechstein-Anhydrite. Geol Jahrb (Hannover) A 85, 85 p

Ricken W (1986) Diagenetic bedding: a model for marl-limestone alternations. Lecture Notes Earth Sci 6, Springer, Berlin Heidelberg New York, 210 p

Ricken W (1991) Time span assessment of sequences, beds, and gaps: an overview. In: Einsele G, Ricken W, Seilacher A (1991) Cycles and events in stratigraphy. Springer, Berlin Heidelberg New York, p 773-794

Ricken W, Eder W (1991) Diagenetic overprint in calcareous rocks: modification of stratification and bedding rhythm - an overview. In: Einsele G, Ricken W, Seilacher A (eds) Cycles and events in stratification. Springer, Berlin Heidelberg New York, p 430-449

Riegel W (1991) Coal cyclothems and some models of their origin. In: Einsele G, Ricken W, Seilacher A (eds) Cycles and events in stratigraphy. Springer, Berlin Heidelberg New York, p 733-750

Robaszynski F, Caron M, Dupuis C, Amédro F, Gonzáles Donoso JM, Linares D, Hardenbol J, Gartner S, Calandra F, Deloffre R (1990) A tentative integrated stratigraphy in the Turonian of central Tunisia: Formations, zones and sequential stratigraphy in the Kalaat Senan area. Bull Cent Rech Explor-Prod Elf-Aquitaine, 14: 213-384

Robert P (1985) Histoire géothermique et diagenèse organique. Bull Cent Rech Explor-Prod Elf-Aquitaine Mem 8, Pau, 345 p

Roberts DG, Backman J, Morton AC, Murray JW, Keene JB (1984) Evolution of volcanic rifted margins: synthesis of Leg 81 results on the west margin of Rockall Plateau. Init Rept DSDP 81: 883-911, US Govt Print Office, Washington

Roberts RG, Sheahan PA (eds) (1988) Ore deposit models. Geosci Canada, Reprint Ser 3, Geol Assoc Canada Publ, St. John's Newfoundland, 194 p

Roberts WH III, Cordell RJ (1980) Problems of petroleum migration: introduction. In: Roberts WH III, Cordell RJ (eds) Problems of petroleum migration. Am Assoc Petrol Geol, Stud Geol 10: VI-VIII

Robinson D (1987) Transition from diagenesis to metamorphism in extensional and collision settings. Geology 15: 866-869

Roden GI (1964) Oceanographic aspects of Gulf of California. In: van Andel TH, Shor GG (eds) Marine geology of the Gulf of California. Am Assoc Petrol Geol Mem 3: 30-58

Rodine JD, Johnson AM (1976) The ability of debris, heavily freighted with coarse clastic materials, to flow on gentle slopes. Sedimentology 23: 213-234

Roedder E (1984) Fluid inclusions. Reviews in mineralogy 12, Bookcrafters, Chelsea, Michigan, 644 p

Roehl U (1990) Parallelisierung des norddeutschen oberen Muschelkalks mit dem süddeutschen Hauptmuschelkalk anhand von Sedimentationszyklen. Geol Rundsch 79: 13-26

Roll A (1979) Versuch einer Volumenbilanz des Oberrheintalgrabens und seiner Schultern. Geol Jahrb (Hannover) A 52: 3-82

Romankevich EA (1984) Geochemistry of organic matter in the ocean. Springer, Berlin Heidelberg New York, 334 pp

Rosendahl BR, Reynolds DJ, Lorber PM, Burgess CF, McGill J, Scott D, Lambiase JJ, Derksen SJ (1986) Structural expression of rifting: lessons from Lake Tanganyika, Africa. In: Frostick LE, Renaut RW, Reid I, Tiercelin JJ (eds) Sedimentation in African rifts. Geol Soc Spec Publ 25: 29-43

Rossignol-Strick M (1985) Mediterranean Quaternary sapropels, an immediate response of the African monsoon to variation in insolation. Palaeogeogr Palaeoclimatol Palaeoecol 49: 237-263

Rouchet R de (1978) Éléments d'une théorie géomechanique de la migration de l'huile en phase constituée. Bull Cent Rech Explor-Prod Elf-Aquitaine 2: 337-373

Royden LH (1988) Late Cenozoic tectonics of the Pannonian basin system. In: Royden LH, Horváth F (eds) The Pannonian basin: a study in basin evolution. Am Assoc Petrol Geol Mem 45: 27-48

Royden LH, Horváth F (eds) (1988) The Pannonian basin: a study in basin evolution. Am Assoc Petrol Geol Mem 45, 394 p

Royden LH, Keen CE (1980) Rifting processes and thermal evolution of the continental margin of eastern Canada determined from subsidence curves. Earth Planet Sci Lett 51: 343-361

Royden LH, Horváth F, Burchfiel BC (1982) Transform faulting, extension and subduction in the Carpathian-Pannonian region. Geol Soc Am Bull 93: 717-727

Royden LH, Sclater JG, Von Herzen RP (1980) Continental margin subsidence and heat flow: Important parameters in formation of petroleum hydrocarbons. Am Assoc Petrol Geol Bull 64: 173-187

Ruiz-Ortiz PA, Bustillo MA, Molina JM (1989) Radiolarite sequences of the Subbetic, Betic Cordillera, southern Spain. In: Hein JR, Obradovic J (eds) Siliceous deposits of the Tethys and Pacific regions. Springer, New York Heidelberg, p 107-127

Rupke NA, Stanley DJ (1974) Distinctive properties of turbiditic and hemipelagic mud layers in the Algero-Balearic Basin, Western Mediterranean Sea. Smithonian Contrib Earth Sci, 13, 40 p

Rust BR (1981) Sedimentation in an arid-zone anastomosing fluvial system: Coopers creek, central Australia. J Sediment Petrol 51: 745-755

Rust BR, Koster EH (1984) Coarse alluvial deposits. In: Walker RG (ed) Facies models, 2nd edn. Geosci Canada Reprint Ser 1: 53-69

Rust BR, Legun AS (1983) Modern anastomosing-fluvial deposits in arid central Australia, and a Carboniferous analogue in New Brunswick, Canada. In: Collinson JD, Lewin J (eds) Modern and ancient fluvial systems. Int Assoc Sedimentol Spec Publ 6: 385-392

Rust BR, Nanson GC (1989) Bedload transport of mud as pedogenic aggregates in modern and ancient rivers. Sedimentology 36: 291-306

Ryer TA (1987) Transgressive-regressive cycles and the occurrence of coal in some Upper Cretaceous strata of Utah. Geology 11: 207-210

Sadler PM, Strauss DJ (1990) Estimation of completeness of stratigraphical sections using empirical data and theoretical models. J Geol Soc London 147: 471-485

Sageman BB, Wignall PB, Kauffman EG (1991) Biofacies models of oxygen-deficient facies in epicontinental seas: tool for paleoenvironmental analysis. In: Einsele G, Ricken W, Seilacher A (eds) Cycles and events in stratigraphy. Springer, Berlin Heidelberg New York, p 542-564

Sahagian DL (1987) Epeirogeny and eustatic sea level changes as inferred from Cretaceous shore line deposits: applications to the central and western United States. J Geophys Res 92: 4895-4904

Saito Y (1989a) Modern storm deposits in the inner shelf and their recurrence intervals, Sendai Bay, northeast Japan. In: Taira A, Masuda F (eds) Sedimentary facies in the active plate margin. Terra Scientific Publ, Tokyo, p 331-344

Saito Y (1989b) Storm-built sand ridges on the inner shelf of Kashima-Nada, Northeast Japan. In: Taira A, Masuda F (eds) Sedimentary facies in the active plate margin. Terra Scientific Publ, Tokyo, p 319-330

Salop LJ (1983) Geological evolution of the Earth during the Precambrian. Springer, Berlin Heidelberg New York, 459 p

Sandberg PA (1985) Nonskeletal aragonite and CO_2 in the Phanerozoic. In: Sundquist ET, Broecker WS (eds) The carbon cycle and atmospheric CO_2: natural variations Archean to Present. Am Geophys Union Monogr 32, Washington DC, p 585-554

Sang L, You Z (1988) The metamorphic petrology of the Susong Group and the origin of the Susong phosphorite deposits, Anhui province. Precambrian Res 39: 65-76

Sarg JF (1988) Carbonate sequence stratigraphy. In: Sea-level changes - an integrated approach. Soc Econ Paleontol Mineral Spec Publ 42: 155-181

Sarnthein M (1972) Sediments and history of the postglacial transgression in the Persian Gulf and Gulf of Oman. Marine Geol 12: 245-266

Sarnthein M (1973) Quantitative Daten über benthische Karbonatsedimentation in mittleren Breiten. Veröffentl Univ Innsbruck (Festschrift Heißel) 86: 267-279

Sarnthein M (1978) Sand deserts during glacial maximum and climatic optimum. Nature 272 (5648): 43-46

Sarnthein M, Diester-Haass L (1977) Eolian-sand turbidites. J Sediment Petrol 47: 868-890

Sarnthein M, Mienert J (1986) Sediment waves in the eastern equatorial Atlantic: sediment record during Late Glacial and Interglacial times. In: Summerhayes CP, Shackleton NJ (eds) North Atlantic palaeoceanography. Geol Soc Spec Publ 21: 119-130

Sarnthein M, Walger E (1974) Der aeolische Sandstrom aus der Westsahara zur Atlantikkueste. Geol Rundsch 63: 1065-1087

Sarnthein M, Winn K, Zahn K (1987) Paleoproductivity of oceanic upwelling and the effect on atmospheric CO_2 and climatic change during deglaciation times. In: Berger WH, Labeyrie LD (eds) Abrupt climate change. Reidel, Dordrecht, p 311-337

Sarnthein M, Winn K, Duplessy JC, Fontugne MR (1988) Global variations of surface water productivity in low- and mid-latutudes on CO_2 reservoirs of the deep ocean and atmosphere during the last 21,000 years. Paleoceanography 3: 361-399

Savelli C (1988) Late Oligocene to Recent episodes of magmatism in and around the Tyrrhenian Sea: implications for the processes of opening a young inter-arc basin of intra-orogenic (Mediterranean type). Tectonophysics 146: 163-191

Savrda CE, Bottjer DJ (1987) Trace fossils as indicators of bottom-water redox conditions in ancient sedimentary environments. In: Bottjer DJ (ed) New concepts in the use of biogenic sedimentary structures for paleoenvironmental interpretation. Soc Econ Paleontol Mineral, Pacific Section, Volume and Guidebook 52, p 3-26

Savrda CE, Bottjer DJ (1989) Anatomy and implications of bioturbated beds in "black shale" sequences: examples from the Jurassic Posidonienschiefer (Southern Germany). Palaios 4: 330-342

Savrda CE, Bottjer DJ, Seilacher A (1991) Redox-related benthic events. In: Einsele G, Ricken W, Seilacher A (eds) Cycles and events in stratigraphy. Springer, Heidelberg New York, p 524-541

Sawyer DS (1988) Thermal evolution. In: Sheridan RE, Grow JA (eds) The Geology of North America, vol 1/2, The Atlantic continental margin. US Geol Soc Am, Boulder, p 417-428

Sawyer DS, Hsui AT, Toksöz MN (1987) Extension, subsidence and thermal evolution of the Los Angeles basin - a two-dimensional model. Tectonophysics 133: 15-32

Sawyer DS, Swift BA, Sclater JG. Tolsöz MN (1982) Extensional model for the subsidence of the northern United States Atlantic continental margin. Geology 10: 134-140

Saxov S, Nieuwenhuis JK (eds)(1982) Marine slides and other mass movements. NATO Conference Series IV, Marine Sciences, Plenum Press, New York London, 353 p

Schaal S, Ziegler W (eds) (1988) Messel - ein Schaufenster in die Geschichte der Erde und des Lebens. Kramer, Frankfurt a.M., 315 p

Scherer M (1987) Parameters influencing porosity in sandstones: a model for sandstone porosity prediction. Am Assoc Petrol Geol Bull 71: 485-491

Schidlowski M (1987) Application of stable carbon isotopes to early biochemical evolution on Earth. Annu Rev Earth Planet Sci 15: 47-72

Schidlowski M, Eichmann R, Junge CE (1975) Precambrian sedimentary carbonates: carbon and oxygen isotope geochemistry and implications for the terrestrial oxygen budget. Precambrian Res 2: 1-69

Schidlowski M, Hays JM, Kaplan JR (1984) Isotopic inferences of ancient biochemistry - carbon, sulfur, hydrogen, and nitrogen. In: Schopf JW (ed) Earth's earliest biosphere. Princeton Univ Press, Princeton, New Jersey, p 149-183

Schindler C (1974) Zur Geologie des Zürichsees. Eclogae Geol Helv 67: 163-196

Schindler E (1990) The late Frasnian (upper Devonian) Kellwasser crisis. In: Kauffman EG, Walliser OH (eds) Extinction events in earth history. Lecture Notes Earth Sci 30, Springer, Berlin Heidelberg New York, p 151-159

Schlager W (1981) The paradox of drowned reefs and carbonate platforms. Geol Soc Am Bull 92: 197-211

Schlager W, Camber O (1986) Submarine slope angles, drowning unconformities, and shelf-erosion on limestone escarpments. Geology 14: 762-765

Schlager W, Philip J (1990) Cretaceous carbonate platforms. In: Ginsburg RN, Beaudoin B (eds) Cretaceous resources, events and rhythms. NATO ASI Ser C 104, Kluwer, Dordrecht, p 173-195

Schlager W, Bourgeois F, Mackenzie G, Smit J (1988) Boreholes at Great Isaac and Site 626 and the history of the Florida Straits. In: Austin JA, Schlager W et al., Proc ODP, Sci Results 101: 425-437, College Station TX (Ocean Drilling Project)

Schlanger SO, Cita MB (eds) (1982) Nature and origin of Cretaceous carbon-rich facies. Academic Press, London, 229 p

Schlanger SO, Douglas RG (1974) Pelagic ooze-chalk limestone transition and its implications for marine stratigraphy. In: Hsü KJ, Jenkyns HC (eds) Pelagic sediments on land and under the sea. Int Assoc Sedimentol Spec Publ 1: 117-148

Schlanger SO, Premoli Silva I (1986) Oligocene sea-level falls recorded in mid-Pacific atoll and archipelagic apron settings. Geology 14: 392-395

Schlanger SO, Jenkyns HC, Premoli-Silva I (1981) Volcanism and vertical tectonics in the Pacific basin related to global Cretaceous transgressions. Earth Planet Sci Lett 52: 435-449

Schlee JS (ed) (1984) Interregional unconformities and hydrocarbon accumulation. Am Assoc Petrol Geol Mem 36, 184 p

Schlische RW, Olsen PE (1990) Quantitative filling model for continental extensional basins with applications to early Mesozoic rifts of eastern North America. J Geol 98: 135-155

Schmalz R (1969) Deep-water evaporite deposition: a genetic model. Am Assoc Petrol Geol Bull 53: 798-823

Schmidt V, McDonald DA (1979) The role of secondary porosity in the course of sandstone diagenesis. In: Scholle PA, Schluger PR (eds) Aspects of diagenesis. Soc Econ Paleontol Mineralog Spec Publ 26: 175-207

Schmidt-Witte H, Einsele G (1986) Rezenter und holozäner Stoffaustrag aus den Keuper-Lias-Einzugsgebieten des Naturparks Schönbuch. In: Einsele G (ed) Das landschaftsökologische Forschungsprojekt Naturpark Schönbuch. VCH-Verlag Weinheim, p 369-391

Schmincke HU (1988) Pyroklastische Gesteine. In: Füchtbauer H (ed) Sedimente und Sedimentgesteine. Schweizerbart, Stuttgart, p 731-778

Schmincke H-U, Bogaard P van den (1991) Tephra layers and tephra events. In: Einsele G, Ricken W, Seilacher A (eds) Cycles and events in stratigraphy. Springer, Heidelberg New York, p 392-429

Schneider HJ, Walter HW (1988) Erzlagerstätten in Sedimenten. In: Füchtbauer H (ed) Sedimente und Sedimentgesteine. Schweizerbart, Stuttgart, p 569-601

Schneider R, Wefer G (1990) Shell horizons in Cenozoic upwelling-facies sediments off Peru: distribution and mollusk fauna in cores from Leg 112. In: Suess E, von Huene R, et al., Proc ODP, Sci Results 112: 335-352, College Station, TX (Ocean Drilling Program)

Scholle PA, Spearing DR (eds) (1982) Sandstone depositional environments. Am Assoc Petrol Geol Mem 31, 410 p

Scholle PA, Bebout DG, Moore CH (eds) (1983a) Carbonate depositional environments. Am Assoc Petrol Geol Mem 33: 345-440

Scholle PA, Arthur MA, Ekdale AA (1983b) Pelagic environment. In: Scholle PA, Bebout DG, Moore CH (eds) Carbonate depositional environments. Am Assoc Petrol Geol Mem 33: 619-691

Scholle PA, James NP, Read JF (eds)(1989) Carbonate sedimentology and petrology. Short Course in Geology, 4, Am Geophys Union, Washington DC, 160 p

Schönenberg R, Neugebauer J (1987) Einführung in die Geologie Europas, 5th edn. Rombach, Freiburg, 294 p

Schopf JW (ed) (1983) Earth's earliest biosphere. Princeton Univ Press, Princeton

Schopf JW, Packer BM (1987) Early Archean (3.3-billion to 3.5-billion-year-old) microfossils from Warrawoona Group, Western Australia. Science 237: 70-73

Schopf TJM (1980) Paleoceanography. Harvard Univ Press, Cambridge, Mass, 341 p

Schreiber HC (1986) Arid shorelines and evaporites. In: Reading HG (ed) Sedimentary facies and environment, 2nd edn, Blackwell, Oxford, p 189-228

Schreiber BC (ed) (1988a) Evaporites and hydrocarbons. Columbia Univ Press, New York, 475 p

Schreiber BC (1988b) Subaqueous evaporite deposition. In: Schreiber BC (ed) Evaporites and hydrocarbons. Columbia Univ Press, New York, p 182-255

Schreiber BC, Hsü KJ (1980) Evaporites. In: Hobson GD (ed) Developments in petroleum geology 2, Applied Science Publishers, London, p 87-138

Schröder B (1982) Entwicklung des Sedimentbeckens und Stratigraphie der klassischen Germanischen Trias. Geol Rundsch 71: 783-794

Schroeder JH (1988) Spatial variations in the porosity development of carbonate cements and rocks. Facies (Erlangen) 18: 181-204

Schroeder JH, Purser BH (1986) Reef diagenesis. Springer, Berlin Heidelberg New York, 455 pp

Schumm SA (1981) Evolution and response of the fluvial system, sedimentologic implications. In: Ethridge FG, Flores RM (eds) Recent and ancient nonmarine depositional systems: models for exploration. Soc Econ Paleontol Mineralog Spec Publ 31: 19-29

Schwab FL (1986) Sedimentary "signatures" of foreland basin assemblages: real or counterfeit? In: Allen PA, Homewood P (eds) Foreland basins. Int Assoc Sedimentol Spec Publ 8, Blackwell, Oxford, p 395-410

Schwartz FW, Longstaffe FJ (1988) Ground water and clastic diagenesis. In: Back W, Rosenhein JS, Seaber PR (eds) Hydrogeology. The geology of North America, vol 0-2, Geol Soc Am, Boulder, Colorado, p 413-434

Schwarz HU (1975) Sedimentary structures and facies analysis of shallow-marine carbonates. Contributions to sedimentology 3, Schweizerbart, Stuttgart, 100 p

Schwarz HU (1982) Subaqueous slope failures - experiments and modern occurrences. Contributions to sedimentology 11, Schweizerbart, Stuttgart, 116 p

Schwarz HU, Einsele G, Herm D (1975) Quartz-sandy, grazing-conoured stromatolites from coastal embayments of Mauritania. West Africa. Sedimentology 22: 539-561

Schwarzacher W (1975) Sedimentation models and quantitative stratigraphy. Developm Sedimentol 19, Elsevier, Amsterdam, 382 p

Schwarzacher W (1987) Astronomically controlled cycles in the lower Tertiary of Gubbio (Italy). Earth Planet Sci Lett 84: 22-26

Schwarzacher W, Fischer AG (1982) Limestone-shale bedding and perturbations of the earth's orbit. In: Einsele G, Seilacher A (eds) Cyclic and event stratification. Springer, Berlin Heidelberg New York, p 72-95

Schwarzbach M (1974) Das Klima der Vorzeit, 3rd edn. Enke, Stuttgart, 380 p

Schweller WJ, Kulm LD (1978) Depositional patterns and channelized sedimentation in active eastern Pacific trenches. In: Stanley DJ, Kelling G (eds) Sedimentation in submarine canyons, fans and trenches. Hutchinson & Ross, Dowden, p 311-324

Sclater JG, Christie PAF (1980) Continental stretching: an explanation of the post Mid-Cretaceous subsidence of the central North Sea basin. J Geophys Res 85: 3711-3739

Sclater JG, Shorey MD (1989) Mid-Jurassic through mid-Cretaceous extension in the Central Graben of the North Sea - part 2: estimates from faulting observed on seismic lines. Basin Res 1: 201-215

Scoffin TP (1987) An introduction to carbonate sediments and rocks. Blackie, Glasgow London, 274 p

Scruton PC (1960) Delta building and the deltaic sequence. In: Shepard FP, Phleger FB, Van Andel TH (eds) Recent sediments, northwest Gulf of Mexico. Am Assoc Petrol Geol, Symp vol, Tulsa, p 82-102

Seibold E (1952) Chemische Untersuchungen zur Bankung im unteren Malm Schwabens. Neues Jahrb Geol Paläontol (Stuttgart) Abh 95: 337-370

Seibold E (1970) Nebenmeere im humiden und ariden Klimabereich. Geol Rundsch 60: 73-105

Seibold E (1978) Deep sea manganese nodules: the challenge since "Challenger". Episodes 4: 3-8

Seibold E, Berger WH (1982) The sea floor. Springer, Berlin Heidelberg New York, 288 p

Seibold E, Exon N, Hartmann M, Kögler FC, Krumm GF, Lutze GF, Newton RS, Werner F (1971) Marine Geology of Kiel Bay. VIIIth Int Sedimentol Congress, Heidelberg, Sedimentology of parts of Central Europe, Guidebook, p 209-235

Seilacher A (1962) Paleontological studies on turbidite sedimentation and erosion. J Geol 70: 227-234

Seilacher A (1991) Events and their signature - an overview. In: Einsele G, Ricken W, Seilacher A (eds) Cycles and events in stratigraphy. Springer, Heidelberg New York, p 222-226

Selby MJ (1974) Rates of denudation. New Zealand J Geogr 56: 1-13

Selley RC (1976) An Introduction to sedimentology. Academic Press, London, 408 p

Selley RC (1985a) Ancient sedimentary environments and their subsurface diagenesis, 3rd edn. Chapman & Hall, London, 317 p

Selley RC (1985b) Elements of petroleum geology. Freeman, New York, 449 p

Sellwood BW (1986) Shallow-marine carbonate environments. In: Reading HG (ed) Sedimentary environment and facies, 2nd edn. Blackwell, Oxford, p 283-342

Selverstone J (1985) Petrologic constraints on imbrication, metamorphism, and uplift in the SW Tauern Window, Eastern Alps. Tectonics 4: 687-704

Sengör AMC, Görür N, Saroglu F (1985) Strike-slip faulting and related basin formation in zones of tectonic escape: Turkey as a case study. In: Biddle KT, Christie-Blick N (eds) Strike-slip deformation, basin formation, and sedimentation. Soc Econ Paleontol Mineral Spec Publ 37: 227-264

Sepkoski JJ Jr (1982) Flat-pebble conglomerates, storm deposits, and the Cambrian bottom fauna. In: Einsele G, Seilacher A (eds) Cyclic and event stratification, Springer, Heidelberg New York, p 371-385

Sepkoski JJ Jr (1989) Periodicity in extinction and the problem of catastrophism in the history of life. J Geol Soc London 146: 7-19

Sepkoski JJ Jr, Bambach RK, Droser ML (1991) Secular changes in Phanerozoic event bedding and the biologic overprint. In: Einsele G, Ricken W, Seilacher A (eds) Cycles and events in stratigraphy. Springer, Heidelberg New York, p 298-312

Sestini G (1989) Nile Delta: a review of depositional environments and geological history. In: Whateley MKG, Pickering KT (eds) Deltas; sites and traps for fossil fuels. Geol Soc Spec Publ 41: 99-127

Shanmugam G, Moiola RJ (1985) Submarine fan models: problems and solutions. In: Bouma AH, Normark WR, Barnes NE (eds) Submarine fans and related turbidite systems. Springer, New York Heidelberg, p 29-34

Shaver RH (1977) Silurian reef geometry - new dimensions to explore. J Sediment Petrol 47: 1409-1424

Shaw J (1985) Subglacial and ice marginal environments. In: Ashley GM, Shaw J and Smith ND (eds) Glacial sedimentary environments. SEPM Short Course No 16, p 7-84

Shearman DJ (1978) Evaporites of coastal sabkhas. Soc Econ Paleontol Mineral, Short Course 4: 6-42

Shearman DJ, Smith AJ (1985) Ikaite, the parent mineral of jarrowite-type pseudomorphs. Proc Geol Assoc London 96: 305-314

Sheldon RP (1981) Ancient marine phosphorites. Annu Rev Earth Planet Sci 9: 251-284

Shepard FP (1973) Submarine geology, 3rd edn. Harper and Row, New York, 551 p

Shepard FP, Dill RF (1966) Submarine canyons and other sea valleys. Rand McNally, Chicago, 381 p

Sheppard RA, Gude 3rd AJ (1986) Magadi-type chert - a distinctive diagenetic variety from lacustrine deposits. In: Mumpton FA (ed) Studies in diagenesis. US Geol Surv Bull 1578: 335-345

Sheridan MF (1979) Emplacement of pyroclastic flows: a review. In: Chapin CE, Elston WE (eds) Ash-fall tuffs. Geol Soc Am Spec Pap 180: 125-136

Sheridan RE (1981) Recent research on passive continental margins. In: Warme JE, Douglas RG, Winterer EL (eds) The Deep Sea Drilling Project: a decade of progress. Soc Econ Paleontol Mineral Spec Publ 32: 39-55

Sheridan RE (1987) Pulsation tectonics as the control of long-term stratigraphic cycles. Paleoceanography 2: 97-118

Sheridan RE, Enos P (1979) Stratigraphic evolution of the Blake Plateau after a decade of scientific drilling. In: Talwani M, Hay W, Ryan WBF (1979) Deep drilling results in the Atlantic Ocean: continental margins and paleoenvironment. Maurice Ewing Ser 3, Am Geophys Union, Washington DC, p 109-122

Sheridan RE, Grow JA (eds) (1988) The Geology of North America, vol 1-2, The Atlantic continental margin. US Geol Soc Am, Boulder

Shinn EA (1983) Tidal flat environment. In: Scholle PA, Bebout DG, Moore CH (eds) Carbonate depositional environments. Am Assoc Petrol Geol Mem 33: 171-210

Shinn AE, Steinen RP, Lidz BH, Swart PK (1989) Whitings, a sedimentologic dilemma. J Sediment Petrol 59: 147-161

Shipboard Scientific Party (1988a) Site 704. In: Gieselski PF, Kristoffersen Y et al. Proc ODP, Init Repts 114: 621-795, College Station, TX (Ocean Drilling Project)

Shipboard Scientific Party (1988b) Synthesis of shipboard results: Leg 110 transect of the northern Barbados Ridge. In: Mascle A, Moore JC, et al., Proc Init Repts (Pt A), ODP 110: 577-587, College Station, TX (Ocean Drilling Project)

Shipboard Scientific Party (1988c) Introduction, objectives, and principal results, Leg 112, Peru contiental margin. In: Suess E, von Huene R, et al., Proc ODP, Init Repts 112: 5-22, College Station, TX (Ocean Drilling Project)

Shoemaker EM (1984) Large body impacts through geologic time. In: Holland HD, Trendall AF (eds) Patterns of change in Earth evolution, Springer, Berlin Heidelberg New York, p 15-40

Shurr GW (1984) Geometry of shelf sandstone bodies in the Shannon sandstone of southeastern Montana. In: Tillman RW, Siemers CT (eds) Siliciclastic shelf sediments. Soc Econ Paleontol Mineral Spec Publ 34:63-83

Siehl A, Thein J (1978) Geochemische Trends in der Minette (Jura, Luxembourg, Lothringen). Geol Rundsch 67: 1054-1977

Simm RW, Kidd RB (1984) Submarine debris flow deposits detected by large-range side-scan sonar 1000 km from source. Geo-Marine Lett 3/1:13-16

Simoneit BRT, Stuermer DH (1982) Organic geochemical indicators for sources of organic matter and paleoenvironmental conditions in Cretaceous oceans. In: Schlanger SO, Cita MB (eds) Nature and origin of Cretaceous carbon-rich facies. Academic Press, London, p 145-163

Simonson B (1985) Sedimentological constraints on the origin of Precambrian iron formations. Geol Soc Am Bull 96: 244-252

Sirocko F, Sarnthein M (1989) Wind-borne deposits in the northwestern Indian Ocean: record of Holocene sediments versus modern satellite data. In: Leinen M, Sarnthein M (eds) Paleoclimatology and paleometeorology: modern and past pattarns of global atmospheric transport. Kluwer, Dordrecht, p 401-433

Skempton AW (1970) The consolidation of clays by gravitational compaction. Quat J Geol Soc London 125: 373-412

Slansky M (1980) Géologie des phosphates sédimentaires. Mém BRGM 114, 94 p

Sloan LC, Barron EJ (1990) Equable climates during Earth history? Geology 18: 489-492

Smith DB (1980) The evolution of the English Zechstein basin. In: Füchtbauer H, Peryt T (eds) The Zechstein basin with emphasis on carbonate sequences. Contributions to sedimentology 9. Schweizerbart, Stuttgart, p 7-37

Smith DG (1983) Anastomosed fluvial deposits, modern examples from Western Canada. In: Collinson JD, Lewin J (eds) Modern and ancient fluvial systems. Int Assoc Sedimentol Spec Publ 6: 155-168

Smith GA (1988) Sedimentology of proximal to distal volcaniclastics dispersed across an active fold belt: Ellensburg Formation (late Miocene), central Washington. Sedimentology 35: 953-977

Smith KLJr, Carlucci AF, Jahnke RA, Craven DB (1987) Organic carbon mineralization in the Santa Catalina Basin: benthic boundary layer metabolism. Deep-Sea Res 34: 185-211

Smith L, Forster C, Woodbury A (1989) Numerical simulation techniques for modeling advectively disturbed thermal regimes. In: Beck AE, Garven G, Stegena L (eds) Hydrogeological regimes and their subsurface thermal effects. Geophys Monogr 47, IUGG vol 2, Am Geophys Union, Washington DC, p 1-5

Smith ND, Ashley GM (1985) Proglacial lacustrine environment. In: Ashley GM, Shaw J, Smith ND (eds) Glacial sedimentary environments. SEPM Short Course 16, p 135-216

Smoot JP (1991) Sedimentary facies and depositional environments of early Mesozoic Newark Supergroup basins, eastern North America. Palaeogeogr Palaeoclimatol Palaeoecol 84: 369-423

Smosna R (1989) Compaction law for Cretaceous sandstones of Alaskas north slope. J Sediment Petrol 59: 572-584

Snedden JW, Nummedal D, Amos AF (1988) Storm- and fair-weather combined flow on the central Texas continental shelf. J Sediment Petrol 58: 580-595

Solle G (1966) Rezente und fossile Wüste. Notizbl Hess Landesamt Bodenforsch (Wiesbaden) 94: 54-121

Song T, Gao J (1985) Tidal sedimentary structures from upper Precambrian rocks of the Ming tombs district, Beijing (Peking), China. Precambrian Res 29: 93-107

Sonnenfeld P (1984) Brines and evaporites. Academic Press, Orlando, 613 p

Sonnenfeld P, Perthuisot J-P (1989) Brines and evaporites. Short course in geology 3, Am Geophys Union, Washington DC, 126 p

Soudry D, Lewy Z (1988) Microbially influenced formation of phosphate nodules and megafollil moulds (Negev, southern Israel). Palaeogeogr Palaeoclimatol Palaeoecol 64: 15-33

Souquet P, Eschard R, Lods H (1987) Facies sequences in large-volume debris- and turbidity-flow deposits from the pyrenees (Cretaceous; France, Spain). Geo-Marine Lett 7: 83-90

Southard JB, Lambie JM, Federico DC, Pile HT, Weidman CR (1990) Experiments on bed configurations in fine sands under bidirectional, purely oscillatory flow, and the origin of hummocky cross-stratification. J Sediment Petrol 60: 1-17

Southgate PN (1986) Cambrian phoscrete profiles, coated grains, and microbial processes in phosphogenesis: Georgiy basin, Australia. J Sediment Petrol 56: 429-441

Southgate PN (1989) Relationship between cyclicity and stromatolite form in the late Proterozoic Bitter Springs Formation, Australia. Sedimentology 36: 323-339

Stach E, Mackowsky M-TH, Teichmüller M, Taylor GH, Chandra D, Teichmüller R (1982) Coal petrology, 3rd edn. Bornträger, Berlin Stuttgart, 535 p

Stallard RF (1985) River chemistry, geology, geomorphology, and soils in the Amazon and Orinoco basins. In: Drever JI (ed) The chemistry of weathering. NATO ASI Series C, Math Phys Sci, Reidel, Dordrecht, p 293-316

Stanley DJ (ed) (1972) The Mediterranean Sea: a natural sedimentation laboratory. Dowden, Hutchinson & Ross, Stroudsburg, 765 p

Stanley DJ (1977) Post-Miocene depositional patterns and structural displacement in the Mediterranean. In: Kanes WH, Stehli FG (eds) The ocean basins and margins, vol 4 A: The eastern Mediterranean. Plenum, New York London, p 77-150

Stanley DJ (1982) Welded slump-graded sand couplets: evidence for slide generated turbidity currents. Geo-Marine Lett 2: 149-155

Stanley DJ (1985) Mud depositional processes as a major influence on Mediterranean margin-basin sedimentation. In: Stanley DJ, Wezel FC (eds) Geological Evolution of the Mediterranean Basin. Springer, New York Heidelberg, p 377-410

Stanley DJ (1986) Turbidity current transport of organic-rich sediments: Alpine and Mediterranean examples. Marine Geol 70: 85-101

Stanley DJ (1988a) Turbidites reworked by bottom currents: Upper Cretaceous examples from St. Croix, U.S. Virgin Islands. Smithonian Contrib Marine Sci 33, Washington DC, 79 p

Stanley DJ (1988b) Subsidence in the northeastern Nile delta: rapid rates, possible causes, and consequences. Sci 240: 497-500

Stanley DJ, Blanpied C (1980) Late Quaternary water exchange between the eastern Mediterranean and the Black Sea. Nature 285 (5766): 537-541

Stanley DJ, Kelling G (eds)(1978) Sedimentation in submarine canyons, fans, and trenches. Dowden, Hutchinson and Ross, Stroudsburg, 382 p

Stanley DJ, Wezel FC (eds) (1985) Geological evolution of the Mediterranean basin. Springer, New York Heidelberg

Stanton RJ, Flügel E (1989) Problems with reef models: the late Triassic Steinplatte "reef" (northern Alps, Salzburg/Tyrol, Austria). Facies (Erlangen) 20: 1-138

Stearly RF, Ekdale AA (1989) Modern marine bioerosion by macroinvertebrates, northern Gulf of California. Palaios 4: 453-467

Steckler MS, Watts AB, Thorne JA (1988) Subsidence and basin modeling at the U.S. Atlantic passive margin. In: Sheridan RE, Grow JA (eds) The geology of North America, vol 1/2, The Atlantic continental margin. US Geol Soc Am, Boulder, p 399-415

Steckler MS, Reynolds DJ, Coakley BJ, Swift BA, Jarrard RD (1990) Dynamics of sedimentary sequence development: The implications of flexural, isostacy, compaction and sediment dynamics for the measurement of sea level change. 13th Intern Sedimentol Congress, Aug 1990, Nottingham, Int Assoc Sedimentol, Abstracts of Papers, p 523-524

Steel R, Gloppen TG (1980) Late Caledonian (Devonian) basin formation, western Norway: signs of strike-slip mobile zones. In: Ballance PL, Reading HG (1980) Sedimentation in oblique-slip mobile zones. Int Assoc Sedimentol Spec Publ 4, Blackwell, Oxford, p 79-103

Stein R (1986) Organic carbon and sedimentation rate - further evidence for anoxic deep water conditions in the Cenomanian/Turonian Atlantic Ocean. Marine Geol 72: 199-209

Stein R (1991) Accumulation of organic carbon in marine sediments. Lecture Notes Earth Sci 34, Springer, Berlin Heidelberg New York, 217 p

Stein R, Rullkötter J, Welte D (1986) Accumulation of organic-carbon-rich sediments in the Late Jurassic and Cretaceous Atlantic Ocean - a synthesis. Chemical Geol 56: 1-32

Stein R, Sarnthein M, Suendermann J (1986) Late Neogene submarine erosion events along the north-east Atlantic continental margin. In: Summerhays CP, Shackleton NJ (eds) North Atlantic palaeogeography. Geol Soc Spec Publ 21: 103-118

Stein R, Rullkötter J, Welte D (1989) Changes in paleoenvironments in the Atlantic Ocean during Cretaceous times: results from black shale studies. Geol Rundsch 78: 883-901

Steininger FF, Rögl F (1984) Paleogeography and palinspastic reconstruction of the Neogene of the Mediterranean and the Paratethys. In: Dixon JE, Robertson AHF (eds) The geological evolution of the eastern Mediterranean. Geol Soc Spec Publ 17: 659-668

Stets J, Wurster P (1982) Atlas and Atlantic - structural relations. In: Rad U von, Hinz K, Sarnthein M, Seibold E (eds) Geology of the Northwest African continental margin. Springer, Berlin Heidelberg New York, p 69-85

Stockmal GS, Beaumont C (1987) Geodynamic models of convergent margin tectonics: the southern Canadian Cordillera and the Swiss Alps. In: Beaumont C, Tankard AJ (eds) Sedimentary basins and basin forming mechanisms. Can Soc Petrol Geol Mem 12: 393-411

Stockmal GS, Beaumont C, Boutilier R (1986) Geodynamic models of convergent margin tectonics, transition from rigid margin to overthrust belt, and consequences for foreland basin development. Bull Am Assoc Petrol Geol 70: 181-190

Stoffers P, Kühn R (1974) Red Sea evaporites: a petrographic and geochemical study. In: Whitmarsh RB, Weser OE, Ross DA et al, Init Repts DSDP 23, US Govt Print Office, Washington, p 821-865

Stoffers P, Ross DA (1974) Sedimentary history of the Red Sea. In: Whitmarsh RB, Weser OE, Ross DA et al, Init Repts DSDP 23, US Govt Print Office, Washington, p 849-865

Stoffers P, Ross DA (1979) Late Pleistocene and Holocene sedimentation in the Persian Gulf - Gulf of Oman. Sediment Geol 23: 181-208

Stoneley R (1981) Petroleum: the sedimentary basin. In: Tarling DH (ed) Economic geology and geotectonics. Wiley, p 51-71, or in: Foster NH, Beaumont EA (eds, 1987) Geologic basins I: classification, modeling, and predictive stratigraphy. Treatise of petroleum geology, Reprint Ser 1, Am Assoc Petrol Geol, Tulsa, p 102-122

Stoneley R (1987) A review of petroleum source rocks in parts of the Middle East. In: Brooks J, Fleet AJ (eds) Marine petroleum source rocks. Geol Soc Spec Publ 26: 263-269

Stow DAV (1980) A physical model for the transport and sorting of fine-grained sediment by turbidity currents. Sedimentology 27: 31-46

Stow DAV (1981) Laurentian fan: morphology, sediments, processes, and growth pattern. Am Assoc Petrol Geol Bull 65 I: 375-393

Stow DAV (1984) Anatomy of debris-flow deposits. In: Hay WW, Sibuet JC et al, Initial Repts DSDP 75, US Govt Printing Office, Washington, p 801-807

Stow DAV (1986) Deep clastic seas. In: Reading HG (ed) Sedimentary Environments and Facies, 2nd edn. Blackwell, Oxford, p 399-444

Stow DAV, Cochran JR, and ODP Shipboard Scientific Party (1989) The Bengal Fan: some preliminary results from ODP drilling. Geo-Marine Lett 9: 1-10

Stow DAV, Holbrook JA (1984) North Atlantic contourites: an overview. In: Stow DAV, Piper DJW (eds) Fine-grained sediments: deep-water processes and facies. Geol Soc Spec Publ 15: 245-256

Stow DAV, Piper DJW (eds) (1984) Fine-grained sediments: deep-water processes and facies. Geol Soc Spec Publ 15, Blackwell, Oxford, 659 p

Stow DAV, Howell DG, Nelson CH (1985) Sedimentary, tectonic, and sea-level controls. In: Bouma AH, Normark WR, Barnes NE (eds) Submarine fans and related turbidite systems. Springer, New York Heidelberg, p 15-34

Stowe K (1979) Ocean science. Wiley, New York, 610 p

Strasser A (1988) Shallowing-upward sequences in Purbeckian peritidal carbonates (lowermost Cretaceous, Swiss and French Jura Mountains). Sedimentology 35: 369-383

Strasser A (1991) Lagoonal-peritidal sequences in carbonate environments: autocyclic and allocyclic processes. In: Einsele G, Ricken W, Seilacher A (eds) Cycles and events in stratigraphy. Springer, Heidelberg New York, p 709-721

Strobel J, Cannon R, Kendall CGSC, Biswas G, Bezdek J (1989) Interactive (SEDPAK) simulation of clastic and carbonate sediments in shelf to basin settings. Computers & Geosci 15: 1279-1290

Stubblefield WL, McGrail DW, Kersey DG (1984) Recognition of transgressive and post-transgressive sand ridges on the New Jersey continental shelf. In: Tillman RW and Siemers CT (eds) Siliciclastic shelf sediments. Soc Econ Paleontol Mineral Spec Publ 34:1-23

Sturm M, Matter A (1978) Turbidites and varves in Lake Brienz (Switzerland): deposition of clastic detritus by density currents. In: Matter A, Tucker ME (eds) Modern and ancient lake sediments. Int Assoc Sedimentol Spec Publ 2: 147-168

Subramanian V (1985) Geochemistry of river basins. Part I: water chemistry, chemical erosion and water-mineral equilibria. Mitt Geol-Paläontol Inst Univ Hamburg, SCOPE/UNEP Sonderb 58: 495-512

Suess E (1980) Particulate organic carbon flux in the oceans: surface productivity and oxygen utilization. Nature 288: 260-263

Suess E, Whiticar MJ (1989) Methane-derived CO_2 in pore fluids expelled from the Oregon subduction zone. Palaeogeogr Palaeoclimatol Palaeoecol 71: 119-136

Suess E, Balzer W, Hesse K-F, Müller PJ, Ungerer CA, Wefer G (1982) Calcium carbonate hexahydrate from organic-rich sediments of the Antarctic shelf: precursors of glendonites. Science 216: 1128-1131

608

Suess E, Kulm LD, Killingley JS (1987) Coastal upwelling and a history of organic-rich mudstone deposition off Peru. In: Brooks J, Fleet AJ (eds) Marine petroleum source rocks. Geol Soc Spec Publ 26: 181-197

Sullivan R (1985) Origin of lacustrine rocks of Wilkins Peak Member, Wyoming. Am Assoc Petrol Geol Bull 69: 913-922

Summerfield MA (1991) Subaerial denudation of passive margins: regional elevation versus local relied models. Earth Planet Sci Lett 102: 460-469

Summerhayes CP (1983) Sedimentation of organic matter in upwelling regimes. In: Thiede J, Suess E (eds) Coastal upwelling - its sediment record. Part B, NATO Conference Ser 4: Marine Sci 10, Plenum Press, New York London, p 29-72

Sun Shuncai (1988) Lakes in China and Chinese lacustrine sedimentology - a brief survey. Mitt Geol-Paläontol Inst, Univ Hamburg, SCOPE/UNEP Sonderb 66: 165-175

Suppe J (1985) Principles of structural geology. Prentice-Hall, Englewood Cliffs, 537 p

Surlyk F, Clemmensen LB, Larsen HC (1981) Post-Paleozoic evolution of the East Greenland continental margin. In: Kerr JW, Fergusson AJ (eds) Geology of the North Atlantic borderlands. Can Soc Petrol Geol Mem 7: 611-646

Suttner LJ, Dutta PK (1986) Alluvial sandstone composition and paleoclimate, I. Framework mineralogy. J Sediment Petrol 56: 329-345

Svendsen JI, Mangerud J, Miller GH (1989) Denudation rates in the Arctic estimated from lake sediments on Spitsbergen, Svalbard. Palaeogeogr Palaeoclimatol Palaeoecol 76: 153-168

Sverjensky DA (1986) Genesis of Mississippi Valley-type lead-zinc deposits. Annu Rev Earth Planet Sci 14: 177-199

Swift DJP (1985) Response of the shelf floor to flow. In: Tillman RW, Swift DJP, Walker RG (eds) Shelf sands and sandstone reservoirs. Soc Econ Paleontol Mineral Short Course 13: 135-241

Swift DJP, Hudelson PM, Brenner RL, Thompson P (1987) Shelf construction in a foreland basin: storm beds, shelf sandbodies, and shelf-slope depositional sequences in the Upper Cretaceous Mesaverde Group, Book Cliffs, Utah. Sedimentology 34: 423-457

Swift DJP, Rice DD (1984) Sand bodies on muddy shelves: a model for sedimentation in the Western Interior Cretaceous seaway, North America. In: Tillman RW and Siemers CT (eds) Siliciclastic shelf sediments. Soc Econ Paleontol Mineral Spec Publ 34: 43-62

Swift JH (1984) The circulation of the Denmark Strait and Iceland-Scotland overflow waters in the North Atlantic. Deep-Sea Res 31: 1339-1355

Tada R (1991) Compaction and cementation in siliceous rocks and their possible effect on bedding enhancement. In: Einsele G, Ricken W, Seilacher A (eds) Cycles and events in stratification. Springer, Heidelberg New York, p 480-491

Tada R, Siever R (1989) Pressure solution during diagenesis. Annu Rev Earth Planet Sci 17: 89-118

Taira A, Katto J, Tashiro M, Okamura M, Kodama K (1988) The Shimanto Belt in Shikoku, Japan - evolution of Cretaceous to Miocene aaccretionary prism. Modern Geol 12: 1-42

Talbot MR, Kelts K (eds) (1989) The Phanerozoic record of lacustrine basins and their environmental signals. Palaeogeogr Palaeoclimatol Palaeoecol 70, 304 p

Talwani M, Hay W, Ryan WBF (eds) (1979) Deep drilling results in the Atlantic ocean: continental margins and paleoenvironment. Maurice Ewing Ser 3, Am Geophys Union, Washington DC

Tamrazyan GP (1989) Global peculiarities and tendencies in river discharge and wash-down of the suspended sediments - the Earth as a whole. J Hydrol 107: 113-131

Tankard AJ, Balkwill HR (eds)(1989) Extensional tectonics and stratigraphy of the North Atlantic margins. Am Assoc Petrol Geol Mem 46, 641 p

Tankard AJ, Welsink HJ (1988) Extensional tectonics, structural styles and stratigraphy of the Mesozoic Grand Banks of Newfoundland. In: Manspeizer W (ed) Triassic-Jurassic rifting, parts A and B. Developm Geotectonics 22, Elsevier, Amsterdam, p 129-165

Tarkian M, Garbe CD (1988) Geochemistry and genesis of sulfide ore deposits in the volcano-sedimentary sequence of the western Grauwackenzone (Eastern Alps, Austria). In: Friedrich GH, Herzig PM (eds) Base metal sulfide deposits in sedimentary and volcanic environments. Soc Geology Applied to Mineral Deposits, Spec Publ 5, Springer, Berlin Heidelberg New York, p 149-168

Terzhagi K (1925) Erdbaumechanik auf bodenphysikalischer Grundlage. Deuticke, Leipzig

Teichert C (1958) Concepts of facies. Am Assoc Petrol Geol Bull 42: 2718-2744

Teichmüller R (1973) Die paläogeographisch-fazielle und tektonische Entwicklung eines Kohlenbeckens am Beispiel des Ruhrkarbons. Z Deutsch Geol Ges 124: 149-165

Teichmüller M, Teichmüller R (1979) Diagenesis of coal (coalification). In: Larsen G, Chilingarian GV (eds) Diagenesis in sediments and sedimentary rocks. Developm Sedimentol 25A, Elsevier, Amsterdam, p 207-246

Teichmüller M, Teichmüller R, Bartenstein H (1984) Inkohlung und Erdgas - eine neue Inkohlungskarte der Karbon-Oberfläche in Nordwestdeutschland. Fortschr Geol Rheinld Westf (Krefeld) 32: 11-34

Teisserene P, Villemin J (1990) Sedimentary basin of Gabon - geology and oil systems. In: Edwards JD, Santogrossi PA (eds) Divergent/passive margin basins. Am Assoc Petrol Geol Mem 48: 117-199

Teller JZ, Last WM (1990) Paleohydrological indicators in playas and salt lakes, with examples from Canada, Australia, and Africa. Palaeogeogr Palaeoclimatol Palaeoecol 76: 215-240

Ten Haven HL, Baas M, De Leeuw JW, Schenck PA (1987) Late Quaternary sapropels, I - on the origin of organic matter in sapropel S_7. Marine Geol 75: 137-156

Ten Haven HL, Littke R, Rullkötter J, Stein R, Welte DH (1990) Accumulation rates and composition of organic matter in late Cenozoic sediments underlying the active upwelling area off Peru. In: Suess E, von Huene R et al. Proc ODP, Sci Results 112: 591-605, College Station, TX (Ocean Drilling Program)

Teng LS, Gorsline DS (1989) Late Cenozoic sedimentation in California continental borderland basins as revealed by seismic facies analysis. Geol Soc Am Bull 101: 27-41

Terwindt JHJ (1988) Palaeotidal reconstructions of inshore tidal depositional environments. In: Boer PL de, van Gelder A, Nio SD (eds) Tide-influenced sedimentary environments and facies. Reidel, Dordrecht, p 233-263

The Leg 116 Shipboard Scientific Party (1987) Himalayan uplift history observed on the equator. Geotimes 33: 9-12

Thein J, von Rad U (1987) Silica diagenesis in continental rise and slope sediments off eastern North America (Sites 603, Leg 93; Sites 612 and 613, Leg 95). In: Poag CW, Watts AB et al. Init Repts DSDP 95: 501-525, US Govt Print Office, Washington

Thiede J (1978) Pelagic sedimentation in immature ocean basins. In: Ramberg IB, Neumann ER (eds) Tectonics and geophysics of continental rifts. Reidel, Dordrecht, p 237-248

Thiede J (1981) Sedimentation und physiographische Entwicklung des Nordatlantiks seit dem mittleren Mesozoikum. Geol Rundsch 70: 316-326

Thiede J, Suess E (eds) (1983) Coastal upwelling - its sediment record. Part B, NATO Conference Ser 4: Marine Sci 10, Plenum Press, New York London, 610 p

Thiede J, Diesen GW, Knudsen B-E (1986) Patterns of Cenozoic sedimentation in the Norwegian-Greenland Sea. Marine Geol 69: 323-352

Thiede J, Eldholm O, Taylor E (1989) Variability of Cenozoic Norwegian-Greenland Sea paleoceanography and northern hemisphere paleoclimate. In: Eldholm O, Thiede J, Taylor E et al. Proc ODP, Sci Results 104: 1067-1118, College Station, TX (Ocean Drilling Program)

Thierstein HR (1979) Paleoceanographic implications of organic carbon and carbonate distribution in Mesozoic deepsea sediments. In: Talwani M, Hay W, Ryan WBF (eds) Deep drilling results in the Atlantic Ocean: continental margins and paleoenvironment. Maurice Ewing Ser 3, Am Geophys Union, Washington DC, p 249-274

Thomson J, Calvert SE, Mukherjee S, Burnett WC, Bremner JM (1984) Further studies on the nature, composition and ages of contemporary phosphorite from the Namibian shelf. Earth Planet Sci Lett 69: 341-353

Thornburg TM, Kulm LD (1987) Sedimentation in the Chile Trench: depositional morphologies, lithofacies, and stratigraphy. Geol Soc Am Bull 98: 33-52

Thornes JB, Brunsden D (1977) Geomorphology and time. Methuen, London

Thornton SE (1984) Basin model for hemipelagic sedimentation in a tectonically active continental margin: Santa Barbara basin, California continental borderland. In: Stow DAV, Piper DJW (eds) Fine-grained sediments: deep-water processes and facies. Geol Soc Spec Publ 15: 377-394

Thunell RC, Locke SM, Williams DF (1988) Glacio-eustatic seal-level control on Red Sea salinity. Nature 334: 601-604

Thiry M, Milnes AR (1991) Pedogenic and groundwater silcretes at Stuart Creek opal field, South Australia. J Sediment Petrol 61: 111-127

Tiercelin JJ, Faure H (1978) Rates of sedimentation and vertical subsidence in neorifts and paleorifts. In: Ramberg B, Neumann ER (eds) Tectonics and geophysics of continental rifts. Reidel, Dordrecht, p 41-47

Tietze K-W (1982) Zur Geometrie einiger Flüsse im Mittleren Buntsandstein (Trias). Geol Rundsch 71: 813-823

Tillman RW, Siemers CT (1984) Siliciclastic shelf sediments. Soc Econ Paleontol Mineral Spec Publ 34

Tillman RW, Swift DJP, Walker RG (eds) (1985) Shelf sands and sandstone reservoirs. Soc Econ Paleontol Mineral Short Course Notes 13, 708 p

Tipper JC (1987) Estimating stratigraphic completeness. J Geology 95: 710-715

Tissot BP, Welte DH (1984) Petroleum formation and occurrence, 2nd edn. Springer, Berlin Heidelberg New York, 699 pp

Tissot BP, Pelet R, Ungerer Ph (1987) Thermal history of sedimentary basins, maturation indices, and kinetics of oil and gas generation. Am Assoc Petrol Geol, Bull 71: 1445-1466

Toomey DF (ed) (1981) European fossil reef models. Soc Econ Paleontol Mineral Spec Publ 30, 546 p

Torrent J, Schwertmann U (1987) Influence of hematite on the color of red beds. J Sediment Petrol 57: 682-686

Toth J (1980) Cross-formational gravity-flow of groundwater: a mechanism of the transport and accumulation of petroleum. In: Roberts WH III, Cordell RJ (eds) Problems of petroleum migration. Am Assoc Petrol Geol, Stud in Geol 10: 121-167

Toy TJ (1982) Accelerated erosion: process, problems, and prognosis. Geology 10: 524-529

Trappe J (1991) Stratigraphy, facies distribution and paleogeography of the marine Paleogene from the western High Atlas. N Jahrb Geol Paläontol Abh (Stuttgart) 180: 279-321

Trendall AF, Morris RC (eds)(1983) Iron-formation: facts and problems. Developm Precambrian Geol 6, Elsevier, Amsterdam, 558 p

Trurnit P, Amstutz GC (1979) Die Bedeutung des Rückstandes von Druck-Lösungsvorgängen für stratigraphische Abfolgen, Wechsellagerung und Lagerstättenbildung. Geol Rundsch 68: 1107-1124

Trusheim F (1960) Mechanics of salt migration in northern Germany. Am Assoc Petrol Geol Bull 44: 1519-1540

Tucholke BE, Embley RW (1984) Cenozoic regional erosion of the abyssal sea floor off South Africa. In: Schlee JS (ed) Interregional unconformities and hydrocarbon accumulation. Am Assoc Petrol Geol Mem 36: 145-164

Tucker ME (1974) Sedimentology of Paleozoic pelagic limestones: the Devonian Griotte (Southern France) and Cephalopodenkalk (Germany). In: Hsü KJ, Jenkyns HC (eds) Pelagic sediments on land and under the sea. Int Assoc Sedimentol Spec Publ 1: 71-92

Tucker ME (1982) Precambrian dolomites: petrographic and isotopic evidence that they differ from Phanerozoic dolomites. Geology 10: 7-12

Tucker ME, Wilson JL, Crevello PD, Sarg JR, Read F (eds) (1990) Carbonate platforms: facies, sequences and evolution. Int Assoc Sedimentol Spec Publ 9, Blackwell, Oxford, 328 p

Tucker ME, Wright VP (1990) Carbonate sedimentology. Blackwell, Oxford, 482 p

Tucker RM, Cann JR (1986) A model to estimate the depositional brine depth of ancient haline rocks: implications for ancient subaqueous depositional environments. Sedimentology 33: 401-412

Turner P (1980) Continental red beds. Developm Sedimentol 29, Elsevier, Amsterdam, 562 p

Twidale CR, Milnes AR (1983) Aspects of the distribution and disintegration of siliceous duricrusts in arid Australia. Geol Mijnbouw 62: 373-382

Tye RS, Coleman JM (1989) Depositional processes and stratigraphy of fluvially dominated lacustrine deltas: Mississippi delta plain. J Sediment Petrol 59: 973-996

Tye RS, Kosters EC (1986) Styles of interdistributary basin sedimentation: Mississippi delta plain, Louisiana. Trans Gulf Coast Assoc Geol Soc 36: 575-588

Uemura T, Mizutani S (1984) Geological structures. Wiley, Chichester, 309 p

Vail PR (1987) Seismic stratigraphy interpretation using sequence statigraphy. Part I: Seismic stratigraphy interpretation procedure. In Bally AW (ed) Atlas of seismic stratigraphy. Am Assoc Petrol Geol, Stud in Geol 27: 1-10

Vail PR, Sangree JB (1988) Sequence stratigraphy interpretation of seismic, well and outcrop data workbook. NATO Adv Res Workshop, Sept 1988, Digne, France, Assoc Sedimentol Fr, 19 parts

Vail PR, Hardenbol J, Todd RG (1984) Jurassic unconformities, chronostratigraphy, and sea level changes from seismic stratigraphy and biostratigraphy. In: Schlee JS (ed) Interregional unconformities and hydrocarbon accumulation. Am Assoc Petrol Geol Mem 36: 129-144

Vail PR, Audemard F, Bowman SA, Eisner PN, Perez-Cruz C (1991) The stratigraphic signatures of tectonics, eustacy and sedimentology. In: Einsele G, Ricken W, Seilacher A (eds) Cycles and Events in stratigraphy. Springer, Heidelberg New York, p 617-659

Vail PR, Mitchum RM Jr, Todd RG, Widmier JM, Thompson S, Sangree JR, Bubb JN, Hatlelid WG (1977) Seismic stratigraphy and global changes of sea level. In: Payton CE (ed) Seismic stratigraphy - application to hydrocarbon exploration. Am Assoc Petrol Geol Mem 26: 49-212

Valeton I (1983) Klimaperioden lateritischer Verwitterung und ihr Abbild in den synchronen Sedimentationsräumen. Z Deutsch Geol Ges 134: 413-452

Valeton I (1988) Sedimentäre Phosphatgesteine. In: Füchtbauer H (ed) Sedimente und Sedimentgesteine. Schweizerbart, Stuttgart, p 543-567

Valloni R, Mezzardi G (1984) Compositional suites of terrigenous deep-sea sands of the present continental margins. Sedimentology 31: 353-364

Van Andel TH (1964) Recent marine sediments of the Gulf of California. In: van Andel TH, Shor GG (eds) Marine geology of the Gulf of California - a symposium. Am Assoc Petrol Geol Mem 3: 216-310

Van Andel TH (1983) Estimation of sedimentation and accumulation rates. In: Heath GR (ed) Sedimentology, physical properties and geochemistry in the Initial Reports of the Deep Sea Drilling Project: an overview. US Dept Commerce, Nat Oceanic and Atmosph Administr, Envrionm Data and Inform Service, Boulder, p 93-101

Van Andel TH (1985) New views on an old planet. Cambridge Univ Press, Cambridge, 324 p

Van Andel TH, Thiede J, Sclater JG, Hay WW (1977): Depositional history of the South Atlantic ocean during the last 125 million years. J Geol 85: 651-698

van Hinte JE (1978) Geohistory analysis - application of micropaleontology in exploration geology. Am Assoc Petrol Geol Bull 62: 201-222

van Hinte JE, Cita MB, van der Weijden CH (eds) (1987) Extant and ancient anoxic basin conditions in the eastern Mediterranean. Marine Geol 75, 1/4 (Spec Iss), 281 p

Van Houten FB (1964) Cyclic lacustrine sedimentation, Upper Triassic Lockatong formation, central New Jersey and adjacent Pennsylvania. Kansas Geol Surv Bull 169: 497-531

Van Houten FB (1973) Origin of red beds: a review - 1961-1972. Annu Rev Earth Planet Sci 1: 39-61

Van Houten FB, Arthur MA (1989) Temporal patterns among Phanerozoic oolitic ironstones and oceanic anoxia. In: Young TP, Taylor EG (eds) Phanerozoic ironstones. Geol Soc Spec Publ 46: 33-49

Van Houten FB, Purucker ME (1984) Glauconitic peloids and chamositic ooids - favorable factors, constraints and problems. Earth-Sci Rev 20: 211-243

Van Straaten LMJU (1954) Composition and structure of recent marine sediments in the Netherlands. Leidse Geol Mededel 19: 1-110

van Wagoner JC, Mitchum RM, Posamentier HW, Vail PR (1987) Seismic stratigraphy using sequence stratigraphy. Part II: Key definitions of sequence stratigraphy. In: Bally AW (ed) Atlas of seismic stratigraphy. Am Assoc Petrol Geol Stud in Geol 27: 11-14

Van Wagoner JC, Posamentier HW, Mitchum RM, Vail PR, Sarg JF, Loutit TS, Hardenbol J (1988) An over-view of the fundamentals of sequence stratigraphy and key definitions. In: Sea-level changes - an integrated approach. Soc Econ Paleontol Mineral Spec Publ 42: 39-45

Vecsei A, Frisch W, Pirzer M, Wetzel A (1989) Origin and tectonic significance of radiolarian chert in the Austroalpine rifted continental margin. In: Hein JR, Obradovic J (eds) Siliceous deposits of the Tethys and Pacific regions. Springer, New York Heidelberg, p 65-80

Veevers JJ (1988) Gondwana facies started when Gondwanaland merged in Pangea. Geology 16: 732-734

Veevers JJ (1990) Tectonic-climatic supercycle in the billion-year plate-tectonic eon: Permian Pangean icehouse alternates with Cretaceous dispersed-continents greenhouse. Sediment Geol 68: 1-16

Veizer J (1988) The evolving exogenic cycle. In: Gregor CB, Garrels RM, Mackenzie FT, Maynard JB (eds) Chemical cycles in the evolution of the Earth. Wiley, New York, p 175-220

Veizer J, Fritz P, Jones B (1986) Geochemistry of brachiopods: oxygen and carbon isotopic records of Paleozoid oceans. Geochim Cosmochim Acta 50: 1679-1696

Velbel MA (1985) Mineralogically mature sandstones in accretionary prisms. J Sediment Petrol 55: 685-690

Vessel RK, Davies DK (1981) Nonmarine sedimentation in an active forearc basin. Soc Econ Paleontol Mineral Spec Publ 31: 31-45

Visser MJ (1980) Neap-spring cycles reflected in Holocene subtidal large-scale bedform deposits: a preliminary note. Geology 8: 543-546

Von der Borch CC, Lock DE (1979) Geological significance of Coorong dolomites. Sedimentology 26: 813-824

Von Huene R (1986) To accrete or not accrete, that is the question. Geol Rundsch 75: 1-15

Von Huene R, Lallemand S (1990) Tectonic erosion along the Japan and Peru convergent margins. Geol Soc Am Bull 102: 704-720

Vorren TO, Lebesbye E, Andreassen K, Larsen KB (1989) Glacigenic sediments on a passive continental margin as exemplified by the Barents Sea. Marine Geol 85: 251-272

Wagenbreth O, Steiner W (1982) Geologische Streifzüge - Landschaft und Erdgeschichte zwischen Kap Arkona und Fichtelberg. VEB Deutscher Verlag für Grundstoffindustrie, Leipzig, 204 p

Walcott RI (1987) Geodetic strain and the deformational history of the North Island of New Zealand during the late Cainozoic. In: Oxburgh ER, Yardley BWD, England PC (eds) Tectonic settings of regional metamorphisms. Philos Trans Royal Soc London A, 321: 163-181

Walger E (1964) Zur Darstellung von Korngrößenverteilungen. Geol Rundsch 54: 976-1002

Walker BH, Francis EH (1987) High-level emplacement of an olivine-dolerite sill into Namurian sediments near Cardenden, Fife. Trans Royal Soc Edinburgh, Earth Sci 77: 295-307

Walker JCG, Drever JI (1988) Geochemical cycles of atmospheric gases. In: Gregor CB, Garrels RM, Mackenzie FT, Maynard JB (eds) Chemical cycles in the evolution of the Earth. Wiley, New York, p 55-76

Walker RG (1973) Mopping up the turbidite mess. In: Ginsburg, RN (ed) Evolving concepts in sedimentology. Johns Hopkins Univ Press, Baltimore London, p 1-37

Walker RG (1978) Deep-water sandstone facies and ancient submarine fans: models for exploration for stratigrphic traps. Am Assoc Petrol Geol Bull 62: 932-966

Walker RG (ed) (1984a) Facies models, 2nd edn. Geosci Canada Reprint Ser 1, Geol Assoc Canada, 317 p

Walker RG (1984b) Shelf and shallow marine sands. In: Walker RG (ed) Facies models, 2nd edn. Geosci Canada Reprint Ser 1, Geol Assoc Canada, p 141-170

Walker RG (1984c) Turbidites and associated coarse-grained clastic deposits. In: Walker R (ed) Facies models, 2nd edn. Geosci Canada Reprint Ser 1, Geol Assoc Canada, p 171-188

Walker RG, Cant DJ (1984) Sandy fluvial systems. In: Walker RG (ed) Facies models, 2nd edn. Geosci Canada Reprint Ser 1, Geol Assoc Canada, p 71-89

Walker RG, Duke WL, Leckie DA (1983) Hummocky stratification: significance of its variable bedding sequences - discussion and reply. Geol Soc Am Bull 94: 1245-1251

Walker TR (1967) Formation of red beds in modern and ancient deserts. Geol Soc Am Bull 78: 353-368

Walker TR (1976) Diagenetic origin of continental red beds. In: Falke H (ed) The continental Permian in Central, West, and South Europe. Reidel, Dordrecht, p 240-282

Walker TR (1979) Red color in dune sand. In: McKee ED (ed)(1979) A study of global sand seas. US Geol Surv Prof Pap 1052: 61-81

Walling DE, Webb, BW (1986) Solutes in river systems. In: Trudgill ST (ed) Solute processes. Wiley, Chichester New York, p 251-327

Walsh JJ (1988) On the nature of continental shelves. Academic Press, 515 p

Wanless HR, Tedesco LP, Tyrell KM (1988) Production of subtidal and surficial tempestites by hurricane Kate, Caicos platform, British West Indies. J Sediment Petrol 58: 739-750

Waples DW (1980) Time and temperature in petroleum formation: Application of Lopatin's method to petroleum exploration. Am Assoc Petrol Geol Bull 65: 916-926

Ward LG, Ashley GM (eds)(1989) Physical processes and sedimentology of siliciclastic-dominated lagoonal systems. Marine Geol (Spec Iss) 88: 181-364

Ward RF, Kendall CGSC, Harris PM (1986) Upper Permian (Guadalupian) facies and their association with hydrocarbons - Permian Basin, West Texas and New Mexico. Am Assoc Petrol Geol 70: 239-262

Warme JE (1988) Jurassic carbonate facies of the central and eastern High Atlas Rift, Morocco. In: Jacobshagen VH (ed) The Atlas system of Morocco. Lecture Notes Earth Sci 15, Springer, Berlin Heidelberg New York, p 169-199

Warren JK (1979) Aeolian processes. In: Embleton C, Thornes J (eds) Process in geomorphology. Arnold, London, p 325-351

Warren JK (1989) Evaporite sedimentology: importance in hydrocarbon accumulation. Prentice-Hall, Englewood Cliffs, New Jersey, 285 p

Warren JK (1990) Sedimentology and mineralogy of dolomitic Coorong lakes, South Australia. J Sediment Petrol 60: 843-858

Wasson RJ (1986) Geomorphology and Quaternary history of the Australian continental dune fields. Geogr Rev Japan 59: 55-67

Watts AB, Steckler MS (1979) Subsidence and eustacy at continental margin of eastern North America. In: Talwani M, Hay W, Ryan WBF (eds) Deep drilling results in the Atlantic Ocean, continental margins and paleoenvironment. Maurice Ewing Ser 3, Am Geophys Union, p 218-234

Watts AB, Thorne J (1984) Tectonics, global changes in sea level and their relationship to stratigraphical sequences at the U.S. Atlantic continental margin. Marine and Petrol Geol 1: 319-339

Watts AB, Karner GD, Steckler Ms (1982) Lithospheric flexure and the evolution of sedimentary basins. Philos Trans Royal Soc A, 305: 249-281

Watts KF, Blome CD (1990) Evolution of carbonate platform margin slope and its response to orogenic closing of a Cretaceous ocean basin, Oman. In: Tucker ME, Wilson JL, Crevello PD, Sarg JR, Read F (eds) Carbonate platforms: facies, sequences and evolution. Int Assoc Sedimentol Spec Publ 9, Blackwell, Oxford, p 291-323

Weaver CE (1989) Clays, muds, and shales. Developm Sedimentol 44, Elsevier, Amsterdam, 820 p

Weedon GP (1991) The spectral analysis of stratigraphic time series. In: Einsele G, Ricken W, Seilacher A (eds) Cycles and events in stratigraphy. Springer, Heidelberg New York, p 840-854

Wefer G, Heinze P, Suess E (1990) Stratigraphy and sedimentation rates from oxygen isotope composition, organic carbon content, and grain-size distribution at the Peru upwelling region: Holes 680B and 686B. In: Suess E, von Huene R et al. Proc ODP, Sci Results 112: 355-362, College Station, TX (Ocean Drilling Program)

Weggen J, Valeton I (1990) Polygenetic lateritic iron ores on BIF's in Minas Gerais/Brazil. Geol Rundsch 79/2: 301-318

Weimer P (1990) Sequence stratigraphy, facies geometries, and depositional history of the Mississippi fan, Gulf of Mexico. Am Assoc Petrol Geol Bull 74: 425-453

Weimer RJ (1960) Upper Cretaceous stratigraphy, Rocky Mountain area. Am Assoc Petrol Geol Bull 44: 1-20

Weimer RJ (1984) Relation of unconformities, tectonics, and sea-level changes: Cretaceous of Western Interior. In: Schlee JS (ed) Interregional unconformities and hydrocarbon accumulation. Am Assoc Petrol Geol Mem 36: 7-35

Weimer RJ (1986) Relationship of unconformities, tectonics, and sea level changes in the Cretaceous of the Western Interior, United States. In: Peterson JA (ed) Paleotectonics and sedimentation in the Rocky Mountain region, United States. Am Assoc Petrol Geol Mem 41: 397-422

Weimer RW (1976) Deltaic and shallow marine sandstones: sedimentation, tectonics, and petroleum occurrences. Am Assoc Petrol Geol, Educ Course Note Ser 2, 167 p

Weissert H, McKenzie J, Hochuli P (1979) Cyclic anoxic events in the early Cretaceous Tethys ocean. Geology 7: 147-151

Weller JM (1964) Development of the concept and interpretation of cyclic sedimentation. In: Merriam DF (ed) Symposium on cyclic sedimentation. Kansas Geol Surv Bull 169: 607-621

Wells JT, Coleman JM (1981) Periodic mudflat progradation, northeastern coast of South America: a hypothesis. J Sediment Petrol 51: 1069-1076

Wells NA, Dorr JA (1987) A reconnaissance of sedimentation on the Kosi alluvial fan of India. In: Ethridge FG, Flores RM, Harvey MD (eds) Recent developments in fluvial sedimentology. Soc Econ Paleontol Mineral Spec Publ 39: 51-61

Welsink HJ, Srivastava SP, Tankard AJ (1989) Basin architecture of the Newfoundland continental margin and its relationship to ocean crust fabric during extension. In: Tankard AJ, Balkwill HR (eds) Extensional tectonics and stratigraphy of the North Atlantic margin. Am Assoc Petrol Geol Mem 46: 197-213

Wendt J (1988) Condensed carbonate sedimentation in the late Devonian of the eastern Anti-Atlas. Eclogae Geol Helv 81: 155-173

Wendt J, Aigner T (1985) Facies patterns and depositional environments of Palaeozoic cephalopod limestones. Sediment Geol 44: 263-300

Werner F (1963) Über den inneren Aufbau von Strandwällen an einem Küstenabschnitt der Eckernförder Bucht. Meyniana (Kiel) 13: 108-121

Wernicke B (1985) Uniform-sense normal simple shear of the continental lithosphere. Can J Earth Sci 22: 108-125

Wernicke B, Burchfiel BC (1982) Modes of extensional tectonics. J Struct Geol 4: 105-115

Wessely G (1988) Structure and development of the Vienna basin in Austria. In: Royden LH, Horváth F (eds) The Pannonian basin: a study in basin evolution. Am Assoc Petrol Geol Mem 45: 333-346

Westgate JW, Gee CT (1990) Paleoecology of a middle Eocene mangrove biota (vertebrates, plants, and invertebrates) from southwest Texas. Palaeogeogr Palaeoclimatol Palaeoecol 78: 163-177

Wetherill GW (1990) Formation of the Earth. Annu Rev Earth Planet Sci 18: 205-256

Wetzel A (1986) Sedimentphysikalische Eigenschaften als Indikatoren für Ablagerung, Diagenese und Verwitterung von Peliten. Habilitationsschrift, Geowiss Fakultät Univ Tübingen, 116 p (unpublished)

Wetzel A (1989) Influence of heat flow on ooze/chalk cementation: quantification from consolidation parameters in DSDP Sites 504 and 505 sediments. J Sediment Petrol 59: 539-547

Wetzel A (1991) Stratification in black shales: depositional models and timing - an overview. In: Einsele G, Ricken W, Seilacher A (eds) Cycles and events in stratigraphy. Springer, Heidelberg New York, p 508-523

Wetzel A, Aigner T (1986) Stratigraphic completeness: tiered trace fossils provide a measuring stick. Geology 14: 234-237

Wetzel A, Kohl B (1986) Accumulation rates of Mississippi fan sediments cored during Deep Sea Drilling Project Leg 96. In: Bouma AH, Coleman JM, Meyer AW, et al., Init Repts DSDP 96: 595-600, US Govt Print Office, Washington

Wezel WC (ed) (1980) Sedimentary basins of Mediterranean margins. Nazionale delle Ricerche, Università degli Studi, Urbino, Italy

Whitaker FF, Smart PL (1990) Active circulation of saline ground waters in carbonate platforms: evidence from the Great Bahama Bank. Geology 18: 200-203

White R, Louden KE (1982) The Makran continental margin: structure of a thickly sedimented convergent plate boundary. In Watkins JS, Drake CL (eds) The geology of continental margins. Am Assoc Petrol Geol Mem 34: 499-518

White RS (1989) Volcanism and igneous underplating in sedimentary basins and at rifted continental margins. In: Price RA (ed) Origin and evolution of sedimentary basins and their energy and mineral resources. Geophys Monogr 48 (IUGG vol 3), Am Geophys Union, Washington DC, p 125-127

Whitmarsh RB, Weser OE, Ross DA et al. Init Reports DSDP 23, US Govt Print Office, Washington, 1180 p

Wildi W, Funk H, Loup B, Amato E (1989) Mesozoic subsidence history of the European marginal shelves of the alpine Tethys (Helvetic realm, Swiss Plateau and Jura). Eclogae Geol Helv 82: 817-840

Wilgus C, Hastings B, Ross C, Posamentier H, Van Wagoner J, Kendall CGSC (eds) (1988) Sea level changes: an integrated approach. Soc Econ Paleontol Mineral Spec Publ 42, 407 p

Williams GE (1989a) Late Precambrian tidal rhythmites in South Australia and the history of the Earth's rotation. J Geol Soc London 146: 97-111

Williams GE (1989b) Tidal rhythmites, geochronometer for the ancient Earth-Moon system. Episodes 12, 3: 162-171

Williams HFL, Roberts MC (1989) Holocene sea-level change and delta growth: Fraser River delta, British Columbia. Can J Earth Sci 26: 1657-1666

Williamson PG, Savage RJG (1986) Early rift sedimentation in the Turkana basin, northern Kenya. In: Frostick LE, Renaut RW, Reid I, Tiercelin JJ (eds) Sedimentation in African rifts. Geol Soc Spec Publ 25: 267-283

Wilson JL (1975) Carbonate facies in geologic history. Springer, Berlin Heidelberg New York, 471 p

Wilson JL, Jordan C (1983) Middle shelf. In: Scholle PA, Bebout DG, Moore CH (eds) Carbonate depositional environments. Am Ass Petrol Geol Mem 33: 297-343

Wilson JT (1965) A new class of faults and their bearing on continental drift. Nature 207: 343-347

Wilson JT (1968) Static or mobile Earth: the current scientific revolution. Am Philos Soc Proc 112: 309-320

Windley BF (1984) The evolving continents, 2nd edn. Wiley, Chichester, 399 p

Winterer EL, Bosellini A (1981) Subsidence and sedimentation on a Jurassic passive continental margin (Southern Alps, Italy). Am Assoc Petrol Geol Bull 65: 394-421

Winterer EL, Hinz K (1984) The evolution of the Mazagan continental margin: a synthesis of geophysical and geological data with results of drilling during Deep Sea Drilling Project Leg 79. In: Hinz K, Winterer EL et al. Init Repts DSDP 79: 893-919, US Govt Print Office, Washington

Witzke BJ (1987) Models for circulation patterns in epicontinental seas applied to Paleozoic facies or North America craton. Paleoceanography 2: 229-248

Wohletz KH, Sheridan MF (1979) A model of pyroclastic surge. In: Chapin CE, Elston WE (eds) Ash-fall tuffs. Geol Soc Am Spec Pap 180: 177-194

Woldegabriel G, Aronson JL, Walter RC (1990) Geology, geochronology, and rift basin development in the central sector of the Main Ethiopian Rift. Geol Soc Am Bull 102: 439-458

Woldstedt P, Duphorn K (1974) Norddeutschland und angrenzende Gebiete im Eiszeitalter. Koehler, Stuttgart, 500 p

Wolff M, Füchtbauer H (1976) Die karbonatische Randfazies der tertiären Süßwasserseen des Nördlinger Ries und des Steinheimer Beckens. Geol Jahrb (Hannover) D 14: 3-53

Wood JR, Hewett TA (1984) Reservoir diagenesis and convective fluid flow. In: McDonald DA, Surdam RC (eds) Clastic diagenesis. Am Assoc Petrol Geol Mem 37: 99-110

Wopfner H, Twidale CR (1988) Formation and age of desert dunes in the Lake Eyre depocenters in central Australia. Geol Rundsch 77: 815-834

Worsley TR, Nance D, Moody JB (1984) Global tectonics and eustacy for the past 2 billion years. Marine Geol 58: 373-400

Wright LD and Coleman JM (1973) Variations in morphology of mayor river deltas as function of wave and river discharge regimes. Bull Am Assoc Petrol Geol 57: 370-398

Wright RF, Matter A, Schweingruber M (1980) Sedimentation in Lake Biel, an eutrophic hard-water lake in northwestern Switzerland. Schweiz Z Hydrol 42: 101-126

Wright VP (1986) Facies sequences on a carbonate ramp: the Carboniferous limestone of South Wales. Sedimentology 33: 221-241

Wurster P (1964) Geologie des Schilfsandsteins. Mitt Geol Staatsinst Hamburg 33, 140 p

Wurster P (1968) Paläogeographie der deutschen Trias und die paläogeographische Orientierung der Lettenkohle in Südwestdeutschland. Eclogae Geol Helv 61: 157-166

Wurster P, Stets J (1982) Sedimentation in the Atlas Gulf II: Mid-Cretaceous events. In: Rad U von, Hinz K, Sarnthein M, Seibold E (eds) Geology of the Northwest African continental margin. Springer, Berlin Heidelberg New York, p 439-458

Wyllie PJ (1971) The dynamic earth. Wiley, New York, 416 p

Yaalon DM, Dan J (1974) Accumulation and distribution of loess-derived deposits in the semi-desert and desert fringe of area of Israel. Z Geomorph NF (Suppl) 20: 91-105

Ying Wang, Mei-e Ren, Dakuei Zhu (1986) Sediment supply to the continental shelf by the major rivers of China. J Geol Soc London 143: 935-944

Yose LA, Heller PL (1989) Sea-level control of mixed-carbonate-siliciclastic, gravity-flow deposition: Lower part of the Keeler Canyon Formation (Pennsylvanian), southeastern California. Geol Soc Am Bull 101: 427-439

Young TP, Taylor EG (eds)(1989) Phanerozoic ironstones. Geol Soc Spec Publ 46, 251 p

Zachar D (1982) Soil erosion. Developm in Soil Sci 10, Elsevier, Amsterdam, 547 p

Zafiriou OC, Gagosian RB Peltzer ET, Alford JB, Loder T (1985) Air-to-sea fluxes of lipids at Enewetak atoll. J Geophys Res 90: 2409-2423

Zankl H (1971) Upper Triassic carbonate facies in the northern Limestone Alps. In: Müller G (ed) Sedimentology of parts of Central Europe, 8th Int Sedimentol Congr Heidelberg, Guidebook. Kramer, Frankfurt, p 147-185

Zbinden EA, Holland HD, Feakes CR (1988) The Sturgeon Falls paleosols and the composition of the atmosphere 1.1 Ga BP. Precambrian Res 42: 141-163

Zhao WI, Morgan WJ (1985) Uplift of the Tibetan plateau. Tectonics 4: 359-369

Zharkov MA (1981) History of Paleozoic salt accumulation. Springer, Berlin Heidelberg New York, 309 p

Zhu J, Zhang F, Xu K (1988) Depositional environment and metamorphism of early Proterozoic iron formation in the Lüliangshan region, Shanxi province, China. Precambrian Res 39: 39-50

Zhu X (ed)(1989) Chinese sedimentary basins. Elsevier, Amsterdam, 238 p

Ziegler MA (1989) North German Zechstein facies patterns in relation to their substrate. Geol Rundsch 78: 109-127

Ziegler MA (1989) North German Zechstein facies patterns in relation to their substrate. Geol Rundsch 78: 109-127

Ziegler PA (1979) Factors controlling North Sea hydrocarbon accumulations. World Oil 189 (6): 111-124

Ziegler PA (1982) Geological atlas of western and central Europe. Shell Internationale Petrolleum Maatschappij B.V., The Hague. Distrib. Elsevier, Amsterdam, 130 p, 40 Encl

Ziegler PA (1988) Evolution of the Arctic-North Atlantic and the Western Tethys. Am Assoc Petrol Geol Mem 43, 198 p, 30 pl

Ziegler PA (1989) Evolution of Laurussia: a study in late Paleozoic plate tectonics. Kluwer, Dordrecht, 102 p, 14 pl

Zimmerman HB, Shackleton HJ, Backman J, Kent DV, Baldauf JG, Kaltenback AJ, Morton AC (1984) History of Plio-Pleistocene climate in the northeastern Atlantic, Deep Sea Drilling Project Hole 552A. In: Roberts DG, Schnitker D et al. Init Repts DSDP 81: 861-875, US Govt Print Off, Washington

Zonenhain LP, LePichon X (1986) Deep basins of the Black Sea and Caspian Sea as remnants of Mesozoic back-arc basins. In: Auboin J, LePichon X, Monin AS (eds) Evolution of the Tethys. Tectonophysics 123: 181-211

Zuffa GG (1987) Unravelling hinterland and offshore paleogeography from deep-water arenites. In: Legget JK, Zuffa GG (eds) Marine clastic sedimentology. Graham and Trotman, London, p 39-61

Subject Index

Italized numbers refer to figures, T to tables